Colorimetric Chemical Analytical Methods

L. C. THOMAS, B.Sc., F.R.I.C.

G. J. CHAMBERLIN

Revised by G. Shute, L.R.I.C., M.I.W.E.S.
(The Tintometer Ltd.)

9th Edition

The Tintometer Ltd., Salisbury, England.
and at:- Tintometer GmbH, Westfalendamm 73, D4600 Dortmund 1, W. Germany

Also Distributed by John Wiley and Sons Ltd., Chichester, New York, Brisbane and Toronto
ISBN No. 0 471 27605 7

LIBRARY OF CONGRESS CATALOGUE PART No. 79-56635

1st Edition 1953
2nd Edition 1953
3rd Edition 1953
4th Edition 1954
5th Edition 1958
6th Edition 1962
7th Edition 1967
8th Edition 1974
9th Edition 1980

Preface to the Ninth Edition

All the tests recorded in this book were developed with a specific application in mind, and, in previous editions, the tests were written up for the original application. However many of the tests have far wider applications and this edition includes details of methods of sample preparation for a variety of these. The sample preparation instructions and references are given in an introduction which covers all the tests which are included for any one radical. Where reference is made to a journal which may not be generally available, reference to an appropriate abstract is also included wherever possible.

Tests in the Inorganic Section include all applications published in Analytical Abstracts up to the end of Volume 19 and references to methods of isolating the ions under test have been included. This has increased the number of references in this section to 1488.

In order to facilitate reference to the appropriate method of sample preparation the instructions have been subdivided into the following groups:—

>   Air
>   Beverages
>   Biological materials
>   Chemicals
>   Dairy products
>   Foods and edible oils
>   Fuels and lubricants
>   Metals and alloys
>   Minerals, ores and rocks
>   Miscellaneous
>   Paper
>   Plants
>   Plastics and polymers
>   Sewage and industrial wastes
>   Soil
>   Water

These headings always appear in alphabetical order as applicable, and the groups are further subdivided as required.

In this edition some new tests have been added and redundant ones omitted. Sections are included which describe the ion exchange resins and the buffer solutions mentioned in the text.

Colorimetric Chemical Analytical Methods

9th Edition

GENERAL INDEX

	Page
Preface	i
Indices	ii
Introduction to analytical methods	vi
Methods of stating concentration	x
Normal and molar solutions	xi
Ion-exchange resins	xv
Buffer solutions	xxix
Lovibond apparatus used in the various tests	xxxi
pH in industry, medicine, and agriculture	1
Chemical Analysis—Organic	21
Chemical Analysis—Inorganic	71
Biochemistry Pathology and Pharmacology	385
Toxic substances in air	459
Colour grading and quality tests	557

Colorimetric Chemical Analytical Methods

9th Edition

Aids to the use of this book

There are 2 indices, to facilitate quick reference:—

General

which gives the over-all layout of the book.

Alphabetical

listing all the tests which are covered by the book.

COLORIMETRIC CHEMICAL ANALYTICAL METHODS

Alphabetical Index of Tests

Test	Page
Acetone in air	467
Acid Wash Test	560
Acrylonitrile	470
Alcohol	23
Aluminium	73
Amines	24
Ammonia	90, 389
α-Amylase (Eggs)	565
α-Amylase (Malt)	562
α-Amylase (Blood)	387
Anionic Detergents	41
Aniline	472
Anti-icing additives in fuel	567
Arsenic	98
Arsine	474
A.S.T.M. (Oil) scale	595
Aschaffenburg & Mullen test	603
Barrett colour scale	570
Beer colours (E.B.C.)	571
Benzene in air	476
Benzene (colours)	573
Benzene hexachloride	29
Bilirubin	391
Bismuth	104
Bromide	110, 393
Bromine	113
Butter colours	575
Cadmium	117
Calcium	123
Carbohydrates	31
Carbon dioxide	478
Carbon disulphide	32, 480
Carotene	33
Chemical Oxygen Demand	576
Chloride	129
Chlorine in water	134
Chlorine in air	483
Chlorine dioxide	148
Chloroform	484
Cholesterol	395
Cholinesterase	396
Chromic acid mist	485
Chromium	150
Coal tar products (acid wash test)	560
Cobalt	159
Coconut oil colours	580
Copper	168
Copper oxide fumes	487
Cresylic acid—colour grading	581

Test	Page
Cyanide	34
Cyanuric Acid	39
Cyclohexanone	490
Detergents	41
Dichlorophen	44
European Brewery Convention (E.B.C.)	571
F.A.C. colour scale	582
Fluoride	184
Fluoride in air	493, 501
Formaldehyde in water	48
" " in air	496
Furfuraldehyde	50
Gardner scale	583
Glucose (see sugar)	
Haemoglobin	400
Hardness of water	584
Hazen scale (colour of water)	586
Honey grading	588
Humidity	589
Hydrazine	192
Hydrogen cyanide	499
Hydrogen fluoride in air	501
Hydrogen peroxide	195
Hydrogen sulphide	196, 503
Iodine	200
Iodine colour scale	592
Iron	201
Iron in blood	405
Iron oxide fumes in air	507
Isocyanates	509
Isophorone	512
Kay and Graham test	606
Ketone vapour	515
Lactate dehydrogenase	407
Lactic acid	593
Lange's colloidal gold reaction	409
Lauryl pentachlorphenol	51
Lead	219, 411
Lead (airborne)	519
Lime requirement in soils	18
Liver function test (sulphobromophthalein)	443
Lubricating oil colour (A.S.T.M.)	595

IV

COLORIMETRIC CHEMICAL ANALYTICAL METHODS

Alphabetical Index of Tests—continued

Test	Page
Magnesium	229
Manganese	239
Mercury in air	523
Methyl α-chloro-acrylate	527
Milk products (colour)	596
Naphtha (colour)	573
Nickel	253
Nicotine	54
Nicotinic acid	413
Nitrate	263
Nitrite	277
Nitrobenzene	529
Nitrogen	285
Nitrogen potentially available in soils	288
Nitrogen (α-amino)	290
Nitrogen dioxide	531
Nitrous fumes	534
Oils (colour) F.A.C.,	582
,, ,, Gardner	583
,, ,, A.S.T.M.	595
Organic acids	56
Oxidants in air	537
Oxygen, dissolved	291
Ozone, dissolved	304
Ozone in air	304, 539
Paint Research Station Colour Scale	622
"Panacide"	44
Pasteurisation—efficiency of	603
Penicillin in milk	597
Pentachlorophenates	58
Pentachlorophenol	51
Permanganate value	601
pH	1
Phenols, monohydric	61
Phenols, total	63
Phosgene	541
Phosphatase	418, 603
Phosphate/Phosphorus	308, 424
Potassium	331
Proteins	426
PSP renal function test	416
Pyridine	543
Quality grading	557
Quaternary Ammonium Compounds	24

Test	Page
Reducing sugars in potatoes	610
Resazurin tests	612, 617
Rubber latex colour	619
Salicylate in blood	433
Sand for road paints	620
Sand quality	621
Shellac colours (P.R.S. scale)	622
Silicon	334
Sodium	344
Styrene in air	545
Sugar	65, 435, 610
Sulphate	348
Sulphetrone	441
Sulphite	353
Sulphobromophthalein test	443
Sulphonamides	445
Sulphur	354
Sulphur dioxide	547
Swimming pools pH and Chlorine	623
Thiocyanate	66
Thiophen	68
Tin	358
Titanium	364
Toluene (colour of)	573
Toluene in air	551
Toxic substances in air	461
Transaminase	447
Trichloracetic acid	450
Trichlorethylene (see trichloracetic)	
Urea	451
Uric acid	457
Urine, pH of	415
Vanadium	370
Varnish colours P.R.S. Scale	622
,, ,, Gardner Scale	583
Water, colour of (Hazen scale)	586
Xylene, colour of	573
Xylene in air	551
Zinc	375
Zinc oxide fumes	553

COLORIMETRIC CHEMICAL ANALYTICAL METHODS

Introduction

COLORIMETRIC CHEMICAL ANALYSIS is defined as the determination of the amount of a particular chemical which is present in a sample, using a chemically developed colour as a measure of the concentration. This method of analysis requires three separate procedures:

1. The isolation of the chemical to be determined from any interfering materials which may be present in the sample
2. The production of a colour by the action of some chemical reagent upon the isolated chemical
3. The measurement of the colour so produced.

These three processes are quite distinct and will therefore be discussed separately.

1. ISOLATION of the material to be estimated depends to a large extent on the nature of the sample in which it occurs and on the particular chemical ions which will interfere with the quantitative method to be used. Each individual determination therefore poses separation problems for which there can be no universal answer. The most that can be done to help the analyst is to indicate the general methods which can be used for getting material into solution; to state in the individual test procedures any solution method which cannot be used; and to refer the reader to published papers on specific applications of determinations, which contain details of the methods used for getting the samples into solution. This has been done in this edition and is especially important in the Inorganic Section.

In many applications of colorimetric chemical analysis the problem of getting the sample into solution does not arise, as the sample itself is a liquid. This is the case in the examination of water samples, the examination of trade effluents, quality control by examination of a liquid product, examination of body fluids, etc., etc. Such applications cover roughly 50% of the analyses for which the tests described in this book were originally devised.

Even in these cases however it may be necessary to remove some interfering substances before developing the test colour. Such removals can be effected by the use of adsorbents such as activated charcoal; by ion exchange resins; by column chromatography; or by developing the analytical colour in the presence of the impurity and then selectively extracting the colour with a solvent before estimation. Where the use of such techniques is required, then specific mention of the fact is made in the test procedure.

An alternative technique which is also widely used in the original tests is that of extracting the material to be determined from the sample matrix by suitable solvent procedures. No single method of extraction is suitable for use in every case and individual reference to the preferred techniques is made in the appropriate test instructions.

In many cases, especially in metallurgical analysis, it is possible to dissolve the sample fairly readily by the use of an acid or a mixture of acids. The acid used and the optimum concentration of acid will vary according to the chemical nature of the sample. However, provided that care is taken not to use an acid the anions from which will interfere with the subsequent colorimetric reaction, or which will destroy the ion to be determined, and provided that the final pH is adjusted to its optimum value for that particular determination, then any procedure which will dissolve the sample can be adopted safely.

An extension of this acid solution technique is that of "wet ashing"[1,2]. This method of solution is particularly useful when small amounts of metal ions have to be determined in organic matter and it is essential to destroy completely the organic material.

In this method of solution the sample is digested with a mixture of sulphuric, nitric and/or perchloric acids, sometimes in the presence of a catalyst, until the solution becomes colourless. The sulphuric acid is then evaporated and the residue heated until copious white fumes are evolved. The residue is subsequently taken up in water, neutralised if necessary, and treated with the appropriate colorimetric reagent.

Dry ashing is also sometimes employed, usually as a method of concentration prior to chemical solution. This technique also is limited to the determination of metal ions, which are not destroyed by ashing. Great care must be taken when using this technique that none of the material is lost by evaporation, or by sputtering. To avoid the loss of the more volatile elements, ashing should be carried out at as low a temperature as possible (400-450°C) and air-flow through the furnace should be limited to the minimum required to complete oxidation. It has recently been demonstrated[3,4] that when ashing biological materials, even at temperatures as low as 110°C, losses of cobalt, copper, manganese and zinc can occur. The use of wet ashing is therefore recommended when the determination of any of these elements is included in the analysis. Where the possible loss of these elements is not important dry ashing at 400°C may be more convenient, especially for large samples. It is sometimes preferable to prepare a 'sulphated ash' and this is done by carefully adding sulphuric acid to the initial ash and reheating, repeating the procedure until all traces of carbon have disappeared from the ash.

A further technique—which is useful for the dissolution of rocks, ores and other refractory samples—is that of fusion followed by acid solution[5]. The sample is weighed into a platinum dish, mixed with a suitable flux such as lithium metaborate[5] ($LiBO_3$), heated for 5–10 minutes and then dropped, while still hot, into dilute acid. Many variants on this technique will be found in the chemical literature. Some are specifically mentioned in the sample preparation sections of Part III of this book.

2. COLOUR PRODUCTION is carried out by chemically coupling the ion to be determined, if it is not already coloured, with a suitable chromophoric group. Many colorimetric reagents are now available for this purpose **and virtually any substance can now be determined colorimetrically by the use of the appropriate reagent.** The reagents used for the determination of specific ions are described in the appropriate test procedures. It must be emphasised that these are only those reagents for which discs have been calibrated in connection with specific procedures. The Tintometer Ltd. are always ready to produce further discs for use with alternative reagents and to advise on the appropriate reagent for use in particular circumstances.

High grade "analytical" reagents must always be employed in these tests. A widely used British description is AnalaR or AR. On the continent of Europe the designation PA is recognised as a suitable grade, and ACS is used in the U.S.A. to denote "conforming to the requirements of the American Chemical Society".

3. MEASUREMENT OF THE COLOUR is carried out by visually matching the test colour against a range of predetermined colour standards. It would obviously be impracticable to have every possible colour in any range of individual standard colours, and a method has therefore been devised whereby any colour can be matched by a mixture of standards of the three primary colours, red, yellow and blue. This method, using the **Lovibond Colour Scale**, was devised 100 years ago by J. W. Lovibond. The scale consists of three series of unfadeable coloured glass slips, one series of red (magenta), one of yellow and one of blue. Each series is a linear scale of the same colour, increasing by equal steps from the palest shade perceptible to a fully saturated (spectral) colour, and is divided into units and decimals of a unit. The colours are permanent and are exactly reproducible by the makers (The Tintometer Limited, of Salisbury, England). By super-imposing suitably chosen glasses, any colour can be matched—there are, in all, nine million possible combinations of these glasses—including a grey series down to black. All possible colours can in this way be recorded in terms of three numbers (e.g. "a" units Red, "b" units Yellow, "c" units Blue) and reproduced from these figures at any laboratory which has access to a Tintometer or to a set of Lovibond glasses.

Lovibond units of colour have received international recognition, and standardising bodies in over 20 different countries quote colours in terms of units of the Lovibond Scale.

The Lovibond Scale has also been related to the **Commission Internationale de L'Eclairage** system of colour specification[6] so that records are interchangeable between the two systems. A specification expressed in C.I.E. chromaticity co-ordinates can thus be converted into Lovibond units, assembled, and looked at as an actual sample of the colour.

In chemical colorimetry it is usually most convenient to match the colour produced in the test against a series of colour standards representing known amounts of the compound being determined. In photoelectric colorimetry this series of standards is frequently reduced to a calibration curve obtained by plotting absorbance at a given wavelength against concentration; it is however basically the same matching procedure, but carried out instrumentally over a more limited range of wavelengths. For the purpose of visual photometry, instead of reducing the series of standards to a series of absorbance figures, the colours are measured in Lovibond units and reproduced as permanent glass standards using the Lovibond glasses. For convenience, these standards are mounted in discs for use either in Lovibond Comparators or in Lovibond Nesslerisers.

These standards are available as an exact match of the standard solutions prepared by skilled workers in the laboratories of specialists, in many different parts of the world, who co-operate with The Tintometer Limited. The standards are then checked and cross-checked in other laboratories. Being permanent, the colours can be relied upon indefinitely and thus not only save time previously spent in preparing fresh chemical standards, but also simplify procedure and give confidence. Any error in carrying out the test is immediately apparent from an inability to match the solution against the standards, and the operator is able to institute an immediate check of the reagents and technique.

It is from considerations of this nature, based on experiences reported from a number of laboratories that our slogan, *"Measure colour with colour and SEE what you are doing"* has developed.

This book contains details of many chemical tests which have been specially developed for use with Lovibond colour standards. Although the tests have frequently been developed with a specific application in mind, it must be stressed that ultimately all colorimetric tests reduce to the measurement of a colour in a solution from which all possible interfering colours have been removed. The application of these tests to other materials is thus merely a question of adequate prior treatment of the sample and it is to assist in this that the introductory notes have been written.

Colorimetric tests are normally used for trace amounts, where this order of accuracy is acceptable. For much higher concentrations much greater accuracy is usually demanded and other methods of analysis are used. However in many control and limit tests an accuracy of $\pm 10\%$ is acceptable, and colorimetric tests can be used if careful dilution is carried out. One dilution rule must be obeyed however. **Never** dilute by more than 1:10 in any single step.

References
1. Society for Analytical Chemistry, *Analyst,* 1960, **85,** 643
2. G. F. Smith, *Talanta,* 1964, **11,** 633
3. G. R. Doshi, C. Sreekumaran, C. D. Mulay and B. Patel, *Current Sci.,* 1969, **38** (9), 206; *Analyt. Abs.,* 1970, **19,** 1474
4. P. Strohal, S. Lulic and O. Jelisavcic, *Analyst,* 1969, **94,** 678
5. C. O. Ingamells, *Talanta,* 1964, **11,** 665
6. R. K. Schofield, *J. Sci. Inst.* 1939, **16,** 74

Methods of stating concentration in analytical tests

Chemists use a variety of ways to report answers, and the following information will be of help in converting results from one form to another.

Parts per million—abbreviated to ppm—is always understood to mean weight to weight ratio. In the case of a solution, the weight of 1 litre of water is 1 kilogram (1,000 grams), so that 1 gram of a substance in 1 litre of water would be 1 part per thousand or 1,000 ppm.

1 milligram (abbreviated as mg) is one thousandth of a gram, so 1 mg/litre is the equivalent of 1 ppm.

1 microgram (abbreviated as μg) is one thousandth of a milligram or one millionth of a gram. Hence 1 microgram in 1 litre is 0.001 ppm, and 1 microgram in 1 millilitre (ml) is 1 ppm.

Summary

Parts per million (ppm) = milligrams per litre (mg/l).
= micrograms per millilitre (μg/ml)

In some of the tests in this book, the answer read from the disc is expressed in micrograms or milligrams in the actual amount of sample solution taken. This widens the range of the test, because by adjusting the volume of sample suitably, different concentrations can be included in the scope of the disc. Thus, if an answer of 5 micrograms is obtained and 10 ml of the sample was used, there would be 0.5 micrograms in 1 ml, or 0.5 ppm. If only 1 ml of the original sample were used, and diluted with the solvent, there would be 5 micrograms in 1 ml or 5 ppm. If 100 ml of the original sample were taken and this was concentrated by evaporation to 10 ml, an answer of 5 micrograms (μg) would represent 5 μg in 100 i.e. 0.05 ppm. If an answer is expressed as parts per 100,000 (as is the custom in some countries when reporting for example oxygen dissolved in water) the answer is multiplied by 10 to bring to ppm. Thus 1.5 parts per 100,000 equals 15 ppm.

COLORIMETRIC CHEMICAL ANALYTICAL METHODS

Normal and Molar Solutions

A *normal* solution (N) contains one gram equivalent of hydrogen per litre. For example a normal solution of nitric acid (HNO_3) requires the gram molecular weight (H = 1.008, N = 14.008, O = 16.00) i.e. 63.02 g per litre in order to get one gram of hydrogen. However for sulphuric acid (H_2SO_4), which has two hydrogen atoms per molecule, only half the gram molecular weight i.e. 49.04 g per litre is required.

In the case of oxidising agents the normal solution contains one gram equivalent of available oxygen per litre. As an example, in the case of potassium permanganate ($KMnO_4$), owing to the different reactions which take place in acid and alkaline solutions, the values for the normal solution vary according as the permanganate is to be used in acid or alkaline media.

A *molar* solution (M) is one which contains the gram molecular weight of the substance per litre. Therefore whereas for nitric acid the N and M solutions are identical, for sulphuric acid $M = 2N$.

Notes referring to following Table

§ The commercial product, syrupy ammonium lactate, usually contains about 60% w/w $NH_4C_3H_5O_3$. Due allowance must be made for this in preparing normal or molar solutions.

† When the commercial product is itself a solution it is difficult to prepare standard solutions unless the concentration of the commercial solution is accurately known. The following approximations are sufficiently accurate for most purposes. For volumetric work the solutions must be rigorously standardised after preparation.

   1. *Ammonium hydroxide* Dilute 67 ml of the concentrated (Sp gr 0.880) solution to 1 litre.
   2. *Hydrochloric acid* The concentrated solution contains about one third of its weight of HCl. 180.15 g of constant boiling point solution per litre gives an exactly normal solution; 120 g of concentrated acid diluted to 1 litre is slightly above normal.
   3. *Nitric acid* 65 ml or 93 g of acid (Sp gr 1.42) diluted to 1 litre gives a solution slightly stronger than normal.
   4. *Sulphuric acid* Take 30 ml of concentrated acid (Sp gr 1.84) and pour this slowly and carefully into 100 – 220 ml of water. Cool, mix thoroughly and dilute to 1 litre.

‡ Sodium and potassium hydroxides absorb carbon dioxide from the air. It is therefore difficult to be certain that the solid is not contaminated with some carbonate. The following will give solutions sufficiently close to normal for most colorimetric purposes.
   1. *Potassium hydroxide* Dissolve 64 g of KOH of assay value 85%, or better, in water and dilute to 1 litre. Standardise against normal acid if an accurately normal solution is required.
   2. *Sodium hydroxide* Dissolve 42 g of NaOH of assay value 95%, or better, in water and dilute to 1 litre. Standardise if an accurate normality is required.

Name	Formula	Molecular Weight	Concentration g/litre Normal	Concentration g/litre Molar
Acetic acid	CH_3COOH	60.05	60.05	60.05
Ammonium acetate	CH_3COONH_4	77.08	77.08	77.08
Ammonium borate	$NH_4HB_4O_7.3H_2O$	228.33	228.33	228.33
Ammonium chloride	NH_4Cl	53.49	53.49	53.49
Ammonium citrate	$(NH_4)_3C_6H_5O_7$	243.22	81.07	243.22
Ammonium fluoride	NH_4F	37.04	37.04	37.04
Ammonium hydrogen phosphate	$(NH_4)_2HPO_4$	132.06	132.06	132.06
Ammonium hydroxide†	NH_4OH	35.05	35.05	35.05
Ammonium lactate§	$CH_3.CHOH.COONH_4$	107.11	107.11	107.11
Ammonium persulphate	$(NH_4)_2S_2O_8$	228.20	114.10	228.20
Ammonium phosphate—see Ammonium hydrogen phosphate				
Ammonium thiocyanate	NH_4SCN	76.12	76.12	76.12
Aqueous ammonia—see Ammonium hydroxide				
Barium chloride	$BaCl_2.2H_2O$	244.31	122.16	244.31
Beryllium acetate	$(CH_3COO)_2Be$	127.10	63.55	127.10
Bis-(2-ethylhexyl)hydrogen phosphate	$(C_8H_{17})_2HPO_4$	322.43	107.48*	322.43
Caicium chloride	$CaCl_2$	110.99	55.50	110.99
	$CaCl_2.2H_2O$	147.02	73.51	147.02
Calcium hydroxide	$Ca(OH)_2$	74.09	37.05	74.09
Caustic potash—see Potassium hydroxide				
Caustic soda—see Sodium hydroxide				
Citric acid	$C(OH)(COOH)(CH_2COOH)_2$	210.14	70.05	210.14
Cobalt sulphate	$CoSO_4$	155.00	77.50	155.00
	$CoSO_4.7H_2O$	281.10	140.55	281.10
Copper sulphate	$CuSO_4.5H_2O$	249.68	124.84	249.68
p-Dimethylaminobenzaldehyde	$(CH_3)_2NC_6H_4CHO$	149.19	—	149.19
Disodium monophenyl phosphate	$C_6H_5Na_2PO_4.2H_2O$	254.09	84.70*	254.09
E.D.T.A. (Ethylenediaminetetra-acetic acid)	$(CH_2.N(CH_2COOH)_2)_2$	292.25	73.06	292.25

* Based on anion

COLORIMETRIC CHEMICAL ANALYTICAL METHODS

Name	Formula	Molecular Weight	Concentration g/litre	
			Normal	Molar
2-Ethylhexyl pyrophosphate	$((C_8H_{17})_2PO_3)_2O$	631.74	—	631.74
Hydrazine	$N_2H_4.H_2O$	50.06	25.03	50.06
Hydrobromic acid	HBr	80.92	80.92	80.92
Hydrochloric acid†	HCl	36.46	36.46	36.46
Hydrofluoric acid	HF	20.01	20.01	20.01
Iodine	I_2	253.81	126.90	253.81
Iodine chloride	ICl	162.36	—	162.36
Lead nitrate	$Pb(NO_3)_2$	331.20	165.60	331.20
Magnesium chloride	$MgCl_2.6H_2O$	203.31	101.66	203.31
Nickel diethyl phosphorodithioate	$((C_2H_5O)_2PS_2)_2Ni$	430.37	215.19	430.37
Nitric acid†	HNO_3	63.01	63.01	63.01
Perchloric acid	$HClO_4$	100.46	100.46	100.46
Batho Phenanthrolene	$C_{24}H_{16}N_2$	332.41	—	332.41
N-Phenylbenzohydroxamic acid	$C_6H_5.CO.N(OH)C_6H_5$	213.24	213.24	213.24
Potassium bromate	$KBrO_3$	167.01	167.01	167.01
Potassium chloride	KCl	74.56	74.56	74.56
Potassium dichromate	$K_2Cr_2O_7$	294.19	49.03**	294.19
Potassium hydrogen phthalate	$C_6H_4.COOH.COOK$	204.22	102.11*	204.22
Potassium hydroxide‡	KOH	56.11	56.11	56.11
Potassium iodate	KIO_3	214.00	214.00	214.00
Potassium iodide	KI	166.01	166.01	166.01
Potassium 2-naphthol-3:6-disulphonic acid	$C_{10}H_5OH(SO_3K)_2$	380.49	190.25	380.49
Potassium nitrate	KNO_3	101.11	101.11	101.11
Potassium permanganate	$KMnO_4$	158.04	—	158.04
acid solution			31.61**	
neutral or alkaline solution			52.67**	
Potassium sulphate	K_2SO_4	174.27	87.14	174.27
Slaked lime—see Calcium hydroxide				

* Based on anion ** Based on available oxygen

COLORIMETRIC CHEMICAL ANALYTICAL METHODS

Name	Formula	Molecular Weight	Concentration g/litre Normal	Concentration g/litre Molar
Sodium acetate	CH_3COONa	82.03	82.03	82.03
Sodium barbitone	$(C_2H_5)_2.C.CO.NH.C(ONa):N.CO$	206.18	206.18	206.18
Sodium carbonate	Na_2CO_3	105.99	53.00	105.99
Sodium chloride	$NaCl$	58.44	58.44	58.44
Sodium citrate	$Na_3C_6H_5O_7.2H_2O$	294.10	98.03	294.10
Sodium hydroxide‡	$NaOH$	40.00	40.00	40.00
Sodium 2-naphthol-3:6-disulphonic acid	$C_{10}H_5OH(SO_3Na)_2$	348.27	174.14	348.27
Sodium nitrate	$NaNO_3$	85.00	85.00	85.00
Sodium nitrite	$NaNO_2$	69.00	69.00	69.00
Sodium thiosulphate	$Na_2S_2O_3.5H_2O$	248.18	248.18**	248.18
Sulphuric acid†	H_2SO_4	98.08	49.04	98.08
Sulphurous acid	H_2SO_3	82.08	41.04	82.08
Tartaric acid	$(CHOH.COOH)_2$	150.09	75.05	150.09
Tin	Sn_4	474.76	118.69	474.76
Titanous sulphate	$Ti_2(SO_4)_3$	383.82	191.91**	383.82
Tributyl phosphate	$(C_4H_9)_3PO_4$	266.32	—	266.32

** Based on available oxygen

Ion-Exchange Resins

Ion-exchange materials have been employed for thousands of years. Aristotle, for example, records that filters of sand or soil were used for the purification of drinking water, although it was not until very much later in history that the chemical nature of the processes involved was recognised.

Early in this century the use of natural and synthetic zeolites for the softening of water became well established. The introduction of synthetic ion-exchange materials, based on insoluble phenol-formaldehyde resins, in the middle 1930's greatly extended the range of application of the ion-exchange technique, and this range was further extended by the replacement of phenol-formaldehyde by the more stable cross-linked polystyrene resins.

These ion-exchange materials consist of an insoluble matrix containing ionic groups incorporated in the lattice as integral parts of the structure of the polymeric matrix. These ionic groups are associated with an equivalent number of labile ions which are capable of exchanging with ions in the surrounding medium. Thus the materials used for water softening act by removing calcium and magnesium ions and replacing them with sodium ions. When all the available sodium ions have been replaced, the ion-exchange material can be regenerated by flushing with sodium chloride solution; whereupon the calcium and magnesium ions are flushed out as soluble chlorides and replaced by sodium.

Ion-exchange has, in recent years, been increasingly applied to the problem of ion separation in analytical chemistry to the extent where it rivals, or even exceeds, in importance the traditional separation techniques of precipitation and distillation. Ion-exchange resins can now be tailored for specific applications to the point where virtually any separation of even closely related ions becomes possible. This has inevitably resulted in a wide proliferation of ion-exchange materials with a baffling variety available to the occasional user. To help in the choice of the correct resin, and to give some guidance to the selection of an available replacement for an unavailable material recommended in separation procedures reported in the literature, the following notes and tables have been prepared. For a fuller discussion of the principles and applications of ion-exchange, reference should be made to one of the many excellent books now available on the subject.

Ion-exchange resins can be divided into two general classifications, cation exchangers and anion exchangers. A cation exchanger is a material having anionic groups incorporated in the matrix, the exchangeable cation being associated with these fixed anions. Conversely the anion exchangers have cationic groups permanently bonded in the matrix. These two general classifications may each be further sub-divided into strongly and weakly acidic cation exchangers, and strongly and weakly basic anion exchangers. It is on this classification of ionic strength that the table of ion-exchange resins has been based.

Cation exchangers have sulphonic acid (SO_3^-) or carboxylic acid (CO_2^-) groups which confer strongly and weakly acid properties respectively to the resin. Many of the older sulphonic ion-exchangers were based on phenol-formaldehyde resin and were thus bi-functional, as they also contained phenolic hydroxyl groups which were also ionised at high pH. These resins were, therefore, only usable at a pH below 8.5. Modern polystyrene-based resins are usable over a wide range of pH. The carboxylic ion-exchangers behave as weak acids and are capable of being buffered. They are thus especially suitable for cation exchange at controlled pH, but only have a useful capacity in neutral or alkaline solutions.

The cation exchange resins can be used either in the free acid, or hydrogen, form (H^+) in which absorbed cations will release an equivalent amount of hydrogen ions into solution, or in the salt, usually sodium, form (Na^+) in which sodium ions are released. In dilute solutions strongly acidic resins absorb multivalent ions in preference to univalent ions whereas in concentrated solutions this effect is reversed. This explains why in water softening the low concentrations of calcium and magnesium ions are strongly absorbed but are readily replaced by sodium from concentrated sodium chloride solution.

Anion exchange resins have amino or substituted amino groups incorporated in the resin. Resins containing amino and mono- and di-substituted amino groups are weakly basic ion-exchangers, whereas those containing quaternary ammonium groups are strongly basic. The weakly basic resins have a high affinity for hydroxyl ions and can only be used in neutral or acid solutions.

The ion-exchange resins listed in Table 1 are a selection of those commonly available. Resins having a similar structure will have similar but not identical properties. When a resin which is specified in the sample preparation procedures is not available, it may be replaced by one having similar characteristics, but the separation should be tested, as experimental conditions may have to be modified slightly.

The details given in the Tables have been abstracted from published sources including the literature from chemical supply companies. While every effort has been made to ensure the accuracy of the information in the Tables, the authors cannot accept responsibility for this accuracy, as in several instances the data from different suppliers of the same resin has been contradictory. The names of the manufacturers, or suppliers, of the various resins are listed in Table 2, which links trade names with manufacturers. Table 3 lists the addresses of the various manufacturers, arranged alphabetically.

TABLE I TYPES OF ION-EXCHANGE RESIN

Type	Strength	Active Group	Resin	Trade Name	Type No.	Remarks
Cation	Weak	Carboxylic acid	Acrylic	Amberlite	IRC–50	
					CG–50	Chromatographic grade of IRC-50
					IRC–84	
					XE 64	
					XE 89	
					CS–101	
				Duolite	C–100	
				Kastel	H–70	
				Permutit	226	
				Zeo-Karb	CN	
			Phenolic	Wolfatit	A–1	
			Polystyrene	Dowex	CP–300	Also sold as Chelex 100
				Wolfatit	CV	
			Unspecified	Imacti	25	Granular
				Ionac	265	Also referred to as Imac
					270	
				Lewatit	C	
					CNO	
					CNP	Macroporous
				Permutit	C	
				Relite	CC	
				Zeo-Karb	227	
	Medium	Mixed carboxylic/sulphonic acids	Acrylic	Allassion	CPM	
			Phenolic	Zeo-Karb	216	
			Unspecified	Dowex	CCR–1	Granular
				Imacti	C–19	Granular
					Dusarit-S	Granular
						Granular
				Konvertat	CMS	
				Lewatit	CNS	
					S–53	
				Permutit	CM–2	
				Relite		

Table I continued

Type	Strength	Active Group	Resin	Trade Name	Type No.	Remarks
Cation	Strong	Phosphonic acid	Polystyrene	Duolite	C-63	
					ES-63	
					X-219	
		Phosphonous acid	Polystyrene	Nalcite	C-62	
		Phosphoric acid	Polystyrene	Duolite	C-65	
		Sulphonic acid	Coal	Duolite	C	Granular
				Allassion	CP	
				Dycatan	C-150	
				Ionac	C-40N	Granular
				Permutit	C-40P	
					H-53	
			Phenolic	Soucol	Na	
				Zeo-Karb	C-131	Argentinian
				Acuolite	IR-100	
				Amberlite	IR-105	Methylene sulphonic acid
					IR-105G	Identical to Duolite C-3
					40	
				Bio-Rex	BK	
				Deionite	30	
				Diaion	C-3	Methylene sulphonic acid
				Dowex	C-10	Methylene sulphonic acid granular
				Duolite	C-11	
					C-200	
				Imacti	KSN	
				Ionac	PN	
				Lewatit		
				Resex	F	
				Wolfatit	K	
					KS	Granular
					P	Granular
					X	Catalyst

Table I continued

Type	Strength	Active Group	Resin	Trade Name	Type No.	Remarks
Cation	Strong	Sulphonic acid	Phenolic	Zeo-Karb	215	Methylene sulphonic acid granular
					315	Methylene sulphonic acid
			Polystyrene	Allassion	CS	Identical to Duolite C20
					CS–AD	Identical to Duolite C20L
				Amberlite	IR–112	
					IR–120	Equivalent to Dowex 50 and Zerolit 225
					CG–120	Chromatographic grade of IR–120
					IR–122	
					IR–124	
					XE–100	
				Amberlyst	15	
				Dowex	50	Identical to Nalcite HCR
					50–W	Colourless, heavy-duty form of 50, identical to Nalcite HDR
					200	
				Duolite	C–20	Identical to Allassion CS
					C–20L	Larger bead size of C–20, identical to Allassion CS–AD
					C–25	Macroporous
					C–27	Porous
					ES–26	Macroporous
				Dusarit	S	
				Imacti	C–8P	Porous
					C–12	
					C–16P	Porous
					C–22	
				Ionac	C–240	
					C–242	
					C–249	
				Kastel	C–300	Identical to Montecatini C–300
					C–300 AGR	Heavy duty form of C–300
				Katex	KP–1	
				K U	2	
				Lewatit	SP–100	Macroporous

Table I continued

Type	Strength	Active Group	Resin	Trade Name	Type No.	Remarks
Cation	Strong	Sulphonic acid	Polystyrene	Montecatini	C–300	Identical to Kastel C–300
					Cation G–300	
					Resina	
					Cationica	
				Nalcite	HCR	Identical to Dowex 50
					HDR	Identical to Dowex 50W
					HGR	High temperature form of HCR
				Permutit	C–50D	
					Q	
					RS	
				Resex	P	
				Wolfatit	KPS–200	
				Zeo-Karb	225	
					225 X4	4% Di-vinyl benzene (DVB)
					225 NC	
					325	
					425	
				Zerolit	Calcium Resin	8% DVB
					225	Equivalent to Amberlite IR–120 and Dowex 50
			Unspecified	Amberlite	IR–200	Macroporous
					IR–200C	Macroporous
				Diaion	SK–1B	
					SK–12	
					SK–110	
				Lewatit	S–100	
					SP–115	
					SP–120	
				Relite	CF	
					CFS	Porous
					CFZ	Granular Macroporous
				Varion	Redex CF	Macroporous catalyst
					KB	Large bead
					KS	Small bead

Table I continued

Type	Strength	Active Group	Resin	Trade Name	Type No.	Remarks
Cation	Unspecified	Unspecified	Phenolic	Diaion	K	
			Unspecified	Espatite	1	
					TM	
				K U	1	Precipitated sodium alumino silicate gel
				Decalso	F	Precipitated sodium alumino silicate gel
					Y	
Anion	Weak	Amine	Polystyrene	Imacti	A-19	
		Polyamine	Acrylic	Bio-Rex	5	
				Imacti	A-17	
				Kastel	A-100	
			Phenolic	Amberlite	IR-4B	
				Deacidite	E	
				Permutit	A-230A	
			Polystyrene	Amberlite	CG-45	Chromatographic grade of IR-45
					IR-45	Matrix chloromethylated polystyrene and DVB
				Deacidite	J	
					M	
					MIP	
				Dowex	3	Identical to Nalcite WBR
				Nalcite	WBR	Identical to Dowex 3, co-polymer with DVB
				Permutit	A-240A	
				Polyaminostyrene		Norwegian
		Secondary amine	Phenolic	Dulite	A-2	Granular similar to A5
					A-5	Granular similar to A2
					A-7	Granular and porous
		Mixed secondary and tertiary amines	Unspecified	Permutit	CCG	

Table I continued

Type	Strength	Active Group	Resin	Trade Name	Type No.	Remarks
Anion	Weak	Tertiary amines	Acrylic	Amberlite	IR-68	Porous
			Hydrocarbon	Duolite	A-114	Similar to A-4 but with difference matrix
			Phenolic	Duolite	A-4	
					A-6	Granular
			Polystyrene	Amberlyst	A-21	
				Deacidite	G	Ethyl substituents on amine group
					H	Methyl substituents on amine group
					HIP	Iso-porous version of H
				Duolite	A-14	Similar to A-4 but with different matrix
			Unspecified	Permutit	W	
				Amberlite	IR-93	Macroporous, methyl amino-substituents
				Lewatit	MP-60	Macroporous, methyl amino-substituents
			Polystyrene	Wolfatit	AK-40	Porous
					EE-2	Granular, porous
		Unspecified			M	
					MD	
					N	Granular
					Y-13	Granular
					44	Identical to Nalcite WGR
				Dowex	ES-15	Granular
				Duolite	A-20	
				Imacti	A-27	
				Ionac	A-260	Granular
					A-310	
					A-315	
				Kaken Kagaku	KK-12	Japanese
				Lewatit	M	
					MP-62	Macroporous
				Merck	II	
				Montecatini	A-20	

Table I continued

Type	Strength	Active Group	Resin	Trade Name	Type No.	Remarks
Anion	Weak	Unspecified	Unspecified	Nalcite	WGR	
				Permutit	E	
				Relite	MS-170	Porous
	Medium	Mixed tertiary and quaternary amine	Polystyrene	Imacti	A-13	
				Wolfatit	L-150	
					L-165	
			Unspecified	Allassion	AWB-3	Identical to Duolite A-30B
					AW-2	Granular
					A3-03	Isoporous
				Duolite	A-4F	
					A-30	Similar to A-41 and A-43
					A-30B	Identical to Allassion AWB-3
					A-41	Similar to A-30 and A-43
					A-43	Similar to A-30 and A-41
					A-53	
					ES-57	Porous
				Ionac	A-300	
					A-302	Granular
				Lewatit	M-1	
					M-1H	
				Permutit	A	
					F	
	Strong	Quaternary amine Type I alkyl	Polystyrene	Resanex	AR-10	Identical to Duolite A-101D
				Allassion	IR-400	3–5% DVB
				Amberlite	IR-400C	
					IR-401	High porosity
					IR-401S	Porous
					IR-402	Intermediate porosity
					IR-900	Macroporous
					IR-904	Macroreticular
				Amberlyst	A-26	Macroreticular
					A-27	
				Anex	AP1	Czechoslovakian

Table I continued

Type	Strength	Active Group	Resin	Trade Name	Type No.	Remarks
Anion	Strong	Quaternary Amine Type I alkyl	Polystyrene	Deacidite	FF	
					FF-510	
					FF-530	
					FFIP	Isoporous
					FFDVP-	
					SRA 133	Special form of FFIP
					FXIP	Special resin for use in mixed beds
					K	
					KMP	Macroporous
				Diaion	SA-10A	
					SA-11A	
				Dowex	1	Identical to Nalcite SBR
					11	
					21K	Porous, identical to Nalcite SBR-P
				Duolite	A-42	Based on trimethylbenzyl ammonium chloride group
					A-101	
					A-101D	Macroporous, identical to Allassion AR-10
				Imacti	ES-111	
					S5-40P	
					S5-50P	
				Ionac	A-540	
					A-580	
					A-590	
					A-935	Porous
					A-936	
				Kastel	NA-35	Very pure OH form of A-935
					A-500	Porous
					A-500P	
				Lewatit	M-500	Macroporous
					MR-500	Identical to Dowex 1
				Nalcite	SBR	Porous, identical to Dowex 21K
					SBR-P	
					SBR-M	
				Permutit	S-1	

Table I continued

Type	Strength	Active Group	Resin	Trade Name	Type No.	Remarks
Anion	Strong	Quaternary amine Type I alkyl	Polystyrene	Relite	3A	Porous
					3A2	Macroporous
					3AZ	
				Resanex	HP	
				Varion	AT	
				Wolfatit	ES	Porous
					SBW	Porous
		Quaternary amine Type II alkyl	Polystyrene	Allassion	AR–20	Identical to Duolite A–102D
				Amberlite	IR–410	
					IR–910	Macroporous
					IR–911	Macroporous
				Amberlyst	A–29	
				Deacidite	N	Isoporous
					NIP	
					NX	Isoporous, intermediate in strength between FFIP and NIP
					PIP	
				Diaion	SA–20A	
					SA–21A	Identical to Nalcite SAR
				Dowex	2	Based on dimethylhydroxy-ethylbenzyl ammonium chloride
				Duolite	A–40	
					A–102	
					A–102D	Porous, identical to Allassion AR–20
				Imacti	S5–52	Porous
				Ionac	A–550	Porous
				Kastel	A–300	
					A–300P	Porous
				Lewatit	M–600	
					MP–600	Macroporous
				Nalcite	SAR	Identical to Dowex 2
				Permutit	S–2	
				Relite	2A	
					2AS	Porous
				Varion	AD	
				Wolfatit	SBK	

Table I continued

Type	Strength	Active Group	Resin	Trade Name	Type No.	Remarks
Anion	Strong	Quaternary amine Type I or II unspecified	Polystyrene	Imacti	A-21	Porous
					S-4	
				Kastel	A-501D	Porous
					A-510	Porous
				Lewatit	MN	
				OAL		Czechoslovakian
				PEK		Russian
				Permutit	SK	Contains pyridinium groups
					SKB	Contains pyridinium groups
					SBT	Porous
				Wolfatit	SBU	
			Unspecified	Lewatit	M-11	
				Permutit	A-300D	
					ES	
	Unspecified	Ammonia epichlorhydrin	Unspecified	Dowex	S	
		Unspecified	Unspecified	Allassion	4	Porous
					ADM	
				AN	AS	Russian
				Diaion	2F	
				EDE	A	Russian
				M	10	Russian
				TM		Russian
Miscellaneous		Nil	Polystyrene	Amberlite	XAD-2	Used to absorb water soluble organic compounds

TABLE II
Manufacturers of Ion-Exchange Resins

Resin	Manufacturer
Acuolite	Kelly
Allassion	Dia Prosim
Amberlite	Rohm & Haas
Amberlyst	Rohm & Haas
Anex	Czechoslovakian
Bio-Rex	Bio-Rad Labs
Deacidite	Permutit
Decalso	Permutit
Deionite	Hodogaya Kagaku
Diaion	Mitsubishi
Dowex	Dow Chemical Co
Duolite	Diamond Alkali Co
Dusarit	Industrial Maatschappi
Dyeatan	Douglas Holt Co
Espatite	Russian
Imac or Imacti	Industrial Maatschappi
Ionac	American Zeolite Corpn
Kaken Kagaku	Japanese
Kastel	Montecatini
Katex	Czechoslovakian
Konvertat	Hoganas Billesholms
Lewatit	Farbenfabriken Bayer
Merck	E. Merck
Montecatini	Montecatini
Nalcite	National Aluminium Co
Permutit	Permutit
Polyaminostyrene	Norsk Elektrisk
Relite	Resin D'ion
Resanex	Joseph Crossfield
Resex	Joseph Crossfield
Soucol	Joseph Crossfield
Varion	Chemolimpex
Wolfatit	Veb. Farkenfabrik Wolfen
Zeo-Karb	Permutit
Zerolit	Permutit

Note: In some catalogues, Zeo-Karb and Deacidite are both referred to under the name Zerolit.

TABLE III

Manufacturers' fuller details

American Zeolite Corporation

Bio-Rad Laboratories, Richmond, California, USA
Chemolimpex, P.O. Box 121, Budapest 5, Hungary
Joseph Crossfield & Sons Ltd, Warrington, Lancs.
Diamond Alkali Co, (Western Division), P.O. Box 829, Redwood City, California, USA
Dia Prosim, 107 Rue Edith Cavell, Vitry s-Seine, France
Dow Chemical Co, Midland, Michigan, USA
Farbenfabriken Bayer A.G., Leverkusen, Germany
Hodogaya Kagaku Co, Japan
Hoganas Billesholms, Sweden
Douglas Holt Co, England
Industrieele Maatschappij Activit N.V., P.O. Box 240C, Amsterdam, Holland
Louis Kelly & Co, Argentina
E. Merck, Darmstadt, West Germany
Mitsubishi Kasei Kogyo Co, Japan, (U.K. agent, Mitsui, 83 Cannon St., London)
Montecatini S.G., Milan, Italy
National Aluminium Co, Chicago, Illinois, USA
Permutit A.G., Berlin–Schmargendorf, Auguste Viktoria St, Germany
The Permutit Co, 50 West 44th St., New York 36, New York, USA
Permutit Co., Ltd, Gunnersbury Avenue, London W4
(Permutit) Philips et Pain–Vermorel, 31 Rue de la Vanne, Montrange (Seine), France
Resin D'ion, Italy
Rohm & Haas Co., Philadelphia, USA
Veb Farbenfabrik Wolfen, Kr. Bitterfeld, Germany

Buffer Solutions

It is possible to prepare a solution which contains a definite proportion of hydrogen ions irrespective of dilution, that is its pH does not vary with dilution. This solution is called a buffer solution. Such solutions are of great importance in analyses in which it is essential that an extraction, precipitation or other separation procedure be carried out at a controlled pH. The following Tables give the composition of the common buffer solutions which are referred to in the text. If buffers are referred to which are not included in these Tables, this is because the abstract of the published method did not contain details of the buffer composition and the editors have not been able to find details of the buffer composition in other reference works. In such cases it is recommended that the Universal buffer solution be used if possible.

Acetate buffer (1) pH range 0.65-5.20 at 18°C

50 ml of normal sodium acetate plus X ml of normal hydrochloric acid, diluted to 250 ml with distilled water.

X ml HCl	pH	X ml HCl	pH
100	0.65	47.5	3.29
90	0.75	45	3.61
80	0.91	42.5	3.79
70	1.09	40	3.95
65	1.24	35	4.19
60	1.42	30	4.39
55	1.71	25	4.58
52.5	1.99	20	4.76
50	2.64	15	4.92
		10	5.20

Acetate buffer (2) pH range 3.42-5.89 at 18°C

Mix $0.2N$ acetic acid and $0.2N$ sodium acetate in the following proportions.

pH	Acetic acid	Sodium Acetate	pH	Acetic acid	Sodium Acetate
3.42	9.5	0.5	4.80	4.0	6.0
3.72	9.0	1.0	4.99	3.0	7.0
4.05	8.0	2.0	5.23	2.0	8.0
4.27	7.0	3.0	5.37	1.5	8.5
4.45	6.0	4.0	5.57	1.0	9.0
4.63	5.0	5.0	5.89	0.5	9.5

Borate buffer pH range 7.8-10.0

Take 50 ml of a solution containing 12.369 g of boric acid (H_3BO_3) and 14.911 g of potassium chloride (KCl) per litre, add the following amounts of $0.2M$ caustic soda (NaOH) and then dilute to 200 ml with distilled water.

pH	ml NaOH	pH	ml NaOH
7.8	2.65	9.0	21.40
8.0	4.00	9.2	26.70
8.2	5.90	9.4	32.00
8.4	8.55	9.6	36.85
8.6	12.00	9.8	40.80
8.8	16.40	10.0	43.90

Phosphate buffer (1) pH range 5.8-8.0 at 20°C

Take 50 ml of $0.2M$ of potassium dihydrogen phosphate (KH_2PO_4), add the following amounts of $0.2M$ caustic soda (NaOH) and dilute to 200 ml with distilled water.

pH	ml NaOH	pH	ml NaOH
5.8	3.66	7.0	29.54
6.0	5.64	7.2	34.90
6.2	8.55	7.4	39.34
6.4	12.60	7.6	42.74
6.6	17.74	7.8	45.17
6.8	23.60	8.0	46.85

Phosphate buffer (2) pH range 5.29-8.04 at 18°C

Mix $M/15$ disodium hydrogen phosphate (Na_2HPO_4) and $M/15$ potassium dihydrogen phosphate (KH_2PO_4) in the following proportions.

pH	Na_2HPO_4	KH_2PO_4	pH	Na_2HPO_4	KH_2PO_4
5.29	0.25	9.75	6.81	5.0	5.0
5.59	0.5	9.5	6.98	6.0	4.0
5.91	1.0	9.0	7.17	7.0	3.0
6.24	2.0	8.0	7.38	8.0	2.0
6.47	3.0	7.0	7.73	9.0	1.0
6.64	4.0	6.0	8.04	9.5	0.5

Universal buffer pH range 2.6-12.0 at 18°C

Dissolve 6.008 g of citric acid ($C(OH)(COOH)(CH_2COOH)_2.H_2O$), 3.893 g of potassium dihydrogen phosphate (KH_2PO_4), 1.769 g of boric acid (HBO_3) and 5.266 g of diethyl barbituric acid (barbitone $(C_2H_5)_2.C.CO.NH.CO.NH.CO$) in distilled water and make the volume up to one litre. All the chemicals should be of analytical reagent quality.

To 100 ml of this solution add the required volume of $0.2N$ caustic soda (NaOH) solution to give the pH as indicated below:

pH	ml NaOH	pH	ml NaOH	pH	ml NaOH
2.6	2.0	5.8	36.5	9.0	72.7
2.8	4.3	6.0	38.9	9.2	74.0
3.0	6.4	6.2	41.2	9.4	75.9
3.2	8.3	6.4	43.5	9.6	77.6
3.4	10.1	6.6	46.0	9.8	79.3
3.6	11.8	6.8	48.3	10.0	80.8
3.8	13.7	7.0	50.6	10.2	82.0
4.0	15.5	7.2	52.9	10.4	82.9
4.2	17.6	7.4	55.8	10.6	83.9
4.4	19.9	7.6	58.6	10.8	84.9
4.6	22.4	7.8	61.7	11.0	86.0
4.8	24.8	8.0	63.7	11.2	87.7
5.0	27.1	8.2	65.6	11.4	89.7
5.2	29.5	8.4	67.5	11.6	92.0
5.4	31.8	8.6	69.3	11.8	95.0
5.6	34.2	8.8	71.0	12.0	99.6

Lovibond Colorimetric Apparatus used in the tests described in this book

All the colorimetric methods described in the various sections of this book finish by producing a coloured solution, the colour of which is specific for that particular radical and proportional to the concentration present. The Tintometer Limited manufactures carefully checked permanent glass colour standards which accurately represent the exact colours for the various concentrations within the range mentioned in the relevant tests, and these colour standards are fitted into one or other of the instruments described in the following pages.

The "Lovibond 2000" Comparator, with its various accessory parts, constitutes the most complete and convenient equipment yet produced for carrying out visual colorimetric chemical analytical tests.

The comparator illustrated below is a black moulded case of modern design into which are fitted a glass cell to hold the sample liquid, and another similar cell for a "blank" solution to provide compensation for inherent colour in sample or reagent. The disc appropriate to the test is slotted into the comparator, and by means of the spring ratchet, each glass colour standard registers exactly in the field of view in turn, while the numerical value of the appropriate colour standard is located at the indicator aperture.

TK 100 "Lovibond 2000" Comparator.

COLORIMETRIC CHEMICAL ANALYTICAL METHODS

The interchangeable discs to fit the Lovibond Comparator contain any number of Lovibond glass permanent colour standards from 1 to 110 according to the test, and there is a selection from many hundreds of different discs to cover a wide range of tests, as described in this book. Special discs can be made to order to contain colours to match customers' own samples. In some cases a "dulling" screen is needed in front of the sample, and in such instances it is built into the disc as an integral part.

Lovibond disc

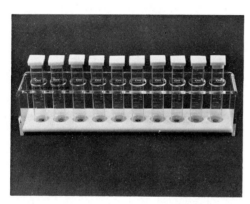

For the majority of the tests in this book, the optical depth of liquid through which one looks has been standardised at 13.5 mm, and moulded glass cells calibrated at 1 ml intervals from 2 to 12 ml are available. This calibration will in many instances obviate the need for pipettes to measure sample and reagents.

DB 424 Moulded 13.5 mm cells, with cell stand (optional).

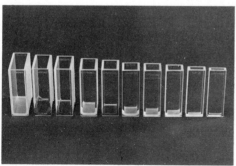

Where some other optical depth is needed, this is clearly stated in the instructions, and there is a range of optically correct fused cells for use in the Lovibond Comparator from 1 mm up to 40.0 mm.

Lovibond Comparator cells

COLORIMETRIC CHEMICAL ANALYTICAL METHODS

The accuracy of visual comparison of colours is greatly dependent upon adequate and suitable lighting conditions. When using natural daylight, a good north daylight is necessary (south in the Southern Hemisphere) to avoid direct sunlight. When this is not available, the **Lovibond White Light Cabinet** should be used.

This same cabinet is used with the whole range of attachments for different tests as described below.

DB 416 White Light Cabinet

Certain tests in the Lovibond range require an optical depth between 13.5 mm and 40 mm, and this is always stated in the instructions. To use such cells, adjust the integral large cell holder to suit the size of cell to be used.

Some of the Lovibond tests, where pale coloured solutions are involved, require the depth of a Nessler cylinder, 113 mm or 250 mm optical path, in order to develop enough colour for easy distinguishing. Where required, this is always stated in the instructions. For such tests the **Nessler Attachment** is fitted to the basic Comparator unit.

DB 412 Nessler attachment

Some of the tests require the sample to be looked at by reflected light—for example toxic gas tests where the colour is developed on a test paper, or for grading the colour of butter or sand. In this case, the Lovibond Comparator is used in conjunction with a **Surface Viewing Stand.**

DB414 Surface viewing stand

For the range of milk tests, such as the Resazurin and the Phosphatase tests, milk in a test tube has to be viewed obliquely by reflected light. For such tests the Lovibond Comparator is used in conjunction with the **Milk Viewing Stand.**

DB 415 Milk Viewing Stand

For all different combinations above, the Lovibond White Light Cabinet is strongly recommended as a valuable assistance.

Some tests in this book relate to the use of the B.D.H. Lovibond Nessleriser MK III and this unit is shown below.

It is a complete instrument and not an attachment for the Lovibond Comparator. It has its own integrated Comparator and prism unit, also a pair of lamps giving a good artificial daylight source. Removal of the blue filter enables the use of normal daylight when electrical power is not available. Two 50 ml. Nessler tubes are supplies as standard.

COLORIMETRIC CHEMICAL ANALYTICAL METHODS

Other Lovibond equipment referred to in this book

The Lovibond Tintometer is a general colorimeter containing a range of Lovibond red, yellow, and blue glass filters which can be combined to match and measure any colour, and reproduce an example of that colour at will. This instrument is universal, but each test requires calibration, whereas this work has already been done for the Comparator. The Tintometer is used for oil rancidity and certain other tests given in this book.

AF 900 Lovibond Tintometer Model E.

Lovibond Grading Strips are used in certain tests such as the pH of the surface of paper, and the Resazurin meat test, where the Comparator is not appropriate.

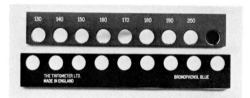

AF 263 Grading Strips

The Seta-Lovibond Colour Comparator has been specially produced to conform to the requirements of the American Society for Testing Materials test D 1500 for Lubricating Oils, and is used for this test only.

Details on page 593

AF 760 Seta-Lovibond Colour Comparator

The Lovibond 3-aperture Comparator, used for colour grading tests to conform to certain official specifications such as, for example, the Gardner Scale for oils and varnishes, and the F.A.C. scale for fats. This enables the colour of the sample to be viewed between colour standards darker and lighter, to make it easier to decide on the colour grade.

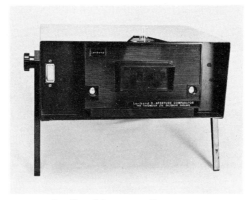

Lovibond 3-aperture Comparator

The Lovibond "Water Baby" Comparator is used for certain tests where only a limited range of colour standards is required (for example pH and chlorine standards for small swimming pools). The standards are not interchangeable for other tests, and this comparator can only be used for the one test for which it is made.

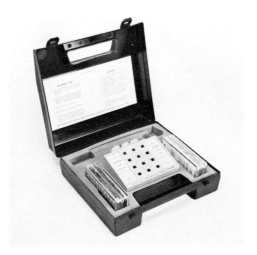

Lovibond "Water Baby" Comparator

The Colorimetric Determination
of
pH

The Determination of pH Values (1)

Introduction

In nearly all branches of research and industrial work, from sugar manufacture to sewage disposal, from brewing to boiler-water control, it has long been realised that the control of the acidity and alkalinity of the various solutions used is of paramount importance, since this factor frequently has a marked effect on yield, quality, stability, etc.

Explanation of pH*

The essential point to be realised is the difference between **total acidity** and **active acidity (or alkalinity)**. The total acidity is measured by the usual titration method, and furnishes the answer as to **how much** of the given acid there is in a given volume of the solution: but this is no guide as to the **intensity** of the acidity, which, of course, is the controlling factor in many processes. For instance, equivalent quantities of hydrochloric acid and boric acid in solution show exactly the same acidity by titration, but there is clearly a large difference in the acidic properties of these solutions: this difference one may call the **intensity** of acidity. In the past, failure to appreciate this distinction, and attempts made to control processes by rough test-papers or by titration, have resulted in failures and wastage in chemicals and time.

Of recent years this matter has been fully investigated and shown to be dependent on the Hydrogen Ion Concentration or pH. According to the theory of electrolytic dissociation, all liquids of which water is a constituent contain free, positively charged, hydrogen ions, and negatively charged hydroxyl ions. When the numbers of these two ions present in a liquid are equal, the liquid is said to be neutral. Thus, a litre of pure freshly-distilled water contains one ten-millionth of a gram (10^{-7}) of ionised hydrogen and an equivalent amount of hydroxyl ion, and the water is neutral. Addition of an acid to water increases the concentration of the hydrogen ions and decreases the concentration of the hydroxyl ions: the water becomes acid in reaction, and, according to the increase in the hydrogen ion concentration, so the **active acidity** increases. The converse effect is obtained by the addition of an alkali to water.

The product of the concentration of these two ions expressed in grams is **constant at a given temperature,** so that if one is known the other is easily calculated, and it is usual to speak only of the Hydrogen Ion Concentration. This concentration is expressed as grams of active or ionised hydrogen per litre of the liquid. Decinormal hydrochloric acid is almost completely ionised, and 1 litre contains approximately 0.1 gram of ionic hydrogen (H^+), and the H^+ concentration is 0.1 or 10^{-1}. At the other extreme, decinormal sodium hydroxide contains 0.000,000,000,000,086 gram of ionic hydrogen per litre, and its H^+ concentration is expressed at 8.6×10^{-14} or $10^{-13.07}$. This method of expression is clumsy, so the term "pH" has been adopted, the letter "p" standing for the German "potenz" or power (mathematical). **The pH value is the logarithm to the base 10 of the reciprocal of the hydrogen ion concentration,** i.e., the index of the hydrogen ion concentration with the negative sign changed to positive. Thus in the examples given above, the pH of decinormal hydrochloric acid is 1.0, of decinormal sodium hydroxide is 13.07, and of pure freshly-distilled water 7.0.

Thus simply, a solution having a pH value of 7 is neutral (neither acid nor alkaline), and as this value decreases below 7 the hydrogen ion concentration increases, and with it the active acidity. Above 7, the reverse is the case, the hydroxyl ion concentration increases, and with it the alkalinity. It must be remembered that this is a logarithmic scale, so that pH 5 represents ten times more ionic hydrogen than pH 6, and pH 4 one hundred times more ionic hydrogen than pH 6, and so on.

---

*British Standard 1647 : 1950 "pH Scale" adopted an operational standard, and defined a primary standard in terms of chemical constitution. It then defined the difference in pH between any two solutions in terms of electromotive force. The above paragraph, however, may be taken as an adequate explanation of pH for all ordinary purposes.

Indicators

In measuring pH values, advantage is taken of the fact that certain dyes, known as indicators, change their colour in a definite and reproducible manner and degree, according to the pH value of the solution with which they are mixed. The simplest and best known of these is litmus, which is red in an acid solution and blue in an alkaline solution. As, however, it has a very wide range between its extreme acid colour (pH 4.6) and its extreme alkaline colour (pH 8.4) and as the actual colours vary with different qualities of litmus, it is unsuitable for the determination of pH values. To ascertain, in addition to the fact that a solution is acid, precisely what degree of acidity it possesses, recourse is had to indicators which show a complete colour change through a short pH range. Thus Bromo Thymol Blue shows a full yellow colour in acid at pH 6.0 and as the pH value increases the colour changes gradually to green at neutral and then to blue as the solution becomes alkaline, so that at pH 7.6 it is deep blue. Other indicators show their full colour change all on one side or the other of neutral; for example, Cresol Red is yellow at pH 7.2 (slightly alkaline) and changes through orange to purplish red at pH 8.8 (more highly alkaline). These colour changes are constant for given concentrations of each indicator. It is only necessary therefore to have sufficient indicators available, spanning the pH scale, and to find some means of recording the colour changes, to be able to ascertain the pH value of any aqueous liquid.

Principle of the method

The colour changes of indicators are most easily identified by visual comparison with Lovibond permanent glass colour standards. For routine use these standards are conveniently mounted in a moulded plastic disc for use with the Lovibond Comparator. Each disc contains glass standards matching the colour produced by a named indicator when added to liquid of pH corresponding to the values stated on the disc. Thus, to revert to the example of Bromo Thymol Blue, if a stated quantity of this indicator solution be added to a liquid whose pH value is between 6.0 and 7.6, the colour produced will be matched by one of the standards in the Bromo Thymol Blue disc. The numerical value corresponding to the matching standard is the pH of the solution. If the pH value were below pH 6.0 or above pH 7.6, this disc would not be suitable for its determination. The colour of an indicator solution changes only within the limits of its stated pH range. Beyond these limits further change of pH has no effect on the colour of the indicator solution. In such a case, another indicator would be employed whose range was more suitable. Some indicators have a double range of usefulness. For example, Cresol Red changes from red to yellow as the pH value drops from 8.8 to 7.2, remains at this colour until the pH value has dropped as low as pH 1.8, and then commences to reverse the process as the pH value drops still lower in the acid direction, regaining a red colour at pH 0.2. There are thus two discs for this indicator. A sufficient number of indicators and discs are available to cover the whole range from pH 0.2 to 14.0 in small steps, with an overlap of readings between each adjacent pair, so that any reading at the end-point of an indicator should be confirmed by repeating the test with the next indicator. If one is dealing with a solution whose pH value is totally unknown, recourse should be had to the Universal Indicator, which has the enormous range of pH 4.0—11.0, changing through all the colours of the spectrum in correct order. This, of course, gives a very approximate answer, which is only intended to serve as a guide to which indicator is appropriate for the more accurate determination of the pH value. Alternatively, "B.D.H. Full Range" Indicator covering the pH range 1.0—13.0 is available.

COLORIMETRIC CHEMICAL ANALYTICAL METHODS   pH Values (1)

Indicators and their appropriate Standard Lovibond Comparator Discs

Discs, containing colours standardised on the following indicators, are available showing colour changes in steps of 0.2 pH except where otherwise stated:—

pH range	Indicator	Standard Disc	Notes
0.2 — 1.8	Cresol Red	2/1Y	
1.0 —13.0	B.D.H. Full Range	2/1ZE	Steps 1, 3, 5, 6, 7, 8, 9, 11 & 13
1.0 — 2.6	m-Cresol Purple	2/1W	
1.2 — 2.8	Tropaeolin 00	2/1 I	
1.2 — 2.8	Thymol Blue†	2/1A	
2.8 — 4.4	Bromo Phenol Blue	2/1B	
2.8 — 4.4	2,4-Dinitrophenol α	KAR	pH of glue and gelatine monochromatic indicator
2.8 — 4.4	Methyl Orange	2/1ZG	
3.0 — 4.6	B.D.H. 3046	2/1V	
3.6 — 5.2	Bromo Cresol Green†	2/1C	
4.0 — 5.6	2,5-Dinitrophenol γ	KAS	Monochromatic indicator pH of glue and gelatine
4.0 — 8.0	B.D.H. 4080	2/1CC	Steps of 0.5
4.0 — 8.0	Palin Soil†	2/1ZD	Steps of 0.5
4.0 — 9.0	N.B.A. Mixed Indicator	2/1ZH	"Neutrality of xyloles and benzoles"
4.0 —11.0	Universal	2/1P	Steps of 1.0
4.4 — 6.0	B.D.H. 4460	2/1D	
4.4 — 6.0	Methyl Red	2/1E	
4.8 — 6.4	Chlorophenol Red	2/1F	
4.8 — 6.4	Ethyl Red	2/1AA	for pH of formaldehyde solutions in the plastics industry
5.2 — 6.8	Bromo Cresol Purple†	2/1G	
5.4 — 7.0	p-Nitrophenol	KAX	Monochromatic indicator pH of glue and gelatine
6.0 — 7.6	Bromo Thymol Blue†	2/1H	
6.8 — 8.4	Phenol Red†	2/1J	
6.8 — 8.4	m-Nitrophenol	KAZ	Monochromatic indicator
7.0 — 8.6	Diphenol Purple†	2/1 O	
7.2 — 8.8	Cresol Red	2/1K	
7.6 — 9.2	m-Cresol Purple	2/1Z	
8.0 — 9.6	Thymol Blue†	2/1L	
8.6 —10.0	B.D.H. 8610	2/1U	8 standards only
8.6 —10.2	Phenolphthalein	2/1ZF	
9.0 —11.0	B.D.H. 9011	2/1M	5 standards only. Steps of 0.5
10.0 —14.0	B.D.H. 1014	2/1BB	In steps of 0.5

†These indicators are also available in tablet form.

Indicators supplied by The Tintometer Ltd. are specially prepared to conform to the colours on the Lovibond discs.

Technique

Fill both Comparator cells to the 10 ml mark with the solution to be tested, and to the right-hand tube, only, add the appropriate quantity of indicator with the pipette provided. Use 0.5 ml of indicator, except in the cases of Universal Indicator, when only 0.1 ml is used (0.2 ml if the solution proves to be acid), and of B.D.H. Full Range Indicator when 0.2 ml is used. Do not immerse the tip of the pipette beneath the surface of the liquid being tested. Carefully mix indicator and liquid, by stirring with a clean glass rod, or by pouring from one cell to another. Insert the appropriate disc in the Lovibond Comparator, revolve until the nearest colour match is obtained, and read the pH value in the indicator recess at the bottom right-hand corner of the comparator. A standard source of white light should be used as the illuminant for colour matching (Note 4).

In tests on water which has been chlorinated, residual chlorine may bleach some indicators. This effect can be avoided, without interfering with the *p*H measurement, by adding a very small crystal of "hypo" (Sodium thiosulphate, $Na_2S_2O_3.5H_2O$) to the water before adding the indicator. When tablets are used, these already contain hypo.

If there is any doubt as to which is the appropriate indicator to use, carry out the test first with the Universal Indicator to obtain an approximate reading, then select the indicator with the mid-point nearest this approximate determination.

Steps on most of the discs are of 0.2 *p*H divisions, and values can usually be interpolated between these. Carefully wash cells and pipette after use. The reason for using a "blank" in the left-hand cell is to give compensation for any inherent colour or turbidity in the liquid under test. If, for instance, a brownish liquid were being tested, the colour resulting from the mixture of liquid and indicator would be influenced by the inherent colour of the sample, and so would the colour as seen through the glass in the left-hand aperture, and the departure from normal would thus be compensated.

Some miscellaneous applications

(a) *Adjustment of the pH value of solutions*

The converse of determination of *p*H, i.e. the adjustment of a solution to any required *p*H value, is an easy matter using the Lovibond Comparator. For this work, a Comparator test tube of the same diameter as the standard cell, but of greater length, can be supplied.

Place in the tube 10 ml of the solution which is to be adjusted, add 0.5 ml of the indicator which covers the required *p*H value, set the disc at the required reading and titrate from a burette with standard acid or alkali, as the case may be, until the appropriate colour is obtained.

The solution must be well mixed between additions from the burette. Note the amount required, and carry out the adjustment on a suitable volume of the solution without adding any indicator.

(b) *Determination of pH of Neutral Solutions—pH 7.0*[1]

One of the most difficult *p*H determination is that of distilled water, or water approaching that state of purity. The reason for this difficulty is the lack of "buffering" action in such water. In most liquids one has to test there are traces of a number of substances or impurities, and the chances are very high that there are present some salts of a weak acid or weak base. The salts act as "buffers"—that is, they strongly resist any change in the *p*H of the solution, and consequently slow down the alteration in *p*H value which occurs when an acid or alkali is added to the solution. For example, if 0.1 ml of decinormal hydrochloric acid be added to 10 ml of distilled water with a *p*H of 7.0, the *p*H will be changed to 3.0; if however, 0.1 ml of decinormal hydrochloric acid be added to 10 ml of decinormal ammonium acetate solution also with a *p*H of 7.0, only a small change in *p*H occurs, and it will require the addition of over 10 ml of the acid to reduce the *p*H to 3.0. Thus the ammonium acetate solution has exerted a strong buffer action.

Now, indicator solutions are adjusted to the mid-point of their *p*H range, so that they may be either acid or alkaline, according to which indicator is being used. Owing, however, to this "buffering" action which most solutions possess, the *p*H value of the indicator itself does not appreciably affect the result. When, however, **unbuffered** (soft) waters are to be tested (e.g., freshly distilled or certain natural waters) the result obtained will be largely influenced by the indicator itself. While there are certain chemical precautions which can be taken to obviate this difficulty, they are not simple, and a much simpler method is available through the use of the Lovibond Nessleriser. In this case, by a suitable re-adjustment of the proportions of test liquid and indicator, the possibility of such errors has been reduced to a minimum.

(c) *Determination of pH of sea water*

The determination of the *p*H value of sea water is of interest because the *p*H has important effects on the viability of marine organisms, although such effects are usually of importance only when moderate *p*H changes are involved. The *p*H value is also a useful indication of the condition of the buffer system of sea water, e.g. photosynthesis by utilising carbon dioxide dissolved in the water changes the pH. Certain precipitation reactions are pH sensitive and the pH thus determines which materials can be precipitated from the water and sedimented.

All indicators can be used to determine the pH value of sea water. The following points should be noted however
    (i) The water sample must be fresh
    (ii) A blank must be used to compensate for any inherent colour in the sample
    (iii) Account must be taken of the errors due to the high and variable salt content. Whereas this is relatively constant at 3.5% in ocean water it falls to 0.6% or lower in estuarine samples. For correction factors see Note 1. The approximate ionic strength of sea water is 0.6.

(d) *Determination of pH of Jam*

Dilute the jam to ten times its volume with distilled water. Filter off the solids and determine the pH on one portion of the filtrate using another portion as a blank.

(e) *Determination of the pH of Milk*[3]

Dilute the milk to twenty times its volume with freshly boiled distilled water. Determine the pH with Bromo Thymol Blue indicator using a tube containing diluted milk without indicator as a blank. To correct for protein and salt errors (Notes 1 and 2) **add** 0.2 pH units to the figure obtained from Disc 2/1H.

(f) *Determination of the pH of Whey*[3]

Use either Bromo Thymol Blue or B.D.H. 4460 indicator, as appropriate, on undiluted whey. Use a tube of whey as a blank. To correct for salt and protein errors (Notes 1 and 2) **subtract** 0.2 pH units from the figure obtained from the appropriate disc.

(g) *Determination of pH of Sewage and Sewage Effluents*[4]

Fill three standard comparator tubes with the sample and to one add 0.1 ml of Universal Indicator. Mix well, place in the right-hand compartment of the Comparator, place another tube in the left-hand compartment and compare the colour of the right-hand tube with the glass standards in Disc 2/1 P, using a standard source of white light (Note 4). Choose an indicator, from the list above, which has its mid-point at, or near, the approximate pH obtained with Universal Indicator. Add 0.5 ml of this indicator to the third tube, mix and compare with the standards in the appropriate disc as before, using the original blank.

With turbid solutions it is difficult to compare colours accurately. In these cases the samples should be filtered before measurement.

In solutions containing residual chlorine certain indicators may be bleached. The chlorine must therefore be removed by adding a very small crystal of "hypo" to the water before adding the indicator. The pH tablets sold by The Tintometer Ltd. already contain hypo.

(h) *Standardisation of Buffer Solutions for Electrical pH Meters*

Electrical pH meters measure differences between the pH of a solution and that of the buffer solution on which they were standardised. This setting requires frequent checking. If the buffer solution itself has not exactly the pH value attributed to it then all readings on the pH meter will be in error by the amount by which the pH value of the buffer differs from its reputed value.

Buffer solutions do not keep indefinitely. Errors may also arise from one or more of the following sources:—
    (i) Deterioration due to take-up of alkali from the container after long storage
    (ii) Errors in making up
    (iii) Contamination from dirty glassware.

To provide a convenient check on standardising buffer solutions a disc containing nine permanent glass colour standards is available (reference 2/1 ZA). These nine standards cover the three usual points at which standardisation is carried out, namely pH 4.0, 7.0 and 9.2. Each point is checked with the appropriate indicator and three standards covering a spread of ± 0.2 pH units are provided for each point.

The test is performed by pouring the buffer solution into a comparator tube up to the 10 ml mark, adding 0.5 ml of the appropriate indicator (Bromo Cresol Green for pH 4.0, Bromo

Thymol Blue for pH 7.0 and Thymol Blue for pH 9.2) and checking against the colour standards in the disc.

(i) Determination of the pH of Brine

Calcium chloride ($CaCl_2$) brine of concentration about 30 g/100 g solution, specific gravity 1.13—1.5 and pH 7.6—8.6, has zero salt error when the pH is determined with Cresol Red.

Notes

1. *"Salt Errors"* The colorimetric method is based on the assumption that if an indicator in two solutions has the same colour, then the pH value of the two solutions is the same. This will only be true if the ionic strengths of the two solutions are identical. At other strengths an error, commonly referred to as the "Salt Error," is introduced. The discs were standardised using buffer solutions with an ionic strength of 0.10. If colorimetric pH measurements are made on solutions having an ionic strength significantly different from 0.1, then a systematic error may be introduced. The magnitude of this error will depend on the particular indicator which is being used, and on the ionic strength of the test solution. In general at ionic strengths below 0.1 the colorimetric pH value will be lower than the electrometric value, while at ionic strengths above 0.1 the reverse will be true. In some cases, errors as high as 0.28 pH units have been reported. It should be remembered however that some indicators, such as Methyl Red, have zero salt errors.

The calculation of the ionic strength of a solution requires an exact knowledge of the concentrations of all ions present in the solution. This information is not normally available. The publication of a list of correction factors for indicators at various ionic strengths has proved to be of little practical value. It is therefore recommended that, in those cases where the absolute pH value is required to an accuracy better than $\pm$ 0.3 units, the pH should be measured both colorimetrically and electrometrically. The difference between the two results can then be applied to other colorimetric pH readings using that indicator, as a constant correction factor, provided that there is no change in the ionic composition of the test solution.

2. *Protein effects* There is some disagreement between pH values determined colorimetrically and those determined electrically if proteins are present in the sample solution. The magnitude of the error depends on the type of protein, its concentration, and on the indicator being used.

3. *Colloid effects* The particles of a colloid solution may absorb some of the ions present in a solution, including indicator ions, and thus alter the activity coefficients of these ions and hence the apparent pH. Fine particles in suspension sometimes have the same effect. For this reason turbid solutions should be filtered before measuring their pH. In the case of colloidal solutions test with more than one indicator. If the results agree then no colloidal effect is operating.

4. A strongly recommended standard source of white light is the Lovibond White Light Cabinet. In the absence of a standard source, north daylight should be used wherever possible.

5. *Advantages of the Comparator* Compared with other methods of measuring pH values, the Lovibond Comparator has the great advantages of extreme simplicity, permanence and strength, and (as compared with electrometric methods) cheapness. An unskilled worker can obtain reliable results without training; such results are consistent over long periods. The colour glasses are permanent and therefore do not require renewing as is the case with tubes of coloured solutions. The comparator is robust in construction, and, complete with disc and the necessary accessories for one indicator, fits into the pocket. By the use of sufficient discs, practically the whole range of pH values is adequately covered. Maintenance is negligible, which justifies the initial outlay.

6. For details of complete kits designed for pH tests, apply to The Tintometer Ltd., Salisbury.

References

1. See *Determination of pH values* (2)
2. The British Standards Institution, B.S. 1427, 1962, *"Routine Control Methods of Testing Water used in Industry"*
3. J. G. Davis and C. C. Thiel, *J. Dairy Res.*, 1940, **11**, 71
4. Min. of Housing and Local Govt., *"Methods of Chemical Analysis as applied to Sewage and Sewage Effluents"*, H.M. Stationery Office, London, 1956

The Determination of pH Values (2)
Nessleriser method

Introduction

By using Nessleriser glasses instead of test tubes for pH tests, and looking down through the column of liquid instead of across it, the proportion of indicator solution needed in the liquid is reduced to a minimum. This avoids the considerable errors the reaction of the indicator solution may otherwise introduce into colorimetric determinations of the pH of soft natural waters and other relatively unbuffered solutions.

Principle of the method

Except in the disc for the Universal Indicator the colour glasses of the Nessleriser series of pH discs are standardised against the colours produced by adding 0.2 ml of the respective indicator solution to 50 ml quantities of standard buffer solutions. Colours in the Universal Indicator disc are standardised on 25 ml quantities of solution containing 0.1 ml of indicator in acid solutions and 0.05 ml in alkaline solutions. When 0.05 ml cannot be measured conveniently, 50 ml of solution containing 0.1 ml of indicator should be prepared and 25 ml of the mixture used for the test. With the Universal Indicator disc the quantity of solution without indicator used as a blank in the left-hand Nessleriser glass should also be 25 instead of 50 ml.

Indicators and their appropriate Standard Lovibond Nessleriser Discs

The discs designed for use with the Lovibond Nessleriser are distinguished from those used in conjunction with the Lovibond Comparators by the word "**Nessleriser**" on the moulding. The following discs are available:—

pH range	Indicator (Note 1)	Standard Disc	Notes
1.2 — 2.8	Thymol Blue	NLM	
2.8 — 4.4	Bromo Phenol Blue	NLN	
3.0 — 4.6	B.D.H. 3046	NLR	
3.6 — 5.2	Bromo Cresol Green	NLA	
4.0 — 8.0	B.D.H. 4080	NLT	
4.4 — 6.0	Methyl Red	NLL	
4.4 — 6.0	B.D.H. 4460	NLO	
5.2 — 6.8	Bromo Cresol Purple	NLB	
6.0 — 7.6	Bromo Thymol Blue	NLC	
6.8 — 8.4	Phenol Red	NLD	
7.0 — 8.6	Diphenol Purple	NLP	
7.2 — 8.8	Cresol Red	NLE	
7.6 — 9.2	m-Cresol Purple	NLK	
8.0 — 9.6	Thymol Blue	NLF	
9.0 —11.0	B.D.H. 9011	NLG	5 standards only. Steps of 0.5
10.0 —14.0	B.D.H. 1014	NLS	Steps of 0.5
4.0 —11.0	Universal	NLH	Steps of 4, 5, 6, 7, 8, 9, 9.4, 10, 11
1.0 —13.0	B.D.H. Full Range	NLU	Steps of 1, 3, 5, 6, 7, 8, 9, 11, 13

Technique

Fill one of the Nessleriser glasses (Note 2) to the 50 ml mark with the fluid under examination (Note 3) and place in the left-hand compartment of the Nessleriser. Place the appropriate quantity of indicator in the other Nessleriser glass and fill to the 50 ml mark with the fluid under examination, mix thoroughly and place in the right-hand compartment. Stand the Nessleriser before a uniform source of white light (Note 4) and compare the colour produced in the test solution with the colours in the standard disc, rotating the disc until a colour match is obtained (Note 5).

The markings on the discs represent the pH values corresponding with the colours in the field of view.

COLORIMETRIC CHEMICAL ANALYTICAL METHODS   *p*H Values (2)

Notes

1. Indicators supplied by The Tintometer Ltd. are specially prepared to conform to the colours on the Lovibond Nessleriser discs.

2. Lovibond Nessleriser discs are standardized on a depth of liquid of 113 ± 3 mm. The height, measured internally, of the 50 ml calibration mark on Nessleriser glasses used with the instrument must be within the same limits. Tests with Nessleriser glasses not conforming to this specification will give inaccurate results.

3. It is important that the liquid being tested should be as clear as possible, as turbidity reduces the amount of light transmitted and consequently dulls down the colours, and that the correct quantities of indicators should be used—see introduction overleaf.

4. A strongly recommended standard source of white light is the Lovibond White Light Cabinet. In the absence of a standard source, north daylight should be used wherever possible.

5. No reliance should be placed on a result indicated by the colours at the extreme ends of the range of an indicator, as a *p*H value outside the range will give only an end colour. In such cases, the test should be repeated using an indicator with an over-lapping range.

The Determination of the *p*H of Nickel Plating Solutions

Introduction

The optimum working conditions for nickel-plating depend on various factors such as the composition and concentration of the plating solutions, the temperature, the degree of acidity (*p*H value), the class of work, the desired finish, and the current density employed. One of the most important of these conditions is the *p*H value, and the character of the deposit is largely dependent upon it.

Principle of the method

The *p*H of plating solutions can be determined colorimetrically by means of a suitable indicator. For the *p*H range 5.2 to 6.8, which is the usual range for nickel plating baths, the indicator recommended is Bromo Cresol Purple. The deep green colour of the nickel solution makes it difficult to match the colour of the indicator when the normal disc for Bromo Cresol Purple is used and two special discs have therefore been prepared for this particular application, one disc for each of the two plating-bath compositions most commonly used.

Reagent required

Bromo Cresol Purple indicator solution.

The Standard Lovibond Comparator Discs 2/3A and 2/3B

These discs both cover the *p*H range 5.2 to 6.8.

Disc A is for use with solutions containing about 120 g of crystalline nickel sulphate per litre.

Disc B is for use with solutions containing from 200 to 250 g of crystalline nickel sulphate per litre.

Technique

Fill two standard comparator tubes to the 10 ml. mark with the plating solution to be tested and add 0.5 ml of the indicator solution to one of the tubes. Mix the contents of this tube and then place it in the right-hand compartment of the comparator. Place the other tube in the left-hand compartment and place the disc appropriate to the plating solution being tested in position in the lid of the comparator. Place the comparator before a standard source of white light, such as the Lovibond White Light Cabinet or, failing this, north daylight. Compare the colour of the solution with the colour standards in the disc and, when a match has been attained, read the *p*H value from the indicator recess in the right-hand bottom corner of the comparator case (Note 2).

Notes

1. *Other pH ranges* For plating solutions with a lower *p*H (higher acidity) than 5.2 use Bromo Cresol Green indicator and Disc 2/1C. This combination covers the *p*H range 3.6 to 5.2.

2. *Salt error* Electrometric methods for the determination of *p*H values of nickel plating solutions give values differing by several decimal points from those obtained by colorimetric methods. This arises from the high concentration of nickel salts in solution affecting the colorimetric method and giving rise to apparent *p*H values which are 0.2 to 0.6 *p*H units higher than the values measured on the same solutions using a hydrogen or quinhydrone electrode. As an approximate correction it is recommended that 0.4 be subtracted from the reading obtained from the disc. For recording absolute *p*H values the electrometric methods are unquestionably the more accurate. However, for all practical purposes, the colorimetric method is more suitable for factory use as it can be performed rapidly by any ordinary workman and produce a direct reading without any need for further calculation. The results should be recorded as "apparent *p*H values by the colorimetric method".

The Determination of the *p*H of Paper

Introduction

The *p*H of a paper, which is the chemist's way of describing its acidity or alkalinity, is a critical factor in its response to printing. Excess acid or alkali has been shown to cause some colours to fade, and set-off and smudging can nearly always be traced to paper acidity (low paper *p*H)[1]. Printers must therefore be able to gauge the *p*H of a paper with reasonable accuracy if they are to avoid drying and fading faults.

An established standard method[2] of measuring the *p*H of paper involves preparing a water extract of the paper, adding a chemical indicator, and comparing the colour of the solution with permanent glass standards in a Lovibond *p*H Comparator. This extraction is time consuming but has the advantage that the extracts correlate with particular aspects of the paper-making process. For instance cold-water extraction correlates with the back-water *p*H at which the paper was made, while hot-water extraction correlates with the tendency of papers to inhibit the drying of printing inks (see also Notes 1-4). These extraction methods will be outlined in Part I of this test procedure.

In an attempt to speed up the standard test, the direct application of the indicator to the paper, followed by comparison of the colour of the "smear" with the standard Lovibond disc glasses for the particular indicator used, has been suggested. This short cut has, unfortunately, been found to be unreliable, as the colour of a stain for any particular *p*H does not correspond to that of the solution of the same *p*H. This is due to the dichroic nature of the indicator solutions.

Research carried out at the Printing Industries Research Association, (PIRA),[3,4] has led to the development of a method of assessing the surface *p*H of paper which will be described below. As this method will often give results which differ from those obtained by the standard hot-extract method, a special index has been devised to avoid confusion between results from the two techniques. This is called the Lovibond-PIRA *p*H Index. The *p*H values in this index range from 10 to 200, each step of 10 in the index representing a *p*H difference of about 0.2 units. Figures below 100 represent increasing acidity while those above 100 represent increasing alkalinity. The value 100 is analagous to a hot extract value of 5.0.

Probably the greatest use of the new *p*H index will be in connection with ink drying problems. It has been shown[3,5] that, as an approximation, uncoated papers should have a Lovibond-PIRA index of 100 or above. Similar specifications can be built up for other paper *p*H requirements. The new test should also prove satisfactory for the routine quality control of paper production, once the individual manufacturer has established his own correlation between smear *p*H index and the hot-extract standard *p*H test results.

Part I Water Extraction Methods

Principle of the tests

A sample of paper is macerated with water and the *p*H of the extract is determined by comparing the colour, produced by the addition of an appropriate indicator to the extract, with the colours of Lovibond permanent glass colour standards.

Reagents required—and the corresponding Standard Lovibond Comparator Discs

1. *Distilled Water* Freshly distilled or deionized water of *p*H 5.9—7.0 which has been proved (Note 5) to be free from alkaline impurities and which meets the requirements of B.S. 3978.

2. Indicator solutions and the appropriate Discs

pH range	Indicator	Standard Disc	Notes
0.2 — 1.8	Cresol Red	2/1Y	
1.0 — 2.6	m-Cresol Purple	2/1W	
1.2 — 2.8	Tropaeolin 00	2/1I	
1.2 — 2.8	Thymol Blue†	2/1A	
2.8 — 4.4	Bromo Phenol Blue	2/1B	
2.8 — 4.4	Methyl Orange	2/1ZG	
3.0 — 4.6	B.D.H. 3046	2/1V	
3.6 — 5.2	Bromo Cresol Green†	2/1C	
4.0 — 8.0	B.D.H. 4080	2/1CC	in steps of 0.5
4.4 — 6.0	B.D.H. 4460	2/1D	
4.4 — 6.0	Methyl Red	2/1E	
4.8 — 6.4	Chloro Phenol Red	2/1F	
5.2 — 6.8	Bromo Cresol Purple†	2/1G	
6.0 — 7.6	Bromo Thymol Blue†	2/1H	
6.8 — 8.4	Phenol Red†	2/1J	
7.0 — 8.6	Diphenol Purple†	2/1 O	
7.2 — 8.8	Cresol Red	2/1K	
7.6 — 9.2	m-Cresol Purple	2/1Z	
8.0 — 9.6	Thymol Blue†	2/1L	
8.6 — 10.0	B.D.H. 8610	2/1U	8 standards only
8.6 — 10.2	Phenolphthalein	2/1ZF	
9.0 — 11.0	B.D.H. 9011	2/1M	5 standards only
10.0 — 14.0	B.D.H. 1014	2/1BB	in steps of 0.5
4.0 — 8.0	B.D.H. Soil	2/1N	in steps of 0.5
4.0 — 11.0	Universal	2/1P	in steps of 1.0
1.0 — 13.0	B.D.H. Full Range	2/1ZE	Steps 1, 3, 5, 6, 7, 8, 9, 11 and 13

† These indicators are available in tablet form.

Indicators supplied by The Tintometer Ltd. are specially prepared to conform to the colours on the Lovibond discs.

Technique

a) Cold Extraction Method[2]

Weigh 2.00 g of the air-dry test sample into a 100 ml beaker made of chemically resistant glass. Measure out 100 ml of distilled water and add about 20 ml of this to the beaker. Macerate the paper with a flattened glass stirring-rod until the sample is uniformly wet, and then add the remainder of the distilled water. Stir for about 30 seconds, cover the beaker with a clean watch-glass and allow it to stand for an hour. At the end of this time stir again and then carefully decant 10 ml of the extract into a standard test tube. Add 0.1 ml of Universal Indicator, mix, and compare the colour with the colours in the appropriate disc (2/1P) using a standard source of white light (Note 6). Choose another indicator whose range spans the pH value determined with the Universal Indicator. Fill a further standard test tube to the 10 ml mark with a further portion of the extract, add 0.5 ml of the indicator, mix, and again compare the colour with the standards in the appropriate disc using the dulling screen if one is provided with the disc.

This entire procedure should be carried out at a room temperature of $20\pm5°C$.

b) Hot Extraction Method[2]

Weigh 2.00 g of the air-dry test sample into a 250 ml flask made of chemically resistant glass. Measure out 100 ml of distilled water and add about 20 ml of this to the flask. Macerate the paper with a flattened glass stirring-rod until the sample is uniformly wet and then add the remainder of the distilled water. Affix to the flask an air condenser of about 1 cm internal diameter and at least 75 cm long. If a bung is used to fix the condenser to the flask then cover the bung with clean metal foil. Place the flask in a boiling water bath so that the water level in the flask is below that in the bath and maintain the flask at $95\pm5°C$, without boiling, for an hour. Cool rapidly to 15–25°C, decant and measure the pH of the extract as in (a) above. If the decanted liquid is not clear, filter through a well washed, sintered glass filter (Note 4).

Part II The Lovibond-PIRA pH Index[3,5]

Principle of the method

A drop of indicator is smeared, by means of a microscope slide, over the surface to be tested. The colour of the centre of the resulting smear is compared with Lovibond permanent glass standards (Note 7).

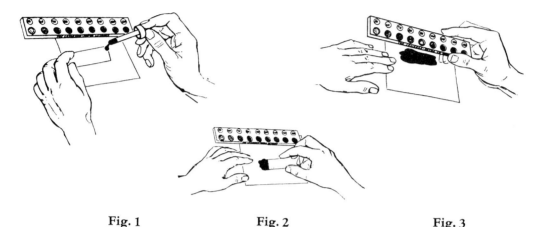

Fig. 1 Fig. 2 Fig. 3

Reagents required

Bromo Phenol Blue	(Standard pH range 2.8–4.4)	Index 10–90
Bromo Cresol Green	(,, ,, ,, 3.6–5.2)	Index 50–130
Bromo Cresol Purple	(,, ,, ,, 5.2–6.8)	Index 130–200

The Standard Lovibond-PIRA pH Colour Slides

The specially matched colour standards of the Lovibond-PIRA Scale are mounted in black anodised aluminium slides. Three slides are available, one corresponding to each of the above indicators and covering the ranges quoted. It will be seen that there are overlaps between these ranges. The scale however is continuous.

Technique

Place a drop of the Bromo Cresol Green indicator on the narrow edge of a clean microscope slide (Figure 1). Draw this across the paper to produce a smear about $3\frac{1}{2}$" long by 1" wide (Figure 2). If the paper is very smooth or non-absorbent, apply only light pressure so that the smear is always of approximately the same size. Allow about a minute for the indicator film to soak in, and then compare the central, uniform area of colour in the smear with the standard glasses in the slide (Figure 3). In order to compensate for any inherent colour in the paper the standards should be viewed against a clear area of the test sheet, using a standard source of white light.

If the colour of the smear is found to be at the end of the Bromo Cresol Green slide the test should be repeated with either Bromo Phenol Blue or Bromo Cresol Purple, depending on whether the colour is below or above the standards in the slide. A reading near the centre of a scale is more accurate than one at either end, as indicators are more sensitive in the mid part of their range.

Notes

1. Hot extraction of papers made with alum usually yields extracts with pH values lower than those obtained by cold extraction.

2. Melamine resins can undergo alkaline hydrolysis during hot extraction thus yielding extracts of higher pH value than those obtained by cold extraction.

3. Papers yielding alkaline extracts owing to the presence of bicarbonate will give higher extract pH values with hot extraction than with cold.

4. The *p*H values of extracts obtained by hot extraction will depend on whether or not the hot extract has been filtered. If the extract is cooled with the paper present a proportion of the hydrogen ions produced during the hot extraction are absorbed by the fibres, the proportion absorbed being determined by the nature of the fibres present.

5. The distilled water should be tested for alkaline impurity by boiling for a few minutes, cooling, and then measuring its *p*H. If this is greater than 7.3 the water must be redistilled, using a double still-head, after the addition of approximately 1 g of potassium permanganate and 4 g of sodium hydroxide per litre of water.

6. A strongly recommended standard source of white light for use with the Lovibond Comparator is the Lovibond White Light Cabinet. In the absence of a standard source of light, then north daylight should be used wherever possible.

7. The Lovibond-PIRA test is designed as a quick test for routine spot checks on the surface *p*H of paper, and for the applications discussed above. It is not intended to replace the standard extraction tests of Part I.

References
1. R. R. Coupe *et al*, Printing, Packaging & Allied Trades Research Association, Interim Reports Nos. 84a, b, c & d
2. The British Standards Institution, B.S. 2924, 1968
3. D. H. Charlesworth and R. R. Coupe, PIRA Lab. Report No. 31, July 1960
4. PIRA Technical Reference Note PRHT 2, September 1969
5. G. J. Chamberlin and R. R. Coupe, *Printing Technology*, 1962, **6**, 7

The Determination of the *p*H of Soils
using Palin tablets

Introduction

Gardeners, nurserymen, and farmers need to know the *p*H of their soil because different crops grow best at different *p*H levels. Most people know that alkaline soil (*p*H above 7.0) is termed "sweet" and that acid soil (*p*H below 7.0) is termed "sour", but these terms are too vague for soil control. An exact knowledge of the *p*H value of the soil will assist in deciding what crops will grow best, or alternatively will indicate what changes are necessary to be made by the addition of suitable chemicals (usually lime to raise the *p*H or sulphate of ammonia to lower it).

In the *p*H scale, which is logarithmic, 7.0 is neutral—neither acid nor alkaline. 5.0 is ten times more acid than 6.0, and 4.0 is one hundred times more acid than 6.0.

Values above 7.0 indicate alkalinity, also on a logarithmic scale.

Soils having a *p*H above 7.0 may cause crop failure, because plant foods such as phosphate and trace elements are "locked up" and are not available to the roots.

Soils with a *p*H too far below 7.0 are acid and may cause failure through calcium deficiency, or because aluminium is freed and becomes toxic to plants, or because other nutrients are "locked up", or because the micro-organisms of the soil are rendered inactive.

Most crops prefer a soil of about 6.5 *p*H, but the following are some of those which prefer an acid soil below 6.5:—

Lawn grasses, potatoes, many flowering shrubs and ornamental trees such as azalea, rhododendron, magnolia, holly and birch, soft fruits such as raspberries and strawberries, and orchids and lilies.

Crops which prefer a soil with a *p*H above 6.5 (neutral or alkaline) include, among others:—

Legumes, asparagus, beet, carnation, clematis, sweet pea, and many of the exotic flowers.

Principle of the method

The method here adopted for measuring *p*H is a colorimetric indicator coupled with a flocculant, so that the need for filtration of the soil solution is eliminated. The indicator is in tablet form, giving greater convenience in the field.

Reagents required

1. Tablets of Palin soil indicator
2. Tablets of Palin flocculant
3. Distilled water

The Standard Lovibond Comparator Disc 2/1ZD

This disc covers the *p*H range 4.0—8.0 in 9 steps of 0.5.

Technique

Sample the soil carefully, taking the soil samples to a depth of about six inches from many different places throughout the plot. At least 10 samples are needed to give a representative answer for a small field.

Method A

Mix these samples thoroughly in a bucket, remove stones and pieces of organic matter, place in a pile, flatten and divide into quarters. Use one of these quarters (which is reasonably representative of the whole) for testing. Place this representative soil sample in the shaking bottle to a depth of about $\frac{1}{2}$″, and add 5 times as much distilled water—that is, up to about 3″ depth—and stopper. Shake vigorously for a minute, allow the coarse matter to settle, and decant into two Comparator test tubes up to the 10 ml mark.

or alternatively

Method B

Take the desired number of soil samples, each in one of the sampling tubes, and test these separately. Use about 1″ height of soil and 5″ distilled water; shake vigorously, and then decant into two Comparator tubes. One tube should be filled to the 10 ml mark, and the other tube (for colour compensation) contains the remains of the solution. The exact amount is not important so long as there is enough height to cover the left-hand Comparator aperture. When using this method, which saves taking a large sample and mixing and quartering, judgement should be based on a study of a fair number of these individual spot tests.

Whichever method is used up to this point, proceed as follows:

To one of the tubes, add one flocculating tablet and shake vigorously to dissolve. This will cause the suspended matter to fall to the bottom, leaving only the extracted soil colour to be seen. If settlement is not rapid on allowing to stand, shake vigorously a second time. Place this tube in the left-hand compartment of the Comparator, behind the glass colour standards, to compensate for inherent colour in the solution. To the other tube, containing 10 ml of solution, add one indicator tablet which contains flocculant and shake vigorously to dissolve, as before, repeating a second time if necessary. Place in the right-hand compartment of the Comparator and revolve the disc containing the colour standards until the nearest match is found. The pH value is then read from the lower right-hand window. In the field, view with your back to the sun. For an accurate standardised light for use indoors, the Lovibond White Light Cabinet should be used.

Notes

1. If a value below 6.5 is found, lime is required for many crops. To ascertain exactly how much, a second test, known as a "Lime Requirement Test" should be applied. Full details are available from The Tintometer Ltd. For a rough approximation, however, use may be made of the following table to calculate the amount of lime required:

pH	Hydrated Lime required					
	Sandy soils		*Loam*		*Clay or Peaty soils*	
	oz/sq yard	grams/m^2	oz/sq yard	grams/m^2	oz/sq yard	grams/m^2
4	12	285	16	380	20	476
5	8	190	12	285	16	380
6	4	95	8	190	12	285

To obtain lime required in tons per acre, divide the ozs/yd^2 figure by 8.
For Quicklime, reduce quantities by one-quarter and for Carbonate of Lime increase by one-third.

Reference

D. W. Gilchrist Shirlaw, "*Soil Fertility and Field Methods of Analysis.*" Cassell, London, 1962.

The Determination of Soil Lime Requirement
using nitrazine yellow

Introduction

Lime is added to soil to neutralise acidity. It also supplies the essential nutrient calcium. Although the object of lime addition is to raise the pH value of the soil, the amount of lime required cannot be determined accurately from a pH test alone, because for any given pH value the lime requirement is affected by the soil texture and by the presence of organic matter. It is therefore desirable to carry out a separate test to determine the lime requirement.

The present test[1] has been devised for this purpose and results in a direct indication of the amount of lime which should be added to the soil to bring the pH up to 6.5.

There are three types of lime which are used to neutralise soil acidity. These are calcium carbonate (ground limestone, chalk, carbonate of lime, $CaCO_3$), calcium oxide (burnt lime or quicklime, CaO) and calcium hydroxide (hydrated or slaked lime, $Ca(OH)_2$). Each type has a different neutralising power and thus the rate of application for a given lime requirement will depend on which substance is used. The figures obtained from the disc refer to calcium oxide. For a lime consisting of calcium carbonate, double the disc readings: for one consisting of calcium hydroxide multiply by 4/3. These factors will give approximate values. See Note 1.

Principle of the method

The soil is shaken with a buffer solution, and the resulting change in the pH value of this buffer, as measured by means of Nitrazine Yellow indicator, has been related directly in the disc readings to the lime requirement per unit area to bring the soil pH up to 6.5.

Reagents required
1. Buffer tablets (Note 2)
2. *Buffer solution* Dissolve two buffer tablets in 300 ml of distilled water.
3. Nitrazine Yellow indicator tablets (Note 2)
4. Activated charcoal

The Standard Comparator Disc 2/1ZB

This disc contains 5 standards corresponding to lime requirement of 0, 8, 16, 32 and 64 ozs of calcium oxide per square yard (Note 3). A conversion table for use with other liming agents and with metric units is included in Note 1.

Technique

Pour 50 ml of the buffer solution (reagent 2) into a stoppered shaking tube. Measure 8 ml of soil (Note 4) into a calibrated test tube and tip this into the buffer solution. Stopper the tube and shake for 1 minute and then shake at frequent intervals during the next 30 minutes. Filter. If the filtrate is coloured add a small amount of activated charcoal, shake the mixture and re-filter. Pour 10 ml of the colourless filtrate into each of the two Comparator tubes and place one of these in the left-hand compartment of the comparator. To the other tube add 1 indicator tablet (reagent 3), shake until this has dissolved and then place the tube in the right-hand compartment of the Comparator. Match the colour of the indicator solution with the standards in the disc using north daylight.

Notes

1.

Standard reading CaO		Approximate Equivalent amounts Ca(OH)$_2$		CaCO$_3$	
oz/sq yd	g/sq m	oz/sq yd	g/sq m	oz/sq yd	g/sq m
0	0	0	0	0	0
8	270	12	400	16	540
16	540	24	800	32	1080
32	1080	48	1600	64	2160
64	2160	96	3200	128	4320

2. Reagent tablets may be obtained from The Tintometer Ltd.

3. If the following crops are to be grown the indicated lime requirement should be reduced by 20 oz/sq yd of calcium oxide (or the equivalent for other liming materials). If this results in a negative answer then no lime should be added. Plants which like a slightly acid soil are:—iris, lily, orchid, blueberries, cranberries, raspberries, strawberries, azalea, broom, camelia, magnolia, rhododendron, birch, holly, laurel, potatoes, and most of the better lawn grasses.

4. The results of any analysis on soil can only be as accurate as is the sample itself. As only a small amount of soil is used in the test, great care must be taken to make this truly representative of the area to be treated. At least 30 small samples, to a depth of about 6 inches, should be taken from any plot, at well-spaced intervals. Mix these samples thoroughly, remove stones and pieces of organic materials, place the soil in a heap, flatten this and divide into quarters. Take two opposite quarters mix together, form in a heap, flatten and again divide into quarters. Repeat this process until one of the quarters is of the size required for the sample.

Reference

1. D. W. G. Shirlaw, *"Soil Fertility and Field Methods of Analysis"* Cassell, London, 1962.

The Determination of pH of Urine
using Universal Indicator

Introduction

One of the main functions of the kidneys is to maintain the acid-base equilibrium of the blood. Acids are constantly being formed as carbon-dioxide is produced by the oxidation of foodstuffs, and as phosphoric and sulphuric acids are produced by the oxidation of phosphorus and sulphur contained in ingested protein. On the other hand vegetable foods contain basic ions such as sodium, potassium and calcium which tend to make the reaction of the blood more alkaline. By the appropriate excretion of acidic or basic radicles the kidneys maintain the pH of the blood within narrow limits. When alkalaemia tends to occur, alkaline urine is excreted; in acidaemia, acid urine is excreted. The extreme range of pH changes in the urine which healthy kidneys can achieve is from 4.8 to 7.9.

Acid Urine

Freshly voided urine is usually slightly acid if the subject is eating an ordinary mixed diet. The degree of acidity is increased by a high protein intake. Marked acidity of the urine occurs in uncontrolled diabetes mellitus when aceto-acetic and hydroxybutyric acids are formed and eliminated in the urine. Very acid urines are also formed in such conditions as emphysema and chronic bronchitis, when the inadequate ventilation gives rise to a respiratory acidosis.

In the treatment of urinary infections, the urine may be made acid by giving ammonium chloride, or calcium or ammonium mandelate. The optimal degree of acidity in a freshly voided urine sample is pH 5.0.

Alkaline Urine

The reaction of urine may become alkaline after a meal—the so-called alkaline tide caused by the secretion of hydrochloric acid into the stomach. The reaction of the urine also tends to be alkaline on a vegetarian diet. A strongly alkaline urine is often found when there is infection of the renal tract with urea-fermenting organisms, such as *B. Proteus,* which produce ammonia. An alkaline urine is excreted in respiratory alkalosis such as occurs in hysterical hyperventilation, in which the patient exhales large amounts of carbon-dioxide. The urine is also alkaline in conditions which cause metabolic alkalosis such as vomiting, in pyloric stenosis, and the ingestion of large amounts of alkali.

Urine is deliberately made alkaline in the treatment of certain urinary infections and during sulphonamide therapy by giving sodium or potassium citrate or bicarbonate. The night urine is the most difficult to make alkaline and when the pH is greater than 7.0 sufficient alkalis are being given.

Principle of the method

The pH of the freshly voided urine is determined with Universal Indicator, the colour produced being compared with a series of Lovibond permanent glass standards representing the colours produced by solutions of known pH value.

Reagent required

Universal Indicator

The Standard Lovibond Comparator Disc 2/1 P

This disc covers the pH range from 4.0 to 11.0 in steps of 1.0.

Technique

Fill both Comparator cells to the 10 ml mark with the freshly voided urine. To the right-hand cell only add 0.2 ml Universal Indicator. Do not immerse the tip of the pipette beneath the surface of the urine. Carefully mix by stirring with a clean glass rod or by pouring into another cell and back again. Hold the Comparator about 18 inches in front of the eye facing a uniform source of white light (Note) and rotate the disc until the colours match. The pH of the specimen is shown in the indicator recess at the bottom of the right-hand corner.

Note

A strongly recommended standard source of white light is the Lovibond White Light Cabinet. In the absence of a standard source north daylight should be used wherever possible.

Chemical Analysis
Organic

The Determination of Alcohol
using chromic acid

Introduction

The alcohol content of soft drinks is an important criterion in judging the market value of the product, and statutory maxima for alcohol contents are imposed on a number of beverages including fruit juices. The present method has been developed to provide a rapid and accurate means of determining the alcohol content of beverages, and of vinegar.

Principle of the method

Alcohol reduces solutions of hexavalent chromium to the trivalent form and this reacts with nitrate to produce a deep blue complex. The intensity of the blue colour, which is proportional to the concentration of trivalent chromium and hence to the alcohol concentration, is measured by comparison with Lovibond permanent glass colour standards.

Reagent required

Chromic acid solution Dissolve 21.28 g of potassium dichromate ($K_2Cr_2O_7$) in 100 ml of distilled water and mix with 585 ml of 65% nitric acid (HNO_3). Dilute the resulting solution to 1,000 ml with distilled water.

The chemicals used should be of analytical reagent quality.

The Standard Lovibond Comparator Disc 3/80

This disc covers the range 1, 2, 3, 4, 4.5, 5 and 6 g of alcohol (EtOH) per litre.

The master disc, against which all subsequent copies are checked, was tested and approved by Dr. L. Jakob, Landes- Lehr- und Forschungs- anstalt für Wein und Gartenbau, 673 Neustadt (Weinstrasse), West Germany.

Technique

Place 5 ml of the sample in a suitable micro-distillation apparatus. Place 5 ml of the chromic acid solution in a flask and insert the delivery tube of the distillation apparatus below the level of the chromic acid solution in the flask. Distil the sample until about 2 ml remains, taking care to avoid sucking the chromic acid back into the still. After distillation remove the flask containing the chromic acid solution, wash the condenser and delivery tube with distilled water and add the washings to the chromic acid solution. Transfer this solution quantitatively to a 50 ml volumetric flask and dilute to the mark with distilled water. Mix thoroughly. Transfer 10 ml of this solution to a standard comparator test tube or 13.5 mm cell and place this in the right-hand compartment of the Comparator. Place the Comparator in front of a standard source of white light, such as the Lovibond White Light Cabinet or, failing this, north daylight, and match the colour of the sample with the standards in the disc, interpolating between standards if necessary. Read off the alcohol content from the indicator window in the comparator.

If the result falls in the region 0 – 2.5 g per litre, the accuracy can be improved by repeating the determination using 10 ml of sample, distilling down to 5 ml and proceeding as above. In this case the final result read from the disc must be divided by 2 to obtain the alcohol content.

Note

Volatile acids do not interfere with this test. Esters however do interfere, and the method is therefore not suitable for the investigation of aroma in beverages.

Reference

P. Böhringer und L. Jakob, *Zeitschr. Flüssiges Obst.*, 1964, **31**, 223

The Determination of Amines (1)
using Bromo Cresol Green

Introduction

This test[1] (Note 1) has been developed primarily for the determination of traces of long-chain aliphatic amines in water. Film-forming amines of this type have been used in water treatment for the prevention of the growth of algae and also for the protection of steam lines. When used in this way, the determination of residual amine in cooling waters and condensates is essential for control purposes. Alternative methods[2-11] which have been proposed for this determination are unsuitable for field use, as they require exacting conditions for the production of accurate results. The present method, on the other hand, is capable of producing accurate results under field conditions, even in the hands of relatively unskilled operators.

Sulphonphthaleins react with primary, secondary, tertiary and quaternary amines (quaternary ammonium compounds-QAC) and this test, although primarily developed for long-chain amines can be applied to the determination of other amines provided that the disc is recalibrated in terms of the particular amine being estimated (see Note 4).

Principle of the method

Amines react with sulphonphthaleins, in acid solution, to form a yellow complex. This complex is extracted from the aqueous solution by shaking with chloroform in a special comparator extraction tube. The intensity of the yellow colour in the lower (chloroform) layer after extraction is proportional to the concentration of amine, and is estimated by comparison with Lovibond permanent glass colour standards.

Reagents required

1. Amine test indicator tablets (equivalent to 0.5 ml of a 0.1% solution of tetra-bromo-*m*-cresol sulphonphthalein i.e. "Bromo Cresol Green," with an acid content equivalent to 15 mg alkali, expressed as $CaCO_3$)
2. Chloroform ($CHCl_3$), analytical reagent grade

The Standard Comparator Disc 3/58

This disc covers the range 1—10 ppm of pure octadecylamine and is designed for use with special extraction tubes, which are available from The Tintometer Ltd.

Other amines may be determined with this disc by using appropriate conversion factors for the disc readings (Note 4).

Technique

Take two of the special calibrated extraction tubes and carefully add chloroform to each, up to the 5 ml mark. Place an indicator tablet in each tube. Into one tube introduce 10 ml of sample, i.e. up to the 15 ml mark. To the other tube add 10 ml of distilled, or untreated, water to form the blank. Stopper both tubes and shake for 30 seconds.

Allow the chloroform layers to separate. If amine is present in the sample the lower (chloroform) layer will be coloured yellow. Place the sample tube in the right-hand compartment of the Lovibond Comparator and the blank tube in the left-hand compartment. The tubes have been designed so that the lower layers will cover the comparator apertures obviating the need to transfer the chloroform layers to separate comparator cells. Compare the colour of the sample with the Lovibond permanent glass standards in the disc, using a standard source of white light, such as the Lovibond White Light Cabinet or, failing this, north daylight.

COLORIMETRIC CHEMICAL ANALYTICAL METHODS — Amines (1)

Notes

1. This method has been developed by Houseman and Thompson Ltd. Permission to quote from the published literature is gratefully acknowledged.

This method is only applicable if the pH of the sample is below 8.5; amine treatment is not suitable for more alkaline waters. The hardness of the water is also critical and should not exceed 2,000 ppm. Beyond this level, which increases with evaporation, amines are ineffective. The chloride concentration gives an indication of the hardness level as the chloride remains soluble while other salts precipitate from the water.

2. If the intensity of the colour in the sample tube is greater than that of the 10 ppm standard, the test should be repeated using a smaller volume of sample and making the total volume of water added up to 10 ml by means of distilled water. The final reading should then be multiplied by the appropriate dilution factor, i.e. by 10 divided by the volume of sample used.

3. If the solution is insufficiently acidified, free ammonia tends to give a false reading. In the 10 ml sample, there is sufficient acid in the tablet to allow up to 1,500 ppm total alkalinity to be tolerated before the yellow colour is interfered with. Ammonium, iron and copper salts and hydrazine do not interfere.

4. Amines other than octadecylamine can be determined using this disc provided that a suitable correction factor is applied to the disc readings. The appropriate correction factor may be calculated in the following manner. The colour which is developed is proportional to the molecular concentration of the amino group, or groups, in the amine, irrespective of whether it is a primary, secondary, tertiary or quaternary amine. This molecular concentration is simply the molecular weight of the amine divided by the number of amino groups present in the molecule.

$$\text{Correction factor} = \frac{\text{Molecular Weight of new amine}}{\text{No. of amino groups} \times \text{Molecular Weight octadecylamine (269)}}$$

A few examples will make this clear:—

Morpholine $\underline{NH.CH_2.CH_2.O.CH_2.CH_2}$ Molecular Weight 87.1
No. of amine groups 1

$$\text{Correction factor} = \frac{87.1}{1 \times 269} = 0.32$$

Ethylene diamine $NH_2.CH_2.CH_2.NH_2$ Molecular Weight 60.1
No. of amine groups 2

$$\text{Correction factor} = \frac{60.1}{2 \times 269} = 0.11$$

Amine 220 $C_{17}H_{33}.\underline{C:NC_2H_4N}.C_2H_4OH$ Molecular Weight 350.6
No. of amine groups 2

$$\text{Correction factor} = \frac{350.6}{2 \times 269} = 0.65$$

Tetraethylammonium bromide $(C_2H_5)_4NBr$ Molecular Weight 210.2
No. of amine groups 1

$$\text{Correction factor} = \frac{210.2}{1 \times 269} = 0.78$$

5. The apparatus supplied by The Tintometer Ltd. for this test consists of:—
   Lovibond Comparator
   2 calibrated extraction tubes
   Disc 3/58
   Bottle for chloroform
   Bottle of amine test tablets

A fitted case to hold the above and other ancillary equipment is available.

References

1. A. S. Pearce, *Chem. & Ind.,* 1961, 825
2. A. Milun and F. Moyer, *Anal. Chem.,* 1956, **28,** 1204
3. E. Carkhuff and W. Boyd, *J. Am. Pharm. Ass. Sci., Ed.,* 1954, **43,** 240
4. P. J. Lloyd and A. D. Carr, *Analyst,* 1961, **86,** 335
5. K. B. Coates, *Corrosion Tech.,* 1960, **7,** 46
6. H. M. Hershenson and D. N. Hume, *Anal. Chem.,* 1957, **29,** 16
7. J. Johnston, *Anal. Chem.,* 1953, **25,** 1764
8. J. Johnston and F. E. Critchfield, *Anal. Chem.,* 1956, **28,** 436
9. *idem ibid,* 1957, **29,** 957
10. J. Johnston and G. L. Fink, *Anal. Chem.,* 1956, **28,** 436
11. A. J. Milun, *Anal. Chem.,* 1957, **29,** 1502

The Determination of Amines (2)
using Methyl Orange

Introduction

This test is based on a method developed by Silverstein[1] for the colorimetric determination of traces of amines in water. It is very simple and also has the advantage of being more sensitive than some alternative methods[2,3]. It is suitable for the determination of octadecylamine, which is used as a corrosion inhibitor in water treatment.

Principle of the method

Certain primary, secondary and tertiary amines react with Methyl Orange at a pH of 3-4 to form a yellow complex. This complex is soluble in organic liquids, such as methyl chloroform, which can be used to extract the complex from the water. The colour intensity of the complex solution is a direct measure of the total amine concentration, and this is determined by comparing the colour with a series of Lovibond permanent glass standards.

Reagents required

1. *Methyl Orange indicator* Dissolve 0.04 g of Methyl Orange in 100 ml of hot distilled water, or in 20 ml of industrial methylated spirit followed by 80 ml of distilled water. Cool, and filter if necessary. This is sold by B.D.H. Chemicals Ltd. as "Alfloc Methyl Orange".

2. *Amine buffer solution* Dissolve 125 g of potassium chloride (KCl) and 70 g of sodium acetate trihydrate ($CH_3COONa.3H_2O$) in 500 ml of distilled water. Add 300 ml of glacial acetic acid (CH_3COOH) and make up to 1 litre with distilled water.

3. *Methyl chloroform* 1,1,1-trichloroethane (CCl_3CH_3).

   Reagents 2 and 3 should be of analytical reagent grade.

The Standard Lovibond Comparator Disc 3/64

This disc covers the range 0-2 ppm of octadecylamine in 5 steps as follows:—

    0   0.25   0.5   1.0   2.0 ppm

Technique

If the water to be sampled is above ambient temperature the sample should be collected through a small stainless-steel cooler.

Run the sample into one of two stoppered 50 ml cylinders which have been silicone treated, (Note 2). When a representative sample has been collected, reject excess sample down to the 50 ml mark.

Into the other 50 ml cylinder add 50 ml of distilled water. To both cylinders add 2 ml of amine buffer solution (reagent 2), 1 ml of Methyl Orange indicator (reagent 1) and 10 ml of methyl chloroform (reagent 3). Shake both cylinders for 5 minutes. Allow to stand for 3 minutes and then decant as much as possible of the upper, aqueous, layer from both cylinders. Transfer the remaining liquid in each cylinder into a 10 ml calibrated test tube, making sure that the methyl chloroform falls to the bottom of the tube and ignoring any aqueous solution that may be lost by overflow from the test tube. Remove any surplus liquid from the outside of the test tubes with a cloth.

Place the test tube containing the treated sample in the right-hand compartment of a Lovibond Comparator containing Disc 3/64, and place the tube containing the blank in the left-hand compartment. Compare the colour of the treated sample with the Lovibond permanent glass standards in the disc, using a standard source of white light, such as the Lovibond White Light Cabinet, to illuminate the comparator. If a White Light Cabinet is not available use north daylight wherever possible.

Notes

1. The adaptation of this test to the comparator is due to the 'Alfloc' Water Treatment Service of I.C.I. Ltd., to whom acknowledgement is made.

2. The 50 ml stoppered cylinders can be silicone coated by completely filling the cylinder with a 2% aqueous solution of I.C.I. Silicone Emulsion M.460 and stoppering the cylinder. After 15 minutes drain out the solution and dry the cylinders at a temperature of about 180°C. The coating will last about 6 months. Retreatment is necessary when a normal water meniscus reappears.

References

1. R. M. Silverstein, *Anal. Chem.*, 1963, **35,** 154
2. K. B. Coates, *Corrosion Tech.*, 1960, **7,** 46
3. A. S. Pearce, *Chem. and Ind.*, 1961, 825
4. Brit. Standards Inst., B.S. 4445: 1969, "*Schedule of tests for gasification and reforming plants using hydrocarbon feedstocks*", London, 1969

The Determination of Benzene Hexachloride

using thiocyanate

Introduction

Benzene hexachloride (BHC; 1,2,3,4,5,6-hexachloro*cyclo*hexane; Gammexane; Lindane) is a widely used insecticide. Analytical procedures are required both to determine the level to which it has been applied, in order to avoid wastage of material, and also to measure the residual level on crops, especially on those which are to be used as foodstuffs.

The present test[1,2] has been devised primarily to measure the BHC level on dressed seeds. However, by suitable modification of the extraction procedure together with subsequent concentration if necessary, it can be used for other applications. The only limitation is the requirement to bring the final concentration of the extract within the range of the disc, i.e. to between 100 and 500 µg per ml.

Principle of the method

The BHC is extracted and the extract is hydrolysed with alkali. The chloride ions released react with mercuric thiocyanate to displace thiocyanate ions which then form a red complex with ferric ions. The intensity of this red colour, which is proportional to the thiocyanate concentration, and thus to the original BHC concentration, is measured by comparison with a series of Lovibond permanent glass colour standards.

Reagents required

1. *Toluene* ($CH_3C_6H_5$)
2. *Sodium hydroxide* ($NaOH$) 0.1 N *alcoholic solution* Dilute 10 ml of a 20% w/v aqueous solution of NaOH to 500 ml with chloride-free 95% alcohol (C_2H_5OH)
3. *Ammonium ferric sulphate* [$(NH_4)_2SO_4.Fe_2(SO_4)_3.24H_2O$] *solution* Dissolve, by heating, 1 g of ammonium ferric sulphate in distilled water containing 15 ml of concentrated nitric acid (HNO_3), and dilute the cooled solution to 100 ml with distilled water
4. *Mercuric thiocyanate* [$Hg(SCN)_2$] 0.1% *alcoholic solution* Dissolve 0.1 g of mercuric thiocyanate, by stirring with a glass rod, in 100 ml of chloride-free 95% alcohol

All chemicals used in the preparation of reagents should be of analytical reagent quality.

The Standard Lovibond Comparator Disc 3/84

This disc covers the range 200-1000 ppm of BHC in steps of 200, 300, 400, 500, 600, 700, 800, 900, and 1,000 ppm, based on a 10 g sample. This is equivalent to 100-500 µg in the 1 ml of extract used in the test.

Technique

Weigh accurately 10 g of cereal grain (or 5 g of sugar beet seed—Note 2) into a 6″ × ¾″ test tube. Add 10 ml of toluene (reagent 1), stopper the tube and shake **vigorously** for 1 minute. Drain off the extract into a second test tube. Add a further 10 ml of toluene to the sample in the first test tube, stopper and shake vigorously for a further 1 minute. Add this extract to the solution already in the second test tube, and mix thoroughly. Pipette 1 ml of these combined extracts into a third test tube. Pipette 10 ml of alcoholic alkali solution (reagent 2) into the third test tube, stopper, mix thoroughly, and then allow to stand for 15 minutes. After this time add to this tube 2 ml of ammonium ferric sulphate solution (reagent 3) and 2 ml of mercuric thiocyanate solution (reagent 4). Mix after each addition. Rinse the comparator test tube with a small amount of this test solution, and then half fill the tube with the test solution. Place the tube in the right-hand compartment of the Comparator and stand the Comparator before a standard source of white light, such as the Lovibond White Light Cabinet, or, failing this, north daylight. Compare the colour of the solution with the standards in the disc and note the reading in the indicator window of the Comparator when the nearest match is obtained. This is equal to the BHC content of the sample in ppm (Note 2).

Notes

1. This test was developed by the Agricultural Division of Imperial Chemical Industries Ltd. and is quoted by permission of the author.

2. If less than 10 g of sample is used the result must be adjusted proportionately, e.g. for 5 g multiply the result obtained by $10/5 = 2$; thus for a disc reading of 400 report 800 ppm.

3. As chloride ion interferes with this test by giving a background colour in the final solution, all apparatus must be rinsed with distilled water before use. After using, wash the pipettes and test tubes with warm water and then rinse with distilled water.

4. During the determination care must be taken to avoid exposing the test solution to sunlight, which causes the colour to fade rapidly.

5. The method can be checked, if desired, by means of a standard BHC solution equivalent to 600 ppm. Prepare a stock solution of BHC (5 mg per ml) by dissolving 0.5 g of Lindane (minimum 99% gamma BHC) in toluene and diluting to 100 ml with toluene. The 600 ppm standard is prepared by diluting 3 ml of this stock solution to 50 ml with toluene. To check the method use 1 ml of this standard solution instead of the 1 ml of extract in the procedure given above and develop the colour. This should match the 600 ppm standard in the disc.

References

1. Plant Protection Ltd. "*Standard Method No. 275, Addendum II/B,*" June 1967
2. S. H. Yuen, *Analyst,* 1964, **89,** 726

The Determination of Carbohydrates
using orcinol

Introduction
The determination of carbohydrates is of importance in the examination of foodstuffs, industrial effluents, boiler feed water, and sewage. The carbohydrate colour reaction of Tollens[1] has been modified by Tillmans and Philippi[2] for application to foodstuffs. This method has been further modified for use with the Lovibond Comparator.

Principle of the method
Acid orcinol produces a yellow to orange red colour when it reacts with carbohydrates (Note 1). The intensity of this colour, which is proportional to the carbohydrate concentration, is determined by comparison with a series of Lovibond permanent glass colour standards.

Reagents required
1. *Orcinol* $(3,5\text{-}dihydroxytoluene, CH_3C_6H_3(OH)_2.H_2O)$
2. *Sulphuric acid* 66% v/v H_2SO_4. Slowly add 66 ml of concentrated sulphuric acid to 20 ml of water. Cool and dilute to 100 ml with distilled water
3. *Acid orcinol reagent* Dissolve 0.2 g of orcinol in 99.8 g of 66% sulphuric acid

All chemicals used should be of analytical reagent quality.

The Standard Lovibond Comparator Disc 3/8
This disc covers the range from 5 to 75 micrograms of carbohydrate (calculated as glucose) with steps of 5, 7, 10, 14, 20, 27, 38, 54 and 75 micrograms. These values refer to the amounts in the 10 ml of final volume. Using 0.5 ml of sample solution, and the technique described below, this is equivalent to the range 50–750 ppm of carbohydrate in the sample solution.

Technique
Dissolve, or suspend, the material to be tested in water. To 0.5 ml of this solution add 10 ml of the acid orcinol reagent (reagent 3). Heat the mixture in a boiling water-bath for 12 minutes. Cool and dilute to 50 ml with water. Cool again and transfer 10 ml of this solution to a 13.5 mm cell or standard test tube. Place this tube in the right-hand compartment of the Comparator and place the Comparator in front of a standard source of white light such as the Lovibond White Light Cabinet or, failing this, north daylight. Rotate the disc until the colour of the sample is matched by one of the colour standards. Read off the carbohydrate concentration from the indicator recess in the bottom right-hand corner of the Comparator.

Notes
1. Maltose, lactose, fructose, xylose, arabinose, sucrose, starch, and glycogen give this colour reaction; glucosamine does not. Lactic acid, tartaric acid and acetone give negative reactions.
2. Formaldehyde and benzaldehyde interfere.

References
1. B. Tollens, "*Les Hydrates de Carbone*", Dunot et Pinat, Paris, 1918
2. J. Tillmans and K. Philippi, *Biochem. Z.,* 1929, **215,** 26

The Determination of Carbon Disulphide in Benzene
using diethylamine and copper acetate

Introduction
This method was developed for determining the carbon disulphide content of benzoles and can be adapted to cover the range 1 to 1,200 mg/kg sulphur as carbon disulphide in pure benzene. This disc is specified in the Standardisation of Tar Products Test Committee (S.T.P.T.C.) Method, Serial No. R.L.B. 17-67[1], in British Standards 135:1963[2], and in Indian Standards 534 and 535[3].

The test may be applied to samples containing unsaturated compounds, provided that, if the sample is coloured, a "blank" is used in compensation. Caution is required because some unsaturated hydrocarbons produce a colour with the reagents when no carbon disulphide is present.

Principle of the Method
The sample is mixed with diethylamine and copper acetate, and the colour produced is compared with the standard colours in the disc. This was prepared in conjunction with the National Benzole Association Research Department[4,5] by carrying out the test on benzene solutions containing known quantities of carbon disulphide treated in the same way.

Reagents required
1. Diethylamine [$(C_2H_5)_2NH$] which has been redistilled within 14 days of the test, and kept in a closely stoppered amber-coloured bottle
2. A solution containing 0.03% w/v cupric acetate [$(CH_3COO)_2Cu.H_2O$] in absolute alcohol
3. Carbon disulphide-free toluene (C_7H_8), tested for carbon disulphide by the following procedure:—mix 23 ml of the toluene with 1 ml reagent 1 and 1 ml of reagent 2; the colour of the resultant mixture must not be darker than the standard disc corresponding to 20 micrograms (0.02 mg) of sulphur as carbon disulphide

The Standard Lovibond Comparator Disc 3/18
The disc covers the range 20, 30, 40, 50, 60, 70, 80, 100 and 120 micrograms (0.02 to 0.12 mg) sulphur (S) present as carbon disulphide (CS_2), and is designed for use with 40 mm cells and the Lovibond Comparator.

Technique
Pipette 1 ml reagent 1 and 1 ml reagent 2 into a 25 ml stoppered volumetric flask; add by pipette the appropriate volume (determined by a preliminary test) of sample containing between 0.04 and 0.12 mg sulphur as carbon disulphide. Dilute the mixture to 25 ml with carbon disulphide-free toluene, and mix well. Maintain the contents of the flask at 20°C ±5°C for 8 minutes after mixing, making use of a water bath if the ambient temperature is outside these limits; transfer the mixture to a 40 mm cell and place this in the right-hand compartment of the Comparator; place a portion of the untreated sample in another 40 mm cell in the left-hand side of the Comparator. Match the colour of the sample as closely as possible by rotating the disc, the colour of the next highest value being taken if an exact match cannot be made. Make the comparison by illumination of a standard source of white light, such as the Lovibond White Light Cabinet or, failing this, north daylight. Complete the comparison within 10 minutes of the mixing of the sample and reagents.

The figure shown on the disc represents the micrograms (0.001 mg) of sulphur in the original volume of sample taken. From these values the concentration of sulphur as carbon disulphide (mg/kg or mg/litre) may be calculated.

References
1. S.T.P.T.C., *Standard Methods for Testing Tar and its Products,* 6th edition, 1967
2. *British Standard,* 135:1963
3. *Indian Standards* 534 and 535 (Benzene), revised 1965
4. T. A. Dick, *National Benzole Association Memorandum,* R.465, 1946
5. T. A. Dick, *J. Soc. Chem. Ind.,* 1947, **66,** 253

The Determination of Carotene

Introduction

It is generally agreed that from the carotene content of a lucerne or grass meal an indication of its general quality may be obtained. It can be said, as a rule, that if the carotene content is high, then the protein and Vitamin C content are high and the fibre is low. There are, of course, exceptions to this rule.

There is a government regulation that certain poultry rations must contain a grass meal of a stated carotene content. Many meals are sold according to their carotene and protein contents.

Previously the Lovibond Tintometer has been largely used for the colorimetric part of the assay for carotene, but this disc offers a more rapid method, and, as it is calibrated directly in mg carotene, it is simpler to use.

Principle of the method

The β-carotene is estimated by means of the yellow colour of its solution in light petroleum. Before the β-carotene can be estimated however it is necessary to

(a) separate β-carotene from the non-active carotenoids
(b) remove all forms of chlorophyll
(c) prevent any possibility of isomerisation taking place[3].

The present method has been found to fulfil the requirements stated above. The carotene is extracted by means of light petroleum. All interfering materials are removed by chromatography on a bone-meal column, and the carotene is estimated in the eluate.

Reagent required

Light petroleum (b.pt. 80°—100°C)

The Standard Lovibond Comparator Disc 3/23

The disc covers the range 0.02—0.1 mg carotene/100 ml of solution, in steps of 0.01 mg and is designed to be used with the Lovibond Comparator and 25 mm cells.

Technique

Boil from 1 to 2 g of grass meal with 50 to 60 ml of light petroleum, of b.p. 80° to 100°C, under reflux for 1 hour on a steam bath. Cool the flask and contents and filter the extract onto a 2-inches by 1-inch column of bone meal. Rinse the flask and residue with small quantities of light petroleum. Apply suction to the column and elute with light petroleum.

Measure the volume of the carotene solution obtained. This solution is normally too concentrated for direct recording in the Comparator, therefore prepare a series of dilutions using light petroleum as the diluent. Take a solution which falls within the colour range of the disc, pour it into a 25 mm cell and place this cell in the right-hand compartment of the Comparator. Fill an identical cell with distilled water and place this cell in the left-hand compartment. Place the Comparator before a standard source of white light such as the Lovibond White Light Cabinet or, failing this, north daylight, and compare the colour of the sample with the standards in the disc.

When a match is obtained the figure shown at the indicator window represents the mg carotene per 100 ml petroleum in the actual solution under test. A simple calculation then gives the carotene content of the meal in milligrams per kilogram.

References

1. Carotene Committee of the Crop Driers Association, *Analyst,* 1941, **66**, 334
2. W. A. G. Nelson, *Analyst,* 1947, **72**, 200
3. L. Zechmeister, *Chem. Revs.,* 1944, **34** (2), 267
4. Anal. Methods Ctte. of the Soc. for Anal. Chem., *Analyst,* 1950, **75**, 568; 1952, **77**, 171

The Determination of Cyanide (1)
using pyridine-pyrazolone reagent

Introduction

The determination of traces of cyanide is of considerable general importance. Cyanides are one of the most toxic of all chemicals towards fish, and great care has to be taken to avoid discharging untreated cyanide wastes into rivers and streams. Such toxic wastes may arise from electroplating and case-hardening processes and also from gas liquors. The destruction of cyanide in such wastes can be carried out by chlorination in alkaline solution. The cyanide content of effluents, after treatment, should not exceed 1-2 ppm (as HCN). This test has been devised primarily for the estimation of cyanide in river waters, sewage and trade effluents.

Principle of the method

The method is based on those described by Epstein[1] and the American Public Health Association[2]. After treatment with chloramine-T cyanide reacts with a pyridine-pyrazolone reagent to give a pink colour which changes through mauve to blue on standing. The blue colour attained after 35 to 40 minutes is sufficiently stable to permit accurate visual comparison with Lovibond permanent glass standards. Chlorate, nitrate, phosphate, borate and sulphate do not interfere. Heavy metals, which might interfere, are removed by prior distillation of the sample.

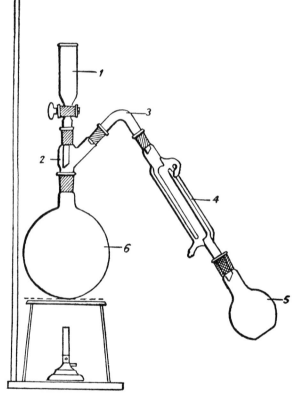

1. Dropping funnel   Cat. No. D. 3/12
2. Multiple adaptor   Cat. No. M.A. 2/3
3. Bend   Cat No. S.H. 1/22
4. Double surface condenser   Cat. No. C. 5/12
5. 250 ml Pyrex, flat bottom flask, medium neck B. 24
6. 1,000 ml Pyrex, round bottom flask, short neck B. 24

    Also required:—   B. 24 stopper
                               B. 19 stopper

Catalogue numbers refer to:—
    Quickfit & Quartz Ltd., Mill Street, Stone, Staffs.

Reagents required

1. *Acetic acid*   0.5N solution of glacial acetic acid (CH_3COOH analytical reagent grade) in distilled water.
2. *Chloramine-T*   0.2% aqueous solution which solution must be prepared freshly each day.
3. *Pyridine-pyrazolone reagent*   Dissolve 0.5 g of 3-methyl-1-phenyl-5-pyrazolone in 250 ml of a 50:50 mixture of ethyl alcohol (absolute, analytical reagent grade), and distilled water. This solution is stable. Dissolve, with shaking, 0.020 g of bis-pyrazolone in 20 ml of pyridine make sure solution is complete and make up to exactly 100 ml with the 3-methyl-1-phenyl-5-pyrazolone solution. The resulting solution is unstable, even when stored in the dark, and must be prepared freshly each day.

The Standard Lovibond Nessleriser Disc NOC

This disc covers the range 0.02—0.2 ppm of cyanide (CN^-), in steps of 0.02 ppm.

Technique

Place 250 ml of the sample in an all glass distillation apparatus (see figure), and distil to completion (Note 2). Cool the distillate and make up to 250 ml with distilled water, in a volumetric flask. Pipette 25 ml of this distillate into a Nessleriser cylinder, and neutralise to pH 6 to 7, as shown by indicator papers, using the dilute acetic acid. Add 1 ml of the chloramine-T solution, mix, and allow to stand for 2 minutes. Add 5 ml of the pyridine-pyrazolone reagent, adjust the final volume to 50 ml with distilled water and mix. Exactly 40 minutes after adding the chloramine-T reagent place the cylinder in the right-hand compartment of the Nessleriser, place an identical cylinder filled with distilled water in the left-hand compartment and match the colour of the sample solution against that of the standards in the disc, using a standard source of white light, such as the Lovibond White Light Cabinet or, failing this, north daylight.

Notes

1. It must be emphasized that the readings obtained by means of the Lovibond Nessleriser and disc are only accurate provided that Nessleriser cylinders are used which conform to the specification employed when the discs were calibrated, namely, that the 50 ml calibration mark shall fall at a height of 113 mm $\pm$ 3 mm measured internally.
2. This method covers cyanides, soluble in water, which hydrolyse and pass over as HCN. The addition of 0.1 g of tartaric acid per 100 ml of sample will ensure full recovery of cyanides. In effluent samples which contain free alkali it is necessary to make the sample acid before distillation.

References

1. J. Epstein, *Anal. Chem.*, 1947, **19**, 272
2. "*Standard methods for the Examination of Water and Wastewater*", American Public Health Association (A.P.H.A.) (11th edition), New York, 1965

COLORIMETRIC CHEMICAL ANALYTICAL METHODS

The Determination of Cyanide (2)
using benzidene-pyridine

Introduction

This[1,2] was the original cyanide test for which Lovibond permanent glass colour standards were prepared. Details of the test, and the corresponding disc, were withdrawn however when benzidene ceased to be available on the British market. Although the use of benzidene, which is a powerful carcinogen, is prohibited in Britain, this test has been reinstated in response to the demands of chemists from countries where no such prohibition is extant. It has the advantage over the pyridine-pyrazolene test[3,4] of not requiring a prior distillation. It is thus much more useful for applications, such as the examination of river water for pollution, where it is inconvenient to have to take samples to a laboratory for examination.

Principle of the method

The cyanide is first converted to cyanogen bromide, which is then coupled with benzidene and pyridine to form a red colour. The intensity of this colour, which is proportional to the concentration of cyanide, is measured by comparison with a series of Lovibond permanent glass colour standards.

Reagents required

1. *2N acetic acid* (CH_3COOH)
2. *Bromine water (saturated)*
3. *Arsenious acid solution* Dissolve 2 g of arsenious oxide (As_2O_3) in 100 ml of distilled water
4. *Benzidene reagent* Dissolve 5 g of benzidene hydrochloride (Note 1) in 100 ml of distilled water containing 2 ml of hydrochloric acid (HCl sp gr 1.18). Prepare this solution freshly as required each day
5. *Pyridine reagent* Approximately 60% v/v solution of pyridine in distilled water (mixture of constant boiling point)

The Standard Lovibond Comparator Disc 3/33

This disc covers the range 0.05 to 1.0 ppm of cyanide (HCN). The steps are 0.05, 0.1, 0.2, 0.3, 0.4, 0.5, 0.6, 0.8 and 1.0 ppm. A special stoppered tube is required.

Technique

Take 5 ml of sample, adjusted by dilution if necessary to contain not more than 2 µg of cyanide, and place this in a glass-stoppered comparator tube calibrated at 10 ml. Add 0.9 ml of $2N$ acetic acid (reagent 1), then add 0.2 ml of bromine water (reagent 2) and mix thoroughly. After 2 minutes add 0.2 ml of arsenious acid (reagent 3) to remove excess bromine, and disperse any released bromine vapour by blowing across the mouth of the tube. Mix 3 ml of pyridine solution (reagent 5) with 0.6 ml of benzidene reagent (reagent 4) and add this mixture to the contents of the tube, dilute to the 10 ml mark with distilled water and mix thoroughly. Stopper the tube and allow the mixture to stand in the dark for 25–30 minutes at a temperature between 15 and 20°C. Prepare a blank using 5 ml of distilled water in the place of the sample but otherwise following the above instructions in every respect. After the colours have developed place the sample tube in the right-hand compartment of the Comparator and place the blank in the left-hand compartment. Illuminate the Comparator with a standard source of white light, such as the Lovibond White Light Cabinet, and match the colour of the sample with that of the standards in the disc. If no standard source of white light is available then north daylight should be used wherever possible.

Notes

1. **Benzidene is a highly carcinogenic compound and its use is forbidden in the U.K. by Statutory Instrument. Great care should be taken not to allow benzidene or its solutions to come into contact with the skin.**
2. Thiocyanate responds equally well to this test in proportion to its molecular weight. There is no interference from ferricyanide, cyanate, ammonium salts or cobalt salts.
3. The blank is likely to fade in time. If a large number of tests is being carried out it is advisable to prepare a fresh blank every hour.
4. If river water is being tested, turbidity and as much colour as possible should be filtered out.
5. If possible the test should be carried out at 20°C and the temperature of all solutions should be kept at, or adjusted to, this level.
6. If the water being tested is decidedly alkaline, an aliquot of 5 ml should be titrated to neutrality to phenolphthalein using $2N$ acetic acid. An equal amount of acid should then be added to both the 5 ml test sample, and to the blank, in addition to the 0.9 ml mentioned in the test procedure. The cyanide concentration found must then be adjusted by multiplying by the factor (10+vol. added in mls)/10 to allow for the additional dilution. Unless the water is very alkaline this factor will be within the experimental error of the method and can be ignored.
7. The test tube in which the solutions are mixed should be scrupulously cleaned and should be scrubbed with a test tube brush between tests to ensure that no benzidene adheres to the tube walls. Mixing should be carried out with a glass rod and **under no circumstances** should the reagent be allowed to come in contact with the skin. The acetic acid used to render the solution slightly acidic should not be added until immediately before the addition of bromine water.

References

1. W. N. Aldridge, *Analyst,* 1944, **69**, 262
2. W. N. Aldridge, *Analyst,* 1945, **70**, 474
3. J. Epstein, *Anal. Chem.*, 1947, **19**, 272
4. Cyanide(1) p. 38

The Determination of Cyanide (3)
using pyridine and *p*-phenylenediamine

Introduction
The reaction of cyanogen bromide with pyridine and an aromatic amine (the König synthesis), which is a standard method for the determination of low concentrations of cyanide ion, has been examined[1] with a view to finding a better heterocyclic amine than pyridine and a less carcinogenic aromatic amine than benzidene.

No heterocyclic amine better than pyridine was found. The most favourable aromatic amine, when carcinogenic hazards were taken into consideration, proved to be *p*-phenylenediamine. The present test[2] was therefore developed for cyanide ion in concentrations of 0.05-1.0 ppm in waters.

Principle of the method
In acid solutions cyanides react with pyridine and *p*-phenylenediamine to form a dark red coloured complex. The intensity of the colour of this complex, which is proportional to the concentration of cyanide ion, is measured by comparison with a series of Lovibond permanent glass colour standards.

Reagents required
1. *Hydrochloric acid* Concentrated
2. *Bromine water, saturated* Add 3 ml of bromine to 100 ml of distilled water. This reagent should be prepared on day of use.
3. *Pyridine* Add 3 ml of reagent 1 to 18 ml of pyridine dissolved in 12 ml of water.
4. *p-Phenylenediamine dihydrochloride* Dissolve 0.36 g of *p*-phenylenediamine dihydrochloride in 100 ml of distilled water. Store the solution in a brown bottle.
5. *Sodium arsenite* Dissolve 1.5 g of sodium arsenite ($NaAsO_2$) in 100 ml of water.
6. *Pyridine-p-phenylenediamine reagent* Mix 3 volumes of reagent 3 with 1 volume of reagent 4.

All chemicals used for the preparation of reagents should be of analytical reagent quality.

The Standard Lovibond Comparator Disc 3/86
This disc contains standards corresponding to 0.05, 0.1, 0.2, 0.3, 0.4, 0.5, 0.6, 0.8 and 1.0 ppm of cyanide (CN^-) ions.

Technique
Place 5 ml of sample solution in one 13.5 mm comparator tube and 5 ml of distilled water in a second, identical, tube. Add 0.1 ml (2 drops) of reagent 1, and 0.2 ml (4 drops) of reagent 2 to each tube and mix thoroughly. After 2 minutes add 0.2 ml (4 drops) of reagent 5 to each tube to remove any excess bromine, add 4 ml of reagent 6 and make up to the 10 ml mark with distilled water. Place the sample tube in the right-hand compartment of the Comparator and the second tube in the left-hand compartment. After 30 minutes place the Comparator before a uniform source of white light such as the Lovibond White Light Cabinet or, failing this, north daylight, and match the colour of the sample solution with the standards in the disc.

Acknowledgement is made to Lancy Laboratories Ltd. for assistance in developing this disc.

Note
All colorimetric methods for cyanide only react to "free" and simple complexes of cyanide, i.e. sodium, potassium, zinc, copper and cadmium cyanides.

Heavy metals, e.g. nickel, iron, silver and gold, form tight complexes with cyanide which can only be broken down by distillation. If a sample containing a complex cyanide is distilled and the distillate collected in caustic soda, then the cyanide is present as sodium cyanide, and thus a figure for total cyanide may be obtained.

References
1. L. S. Bark and H. G. Higson, *Talanta*, 1964. **11**, 471
2. L. S. Bark and H. G. Higson, *Talanta*, 1964, **11**, 621

The Determination of Cyanuric Acid in Swimming Pool Water
using Palin CNA Reagents

Introduction
The addition of cyanurates to swimming pool water enables a reserve of "free available chlorine" to be built up in the water, and this is now increasingly being used. The recommended level is a minimum of 30 ppm cyanuric acid and the top level is still a matter of debate.

Too low a level is ineffective, and too high a level is inadvisable. The new colorimetric testing method, devised by Dr. A. T. Palin, is simple to perform and is accurate.

Principle
Any possible errors in the test caused by the presence of interfering substances in the water are virtually eliminated by first diluting the sample with a considerable excess of distilled water. If it is necessary to use tap water see Note 2. At a suitable pH, obtained by adding a buffer solution, the cyanuric acid in the diluted sample is precipitated by adding an appropriate amount of a very dilute solution of a mercuric salt. After filtering, a suitable indicator is added to the filtrate to produce a colour which provides a measure of the original concentration of cyanuric acid in the water.

Reagents required
1. Buffer Solution CNA–1
2. Precipitating Solution CNA–2
3. Indicator Solution CNA–3

The Standard Lovibond Comparator Disc 3/92
This disc provides standards 20, 25, 30, 35, 40, 50, 60, 70 and 80 ppm cyanuric acid based on a 2 ml sample.

Technique
Pipette 2 ml sample of the pool water into a measuring cylinder or Nessler tube graduated at 50 ml Make up to this mark with distilled water. (If using tap water see Note 2.) Then add 0.5 ml of Buffer solution CNA–1 and exactly 0.5 ml of Precipitating Solution CNA–2. Mix thoroughly.

Filter through Whatman No. 1 paper into 10 ml Comparator cell using the first runnings to rinse the cell. Then fill to 10 ml mark. Next add 0.2 ml of Indicator Solution CNA–3 to this 10 ml cell. Mix *gently* and immediately match against the disc using a distilled water blank in the left hand compartment of the Comparator.

Notes
1. The colours produced gradually precipitate on standing or if mixed vigorously. It is desirable, therefore, that colour matching should be completed within 1 or 2 minutes.
2. While the use of distilled water for dilution is preferred, it has been shown that tap water may be used instead. The resultant colours may, however, be too red for matching against the disc. To correct this it is necessary to use a different buffer, which is available from The Tintometer Limited in crystal form. The above standard procedure must then be modified as follows:

   (a) Omit the use of Buffer Solution CNA–1
   (b) If the final colour in the Comparator tube is too red add a pinch (0.1 to 0.2 g) of Buffer Crystals CNA–4 and mix *gently* until the red tint disappears and a match can be made.

3. For lower concentrations of cyanuric acid, 5 ml instead of 2 ml of sample may be taken, in which case the disc readings are multiplied by 0.4 thus giving a range of values from 8 to 32 ppm. For higher concentrations use 1 ml of sample, when the readings are doubled to give a range from 40 to 160 ppm.

4. Reagents should be stored in the amber bottles in a cool place. The Indicator Solution CNA–3 will not keep indefinitely (shelf life about 12 months) and is particularly susceptible to exposure to strong light. The standard Tintometer amber bottle with small pipette should be used with all reagents. A mouth-pipette must not be used with the Precipitating Solution CNA–2.

5. A complete kit consisting of a fitted wood case is available containing:
Lovibond Comparator
3 square moulded 10 ml test tubes
Disc 3/92 (colour standards for 20, 25, 30, 35, 40, 50, 60, 70, 80 ppm)
3 bottles of reagents with fitted pipettes
1 Nessler cylinder for measuring 50 ml sample and a stirring rod funnel and filter paper
1 pipette 2 ml
Code AF121

If reagent CNA–4 (see Note 2) is required, this must be ordered as an accessory.
A complete kit incorporating this test, and also tests for pH, Residual Chlorine and Alkalinity is also available. See leaflet available from The Tintometer Limited, Salisbury, England.

The Determination of Anionic Synthetic Detergents

using Methylene Blue

Introduction

This test supercedes an earlier method (based on "Teepol", an alkyl sulphate) which used disc 3/24. The present method has been developed to enable small amounts of any anionic detergent to be estimated. This test differs from previous tests in the method of extraction of the detergent. This has been the subject of several papers[1-5] and, of the various methods proposed, that of Longwell and Maniece[5] has been accepted as the 'official' test in Great Britian. The procedure is however still not completely satisfactory, particularly in comparison with the 'referee' test[6] which uses infrared absorption spectroscopy. An alternative extraction procedure developed by Webster and Halliday[7] is also described. This is claimed to give much better agreement with results obtained using the 'referee' test.

The basis of calibration has been changed from "Teepol" to "Manoxol O.T.", as extensive investigation by Longwell and Maniece[5], showed that this material was a more stable and reproducible basis for standardisation.

Principle of the method

Alkylbenzenesulphonates combine with Methylene Blue to form a complex which is soluble in chloroform. The intensity of the colour in the chloroform layer is proportional to the amount of detergent present, which is estimated by comparison of the colour with that of the Lovibond permanent glass standards.

Other substances found in sewage and river waters also form a chloroform-soluble complex with Methylene Blue. In the original test this interference was overcome, at least in part, by a differential extraction at two different pH's. In the 'official" method of Longwell and Maniece[5] this extraction technique is replaced by an alkaline extraction followed by an acid wash. In the Webster and Halliday[7] method the sample is first subjected to acid hydrolysis, then to extraction with an amine solution and then finally to alkaline chloroform extraction and acid washing, as in the 'official' method.

Reagents required

1. *Sulphuric acid* H_2SO_4 approximately $1.0N$
2. *Sodium hydroxide solution* NaOH 10% w/v, aqueous
3. *Chloroform* $CHCl_3$ analytical reagent grade
4. *Light petroleum,* boiling range 40° to 60°C
5. *1-Methylheptylamine* (available from Eastman Chemicals) or 2-amino octane
6. *Neutral buffer solution* Dissolve 10 g of potassium dihydrogen orthophosphate (KH_2PO_4) in 800 ml of distilled water, adjust to pH 7.5 with sodium hydroxide solution (reagent 2) and dilute to 1 litre with distilled water.
7. *Alkaline buffer solution* Dissolve 10 g of disodium hydrogen orthophosphate (Na_2HPO_4) in 800 ml of distilled water, adjust to pH 10.5 with sodium hydroxide solution (reagent 2) and dilute to 1 litre.
8. *Methylene Blue* Dissolve 0.2 g B.P. quality in 1000 ml of distilled water. To reduce the blank value, extract sufficient of this solution for one day's use three times with chloroform, discarding the extracts (See Note 1).

The Standard Lovibond Comparator Disc 3/48

The disc covers the range 1 to 16 ppm of active material, (1, 2, 3, 4, 6, 8, 10, 12, 16) calculated as Manoxol O.T. (sodium dioctylsulphosuccinate) and based on a 10 ml sample.

Disc 3/24, which was calibrated against 'Teepol' in the original test, may also be used with these methods provided that the figure obtained from the disc is multiplied by 1.4 to convert it to the equivalent value for Manoxol O.T.

Technique

(a) "Official" method[5]

Take sufficient sample to contain 20 to 150 μg of anion active material (Note 2). Place sample in a separating funnel and make up to 100 ml with distilled water*.

Add 10 ml of the alkaline phosphate solution (Reagent 7), 5 ml of Methylene Blue solution (reagent 8) and 15 ml of chloroform. Shake gently, twice a second, for 1 minute. Allow to separate. Break up any emulsion by gentle agitation with a polythene rod. Run the clear chloroform layer into a second separating funnel containing 110 ml of distilled water, 5 ml of Methylene Blue solution and 1 ml of 1.0N sulphuric acid (reagent 1). Rinse the first separating funnel with 2 ml of chloroform and run this into the second separating funnel. Shake the second separating funnel and allow the layers to separate. Run the chloroform layer through a small funnel, plugged with glass wool which has been moistened with chloroform, into a 50 ml volumetric flask. Rinse the second separating funnel with 2 ml of chloroform and add this to the contents of the flask. Repeat the extraction of the contents of the two funnels twice more with further 10 ml portions of chloroform, and combine the extracts in the volumetric flask. Make up to the 50 ml mark with chloroform.

Thoroughly mix the contents of the volumetric flask by repeated inversion, and transfer an aliquot to a standard test tube. Place this tube in the right-hand compartment of the Comparator. Repeat the above extraction procedure on reagents alone and place an aliquot of the resulting chloroform solution in an identical test tube in the left-hand compartment of the Comparator. Compare the colour of the test solution with that of the permanent glass standards in the disc, using a standard source of white light, such as the Lovibond White Light Cabinet or, failing this, north daylight. For interpretation, see Note 2.

(b) Webster and Halliday modification[7]

Place a volume of the sample containing 25 to 150 μg of anionic detergent in a 400 ml beaker, and adjust the volume to 100 ml by dilution or evaporation as necessary. Use 100 ml of distilled water as a blank and treat this in exactly the same manner as the sample.

Add 50 ml of 1.0N sulphuric acid (reagent 1), cover with a clock-glass and boil for 1 hour, adjusting the rate of boiling so that the volume is about 50 ml at the end of this time. Add water if necessary to prevent the volume falling below this level. Neutralise the solution to phenolphthalein with sodium hydroxide solution (reagent 2), taking care not to overshoot the endpoint, and transfer to a 250 ml separating funnel with sufficient water to make a total volume of 100 ml. Use this same beaker for the subsequent collection of chloroform extracts.

Add 10 ml of the neutral buffer solution (reagent 6), and extract four times with separate 25 ml portions of chloroform, to each of which has been added 1 drop of 1-methylheptylamine from a pipette dropper. (For samples that form persistent emulsions, increase the number of extractions to five; most emulsions can be readily broken by rubbing with a paddle-shaped polythene rod). For each extraction shake gently and evenly twice a second for two minutes. Combine the chloroform extracts in the 400 ml beaker, add 5 ml of water, and evaporate the chloroform and excess of amine on a steam or hot-water bath. Discard the aqueous layer from the separating funnel, which can be used, without washing, for the extraction with light petroleum.

Wash the sides of the beaker with 10 to 15 ml of water, warm for a few minutes on the bath to ensure complete dissolution, and transfer to the separating funnel with sufficient water to make a total volume of 50 ml. Reserve the beaker for the subsequent collection of light petroleum extracts. Add 10 ml of neutral buffer solution, and extract four times with 25 ml portions of light petroleum, to each of which has been added 1 drop of 1-methylheptylamine. Shake for 2 minutes for each extraction. Run the aqueous layer, as necessary, into a second separating funnel, and combine the light petroleum extracts in the 400 ml beaker.

Discard the aqueous layer after the fourth extraction, place 15 ml of methanol containing 1 drop of 1-methylheptylamine in the separating funnel, and shake for 30 seconds. Run the methanol into the second separating funnel, and again shake for 30 seconds. Run the methanol into the beaker containing the light petroleum extracts. Repeat the rinsing procedure with 25 ml of water, and add this to the methanol-light petroleum mixture.

Evaporate the light petroleum on the steam or hot-water bath, and then boil on a hot-plate until free from excess of amine, as indicated by odour and by the solution becoming colourless to phenolphthalein. Maintain the volume at approximately 50 ml by adding water as necessary. Add 5 ml of sodium hydroxide solution (reagent 2), and boil gently for 15 minutes. Occasionally rinse the sides of the beaker with water to maintain the volume at approximately 50 ml.

Remove all traces of 1-methylheptylamine from the separating funnel by rinsing with alcohol-hydrochloric acid mixture and then with water, and discard the rinsings. Transfer the sample solution to the separating funnel, neutralise to phenolphthalein with dilute sulphuric acid (reagent 1) and add 1 drop of dilute sodium hydroxide solution (reagent 2 diluted 1-250, i.e. 1 drop in 10 ml of distilled water) to adjust the pH to between 8 and 10.

Cool and proceed as from * in the 'official' method.

Notes

1. The extracted methylene blue solution should not be stored for more than 1 day, as any traces of chloroform retained may slowly decompose and eventually lead to high blanks.

2. It is generally impracticable to take more than 10 ml of raw or settled sewage, owing to the degree of emulsion formation on shaking with chloroform, but it is possible to take up to 100 ml of good quality effluent, river or drinking water when the detergent content is very low. If a sample volume other than 10 ml is taken the final result is calculated from the reading obtained from the disc by means of the following formula :—

$$\text{Detergent concentration (as Manoxol O.T.} =) \frac{\text{Disc reading} \times 10}{\text{Sample volume in ml}} \text{ppm}$$

References

1. J. H. Jones, *J. Ass. Off. Agric. Chem.*, 1945, **28**, 398
2. H. C. Evans, *J. Soc. Chem. Ind.*, 1950, **69**, S76
3. W. F. Lester and R. D. Raybould, *J. Inst. Sew. Purif.*, 1950, 393
4. P. N. Degens, H. C. Evans, J. D. Kommer and P. A. Winsor, *J. Appl. Chem.*, 1953, **3**, 54
5. J. Longwell and W. D. Maniece, *Analyst*, 1955, **80**, 167
6. E. M. Sallee et al, *Anal. Chem.*, 1956, **28**, 1822
7. H. L. Webster and J. Halliday, *Analyst*, 1959, **84**, 552

The Determination of Dichlorophen (1)

(Panacide*; 5,5′-dichloro-2,2′-dihydroxy-diphenyl-methane)

*Registered Trade Mark of BDH Chemicals Ltd., Poole, Dorset.

using 4-amino-antipyrine

Introduction

Phenolic fungicides are extensively used as mildew preventatives for fabrics destined to be used in tropical climates. One of the compounds most extensively used for this purpose is dichlorophen, as an algicide, bactericide, and fungicide.

The method which has been most generally applied to the determination of dichlorophen in fabrics is that of Gottlieb and Marsh[1]. Ashton[2] modified the oxidation stage of the original method, in order to improve the stability of the reagent blanks, and the present test is an adaptation of Ashton's method. This test was developed at the request of the British Standards Institution Committee formed for the revision of B.S. 2087.

Principle of the method

This method is based on the observation by Emerson[3], that 4-amino-antipyrine condenses with phenols, in the presence of an alkaline oxidising agent, to form an antipyrine dye. The condensation of dichlorophen and 4-amino-antipyrine in the presence of alkaline potassium periodate yields a red colour, the intensity of which is proportional to the concentration of dichlorophen. The intensity of the colour is estimated by comparison with Lovibond permanent glass standards.

Reagents required

1. *Buffer solution   pH 9.68*

Borax ($Na_2B_4O_7.10H_2O$), analytical reagent grade	13.75 g
Sodium hydroxide (NaOH), analytical reagent grade	1.125 g *
Distilled water	1,000 ml

     * Alternatively, use 28.2 ml of N NaOH, and reduce the amount of water accordingly.

2. *1% 4-amino-antipyrine solution*   Dissolve 1 g of 4-amino-antipyrine in 100 ml of reagent 1

3. *0.25% Potassium periodate solution*   Dissolve 1.25 g of analytical reagent grade potassium periodate (KIO_4) dissolved in 500 ml of reagent 1

The Standard Lovibond Comparator Disc 3/57

This disc covers the range 0.4 to 2.0% of dichlorophen in steps of 0.2%, based on a 1 g sample, and is calibrated for use with a 5 mm cell.

Technique

Extract a 1 g sample by boiling with four successive 25 ml portions of reagent 1. Allow each individual portion to boil for 7 minutes before decanting. Combine the four extracts cool, add a further 100 ml of reagent 1 and make up to 200 ml with distilled water. Filter a portion and, from the filtrate pipette 10 ml into a 50 ml volumetric flask. Add roughly 30 ml of reagent 1; accurately pipette in 1.5 ml of reagent 2 *followed by* 10 ml of reagent 3 (Note 2).

Dilute the solution to the 50 ml mark with reagent 1 and mix well. Transfer a portion of the diluted solution to a 5 mm cell and place in the right-hand compartment of the Comparator.

Fill an identical 5 mm cell with a blank solution, prepared by mixing 38.5 ml of reagent 1, 1.5 ml of reagent 2 and 10 ml of reagent 3, ***mixed in that order*** (Note 2), and place the cell in the left-hand compartment. Five minutes after placing the solutions in the Comparator match the colour of the test solution against the Lovibond permanent glass standards in the disc, using a standard source of white light, such as the Lovibond White Light Cabinet or, failing this, north daylight.

Notes

1. This modification of Ashton's[2] method has been developed by B.D.H. Chemicals Ltd. Permission to quote this method is gratefully acknowledged.

2. The amino-antipyrine solution (reagent 2) must always be added and well mixed before the periodate solution (reagent 3) is added.

It is also important to ensure that the temperature of this reaction remains within the range 17–20°C if accurate results are to be obtained.

References

1. S. Gottlieb and P. B. Marsh, *Ind. Eng. Chem. Anal. Edn.,* 1946, **18,** 16
2. F. C. Ashton, *Analyst,* 1960, **85,** 685
3. E. I. Emerson, *J. Org, Chem.,* 1943, **8,** 417

The Determination of Dichlorophen (2)

(Panacide*, 5,5' - dichloro - 2,2' dihydroxy-diphenyl-methane)

*Registered Trade Mark of BDH Chemicals Ltd., Poole, Dorset.

using 4-amino-phenazone (4-amino-antipyrine)

Introduction

Dichlorophen is widely used as a mildew preventive for fabrics, for slime prevention in paper-making machinery, and for size preservation and impregnation of paper against rot.

Dichlorophen also has many applications in the prevention of algae, bacterial slime and fungal growth in water, and hence is used in cooling towers and in refrigeration, humidifying and air-conditioning plant.

Principle of the method

The method is based on the production of a red colour when dichlorophen condenses with 4-aminophenazone in the presence of alkaline potassium periodate. Emerson[1] originally observed that phenols condensed with 4-aminophenazone to give phenazone dyes, and Gottlieb and Marsh[2] adopted this principle for the estimation of dichlorophen in textiles. Ashton[3] modified this method by replacing the potassium ferricyanide oxidising agent with potassium periodate, which gives more stable reagent blanks.

Reagents required

1. *Buffer solution pH9.68*

Borax ($Na_2B_4O_7.10H_2O$), analytical reagent grade	13.75 g
Sodium hydroxide (NaOH), analytical reagent grade	1.125 g *
Distilled water to	1,000 ml

    * Alternatively, use 28.2 ml of N NaOH, and reduce the amount of water accordingly.

2. *1% 4-aminophenazone solution*
    1 g of 4-amino-phenazone in 100 ml of reagent 1.

3. *0.25% potassium periodate solution*
    1.25 g of analytical reagent grade potassium periodate (KIO_4) dissolved in 500 ml of reagent 1. (This reagent should be freshly prepared every few days).

The Standard Lovibond Comparator disc 3/57B

This disc covers the range 20 to 100 ppm of dichlorophen in steps of 10 ppm based on a 10 ml water sample and is calibrated for use with a 5 mm cell.

Technique

To a 10 ml sample in a 50 ml Nessler cylinder or volumetric flask, add 25 ml of reagent 1. Accurately pipette in 1.5 ml of reagent 2, mix well, and then add 10 ml of reagent 3. Dilute to 50 ml with reagent 1 and mix well. Transfer a portion of the solution to a 5 mm cell and place it in the right hand compartment of the comparator.

A blank solution is prepared by adding 1.5 ml of reagent 2 to approximately 35 ml of reagent 1 in a Nessler cylinder. The solution is mixed and 10 ml of reagent 3 is added. The volume is made up to 50 ml with reagent 1 and the solution mixed well. A portion is transferred to a 5 mm cell and placed in the left hand compartment of the comparator.

The colour is matched against the disc 10 minutes after the addition of reagent 3, using North daylight or failing this the Lovibond white light equipment.

Notes

1. This modification of Ashton's method has been developed by BDH Chemicals Limited and permission to quote it is gratefully acknowledged.

2. The 4-amino-phenazone solution (reagent 2) must always be added and well mixed before the periodate (reagent 3) is added. If accurate results are to be obtained the temperature of the reaction must be maintained within the range 17 – 20°C.

3. It must be emphasized that the reaction is not specific for dichlorophen. If other phenolic compounds are present a similar colour will be produced under the above conditions.

4. If the colour produced is paler than the 20 ppm step on the disc it is possible to obtain a reading by using a cell of greater optical depth, e.g. if a 20 mm cell is used the disc readings are divided by 4 to give a range of 5 to 25 ppm.

References

1. E. I. Emerson, *J. Org. Chem.*, 1943, **8,** 417
2. S. Gottlieb and P. B. Marsh, *Ind. Eng. Chem. Anal. Ed.*, 1946, **18,** 16
3. F. C. Ashton, *Analyst*, 1960, **85,** 685.

The Determination of Formaldehyde

using chromotropic acid
(1,8-dihydroxynaphthalene–3,6-disulphonic acid)

Introduction

Formaldehyde is one of the most important industrial chemicals, being used in the manufacture of synthetic resins and lacquers; as an intermediate in the manufacture of dyestuffs, explosives and other chemicals; and in dyeing, bleaching, calico printing, etc. Its ability to render glue and gelatin almost completely insoluble has resulted in its use in the textile and printing industries as a waterproofing agent for fabrics, for fixing glues and sizes and for increasing the fastness of dyes to washing.

The antiseptic properties of formaldehyde are used in such pharmaceutical preparations as medicines, deodorants and general and internal antiseptics. It has been prohibited for use as a food preservative but may occasionally be found in both milk and meat. It is extensively used in the preservation of histological and anatomical specimens.

Principle of the method

When warmed with chromotropic acid, in concentrated sulphuric acid solution, formaldehyde gives a violet-red colour. The exact chemistry of this reaction is, as yet, unknown. It has been suggested[1] that the first step is probably a condensation of formaldehyde with an aromatic hydroxyl group of the chromotropic acid producing a hydroxydiphenylmethane derivative. This would be soluble in the sulphuric acid and, on contact with air, is assumed to lead to a coloured oxidation product. Alternatively it is suggested[2] that the action of the concentrated sulphuric acid causes a phenol-aldehyde condensation followed by oxidation to a quinoid compound.

The chromotropic acid solution does not react with acetaldehyde, proprionaldehyde, n- and iso-butyraldehydes, iso-valeraldehyde, œnanthol, crotonaldehyde, chloralhydrate, glyoxal or aromatic aldehydes. A yellow colour is produced with glyceraldehyde, furfural, arabinose, fructose and sucrose. If, however, the blank solution is made up with an equal amount of any of these materials, then 0.5 micrograms of formaldehyde may be detected in the presence of 100 times the quantity of sucrose or fructose and of more than 300 times the quantity of the other compounds. Other sugars, acetone and carboxylic acids do not react with the chromotropic acid reagent, neither do aromatic acids[2]. The halogen derivatives of aromatic acids, and also phenoxy-acetic acid, do react to produce a wine colour[3,4].

Certain organic compounds split off formaldehyde when treated with acids and therefore develop a colour with the acid solution of chromotropic acid; hexamethylene-tetramine, formaldoxime, cellulose formals, etc.[2,5] are typical of such compounds.

This method may be adapted for other estimations. Formic acid may be determined in small quantities by reduction to formaldehyde by nascent hydrogen[1,2]. Glucose interferes with this determination because it is partially reduced to formic acid and thence to formaldehyde during the reduction. Similarly methyl alcohol may be estimated in the presence of ethyl and higher alcohols, sugar and other aldehydes, etc., if the oxidation of methyl alcohol to formaldehyde, with potassium permanganate, is carried out under specific conditions[2,6].

Some metals, such as titanium, hexavalent chromium, and iron, will interfere with this determination. Oxidising agents, such as chlorates and nitrates, also interfere[7].

Reagents required

1. *Chromotropic acid sodium salt* A special reagent prepared by B.D.H. Chemicals Ltd., for formaldehyde determinations, and should be specified as such when ordering
2. *Sulphuric acid, concentrated* H_2SO_4 analytical reagent grade
3. *Reagent solution* Add 98 ml of sulphuric acid to 2 ml of a 5% aqueous solution of the chromotropic acid salt. This solution must be prepared freshly each day. Great care must be exercised when adding the concentrated sulphuric acid to the aqueous solution, especially in the initial stages of the addition

The Standard Lovibond Nesseleriser Disc NOD

The disc contains 9 standards corresponding to 1, 2, 3, 4, 5, 6, 7, 8 and 10 micrograms of formaldehyde (HCHO).

Technique

Take 4 ml of the sample, or adjust the amount taken to exactly 4 ml, with distilled water, add 4 ml of the chromotropic acid reagent (reagent 3) rapidly, but with caution, mix, and allow to stand for 30 minutes. The heat developed accelerates the reaction and full colour development. Transfer to a standard Nessleriser glass and make up to 50 ml with distilled water. Place the glass in the right-hand compartment of the Nessleriser. Fill an identical glass to the mark with distilled water and place in the left-hand compartment. Match the colour of the sample against the permanent glass standards in the disc using a standard source of white light, such as the Lovibond White Light Cabinet or, failing this, north daylight. The reading on the disc gives the number of micrograms of formaldehyde in the amount of sample used.

Notes

1. Great care must always be taken in handling concentrated acids, and it must be remembered that the chromotropic acid reagent is still virtually concentrated acid.

2. It must be emphasized that the readings obtained by means of the Lovibond Nessleriser and disc are only accurate provided that the Nessleriser glasses used conform to the specification employed when the disc was calibrated, namely that the 50 ml calibration mark falls at a height of 113 ± 3 mm measured internally.

References

1. E. Eegriwe, *Z. Anal. Chem.*, 1937, **117**, 22
2. F. Feigl, *"Spot Tests"*, 4th edition, Elsevier Publishing Co., London, 1954
3. V. H. Freed, *Science*, 1948, **107**, 98
4. R. P. Marquardt and E. N. Luce, *Anal. Chem.*, 1951, **23**, 1484
5. C. L. Hoaffpavir, *Ind. Eng. Chem. (Anal. Ed.)*, 1943, **15**, 605
6. E. Eegriwe, *Mikrochim. Acta.*, 1937, **2**, 329
7. *"The BDH Book of Organic Reagents"*, B.D.H. Chemicals Ltd., Poole

The Determination of Furfuraldehyde
using aniline acetate

Introduction

This test has been developed primarily for the determination of furfuraldehyde in whisky, as this gives useful evidence of the character of a whisky.

It has been shown that the test is of wider application, however, and a modification has been recommended by Lampitt, Hughes and Trace[2], for the estimation of furfuraldehyde in vinegar. Garratt[3] has suggested its application to the detection, and approximate estimation, of Japanese mint oil in peppermint oils. Garratt[4] has also shown that the test may be applied to the detection of light camphor oil (9.2 Lovibond red) in rosemary oil (0.8 red) or clove oil (23.0 red) in bay (1.4 red) and in pimento berry oil (1.1 red), using a Lovibond Tintometer.

Principle of the method

The test is based on the red colour produced when furfuraldehyde reacts with aniline salts. Methyl-furfuraldehyde, hydroxymethyl-furfuraldehyde and formaldehyde do not produce the red colour.

Reagent required

Aniline acetate Prepare by mixing equal volumes of
Aniline ($C_6H_5NH_2$), freshly redistilled
Glacial acetic acid (CH_3COOH)
Distilled water

This **must be** freshly prepared, using analytical reagent grade chemicals.

The Standard Lovibond Comparator Disc 3/30

The disc contains colour standards for
0.25, 0.5, 0.7, 0.8, 0.9, 1.0, 1.1, 1.2, 1.3
grams furfuraldehyde per 100,000 ml absolute alcohol.

Technique

Distil 250 ml of the whisky to be tested, very gently to avoid charring. When the contents of the distillation flask are down to about 20 ml, add about 10 ml distilled water, and continue the distillation until about 240 ml are collected. Make volume up to 250 ml with distilled water. If the whisky is 30 under proof, then take about 275 ml and distil 250 ml as above, to ensure that the distillate for analysis is always about 43.4% by volume absolute alcohol.

To 10 ml of the distillate add 0.5 ml of the reagent, mix well, and shield from strong light for 15 minutes. The temperature of the reaction should be controlled to about 15°–20°C.

Place a "blank" of distilled water in a 13.5 mm cell in the left-hand compartment of the Comparator, and the test solution in a similar cell in the right-hand compartment, and match against the disc after **exactly** 15 minutes, using a standard source of white light, such as the Lovibond White Light Cabinet or, failing this, north daylight. The answers read from the indicator recess represent grams furfuraldehyde per 100,000 ml of absolute alcohol.

References

1. Simmonds, *"Alcohol"*, MacMillan, London, 1919
2. L. H. Lampitt, E. B. Hughes and L. H. Trace, *Analyst*, 1927, **52**, 260
3. D. C. Garratt, *Analyst*, 1935, **60**, 369
4. D. C. Garratt, *Analyst*, 1935, **60**, 595

The Determination of Lauryl Pentachlorophenol and Pentachlorophenol
using copper pyridine reagent

Introduction

Lauryl pentachlorophenol and pentachlorophenol are used as rot-proofing agents and as fungicides on rubber latex, flax and jute textiles, fire hose and canvas. Certain official specifications (e.g. British Standards 2087:1954 and 1672:1954) lay down the minimum amount of the agent which will be acceptable or adequate for the purpose. In such cases, a chemical test is required to ascertain that these conditions have been fulfilled. The following test has been designed for this type of control.

Principle of the method

The lauryl pentachlorophenol is saponified to liberate pentachlorophenol and lauric acid as their sodium salts. The solution is then acidified in order to liberate the free phenol, which is separated from other organic debris by steam distillation and concentrated by extracting the distillate with chloroform.

The chloroform extract is treated with a copper-pyridine reagent, which forms a copper-pyridine-pentachlorophenol complex soluble in chloroform, and imparts an intense brown colour to the solution.

The intensity of this colour is matched against Lovibond permanent glass standards.

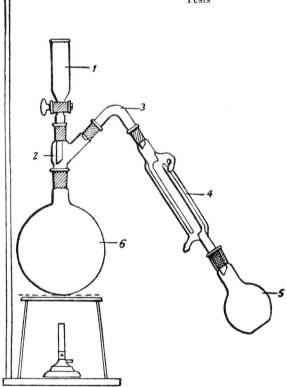

Apparatus used for Lauryl Pentachlorophenol Tests

1. Dropping funnel   Cat. No. D. 3/12
2. Multiple adaptor   Cat. No. M.A. 2/3
3. Bend   Cat. No. S.H. 1/22
4. Double surface condenser   Cat. No. C. 5/12
5. 250 ml Pyrex, flat bottom flask, medium neck B. 24
6. 1,000 ml Pyrex, round bottom flask, short neck B. 24

Also required:—   B. 24 stopper   B. 19 stopper

Catalogue numbers refer to:—
    Quickfit & Quartz Ltd., Mill Street, Stone, Staffs.

Reagents required

1. *4N sodium hydroxide (NaOH) solution*
2. *Concentrated hydrochloric acid (HCl)*   35—36% sp gr 1.18
3. *Chloroform* $CHCl_3$ analytical reagent grade
4. *Copper sulphate solution*   Dissolve 3 g of copper sulphate (analytical reagent grade, $CuSO_4 5H_2O$) in 60 ml of distilled water
5. *Pyridine (C_5H_5N)*   Redistilled

COLORIMETRIC CHEMICAL ANALYTICAL METHODS — Lauryl Pentachlorophenol and Pentachlorophenol

The Standard Lovibond Comparator Discs, 3/37A and 3/37B

Disc A Pentachlorophenol 0.1% to 0.9%, in steps of 0.1%
Disc B Lauryl Pentachlorophenol 0.2% to 1.0%, in steps of 0.1%.

These discs must be used in conjunction with a 5 mm cell.

Apparatus

The distillation apparatus used is as detailed in the accompanying diagram. During refluxing the double surface condenser can function as a reflux condenser.

It is preferable to use glassware with ground glass joints, such as is supplied by Quickfit & Quartz Limited, Stone, Staffordshire.

Technique

The determination of lauryl pentachlorophenol on all materials other than wool

Weigh 5 g of material, cut up and place in a litre flask together with 100 ml of $4N$ sodium hydroxide. A few pieces of unglazed porcelain are also introduced into the flask to prevent bumping. Boil the mass under reflux for two hours.

At the end of this period, fit up the flask for distillation. Add 250 ml of distilled water and 80 ml of concentrated hydrochloric acid to the flask through the dropping funnel and distil the mass until approximately 200 ml of distillate is collected. The distillate is then extracted with 20 ml of chloroform, the condenser being washed out by the chloroform on its way to extract the distillate.

Separate the chloroform layer in a 250 ml separating funnel and transfer it to a 50 ml separating funnel; add 30 drops of copper sulphate solution and 30 drops of pyridine to this chloroform solution, shake well and allow to separate.

Filter the coloured chloroform solution into a test tube and from this fill a 5 mm cell. Place in a Lovibond Comparator. Hold the Comparator about 18 inches from the eye, facing a standard source of white light such as the Lovibond White Light Cabinet or, failing this, north daylight and compare the intensity of the colour with the colours on disc B. The percentage of true lauryl pentachlorophenol contained in the fabric is read off direct. (See—Note 1).

The determination of lauryl pentachlorophenol on wool

Owing to the presence of sulphur compounds, produced during saponification when wool is present, trouble may be encountered when carrying out lauryl pentachlorophenol estimations if hydrochloric acid is used for acidification prior to distillation.

This may be obviated by using 100 ml of $8N$ formic acid solution (containing zinc sulphate) in place of the 80 ml of concentrated hydrochloric acid. Prepare this $8N$ formic acid solution as follows:—

Dissolve 100 g of zinc sulphate ($ZnSO_2.7H_2O$) in 500 ml of distilled water, add 433 g 85% formic acid and make up to one litre with distilled water.

British Standards Institution method

B.S. 2087:1954 gives a standard method for determination of lauryl pentachlorophenol on textile materials, but it has been found that, when using this method on some fabrics, a pronounced reddening of the chloroform solution is obtained after the addition of the copper pyridine indicator due, presumably, to the action of the residual concentrated hydrochloric acid on the textile material.

The determination of pentachlorophenol on fabrics

The estimations are carried out substantially as directed for the estimation of lauryl pentachlorophenol. The only modification is that, as the pentachlorophenol is present as a protein complex in wool, it will only be necessary to reflux with sodium hydroxide for a short period in order to disintegrate the wool before acidification and distillation.

When estimating pentachlorophenol, **use Disc A,** as the colours are slightly different from those developed in the other test, and matching would be difficult. The percentage of true pentachlorophenol is read direct from this disc, as these colours were matched against actual solutions of pentachlorophenol.

Notes

1. The word "true" is used because these glass colour standards were matched against actual solutions of lauryl pentachlorophenol and not against secondary standards.

If the colour intensity obtained is darker than the darkest colour in the disc, it is only necessary to dilute the chloroform solution with more chloroform and then, after comparing the colour, multiply by the dilution factor the % lauryl pentachlorophenol figure obtained.

2. These discs were prepared in conjunction with the laboratories of Catomance Limited, Welwyn Garden City, and thanks are expressed to them for their skilled assistance in modifying the technique for these special tests.

Reference

R. W. Moncrieff, "*Mothproofing,*" Leonard Hill Limited, London, 1952

The Determination of Nicotine
using silicomolybdic acid

Introduction

Nicotine is widely used as an insecticide in horticulture, and the following method was developed by Sutherland, Daroga and Pollard[1] primarily for the determination of small amounts of nicotine in soils. The method is a modification of that proposed by Hofmann[2], and may also be used for the determination of nicotine in tobacco provided that the nicotine is removed from the non-volatile alkaloids and the proteins by steam distillation.

Principle of the method

Nicotine is precipitated as its complex with silicomolybdic acid. This precipitate is then reduced with glycine and sodium sulphite. The blue colour produced is compared with permanent glass standards, using a Lovibond Nessleriser.

Reagents required

1. *Silicomolybdic acid*

Molybdic anhydride (MoO_3)	14.4 g
Sodium hydroxide (NaOH Normal solution)	100 ml
Silica (as sodium silicate solution)	0.7 g
Hydrochloric acid (HCl 10% solution)	

   The molybdic anhydride is dissolved in the sodium hydroxide by warming, and the sodium silicate solution added. Hydrochloric acid is then added slowly, with stirring, until the solution becomes green. The solution is diluted to 900 ml and heated on a boiling water-bath for 3 hours. After cooling, the solution is stored at ambient temperature for 24 hours, the excess silica is filtered off, and the filtrate is diluted to 1 litre.

2. *Wash liquid*

   10% of sodium chloride (NaCl) in 1% hydrochloric acid (HCl)

3. *Reducing solution*

Glycine (NH_2CH_2COOH)	1 g
Sodium sulphide (Na_2S 20% aqueous soln.)	15 ml
Ammonia solution (NH_4OH d 0.880)	5 ml

   Dissolve the glycine in the sodium sulphide solution, add the ammonia slowly, while stirring. Dilute the solution to 100 ml.

The Standard Lovibond Nessleriser Disc NOF

This disc covers the range 0.2 to 1.8 mg of nicotine, in steps of 0.2.

Technique

Extract a suitable quantity of sample in 1% hydrochloric acid (HCl). Filter, and make up to a known volume with 1% hydrochloric acid. Take 14 ml of this solution (called "original volume" in formula below) and add 2 ml of reagent 1. Warm, shake, and allow to stand at ambient temperature for at least 2 hours. Filter through a 42 Whatman filter paper in a small (OO) Gooch crucible, and collect the precipitate. Wash this and the beaker with successive quantities of 1.0, 0.5 and 0.5 ml of reagent 2. Pipette 2 to 3 ml of reagent 3 into the beaker to dissolve any residual precipitate, and then pour this into the Gooch crucible and dissolve the precipitate there by stirring with a glass rod. Suck this solution into the filter flask, and wash the beaker and filter with further quantities of reagent 3.

Combine the filtrate and washings and make up to 50 ml in a graduated flask with distilled water. Heat for 25 minutes in a water bath at 45°C to develop the colour.

Transfer the coloured solution to a standard Nessleriser cylinder and place this in the right-hand compartment of a Lovibond Nessleriser.

Compare the colour against the disc, using a standard source of white light such as the Lovibond White Light Cabinet.

Calculate results as follows:—

$$\frac{\text{Reading in mg} \times \text{original volume in ml}}{15 \times \text{gms sample taken}} \times 1000 = \text{ppm nicotine}$$

Notes

1. It is important that the instructions are followed exactly in all particulars as any variation will affect the final colour.

2. The Nessleriser tube used must be identical with the standard tube used in calibrating the disc, if accurate results are to be obtained, i.e. the calibration mark must fall at a height of 113 ± 3 mm measured internally.

References

1. G. L. Sutherland, R. P. Daroga and A. G. Pollard, *J. Soc. Chem. Ind.*, 1939, **58,** 284
2. R. Hofmann, *Biochem, Z.*, 1933, **260,** 26

The Determination of Organic Acids and their Salts

using ferric hydroxamate

Introduction

During the anaerobic digestion of sewage sludge and other organic wastes, significant concentrations of salts of the lower ("volatile") fatty acids are formed. These are converted, under normal operating conditions in sewage plants, into methane and carbon dioxide. A breakdown in the normal digestion process is often indicated by a sudden rise in the concentration of "volatile" acids, and frequent checks on this concentration are used for control purposes.

Of the methods which have been used for the determination of organic acids in sewage sludge the distillation method[1] is tedious, requires large samples and the results are markedly affected by slight variations in the experimental conditions. Paper chromatographic methods[2,3] are fairly rapid but are either non-reproducible or at best only semi-quantitative. The ether extraction method[4] is also time consuming. Potentiometric titration[5] is both rapid and accurate but is not readily adaptable to multiple analysis.

The present method[6] has been developed to overcome the drawbacks in previous methods and to provide a technique more suitable for plant control purposes. It is rapid, a single determination taking only 25 minutes including time for sample preparation, and is readily adaptable to multiple determinations, it being possible to complete 16 analyses in 2 hours. Although this method was developed primarily for the examination of sewage sludge liquors it has been demonstrated[6] that it is also applicable to other organic industrial wastes.

Principle of the method

After the removal of suspended matter the sample of liquor is treated with ethylene glycol in the presence of sulphuric acid. This treatment converts carboxylic acids and their salts to esters which are, in turn, converted into hydroxamic acids by reaction with hydroxylamine. The colour of the complexes formed by reaction of the hydroxamic acids with ferric chloride[7] is a measure of the hydroxamic acid concentration and thus of the original organic acid concentration. This is estimated by comparing the colour with a range of Lovibond permanent glass colour standards.

Reagents required

1. *Diluted sulphuric acid* Mix equal volumes of sulphuric acid (H_2SO_4 sp gr 1.84) and distilled water (Note 1)
2. *Acidic ethylene glycol* Mix 30 ml of ethylene glycol (HOC_2H_4OH) with 4 ml of reagent 1 (Note 2). **This reagent must be freshly prepared daily**
3. *Sodium hydroxide* $4.5N$ Dissolve 180 g of sodium hydroxide (NaOH), from a freshly opened bottle, in distilled water and dilute to 1 litre
4. *Hydroxyammonium sulphate solution* 10% $(NH_2OH)_2H_2SO_4$ w/v in distilled water
5. *Hydroxylamine reagent* Mix 20 ml of reagent 3 with 5 ml of reagent 4 **immediately before use**
6. *Acidic ferric chloride* Dissolve 20 g of ferric chloride hexahydrate ($FeCl_3.6H_2O$) in 500 ml of distilled water, add exactly 40 ml of reagent 1, and dilute to 1 litre.

All chemicals should be of analytical reagent quality.

The Standard Lovibond Comparator Disc 3/62

This disc (Note 3) covers the range 200—5,000 ppm of acetic acid (Note 4) in 9 steps, (200, 400, 600, 800, 1000, 1500, 2000, 3000, 5000) and is calibrated for use with 25 mm cells in a Lovibond Comparator.

Technique

Clarify the sample, either by filtration using a filter-aid, or by sedimentation in a centrifuge. Measure 0.5 ml of the clarified liquor into a dry 12.5 × 1.5 cm test tube. Into a similar tube measure 0.5 ml of distilled water to provide a blank. To both tubes add 1.7 ml of acidic ethylene glycol (reagent 2) from a burette, and mix thoroughly. Heat in a boiling-water bath for 3 minutes. Immediately cool the test tubes in cold water. Add 2.5 ml of hydroxylamine reagent (reagent 5) and mix. Measure 10 ml of acidic ferric chloride solution (reagent 6) into each of a pair of 25 ml volumetric flasks and add the solutions from the test tubes, using distilled water to rinse out the last traces from the tubes into the flasks. Make up to the mark with distilled water and **mix thoroughly by shaking the flasks vigorously.** Remove the stoppers from the flasks, to facilitate the escape of evolved gases and set the flasks aside for 5 minutes. After this interval has elapsed transfer the sample solution to a 25 mm comparator cell and place this in the right-hand compartment of the Comparator. Fill an identical cell with the solution from the blank determination (Note 2) and place this in the left-hand compartment. Compare the colour of the sample solution with the Lovibond permanent glass colour standards in the disc, using a Lovibond White Light Cabinet to illuminate the Comparator. Where this standardised light source is not available, north daylight should be used as the illuminant.

Notes

1. When diluting sulphuric acid it is important to remember that the **acid** must be **added** slowly **to the water and not vice versa.**

2. If the colour of the blank performed on 0.5 ml of distilled water corresponds to an organic acid concentration greater than 200 ppm, as shown by the lowest standard in the disc, then the ethylene glycol must be purified by reduced-pressure distillation from sodium hydroxide. When the ethylene glycol is pure it is claimed[6] that it is possible to detect the difference between the colour of the blank and the colour from a solution containing 13 ppm of acetic acid.

3. This disc was prepared in collaboration with, and has been checked by, the Water Pollution Research Laboratory.

4. The individual acids are seldom determined separately in current practice. It is usual for a composite figure for the concentration of organic acids to be determined by any convenient method and for this composite figure to be quoted in terms of the equivalent concentration of acetic acid.

References

1. J. E. Frook, *Sewage Ind. Wastes*, 1957, **29**, 18
2. A. M. Buswell, J. R. Boring and J. R. Milam, *J. Water Pollut. Control Fed.*, 1960, **32**, 721
3. H. F. Mueller, A. M. Buswell and T. E. Larson, *Sewage Ind. Wastes*, 1956, **28**, 255
4. J. F. Thomas, C. R. Wherry and E. A. Pearson, *Proc. 10th Ind. Wastes Conf. Purdue Univ.*, 1955, 267
5. R. Di Lallo and O. E. Albertson, *J. Water Pollut. Control Fed.* 1961, **33**, 356
6. H. A. C. Montgomery, Joan F. Dymock and N. S. Thorn, *Analyst*, 1962, **87**, 949
7. U. T. Hill, *Ind. Eng. Chem. (Anal. Ed.)*, 1946, **18**, 317

The Determination of Pentachlorophenates
using Methylene Blue

Introduction

Pentachlorophenates are widely used as industrial fungicides and bacteriocides for slime control in water systems, for the prevention of sap stain in green timber during seasoning, and for protection against bacteria and fungi in the paper, leather and textile industries. They are also used as molluscicides for the destruction of snails which act as intermediate hosts of human schistosomes. These snails are found in streams, irrigation ditches, swamps and lakes and are responsible for the wide incidence of bilharziosis in the Middle East, particularly Egypt.

This test has been developed especially for the field control of pentachlorophenates used for the destruction of these snails. It is necessary to determine the pentachlorophenate concentration in water at various localities. As these materials are effective at concentrations of 100 ppm, or less, the method has been developed to cover the range 5—100 ppm. It can, of course, be applied to higher concentrations provided that the sample is diluted prior to estimation. The equipment is portable, robust, and easy to use. After a short period of instruction even unskilled personnel should be able to analyse a water sample in about 5 minutes.

Principle of the method

It has been reported by Wallin[1], and by Haskins[2,3], that sodium and copper pentachlorophenates, in alkaline solution, will combine with Methylene Blue to form a chloroform-soluble blue complex. Over a limited range of concentrations it is possible to estimate the pentachlorophenate content of a water sample by forming this complex, extracting it with chloroform, and comparing the colour intensity of the chloroform extract with permanent glass standards.

Reagents required

1. *Chloroform* $CHCl_3$ analytical reagent grade
2. *Sodium bicarbonate solution* 20 g $NaHCO_3$ analytical reagent grade in 100 ml distilled water
3. *Sodium citrate* $Na_3C_6H_5O_7.2H_2O$ analytical reagent grade
4. *Methylene Blue—sodium bicarbonate reagent*
   Dissolve three Methylene Blue tablets, each containing 19 mg of methylene blue chloride (note 5), or about 0.05 g of methylene blue crystals (BDH code 26132), in 100 ml of distilled water using a volumetric flask. Place 25 ml of this solution in a 250 ml separating funnel, add 50 ml of saturated sodium bicarbonate solution followed by 50 ml of chloroform. Shake the separator, allow the layers to separate, and reject the pink lower layer. Repeat the extraction, using 50 ml portions of chloroform until the rejected layer is colourless. This usually requires about six additions of chloroform. Transfer the upper (aqueous) layer to an amber 125 ml reagent bottle. This blue solution must be prepared freshly each day.

The Standard Lovibond Comparator Disc B.T. 400

The disc covers the range 5, 7.5, 10, 12.5, 15, 20, 25, 50, 100 ppm (mg/l) of sodium pentachlorophenate. As the pentachlorophenate content of copper pentachlorophenate differs from that of the sodium salt by less than 3%, this disc may also be used for the determination of the copper salt without any significant loss of accuracy.

Technique

Take two clean, dry, stoppered graduated tubes (Lovibond 12×1.5 cm, AF223 acid wash tubes) and mark these A & B on the necks. Fill each tube to the first graduation with chloroform. Using a 5 ml pipette fitted with a rubber bulb, obtain a sample for analysis by immersing the tip of the pipette into the water to be tested and squeezing and releasing the bulb three times before withdrawing the pipette. Add this sample to tube A, forming an aqueous layer above the chloroform and up to the second graduation.

Fill tube B in the same manner using water from the same source as the analytical sample to be tested, but taken before pentachlorophenate treatment. If this is not available then distilled water may be used. Add 1 ml of the methylene blue sodium bicarbonate reagent, using a 1 ml pipette fitted with bulb, to tubes A and B. Replace the glass stoppers in the tubes, shake both tubes for 30 seconds and then stand them upright for two minutes to allow the layers to separate. Eliminate any bubbles which are adhering to the inside of the tube by holding the tube almost horizontal and rotating it about its axis, (Figure 1).

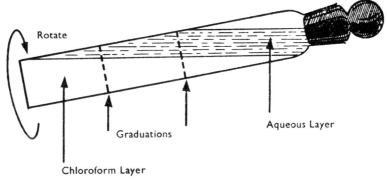

Figure 1

Place tube B in the left-hand compartment of the Comparator. Examine the upper (aqueous) layer in tube A. If this is colourless, or only pale blue, add a further 1 ml of the Methylene Blue reagent, shake for 30 seconds and allow to stand for two minutes. Repeat this operation until the upper layer is a definite blue. Having removed any bubbles, in the manner already described, place tube A in the right-hand compartment of the Comparator. Match the colour of the chloroform layer in tube A with that of the permanent glass standards in the disc using a standard source of white light such as the Lovibond White Light Cabinet or, failing this, north daylight. Note the disc reading. This is the sodium pentachlorophenate concentration in the sample in parts per million.

Notes

1. It will be noticed that the disc calibration ensures the greatest accuracy in the lower part of the range. A reading observed to be between 25 and 100 ppm should always be confirmed by diluting the sample and repeating the test.

2. If the initial procedure yields a blue colour deeper than the 100 ppm standard in the disc, the sample should be diluted with a known volume of distilled water, to bring the pentachlorophenate solution within the range of concentration covered by the disc. The diluted solution is used for the test and the answer obtained should be multiplied by the degree of dilution, (Table 1).

TABLE 1

Volume of water sample—ml	Volume of distilled water added—ml	Degree of dilution
5	5	2
5	10	3
5	15	4
5	20	5
5	25	6
5	30	7
5	35	8
5	40	9
5	45	10

3. Water containing appreciable amounts of dissolved copper or iron may produce cloudy precipitates which are carried into the chloroform extract. This can be prevented by adding a few small crystals of analytical reagent grade sodium citrate to each sample of water before commencing the test.

4. The disc is calibrated with pure pentachlorophenate. Commercial grades of this material, e.g. Santobrite, contain small amounts of the salts of lower chlorinated phenols. These may also form blue complexes with the reagent but they do not significantly affect the accuracy of the test.

5. Methylene Blue tablets are available from B.D.H. Chemicals Ltd., Poole, Dorset.

References

1. G. R. Wallin, *Anal. Chem.*, 1959, **22**, 1208
2. W. T. Haskins, *Pub. Health. Rep. (Washington)*, 1951, **66**, 33, 1047
3. W. T. Haskins, *Anal. Chem.*, 1951, **23**, 1672

The Determination of Monohydric Phenols
using 4-amino-antipyrine

Introduction

The diazotised sulphanilic acid method described in this book for the estimation of phenol determines total phenols. The present method determines monohydric phenols only. It is thus possible, using a combination of both methods, to determine polyhydric phenols by difference.

The use of amino-antipyrine as a reagent for phenol was suggested by Emerson[1] and extensively investigated by Ettinger *et al*[2]. Procedures have been developed for the application of the test to fungicides[3], water and brine[4], and aqueous wastes from coke plants[5]. It has been found to be particularly suitable for trade effluent analysis, especially for gas-works liquors and for effluents in the phenolic plastics industry.

Principle of the method

The phenolic material is mixed with 4-amino-antipyrine and an alkaline oxidant in a solution of sufficiently high pH to prevent the formation of antipyrine-red. The oxidant normally used is potassium ferricyanide. The colour produced is compared with Lovibond permanent glass standards.

Reagents required

1. *Hydrochloric acid*   HCl analytical reagent grade, diluted with an equal volume of water

2. *Ammonia solution* NH_4OH (0.880) diluted with 8 volumes of water

3. *4-amino-antipyrine*   2% solution in water

4. *Potassium ferricyanide* $K_3Fe(CN)_6$ 8% solution in water

5. *Buffer solution:*—pH 10.0   12.370 g of boric acid (H_3BO_3) and 14.910 g of potassium chloride (KCl) dissolved and made up to 1,000 ml with water. 250 ml of this solution and 44 ml of N sodium hydroxide (NaOH) are made up to 1,000 ml with water

The Standard Lovibond Comparator Disc 3/43

The disc covers the range 0.5-4.0 ppm monohydric phenols, in 8 steps of 0.5 ppm.

Technique

A suitable volume of the sample, containing not more than 400 micrograms of monohydric phenol, is placed in a 100 ml Nessler tube or volumetric flask and made up to about 60 ml with water. 10 ml of buffer solution are added and the volume made up to about 90 ml with water. The pH is checked and adjusted to between 9.6 and 10.0, if necessary, with the acid or alkali (reagents 1 and 2). Close range test papers are accurate enough for this check. The final volume is now adjusted to 100 ml with water

The solution is divided into two 50 ml portions and 1 ml of the antipyrine reagent is added to each portion. Each of the solutions is mixed and a part of one is poured into a 13.5 mm cell (or standard test tube) and placed in the left-hand compartment of the Comparator to act as a blank. To the other 50 ml of solution, 1 ml of the ferricyanide reagent is added and, after mixing, a portion of the resulting solution is placed in a 13.5 mm cell (or standard test tube) in the right-hand compartment of the Comparator. Exactly 10 minutes after the addition of this reagent, the colour is matched against the disc, using a standard source of white light, such as the Lovibond White Light Cabinet or, failing this, north daylight.

The figure read from the indicator window represents ppm of monohydric phenol in the final 100 ml volume. From this figure the concentration in the original volume is calculated :—

$$\text{Concentration} = \frac{\text{Reading} \times 100}{\text{Original volume}} \text{ ppm}$$

References
1. E. Emerson, *J. Org. Chem.,* 1943, **8,** 417
2. M. B. Ettinger *et al., Anal. Chem.,* 1951, **23,** 1783
3. S. Gottleib and P. B. Marsh, *Ind. Eng. Chem. Anal. Ed.,* 1946, **18,** 16
4. R. W. Martin, *Anal. Chem.,* 1949, **21,** 1419
5. J. A. Shaw, *Anal. Chem.,* 1951, **23,** 1788

The Determination of Total Phenols (Tar Acids)
using sulphanilic acid

Introduction

Phenols are found in gas liquor, coke-oven effluents, wastes from the production of phenol-formaldehyde resins, drainage from tarred roads, and in many chemical wastes. They are bactericidal and also very toxic to fish, and are therefore most undesirable in effluents discharged to rivers. If present in high concentrations, phenols may also produce a serious problem to sewage works using biological methods of purification.

In view of their presence in coal tar, phenols are often referred to as 'tar acids.'

Principle of the method

This test is based on the reaction between diazotised sulphanilic acid and phenols. The method adopted is that of Fox and Gauge[1] for the determination of tar acids in road dressings As the tar acids present in road tar consist chiefly of higher phenols and xylenols which produce colours with diazotised sulphanilic acid ranging from yellow to red, it is not possible to make glass colour standards that will be perfect colour matches for the colours produced by different kinds of phenols. An arbitrary mixture of phenols has been used for making the solutions on which the glass colour standards are based, and generally these closely match the colours produced in testing aqueous extracts of road dressings and road drainage. This method determines total phenols. A different method is described in this book for monohydric phenols.

Reagents required

1. *Sulphuric acid* (H_2SO_4, *1:3*)   Mix 1 volume of sulphuric acid with 3 volumes of water, taking care to add the acid to the water slowly while stirring
2. *Sodium hydroxide* ($NaOH$, *8%*)   Dissolve 8 g of sodium hydroxide in 100 ml of water
3. *Sodium hydroxide* ($NaOH$, *20%*)   Dissolve 20 g of sodium hydroxide in 100 ml of water
4. *Chloroform* ($CHCl_3$)
5. *Sulphanilic acid* ($NH_2.C_6H_4.SO_3H$) *solution*   Dissolve 1.91 g of sulphanilic acid in 250 ml of water
6. *Sodium nitrite* ($NaNO_2$) *solution*   Dissolve 0.85 g of sodium nitrite in 250 ml of water
7. *Diazotised sulphanilic acid reagent*   Immediately before use mix 5 volumes of reagent 5 with 1 volume of reagent 1, cool to 4°C and then slowly add 5 volumes of reagent 6.

All chemicals used in the preparation of reagents should be of analytical reagent quality.

The Standard Lovibond Nessleriser Disc NP

The disc covers the range from 0.01 to 0.09 part per 100,000, (i.e. 0.1 to 0.9 ppm) of phenols in the 50 ml diluted solution finally used for matching, in steps of 0.01.

Technique

Acidify a measured volume (A) of the solution under examination—in the case of aqueous extracts of road dressings, 250 ml is a suitable quantity—with 10 ml of sulphuric acid (1 in 4) and extract three times with a total of 100 ml of chloroform. Shake the combined chloroform extracts twice with 10 ml of 20% sodium hydroxide solution diluted with an equal volume of water. Mix the alkaline solutions and dilute with water to 100 ml.

Transfer an aliquot volume (B) of the alkaline solution to a Nessleriser glass, neutralize to litmus paper with sulphuric acid (1 in 4) and dilute to 40 ml with water. Add 5 ml of the diazotised sulphanilic acid and 5 ml of 8% sodium hydroxide solution, mix thoroughly, allow to stand for five minutes and place in the right-hand compartment of the Nessleriser. In the left-hand compartment of the instrument place another Nessleriser glass containing the same quantities of the alkaline solution and sulphuric acid, together with sufficient water to produce 50 ml. Stand the Nessleriser before a standard source of white light—such as the Lovibond White Light Cabinet or, failing this, north daylight and compare the colour produced in the test solution with the colours in the standard disc, rotating the disc until the nearest colour match is obtained. Should the colour in the test solution be deeper than the standard colours, a fresh test should be conducted, using a smaller quantity of the alkaline solution.

The markings on the disc indicate the concentration of phenols in parts per 100,000 in the 50 ml of solution contained in the Nessleriser glass, from which, by simple arithmetic, the concentration in the original solution may be ascertained (Note 3).

Notes

1. It must be emphasized that the readings obtained by means of the Lovibond Nessleriser and disc are only accurate provided that Nessleriser glasses are used which conform to the specification employed when the discs were being calibrated, namely that the 50 ml calibration mark shall fall at a height of 113 mm, plus or minus 3 mm measured internally.

2. The preparation of the necessary disc for the above test was suggested by members of the staff of the Ministry of Agriculture and Fisheries at Alresford, Hampshire, and the method was worked out by them in conjunction with The Tintometer Ltd.

3. To convert to parts per million, use formula $$\frac{\text{Disc reading} \times 50{,}000}{\text{Volume used (A)} \times \text{Aliquot volume (B)}}$$

Reference

1. J. J. Fox and A. G. H. Gauge, *J.S.C.I.,* 1920, **39**, 260T; 1922, **41**, 173T

The Determination of Sugar in Water
using α-naphthol

Introduction
The quantitative determination of small traces of sugar in water is of considerable importance to sugar manufacturers and in all factories where sugar is handled. If the sugar content in boiler water exceeds 50 ppm, corrosion of the boiler is almost certain to ensue. This contamination of the boiler water may be the result of returned condensates, or wash water used in boiler feed, which gives a zero polarimetric reading.

Sugar losses in effluent waters may also be determined by this same method.

Alternative methods, for the determination of sugar in blood, are described in Section E.

Principle of the method
Sugar reacts with α-naphthol and sulphuric acid to give a violet colour, which may be used for the estimation of sugar in dilute solutions[1]. The intensity of this violet colour, which is proportional to the sugar concentration, is measured by comparison with a series of Lovibond permanent glass colour standards. The method has also been applied to the determination of sugar in other dilute solutions[2].

Reagents required
1. *α-Naphthol* $(C_{10}H_7OH)$ *solution*   Dissolve 20 g of α-naphthol in 100 ml of 95% ethanol (C_2H_5OH). Store in an amber glass bottle.
2. *Sulphuric acid* (H_2SO_4) Concentrated (98%)

All chemicals should be of analytical reagent quality.

The Standard Lovibond Comparator Disc 3/29A
This disc covers the range 0, 5, 10, 15, 30, 45, 60, 75, 100 ppm of sugar and is calibrated for use with a 5 mm cell.

Technique
Filter and/or dilute the sample as necessary. Pipette 2 ml into a 6″ × 1″ hard glass test tube. Add 5 drops of reagent 1 directly to the liquid without touching the sides of the tube with the reagent. Mix. Run 5 ml of sulphuric acid (reagent 2) gently down the side of the tube to form a separate lower liquid layer. When addition is complete mix well, taking the precautions usual when dealing with concentrated sulphuric acid. Stand the tube for 3 minutes and then cool under running water for 2 minutes. Pour the solution into a 5 mm comparator cell, stand for a further 2 minutes and then place the cell in the right-hand compartment of the Comparator. Match the colour of the sample against the standards in the disc using a standard source of white light, such as the Lovibond White Light Cabinet or, failing this, north daylight.

Dilutions to the sample, when required, should be made with cold hard tap water and the reagents should be checked weekly by preparing known sugar standards with tap water.

Notes
1. This test procedure was devised in the laboratories of Tate & Lyle Ltd., and permission to reproduce it is gratefully acknowledged.
2. Tests should be made on fresh or recently obtained samples only, as the sugar content of samples several hours old would be greatly reduced by bacterial action.

References
1. R. Frailong, *Bull. soc. chim. sucr. dist.,* 1910, **27,** 1188
2. Forbach and Severin, *Zentra-ges. Physiol. Path. Stoffin.,* 1911, **6,** 177

The Determination of Thiocyanate in Effluents
using ferric chloride

Introduction

The determination of thiocyanate is of great importance in the carbonising industries, since the thiocyanate concentration is used as a basis for the assessment of pollution by carbonising plant effluents. It is also used to assess the efficiency of treatment of these effluents before discharge to sewers or rivers.

The present method[1] has been developed to cover two of the three ranges of thiocyanate concentration met in control work of this nature. These three ranges of concentration are:—

(a) gas-works effluents with thiocyanate concentrations of 400 ppm and above

(b) partially treated gas-works effluents with thiocyanate concentrations of 50 to 400 ppm

(c) sewage works influents and effluents and polluted waters with thiocyanate concentrations of 0 to 20 ppm.

This method will measure thiocyanate concentrations from 5 ppm upwards and will thus cover the control of (a) and (b). For control of (c) the pyridine-pyrazolone method[2,3] is recommended[1] with prior dilution of the sample as required. In both methods the presence of phenolic compounds would interfere with the determination and these are therefore removed from the sample before colour development.

Principle of the method

Phenolic substances are removed from the sample by absorbing the thiocyanate onto De-Acidite E anion-exchange resin (chloride form). The thiocyanate is recovered by elution with ammonia, the ammonia is removed by boiling and the solution remaining is reacted with ferric chloride. The thiocyanate is determined by comparing the red colour of the ferric thiocyanate which is formed with Lovibond permanent glass colour standards.

Reagents required

1. *Ferric chloride reagent solution* Dissolve 100 g of hydrated ferric choride ($FeCl_3.6H_2O$) in water containing 50 ml of concentrated hydrochloric acid (HCl) and make up to 500 ml with distilled water

2. *Ammonium hydroxide solution* (3N) Dilute 170 ml of ammonium hydroxide solution (NH_4OH sp gr 0.880) to 1 litre with distilled water

3. *De-Acidite E anion-exchange resin (chloride form)* Wash well with water and decant the fines. Prepare a 1-inch by 4-inch column in the usual manner and wash with water until only a trace of chloride is present in the washings

    To reactivate the column after use pass 200 ml of dilute (1:9) hydrochloric acid through the column at approximately 10 ml per minute. Wash at a rapid rate with water until the washings contain only a trace of chloride.
    Always keep the column of resin covered with water.

All chemicals should be of analytical reagent quality.

The Standard Lovibond Comparator Disc 3/63

This disc covers the range 5, 10, 15, 20, 30, 40, 50, 60, 70 ppm of thiocyanate ion (SCN^-) and is calibrated for use with 10 mm cells.

Technique

Measure 100 ml of sample into a beaker. Pour it through the anion-exchange resin column at the rate of about 10 ml per minute. When the last of the sample has been added to the column wash it through with 100 ml of distilled water in a rapid stream. Reject the solution which has passed through the column. Elute the thiocyanate which has been retained on the column by washing the column with 200 ml of ammonia solution (reagent 2) at about 10 ml per minute. If the concentration of thiocyanate is known to be more than 300 ppm, use a further 100 ml of ammonia to wash the column. Collect the washings in a 600 ml beaker. Complete the washing with 100 ml of distilled water in a rapid stream. Concentrate the combined eluate and washings by boiling until the volume is decreased to about half and all the ammonia has been removed. Cool, transfer to a volumetric flask, and make up to 200 ml with distilled water. Take an aliquot of this solution and make up to 200 ml in a further volumetric flask, adjusting the size of the aliquot so that the final thiocyanate concentration is in the range covered by the disc (5—70 ppm). Measure 100 ml of this diluted solution into a beaker and add 10 ml of the ferric chloride reagent solution (reagent 1). Mix and transfer a proportion of this coloured solution to a 10 mm cell. Place the cell in the right-hand compartment of the Comparator. Fill an identical cell with a "blank" solution consisting of 100 ml distilled water plus 10 ml of the ferric chloride reagent, and place this in the left-hand compartment. Place the Comparator in a Lovibond White Light Cabinet and compare the colour of the sample with that of the standards in the disc. Read off the thiocyanate concentration of the matching standard and multiply it by the dilution factor to obtain the thiocyanate concentration in the original sample.

If a Lovibond White Light Cabinet is not available, north daylight should be used for colour matching wherever possible.

References

1. T. G. Whiston and G. W. Cherry, *Analyst,* 1962, **87,** 819
2. J. Epstein, *Anal. Chem.,* 1947, **19,** 272
3. "The Determination of Cyanide (1)", page 38

COLORIMETRIC CHEMICAL ANALYTICAL METHODS

The Determination of Thiophen
using alloxan

Introduction

This method is directly applicable for the determination of 2.5 mg/kg to 10 mg/kg sulphur as thiophen in benzene samples that are free from unsaturated compounds. This disc has been adopted by the Standardization of Tar Products Tests Committee (S.T.P.T.C.) under Serial No. RLB 28–67[1], BS 135:1963[2] and Indian Standards 534 and 535 (Benzene)[3]. Higher concentrations of thiophen may be determined by dilution of the sample with thiophen-free benzene.

Principle of the Method

The sample is shaken with a solution of alloxan in sulphuric acid, and the colour produced is compared with the standard colours in the disc which was prepared, in conjunction with the National Benzole Association Research Department, by carrying out the test on benzene solutions containing known quantities of thiophen.

Reagents Required

1. A solution containing 0.01% w/v alloxan ($C_4H_2N_2O_4.H_2O$) dissolved in sulphuric acid containing 90.5% w/w ± 0.5 pure sulphuric acid checked by titration. This reagent must be used within 7 days of its preparation
2. Thiophen-free benzene: See Note 1

The Standard Lovibond Comparator Disc 3/19.

The disc covers the range (in steps of 2 µg) of 4 micrograms to 20 micrograms (0.004 to 0.02 mg) of sulphur (S), present as thiophen in the quantity of material tested. From this value the concentration may be calculated by allowing for any dilution of the original sample (Note 2).

Technique

Prior to the test, the sample and reagent must be brought to a temperature between 15°C and 20°C (Note 3).

Pipette 5 ml of the reagent into a stoppered test tube (Note 4), and then add 2 ml sample by pipette. Stopper firmly and shake for exactly one minute at a rate of 100 to 120 shakes per minute. Place the tube in a water bath at 15° to 20°C for exactly 30 minutes, at the end of which time transfer the tube to the right-hand compartment of the Comparator: place a second tube filled with alloxan reagent only in the left hand side of the Comparator. Match the colour of the sample as closely and as quickly as possible against the disc; take the colour of the next highest value if an exact match cannot be made. Make the comparison using a standard source of white light, such as the Lovibond White Light Cabinet, or, failing this, north daylight.

The figure shown on the disc represents micrograms (0.001 mg) of sulphur in the 2 ml sample taken. From these values the concentration of sulphur as thiophen (in mg/kg or mg/litre) may be calculated.

If the colour produced is greater than the highest standard, a smaller quantity of sample must be taken and diluted to 2 ml with the thiophen-free benzene.

If the colour produced is much pinker than the standards but of a similar intensity, the sample should be diluted one hundred times with thiophen-free benzene and the test repeated on the diluted sample.

Notes

1. Thiophen-free benzene may be prepared by giving nitration grade benzene[2] three thorough washes of 30 minutes each with 5% v/v of 98% w/w sulphuric acid, followed by a water wash, a wash with 5% v/v of 10% w/v sodium hydroxide solution, and a further water wash. Then dry the benzene, distil 95% through an 8-bulb pear column[4] rejecting the first 5% of distillate. When the distilled benzene is tested as given in Technique, the colour produced in the acid layer must be less than a disc reading of 4 µg of sulphur.

2. The lower limit of the test cannot be reduced by using a greater proportion of sample to reagent.

3. No temperature control is necessary when the laboratory temperature is between 15°C and 20°C.

4. The stoppered test tube must conform to the specification of S.T.P.T.C., and British Standards Institution[2] for Acid Wash Test, except that it does not require any calibration mark. Suitable test tubes may be obtained from The Tintometer Ltd, reference AF223.

References
1. S.T.P.T.C., *Standard Methods for Testing Tar and its Products,* 6th edition 1967, RLB 28–67
2. *British Standard,* 135:1963
3. *Indian Standards* 534 and 535 (Benzene)
4. S.T.P.T.C., *Standard Methods for Testing Tar and its Products,* 6th edition, 1967, 553, Schedule No. A1a
5. S.T.P.T.C., *Standard Methods for Testing Tar and its Products,* 6th edition, 1967, RLB 10–67

Chemical Analysis
Inorganic

The Determination of Aluminium (Al)

Introduction

Colorimetric methods for the determination of aluminium using Lovibond permanent glass colour standards are widely used for the examination of treated waters[1]. However aluminium estimations are of importance in many other industries and, provided that the aluminium can be obtained in a suitable solution, these methods can be applied to a wide variety of materials.

The method of sample preparation used to bring the aluminium into solution will of necessity vary with the nature of the sample. Some general methods of sample preparation are discussed in the Introduction to this book, others may be found in the chemical literature. A few illustrative examples from the literature are given below.

Methods of sample preparation

Air[2]

Draw a known volume of air at 10 litres a minute through ashless filter paper. Ash the paper in a porcelain crucible and ignite at 600°C. Fuse the residue with 0.5 g of potassium pyrosulphate ($K_2S_2O_7$) and slowly increase the temperature from 200° to 600°C. Dissolve the cooled melt in 50 ml of water.

Beverages[3]

Beer To 10 ml of a degassed sample add 50 ml of water. Adjust pH to be between 4 and 5 and determine by *Method 3*.

Biological materials[4,5]

Remove the organic matter by wet ashing (See General Introduction) and separate aluminium by the acetate method using ferric phosphate ($FePO_4$) as carrier. Separate aluminium from iron as sodium aluminate ($NaAlO_2$).

Serum[5] Remove organic matter by ashing. Dissolve the ash in hydrochloric acid and remove iron, copper and manganese by ion exchange of their ionic chloride complexes. Aluminium range 5—50 μg/100 ml.

Chemicals[6]

Sodium silicate solutions Evaporate a sample to dryness in a platinum dish. Treat the residue with hydrofluoric (HF) and sulphuric (H_2SO_4) acids, to remove the silica, and dilute the solution to known volume.

Foods and edible oils[7-11]

Remove organic matter by either wet oxidation[8] or dry ashing[7]. For wet oxidation[8] take 10 g of food in a silica flask, add 7 ml of sulphuric acid and sufficient nitric acid to destroy the organic matter. Heat until colourless. Cool. Dilute the residue to 50 ml with distilled water, taking care to dissolve all the aluminium sulphate ($Al_2(SO_4)_3$). Add a slight excess of ammonium hydroxide (NH_4OH) and then boil until the solution no longer smells of ammonia (neutral to methyl orange). Filter, and wash the precipitate in 10 ml of distilled water. Dissolve the precipitate in 15 ml of hot dilute hydrochloric acid (5 ml of $5N$ HCl + 10 ml water), collect the solution in a silica flask, add 10 ml of $5N$ caustic soda (NaOH) solution and boil. Allow the solution to cool, filter, and collect the filtrate in a 50 ml volumetric flask. Wash out the silica flask and the filter with 20 ml of distilled water and add the washings to the filtrate. Make up to 50 ml.

Dry ashing is claimed by Shtenberg et al[9] to interfere with the hæmatoxylin colours. These authors recommend the following procedure:- ash the food by fusion with sodium carbonate (Na_2CO_3) dissolve the melt and remove iron as capronate, precipitate the aluminium with 8-hydroxyquinoline, filter, re-dissolve the precipitate and make up to known volume.

Onna et al[11] recommend the use of an ion exchange column for the separation of aluminium from sodium potassium, magnesium, calcium and iron, with special reference to the analysis of sugar solutions.

Fuels and lubricants[12]

Coal ash Mix 0.5 g of sample, ground to pass 250 mesh, with 2 g of lithium metaborate ($LiBO_2$) in a platinum crucible and heat in a muffle at 900°C until the melt is clear. If the sample contains more than 15% of iron add 30—50 mg of ammonium vanadate (NH_4VO_3) to act as flux. Cool the melt and transfer it to a beaker containing 8 ml of nitric acid and 150 ml of water. Stir vigorously until dissolution is complete. Add 2.5 g of tartaric acid (($CHOH.COOH)_2$) and make up to 250 ml.

Metals, alloys and plating baths[13-31]

Steels and ferrous alloys[13-22] Dissolve the sample in 1:5 hydrochloric acid (HCl), fuse any insoluble matter with potassium pyrosulphate ($K_2S_2O_7$) and extract the melt with hot water. Add this extract to the original solution. Complex the interfering elements with the thioglycollic ($HSCH_2COOH$) and ascorbic ($O.CO.C(OH):C(OH).CH.CH(OH).CH_2OH$) acids and determine aluminium on this solution[13].

Other authors suggest variations both in the method of solution and in the method of removing interfering elements. The alternative acids suggested are mixed hydrochloric and nitric[15, 21], dilute sulphuric[17, 19, 20], or 1:1 nitric[22]. Some authors[17, 18] suggest that the solution obtained after acid treatment should be filtered and used directly for the aluminium determination. Others[19, 20] recommend potassium hydrogen sulphate ($KHSO_4$) as an alternative flux for solution of the residue. Methods suggested for the removal of iron include electrolysis[14, 19]; extraction of ferric chloride with butyl acetate ($CH_3COO.(CH_2)_3CH_3$)[14]; isoamyl acetate ($CH_3COO(CH_2)_2CH(CH_3)_2$)[15, 16] or diethyl ether ($C_2H_5OC_2H_5$)[21]; precipitation of mixed hydroxides[20] and separation of sulphates by means of Wofatit L-150 ion exchange resin. Separation from other interfering metals is usually achieved by extraction of aluminium into chloroform ($CHCl_3$) as its 8-hydroxyquinoline ($N:CH.CH:CH.C_6H_3.OH$) complex[14-16].

Non-ferrous alloys[23-29] Dissolve the sample in a suitable acid and then separate the aluminium from interfering metals by one of the following techniques:–

(i) Make the solution alkaline by the addition of 30 ml of 8% aqueous caustic soda (NaOH). Boil for one minute, add 20 ml of an 8% solution of sodium sulphide (Na_2S). Filter. Acidify the filtrate with $5N$ hydrochloric acid. Add distilled water to bring the total volume up to about 60 ml and allow the solution to stand until any precipitate has settled. Add a little filter-paper pulp as a filter aid and refilter. Evaporate the filtrate to about 30 ml. Transfer to a volumetric flask and make up to 50 ml[23].

(ii) It is suggested by some authors that method (i) may result in the loss of some aluminium by adsorption. Melaven[26] recommends an electrolytic method of separating the aluminium. Raine[27] claims that even when using electrolytic separation prior chemical separation of iron is necessary.

(iii) After dissolving the sample in acid, chromatograph the solution on paper, dissolve the aluminium spot in hydrochloric acid and determine aluminium on this solution[24].

(iv) When using aluminon (sodium salt of aurine tricarboxylic acid) as the colorimetric reagent, prior removal of other elements is less important as this reagent is less susceptible to interference. Suitable procedures for sample treatment are those described by Dufek and Kopa[25] and by Luke and Braun[28].

(v) For the determination of aluminium in the presence of large quantities of iron the method due to Short[29] may be used. Dissolve the sample in hydrochloric acid and add the minimum quantity of nitric acid which will convert all the iron to the ferric state. Evaporate the solution to dryness and redissolve the residue in hydrochloric acid. Extract the ferric chloride ($FeCl_3$) with either ethyl or isopropyl ether (ROR, R=Et or iPr). Separate the aqueous portion, warm to remove the ether and treat with a few drops of nitric acid. Evaporate to low volume. Cool and dilute the solution to 50 ml. Add 3 g of ammonium chloride (NH_4Cl) and adjust the pH to 0.3-0.4. Transfer to a separating funnel, add 0.1 g of cupferron reagent (nitrosophenylhydroxylamine, $C_6H_5N(NO)OH$) dissolved in distilled water. Stand for 5 minutes and then extract with chloroform ($CHCl_3$) three times. Warm the aqueous layer to expel the chloroform, dilute to known volume and determine aluminium.

High purity copper[30] Dissolve 1 g of sample in aqua regia (1 vol, HNO_3+3 vols HCl) and evaporate excess acid. Dissolve the residue in $7M$ hydrochloric acid and extract interfering elements into two 10 ml portions of isobutyl methyl ketone $((CH_3)_2CHCH_2COCH_3)$. Adjust the pH of the aqueous phase to 5.6-5.9, mask the copper by adding sodium thiosulphate $(Na_2S_2O_3.5H_2O)$ until its concentration in the solution is molar (0.25 g per ml) and extract the aluminium from the aqueous phase into four successive 5 ml portions of carbon tetrachloride (CCl_4) as the aluminium complex of 8-hydroxyquinoline.

Metallic chromium[31] Dissolve the sample in sulphuric acid, evaporate to dryness and ignite the residue to 1000°C to form the oxides. Then heat a 20-50 mg aliquot of the oxides at 900°C in a stream of chlorine for 30-40 minutes by which time all the chromium will have evaporated as its chloride and aluminium oxide (Al_2O_3) will remain. Fuse the residue with potassium pyrosulphate $(K_2S_2O_7)$, dissolve the melt in dilute acid and make the solution up to volume.

Minerals, ores and rocks[12, 31-34]

Treat 0.5 g of sample in the manner described for coal ash above[12]. Alternatively[31] heat powdered oxides in a stream of chlorine at 700°C for 30-40 minutes. Aluminium and silicon are selectively volatilised as their chlorides. Aluminium chloride $(AlCl_3)$ condenses on the cold portion of the combustion tube. Wash this out with $0.1N$ hydrochloric acid and make up to volume.

Van Loon and Parissis[33] recommend the use of a pre-ignited graphite crucible rather than platinum for the metaborate fusion. Lebedev and Vlasova[32] and also Osborn[34] prefer fusion with sodium carbonate (Na_2CO_3).

Miscellaneous[35-42]

Cement[35-37] Dissolve the sample in acid, convert the Al to sodium aluminate $(NaAlO_2)$ by the addition of caustic soda and sodium carbonate (Na_2CO_3). which will precipitate iron, titanium, manganese, magnesium and calcium. Precipitate phosphate (PO_4^{3-}) as either the iron or the calcium salt. Filter and determine aluminium on the filtrate.

Dusts and bonded deposits[38] Fuse the sample with caustic soda (NaOH) and extract the cooled melt with water, sulphuric acid and hydrogen peroxide (H_2O_2) solution at 60°C for 1 hour. Remove silica by dehydration with sulphuric acid. Remove interfering ions by electrolysis at a mercury cathode, extract titanium and vanadium, as their cupferrates, into chloroform from sulphuric acid solution at pH less than 1 and determine aluminium on an aliquot of the remaining solution.

Leather[39] Ash the sample, dissolve the residue either by alkali fusion or by wet oxidation and determine aluminium on this solution.

Liming materials and fertilizers[40] Grind sample in an agate mortar to pass 100 mesh. Place 0.5 g of sample, if of limestone, or 0.2 g if a silicate, in a 75 ml nickel crucible. Remove any organic matter by heating at 900°C for 15 minutes without the lid. Cool. Mix 0.3 g of potassium nitrate (KNO_3) with the sample and add 1.5 g of caustic soda (NaOH) pellets. Cover the crucible with its lid and heat to a dull red over a gas flame for 5 minutes. Remove the crucible from the flame and swirl the fused melt around the sides. Cool. Add about 50 ml of water and warm to disintegrate the fused cake. Transfer the contents of the crucible to a 150 ml beaker containing 15 ml of $5N$ perchloric acid $(HClO_4)$. Scrub the crucible and lid with a rubber tipped glass rod and wash any residue into the beaker. Transfer to a 100 ml volumetric flask and dilute to volume.

Refractories[34, 41] Fuse 0.025-0.05 g of sample with 1 g of a mixture of 1 part of sodium tetraborate $(Na_2B_4O_7)$ and 3 parts of sodium carbonate (Na_2CO_3). Extract the cooled melt with 100 ml of 1:3 hydrochloric acid and dilute to 250 ml.

Slime[42] Fuse 0.2 g of dried slime with 2 g of sodium carbonate (Na_2CO_3). Extract the melt with 50 ml of hot water, rapidly heat the solution to the boiling point and filter. Dilute filtrate to 200 ml.

Plants[43-45]

Take 0.5 g of ground dried leaves, ash by digestion with 5 ml of nitric acid and 2 ml of sulphuric acid. Heat until white fumes are evolved and then dilute to known volume. Middleton[45] advocates the removal of other ions by evaporating the solution from wet oxidation to a volume of about 5 ml and transfering this with 10 ml of water to a separating funnel. Add 1 ml of a 6% solution of 8-hydroxyquinoline in 2% acetic acid (CH_3COOH) swirl, neutralise with $5N$ aqueous ammonia (NH_4OH), and add 5 ml of acetate buffer (pH 2.85). Add 10 ml of a 0.3% solution of 8-hydroxyquinoline in chloroform ($CHCl_3$). Shake, separate the layers and determine aluminium on the aqueous layer.

Plastics and polymers[46, 47]

Decompose the sample by heating with sulphuric acid and destroy organic matter by oxidation either with nitric acid[47] or hydrogen peroxide[46]. Filter and make up to volume.

Soil[44, 48-51]

Extract the soil with water[49], $0.01M$ calcium chloride ($CaCl_2$) solution[48], $0.5N$ acetic acid[44], Tamm's reagent (24.9 g of hydrated ammonium oxalate (($COONH_4)_2.H_2O$) and 12.6 g of hydrated oxalic acid (($COOH)_2.2H_2O$) per litre)[50] or $0.2N$ neutral ammonium chloride[51]. Skeen and Sumner[51] claim that the ammonium chloride extractant gives the best measure of extractable aluminium for acid soils. Determine aluminium on extract with aluminon (*Method 3*). Interference from iron can be avoided by reducing it to ferrous state with hydroxylamine (NH_2OH) and complexing with cyanide[49].

Water[52-59]

No sample preparation is required in most cases.

The various colorimetric reagents available for the determination of aluminium have been reviewed by Giebler[60] who recommends aluminon as being least susceptible to interference from other ions. Bogova[61] also recommends aluminon, especially for the determination of aluminium in refractory materials, and finds that the optimum pH for its use is 4.8–5.0.

Separation of aluminium

From calcium and magnesium[62] Absorb the sample solution on a 14×1 cm column of Dowex 50-X8 (H^+ form). Elute the magnesium with 200 ml of $0.7N$ hydrochloric acid, calcium with a further 200 ml of $0.7N$ hydrochloric acid and the aluminium with 300 ml of N hydrochloric acid.

From iron[63] Amounts up to the mg level may be separated by reducing all the iron to the ferrous state; forming a complex with 2,2'-bipyridyl and passing the solution at pH 5.5 through a column of Lewatit MN ion exchange resin saturated with neutral sodium arsenate (Na_2HAsO_4) solution. The iron complex passes through and is quantitatively removed from the column with 1% ammonium acetate (CH_3COONH_4) solution. The adsorbed aluminium is then removed with $0.5N$ hydrochloric acid.

From chromium[64] To the test solution add an excess of 5% ammonium thiocyanate (NH_4SCN) solution. Evaporate to dryness on a water bath, add water and pass the solution through a column of KPS-200 cationite resin (H^+ form). Wash the column with 30 ml of water, combine the washings and the percolate for the determination of chromium. Elute the aluminium with $2M$ hydrochloric acid.

The individual tests using Lovibond colour standards

Method 1 was developed for the determination of Al in water after flocculation with Al salts.

Method 2 is a more sensitive modification of the test and was developed by Strafford and Wyatt[52]. This is the standard method for determining Al in water for industrial use[1]. Both *Method 1* and *Method 2* use hæmatoxylin.

Method 3 is the modification devised by Packham[54] of the method originated by Poole and Segrove[55] using aluminon. It is claimed[54] to be an improvement on both the American[56] and British[57] standard methods for the determination of Al in water. It was pointed out by Palin[55] that the colours developed in Packham's method were identical with those developed in the Palin—DPD method for the determination of residual chlorine in water[59]. The same discs (3/40A, 3/40B) can therefore be used for both chlorine and aluminium tests if appropriate correction factors are applied.

Method 4 is the Water Research Centre method devised by Dougan and Wilson[65] using catechol violet.

References

1. British Standards Institution, *"Routine Control Methods of Testing Water Used in Industry"*, B.S. 1427, London, 1962
2. A. F. Kosternaya and N. A. Zavorovskaya, *Sb. Nauch. Rabot Inst. Okhrany Trada VTSSPS.*, 1962, (1), 82; *Analyt. Abs.* 1964, **11**, 383
3. H. O. Etian and M. A. Rovella, *Proc. Amer. Soc. Brewing Chemists,* 1966, 177; *Analyt. Abs.,* 1967, **14**, 3595
4. W. Oelschlager, *Z. analyt. Chem.,* 1957, **154**, 321; *Analyt. Abs.,* 1957, **4**, 2290
5. M. Siebold, *Z. analyt. Chem.,* 1960, **173**, 388; *Analyt. Abs.,* 1960, **7**, 4903, 4904
6. A. B. Kiss and E. D. Walko, *Magyar Kem. Lapja,* 1968, **23**, (12), 720; *Analyt. Abs.,* 1970, **18**, 2319
7. G. W. Monier-Williams, *"Reports on Public Health and Medical Subjects No. 78, Aluminium in Food",* 1935
8. L. H. Lampitt and N. D. Sylvester, *Analyst,* 1932, **57**, 418
9. A. I. Shtenberg, E. B. Pustylnikova and N. V. Orlova, *Trudy Kom. analit. Khim., Akad. Nauk. S.S.S.R., Otdel Khim. Nauk.,* 1951, **3**, 252; *Chem. Abs.,* 1953, **47**, 2388h
10. A. R. Deschreider, *Congr. internat. ind. agr. 9th Congr. Rome,* 1952; *Chem. Abs.,* 1953, **47**, 2389d
11. K. Onna, G. Akatsuka and C. C. Tu, *J. agric. Food Chem.,* 1963, **11**, 332; *Chem. Abs.,* 1963, **59**, 5701g
12. P. L. Boar and L. K. Ingram, *Analyst,* 1970, **95**, 124
13. A. A. Fedorov and G. P. Sokolova, *Sb. Trud. Tsents. Nauch-Issled. Inst. Chern. Metallurg.,* 1962, (24), 128; *Analyt. Abs.,* 1963, **10**, 3247; 1963 (31), 162; *Analyt. Abs.,* 1964, **11**, 4281
14. E. de la Torre Gonzalez, V. T. Carilla and R. S. Acosta, *Inst. Hierro Acero,* 1964, **17**, 192; *Analyt. Abs.,* 1965, **12**, 2801
15. J. B. Nievas and M. H. Laucina, *Inst. Hierro Acero,* 1964, **17**, 517; *Analyt. Abs.,* 1965, **12**, 5813
16. R. M. Dagnall, T. S. West and P. Young, *Analyst,* 1965, **90**, 13
17. L. Buck, *Chim. analyt.,* 1965, **47** (1), 10; *Analyt, Abs.,* 1966, **13**, 2356
18. V. A. Verbitskaya, V. V. Stepin, I. A. Onorina and L. S. Studenskaya, *Trudy vses Nauch.–Issled. Inst. Stand. Obrazt. spektr. Etalon.,* 1965, **2**, 52; *Analyt. Abs.,* 1967, **14**, 3188
19. British Iron and Steel Res. Ass., *Open Report Mg/D/561/67,* 1967; *Analyt. Abs.,* 1968, **15**, 7271
20. T. V. Sinitsyna, *Trudy vses. Nauch–Issled. Konstr.–Tekhnol.-Inst. podshipnik. Prom.,* 1968, (2(54)), 60; *Analyt. Abs.,* 1970, **18**, 1658
21. D. Filipov and I. Nachev, *Compt. rend. Acad. bulg. Sci.,* 1969, **22** (6), 687; *Analyt. Abs.,* 1970, **19**, 1091
22. J. Musil, *Hutn. Listy,* 1968, **23** (9), 649; *Analyt. Abs.,* 1970, **18**, 204
23. G. E. F. Lundell and H. B. Knowles, *Ind. Eng. Chem.,* 1926, **18**, 60
24. K. N. Munshi and A. K. Dey, *J. prakt. Chem.,* 1962, **18**, 233; *Analyt. Abs.,* 1963, **10**, 2625
25. R. Dufek and L. Kopa, *Hutn. Listy,* 1959, **14**, 620; *Analyt. Abs.,* 1960, **7**, 1690
26. A. D. Melaven, *Ind. Eng. Chem., Analyt.,* 1930, **2**, 180
27. P. A. Raine, *Analyst,* 1949, **74**, 364
28. C. L. Luke and K. C. Braun, *Analyt. Chem.,* 1952, **24**, 1120
29. H. G. Short, *Analyst,* 1950, **75**, 420
30. B. Kassner and W. Angermann, *Z. Chem. Lpz.,* 1967, **7** (11), 438; *Analyt. Abs.,* 1969, **16**, 569
31. A. A. Tumanov and V. G. Petukhova, *Zavodskaya Lab.,* 1969, **35** (6), 654; *Analyt. Abs.,* 1970, **19**, 1089

32. O. P. Lebedev and N. I. Vlasova, *Sb. Nauch. Trudy Nauch.-Issled. Gornorudn Inst. Ukr SSR.*, 1963, **7**, 293; *Analyt. Abs.*, 1965, **12**, 3240
33. J. C. Van Loon and C. M. Parissis, *Analyst*, 1969, **94**, 1057
34. W. O. Osborn, *J. Amer. Ceram. Soc.*, 1961, **44**, 527; *Analyt. Abs.*, 1963, **10**, 154
35. H. Sopora, *Silikat Tech.*, 1962, **13**, 398; *Chem. Abs.*, 1963, **58**, 9943d
36. L. Burglen and P. Longuet, *Rev. Mater. Constr.*, C, 1960, 327; *Analyt. Abs.*, 1961, **8**, 4095
37. K. Kawagaki, M. Saito and K. Hirokawa, *J. Chem. Soc. Japan, ind. Chem. Sect.*, 1965, **68** (3), 465; *Analyt. Abs.*, 1967, **14**, 5434
38. J. Grant, *J. appl. Chem. Lond.*, 1964, **14** (12), 525; *Analyt. Abs.*, 1966, **13**, 1782
39. Int. Union of Leather Chemists' Societies, Leather Analysis Commission, *J. Soc. Leath. Trades Chem.*, 1969, **53** (10), 389; *Analyt. Abs.*, 1970, **19**, 4084
40. P. Chichilo, *J. Assoc. off. agric. Chem.*, 1964, **47**, 1019
41. L. V. Bogova, *Ogneupory*, 1969, (6), 52; *Analyt. Abs.*, 1970, **18**, 3802
42. T. I. Badeeva, L. A. Molot, N. S. Frumina and K. G. Petrikova, *Uch. Zap. Saratovsk. Univ.*, 1962, **75**, 100; *Analyt. Abs.*, 1964, **11**, 875
43. J. R. Gallo, *Bragantia*, 1962, **21**, 411; *Chem. Abs.*, 1963, **59**, 15588c
44. S. A. Harris, *J. Sci. Food Agric.*, 1963, **14**, 259; *Analyt. Abs.*, 1964, **11**, 2409
45. K. R. Middleton, *Analyst*, 1964, **89**, 421
46. K. Novak and V. Mika, *Chem. Prumysl*, 1963, **13**, 360; *Analyt. Abs.*, 1964, **11**, 4414
47. S. Fujiwara and H. Narasaki, *Analyt. Chem.*, 1964, **36**, 206
48. C. R. Frink and M. Peech, *Soil Sci.*, 1962, **93**, 317; *Analyt. Abs.*, 1963, **10**, 1229
49. M. L. Tsap, *Vestn. Sel'sk.-Khoz. Nauk.*, 1963, (2), 122; *Analyt. Abs.*, 1964, **11**, 1133
50. D. T. Pritchard, *Analyst*, 1967, **92**, 103
51. J. B. Skeen and M. E. Sumner, *S. Afr. J. agric. Sci.*, 1967, **10** (1), 3; (2), 303; *Analyt. Abs.*, 1968, **15**, 4334, 5698
52. N. Strafford and P. F. Wyatt, *Analyst*, 1943, **68**, 319
53. British Standards Institution, "Routine Control Methods of Testing Water Used in Industry", B.S. 1427, London, 1962
54. R. F. Packham, *Proc. Soc. Water Treatment Exam.*, 1958, **7**, 102
55. P. Poole and H. D. Segrove, *Trans. Soc. Glass Tech.*, 1955, **39**, 205. See also W. C. Johnson "Organic Reagents for Metals, and Other Recent Monographs", Vol 1, 5th Edn., Hopkin and Williams, London, 1955
56. "Standard Methods for the Examination of Water, Sewage and Industrial Wastes", 10th Edn., Amer. Pub. Health Assoc., New York, 1955
57. British Standards Institution, "Methods of Testing Waters Used in Industry", B.S. 2690, Part 4, London, 1967
58. A. T. Palin, *Proc. Soc. Water Treatment Exam.*, 1960, **9**, 82
59. A. T. Palin, *Analyst*, 1945, **70**, 203
60. G. Giebler, *Z. analyt. Chem.*, 1961, **184**, 401; *Analyt. Abs.*, 1962, **9**, 2536
61. L. V. Bogova, *Trudy. vses. Inst. nauchno-issled. proekt. Rab. ogneup. Prom.*, 1968, (40), 131; *Analyt. Abs.*, 1970, **18**, 809
62. M. Tanaka, *J. Chem. Soc. Japan, Pure Chem. Sec.*, 1963, **84**, 582; *Analyt. Abs.*, 1965, **12**, 5045
63. J. Gera, *Chem. Anal. Warsaw*, 1964, **9**, 541; *Analyt. Abs.*, 1965, **12**, 4542
64. N. P. Panchev and B. Evtimova, *Compt. rend. Acad. bulg. Sci.*, 1965, **18** (12), 1127; *Analyt. Abs.*, 1967, **14**, 1907
65. W. K. Dougan and A. L. Wilson, *Analyst*, 1974, **99**, 413–430

Aluminium Method 1

using hæmatoxylin

Principle of the method

Hæmatoxylin gives a violet-purple lake with aluminium in slightly basic solution. On acidification the lake is stabilised and the colour of the excess dye changes from red to yellow.

Reagents required

1. *Hæmatoxylin (microscopic stain 0.1 g)* To 20 ml freshly boiled distilled water in a suitable flask, add the hæmatoxylin while the water is still hot, swirl gently to dissolve, cool, and make up to 100 ml with distilled water. This solution should be kept in a dark bottle and not used when more than 3 days old. Tablets of 0.1 g may be obtained from Tintometer Ltd.

2. *Ammonium carbonate* Dissolve 15 g of analytical reagent grade ammonium carbonate ($NH_4HCO_3.NH_2.COO.NH_4$) in 100 ml distilled water. The loss of ammonia from this solution, or from the solid ammonium carbonate used in its preparation, must be avoided by keeping the bottle containing the ammonium carbonate tightly stoppered.

3. *Acetic acid ($CH_3 COOH$) 30%* Make up 30 ml of glacial acetic acid to 100 ml with distilled water.

The Standard Lovibond Comparator Disc 3/15

This disc covers the range 0.1 to 0.8 parts of aluminium per million, in steps of 0.1, while the ninth standard, labelled "blank," matches the colour of the reagents alone i.e. equals 0 parts of aluminium per million.

Technique

The details must be followed exactly as departures therefrom will lead to inaccuracies.

To 50 ml of the water under examination add 2 ml of the ammonium carbonate solution (reagent 2), followed by 1 ml of the hæmatoxylin solution (reagent 1), mix thoroughly and allow to stand for 10 minutes. Add 2 ml of the acetic acid solution (reagent 3), mix thoroughly and stand for 5 minutes. Transfer a 10 ml portion of the solution to a standard comparator-tube, and place this tube in the right-hand compartment of the comparator. Fill an identical tube with the water being tested and place this in the left-hand compartment to act as a blank.

Hold the comparator facing a standard source of white light (Note 5) and compare the colour in the test solution with the colours in the standard disc, rotating the latter until a match is obtained. If the colour produced is above the highest colour value in the disc, the test should be repeated with a fresh sample, suitably diluted with distilled water. The result obtained should then be multiplied by the dilution factor.

The exact quantities and times above specified must be rigidly adhered to, in order to obtain accurate results. Bubbles in the solution which cling to the side of the tube render matching difficult, and should be removed by gentle tapping.

The figures on the disc represent parts of aluminium per million (= mg per litre) when 50 ml of the original sample are taken.

Notes

1. Treated waters with high hardness values may show interference, but it has been found that satisfactory results are obtained with hardness figures due to calcium salts up to 10 grains per gallon (approximately 143 ppm) and due to magnesium salts up to 20 grains per gallon, (approximately 286 ppm) both expressed in terms of $CaCO_3$. However it is seldom necessary in practice to know the aluminium content of such water.

2. In order to convert the readings of the disc into parts of alumina (Al_2O_3) per million, multiply the figure obtained by the appropriate factor i.e. parts of aluminium per million $\times$ 1.89 = parts of alumina per million.

3. Interference from manganese causes bleaching, and from iron causes "off shades" (usually greenish). If more than 0.2 ppm of manganese is present, the colour of the hæmatoxylin is bleached within 2 minutes i.e. before the addition of the acetic acid. Iron less than 0.4 ppm or manganese less than 0.2 ppm will allow satisfactory aluminium values to be obtained.

4. This method was developed by the Alfloc Water Treatment Service of I.C.I. Ltd., to whom acknowledgment is made.

5. A strongly recommended standard source of white light is the Lovibond White Light Cabinet. In the absence of a standard source north daylight should be used wherever possible.

Aluminium Method 2
using hæmatoxylin

Principle of the method

In this modification excess hæmatoxylin is removed by means of ammonium borate, the lake is stabilised by sodium alginate solution and the colour is measured at pH 7.5.

Reagents required

1. *Hydrochloric acid (HCl) 5N*
2. *Ammonium carbonate ($NH_4HCO_3 + NH_2COONH_4$) solution 2N (157 g/l)*
3. *Sodium alginate 0.1% solution*
4. *Hæmatoxylin solution 0.1%.* This solution is prepared by dissolving 0.1 g of hæmatoxylin microscopic stain in 100 ml of distilled water containing 0.1 ml of $5N$ hydrochloric acid Tablets of 0.1 g may be obtained from Tintometer Ltd.
5. *Ammonium borate ($NH_4HB_4O_7.3H_2O$) saturated solution*
6. *Distilled water*

The Standard Lovibond Nessleriser Disc NX

The disc covers the range 0.5 to 4.5μg (0.0005 to 0.0045 mg) of aluminium, (Al), in steps of 0.5μg. To convert the disc readings to parts per million (ppm) use the appended Conversion Table.

Technique

Measure a suitable volume, not exceeding 30 ml, of the water under examination, which should be approximately neutral, into a Nessleriser glass and dilute to 30 ml with distilled water. Add 1.0 ml of $5N$ hydrochloric acid (reagent 1), mix well, and add $2N$ ammonium carbonate (reagent 2), until the pH is 7.5 ± 0.2 as shown by a test paper (usually about 5.7 ml), swirling the liquid vigorously round the glass to assist the evolution of carbon dioxide. Add 1 ml of sodium alginate solution (reagent 3), mix, add 5.0 ml of 0.1% hæmatoxylin solution (reagent 4), mix again and cover.

At the same time place 30 ml of distilled water in another Nessleriser glass and treat it exactly as described above for the test solution. After allowing the blank and the test solution to stand for 15 minutes, add to each 5.0 ml of $0.8N$ ammonium borate (reagent 5) while swirling the solutions, dilute each to 50 ml, mixing well.

Place the blank in the left-hand compartment and the test in the right-hand compartment of the Nessleriser and allow to stand for not less than two and not more than five minutes. Stand the Nessleriser before a standard source of white light (Note 5) and compare the colour produced in the test solution with the colours of the standard disc, rotating the disc until the nearest match is obtained. The value shown is in micrograms of aluminium in the original volume taken, and can be converted to ppm by means of the Conversion Table.

Notes

1. The ammonium borate removes excess hæmatoxylin.
2. Iron interferes quantitatively, giving a similarly coloured lake; copper, which can be tolerated in quantities up to 10 mg, produces a purple lake which is almost completely decomposed by ammonium borate. Iron and copper in the amounts found in water are best extracted as their thiocyanates by shaking out into amyl alcohol-ether mixture. Tin also interferes, and therefore all water used in the preparation of reagents etc. should be glass distilled.
3. Hydrochloric acid is used initially to break down the aluminium which may be in a colloidal or complex form. As the pH (7.5) of the final solution is critical the amounts of hydrochloric acid and ammonium carbonate solution should be measured accurately. The sodium alginate acts as a protective colloid but slow fading occurs on standing.

COLORIMETRIC CHEMICAL ANALYTICAL METHODS　　　　　　　　　Aluminium Method 2

    4. It must be emphasised that the readings obtained by means of the Nessleriser and disc are only accurate provided that Nessleriser glasses are used which conform to the specification employed when the discs were calibrated, namely that the 50 ml calibration mark shall fall at a height of 113 mm ± 3 mm, measured internally.

    5. A strongly recommended standard source of white light is the Lovibond White Light Cabinet. In the absence of a standard source, north daylight should be used wherever possible.

Table for Conversion of Disc Readings (μg) to ppm

ml of sample used	\multicolumn{9}{c}{Disc Reading}								
	0.5	1.0	1.5	2.0	2.5	3.0	3.5	4.0	4.5
1	0.5	1.0	1.5	2.0	2.5	3.0	3.5	4.0	4.5
2	0.25	0.5	0.75	1.0	1.25	1.5	1.75	2.0	2.25
5	0.1	0.2	0.3	0.4	0.5	0.6	0.7	0.8	0.9
10	0.05	0.1	0.15	0.2	0.25	0.3	0.35	0.4	0.45
15	0.03	0.07	0.1	0.13	0.17	0.2	0.23	0.27	0.3
20	0.025	0.05	0.075	0.1	0.125	0.15	0.175	0.20	0.225
25	0.02	0.04	0.06	0.08	0.10	0.12	0.14	0.16	0.18
30	0.017	0.033	0.05	0.067	0.083	0.10	0.117	0.13	0.15

Aluminium Method 3
using Aluminon (aurin tricarboxylic acid)

Principle of the method

Aluminon reacts with aluminium in the presence of an ammonium acetate buffer to form a bright red lake. The intensity of this colour, which is proportional to the aluminium concentration, is measured by comparison with a series of Lovibond permanent glass colour standards, using discs designed for the determination of chlorine.

Reagents required

1. *p-Nitrophenol* ($HOC_6H_4pNO_2$) — 1% solution
2. 0.5N *Hydrochloric acid* (HCl) — Dilute 4.5 ml of concentrated hydrochloric acid to 100 ml with distilled water.
3. 5N *Aqueous ammonia* (NH_4OH) — Dilute 33 ml of 0.880 aqueous ammonia to 100 ml with distilled water.
4. *Thioglycollic acid* ($HSCH_2COOH$) — 1% solution
5. *Aluminon reagent* — Dissolve 0.25 g of aluminon in 250 ml of water, warming the solution. Add 5 g of gum acacia followed by 87 g of ammonium acetate ($NH_4C_2H_3O_2$) and 126 ml of 5N hydrochloric acid. Dilute to 500 ml and then filter under suction.

All chemicals used in the preparation of reagents should be of analytical reagent quality.

The Standard Lovibond Comparator Discs 3/40A and 3/40B

Disc 3/40A covers the range 0.1 to 1.0 ppm of chlorine
Disc 3/40B covers the range 0.2 to 4.0 ppm of chlorine
The aluminium equivalent of the chlorine standards must be established by the calibration of each batch of aluminon reagent against samples of known aluminium content (Note 6).

Technique

Pipette 50 ml of sample, which should not contain more than 4 ppm of aluminium into a 100 ml volumetric flask (Note 1). Add 1 drop of *p*-nitrophenol (reagent 1) and, if necessary add 5N aqueous ammonia (reagent 3) dropwise until the solution turns yellow. Add 0.5N hydrochloric acid drop by drop until this yellow colour just disappears. Add 16 drops of thioglycollic acid (reagent 4 Note 2) and 12 ml of aluminon (reagent 5). Make up the volume to within 5 ml of the mark and immerse the flask in boiling water for 15 minutes (Note 3). Cool rapidly in running water to 18–20°C. (Note 4). Make up to 100 ml with distilled water and compare the colour in a 13.5 mm tube with the standards in the disc, using a standard source of white light (Note 5). Place a blank, in the left hand field, of the reagents with distilled water instead of the sample. Read off the aluminium content from the calibration curve prepared from samples of known aluminium content (Note 6). A fresh calibration curve should be plotted for each new batch of aluminon reagent.

If the sample contains fluoride then proceed as follows:—
Pipette 50 ml into a 100 ml platinum evaporating basin and evaporate to dryness on a steambath or hot plate at 100°C. Add 5 ml of concentrated sulphuric acid to the residue and carefully evaporate to dryness on a sandbath. Fuse the residue with 0.5 g of fusion mixture AR (Note 7) and extract the cooled cake with 10 ml of hot water. Add 5 ml of concentrated hydrochloric acid to the basin, heat gently for a few minutes, add 10 ml of distilled water and wash the contents of the basin into the beaker containing the dissolved cake. Boil to remove carbon dioxide, wash into a volumetric flask and continue as above. Carry out a blank determination on a solution prepared by dissolving 0.5 g of fusion mixture in 40 ml distilled water and adding 5 ml concentrated hydrochloric acid and 5 ml concentrated sulphuric acid. Use this blank in the left-hand compartment of the comparator.

COLORIMETRIC CHEMICAL ANALYTICAL METHODS — Aluminium Method 3

Notes

1. New flasks should be treated with caustic soda (NaOH) followed by hot concentrated hydrochloric acid. These flasks should be set aside for aluminium determinations and a few ml of concentrated hydrochloric acid left in them between tests. Pyrex flasks are the most suitable.

2. If the sample contains more than 20 ppm of iron the quantity of thioglycollic acid should be increased.

3. The water-bath should be boiling vigorously when the flask is immersed, it should be large enough not to be cooled below the boiling point by the introduction of the flask and the flask should be positioned at least an inch from the bottom and sides of the bath.

4. For work of the highest accuracy it is recommended that the flask be immersed in a thermostat at $20 \pm 0.5°C$ for 30 minutes after cooling.

5. A strongly recommended standard source of white light is the Lovibond White Light Cabinet. In the absence of a standard source, north daylight should be used wherever possible.

6. The disc readings may be taken as giving the aluminium content directly, up to 0.6 ppm.

7. Fusion mixture AR consists of an equimolar mixture of pure anhydrous potassium and sodium carbonates. Such a mixture, with controlled maximum impurity limits, may be obtained from BDH Chemicals Ltd.

Aluminium Method 4
using Catechol Violet (pyrocatechol sulphonphthalein)

Introduction

A new absorptiometric method for the determination of aluminium in treated waters is now available as a colorimetric method using Lovibond permanent glass colour standards. Catechol violet has advantages over the other reagents used in the determination of aluminium[1,2] Application of the method to other types of samples has not been studied, but if appropriate methods of sample preparation are used the method probably has much wider applications.

Principle of the method

Catechol violet forms a blue complex with aluminium at pH 6.0 to 6.2. Since the blank solution is pale yellow the sample solution may range in colour from yellow green to blue depending on the amount of aluminium present. Iron also reacts with this reagent, but this interference is minimised by complexing the iron with 1, 10- phenanthroline.

Apparatus

Do not use detergents or chromic acid for cleaning the apparatus or polyethylene bottles: the following treatment is satisfactory. Fill or soak the apparatus and bottles with 10%(v/v) hydrochloric acid overnight and wash thoroughly with distilled or deionised water.

Polyethylene bottles: Two types will be required (a) thick-walled sample bottles of a convenient size, and (b) small bottles of approximately 60ml capacity for the analysis of samples. In all cases use bottles whose caps are made solely of polyethylene and which are free of insert rings. Make a mark on the side of each sample bottle to indicate the volume to be collected. (See sample collection).

Reagents required

Whenever possible use analytical reagent grade materials, and use distilled or deionised water for preparing reagent and blank solutions.

1. *Hydrochloric acid, 5N.* Accurately standardised 5N hydrochloric acid can be obtained from B.D.H. Chemicals Ltd. in their A.V.S. range of products. Alternatively, prepare the acid by the dilution of hydrochloric acid (sp. gr. 1. 18) making sure that its normality is 5N (± 0.02N).

2. *Hydrochloric acid* 0.1N. Add 20 ml (± 0.1 ml) of 5N hydrochloric acid to a 1-litre calibrated flask and dilute to the mark with water.

3. *1, 10-phenanthroline reagent solution.* Hydroxylammonium chloride may cause harmful effects and precautions should be taken while dispensing and preparing this solution. Dissolve 50 g (± 0.5 g) of hydroxylammonium chloride ($NH_2OH \cdot HCl$) in approximately 400 ml of water. To this solution add 0.5 g (± 0.005 g) of 1, 10-phenanthroline hydrate ($C_{12}H_8N_2H_2O$) and dissolve. (This solution may be slightly pink due to traces of iron but this is unimportant.) Transfer the solution to a 500 ml calibrated flask, and dilute to the mark with water. Store the solution in a polyethylene bottle; it is stable for at least eleven weeks.

4. *Catechol violet solution.* Dissolve 0.094 g (± 0.001 g) of catechol violet (pyrocatechol sulphonphthalein) in approximately 15 ml of water, allow to stand for 3 to 4 minutes, mixing occasionally. Transfer the solution to a 250 ml calibrated flask, and dilute to the mark with water. Store the solution in a borosilicate glass flask or bottle. The solution is stable for at least eleven weeks. Growths of spores may occur in this solution but if avoided when pipetting they are without effect. Such growths are less when distilled water rather than deionsed water is used, and they can also be decreased by storing the solution in a cool dark place.

5. *Hexamine buffer solution.* Dissolve 150 g (± 0.05 g) of hexamine [$(CH_2)_6.N_4$] in approximately 350 ml of water. Transfer this solution to a 500 ml calibrated flask. Add by a suitable burette 8.4 ml (± 0.1 ml) of ammonia solution (sp. gr. 0.880), dilute to the mark with water, and mix well. (NB. Ensure that the ammonia solution is taken from a fresh stock). Store this solution in a polyethylene bottle, and replace the cap immediately after use; the solution is stable for at least eleven weeks.

The Standard Lovibond Comparator disc 3/108

The disc covers the following range of concentrations:
0.00, 0.05, 0.10, 0.15, 0.20, 0.25, 0.30, 0.40 mg/l (p.p.m.) of aluminium.

Sample collection

Add by pipette sufficient 5N hydrochloric acid to the sample bottle so that when the sample has been collected the final acidity is 0.1N (± 0.002N). This is normally achieved by adding 20 ml of 5N hydrochloric acid in the case of a one litre sample or 10 ml of 5N hydrochloric acid when collecting a 500 ml sample. For very alkaline waters, however, it will be necessary to add more hydrochloric acid to ensure that the final acidity is 0.1N (± 0.002N). The polyethylene bottle should have a mark on its side to indicate the volume of sample to be collected. During sampling, care should be taken to avoid distorting the bottle when holding it and also to prevent contamination of the sample.

Technique

The procedure given below must be followed exactly as departures therefrom will lead to inaccuracies. A blank solution must be prepared at the same time as the first sample to be measured.

Blank solution

(1) Using a 50 ml measuring cylinder, add 35 ml (± 0.5 ml) of 0.1N hydrochloric acid (reagent 2) to a small polyethylene bottle.

(2) Add 1 ml (± 0.1 ml) of the 1, 10-phenanthroline solution (reagent 3) and mix by swirling.

(3) Add 2 ml (± 0.05 ml) of the catechol violet solution (reagent 4) and mix by swirling.

(4) Add 10 ml (± 0.1 ml) of the hexamine buffer solution (reagent 5) recap the bottle, and shake well.

(5) Between 10 and 20 minutes after the end of stage (4), transfer a 10 ml portion of the solution to a standard 13.5 mm comparator cell, and place in the left hand compartment of the comparator. N.B. Prior to placing in the comparator this solution should be pale yellow. A blue or greenish colour signifies either aluminium contamination or incorrect *p*H.

Sample analysis

(6) Using a 50 ml measuring cylinder, add 35 ml (± 0.5 ml) of the acidified sample to a small polyethylene bottle. Repeat stages (2) to (5) inclusive, placing a 10 ml portion of this solution in a comparator cell in the right hand compartment of the comparator.

(7) Hold the comparator facing a standard source of white light,* and compare the sample solution colour with the colours in the standard disc, rotating the latter until a colour match is obtained. Should the sample concentration exceed the top standard on the disc, the determination should be repeated with a fresh smaller sample, V ml, suitably diluted with 0.1N hydrochloric acid (reagent 2) to 35 ml (± 0.5 ml). The aluminium concentration in the original sample, C_t, is then calculated from the equation

$$C_t = \left(\frac{35}{V}\right) Ca$$

where Ca is the concentration of aluminium in the diluted sample.

* See Note 5.

Notes

1. The water used for the blank and reagent solutions should preferably have an aluminium content which is negligible compared with the smallest concentration to be determined in the samples. A method of determining the aluminium in the blank water and also a method for correcting for colour and turbidity is given in references 1 and 2.

2. In a batch of samples, each reagent may be added to all samples before adding the next reagent.

3. After stage (4), the pH of the solution should be 6.0 – 6.2. Check that this pH is achieved each time a batch of buffer solution is prepared, by carrying out the blank solution procedure Stages (1) to (4) inclusive, and by then measuring the pH of this solution using a pH meter.

4. The same blank solution may be used for at least one hour.

5. A strongly recommended source of light is the Lovibond White Light equipment. In the absence of a standard source, north daylight should be used whenever possible.

6. Interference effects. In the absorptiometric method, the interference effects of a wide range of substances were studied (see tables on pages 88, 89). With the exceptions of fluoride and condensed inorganic phosphates, the results showed no serious interference effects at the concentrations of the substances used in the trials. The present colorimetric method has not been studied with respect to interferences, but it is expected that such effects will not be markedly different.

The effect of 1 mgF$^-$/l should be tolerable in many applications but greater concentrations cause much larger errors. If the sample contains condensed inorganic phosphates, place the small polyethylene bottle (with cap fitting loosely) containing the 35 ml sample in a boiling water bath for two hours.

Cool, and then proceed with the reagent additions as described under 'Sample Analysis'.

7. All precautions must be taken to prevent contamination of the samples and the apparatus.

References

1. W. K. Dougan, *and* A. L. Wilson, *"The Absorptiometric Determination of Aluminium in Water. A Comparison of Some Chromogenic Reagents and the Development of an Improved Method". Analyst*, 1974, **99,** pp 413–430.

2. *The Water Research Association Technical Paper* 103, *The Absorptiometric Determination of Aluminium in Treated Water Using the Reagent Catechol Violet*. W. K. Dougan, A. L. Wilson, Medmenham, The Association, 1973.

TABLE X

EFFECTS OF OTHER SUBSTANCES ON THE DETERMINATION OF ALUMINIUM

Other substance	Concentration of other substance/ mg l^{-1}	Effect* of other substance, mg l^{-1} of Al, at an aluminium concentration of	
		0·000 mg l^{-1}	0·300 mg l^{-1}
Ca^{2+}	500	+0·006	+0·003
Ca^{2+}	100	+0·003	+0·004
Mg^{2+}	100	+0·004	+0·006
Na^+	100	+0·002	+0·004
K^+	50	+0·005	+0·003
Alkalinity	300 (as $CaCO_3$)	+0·006	+0·007
Alkalinity	200 (as $CaCO_3$)	0·000	—0·002
SO_4^{2-}	100	+0·001	0·000
NO_3^-	80	+0·005	+0·003
NO_2^-	10	+0·002	+0·005
PO_4^{3-}	8·2 (as P)	—0·005	—0·050
PO_4^{3-}	4·1 (as P)	—0·005	—0·038
PO_4^{3-}	1·7 (as P)	0·000	—0·002
SiO_3^{2-}	50 (as SiO_2)	+0·009	+0·005
Humic acid	10	+0·001	—0·004
Fulvic acid	10		
Detergents†	5	—0·003	—0·042
Chlorine	5	+0·001	0·000
Chlorine	5	+0·001	0·000
Ammonia	0·5 (as N)		
Co^{2+}	2	—0·002	—0·009
Ni^{2+}	2	—0·002	—0·006
Cd^{2+}	2	—0·003	—0·003
Cu^{2+}	2	+0·034	+0·006
Cu^{2+}	1	0·000	+0·003
Zn^{2+}	2	0·000	+0·010
Pb^{2+}	2	+0·052	+0·007
Pb^{2+}	1	+0·004	0·000
Fe^{3+}	1	+0·011	+0·014
Fe^{3+}	0·3	+0·006	+0·004
Mn^{2+}	2	+0·012	+0·006
Mn^{2+}	1	0·000	—0·006
Cr^{3+}	0·5	+0·004	—0·028
Cr^{3+}	0·25	+0·002	—0·015
Cr^{3+}	0·025	+0·005	—0·002
Sn^{2+}	2	+0·002	—0·015
Sn^{2+}	1	+0·002	—0·007
Coagulant aids—			
B.T.I. A100	0·5	+0·005	+0·003
B.T.I. A150	0·5	+0·002	+0·002
Magnafloc LT24	0·5	+0·002	+0·001
Wisprofloc 20	3	+0·002	+0·003
Wisprofloc P	3	+0·003	+0·003

*If the other substances had no effect, results would be expected to lie (95 per cent. confidence limits) within the following ranges:

    0·000 ± 0·003 for 0·000 mg l^{-1} of Al
    0·000 ± 0·006 for 0·300 mg l^{-1} of Al

†Daz, Dreft, Omo, Quix, Surf and Tide (equal proportions by mass were used).

TABLE XI

Effects of Condensed Inorganic Phosphates after Heating Acidified Samples

Substance	Concentration of substance/ mg l^{-1}	Effect* of substance, mg l^{-1} of Al, at an aluminium concentration of	
		0·000 mg l^{-1}	0·300 mg l^{-1}
Pyrophosphate	1 (as P)	+0·003	—0·005
Hexametaphosphate	1 (as P)	+0·001	—0·002
Tripolyphosphate	1 (as P)	0·000	+0·004
Calgon S†	10	+0·001	—0·008
Calgon S	20‡	+0·001	—0·037

*95 per cent. confidence limits for the results are the same as those in Table X.

†Calgon S is a glassy sodium polyphosphate used in water treatment; 10 mg l^{-1} of Calgon S is approximately equivalent to 4·1 mg l^{-1} of phosphorus.

‡Heating period, 3 hours.

TABLE XII

Effect of Fluoride

Concentration of fluoride/ mg l^{-1}	Effect* of fluoride, mg l^{-1} of Al, at an aluminium concentration of			
	0·000 mg l^{-1}	0·043 mg l^{-1}	0·130 mg l^{-1}	0·300 mg l^{-1}
0·5	—0·002	—0·002	—0·002	—0·011
1·0	—0·006	—0·009	—0·011	—0·023
4·0	—0·006	—	—	—0·192

*95 per cent. confidence limits for the results are the same as those in Table X.

The above tables apply only to the absorptiometric method and are reproduced by permission of the 'Analyst'.

The Determination of Ammonia

Introduction

The determination of ammonia (NH_3) using Nessler's reagent (alkaline potassio-mercuric iodide, $KI.HgI_2$), is one of the most widely used colorimetric techniques. Ammonia is extensively used as a refrigerant and also in the chemical industry. Its determination in air samples [1-3] is used in the control of industrial atmospheres. Proteins, amino-acids etc., yield ammonia as one of their breakdown products. The estimation of ammonia in blood [6-14] and urine [15-18] is therefore used in biochemistry as a measure of the body's metabolic efficiency. Ammonia is an essential building-block in the synthesis of nitrogen compounds by both plants and animals. Methods have therefore been developed for its determination in soils [38-41], manures and fertilizers [27-29], plants [31-33] and animal feedstuffs [26], either directly or, after reduction, as a measure of the total nitrogen content. As a result of its presence in protein decomposition products, ammonia is inevitably present in sewage. The determination of its concentration is used as an indication of the degree of contamination of water supplies [42-47], and is an essential part of the analysis of sewage and industrial effluents [34-37]. Methods have also been developed for the determination of ammonia in steels [23,24] and ores [25]. As an indicator for the decomposition of proteins etc. the concentration of ammonia can be used as a measure of the edibility of foods and methods are available for its determination in beverages [4,5], cheese [19], eggs [17,21,22] or food in general [20].

The Nessler reaction has been critically examined [50,51], Leonard [50], having examined all the variables, claims that the maximum colour is obtained from a reagent in which the molar ratio of potassium iodide (KI) to mercuric iodide (HgI_2) is about 2.5:1. Moeller [51] has shown that the reagent can be prepared equally well from a mixture of elemental mercury and iodine as from mercuric iodide. He found that the colour intensity was constant from 5 minutes to 6 hours after adding the reagent and that within the range 20—40°C temperature had no effect on the colour.

Methods of sample preparation

Air [1-3]

The atmosphere to be tested is drawn through an absorption train containing dilute acid and the ammonia is determined directly on the solution so obtained.

Beverages [4,5]

Apple juice and cider [4] Destroy organic matter by Kjeldahl digestion, make the digest alkaline distil the ammonia into dilute acid and determine by *Method 1*.

Fruit drinks [5] Dilute a 5 ml sample to 100 ml. Place a 20 ml aliquot in a distillation flask, add 230 ml of water, a small amount of sodium carbonate (Na_2CO_3), 50 ml of alkaline potassium permanganate ($KMnO_4$) solution and a little solid $KMnO_4$. Distil off 200 ml and determine ammonia on the distillate.

Biological materials

Blood and plasma [6-14] Several different methods of removing the ammonia from the sample have been suggested. These range from distillation [6,12,14] to ion exchange [7,8,9,11,13]. For distillation [6,14] mix the sample with an alkaline buffer solution; distil out the ammonia and water at room temperature under reduced pressure; collect the ammonia in dilute sulphuric acid (H_2SO_4) and Nesslerize. For ion exchange adsorb the ammonia from 2—3 ml of sample onto a suitable resin; wash the resin with water and then desorb the ammonia either by $4M$ sodium chloride (NaCl) [7], $0.1N$ caustic soda (NaOH) [8,11] or directly with Nessler reagent [9]. The use of diffusion to separate the ammonia from blood has also been described [10].

The available methods have been critically reviewed by Provini and Secchi [13] who stress the advantages of methods using ion exchange resins [6,8,9,11]. The following is a simple method [12] which avoids the need for deproteinisation. Place 1—2 ml of blood, containing 0.12—1.2 μ-moles of ammonia, in a micro distillation apparatus with 2—3 ml of water and 2—3 ml of $0.33M$ phosphate buffer (pH 7.5). Collect 5 ml of distillate.

Urine[15-18] Separation of the interfering materials by iodine treatment followed by direct Nesslerization has been suggested by Connerty et al[15]. A Cavett microdiffusion apparatus has been used by Milton and Duffield[16] to determine free ammonia, total nitrogen (N) and urea (CH_4N_2O). Absorption onto Permutit ion exchange resin followed by desorption with 10% caustic soda (NaOH) and Nesslerization has also been described[17, 18] (see *Method 2*).

Dairy products[19]

To determine ammonia in cheese[19] mix 10 g of sample into a smooth paste with $0.5N$ sodium citrate ($Na_3C_6H_5O_7$) at 45°C. Add 50 ml water and stir for 30 minutes. Make up to 200 ml. To a 25 ml aliquot add 25 ml of 20% trichloracetic acid (CCl_3COOH) dropwise, shake and filter. Neutralise 25 ml of clear filtrate with 10% caustic soda, make up to 100 ml and add 2 g Permutit ion exchange resin. Filter. Wash the Permutit with water and transfer to a 100 ml volumetric flask. Add 2 ml of 10% caustic soda, 68 ml of water and 2 ml of Nessler reagent and make up to the mark. Stand for exactly 15 minutes, decant the clear liquid and measure the colour.

Foods and edible oils[17, 20-22]

To determine free ammonia in foods[20] homogenize a 10 g sample, add 250 ml of water and 20 drops of kerosene. Add 5 ml of 40% caustic soda and blow liberated ammonia into dilute sulphuric acid with a stream of air or nitrogen.

To determine ammonia in eggs either dissolve 2 g of sample in 100 ml of water, filter and Nesslerize 50 ml of the filtrate [21], or use[22] the method of Folin and Bell[17], separating the ammonia on Permutit resin.

Metals, alloys and plating baths[23, 24]

Steel Dissolve 3 g of the sample in 30 ml of $6N$ hydrochloric acid. Make alkaline with 50 ml of ammonia-free caustic soda solution. Distil about 50 ml and Nesslerize[25]. Alternatively[24] follow the above procedure except to collect 150 ml of distillate, dilute to 250 ml and use an aliquot of this as the sample.

Minerals, ores and rocks[25]

Dissolve 2 g of ore in 1:1 sulphuric acid, make alkaline with caustic soda and distil ammonia. Reduce nitrates left in the flask with Devarda's alloy (see below under fertilizers) and steam distil the ammonia formed.

Miscellaneous[26-30]

Animal feedstuffs[26] Destroy the organic matter in the sample by Kjeldahl digestion and follow up by direct Nesslerization for total nitrogen.

Fertilizers[27-29] Potrafke et al[27] describe a rapid micro method for various forms of nitrogen in fertilizers. Distillation of ammonia takes only one minute when small volumes are used and the ammonia is swept out by an inert gas. The sum of the ammoniacal and nitrate (NO_3^-) nitrogen is obtained by the use of caustic soda and Devarda's alloy (45% aluminium, 50% copper, 5% zinc). To determine nitrate only, remove ammoniacal nitrogen by oxidation with hypobromite. Applications to amides and organic nitrates are also discussed. Gonnet [28] also discusses the determination of various forms of nitrogen in fertilizers while Ducet[24] is concerned mainly with the analysis of manures.

Vinegar[30] To determine albuminoid ammonia in vinegar use method of Mitra[30]. Neutralise 5 ml sample with $0.2N$ caustic soda, dilute to 100 ml and transfer 20 ml to a distillation flask, add 230 ml of water and a small amount of sodium carbonate (Na_2CO_3), distil off 100 ml and reject, add 50 ml of alkaline potassium permanganate ($KMnO_4$), distil a further 100 ml and Nesslerize. Malt vinegar has albuminoid nitrogen value of 230—400 ppm compared with 0—4 ppm for artificial samples.

COLORIMETRIC CHEMICAL ANALYTICAL METHODS — Ammonia

Plants[31-33]

Barker and Volk[31] discuss the method of successive distillation to separate ammonium, amide, amine and nitrate nitrogen. The loss of nitrate during distillation can be prevented[32] by the addition of salicylic acid (o-$C_7H_6O_3$) and thiosulphate ($S_2O_3^{2-}$) to the sulphuric acid used in the Kjeldahl digestion of plants, especially of grasses. The use of sulphuric acid plus hydrogen peroxide (H_2O_2) is also shown [32] to be both simpler and faster than normal digestion procedures. The interference of tannins, which can lead to anomalously high ammonia returns, can be removed by the prevention of oxidation, either by distillation under nitrogen or by the use of tannin anti-oxidants such as ascorbic acid (O.CO.C(OH):C(OH).CH.CH(OH).CH$_2$OH)[33].

Sewage and industrial wastes [34-36]

Sewage[34,35] In the official method ammoniacal nitrogen is determined by distillation and Nesslerization; albuminoid nitrogen by addition of alkaline potassium permanganate ($KMnO_4$) to the residue from the previous distillation, and redistilling; total organic nitrogen by reducing the sample with Devarda's alloy* in alkaline solution and then distillation; total oxidised nitrogen and nitrate by removing the ammonia by evaporating the sample to dryness in the presence of sodium carbonate, redissolving the residue in ammonia-free distilled water and reducing the oxidised nitrogen with either aluminium foil in caustic soda or by use of a zinc-copper couple, or by Devarda's alloy.

A simple method, for use in small sewage works where laboratory facilities are not readily available, has been developed by the Water Pollution Research Laboratory[35]. This uses a complexing agent to prevent precipitation of temporary hardness. A sample of the effluent is allowed to settle and 2 ml are pipetted into a 50 ml Nessler cylinder. About 25 ml of ammonia free distilled water are added followed by 1 ml of 10% sodium hexametaphosphate (($NaPO_3)_6$) solution. The contents of the cylinder are made up to the 50 ml mark with distilled water and thoroughly mixed.

*See "Fertilisers" above.

Plating bath effluents[36,37] Make sample basic and distil off the ammonia[36]. Reduce nitrate and nitrite in the residue, add sodium sulphide, and distil off ammonia[37].

Soil[38-41]

Shake 20 g of soil in a 500 ml flask with 1 ml of toluene ($PhCH_3$) and 200 ml of $0.05N$ hydrochloric acid for 30 minutes. Filter, make 20 ml of the filtrate up to 200 ml and Nesslerize[38]. Marciszewska[39] however has reviewed five alternative methods and recommends extraction of the soil with $2N$ potassium sulphate (K_2SO_4) at pH 1.5—2.0 followed by the addition of magnesium oxide (MgO) and distillation of the ammonia before Nesslerization. Lynderson and Opem[40] also favour extraction followed by steam distillation. Keay and Menage[41] recommend shaking 2 g of soil with 20 ml of $2N$ potassium chloride (KCl) for 60 minutes, filtering through a Whatman No. 42 paper and determining the ammonia directly on an aliquot of the filtrate.

Water[42-49]

The British standard method[42] for water used in industry recommends the addition of 1 ml of potassium sodium tartrate solution (50 g $KNaC_4H_4O_6.4H_2O$ in 100 ml of H_2O) to 50 ml of sample to prevent turbidity developing during direct Nesslerization. If the sample needs clarifying prior to the test add 1 ml of cupric sulphate solution (10 g of $CuSO_4.5H_2O$ in 100 ml) to 100 ml of sample, mix, add 1 ml of 50% w/v caustic soda mix and allow to settle. If this does not satisfactorily clarify the sample repeat the procedure adding the caustic soda first and then the copper sulphate. If the sample contains hydrogen sulphide, substitute 1 ml of zinc sulphate solution (5.42 g of $ZnSO_4.7H_2O$ in 100 ml) for the copper sulphate solution. Determine ammonia on the filtrate by direct Nesslerization.

When it is necessary to determine both free and albuminoid ammonia the standard method[43] recommends the use of a distillation procedure. After all free ammonia has been distilled out of the sample the albuminoid ammonia is determined by adding alkaline permanganate (6 g $KMnO_4$ + 200 g NaOH per litre) to the distillation flask, and continuing the distillation until no more ammonia is found in the distillate.

For sea water, ammonia can either be distilled in a current of air at 4—5 cm Hg pressure after adjusting the *p*H to about 9.15 with a sodium metaborate-sodium hydroxide buffer ($NaBO_2+NaOH$)[44], or it can be separated by micro-diffusion[45]. In the latter method nitrite and nitrate are also determined after reduction with Raney nickel. Grasshoff[46] describes an indirect method for the determination of ammonia in sea water. The ammonia is oxidised with hypobromite (HBrO) to nitrogen tribromide (NBr_3). The unconsumed hypobromite is destroyed by the addition of sodium nitrite ($NaNO_2$), excess potassium iodide (KI) is added and the liberated iodine is determined.

To determine ammonia in boiler water which also contains hydrazine (N_2H_4)[47] take a 50 ml sample, add a few crystals of potassium iodate (KIO_3) and 0.5 ml of 1:1 hydrochloric acid, stir vigorously allow to stand for not less than 10 minutes and then add the Nessler reagent. To remove interference from additives such as cyclohexylamine, morpholine or octadecylamine[46], take 500 ml of sample add 1:1 hydrochloric acid until the solution is just acid and then add 0.5 ml in excess. Add a few crystals of potassium iodate and then, after not less than 10 minutes, solid sodium carbonate (Na_2CO_3) until the solution is alkaline. Remove any residual iodine colour with sodium thiosulphate ($Na_2S_2O_3$) and distil. Collect 50 ml of distillate in dilute acid.

The individual tests

Method 1 covers the determination of NH_3 by Nesslerization of solutions.

Method 2 is the Folin and Bell[17] technique using Permutit resin to separate the NH_3 from interfering substances and was specifically developed for the determination of metabolized Mepacrine in blood plasma[18], using a formula relating urinary mepacrine and urinary ammonia to plasma mepacrine.

References

1. P. L. Magill, *Amer.Ind.Hyg.Assoc.Quart.*, 1950, **11**, 55; *Chem.Abs.*, 1951, **45**, 2612f
2. G. P. Larson, *Proc.Air Pollution and Smoke Prevention Assoc.Amer.*, 1951, **44**, 127; *Chem.Abs.*, 1952, **46**, 5759a
3. G. A. Lugg and S. A. Wright, *Australia Dept. Supply, Defence Research Labs. Circular No. 14 1953, "Determination of Toxic Gases and Vapours in Air"*; *Chem.Abs.*, 1953, **47**, 9857c
4. C. Macfarlane and J. F. Mears, *Analyst*, 1964, **89**, 428
5. S. N. Mitra and S. C. Roy, *Analyst*, 1953, **78**, 681
6. P. V. D. Burg and H. W. Mook, *Clin.Chim.Acta*, 1963, **8**, 162; *Analyt.Abs.*, 1963, **10**, 5283
7. J. C. B. Fenton, *Clin.Chim.Acta*, 1962, **7**, 163
8. G. E. Miller and J. D. Rice, *Amer.J.Clin.Path.*, 1963, **39**, 97; *Analyt.Abs.*, 1963, **10**, 5285
9. J. H. Hutchison and D. H. Labby, *J.Lab.Clin.Med.*, 1962, **60**, 170; *Analyt.Abs.*, 1963, **10**, 1122
10. D. Seligson and K. Hirahara, *J.Lab.Clin.Med.*, 1957, **49**, 962; *Analyt.Abs.*, 1958, **5**, 617
11. S. G. Dienst and B. Morris, *J.Lab.Clin.Med.*, 1964, **64** (3), 495; *Analyt.Abs.*, 1966, **13**, 298
12. E. D. Wachsmuth and I. Fritze, *Klin.Wschr.*, 1965, **43** (1), 53; *Analyt.Abs.*, 1966, **13**, 2545
13. L. Provini and G. C. Sechi, *Diagnosi.Napoli*, 1963, **19** (7), 289; *Analyt.Abs.*, 1966, **13**, 3057
14. S. Gangolli and T. F. Nicholson, *Clinica Chim.Acta.*, 1966, **14** (5), 585; *Analyt.Abs.*, 1968, **15**, 879
15. H. V. Connerty, A. R. Briggs and E. H. Eaton, *Amer.J.Clin.Path.*, 1957, **28**, 634; *Analyt.Abs.*, 1959, **6**, 1007
16. R. F. Milton and W. D. Duffield, *Lab.Practise*, 1954, **3**, 318; *Analyt.Abs.*, 1954, **1**, 2738
17. O. Folin and R. D. Bell, *J.Biol.Chem.*, 1917, **29**, 329
18. Army Malaria Unit, "*Report to Chemical and Pharmacological Sub-committee, M.R.C. Malaria Committee*" M.L.A. 58 and 83, 1944
19. E. Bernhard, *Mitt.Lebensm.Hyg.Bern.*, 1954, **45**, 115; *Analyt.Abs.*, 1954, **1**, 2812
20. D. S. Ortiz, *Rec.Soc.Venezol.Quim.*, 1953, **5**, 39; *Analyt.Abs.*, 1954, **1**, 3112

21. G. Wendland, *Z.Lebensm.Untersuch.u.Forsch.*, 1948, **88**, 50; *Chem.Abs.*, 1948, **42**, 4286c
22. E. B. Boyce, *J.Assoc.Off.Agr.Chemists*, 1950, **33**, 703; *Chem.Abs.*, 1951, **45**, 780e
23. W. T. Hall and R. S. Williams, *"Iron, Steel and Brass"*, McGraw Hill, New York, 1921
24. L. E. Barton, *J.Ind.Eng.Chem.*, 1914, **6**, 1012
25. U.K.A.E.A. Report PG225 (S) 1962; *Analyt.Abs.*, 1962, **9**, 4142
26. P. C. Williams, *Analyst*, 1964, **89**, 276
27. K. A. Potrafke, M. Kroll and L. Blom, *Analyt.Chim.Acta.*, 1964, **30**, 128
28. P. Gonnet, *Ann.fals et fraudes*, 1951, **44**, 103, *Chem.Abs.*, 1952, **46**, 1692a
29. G. Ducet, *Ann.agron.*, 1946, **16**, 219; *Chem.Abs.*, 1947, **41**, 4878c
30. S. N. Mitra, *Analyst*, 1953, **78**, 499
31. A. V. Barker and R. J. Volk, *Analyt.Chem.*, 1964, **36**, 439
32. D. M. Ekpete and A. H. Cornfield, *Analyst*, 1964, **89**, 670
33. V. R. Popov, *Biokhimaya*, 1958, **23**, 37; *Analyt.Abs.*, 1958, **5**, 3547
34. Min. of Housing and Local Govt., *"Methods of Chemical Analysis as Applied to Sewage and Sewage Effluents"*. H.M. Stationery Office, London, 1956
35. Min. of Technology, *"Notes on Water Pollution"*, No. 44, March 1969
36. E. J. Serfass and R. F. Muraca, *Plating*, 1955, **42**, 265; *Chem.Abs.*, 1956, **50**, 6250i
37. idem.ibid., 1956, **43**, 233; *Chem.Abs.*, 1956, **50**, 7009d
38. N. M. Maihoroda, *Pochvovedenie*, 1961, 100; *Analyt.Abs.*, 1962, **9**, 1709
39. M. Marciszewska, *Szoplik.Zabl.Hig.Warsaw*, 1961, **12**, 235; *Analyt.Abs.*, 1962, **9**, 405
40. D. L. Lyndersen and M. Opem, *Z.analyt.Chem.*, 1958, **159**, 339; *Analyt.Abs.*, 1958, **5**, 3551
41. J. Keay and P. M. A. Menage, *Analyst*, 1970, **95**, 379
42. Brit. Standards Inst., *"Routine Methods of Testing Water Used in Industry"*, B.S. 1427, London, 1962
43. Brit. Standards Inst., *"Methods of Testing Water Used in Industry"*, B.S: 2690, Part 7, London, 1968
44. J. P. Riley, *Analyt.Chim.Acta*, 1953, **9**, 355; *Analyt.Abs.*, 1954, **1**, 825
45. J. P. Riley and P. Sinhaseni, *J.Mar.Biol.Assoc. U.K.*, 1957, **36**, 161; *Analyt.Abs.*, 1958, **5**, 280
46. K. Grasshoff, *Z.analyt.Chem.*, 1968, **234** (1), 13; *Analyt.Abs.*, 1969, **16**, 2764
47. N. T. Crosby, *Analyst*, 1968, **93**, 406
48. Brit. Standards Inst., *"Treatment of Water for Marine Boilers"*, B.S. 1170, London, 1968
49. T. A. Steward, *"Record of Agricultural Res."* Min. of Ag. N. Ireland, 1968, **17**, Part 1, 91
50. R. H. Leonard, *Clin.Chem.*, 1963, **9**, 417; *Analyt.Abs.*, 1964, **11**, 1385
51. G. Moeller, *Z.analyt.Chem.*, 1969, **245** (3), 155; *Analyt.Abs.*, 1970, **19**, 69

Ammonia Method 1
using Nessler's Reagent

Principle of the method

This test is based on Nessler's reagent, a strongly alkaline solution of potassium mercuric iodide. In the presence of ammonia (NH_3) a reddish-brown colour is formed. The intensity of this colour, which is proportional to the ammonia concentration is measured by comparison with Lovibond permanent glass colour standards.

Reagent required

The published formulae for Nessler's reagent vary considerably and when using the Lovibond Nessleriser discs it is important that the reagent employed should correspond with that used for standardising the colours. The following is the formula used for this purpose.

Dissolve 35 g of potassium iodide (KI) and 12.5 g of mercuric chloride ($HgCl_2$) in 800 ml of water and add a cold saturated solution of mercuric chloride until, after repeated shaking, a slight red precipitate remains, then add 120 g of sodium hydroxide (NaOH), shake until dissolved, and finally add a little more of the saturated solution of mercuric chloride and sufficient water to produce 1 litre. Shake occasionally during several days, allow to stand, and use the clear supernatant liquid for the tests.

The Standard Lovibond Nessleriser Discs, NAA, NAB, NAC and NAD

Disc NAA covers the range 1µg to 10µg (0.001 to 0.01 mg) of ammonia, i.e. 0.02 to 0.2 ppm on a 50 ml sample.

Disc NAB covers the range 10µg to 26µg (0.01 to 0.026 mg) of ammonia, i.e. 0.2 to 0.52 ppm on a 50 ml sample.

Disc NAC covers the range 28µg to 60µg (0.028 to 0.06 mg) of ammonia, i.e. 0.56 to 1.2 ppm on a 50 ml sample.

Disc NAD covers the range 60µg to 100µg (0.06 to 0.1 mg) of ammonia, i.e. 1.2 to 2.0 ppm on a 50 ml sample.

Technique

Fill one of the Nessleriser glasses to the 50 ml mark with the solution under examination, which must be free from haze (or with distilled water) and place it in the left-hand compartment of the Nessleriser. Fill the other Nessleriser glass to the 50 ml mark with the solution to be tested, which must not be acid, add 2 ml of Nessler's reagent, mix, and place it in the right-hand compartment of the instrument. Stand the Nessleriser before a standard source of white light (Note 5) and compare the colour produced in the test solution with the colours in the standard disc, rotating the disc until a colour match is obtained.

Notes

1. If turbidity occurs upon addition of the Nesslers' Reagent, repeat the test and add 1 ml of Potassium Sodium Tartrate Reagent (R221) to the sample before the Nesslers' and mix thoroughly.

2. With small quantities of NH_3, from 1µg to 5µg, the colour develops slowly and in order to obtain accurate results, fifteen minutes should be allowed to elapse before matching the colour. When more than 5 µg of ammonia is present, the colour should be matched five minutes after adding the reagent.

3. The markings on the discs represent the actual amounts of ammonia producing the colour in the test. To convert this figure to parts per million (ppm) *divide* by 50. Thus, if on adding the reagent to 50 ml of water a colour equivalent to 5µg is produced, the amount of ammonia present in the water will be 0.1 ppm

4. It must be emphasized that the readings obtained by means of the Lovibond Nessleriser and disc are only accurate provided that Nessleriser glasses are used which conform to the specification employed when the discs were being calibrated, namely that the 50 ml calibration mark shall fall at a height of 113 mm, plus or minus 3 mm, measured internally.

5. A strongly recommended standard source of white light is the Lovibond White Light Cabinet. In the absence of a standard source, north daylight should be used wherever possible.

Ammonia Method 2
using Nessler's Reagent

Principle of the method

Urine contains substances (e.g. creatinine) which interfere with direct Nesslerisation for determining the ammonia (NH_3) content. The ammonia is therefore removed by "Permutit-Decalso" (sodium aluminium silicate). The supernatant urine is discarded, and the "Permutit" washed. The "Permutit" is then made alkaline to liberate the ammonia which is estimated colorimetrically by means of the Lovibond Comparator and the standard blood urea discs.

Reagents required
1. 10% *w/v sodium hydroxide* ($NaOH$)
2. *Nessler's solution* Dissolve 150 g potassium iodide (KI) in 100 ml distilled water. Transfer this solution to a litre flask, add 200 g mercuric iodide ($Hg\ I_2$) and allow to dissolve. When solution is complete, make up to the litre mark with distilled water, and filter. Add 1 litre distilled water to this filtrate. Dilute 15 ml of this stock solution immediately before use with 85 ml of distilled water.
3. *"Permutit."* A convenient fineness is that which will pass through a 60 mesh but not an 80 mesh sieve. If the "Permutit" as obtained is too coarse, it may be ground gently in a mortar. Fine particles should then be removed by repeated shaking with water, and decantation of the turbid supernatant liquid. The "Permutit" can be used repeatedly. The "Permutit" collected from a day's work should be freed from any traces of ammonia by washing with 10% w/v sodium hydroxide. The "Permutit" is then washed in 20% v/v acetic acid (CH_3COOH) and finally with water. It is dried in air without heat; if it is oven-dried the activity is greatly reduced.

The activity of each new batch of "Permutit" must be established before use, and determined against the Lovibond Blood Urea discs. The method is as follows:—

Prepare three standard ammonia-nitrogen solutions
1. 10 *mg ammonia nitrogen per* 100 *ml* Weigh out 94.4 mg of ammonium sulphate (($NH_4)_2\ SO_4$) and make up to 200 ml in a graduated flask with distilled water.
2. 5 *mg ammonia nitrogen per* 100 *ml* Solution 1 diluted 1 in 2 with distilled water
3. 2 *mg ammonia nitrogen per* 100 *ml* 20 ml of solution 1 diluted to 100 ml with distilled water

Take 1 ml of each of these standard solutions through the procedure described below for urine, and draw a graph showing Lovibond Comparator readings plotted against mg ammonia-nitrogen per 100 ml.

The Standard Lovibond Comparator Discs 5/9A and 5/9B

Disc 5/9A. covers the range 20 mg to 100 mg urea per 100 ml in 9 steps of 10 mg.
Disc 5/9B. covers the range 110 mg to 220 mg urea per 100 ml in steps of 10 mg up to 160 and then in steps of 20 mg.

Technique

In a test tube place 100 mg "Permutit." Add 1 ml urine (diluted if necessary) and shake for 5 minutes. Add approximately 10 ml distilled water, shake, and allow to settle. Pour off the supernatant fluid carefully, and add another 10 ml of water. Again allow to settle and decant. Add 1 ml of 10% sodium hydroxide (reagent 1) to the "Permutit," shake, and allow to stand for 5 minutes. Add 8 ml of water and 1 ml of Nessler's solution. Shake, place in a comparator test tube or 13.5 mm cell in the right-hand compartment of the comparator, a "blank" of distilled water in the left-hand compartment behind the glass standards, and compare the colours by revolving the disc until a match is obtained (Note 2). Estimate the ammonia present by reference to the graph prepared for that batch of "Permutit."

With dilute urines, the estimation can be done directly on 1 ml of the urine, but with more concentrated specimens the urine will have to be diluted. The usual dilution is 1 in 2, but specimens have been met in which dilutions of 1 in 5 or even 1 in 10 have had to be made. If a dilution is used, the ammonia-nitrogen concentration as read from the graph must, naturally, be multiplied by the dilution factor.

Notes

1. "Permutit" is the Registered Trade Mark of the Permutit Co. Ltd., London. "Permutit-Decalso" is specially manufactured and graded for biochemical purposes.

2. A standard source of white light should be used for colour matching. A strongly recommended standard source is the Lovibond White Light Cabinet. In the absence of a standard source use north daylight wherever possible.

COLORIMETRIC CHEMICAL ANALYTICAL METHODS

The Determination of Arsenic

Introduction

Compounds of arsenic are widely distributed in nature and deposits of elemental arsenic are also found. Traces of arsenic are always present in both vegetable and animal organisms, and these may play an essential part in the metabolism. Apart from its occurrence in the effluents and residues from mineral treatment plants, compounds of arsenic are used industrially in weed killers, rat killers, fruit-sprays and as a pigment in paint. In small doses arsenic is used in medicine as a tonic and is also used for the treatment of various tropical diseases and of syphilis.

Most arsenic compounds are extremely poisonous and for this reason great care must be taken to exercise control of the concentration of arsenic in foodstuffs, especially in the form of spray residues on fruits.

Methods of Sample Preparation

Biological Materials[1,2]

Bile and Blood[1] Homogenise the sample with a mixture of magnesium nitrate ($Mg(NO_3)_2$) and magnesium oxide (MgO). Heat gently until dry and then ash at 550—600°C. Cool, moisten ash with 10 ml of water and transfer it quantitatively to a 250 ml conical flask with 90 ml of $6N$ hydrochloric acid and proceed by *Method 1*.

Tissue[2] Wet ash sample and use resulting solution.

Urine[2] Before wet ashing treat sample for 12 hours with nitric acid in the cold.

Chemicals[3,4]

Herbicides[3] To about 100 mg of sample add 10 ml of $15N$ nitric acid and 3 ml of $2N$ potassium bromate ($KBrO_3$). Evaporate until the total volume is less than 5 ml, add a further 3 ml of $2N$ potassium bromate, evaporate to dryness and heat on hot plate at about 300° until fumes of bromine have disappeared. Cool, dissolve the residue in $4N$ hydrochloric acid and determine arsenic on this solution.

Elemental Sulphur[4] Heat a 10 g sample in a polypropylene beaker with 54 g of 50% caustic soda (NaOH) solution. Cool, dilute with water add hydrogen peroxide (H_2O_2) and boil. Repeat the peroxide treatment, neutralise with normal sulphuric acid and boil with further addition of peroxide until the solution is clear. Evaporate to dryness, dissolve the residue in water and use this solution.

Foods and edible oils[5-7]

Fruit[5,6] Weigh 1 g of sample into a 100 ml conical flask, add 5 g of potassium chlorate ($KClO_3$), 35 ml of distilled water and 18 ml of hydrochloric acid. Boil gently in a fume cupboard until the reaction ceases and most of the chlorine has been expelled, cool the flask in a stream of cold water, add a further 15 ml of distilled water and use this solution.

Poultry[7] Weigh 3 mg of magnesium oxide (MgO) and 10 g of ground tissue into a crucible. Add 20 ml of Whatman CF 11 fibrous cellulose powder (use only 10 ml for muscle tissue). Mix and then char the mixture over an open flame at high heat until no further smoke is evolved. Cool, add 3 g of magnesium nitrate ($Mg(NO_3)_2.6H_2O$), place the crucible in a cold muffle furnace, heat up to 555°C and ash for two hours at this temperature. Treat the ash in the same manner as described for bile and blood above.

Metals, alloys and plating baths[8-11]

General[8] Dissolve the sample in hydrochloric acid, complex with sodium molybdate (Na_2MoO_4) and extract the yellow complex into ethyl ether ($C_2H_5OC_2H_5$)—butanol (C_4H_9OH) (1:2) mixture. Wet ash the complex and determine arsenic on the solution obtained. Milligram amounts of silica, germanium, iron, aluminium, manganese and nickel do not interfere with the determination of 20 µg amounts of arsenic.

Selenium[9] Dissolve a 2 g sample in 10—15 ml of nitric acid, evaporate to dryness, dissolve the residue in 2—3 ml of water and repeat the evaporation. Dissolve the residue in 50 ml of $10N$ hydrochloric acid, cool to 0°C, precipitate the selenium by passing sulphur dioxide for 90 minutes and set the solution aside for 2 hours. Filter, bubble a stream of air or nitrogen through the filtrate to remove any residual sulphur dioxide. Extract the arsenic into a chloroform ($CHCl_3$) solution of sodium diethyl dithiocarbamate ($(C_2H_5)_2N.CS.SNa.3H_2O$). Wet ash the chloroform extract and determine arsenic on the solution obtained.

Bismuth[10] Heat a 1 g sample to dryness first with 5 ml of nitric acid and then with 4 ml of sulphuric acid. Add 0.3 g of hydrazine sulphate ($N_2H_4.H_2SO_4$), evaporate to a syrup, add 4 ml of water and 13 ml of a 2% solution of hydrogen iodide (HI) in hydrochloric acid. After any salts have dissolved shake the solution with 5 ml of carbon tetrachloride (CCl_4) for 2 minutes. Separate the organic layer, wash with 2 x 5 ml portions of $9N$ hydrochloric acid and re-extract the arsenic into 5 ml of water. Shake for 5 minutes. To the aqueous layer add 5 drops of nitric acid, evaporate to dryness and dissolve the residue in 5 ml of water. Determine arsenic on this solution.

Zinc and Zinc Alloys[11] Dissolve the sample in nitric acid, remove excess nitric acid by heating to white fumes with sulphuric acid. Dilute with water and use this solution.

Minerals, ores and rocks[12-15]

Silicate Rocks[12] Fuse sample with caustic soda (NaOH) and after leaching with water separate the arsenic from the filtrate by adding 2% thionalide (α-mercapto-N,2-naphthyl acetamide, $C_{10}H_7.NH.CO.CH_2SH$) solution in acetone (CH_3COCH_3). Stir for 5 minutes, stand for 10 minutes, boil for 10 minutes to remove the acetone, cool with stirring and set aside to stand overnight. Filter, with the aid of suction, through a small filter paper and wash the precipitate with water. Heat the precipitate and the paper with 7.5 ml of nitric acid, in a small flask fitted with a bulb stopper, until the solution is a pale yellow colour, then remove the stopper and evaporate the nitric acid. Carry out the final stages of the evaporation under a stream of carbon dioxide. Be careful to avoid charring. Dissolve the pale yellow residue in 1 ml of normal sulphuric acid and determine arsenic on this solution.

Ores[13-15] Dissolve sample in 20—30 ml of nitric acid and 5—10 ml of hydrochloric acid, evaporate the solution to small volume, add 10—15 ml of 1:1 sulphuric acid, evaporate to the point where copious fumes appear, add a little more water and again evaporate. Dissolve the residue in 5—10 ml of water, add 30 ml of hydrochloric acid, 1—2 g of cuprous chloride (CuCl) and 0.5 g of potassium bromide (KBr). Extract the arsenious chloride ($AsCl_3$) formed into toluene ($CH_3.C_6H_5$) or benzene (C_6H_6) using 3×15 ml portions. Re-extract the arsenic into water and analyse this solution.

Miscellaneous[16,17]

Animal Feedstuffs[16] To differentiate between naturally occurring arsenic and that added for general growth purposes, extract the sample with 5 ml of hydrochloric acid, dilute to 100 ml with water and use this solution. Wet ashing is required to extract the naturally occurring arsenic.

Wood[17] Take 1 g of dried sawdust and place in a 250 ml graduated flask. Add 50 ml of $5N$ sulphuric acid and 10 ml of 100 volumes hydrogen peroxide (H_2O_2). Heat on a water bath at 70—75°C for 20 minutes and occassionally swirl the flask to keep the contents well mixed. Remove from the bath, add 150 ml of water, swirl, cool to room temperature, dilute to the mark and mix. Mix twice more at 5 minute intervals and then allow the contents to settle for 10 minutes. Filter 10 ml of the supernatant liquid through a dry 7 cm disc of Whatman No. 44 filter paper and discard the first 5 ml of filtrate. Determine arsenic on a 1 ml aliquot of remaining filtrate.

Plants[12]

Oxidise the organic matter by digestion with nitric acid, evaporate to dryness to remove excess acid, dissolve the residue in dilute sulphuric acid and follow the procedure outlined for silicate rocks above.

Soils[18,19]

Weigh 0.25 g of sample into a borosilicate glass test-tube, add 1 g of potassium hydrogen sulphate ($KHSO_4$) and fuse until the melt becomes quiescent. Cool, leach the melt with 5 ml of $0.5N$ hydrochloric acid on a water bath. Cool, filter and make up to a known volume.

Water[12,20]

Take 980 ml of sample[20], containing less than 15 μg of arsenic, add 10 ml of $0.18M$ barium chloride ($BaCl_2$), adjust the pH to 7.2—8.0 with $0.5N$ aqueous ammonia (NH_4OH) and add

1.5 g of barium sulphate ($BaSO_4$). Stir, set aside for 15 minutes, stir again, cover, and set aside overnight. Mix the supernatant liquid with a further 1.5 g of barium sulphate, set aside for 10 hours and siphon off and discard the supernatant liquid. Centrifuge the combined suspensions and wash the residue with water at pH 8.5. Leach arsenic from the barium sulphate carrier with 3×7 ml portions of $0.15N$ hydrochloric acid.

For sea water[12] add 4 ml of a 5% aqueous solution of ascorbic acid (O.CO.C(OH): C(OH).CH.CH(OH)CH_2OH) to 1 litre of sample and heat to boiling point to reduce the arsenic to the trivalent state. Cool for 10 minutes, add a further 2 ml of ascorbic acid solution and cool to room temperature. Proceed as for silicate rocks above.

The individual tests

Method 1 is the silver diethyldithiocarbamate method[5] developed as the standard method for the determination of arsenic. This method has been extensively studied [21-25] and it has been shown [21] that the optimum rate of hydrogen generation is achieved by using 1.6–$2.2N$ hydrochloric acid, or 4–$6N$ sulphuric acid, together with 20 mesh zinc granules. For complete recovery of arsenic the dead-space in the apparatus must be less than 100 ml[21]. The use of hydrochloric acid ensures optimum yield claimed as over 95%[22]. To ensure quantitative yields more than 20 mg of the sodium diethyldithiocarbamate per 10 ml of pyridine should be used[23] and care should be taken to remove silver nitrate ($AgNO_3$) from the reagent as this interferes with the determination[24].

References

1. L. R. Stone, *J.Assoc.Off.Analyt.Chem.*, 1967, **50** (6) 1361; *Analyt.Abs.*, 1969, **16**, 1410
2. M. del Pilar, P. M. de Neuman and A. Singerman, *Revta Asoc. bioquim. Argent.*, 1966, **31**, 10; *Analyt.Abs.*, 1967, **14**, 2680
3. W. F. Carey, *J.Assoc.Off.Analyt.Chem.*, 1968, **51** (6), 1300; *Analyt.Abs.*, 1970, **18**, 2044
4. E. Heinerth, *Revta port.Quim.*, 1967, **9** (4), 202; *Analyt.Abs.*, 1969, **17**, 2704
5. Brit. Standards Inst., "*Method for the determination of Arsenic*", B.S. 4404: 1968, London, 1968
6. Reckitt's (Colours) Ltd., "*Standard Analytical Test Method No. 1.3.10.8*" 1969
7. J. L. Morrison and G. M. George, *J.Assoc.Off.Analyt.Chem.*, 1969, **52** (5), 930; *Analyt.Abs.*, 1970, **19**, 3379
8. R. A. Karanov and A. N. Karolev, *Zh.analit.Khim.*, 1965, **20** (5) 639; *Analyt.Abs.*, 1967, **14**, 112
9. E. Ebner, *Z.analyt.Chem.*, 1964, **206** (2), 106; *Analyt.Abs.*, 1966, **13**, 631
10. S. M. Milaev and T. V. Lyashenko *Sb.nauch.Trudy vses.nauchno-issled.gornometallurg. Inst.tsvet.Metall.*, 1965, (9), 34; *Analyt.Abs.*, 1967, **14**, 669
11. Brit. Standards Inst., B.S. 3630, Part 12, 1970, London, 1970
12. J. E. Portmann and J. P. Riley, *Analyt.Chim.Acta*, 1964, **31** (6), 509; *Analyt.Abs.*, 1966, **13**, 2049
13. G. Atterescu, *Revta Chim.*, 1965, **16** (4), 221; *Analyt.Abs.*, 1966, **13**, 4107
14. L. V. Minasyan, *Zavodskaya Lab.*, 1965, **31** (11), 1326; *Analyt.Abs.*, 1967, **14**, 1377
15. K. Minasyan, *Prom.Armenii*, 1966, (11), 40; *Analyt.Abs.*, 1968, **15**, 1353
16. A. Pugliese, O. Zavattiero Castagnoli, P. F. Bianchi and A. M. Russo, *Veterinaria ital.*, 1964, **15** (9), 703; *Analyt.Abs.*, 1966, **13**, 4503
17. A. I. Williams, *Analyst*, 1970, **95** (7), 670
18. R. E. Stanton, *Econ.Geol.*, 1964, **59** (8), 1599; *Analyt.Abs.*, 1966, **13**, 2054
19. Appl.Geochem.Res.Group, *Tech.Comm.R.Sch.Mines*, 1965, No. 49
20. K. R. Kar and Gurbir Singh, *Mikrochim.Acta*, 1968, (3), 560; *Analyt.Abs.*, 1969, **17**, 1400
21. L. Dubois, T. Teichman and J. L. Monkman, *Mikrochim.Acta.*, 1966, (3), 415; *Analyt.Abs.*, 1967, **14**, 4654
22. P. Blanquet, J. Croizet, M. Croizet and R. Castagnou, *Annls.pharm.fran.*, 1968, **26** (2), 159; *Analyt.Abs.*, 1969, **16**, 3130
23. H. Bode and K. Hachmann, *Z.analyt.Chem.*, 1968, **241** (1), 18; *Analyt.Abs.*, 1970, **18**, 877
24. L. Dubois, T. Teichmann, C. J. Baker, A. Zdrojewski and J. L. Monkman, *Mikrochim.Acta*, 1969, (1), 185; *Analyt.Abs.*, 1970, **18**, 2348

Arsenic

using silver diethyldithiocarbamate

Principle of the method

The arsenic in the sample must first be converted to a soluble form. It is then reduced to the trivalent state using an acid solution of stannous chloride and potassium iodide. After this treatment the addition of zinc to the acid solution liberates nascent hydrogen which reduces the arsenious salts to gaseous arsine which is evolved and absorbed in a pyridine solution of silver diethyldithiocarbamate. The reaction between the arsine and the silver diethyldithiocarbamate produces a pink coloration. The intensity of this colour, which is proportional to the arsenic concentration, is measured by comparison with a series of Lovibond permanent glass colour standards. Some other metals also produce similar colours (Notes 1 & 2)

Reagents required

1. *Hydrochloric acid*   Concentrated (d = 1.18)
2. *Potassium iodide solution*   Dissolve 150 g of potassium iodide in distilled water, make the total volume up to 1 litre with distilled water and store the solution in the dark.
3. *Stannous chloride solution*   Dissolve 40 g of stannous chloride dihydrate ($SnCl_2.2H_2O$) in a mixture of 25 ml of distilled water and 75 ml of concentrated hydrochloric acid. If an appreciable amount of deposit forms in this solution after it has been prepared for some time, discard the old solution and prepare a fresh batch.
4. *Silver diethyldithiocarbamate solution*   Dissolve 1 g of silver diethyldithiocarbamate in water-white pyridine (d = 0.980 approx.) and make up the total volume to 200 ml with this pyridine. **Because of the toxicity of pyridine vapour these operations should be carried out in a fume cupboard.** Store the solution in a well stoppered glass bottle protected from the light. This solution is stable for about two months.

   Silver diethyldithiocarbamate of suitable quality is available commercially. However if difficulties are experienced in obtaining this reagent it may be prepared from the more readily available sodium salt in the following manner:

   Dissolve 10 g of sodium diethyldithiocarbamate in 35 ml of ethanol (95% v/v) and filter. Add to this solution, with continual stirring, 100 ml of diethyl ether. Filter with suction, wash the precipitate with ether and dry in air. Dissolve 2.2 g of this purified sodium diethyldithiocarbamate in 100 ml of distilled water. Dissolve 1.7 g of silver nitrate in another 100 ml of distilled water. Mix these two solutions slowly, with continuous stirring, keeping the mixture at a temperature below 10°C. When mixing is complete allow the precipitate to settle, decant the clear supernatant solution, and wash the precipitate three times with water at a temperature below 10°C. Filter and dry the product under vacuum at room temperature. Preserve in a cool place protected from light.
5. *Zinc*   Granular or shot, 0.5 to 1.0 mm (16—30 mesh)
6. *Impregnated cotton wool*   Dissolve 50 g of lead acetate trihydrate in 250 ml of distilled water. Saturate the cotton wool with this solution, drain, tightly press, and dry under vacuum at room temperature.

   All chemicals which are used in the preparation of these reagents must be of an analytical reagent grade which has been specially approved for arsenic testing.

The Standard Lovibond Comparator Discs 3/87A and 3/87B

Disc 3/87A contains standards corresponding to 2, 3, 4, 5, 6, 7, 8, 9, and 10 micrograms of arsenic (As).

Disc 3/87B contains standards corresponding to 4, 6, 8, 10, 12, 14, 16, 18, and 20 micrograms of arsenic (As).

COLORIMETRIC CHEMICAL ANALYTICAL METHODS — Arsenic

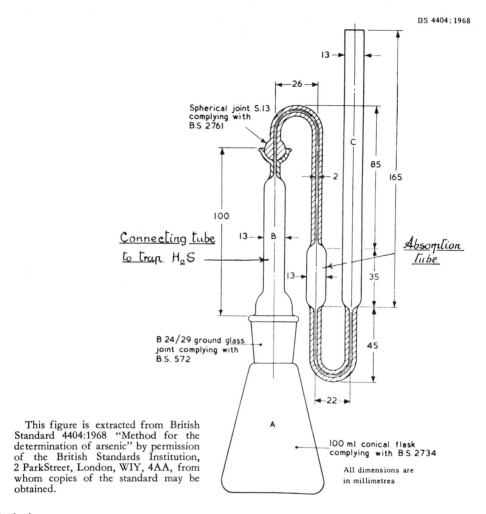

This figure is extracted from British Standard 4404:1968 "Method for the determination of arsenic" by permission of the British Standards Institution, 2 ParkStreet, London, WIY, 4AA, from whom copies of the standard may be obtained.

Technique

Assemble the apparatus components shown in Figure which are commercially available (Note 3). Place the sample solution in the 100 ml conical flask, add 10 ml of the concentrated hydrochloric acid (reagent 1) and dilute, if necessary, to about 55 ml with distilled water. Add 2.0 ml of potassium iodide solution (reagent 2) and 2.0 ml of stannous chloride solution (reagent 3). Mix and allow to stand for 15 to 20 minutes.

During this time lightly pack the top third of the connecting tube, as shown in Figure, with some of the impregnated cotton wool (reagent 6). This will serve to absorb any hydrogen sulphide which is evolved together with the arsine. Assemble the absorption train and transfer 5 ml of silver diethyldithiocarbamate solution (reagent 4) to the absorption tube.

At the end of the 20 minutes standing period add 5 g of zinc (reagent 5) to the solution in the conical flask and rapidly reassemble the apparatus. Allow the gas evolution to proceed for 45—60 minutes at room temperature. After this time disconnect the absorber tube and tilt the absorber so that the reagent solution flows back and forth between the absorber and the bulb to dissolve any red solid and to mix the solution thoroughly. Transfer the solution to a standard 13.5 mm comparator tube and place this tube in the right-hand compartment of the comparator.

Simultaneously with the analysis of the sample carry out a parallel blank determination, using distilled water in place of the sample solution, but following all the other instructions exactly. Place the liquid from the blank determination in an identical comparator tube and place this tube in the left-hand compartment of the comparator.

Place the comparator before a standard source of white light, such as the Lovibond White Light Cabinet or, failing this, north daylight, and match the colour of the sample with the standard colours in the disc. Calculate the arsenic content of the sample, in parts per million, by dividing the number of micrograms, read from the matching standard of the disc, by the number of grams of sample used in the solution tested.

Notes

1. Under the conditions described antimony and germanium will also form colours with the silver diethyldithiocarbamate and their presence will lead to erroneously high results.

2. Salts of chromium, cobalt, copper, mercury, molybdenum, nickel, palladium and silver may interfere with the quantitative evolution and absorption of arsine. However it has been shown that quantitative recoveries of arsenic are obtained when the test solution contains:

less than 50 mg of copper or silver,
less than 5 mg of cobalt or nickel,
less than 1 mg of chromium or mercury,
less than 0.05 ml of concentrated nitric acid.

If more than these quantities are known or suspected to be present, then suitable steps to remove them must be taken during the preliminary preparation of the sample solution.

3. The apparatus shown in Figure may be obtained from:

Fisons Scientific Ltd. .. .. Catalogue No. DOA
Quickfit & Quartz Ltd. .. .. Catalogue No. 1AD

The Determination of Bismuth

Introduction

The determination of bismuth (Bi) is of importance in the pharmaceutical and cosmetic industries and in metallurgy, especially in connection with the refining of copper (Cu) for electrical purposes. The reaction between Bi salts and potassium iodide (KI) in acid solution[1] forms the basis of the most satisfactory method for the colorimetric determination of Bi. Many methods of separating Bi from other radicals which interfere with this reaction have been proposed and a selection of these is included below.

Methods of sample preparation

Biological materials[2-6]

General[2] Take 100 g of sample (containing not more than 0.9 mg of bismuth), wet ash in a Kjeldahl flask to remove the organic matter, add 0.5 g E.D.T.A. (Ethylene diamine tetra-acetic acid) to 100 ml of the resulting solution, neutralise to litmus with 25% aqueous ammonia (NH_4OH) and add 5 ml in excess. Extract the bismuth with 1% iodine in chloroform ($CHCl_3$), wash the extract with water and re-extract the bismuth into nitric acid.

Blood[3] Kjeldahl 50 ml of blood and evaporate the residue to fumes of sulphur trioxide, add 20 ml of concentrated nitric acid, and again heat until sulphur trioxide is evolved. Cool, add 20 ml of water and repeat heating. Cool and dilute to a suitable volume.

Bone[4] Dissolve the sample in hydrochloric acid and precipitate the bismuth by passing hydrogen sulphide. Filter onto a Gooch crucible, dissolve the precipitate in 5 ml of hydrochloric acid, filter and dilute to volume.

Faeces[5] Take 10—50 g and ash with nitric acid as for urine. Take up the residue in hydrochloric acid and evaporate to dryness. Treat with 0.5 ml hydrochloric acid and 5 ml of water. Filter, add 0.5 ml of sodium bisulphite ($Na_2S_2O_5$) solution and dilute to a convenient volume.

Tissue[6] Macerate 30 g of sample and dilute with water if necessary. Add 5 ml of concentrated hydrochloric acid and 5 ml of a 1% cupric chloride ($CuCl_2$) solution. Boil. Immerse a sheet of clean copper foil (4 sq cm per mg of bismuth) in the solution and boil for 1 hour with the copper completely immersed, when the bismuth deposits on the copper. Remove the copper and dissolve in the minimum amount of 1:1 nitric acid. Add 1:1 aqueous ammonia until the solution is faintly turbid. Add 12 drops of 1:1 hydrochloric acid to clarify the solution and dilute 50:50 with boiling water. Heat for an hour on a boiling water bath to complete the precipitation of bismuth oxychloride (BiOCl). Filter and wash. Dissolve the precipitate by washing the filter with 5 ml of 1:3 sulphuric acid. Add 0.5 ml of 5% sulphur dioxide solution and dilute so that the final bismuth concentration is less than 10 mg/100 ml.

Urine[5] Concentrate 5—50 ml of urine to 5 ml of nitric acid, evaporate to dryness and heat but do not char. When nitric acid has been driven off, add further nitric acid and repeat the heating and addition until a white residue remains. Cool, add 2 ml of nitric acid and 25 ml of water and boil. Cool and determine bismuth.

Chemicals[7-9]

Organic[7,8] The method described for the determination of traces of bismuth in copper is also recommended for the determination of bismuth in organic materials. Alternatively[8], remove the organic matter by wet or dry ashing, dissolve the residue in acid, extract with iodine, extract the complex with $0.1N$ hydrochloric acid to remove the lead and then with $1.75N$ hydrochloric acid to remove the bismuth. Determine the bismuth on this latter extract.

Inorganic[9] Dissolve the sample by evaporating to fumes with 20 ml of 57% perchloric acid ($HClO_4$). Dissolve the residue in a little water, evaporate to dryness and repeat the treatment of solution and evaporation three or four times. Add ammonium ferric sulphate ($NH_4Fe(SO_4)_2.12H_2O$) precipitate hydroxides by addition of aqueous ammonia (NH_4OH), collect the precipitate, wash it with hot water, dissolve it in 30 ml of $4N$ hydrochloric acid, dilute with an equal volume of water and pass the solution through a column of anionite AN-31 (Cl^- form). Wash the column with $2N$ hydrochloric acid, desorb the lead with $0.02N$ hydrochloric acid and then desorb the bismuth with 250 ml of N nitric acid.

Metals, alloys and plating baths[7,10-21]

Iron and steel[10,11] Dissolve the sample in acid and extract the bismuth with iodine at pH 8.5. For stainless steel[8,11] dissolve in acid, precipitate bismuth with dithizone in the presence of citrate ($C_6H_5O_7^{3-}$) and potassium cyanide (KCN) at pH 7—10.

Gold and silver[12] Dissolve the sample in acid, remove interfering radicals by complexing with a solution of sodium diethyldithiocarbamate (($C_2H_5)_2$N.CS.S.Na.3H$_2$O). Extract with chloroform (CHCl$_3$). Evaporate the extract to dryness, dissolve in nitric acid and evaporate once more. Repeat this process several times, finally dissolving the residue in hydrochloric acid. Evaporate to dryness and make up the final solution in hydrochloric acid.

Copper and copper alloys[7,13-15] Dissolve a 10 g sample in 60 ml of 1:1 nitric acid. Dilute with 150 ml of water and add 10% sodium carbonate (Na$_2$CO$_3$) solution until a slight permanent precipitate is formed. Add 1 ml excess sodium carbonate, boil for 5 minutes and allow to settle. All the bismuth and some copper is precipitated. Filter, wash the filter with distilled water and re-dissolve the precipitate in the minimum amount of 1:1 nitric acid. Repeat the precipitation to remove any copper, re-dissolve the precipitate in 10 ml of 1:1 nitric acid and determine bismuth.

Alternatively[13] after dissolving the sample in acid and boiling off any nitrogen oxides, co-precipitate the bismuth with aluminium, after addition of aluminium nitrate (Al$_2$(NO$_3$)$_3$), by the addition of aqueous ammonia (NH$_4$OH). Filter, dissolve the precipitate in dilute nitric acid and determine bismuth.

If only traces of bismuth are present in the copper the method recommended by the Fiscal Policy Joint Committee of the Brass, Copper and Nickel Silver Industries[7] should be used. For samples containing up to 0.005% bismuth, dissolve 2 g of drillings in 20 ml of 50% nitric acid, boil until the solution is viscous, add 3 ml of concentrated sulphuric acid and continue heating until copious white fumes are evolved. Cool. Add 30 ml of water and 20 ml of 5% citric acid ($C_6H_8O_7$) solution and make slightly alkaline with aqueous ammonia (0.880 NH$_4$OH). Cool and transfer to a separating funnel. Add 50 ml of fresh 20% potassium cyanide (KCN), dilute to 200 ml and extract the bismuth by shaking with three successive 10 ml portions of a 0.1% solution of diphenylthiocarbazone (dithizone $C_{13}H_{12}N_4S$) in chloroform (CHCl$_3$). Wash the chloroform extracts with 50 ml of water in a second separator. Transfer the chloroform to a beaker and evaporate off nearly all the solvent. Add 1 ml of concentrated sulphuric acid and heat until white fumes are evolved. Oxidise any organic matter which remains by the cautious addition of 30% hydrogen peroxide (H$_2$O$_2$). Cool and add 15 ml of water. Filter off any precipitated lead and estimate bismuth on the solution.

A rapid method for gunmetal brass and bronze has been described by Fitter[14]. Dissolve 2 g of sample in a mixture of 14 ml of $5N$ hydrochloric acid and 6 ml of diluted nitric acid (2:3). Warm if necessary. Make alkaline with N aqueous ammonia and then clarify by the addition of 9N sulphuric acid. Add 10 ml of excess sulphuric acid. Cool and add sufficient 60% potassium iodide (KI) solution to precipitate all the copper. Add 20 ml of 25% sodium hypophosphite (NaH$_2$PO$_2$.H$_2$O) solution and stand for 10 minutes or until completely bleached, then dilute to 200 ml and filter. Analyse 50 ml of the filtrate using *Method 1*. This method is claimed to be valuable for sorting test samples of gunmetal.

Feik[15] uses ion exchange separation in the following method. Dissolve 2 g of alloy in 10 ml of $5M$ nitric acid. Dilute 2 ml of this solution to 20 ml with propane-1,2-diol (CH$_3$.CH(OH).CH$_2$OH) and pass this solution through a column (10 cm × 0.6 cm) of Dowex 1-X8 resin (NO$_3^-$ form). Wash the column with 150 ml of a 9:1 mixture of propane diol and $5M$ nitric acid, to remove lead, and then elute the bismuth with 100 ml of molar nitric acid.

Lead and lead alloys[13,16-19] Dissolve 10 g of sample in 100 ml of 1:4 nitric acid and boil. Add 0.5% sodium chloride (NaCl) solution in excess, to precipitate silver, and 60 ml of boiling 1:20 sulphuric acid with constant stirring, to precipitate lead. Add 30 ml of 1:3 sulphuric acid and allow to cool for at least an hour. Filter, wash by decantation with 1:20 sulphuric acid. Add 5 ml of concentrated hydrochloric acid to the filtrate and a slight excess of aqueous ammonia. Add 1:5 hydrochloric acid until the solution is just acid to methyl orange, boil for 1 minute and stand for 1 hour. Filter and test the filtrate for the absence of bismuth. Pulp the filter paper and precipitate with 10 ml of 1:3 sulphuric acid, add 30 ml of water, boil, cool and filter. Dilute the filtrate to 50 ml and proceed by *Method 1*[13].

Alternative methods of separating the bismuth from an acid solution of the lead are:—

(a)[16] adjust pH to 1.0—1.2 with aqueous ammonia (NH_4OH), add 1 ml of ferric nitrate ($Fe(NO_3)_3$) solution (20 μg per ml) and 2 ml of 1% cupferron ($C_6H_5.N(NO)ONH_4$). Shake with 5 ml of chloroform ($CHCl_3$). Separate the chloroform layer, add a further 1 ml of ferric nitrate solution and 2 ml of cupferron and extract with a further 5 ml of chloroform. Wash the combined chloroform extracts with three 20 ml portions of water then extract the bismuth with two 10 ml portions of 1:7 sulphuric acid.

(b)[17] Extract the sample solution with acetone (CH_3COCH_3) or acetone/isobutyl methyl ketone ($CH_3CH(CH_3)CH_2COCH_3$). Extract bismuth from the solvent with dilute acid.

(c)[18] To 5 ml of sample solution, containing 50—200 μg of bismuth, add 75 ml of water and 2 ml of 2% aluminium nitrate ($Al(NO_3)_3$) solution. Heat to near the boiling point and then add aqueous ammonia (NH_4OH) drop by drop until a dark blue colour appears. Set aside for 20 minutes and then filter. Wash the precipitate with dilute aqueous ammonia, then dissolve the precipitate in 20 ml of 1:1 nitric acid and dilute with water to 50 ml. To 10 ml of this solution add aqueous ammonia to bring the pH to 2.5 and extract the bismuth into a 0.001% solution of dithizone ($C_6H_5.N:N.CS.NH.NH.C_6H_5$) in chloroform. Extract bismuth from the chloroform with 20 ml of 1:100 nitric acid.

(d)[19] Adjust the pH of an aliquot of the sample solution, containing not more than 10 μg of bismuth, to pH 2 and extract the bismuth with 0.01M N-phenylbenzohydroxamic acid ($C_6H_5.CO.N(OH).C_6H_5$) in chloroform by shaking for 3 minutes with each of successive 5 ml, 2.5 ml and 2.5 ml portions. Combine the extracts and re-extract the bismuth by shaking with 5 ml of 5N nitric acid for 5 minutes.

Tin and tinning baths[20,21] *Metals* Dissolve 1—2 g sample in either aqua regia (HNO_3 + 3HCl)[20] or a mixture of 10 ml of 57% perchloric acid ($HClO_4$) and 3 ml of nitric acid[21]. Boil off the oxides of nitrogen and extract the bismuth either with an equal volume of normal 2-ethylhexylpyrophosphate ((($CH_3(CH_2)_3CH(C_2H_5)CH_2O)_2PO)_2O$) and re-extract into 10 ml of a 2% solution of potassium iodide (KI) containing 5—10 drops of 1:4 sulphuric acid[21]; or make alkaline to phenolphthalein indicator ($O.CO.C_6H_4.C(C_6H_4OH)_2$) with aqueous ammonia and treat with 20 ml of aqueous ammonia and 3 ml of 1% solution of sodium diethyldithiocarbamate (($C_2H_5)_2N.CS.S.Na.3H_2O$) and shake with 15 ml of ethylacetate ($CH_3COOC_2H_5$). Repeat extraction two or three times, wash the extract with water several times and then re-extract the bismuth into 10 ml of 5N hydrochloric acid. *Electrolytes*[21] Treat 1—2 ml of sample with 3—4 ml of 3N sulphuric acid and 0.5 ml of 1.5N ammonium fluoride (NH_4F). Extract the solution with 5 ml of normal 2-ethylhexylpyrophosphate in heptane ($CH_3(CH_2)_5CH_3$). Separate the heptane layer and re-extract the bismuth into 10 ml of 2% potassium iodide solution containing 5—10 drops of 1:4 sulphuric acid.

Solder alloys[22] Dissolve 0.1 g of sample in 2.5 ml of 50% nitric acid and 7.5 ml of 50% hydrochloric acid. Add a few drops of bromine water and evaporate to dryness. Dissolve the residue in 20 ml of 8M hydrochloric acid containing a few drops of bromine water and pass the solution through a column of Diaion SA-100 ion exchange resin (Cl⁻ form) at 1 ml a minute. Wash the column with 40 ml of 8M hydrochloric acid and 50 ml of 2M nitric acid. Elute the bismuth with 200 ml of 0.5M nitric acid. Concentrate this eluate and make up to 25 ml with distilled water.

Minerals, ores and rocks[20,23-25]

Copper, silver and lead ores[23] Dissolve a 10 g sample in a mixture of concentrated nitric and hydrochloric acids (3 HNO_3:1 HCl). Evaporate to dryness, heat to 200—250°C and dissolve in 100 ml of 1:10 hydrochloric acid. Filter, add potassium iodide (KI) solution to precipitate copper, filter, add 2 ml of concentrated sulphuric acid and evaporate to dryness. Dissolve in 75 ml of 1% sulphuric acid and filter. Determine bismuth on this filtrate.

The separation of bismuth from interfering radicals by paper chromatography has also been recommended[23] for ore analysis.

Arsenic, tin and antimony ores Fuse a 10 g sample with 50 g of sodium carbonate (Na_2CO_3) and 5 g of sulphur (S). Dissolve the melt in water, boil and filter off the precipitated bismuth. Dissolve the precipitate in 10 ml of 1:1 nitric acid and determine bismuth on the solution.

Ferrochrome, chromium, iron and copper ores[20,24] Dissolve 1—2 g of sample in 20 ml of aqua regia (HNO_3+3HCl), evaporate until about 7 ml of solution remain, cool, make alkaline to phenolphthalein with aqueous ammonia add a further 20 ml of aqueous ammonia and 3 ml of a 1% solution of sodium diethyldithiocarbamate and shake with 15 ml of ethyl acetate. Separate the layers and repeat the extraction of the aqueous layer two or three times. Combine the organic extracts, wash them several times with water and then extract the bismuth into 10 ml of 5N hydrochloric acid.

General method for ores and minerals[25] Weigh 0.5 g of the powdered sample into a test tube graduated at 10 ml. Add 5 ml of nitric acid and boil gently for 30 minutes while stirring. Cool, dilute to 10 ml with water and centrifuge. Determine bismuth on an aliquot portion of the supernatant liquid.

Miscellaneous[26,27]

Pharmaceuticals[26] Dissolve the sample in acid and proceed without need for further treatment.

Sludges[27] Dissolve 1—2 g in aqua regia (1:3 HNO_3:HCl). Evaporate to dryness. Dissolve the residue and evaporate twice with hydrochloric acid. Heat with 10 ml hydrochloric acid and 40 ml of water, filter and treat the filtrate with aqueous ammonia until iron starts to precipitate. Add 1 g of sodium nitrite ($NaNO_2$) and aqueous ammonia until the smell of ammonia persists. Filter, wash with 1% ammonium chloride (NH_4Cl) solution, dissolve in hydrochloric acid and treat the solution with 25 ml of sodium tartrate ($Na_2C_4H_6O_6.2H_2O$) solution. Add 2N caustic soda (NaOH) until neutral to phenol red and then add a further 2 ml in excess and 5 ml of 10% potassium chloride (KCl) solution. Extract the bismuth with chloroform ($CHCl_3$) in the presence of 5 ml of 1% iodine solution. Evaporate the extract to dryness. Take up the residue and evaporate to dryness several times with nitric acid then evaporate with hydrochloric acid and finally dissolve the residue in hydrochloric acid.

Separation from other ions[28-30]

In addition to the methods mentioned above a specific ion exchange resin for bismuth has been described[28]. It has also been claimed[29] that complexing with dibutyl arsenic acid ($AsBu_2$ OOH) and extraction into chloroform at pH 1.8—2.4, results in complete separation of bismuth from lead, copper, cadmium, cobalt, beryllium, chromium, manganese, calcium, magnesium, aluminium and nickel.

Alternatively[30] pass the solution in methanol (CH_3OH)—5N nitric acid (19:1) through a column of Dowex 1-X8 ion exchange resin. Wash the column with methanol 6N hydrochloric acid (9:1), elute lead with normal hydrochloric acid and then bismuth with normal nitric acid.

References

1. F. B. Stone, *J.Soc.Chem.Ind.,* 1887, **6,** 416
1. A. N. Krylova, *Sb.Trud.Sudeb.Med.i Sudeb.Khim Perm.* 1961, 226; *Analyt.Abs.* 1962, 9, 5320
3. E. H. Marchling, *J.Lab.Clin.Med.,* 1933, **18,** 1058
4. J. A. Sultzaberger, *J.Amer.Pharm.Assoc.,* 1927, **16,** 218
5. C. A. Hill, *Lancet,* 1925, **II,** 1281
6. N. A. Valyashko and P. K. Virup, *Ukrain.Khem.Zhur.,* 1930, **5,** Sci.Pt. 275
7. Report of The Chemical Sub-Committee, Fiscal Policy, Joint Committee, Brass, Copper and Nickel Silver Industries, 1935; abstracted in *Analyst,* 1935, **60,** 554
8. J. C. Gage, *Analyst,* 1958, **83,** 672
9. V. V. Stepin, V. I. Ponosov, G. N. Emasheva and N. A. Zobnina, *Trudy vses.nauchno issled.Inst.standart.Obraztsov.,* 1968, **4,** 100; *Analyst.Abs.,* 1970, **18,** 900
10. S. Maekawa, Y. Yoneyama and E. Fujimori, *Japan Analyst.* 1961, **10,** 345; *Analyst.Abs.,* 1963, **10,** 2289
11. C. G. Carlstrom and V. Palvarinne, *Jernkontor Ann.,* 1962, **146,** 453; *Analyt.Abs.,* 1963, **10,** 2292

12. M. Miyamoto, *Japan Analyst,* 1961, **10,** 217; *Analyst.Abs.,* 1963, **10,** 2166
13. V. N. Danilova and P. V. Marchenko, *Zavodskaya Lab.,* 1962, **28,** 654; *Analyt.Abs.* 1963, **10,** 104
14. H. R. Fitter, *Analyst,* 1938, **63,** 107
15. F. Feik and J. Korkisch, *Talanta,* 1964, **11** (12), 1585; *Analyt.Abs.,* 1966, **13,** 1742
16. Y. Ishihara, K. Shibata, H. Kishi and T. Hori, *Japan Analyst,* 1962, **11,** 91; *Analyt.Abs.,* 1964, **11,** 90
17. M. Steffek, *Chem.Listy,* 1964, **58,** 957; *Analyt.Abs.,* 1965, **12,** 6371
18. Z. Skorko-Trybula and J. Chivastowska, *Chem.Analyt. Warsaw,* 1963, **8,** 859; *Analyt.Abs.,* 1965, **12,** 1091
19. J. Chivastowska and J. Bruginska, *Chemia analit.,* 1966, **11** (1) 169; *Analyt.Abs.,* 1967, **14,** 2527
20. Yu. I. Usatenko, A. M. Arishkevich and A. G. Akhmetshin, *Zavodskaya Lab.,* 1965, **31** (7), 788; *Analyt.Abs.,* 1966, **13,** 6174
21. A. A. Shatalova, I. S. Levin, I. I. Shmargolina, Z. S. Komashko and M. V. Solov'eva, *Zavodskaya Lab.,* 1966, **32** (11), 1320; *Analyt.Abs.,* 1968, **15,** 728
22. S. Onuki, *Japan Analyst,* 1963, **12,** 844; *Analyt.Abs.,* 1965, **12,** 625
23. H. Agrinier, *Compt.Rend.,* 1961, **253,** 280; *Analyt.Abs.,* 1962, **9,** 4646
24. S. A. Dekhtrikyan, *Izvest.Akad.Nauk.armyan S.S.R. Nauki o Zemle,* 1964, (6), 53; *Analyt.Abs.,* 1966, **13,** 1274
25. F. N. Ward and H. M. Nakagawa, *Prof.Papers U.S. geol.Surv.,* **575** *D,* 1967, D 239 – D 241; *Analyt.Abs.,* 1969, **16,** 184
26. L. Krowezynski and A. Banaszek, *Acta Polon.Pharm.,* 1957, **14,** 337; *Analyt.Abs.,* 1959, **6,** 1092
27. N. P. Strelnikova and G. G. Lystsova, *Zavodskaya Lab.,* 1962, **28,** 659; *Analyt.Abs.,* 1963, **10,** 103
28. A. Lewandowski and W. Szczepaniak, *Chem.Anal. Warsaw,* 1962, **7,** 593; *Analyt.Abs.,* 1963, **10,** 2230
29. R. Pietsch and E. Pichler, *Mikrochim.Acta,* 1962, 954; *Analyt.Abs.,* 1963, **10,** 594
30. S. S. Ahluwalia and J. Korkisch, *Z.analyt.Chem.,* 1965, **208** (6), 414; *Analyt.Abs.,* 1966, **13,** 3521

Bismuth Method 1
using potassium iodide

Principle of the method

The test is based on the yellow colours produced by the reaction between a bismuth salt and potassium iodide in acid solution to form potassium bismuth iodide (KB_2I_4). Under these conditions, many substances form precipitates or liberate free iodine, producing yellow solutions. Such interfering substances must consequently be removed from the solution containing the bismuth before the quantitative test can be applied. In general, chlorides, nitrates, aluminium, magnesium, zinc and small quantities of copper, lead, ferrous iron, manganese, arsenic and cadmium do not interfere. Silver, tin, antimony, mercury, nitrites and oxidising agents must be absent. In order to ensure that the yellow colour produced in the test is not wholly or in part due to the presence of an oxidizing agent liberating iodine from the acid solution of potassium iodide, a small quantity of dilute sulphurous acid is used in each test.

Reagents required

1. *Sulphuric acid* (H_2SO_4) Approximately $5N$
2. *Potassium iodide* (KI) 10% w/v aqueous solution
3. *Dilute sulphurous acid* (0.5% SO_2) Dilute sulphur dioxide solution (5% w/v SO_2) with distilled water.

All chemicals used in the preparation of these reagents should be of analytical reagent grade.

The Standard Lovibond Nessleriser Discs NBA and NBB

Disc NBA covers the range 0.01 to 0.05 mg of bismuth (Bi) in steps of .0005. This is equivalent to a concentration of 0.4 to 2.0 ppm in 25 ml of solution.

Disc NBB covers the range 0.03 to 0.2 mg of bismuth in steps .03, .04, .05, .06, .08, .1, .12, .16, .2. This is equivalent to a concentration of 1.2 to 8.0 ppm in 25 ml of solution.

Technique

To 25 ml of the solution under examination (or other suitable quantity) free from interfering metals, contained in a Nessleriser glass, add 3 ml of $5N$ sulphuric acid (reagent 1) and sufficient water to produce 50 ml. Then add 2 ml of potassium iodide solution (reagent 2) and 0.2 ml of dilute sulphurous acid (reagent 3), mix and place in the right-hand compartment of the Nessleriser. In the left-hand compartment place a Nessleriser glass containing 3 ml of $5N$ sulphuric acid diluted to 50 ml with distilled water, 2 ml of potassium iodide solution, and 0.2 ml of dilute sulphurous acid. Stand the Nessleriser before a standard source of white light (Note 3) and compare the colour produced in the test solution with the colours in one of the standard discs, rotating the disc until a colour match is obtained.

If the colour obtained is deeper than the standard colours, the test should be repeated on a smaller quantity of the solution.

Notes

1. The application of the Lovibond Nessleriser to the determination of bismuth was suggested by B. Drinkwater, A.R.S.M., of the British Non-Ferrous Metals Research Association. The colour standards were prepared in the laboratories of the Association.

2. It must be emphasized that the readings obtained by means of the Lovibond Nessleriser and disc are only accurate provided that Nessleriser glasses are used which conform to the specification employed when the discs were being calibrated, namely that the 50 ml calibration mark shall fall at a height of 113 mm, plus or minus 3 mm measured internally.

3. A strongly recommended standard source of white light is the Lovibond White Light Cabinet. In the absence of a standard source, north daylight should be used wherever possible.

The Determination of Bromide

Introduction
The determination of bromide in blood is important in the confirmation of bromism as manifest by a skin eruption or by mental changes. Bromide is also determined in the examination of brines and in the examination of fruits for traces of pesticides and fungicides containing bromide. Its determination in molasses has also been reported.

Methods of sample preparation
Biological materials[1-3]
Blood, serum, plasma[1-3] Take a 5 ml sample of venous blood and place it in a clean tube containing a few crystals of potassium oxalate ($K_2C_2O_4.H_2O$). Allow the blood to clot and then centrifuge to separate the cells from the serum. Take 2 ml of the serum, or plasma, add 6 ml of distilled water and 2 ml of 20% trichloracetic acid ($CCl_3.COOH$). Filter and use 5 ml of the filtrate (=1 ml of serum) instead of 1 ml of sample and 4 ml of water in the determination[1,2]

Alternatively[3] remove the protein from a sample of blood, or urine, by treatment with tungstic acid (H_2WO_4). Precipitate the bromide as silver bromide (AgBr) with silver nitrate ($AgNO_3$). Filter. Reduce the precipitate with zinc and dilute hydrochloric acid and filter. Estimate bromide on the filtrate.

Urine[3] Use the second method described above for blood.

Chemicals
Inorganic Dissolve a suitable amount of the sample so that 1 ml of the solution contains about 1 mg of bromide. If fusion be necessary then alkali fusion techniques should be used. No oxidising agents should be used.

Organic Ash the sample without the addition of oxidising agents. Dissolve the ash in hydrochloric acid.

Food and edible oils[4,5]
Citrus fruits[4] Extract the organic bromide by stirring the macerated sample with a solvent such as Skellysolve B.

Molasses[5] Dry ash the sample, after treatment with alkali, at 480—500°C. Extract the ash with hydrochloric acid and determine bromide on the extract.

The individual test
This is the standard gold chloride ($AuCl_3$) test[6,7]. Iodides also produce a colour with $AuCl_3$ and should be removed from the sample solution, if present, in the following manner. To 10—15 ml of the sample solution add 2 ml of N H_2SO_4 and 1 ml of $0.5M$ sodium nitrate ($NaNO_2$) for each 60 mg of iodide present. Boil gently until the solution becomes colourless. Wash down the sides of the flask with distilled water to restore the lost volume and add a few drops of H_2SO_4 and $NaNO_2$. If a colour appears add more H_2SO_4 and $NaNO_2$ and repeat boiling. When solution finally remains colourless boil for a further two minutes, and cool before estimating bromide.

References
1. G. A. Harrison, *"Chemical Methods in Clinical Medicine"*, 4th Editn., J. & A. Churchill, London, 1957
2. H. V. Street, *Clin. Chim. Acta*, 1960, **5**, 938; *Analyt. Abs.*, 1961, **8**, 5125
3. D. Kaplan and E. Schnerb, *Analyt. Chem.*, 1958, **30**, 1703
4. C. E. Castro and R. A. Schmitt, *J. Agric. Food Chem.*, 1962, **10**, 236; *Analyt. Abs.*, 1963, **10**, 349
5. A. R. Deschreider and J. Frateur, *Ind. Aliment. Agric.*, 1957, **74**, 541; *Analyt. Abs.*, 1958, **5**, 2776
6. O. Wuth, *J. Amer. Med. Assoc.*, 1927, **88**, 2013
7. W. A. Taylor, *J. Lab. Clin. Med.*, 1927, **13**, 495

Bromide Method 1
using gold chloride

Principle of the method

The test solution is treated with gold chloride solution, which combines with bromides to give a double salt, the colour of which is a deeper yellow-brown than that of the added gold solution: the colour is compared with that of a series of glass standards prepared to match bromide standard solutions similarly treated.

Reagent required

0.5% *gold chloride* Dissolve 0.5 g of gold chloride ($AuCl_3HCl.3H_2O$), in distilled water, and make up to 100 ml.

The Standard Lovibond Comparator Disc 5/23

The disc covers the range 0, 10, 25, 50, 75, 100, 125, 150, 175 mg of sodium bromide per 100 ml of test solution, i.e. 0—1750 ppm.

Technique

To 1 ml of test solution add 1 ml of 0.5% gold chloride solution and 4 ml of water. Mix, transfer to a comparator test tube and place in the right-hand compartment of the comparator. Hold facing a standard source of white light (Note 3) and rotate the disc until the colour of the liquid is matched by one of the glass standards. The value (expressed in mg sodium bromide per 100 ml of solution) is then read from the indicator recess at the bottom right-hand corner of the comparator.

Notes

1. It is important to check each batch of reagents against the 0 of the disc. Prepare a blank from:—

   5 ml distilled water
   1 ml gold chloride solution

and mix well.

If this does not match the 0 of the disc, it is useless to proceed; the gold chloride may be different from that used in standardising the disc.

2. Iodides, as well as bromides, give double salts with gold chloride. It is therefore essential to make sure that the sample is free from iodides (see previous page).

3. A strongly recommended standard source of white light is the Lovibond White Light Cabinet. In the absence of a standard source north daylight should be used wherever possible.

Confirmatory test for bromide in serum or plasma

If the patient is suffering from skin lesions thought to be due to bromism, and the bromide found by the above test is low (25 mg or less), the fluorescein confirmatory test should be performed, to ascertain that increased bromide is in fact present.

Bromides are oxidised to bromine by chloramine T. The bromine is taken up by fluorescein to yield red tetrabromofluorescein (eosin).

Reagents required for confirmatory test

1. 0.4% *Chloramine-T*
2. 0.125% *fluorescein* Dissolve 125 mg. of fluorescein in 25 ml of $N/10$ sodium hydroxide (NaOH) and add distilled water to 100 ml.
3. *Buffer solution pH 5.3 to 5.4* Dissolve 6.6 g of anhydrous sodium acetate (or 10.9 g of $CH_3COONa.3H_2O$ sodium acetate trihydrate), and 1.2 ml of glacial acetate acid in water and dilute to 100 ml.

Technique

Mix in a small test tube:—

Serum filtrate, as prepared for gold chloride test	1 ml
Buffer solution	0.5 ml
Fluorescein solution	1 drop
Chloramine-T	1 drop

Treat 1 ml of distilled water, or 1 ml of protein-free filtrate from normal serum, in the same way in parallel.

Observe the colour of the contents of the two tubes: the filtrate from the patient's serum will turn orange-red or red, in a few seconds, if more than 5 mg of sodium bromide (NaBr) is present. Observe the tubes for five minutes before declaring the reaction negative. If bromides are not in excess of normal the treated filtrate of the patient's serum will not alter in colour, but will remain like the treated water or treated normal filtrate.

Notes

Iodide, in this test, will give tetraiodofluorescein, which is similar in colour. Iodides should be removed by treating the protein-free filtrate with sodium nitrite solution (see page 110).

The Determination of Bromine

Introduction

The purification, by chlorination, of water containing both bromide and ammonia leads to the formation of both monobromamine and monochloramine. These conditions may arise in the chlorination of sea-water swimming pools or of water contaminated with sea-water. Monobromamine may also be formed as a result of water purification using either bromine itself[1] or a brominating agent such as 1,3-dibromo-5,5-dimethyl hydantoin (DBDMH)[2]. In all these cases the estimation of free residual bromine is a necessary part of the control of the disinfection process.

The main application of the bromine determination is undoubtedly to water testing[6-19]. However the bromine content of wines[3] and of selenium, for use in rectifiers,[4] is also of importance.

Methods of sample preparation

Beverages[3]

Wine To a 50 ml sample add 0.5 ml of 50% caustic soda (NaOH) solution and 1 ml of 4N slaked lime (Ca(OH)$_2$). Stand for 24 hours. Evaporate the solution to dryness and incinerate at 550°C for 15 minutes. Cool, add 1 ml of water and again incinerate after evaporation. Add 5 ml of hot water and then 0.1 N sulphuric acid followed by 0.01 N sulphuric acid until the pH is 4–5.

Then make up to 10 ml and estimate bromine.

Chemicals[4,5]

Selenium[4] Dissolve 0.25–2.5 g of sample in 50% caustic soda, oxidise with hydrogen peroxide (H$_2$O$_2$) and neutralise to pH 7.4 with sulphuric acid.

Solids[5] Fuse 2 g of sample with fusion mixture (Na$_2$CO$_3$ Na$_2$O$_2$), decompose the melt with sulphuric acid and extract with chloroform (CHCl$_3$). Add 10 ml of a solution of sodium carbonate (Na$_2$CO$_3$), boil to remove chloroform and make up to 10 ml.

Water[6-19]

In the majority of cases no special treatment of the sample is required. However for the determination of very low concentrations of bromine some form of sample concentration is required. Uzumasa et al[17] recommend concentration by evaporation in the presence of excess sodium carbonate (Na$_2$CO$_3$). Koga[19] recommends the following method—make a 10–100 ml sample alkaline and then concentrate to less than 10 ml. Neutralise to phenolphthalein with ammonium sulphate ((NH$_4$)$_2$SO$_4$), mix with 1 ml of ammonium sulphate and 2 g of potassium dichromate (K$_2$Cr$_2$O$_7$) and separate bromine by microdiffusion in a Conway apparatus.

The individual test

This is the method devised by Palin[6].

References

1. *Chemical Week,* 1959 (Dec 12), 90
2. R. A. Reed, *Chemical Products,* July 1960
3. P. Jaulmes, S. Brun and J. C. Cabanis, *Chim. Anal.,* 1962, **44,** 327; *Analyt. Abs.,* 1963, **10,** 2493
4. J. Siwecka, *Chem. Anal. Warsaw,* 1958, **3,** 1001; *Analyt. Abs.,* 1960, **7,** 474
5. E. Cogan, *Analyt. Chem.,* 1962, **34,** 716
6. A. T. Palin, *Water and Sewage Works,* 1961, **108,** 461
7. N. D. R. Schaafsma, *Water Pollution Abstract (D.S.I.R.),* 1949, **22,** 196
8. J. K. Johannesson, *Analyst,* 1958, **83,** 155
9. A. T. Palin, *Analyst,* 1945, **70,** 203
10. A. T. Palin, *J. Amer. Water Wks. Assoc.,* 1957, **49,** 873

11. A. T. Palin, *Proc.Soc.Water Treatment & Exam.*, 1957, **6,** 133
12. A. T. Palin, *Baths Service,* 1958, **17,** 21
13. A. T. Palin, *Water & Water Eng.,* 1958, **62,** 30
14. A. T. Palin, *Archiv.d.Badewesens.,* 1958 (August), 216
15. A. T. Palin, *Gas und Wasserfach,* 1958 (October), 1091
16. L. N. Podgornyi and F. I. Bezler, *Byul.Inst.Biol.Vodokhrarilishch Akad.Nauk.SSSR.,* 1959 (2), 56; *Analyt.Abs.,* 1960, **7,** 1187
17. Y. Uzumasa, M. Nishimura and Y. Nasu, *Japan Analyst,* 1959, **8,** 231; *Analyt.Abs.,* 1960, **7,** 1954
18. E. Goldman and D. Byles, *J.Amer.Water Wks.Assoc.,* 1959, **51,** 1051; *Analyt.Abs.,* 1960, **7,** 2494
19. A. Koga, *J.Chem.Soc.Japan,* 1958, **79,** 1026; *Analyt.Abs.,* 1959, **6,** 2795

Bromine Method 1

using diethyl-*p*-phenylene diamine (Palin-DPD)

Principle of the method

Both bromine and monobromamine react with diethyl-*p*-phenylene diamine (DPD) to give a red colour. Chlorine is prevented from reacting by the addition of Glycine, which, at the appropriate *p*H, converts the chlorine to Chloraminoacetic acid, which does not give an immediate colour with DPD. The intensity of the colour which develops on the addition of DPD to the sample in the presence of the Glycine and a buffer, is thus proportional to the concentration of free and combined bromine in the sample. The intensity of the colour is measured by comparison with Lovibond permanent glass standards.

The method does not differentiate between free and combined bromine, but as there does not appear to be the same difference between the chemical and bacteriological action of free and combined bromine as there is between chlorine and the chloramines, their separate determination has little, if any, practical importance.

By a simple supplementary procedure it is possible to adapt the method for the determination of bromine (free plus combined), chlorine and chloramines in the same sample.

Reagents required
 DPD No. 1 tablets.
 DPD Glycine tablets.
 DPD No. 3 tablets.

The Standard Lovibond Comparator Discs 3/53A and 3/53B

 3/53A covers the range 0.2 to 2.0 parts per million bromine in steps of 0.2.
 3/53B covers the range 1.0 to 10.0 parts per million bromine in steps of 1.0.
 These discs are calibrated for use with 13.5 mm cells or test tubes.
 The master discs, against which all reproductions are checked, were tested and approved by Dr. Palin.

Technique

 For total free bromine in the absence of free chlorine (Note 1).

Rinse out a 13.5 mm cell, or test tube, with the sample to be tested and leave in the cell just enough of the liquid to cover a tablet. Drop one DPD No. 1 tablet into this prepared tube and disintegrate. Add the water sample up to the 10 ml mark, mix rapidly to dissolve the remains of the tablet, and place the cell in the right-hand compartment of the comparator. Match immediately against the standards in the disc, using standard white light (Note 2). To compensate for any inherent colour in the sample an identical cell, containing the sample only, should be placed in the left-hand compartment of the comparator. The figure shown in the indicator window of the comparator represents parts per million of free residual bromine present in the sample.

 For free bromine and free and combined chlorine.

The reading obtained in the preceding procedure now corresponds to free bromine plus free chlorine. This is Reading No. 1. After taking this reading add to the right-hand cell one DPD No. 3 tablet. Mix the contents of the cell by shaking until the tablet has dissolved, allow the cell to stand for 2 minutes and then match against the standards as before. This is Reading No. 2. The difference between Reading No. 2 and Reading No. 1 represents the concentration of combined chlorine (Note 3).

COLORIMETRIC CHEMICAL ANALYTICAL METHODS — Bromine (1)

Take an identical cell, rinse out with the water sample as before, add one DPD Glycine tablet and make up to 10 ml with the water sample. Crush tablet and mix to dissolve. Prepare a further cell containing one DPD No. 1 tablet, as in the first procedure, add the 10 ml of sample treated with the DPD Glycine tablet, mix rapidly to dissolve the remains of the DPD No. 1 tablet and match immediately. This is Reading No. 3 and gives the concentration of free bromine. To obtain the concentration of free chlorine deduct Reading No. 3 from Reading No. 1 (Note 3).

Notes

1. If free chlorine is known, or suspected to be present, but only the concentration of free bromine is required, it is necessary only to obtain Reading No. 3 using DPD No. 1 and DPD Glycine tablets.

2. A strongly recommended standard source of white light is the Lovibond White Light Cabinet. If a standard source of white light is not available then north daylight should be used wherever possible.

3. The comparator discs used in these tests are calibrated in terms of bromine. To obtain the concentrations of free and combined chlorine in the differential procedure the difference obtained must be multiplied by 0.44 thus:—

   Free chlorine = $0.44 \times$ (Reading No. 1 — Reading No. 3).
   Combined chlorine = $0.44 \times$ (Reading No. 2 — Reading No. 1).

4. Dissolved oxygen in water can produce a faint colour with the reagent if the solution is allowed to stand. Provided the instructions are followed exactly there is no interference within the period of the test.

5. The only potential interference likely to be present in water is that due to oxidised manganese. This interference can be compensated by developing the manganese colour, in the blank, as follows:—

To a 10 ml of sample in a separate test tube add 1 drop of 0.5% sodium arsenite ($NaAsO_2$) solution and mix. Rinse a comparator cell with sample and add one DPD No. 1 tablet. Crush or allow to disintegrate then add the 10 ml of arsenite treated sample. Mix to dissolve the remains of the tablet and then place this cell in the left-hand compartment of the comparator. In this way the colour due to manganese will have been developed equally in both test and blank and will cancel out.

6. All glassware used must be thoroughly rinsed after making a test. Handling the tablets and unnecessary exposure to air should be avoided. The tablets should be transferred to the comparator cell by shaking into the bottle cap rather than by the use of the fingers.

7. If the colour developed is equal to, or greater than the top standard, the test should be repeated after diluting the sample with distilled water, the final reading being multiplied by the dilution factor.

The Determination of Cadmium

Introduction

Cadmium is the rarest of the common metals and is normally found associated with the sulphide ores of lead and zinc, although it does occur as the rare mineral greenockite (cadmium sulphide). It resembles zinc in its properties but is more volatile and much softer. Cadmium is most commonly used for electroplating steel articles, the coating produced being corrosion resistant and readily soldered. It is also used in glass, in pigments—cadmium sulphide being brilliantly yellow in colour—in alloys and in storage batteries. Its iodide and bromide are used in photography. Salts of cadmium are poisonous.

Methods of sample preparation

Biological materials[1-3]

Blood[1,3] Destroy organic matter by digestion with sulphuric acid until the solution is pale in colour. Complete the oxidation either with hydrogen peroxide (H_2O_2)[1] or aqueous bromine[2], make the solution alkaline ($pH \sim 10$) and extract cadmium with dithizone ($C_6H_5.N:N.CS.NH.NH.C_6H_5$).

Tissue[2] Digest 1—2 g of tissue with sulphuric acid as for blood.

Urine[1,2,3] Digest 50—100 ml sample either with sulphuric acid[1,2] or with a mixture of nitric and perchloric ($HClO_4$) acids[3]. Either proceed as for blood[1,2] or[3] neutralise the acid solution and precipitate lead and cadmium with sodium sulphate (Na_2SO_4) in the presence of ethanol (C_2H_5OH) with strontium as a carrier. Filter, wash the precipitate with sodium sulphate solution, dissolve the precipitate in nitric acid, make alkaline with aqueous ammonia (NH_4OH) and extract cadmium with dithizone.

Chemicals[4]

Perhydrol (H_2O_2) and hydrogen fluoride Decompose the sample by heating with sulphuric acid until dense white fumes are evolved, add water and make the solution alkaline with caustic soda (NaOH). Add potassium cyanide (KCN) and sodium potassium tartrate ($COOK.CHOH.CHOH.COONa.4H_2O$) and extract cadmium into a chloroform solution of dithizone.

Foods and edible oils[5,6]

Food colouring materials[5] Extract the dye with water, make the solution alkaline and extract the cadmium with dithizone.

Sea food[6] Homogenise 500 g sample with the addition of water as required. Weigh 25 g of this slurry into a beaker which has previously been thoroughly washed in nitric acid. Dry in an oven at 100°C for 4 hours and then heat overnight in a muffle furnace at 450°C, raising the temperature slowly from 200°C. Dissolve the ash in 2 ml of nitric acid, dilute with 25 ml of water, heat to boiling, filter and make up to 100 ml.

Metals, alloys and plating baths[7-18]

Aluminium[7] Dissolve 1 g of sample in 28 ml of 20% caustic soda (NaOH), 10 ml of 20% tartaric acid (($CHOH.COOH)_2$) and 5 ml of 1% potassium cyanide (KCN). Add 2 ml of 10% hydroxyammonium chloride ($HO.NH_3Cl$) and extract cadmium with a chloroform solution of dithizone.

Cast Iron[8] Heat 0.5 g of sample with 5 ml of nitric acid and 12 ml of 60% perchloric acid ($HClO_4$) until white fumes are evolved. Add 50 ml of water and filter. Add 10 ml of 50% citric acid ($C(OH)(COOH)(CH_2.COOH)_2.H_2O$) solution, adjust the pH to 9 with aqueous ammonia and extract with dithizone.

Cobalt and nickel[9] Dissolve sample in $2N$ hydrochloric acid, pass solution through a column of EDE-10P resin. Elute cadmium with water.

Magnesium alloys[10] Dissolve 0.2—0.5 g of alloy in 15 ml of 1:5 sulphuric acid containing a few drops of nitric acid. Cool, add 15 ml of $0.05N$ nickel diethyl phosphorodithioate ($((C_2H_5O)_2PS_2)_2Ni$). Set aside for 1 hour then filter. Wash the precipitate, dissolve it in aqueous ammonia and extract the cadmium with dithizone.

Nickel-chromium alloys[11] Dissolve the sample in sulphuric acid, add a little nitric acid, evaporate until dense white fumes are formed, dilute with water, make alkaline and extract the cadmium with dithizone.

Non-ferrous metals[12,13] Dissolve the sample in a mixture of hydrochloric and nitric acids to which a little aqueous bromine has been added. Remove the excess nitric acid and oxides of nitrogen by evaporating three times, adding hydrochloric acid after each evaporation. Dissolve the residue in hydrochloric acid, dilute until the solution is $0.5N$ in hydrochloric acid and then pass it down a column of EDE-10P resin (Cl^- form). Elute cadmium with $0.005N$ hydrochloric acid or water.

Silver alloys[14] Dissolve the sample in a mixture of nitric and fluoroboric (HBF_4) acids, make solution alkaline and extract cadmium with dithizone.

Zinc[15-18] Dissolve the sample in sulphuric acid[15,16] or $2M$ sodium acetate (CH_3COONa) solution [18]. Separate the cadmium from the zinc either by precipitation of cadmium sulphide with hydrogen sulphide, washing out any co-precipitated zinc sulphide with a mixture of ammonium chloride (NH_4Cl) and ammonium phosphate (($NH_4)_3PO_4$) solutions, and redissolving the cadmium sulphide[15]; or by forming a complex with potassium iodide (KI) in $0.75N$ sulphuric acid, shaking this solution with Amberlite LA-2 resin dispersed in xylene and stripping out the cadmium with M sodium carbonate (Na_2CO_3)[16]; or by extracting the cadmium with sodium diethyl phosphorodithioate (($C_2H_2O)_2PS_2Na$) solution in carbon tetrachloride (CCl_4), treating the organic phase with a dithizone solution in chloroform ($CHCl_3$) and shaking with $0.5M$ caustic soda (NaOH)[18].

For the analysis of zinc oxide and other relatively involatile materials [17] heat a sample of the finely ground powder in a stream of hydrogen at 300°C to reduce cadmium oxide to cadmium metal and then volatilise this in a stream of nitrogen at 800°C. Condense the cadmium vapour, dissolve the deposit in hydrochloric acid and make up to volume.

Minerals, ores and rocks[19-21]

Zinc ores and concentrates[19,20] Dissolve 1—5 g of sample in a mixture of hydrochloric and nitric acids, remove any lead as lead sulphate, copper by deposition on metallic iron and separate cadmium from the zinc by forming a complex with iodine in $0.75N$ sulphuric acid containing 5% (w/v) potassium iodide (KI). Pass the solution through a 12 × 1 inch column of Dowex-1 resin (200—400 mesh), wash with $0.1N$ hydrochloric acid and elute cadmium with $3N$ nitric acid.

Silicates[21] Grind sample to pass through 100 mesh and then digest 1 g of the fine powder with a mixture of hydrofluoric (HF), nitric and perchloric ($HClO_4$) acids. Evaporate to the point at which white fumes of perchloric acid appear and then heat more strongly to dryness. Dissolve the residue in the minimum quantity of $2.5M$ hydrochloric acid, add a 20% solution of potassium sodium tartrate ($COOK(CHOH)_2COONa.4H_2O$) to prevent the precipitation of hydrated oxides, and make just alkaline to thymol blue with aqueous caustic soda (NaOH). Add an equal volume of $2.5M$ caustic soda solution and extract the cadmium with dithizone.

Miscellaneous[22]

Inorganic pigments Dissolve the sample in hydrochloric acid, make the solution alkaline and extract the cadmium with dithizone.

Sewage and industrial wastes[23,24]

Destroy organic matter by wet ashing with sulphuric and nitric acids. Treat the residue with a mixture of bromine and hydrogen bromide, add dilute hydrochloric acid, remove excess bromine by boiling, filter and make solution to known volume.

Water[25-28]

Potable water[25-27] No sample preparation is normally required, and for a direct screening test the dithizone colour can be formed directly in the water sample[26]. However under these conditions copper, lead, nickel and zinc will interfere. However the standard method (*Method 1*) using extraction into chloroform will overcome these interferences.

Sea water[28] To ensure freedom from interference Mullin recommends that after the first extraction into chloroform the cadmium should be extracted from the organic layer with excess tartaric acid (($CHOH.COOH)_2$), the solution made strongly alkaline with caustic soda (NaOH) and the cadmium re-extracted with dithizone in the presence of hydroxylamine ($HO.NH_2$) and a small amount of cyanide (KCN).

Concentration of cadmium

For trace analysis cadmium can be concentrated by co-precipitation with the hydroxides of magnesium, iron or bismuth, either by forming the hydroxides in solution or by adding a slurry of the hydroxide. This precipitation has been shown[29] to be quantitative if the pH of the solution is higher than the pH of formation of cadmium hydroxide.

Separation from other ions[30-36]

In addition to the methods of separation from interfering ions mentioned in the individual methods of sample preparation described above the following techniques have appeared in the literature.

From aluminium and zinc[30] Pass sample in a 1:9 mixture of $5N$ nitric acid and ethanol (C_2H_5OH) through a column of Dowex 1—X8 resin. Elute aluminium and zinc with a further portion of the same solvent solution and then elute the cadmium with N nitric acid.

From bismuth, copper, lead, silver and zinc[31] Apply a solution of the mixed chlorides to a 10—13 mm diameter column, containing 10 g of anion exchange resin EDE-10P (0.25—0.5 mm), at a flow rate of 1—2 ml a minute. Elute the copper with $2M$ hydrochloric acid, zinc with $0.65M$ hydrochloric acid, cadmium with $0.25M$ hydrochloric acid, lead with water, bismuth with $5M$ sulphuric acid and silver with $3M$ aqueous ammonia.

From copper[32-34] The cadmium is separated either by selective precipitation of the cadmium as carbonate[32], by means of a 1.5% aqueous solution of ethanolamine carbonate (($HOC_2H_4NH_3)_2CO_3$), or of the copper as sulphide[33] by passing hydrogen sulphide (H_2S) into a $3.5N$ hydrochloric acid solution. Alternatively[34] pass the solution through a column of Dowex 50—X8 resin and elute the cadmium with water and the copper with $4N$ hydrochloric acid.

From zinc[35,36] Either[35] pass mixed chlorides through a column of EDE-10P resin as described under bismuth above[31] or[36] form iodine complex with a solution of $0.75N$ sulphuric acid and 5% potassium iodide (KI), pass this through a column of the same resin, elute the zinc with 30 ml of the sulphuric acid/potassium iodide solution, remove the excess potassium iodide with 50 ml of $0.1N$ hydrochloric acid and desorb the cadmium with 100—200 ml of $3N$ nitric acid.

The individual test

Method 1 is based on a modification of the original method by Saltzman[25].

References

1. P. Sanz-Pedrero and M. D. Hermoso, *Med. Seguridad Trabajo,* 1963, **11** (41), 28; *Analyt. Abs.,* 1964, **11**, 3808
2. L. Truffert, M. Favert and Y. Le Gall, *Annls. Falsif. Expert. chim.,* 1967, **60**. 275; *Analyt. Abs.,* 1969, **16**, 703
3. O. E. Knockaert, G. L. Maes and M. H. Faes, *Am. ind. Hyg. Ass. J.,* 1967, **28** (9), 501; *Analyt. Abs.,* 1969, **16**, 2002
4. W. Kemula, W. Brachaczek and A. Hulanicki, *Chem. Anal. Warsaw,* 1958, **3** (5-6), 923; *Analyt. Abs.,* 1960, **7**, 357
5. E. Kroller, *Z. Lebensmittelunters. u. -Forsch.,* 1964, **125** (5), 401; *Analyt. Abs.,* 1966, **13**, 405
6. J. C. Meranger and E. Somers, *Bull. envir. Contam. Toxicol.,* 1968, **3** (6), 360; *Analyt. Abs.,* 1970, **18**, 2758
7. S. Hashimoto and R. Tanaka, *Japan Analyst,* 1959, **8** (9), 564; *Analyt. Abs.,* 1960, **7**, 4161
8. S. Maekawa and Y. Yoneyama, *Japan Analyst,* 1961, **10** (7), 736; *Analyt. Abs.,* 1963, **10**, 3245
9. V. N. Pavlova and N. P. Strel'nikova, *Zavodskaya Lab.,* 1960, **26** (5), 536; *Analyt. Abs.* 1961, **8**, 136
10. A. I. Busev and L. Ya. Polyak, *Zavodskaya Lab.,* 1959, **25** (6), 668; *Analyt. Abs.* 1960, **7**, 1685
11. V. P. Raumova and L. S. Nadezhina, *Trudy Leningr. Politekh. Inst.,* 1959, (201), 158; *Analyt. Abs.,* 1960, **7**, 3652

12. E. P. Shkrobot and L. M. Bakinovskaya, *Zavodskaya Lab.*, 1966, **32** (12), 1452; *Analyt. Abs.*, 1968, **15**, 1341
13. Brit. Standards Inst., B.S. 3338 : 1965, "*Methods for the sampling of tin and tin alloys, Part 17, Cadmium in solders and metal bearing alloys (photometric method)*"; *Analyt. Abs.*, 1965, **12**, 4480
14. D. P. Stricos and J. T. Porter, *U.S. Atomic Energy Comm., Rep. KAPL-M-DPS-1*, July 17, 1958; *Analyt. Abs.*, 1959, **6**, 2845
15. W. C. G. Sheeler and D. Durant, *Metallurgia Manchr.*, 1959, **59**, 53; *Analyt. Abs.*, 1959, **6**, 3422
16. J. R. Knapp, R. E. Van Aman and J. H. Kanzelmeyer, *Analyt. Chem.*, 1962, **34** (11), 1374; *Analyt. Abs.*, 1963, **10**, 2175
17. W. Geilmann and H. Hepp, *Z. analyt. Chem.*, 1964, **200** (4), 241; *Analyt. Abs.*, 1965, **12**, 2157
18. H. Bode and K. Wulff, *Z. analyt. Chem.*, 1966, **219** (1), 32; *Analyt. Abs.*, 1967, **14**, 4573
19. S. Kallmann, H. Oberthin and R. Liu, *Analyt. Chem.*, 1958, **30** (11), 1846; *Analyt. Abs.* 1959, **6**, 2061
20. S. Kallmann, H. Oberthin and R. Liu, *Analyt. Chem.*, 1960, **32** (1), 58; *Analyt. Abs.*, 1960, **7**, 3651
21. J. R. Butler and A. J. Thompson, *Geochim. cosmochim, Acta*, 1967, **31** (2), 97; *Analyt. Abs.*, 1968, **15**, 2517
22. E. Hezel, *Farbe. u. Lack*, 1963, **69** (11), 828; *Analyt. Abs.*, 1965, **12**, 1098
23. A. A. Christie, J. R. W. Kerr, G. Knowles and G. F. Lowden, *Analyst*, 1957, **82**, 336
24. Joint A.B.C.M. – S.A.C. Cttee. on Methods for the Analysis of Trade Effluents, *Analyst*, 1957, **82**, 764
25. B. E. Saltzman, *Analyt. Chem.*, 1953, **25**, 493
26. M. Lieber, *Wat. & Sewage Wks.*, 1958, **105** (9), 374; *Analyt. Abs.*, 1959, **6**, 2372
27. J. Ganotes, E. Larson and R. Navone, *J. Amer. Wat. Wks. Ass.*, 1962, **54** (7), 852; *Analyt. Abs.*, 1963, **10**, 1224
28. J. B. Mullin and J. P. Riley, *J. Mar. Res.*, 1956, **15**, 103; *Analyt. Abs.*, 1958, **5**, 1388
29. V. T. Chuiko, *Zhur. Neorg. Khim.*, 1957, **2** (3), 685; *Analyt. Abs.*, 1958, **5**, 2510
30. J. Korkisch and F. Feik, *Analyt. chim. Acta*, 1965, **32** (2), 110; *Analyt. Abs.*, 1966, **13**, 2869
31. M. N. Zvereva and N. I. Vinogradova, *Vestn. Leningr. Univ.*, 1961, (10), *Ser. Fizi. Khim.*, (2), 142; *Analyt. Abs.*, 1961, **8**, 4470
32. I. P. Ryazanov, *Sb. Nauch. Trud. Magnitogorsk. Gornometallurg. Inst.*, 1957, (13), 22; *Analyt. Abs.*, 1959, **6**, 849
33. M. Ziegler and K. Suffenplan, *Z. analyt. Chem.*, 1963, **197** (4), 401; *Analyt. Abs.*, 1964, **11**, 4739
34. F. Burriel-Marti and C. Alvarez-Herrero, *An. R. Soc. esp. Fis. Quim. B*, 1966, **62** (1), 45; *Analyt. Abs.*, 1967, **14**, 2436
35. M. N. Zvereva and N. I. Vinogradova, *Uch. Zap. Leningr. Gos. Univ.*, 1960, (297), 46; *Analyt. Abs.*, 1961, **8**, 4509
36. V. P. Razumova, *Izv. Vyssh. Ucheb. Zavedenii, Khim. i. Khim. Tekhnol.*, 1962, **5** (5), 709; *Analyt. Abs.*, 1964, **11**, 902

Cadmium Method 1
using dithizone

Principle of the method

Of all the complexes formed between dithizone and metals only those of tin, indium, cadmium, silver and palladium are soluble in chloroform. The cadmium complex can be separated from all but the silver complex by extraction into chloroform from alkaline solution in the presence of tartrate, which is added to prevent precipitation of hydroxides. The interference of silver with the cadmium determination can then be prevented by the addition of potassium cyanide which holds up the silver but allows the cadmium complex to be extracted. In solutions stabilised by the addition of hydroxylamine cadmium recoveries are better than 95% with decontamination factors of 10^4 for tin, indium, silver and palladium. The amount of cadmium in the chloroform extract is measured by comparison of the colour of the chloroform with a range of Lovibond permanent glass colour standards.

Reagents required

1. *Tartrate solution* Dissolve 25 g of sodium potassium tartrate (COONa.CHOH.CHOH.COOK.4H$_2$O) in 100 ml of distilled water.
2. *Cyanide solution A* Dissolve 40 g of sodium hydroxide (NaOH) and 1.0 g of potassium cyanide (KCN) in 100 ml of distilled water. Store in a polythene bottle.
3. *Cyanide solution B* Dissolve 40 g of sodium hydroxide (NaOH) and 0.05 g of potassium cyanide (KCN) in 100 ml of distilled water. Store in a polythene bottle.
4. *Hydroxylammonium chloride solution* Dissolve 80 g of hydroxylammonium chloride (HO.NH$_3$Cl) in 100 ml of distilled water.
5. *Extraction dithizone solution* Dissolve 80 mg of dithizone (C$_6$H$_5$.N:N.CS.NH.NH.C$_6$H$_5$) in 1 litre of chloroform (CHCl$_3$). Store in an amber glass bottle in a cool place.
6. *Standard dithizone solution* Dissolve 8 mg of dithizone (C$_6$H$_5$, N:N.CS.NH.NH.C$_6$H$_5$) in 1 litre of chloroform (CHCl$_3$). Store in an amber glass bottle in a cool place.
7. *Tartaric acid solution* Dissolve 20 g of tartaric acid ((CHOH.COOH)$_2$) in 1 litre of distilled water.
8. *Thymol blue indicator* 0.04% aqueous solution.
9. *Sodium hydroxide solution* Dissolve 10 g of sodium hydroxide (NaOH) in 100 ml of distilled water.
10. *Chloroform (CHCl$_3$)*

All chemicals used in the preparation of reagents should be of analytical reagent quality.

The Standard Lovibond Comparator Disc 3/89

This disc covers the range 0.5 to 5.0 micrograms of cadmium (Cd) with standards corresponding to 0.5, 1.0, 1.5, 2.0, 2.5, 3.0, 4.0, and 5.0 micrograms. If a 5 ml sample is used for the determination this corresponds to a range 0.1 to 1.0 ppm of cadmium (Note 1).

Technique

Measure a suitable volume (Note 1) of the sample solution into a 50 ml separating funnel and add distilled water to bring the total volume up to about 25 ml. Add 1 drop of indicator (reagent 8). If the solution is not already alkaline, that is if the colour of the indicator is not blue, then add sodium hydroxide solution (reagent 9) until a blue colour appears and persists.

To the alkaline solution add the following reagents in the stated order and mix the solution well after each successive addition:—

1. 1 ml of tartrate solution (reagent 1)
2. 5 ml cyanide solution A (reagent 2)
3. 1 ml of hydroxylammonium chloride solution (reagent 4)

Then add 15 ml of extraction dithizone solution (reagent 5). Shake the mixture for one minute and drain the lower, chloroform, layer into a second 50 ml separating funnel containing 25 ml of tartaric acid solution (reagent 7). Add 10 ml of chloroform (reagent 10) to the aqueous layer remaining in the first separating funnel and shake for one minute. Add the lower layer to the solution in the second separating funnel, taking care not to transfer any of the aqueous layer. Any pressure which is generated in the funnel by shaking should be released by venting through the stopper and not through the stopcock. If the aqueous layer loses the orange colour due to excess dithizone this indicates that too large a volume of sample solution has been used. In this case the determination should be repeated using a smaller sample.

Shake the combined chloroform extracts with the tartaric acid solution in the second separating funnel for two minutes. Allow the layers to separate and discard the lower, chloroform, layer. (As this extraction is being carried out from an acid solution the cadmium in this case will be in the aqueous layer). Add a further 5 ml of chloroform (reagent 10) and shake the mixture for one minute. Again allow the layers to separate and discard the lower layer. Add 0.25 ml (5 drops) of hydroxylammonium chloride solution (reagent 4) and 15 ml of standard dithizone solution (reagent 6) followed by 5 ml of cyanide solution b (reagent 3). Shake for one minute, allow the layers to separate and filter the chloroform layer into a 13.5 mm comparator cell through a small plug of cotton wool to remove any dispersed water globules. Place this cell in the right-hand compartment of the comparator. Compare the colour with the standards in the disc using a standard source of white light such as the Lovibond White Light Cabinet, or failing this, north daylight.

Note

1. The figure shown in the indicator window when a match is achieved represents the micrograms of cadmium present in the volume of the sample used for the determination. To convert this to parts per million, which is equivalent to micrograms per millilitre or milligrams per litre, divide the number of micrograms by the number of millilitres of sample which was used.

The Determination of Calcium

Introduction

Calcium is an important consituent of biological materials, natural waters, plants, pharmaceuticals, ores, slags, fertilisers, cement, glass, lubricating oils, milk, etc., and methods for its determination in all of these materials have been developed. Most determinations are nowadays carried out by flame photometry or by atomic absorption spectroscopy. It is however possible to carry out an indirect colorimetric determination of calcium, and this method has been devised [1-3] for the benefit of those laboratories which do not possess the expensive equipment required for the more sophisticated tests.

Methods of sample preparation

Biological materials [4-6]

Bone[4] Ash the sample. Dissolve the ash in $5N$ hydrochloric acid and dilute to 100 ml with water. Treat an aliquot with 1 – 2 ml of a 2% aqueous solution of sodium diethyldidithio-carbamate $((C_2H_5)_2N.CS.S.Na.3H_2O)$ and extract with three 10 ml portions of chloroform $(CHCl_3)$ to remove heavy metals. Treat the aqueous layer with 0.5 ml of hydrochloric acid, 10 ml of a 1:1 mixture of butanol $(CH_3(CH_2)_2CH_2OH)$ and chloroform and 2.5 ml of a 20% solution of sodium molybdate $(Na_2MoO_4.2H_2O)$. Repeat this last extraction process until a clear organic solution is obtained and then determine calcium on the residual aqueous layer.

Serum See under *Miscellaneous – hair*.

Tissue[5] Ash about 150 mg of tissue at 600 – 650°C. for 9 hours, cool and dissolve the ash in 0.1 ml of normal hydrochloric acid. Add the solution to a column (90 × 5 mm) of Dowex 50-X8 resin (H^+ form) and wash the column with 5 ml of normal hydrochloric acid. Elute magnesium with 5 ml of $2N$ hydrochloric acid and then elute calcium with 5 ml of $3N$ hydrochloric acid.

Urine[6] Acidify the sample with a 1:1 mixture of hydrochloric and acetic (CH_3COOH) acids and extract with chloroform $(CHCl_3)$. Determine calcium on the aqueous phase. Separation is claimed to be quantitative for 0.1 – 0.5 mg of calcium.

Chemicals[7]

Alkali metal compounds Place 20 ml of a 10% solution of the sample in a 100 ml separating funnel. If the sample is a salt, rather than a hydroxide, then add 1 ml of 20% caustic soda (NaOH) solution. Extract the calcium by shaking for 1 minute with 25 ml of a 0.01% solution of 2-hydroxy-2'-(2-hydroxy-1-naphthylazo)-5-methylazoxybenzene in a 1:4 mixture of tributyl phosphate $((C_4H_9O)_3PO))$ and carbon tetrachloride (CCl_4). Re-extract the calcium from the organic phase into 25 ml of $0.01N$ hydrochloric acid.

Foods and edible oils[8]

Potato starch Digest 5 g of the starch with 50 ml of 1:1 nitric acid and 25 ml of perchloric acid $(HClO_4)$. Dilute the final solution to 200 ml with water.

Fuels and lubricants[9]

Coal ash Grind the sample to pass through 250 mesh and then mix 0.5 g with 2 g of lithium metaborate $(LiBO_2)$ in a platinum crucible. Heat in a muffle furnace at 900°C until the melt is clear. If the sample contains more than 15% iron add 30 – 50 mg of ammonium vanadate (NH_4VO_3) to act as a flux. Cool the melt and transfer it to a beaker containing 8 ml of nitric acid and 150 ml of water. Stir vigorously until dissolution is complete. Add 2.5 g of tartaric acid $((CHOH.COOH)_2)$ and make up to 250 ml.

Minerals, ores and rocks[9,10]

Proceed as described for coal ash[9] or alternatively[10] pre-ignite an empty graphite crucible for 30 minutes at 950°C and cool. Take care not to disturb the resulting powdery inside surface. Mix 0.2 g of rock sample powder with 1.4 g of lithium metaborate $(LiBO_2)$ in a porcelain crucible and then transfer the mixture to the pre-ignited graphite crucible. Fuse at 900°C in a muffle furnace for exactly 15 minutes. Pour the melt into 100 ml of 1:24 nitric acid contained in a plastic beaker with a flat bottom. Stir with a Teflon coated stirring rod. Wash the solution into a 250 ml volumetric flask and dilute to volume with 1:24 nitric acid.

Miscellaneous[11-16]

Agricultural feeds[11] Weigh 0.5 – 2.0 g of the sample into a 150 ml borosilicate glass beaker and ash at 600°C. Cool, add 40 ml of 1:3 hydrochloric acid containing a few drops of nitric acid. Bring to the boil on a hot plate. Cool and make up to 200 ml with distilled water. Filter and estimate calcium on an aliquot of the filtrate.

Cement[12] Transfer 1 g of sample into a 200 ml platinum or porcelain evaporating dish, disperse with 15 ml of water, add 10 ml of hydrochloric acid, heat and stir until decomposition is complete. Evaporate nearly to dryness on a steam bath, cool, add 8 ml of hydrochloric acid and after 2 minutes add 50 ml of hot water. Stir, cover the evaporating basin, and digest for 5 minutes. Filter the digest, through a Whatman No. 40 filter paper, into a 500 ml flask. Wash the precipitate with 10 portions of hot 1:99 hydrochloric acid and then with 5 portions of hot water. Add the washings to the contents of the flask. Fuse the precipitate with 0.1 g of lithium chloride (LiCl), cool, dissolve in water, filter and add the filtrate to the contents of the flask. Make up to volume and determine calcium on an aliquot.

Dusts and bonded deposits[13] Fuse the sample with caustic soda (NaOH) pellets, cool the melt and extract it with a mixture of water, sulphuric acid and hydrogen peroxide (H_2O_2) at 60°C for 1 hour. Remove silica by dehydration with sulphuric acid and remove interfering ions by electrolysis at a mecury cathode. Extract titanium and vanadium from a sulphuric acid solution of pH less than 1.0 into a chloroform ($CHCl_3$) solution of cupferron ('copperone', $C_6H_5.N(NO).ONH_4$). Determine on an aliquot of the aqueous layer.

Hair[14] Kamel suggests that the levels of calcium and phosphorus in 1 g of hair are similar to those in 100 ml of serum and that serum levels can be monitored by hair analysis.

Wash the hair sample with water and then with light petroleum and dry. Ash for 12 hours at 500°C. Treat the ash with normal hydrochloric acid and water, filter and make up the solution to a known volume.

Slag[15] Finely grind 0.2 g of sample, moisten with 20 mg of water and heat to dissolution with 10 ml of hydrochloric acid. Boil the solution with 1 – 2 drops of nitric acid and 20 mls of water, precipitate sesquioxides with aqueous ammonia (NH_4OH) and 10 – 20 ml of 10% ammonium persulphate (($NH_4)_2S_2O_8$). Filter and wash precipitate. Add the washings to the filtrate, boil for 10 – 12 minutes, cool and dilute to 250 ml. Determine calcium on an aliquot.

Slime[16] Fuse 0.2 g of dried slime with 2 g of sodium carbonate (Na_2CO_3). Extract the melt with 50 ml of hot water, rapidly heat the solution to the boiling point and filter. Dilute the filtrate to 200 ml.

Water[17]

To separate calcium from magnesium and strontium pass the sample through a column of Dowex 50W-X8 ion exchange resin (H^+ form) and elute the calcium with $2N$ hydrochloric acid.

Concentration of calcium[18]

Wash a 10 – 12 mm column containing about 5 g of oxidised charcoal with $0.1N$ hydrochloric acid at a rate of 60 drops a minute and then follow with water until the percolate is only slightly acid (pH 5 – 6) and gives a negative reaction for calcium with murexide (ammonium purpurate – $C_8H_4N_5O_6.NH_4$). Then pass 100 ml of $2N$ sodium, potassium or ammonium hydroxide (NaOH, KOH or NH_4OH) through the column followed by water until the percolate is neutral. Pass 500 ml of sample solution, containing 20 – 100 μg of calcium, through the column at a rate of 60 drops a minute, wash the column with 100 ml of water and elute the calcium with 50 ml of $0.1N$ hydrochloric acid.

Separation from other ions[7,19-27]

In addition to the methods mentioned for specific applications above, the following general methods have been described:—

Separaration from magnesium[19-24] For 0.25 mM of calcium in 10 ml of 90% isopropanol (($CH_3)_2CHOH$) containing 0.5M nitric acid[19], pass solution through a column of Amberlyst XN-1002 resin at about 0.3 ml a minute. Elute magnesium with 40 ml of a 1:9 mixture of 0.5M nitric acid and 90% isopropanol at 0.5 – 0.6 ml a minute. Then elute calcium with 30 – 40 ml of 0.02M nitric acid at about 2 ml a minute. Alternatively[20] use Dowex 50-X4 resin, elute magnesium with a 6.4M perchloric acid ($HClO_4$):2.6M hydrochloric acid mixture and then elute calcium with 5M hydrochloric acid.

Winowski[24] recommends the use of Lewatit MN anionite (oxalate form). Elute magnesium with 5–20 ml of 0.018% ammonium oxalate (($COO.NH_4)_2.H_2O$) and then elute calcium with 20 ml of 2N hydrochloric acid followed by five 20 ml portions of water.

When aluminium is also present[22] absorb the solution on a 14 × 1 cm column of Dowex 50-X8 (H^+ form). Elute magnesium with 200 ml of 0.7N hydrochloric acid, calcium with a further 200ml of 0.7N hydrochloric acid aluminium with 300 ml of normal hydrochloric acid.

If strontium and magnesium are both present[21] pass the solution through a 22 cm column of Amberlyst XN-1002 resin (NO_3^- form) at 24°C and 0.5 – 0.6 ml a minute. Elute magnesium with 70 ml of 0.25N nitric acid in 95% ethanol (C_2H_5OH), then calcium with 80 – 100 ml of 0.25N nitric acid in 95% methanol (CH_3OH) and strontium with 50 ml of water or 95% methanol (CH_3OH) all at a rate of 0.5 – 0.6 ml a minute.

When barium is also present[23] absorb the solution on a 15 × 1 cm column of Dowex 50-X8 resin (H^+ form) and elute magnesium with 0.7N hydrochloric acid, calcium with 0.8N hydrochloric acid in 10% methanol (CH_3OH), strontium with 1.2N hydrochloric acid and barium with 1.6N hydrochloric acid.

Separation from strontium[25,26] Either use Wofatit KPS-200 resin[25] and elute calcium with 320 ml of 3.4% ammonium sulphate (($NH_4)_2SO_4$) solution, rejecting the first 70 ml, or[26] pass neutral solution, at 10 ml an hour, through a column (24 × 2.1 cm) of Kel-F powder impregnated with bis-(2-ethylhexyl) phosphate (($CH_3CH(C_2H_5)(CH_2)_4O)_2POOH$). Elute strontium with 0.03M nitric acid and then calcium with molar nitric acid.

General[7,27] Use method described above for the sample preparation of chemicals.

The individual test

Method 1 is the indirect method, using sodium phosphate, which has been devised[1-3] for laboratories which are not equipped to perform the more sophisticated tests.

References

1. A. B. Briggs, *J. Biol. Chem.*, 1925, **59**, 255
2. J. H. Roe and B. S. Kahn, *J. Biol. Chem.*, 1929, **81**, 1
3. T. Kuttner and H. R. Cohen, *J. Biol. Chem.*, 1927, **75**, 517
4. A. F. Grishilo and M. V. Shubinets, *Lab. Delo*, 1968, (10), 621; *Analyt. Abs.*, 1970, **18**, 1076
5. M. Hofer, *Experientia*, 1963, **19**(7), 367; *Anaylt. Abs.*, 1964, **11**, 3230
6. H. H. Taussky, *Microchem. J.*, 1963, **7**(1), 89; *Analyt. Abs.*, 1964, **11**, 2712
7. F. P. Gorbenko and V. V. Sachko, *Zhur. analit. Khim.*, 1963, **18**, 1497; *Analyt. Abs.*, 1965, **12**, 1083
8. B. Mica, *Prum. Potravin*, 1964, **15**(10), 526; *Analyt. Abs.*, 1966, **13**, 960
9. P. L. Boar and L. K. Ingram, *Analyst*, 1970, **95**, 124
10. J. C. Van Loon and C. M. Parissis, *Analyst*, 1969, **94**, 1057
11. P. F. Parks and Dorothy E. Dunn, *J. Assoc. off. Agric. Chem.*, 1963, **46**(5), 836; *Analyt. Abs.*, 1965, **12**, 395
12. A. Nestoridis, *Analyst*, 1970, **95**, 51
13. J. Grant, *J. appl. Chem. Lond.*, 1964, **14**(12), 525; *Analyt. Abs.*, 1966, **13**, 1782
14. S. H. Kamel, *J. vet. Sci. U.A.R.*, 1965, **2**(2), 103; *Analyt. Abs.*, 1967, **14**, 1538
15. Z. I. Kuzmina, *Zavodskaya, Lab.* 1964, **30**(10), 1215; *Analyt. Abs.*, 1966, **13**, 653
16. T. I. Badeeva, L. A. Molot, N. S. Frumina and K. G. Petrikova, *Uch. Zap. Saratovsk. Univ.*, 1962, **75**, 100; *Analyt. Abs.*, 1966, **11**, 875

17. J. T. Corless, *J. Chem. Educ.*, 1965, **42** (8), 421; *Analyt. Abs.*, 1966, **13,** 7159
18. F. P. Gorbenko, I. A. Tarkovskaya and M. I. Olevinskii, *Ukr. Khim.Zhur.*, 1964, **30,** 640; *Analyt. Abs.*, 1965, **12,** 5032
19. J. S. Fritz and H. Waki, *Analyt. Chem.*, 1963, **35,** 1079
20. F. Nelson, J. H. Holloway and K. A. Kraus, *J. Chromatog.*, 1963, **11,** 258; *Analyt. Abs.*, 1964, **11,** 3008
21. J. S. Fritz, H. Waki and B. B. Garralda, *Analyt. Chem.*, 1964, **36,** 900
22. M. Tanaka, *J. Chem. Soc. Japan, Pure Chem. Sec.*, 1963, **84,** 582; *Analyt. Abs.*, 1965, **12,** 5045
23. M. Tanaka, *J. Chem. Soc. Japan, Pure Chem. Soc.*, 1964, **85,** 117; *Analyt. Abs.*, 1965, **12,** 5670
24. Z. Winowski, *Chemia analit.*, 1967, **12** (6), 1271; *Analyt, Abs.*, 1969, **16,** 579
25. R. Khristova and P. Ilkova, *God. sof. Univ. Khim.*, 1963/64, **58,** 97; *Analyt. Chem. Acta.* 1965, **33** (4), 434; *Analyt. Abs.*, 1966, **13,** 4704
26. K. H. Lieser and H. Bernhard, *Z. analyt. Chem.*, 166, **219** (5), 401; *Analyt Abs.*, 1967, **14,** 6695
27. F. P. Gorbenko and V. V. Sachko, *Ukr. Khim. Zhur.*, 1964, **30,** 402; *Analyt. Abs.*, 1965, **12,** 3759

Calcium 1
using sodium phosphate

Principle of the method

Calcium is separated from the sample solution as calcium oxalate, and the oxalate is subsequently converted to calcium phosphate by reaction with sodium phosphate. The precipitate of calcium phosphate is dissolved in acid and the phosphate content of the resulting solution is determined by one of the standard methods using the appropriate disc. From this phosphate figure the corresponding calcium concentration is calculated (Note 1).

Reagents required

1. *Ammonium oxalate*   4 g of $(COONH_4)_2.H_2O$ in 100 ml of distilled water.
2. *0.1% ammonium oxalate*   Make up 2.5 ml of reagent 1 to 100 ml with distilled water.
3. *Ammonium hydroxide*   1:1 $NH_4OH:H_2O$
4. *Hydrochloric acid*   1:1 $HCl:H_2O$
5. *Hydrogen peroxide*   30% H_2O_2
6. *Sodium hydroxide* 6N   Dissolve 50 g of NaOH in 135 ml of distilled water. Filter and store in a plastic bottle protected from atmospheric carbon dioxide.
7. *Trisodium phosphate*   1% solution of $Na_3PO_4.12H_2O$
8. *Alcoholic wash solution*   Mix 58 ml of ethyl alcohol (C_2H_5OH) with 10 ml of amyl alcohol $(C_5H_{11}OH)$ and dilute to 100 ml with distilled water. Add 1 drop of reagent 6.

In addition, the reagents for the chosen method of phosphate estimation will be required. All chemicals used should be of analytical reagent quality.

The Standard Lovibond Phosphate Discs

The following Comparator and Nessleriser discs are available for the determination of phosphate:—

Phosphate Method	Disc No. Comparator	Nessleriser	Range	Note
1	3/7		20–220 µg of P_2O_5	
1		NMB	2–20 µg of P_2O_5	
2	3/4		5–100 µg of P_2O_5	
2		NMC	5–60 µg of P_2O_5	
3	5/14		20–160 µg of P	
4	3/12		0–80 ppm of PO_4	using 5 ml sample
5	3/38		10–100 ppm of PO_4	using 5 ml sample
6	3/51		10–400 µ of PO_4	
7	3/60		10–100 ppm of PO_4	using 20–25 ml sample
8	3/70	NOP	0–100 ppm of PO_4	using 50 ml sample
8		NOX	1–10 ppm of PO_4	using 50 ml sample

The equivalent ranges for calcium may be calculated by use of the appropriate conversion factor (Note 1).

Technique

Take 5–10 ml of sample solution, which should contain 0.2–0.5 mg of calcium, add a drop of methyl orange indicator, render the solution just acid with hydrochloric acid (reagent 4) and add two or three drops of excess acid. Heat until nearly boiling, add 1 ml of ammonium oxalate solution (reagent 1) and then ammonium hydroxide (reagent 3) until the solution becomes yellow. Cool with occasional stirring, stand for 2 hours and then filter or centrifuge. Wash the precipitate with cold 0.1% ammonium oxalate solution (reagent 2) and then dissolve the precipitate in a few drops of hot hydrochloric acid (reagent 4). Wash the filter with water and evaporate the acid solution and washings to a small volume – a few drops. Destroy the oxalate with 0.5 ml of hydrogen peroxide (reagent 5) and heat for 30 minutes on a steam-bath.

Dilute the solution to 3–5 ml with distilled water and add 0.5 ml of sodium hydroxide solution (reagent 6). Allow the solution to stand for a few minutes and then add 0.5 ml of sodium phosphate solution (reagent 7). Allow the precipitate to settle for 60 minutes and then collect the precipitate either by centrifuging or by filtration through a sintered glass filter. Wash carefully with 3×2 ml portions of wash solution (reagent 8). Allow most of the alcohol to evaporate and then dissolve the precipitate in acid molybdate solution and proceed according to the method chosen for phosphate.

Notes

1. The concentration of calcium can be calculated from the corresponding phosphate figure read from the disc by use of the appropriate conversion factor:—

Phosphate determined as	For calcium multiply by
P	3.88
PO_4	1.26 (equals approx. $\frac{5}{4}$)
P_2O_5	1.7

2. Readings obtained by the use of the Nessleriser and disc are accurate only if Nessler tubes are used which correspond with those used in calibration of the disc, i.e. tubes in which the 50 ml calibration mark falls at a height of 113 ± 3 mm from the bottom of the tube when measured internally.

The Determination of Chloride

Introduction

Chloride ions are one of the major electrolytes in physiological systems and their determination is of importance in biochemical investigations of body fluids and tissues. The chloride concentration is also used as a measure of the salinity of water, beverages and food.

Methods of sample preparation

Beverages[1]

Beer Prepare a 12 inch column of Amberlite IR-120 ion exchange resin (H^+ form) and pass 50 ml of 6% nitric acid through the column at 1 drop per second. Follow with 50 ml of water and then backwash the column with 250 ml of water. Bring the water level to 0.5 inches above the resin bed. Pass 20 ml of degassed beer through the column at 1 drop per second and reject the first 10 ml of percolate, which is water. When the beer level has fallen to 0.5 inches above the resin bed, maintain this level by the addition of water until 75 ml of percolate has been collected. Determine chloride on this percolate.

Biological materials[2-6]

Blood and plasma[2-4] Transfer 5 ml of blood containing not more than 10 mg of potassium oxalate and no citrate to a 100 ml volumetric flask, dilute with 35 ml of water, add 5 ml of 10% sodium tungstate (Na_2WO_4) solution and mix. Add 3.5 ml of N sulphuric acid, stopper and shake vigorously. When coagulation is complete, as indicated by the precipitate turning dark brown, pour the contents of the flask onto a filter and cover to avoid evaporation. Determine chloride on the filtrate.

Muscle[5] Dry 10–14 mg of tissue in an aluminium pan for 30 minutes at 105°C. Extract the fat by treating the dried tissue for 10 minutes with light petroleum. Dry the tissue and extract the electrolytes by standing the dry fat-free tissue for 1 hour in 5 ml of $0.1 N$ nitric acid. Determine chloride on this solution.

Urine[6] No sample treatment is necessary.

Chemicals[7]

Hydrogen peroxide Decompose 50 ml of sample by heating with deionised water and then with 1 ml of $6 N$ caustic soda ($NaOH$) in a 500 ml beaker at 75–90°C. Evaporate the solution to dryness on a steam-bath and then dissolve the residue in 1–2 ml of deionised water. Determine chloride on this solution.

Foods and edible oils[8]

Meat or fish products Take a 2–5 g sample and dissolve by heating in an Erlenmeyer flask with 10 ml of nitric acid, and 0.1 g of silver nitrate ($AgNO_3$). Filter off the precipitate and dissolve it in ammonium hydroxide (NH_4OH). Pass the solution through Amberlite IR–120 ion exchange resin (NH_4^+ form) and determine chloride in the eluate.

Metals, alloys and plating baths[9]

Zinc plating electrolyte Take 40 ml of electrolyte, precipitate the chloride with 5 ml of N silver nitrate, filter off the precipitate and dissolve it in aqueous ammonia (NH_4OH). Pass this solution through a column of Amberlite IR–120 ion exchange resin (NH_4^+ form) and determine chloride on the eluate.

Miscellaneous[10]

Glass Fuse a 0.25 g sample of the glass with sodium carbonate (Na_2CO_3) and dissolve the melt in water. Just acidify this solution with nitric acid, make up to known volume and determine chloride on an aliquot.

Plastics and polymers[11]

Polythene Heat 3 g of sample under reflux with 95 ml of water and 550 ml of xylene ($C_6H_4(CH_3)_2$). Separate the aqueous phase, add 0.5 ml of ethyl alcohol (C_2H_5OH) and make up to 100 ml with water.

Sewage and industrial wastes[12]

To determine chloride in sulphite waste liquor, pass 50 ml of sample through a 150 mm × 10 mm column of Dowex-2 ion exchange resin (NO_3^- form) at 2.5 ml per minute. Wash the resin with 50 ml of water and elute the chloride with 100 ml of $5M$ ammonium nitrate (NH_4NO_3). Determine chloride on this eluate.

Water[13-16]

No sample preparation is normally required. Where it is necessary to determine both chloride and nitrate on a single sample then these ions may be separated by ion exchange[1-4] as the chloride interferes with the nitrate estimation.

Separation from other ions

Bromide and iodide[17] Pass the solution, in $0.06N$ acetic acid (CH_3COOH) through a column of MG-36 anionite resin (OH^- form). Elute the chloride with $0.01N$ sodium acetate (CH_3COONa) at pH 5.45 at 25 ml per minute.

The individual methods

Method 1 is the standard mercuric thiocyanate method[16,18,19]. Bromides, iodides, cyanides, and thiosulphates interfere and, if present in the sample, should be removed by suitable pre-treatment as described above. Fluorides at the level usually found in water will not interfere.

Method 2 is an indirect method using silver phosphate and measuring the phosphate released by reaction with chloride. It uses any of the standard phosphate discs.

References

1. D. B. West and A. F. Lautenbach, *Proc. Amer. Soc. Brew. Chem.*, 1959, 87; *Analyt. Abs.*, 1960, **7**, 2984
2. O Folin and H. Wu, *J. Biol. Chem.*, 1919, **38**, 84
3. F. L. Rodbey and J. Sendroy, *Clin. Chem.*, 1963, **9**, 668; *Analyt. Abs.*, 1964, **11**, 3238
4. V. Kulhanek and C. Fiser, *Colln. Czech. chem. Commun.*, 1966, **31** (4), 1890; *Analyt.Abs.*, 1967, **14**, 4871
5. C. T. G. Flear and I. Florence, *Clin. Chim. Acta*, 1961, **6**, 129; *Analyt. Abs.*, 1961, **8**, 5113
6. R. A. Crockson, *J. clin. Path.*, 1963, **16**, 473
7. I. Geld and I. Sternman, *Analyt. Chem.*, 1959, **31**, 1662
8. Z. Jedlinski, *Przem. Rol. Spoz.*, 1953, **7**, 365; *Analyt. Abs.*, 1954, **1**, 1661
9. M. Kojima, *Japan Analyst*, 1957, **6**, 309; *Analyt. Abs.*, 1958, **5**, 506
10. F. Plesnivy, *Sklar a Keram*, 1959, **9**, 214; *Analyt. Abs.*, 1960, **7**, 2250
11. H. Narasaki and S. Fujiwara, *Japan Analyst*, 1962, **11**, 1060; *Analyt. Abs.*, 1964, **11**, 1823
12. O. Lagerstrom and O. Samuelson, *Svensk. Papp-Tidn.*, 1959, **62**, 679; *Analyt. Abs.*, 1960, **7**, 3242
13. H. Ambuhl, *Mitt. Lebensmitt. Hyg. Bern.*, 1958, **49**, 241; *Analyt. Abs.*, 1959, **6**, 1549
14. D. Ceausescu, *Z. Analyt. Chem.*, 1959, **165**, 424; *Analyt. Abs.*, 1959, **6**, 3794
15. J. S. Swain, *Chem. & Ind.*, 1956, **20**, 418
16. Brit. Standards Inst., "*Methods of Testing Water used in Industry*", B.S. 2690: Part 6: 1968, London, 1968
17. B. I. Nabivanets, *Ukr., Khim. Zhur.*, 1956, **22**, 816; *Analyt. Abs.*, 1958, **5**, 508
18. I. Iwasaki, S. Uttsumi, K. Hagino and T. Ozawa, *Bull. Chem. Soc. Japan*, 1956, **29**, 860; *Analyt. Abs.*, 1957, **4**, 3954
19. "*Manual on Industrial Water and Industrial Waste Water*", 2nd edn. 1962, p.237

Chloride Method 1
using mercuric thiocyanate

Principle of the method

Chloride ions react with mercuric thiocyanate to produce mercuric chloride and to liberate thiocyanate ions. In the presence of ferric salts these thiocyanate ions produce a characteristic orange colour. The intensity of this colour is proportional to the concentration of thiocyanate ions and hence to the concentration of chloride ions. The intensity of the orange colour is measured by comparison with a series of Lovibond permanent glass colour standards.

Reagents required

1. *Mercuric thiocyanate reagent* A 0.150% solution of mercuric thiocyanate ($Hg(SCN)_2$) in methanol (CH_3OH)
2. *Ammonium ferric sulphate reagent* Dissolve 6 g of ammonium ferric sulphate (iron alum – $(NH_4)_2SO_4.Fe_2(SO_4)_3.24H_2O$) in $5N$ nitric acid (HNO_3) and make up to 100 ml with $5N$ HNO_3.

All chemicals used in the preparation of reagents should be of reagent grade quality.

The Standard Lovibond Comparator Discs 3/71 and 3/72 and Nessleriser Disc NOV.

Disc NOV covers the range 1–10 ppm of chloride ion with steps of 1, 2, 3, 4, 5, 6, 7, 8 and 10 ppm.

Disc 3/71 covers the range 10–200 ppm of chloride ion with steps of 10, 25, 50, 75, 100, 150 and 200 ppm.

Disc 3/72 covers the range 2.5–80 ppm of chloride ion with steps of 2.5, 5, 7.5, 10, 15, 20, 40, 60 and 80 ppm and must be used with 40 mm comparator cells.

Technique

To a 50 ml sample add 2.0 ml of mercuric thiocyanate reagent (reagent 1) and mix thoroughly. Add 5 ml of ammonium ferric sulphate reagent (reagent 2), mix again and allow to stand for 5 minutes. At the same time prepare a blank in an identical manner but using 50 ml of distilled water in the place of the sample.

(a) *Nessleriser* After 5 minutes transfer all the sample mixture to a Nessleriser tube and place this tube in the right-hand compartment of the Nessleriser. Transfer the blank solution to an identical Nessleriser cylinder and place this in the left-hand compartment of the Nessleriser. Place the Nessleriser before a standard source of white light, such as the Lovibond White Light Cabinet or, failing this, north daylight, and match the colour of the sample against the standards in the disc (Note 1).

(b) *Disc 3/72* After 5 minutes fill a 40 mm comparator cell with the sample solution and place this cell in the right-hand compartment of the comparator. Fill an identical cell with the blank solution and place this cell in the left-hand compartment of the comparator. Place the comparator before a standard source of white light and match the colour of the sample against the standard in the disc (Note 1).

(c) *Disc 3/71* After 5 minutes fill a standard test tube or 13.5 mm comparator cell with the sample and place this cell in the right-hand compartment of the comparator. Fill an identical tube or cell with the blank solution and place this in the left-hand compartment. Make sure that the dulling screen is in place with the disc. Place the comparator before a standard source of white light and match the colour of the sample against the standards in the disc (Note 1).

Notes

1. If the colour of the sample is more intense than that of the highest standard in the disc, repeat the test after diluting the sample with a known amount of distilled water. Then multiply the reading obtained from the disc by the dilution factor.

2. The results obtained by the use of the Nessleriser and disc are accurate only if the Nessleriser tubes used conform to the specification used when the disc was standardised i.e. that the 50 ml mark falls at a height of 113 ± 3 mm measured internally.

3. The test should be carried out at 20°C or above. Below this temperature, colour development is very slow, and low readings may be obtained.

Chloride Method 2
using silver phosphate

Principle of the method

The action of chloride ions, in neutral solution, on solid silver orthophosphate results in the precipitation of silver chloride and the passage into solution of phosphate ions. The concentration of phosphate ions, which is proportional to the original concentration of chloride ions, is determined by one of the standard phosphate methods using the appropriate Lovibond permanent glass colour standards.

Reagents required

1. *Silver orthophosphate* Grind silver orthophosphate (Ag_3PO_4) to pass through an 80-mesh sieve, and store the powder in an amber bottle.

2. Reagents for the chosen method of determining phosphate.

The Standard Lovibond Comparator and Nessleriser Discs

The following discs are available for the determination of phosphate:—

Phosphate Method	Disc No. Comparator	Nessleriser	Range	Note
1	3/7		20–220 µg of P_2O_5	2–22 ppm using a 10 ml sample
1		NMB	2–20 µg of P_2O_5	0.2–2 ppm using a 10 ml sample
2	3/4		5–100 µg of P_2O_5	1–20 ppm using a 5 ml sample
2		NMC	5–60 µg of P_2O_5	0.2–2.4 ppm using a 25 ml sample
3	5/14		20–160 µg of P	2–16%
4	3/12		0–80 ppm of PO_4	using 5 ml sample
5	3/38		10–100 ppm of PO_4	using 5 ml sample
6	3/51		10–400 µg of PO_4	2–80 ppm using a 20 ml sample
7	3/60		10–100 ppm of PO_4	using 20–25 ml sample
8	3/70	NOP	0–100 ppm of PO_4	using 50 ml sample
8		NOX	1–10 ppm of PO_4	using 50 ml sample

The equivalent ranges for chloride may be calculated using the appropriate conversion factor (See Notes).

Technique

Weigh exactly 50 mg of silver phosphate powder into each of two ground-glass stoppered centrifuge tubes. To one add 10 ml of carbon dioxide free water (pH 7) to act as a blank. To the other add 10 ml of the sample solution adjusted to pH 5. Invert the tubes every 5 seconds for 7 minutes. Centrifuge for 10 minutes. Pipette 5 ml of the supernatant solution and filter through a fine filter paper. Collect the filtrates in 50 ml volumetric flasks. Wash the paper with about 30 ml of water and collect the washings in the volumetric flasks. Make up to the mark with distilled water. Determine phosphate on an aliquot of this solution, by the chosen method.

Notes

1. The concentration of chloride corresponding to the concentration of phosphate read from the disc can be calculated by means of the following conversion factors:—

Phosphate determined as	For chloride multiply by
P	3.43
PO_4	1.12
P_2O_5	1.50

2. To calculate the chloride concentration in the original sample solution allowance must be made for the dilution during the chloride to phosphate conversion, where 5 ml of solution is diluted to 50 ml. For example if phosphate Method 4 was used and a disc reading of 50 ppm was obtained from the disc then:—

Equivalent chloride concentration $= 50 \times 1.50 = 75$ ppm

1 ppm is equal to 1 μg per ml.

∴ chloride concentration $= 75\,\mu$g per ml of solution used for phosphate determination

∴ chloride concentration of sample solution $= 75 \times 10\,\mu$g per ml
$= 750\,\mu$g per ml

The Determination of Residual Chlorine

Introduction

The administration of chlorine to reservoirs, storage tanks and swimming pools, and the emergency chlorination of water mains, is an accepted practice. In correct proportions it ensures sterility of the water by destroying harmful organisms, such as *E. Coli,* removes contamination (e.g. by sewage) and prevents the growth of algae. It is necessary, however, to keep a close check on the amount of residual chlorine left in the water after its "chlorine demand" has been fulfilled. An excess of chlorine beyond a certain point could be harmful, would give a very unpleasant flavour and smell to potable waters and, in the case of swimming baths, could cause discomfort to bathers through smarting eyes. The Ministry of Housing and Local Government[1,2], recommend a "free" chlorine content of between 0.2 and 0.5 ppm with marginal chlorination, and more than double this with "break-point" chlorination.

Some trade wastes also contain chlorine and its determination is important in the control of effluents.

Methods of sample preparation

Sewage and industrial wastes[3-4]

The standard method of analysis[3] uses *o*-tolidine. Interfering substances are determined by the preliminary removal of chlorine with sodium arsenite ($NaAsO_2$) and the chlorine content is determined by difference. This is automatically performed in these tests by the development of the manganese colour in the blank.

Although it has been claimed[4] that the acid *o*-tolidine test is not applicable to wastes, Rand and Hunter[5] have shown that it gives results which are in good agreement with those obtained by alternative methods.

Water[1,2,6-10,22-27]

No pre-treatment is normally required. The only interfering substance likely to be present in water is oxidised manganese. The effect of this may be compensated by developing the manganese colour in the "blank" as follows: — Place 0.5 ml of buffer (*Method 1* reagent 1) in a clean cell, add a crystal of potassium iodide (KI) and 1 drop of a 0.5% solution of sodium arsenite ($NaAsO_2$). Make up to the 10 ml mark with the sample and mix. Add 0.5 ml of neutral *o*-tolidine solution (*Method 1* reagent 2) and mix. Place this cell in the left-hand compartment of the comparator to act as a blank. A similar procedure is given in *Method 3*.

To prevent interference from large amounts of iron (Fe)[19] add 1 ml of 5% potassium fluoride (KF) to 70 ml of the sample. Next add the *o*-tolidine solution and 1–2 ml of concentrated hydrochloric acid (HCl). Measure the colour after 5 minutes (*Method 2*). Alternatively Johnson[27] suggests that if the *o*-tolidine colour is developed at pH 7, using a phosphate buffer, at 25°C in the presence of sodium bis(2-ethylhexyl) sulphosuccinate ($CH_3(CH_2)_3CH(C_2H_5)$ $CH_2O_2CCH_2CH(SO_3Na)CO_2CH(C_2H_5)(CH_2)_3CH_3$) as a stabiliser, interferences from chloramines, iron, nitrate and manganese are minimised.

For control of the residual chlorine level of water in swimming baths, Williams[24] has compared results obtained using *o*-tolidine with those using iodimetric titration. He suggests that discrepancies between the methods arise from their different sensitivities to combined residual chlorine, and recommends that chemical tests for adequacy of chlorination should be restricted to measurement of pH and of free residual chlorine only, and that when discrepancies arise the *o*-tolidine figure should be preferred.

Palin[25] strongly criticises Williams' conclusions, claims that *o*-tolidine is liable to overestimate free chlorine (although Williams reports that it gives the lowest results) and claims that the best control of swimming bath water is achieved by using the DPD method (*Method 3*) combined with tests for pH and alkalinity. Nicholson[26] confirms that DPD gives the most reliable result in the presence of combined chlorine but makes the proviso that the colour must be measured immediately after the reagents are added. See also below under *Method 3*.

The individual tests

Method 1, is the neutral *o*-tolidine test and is used in preference to the acid version when it is necessary to determine both "free" chlorine and the individual forms of combined chlorine.

Method 2, the acid *o*-tolidine test is mainly used when only "free" chlorine is to be determined, although modifications have been described to extend the test to the determination of total combined chlorine.

Method 3, the Palin-DPD[9,11-15], is considered by many authorities to be the best colorimetric test for chlorine, and was so adjudged by the Water Research Association (U.K.) in Technical Paper 53, 1966 and by Hässelbarth[30]. The method is simplified by the use of tablets, and it assesses separately not only free and combined chlorine, but also the different forms of combined chlorine. Being free of interference from bromides, this method is suitable for use with sea-water baths. It is specified in B.S. 1427[9].

Method 4, the potassium iodide test is used for high chlorine concentrations.

Method 5, is the American Public Health Association version of Method 2. A modification is described for the determination of both free and combined chlorine.

Methods 1, 2 and *5* have the disadvantage that they employ *o*-tolidine, which is considered carcinogenic[21].

Other applications of the chlorine discs

It has been shown that the DPD Tablet No. 1 may also be used as a test for iodine. This test is described later. To convert the disc reading to ppm. iodine multiply by 3.6. Modifications of the DPD technique are also used as tests for bromine and chlorine dioxide, as described in this book.

Palin[28] claims that the colours in the DPD discs are an exact match for the colours developed in Packham's aluminon test[29] for aluminium (see Aluminium—*Method 3*).

References

1. Min. of Housing and Local Govt., *"The Purification of the Water of Swimming Baths"*, H.M. Stationery Office, London, 1957
2. Min. of Housing and Local Govt., *"The Bacteriological Examination of Water Supplies, Report No. 71"*, H.M. Stationery Office, London, 1957
3. Joint A.B.C.M.-S.A.C. Cttee. on Methods for the Analysis of Trade Effluents, *Analyst*, 1958, **83**, 230
4. S. Katz and H. Heukelekian, *Sewage Ind. Waste*, 1959, **31**, 1022; *Analyt. Abs.*, 1960, **7**, 2493
5. M. C. Rand and J. V. Hunter, *J. Water Poll. Control Fed.*, 1961, **33**, 393; *Analyt. Abs.*, 1961, **8**, 5282
6. A. T. Palin, *J. Inst. Water Eng.*, 1949, **3**, 100
7. R. W. Aitken and D. Mercer, *J. Inst. Water Eng.*, 1957, **5**, 321
8. A. T. Palin, Brit. Pat. No. 680,427
9. Brit. Standards Inst., *"Routine Control Methods of Testing Water Used in Industry"*, B.S. 1427, London, Amendment 3, 1968
10. Brit. Standards Inst., *"Methods of Testing Water Used in Industry"*, B.S. 2690, London, 1968
11. A. T. Palin, *Analyst*, 1945, **70**, 203
12. A. T. Palin, *J. Amer. Water Wks. Assoc.*, 1957, **49**, 873; *Analyt. Abs.*, 1959, **6**, 757
13. A. T. Palin, *Baths Service*, 1958, **17**, 21
14. A. T. Palin, *Water and Water Eng.*, 1958, **62**, 30
15. A. T. Palin, *Archiv. des Badewesens*, 1958, August, 216
16. N. S. Chamberlin and J. R. Glass, *J. Amer. Water Wks. Assoc.*, 1943, **35**, 1065
17. Amer. Pub. Hlth. Assoc., *"Standard Methods for the Examination of Water and Waste Water"*, 11th edit., New York, 1960

18. F. J. Hallinan, *J. Amer. Water Wks. Assoc.,* 1944, **36,** 296
19. J. Holluta and H. Meissner, *Z. analyt. Chem.,* 1956, **152,** 112; *Analyt. Abs.,* 1957, **4,** 709
20. U. Hässelbarth, *Chem. Rundschau,* 1967, **48,** November 29
21. Statutory Instrument 1967/879, *"Carcinogenic Substances Regulations 1967",* H. M. Stationery Office, London, 1967
22. A. T. Palin, *J. Inst. Water Eng.,* 1967, **21,** 537
23. A. T. Palin, *Wat. Sewage Wks.,* 1969, **116** (Ref. No.), R51–52; *Analyt. Abs.,* 1970, **19,** 5259
24. H. A. Williams, *J. Assoc. publ. Analysts,* 1964, **2** (2), 34; *Analyt. Abs.,* 1966, **13,** 2050
25. A. T. Palin, *J. Assoc. publ. Analysts,* 1964, **2** (4), 106; *Analyt. Abs.,* 1966, **13,** 2051
26. N. J. Nicholson, *Analyst,* 1965, **90,** 187
27. J. D. Johnson and R. Oeverby, *Analyt. Chem.,* 1969, **41** (13), 1744; *Analyt. Abs.,* 1970, **19,** 4861
28. A. T. Palin, *Proc. Soc. Water Treatment Exam.,* 1960, **9,** 82
29. R. F. Packham, *Proc. Soc. Water Treatment Exam.,* 1958, **7,** 102
30. U. Hässelbarth, *Z. analyt. Chem.,* 1968, **22,** (1), 234

Chlorine Method 1
using neutral *o*-tolidine (Palin)

Principle of the method

The procedure consists in the treatment of the chlorine-containing solution with neutral *o*-tolidine, in the presence of stabilising agents, for the production of a blue coloration. Reaction with chloramine is activated by the addition of potassium iodide. The only interfering substance is manganese, and a method is given for making due allowance for this. Unlike the acid *o*-tolidine method, nitrites in this method are without effect.

Reagents required

1. *Buffer stabilising solution* Dissolve 4 g sodium hexametaphosphate ("Calgon") in about 80 ml distilled water, add 0.4 ml "Teepol 610" (Shell Chemicals Ltd.), mix, and make up to 100 ml with distilled water.

   Note—If subjected to low temperatures, this solution may become turbid, but may be clarified by warming. Cool afterwards to room temperature for use. Alternatively, the inclusion of 10% v/v acetone in the reagent will inhibit precipitation and not influence the reaction.

2. *Neutral o-tolidine solution* 0.1% Place 1 g *o*-tolidine (specially purified Reagent quality B.D.H.) in a mortar with 5 ml of 20% v/v HCl. Grind to a thin paste. Dissolve in about 500 ml distilled water, and then dilute with more distilled water to 1 litre. Store in an amber bottle. The water must be chlorine free.

   ***o*-tolidine is considered carcinogenic, and must be handled with all due precautions.**

3. *Potassium iodide (KI)* Crystals (analytical reagent grade)

The Standard Lovibond Comparator Discs 3/25A and 3/25B

3/25A   covers the range 0.1 to 1.0 parts per million chlorine in steps of 0.1, omitting 0.9. (This disc must be used in conjunction with 40 mm cells).

3/25B   covers the range 0.25 to 5.0 parts per million chlorine in steps 0.25, 0.5, 1.0, 1.5, 2.0, 2.5, 3.0, 4.0, 5.0. (This disc requires 13.5 mm cells or test tubes.)

The chlorine values obtained by the use of these discs are identical with those obtained by the F.A.S. titration method of Palin. The master discs against which all reproductions are checked were tested and approved by Dr. Palin.

Technique

Disc 3/25A Measure into the 40 mm cell 1.0 ml of reagent 1 followed by 1.0 ml of reagent 2 (*in that order*). Add the water sample up to 20 ml, mix and place at once in the right-hand compartment of the comparator, so that it comes behind the centre of the disc. In the left-hand compartment place a similar cell containing the water sample only. Hold the comparator facing a standard source of white light such as the Lovibond White Light Cabinet, or failing this north daylight, and revolve the disc until the nearest match is obtained. The figure then shown in the indicator window represents the parts per million of free chlorien present in the sample. The matching must be made immediately after mixing the solutions.

Disc 3/25B Measure into the 13.5 mm cell or test tube 0.5 ml of reagent 1 followed by 0.5 ml of reagent 2 (*in that order*). Add the water sample up to the 10 ml mark, mix, and place at once in the right-hand compartment of the comparator, so that it comes behind the centre of the disc. In the left-hand compartment place a similar cell or tube filled with the water sample. Match at once as above. The figure shown represents parts per million free chlorine as before.

Determination of Chloramines by these Discs

Monochloramine may be determined as follows:— continue the above test by adding a small crystal of potassium iodide (analytical reagent grade) to the right-hand cell and mix well. Evaluate the colour as previously, and deduct from this the reading first obtained as above. The result represents monochloramine in parts per million chlorine.

Dichloramine is not usually present in significant quantities if the pH of the water is above 7. If it is required to estimate it, proceed as follows:—

To the solution which has already been treated with reagents 1 and 2 and potassium iodide (see above) add 0.5 ml* of 0.5% v/v sulphuric acid (analytical reagent grade) and mix. Then add 0.5 ml* of 5% w/v sodium bicarbonate solution and again mix.

Evaluate this colour as previously, multiply this result by 1.1 to allow for the volume change, and deduct from this answer the result previously obtained with potassium iodide. The result represents parts per million chlorine as dichloramine.

Trichloramine. If nitrogen trichloride is present in the sample, which would be indicated by its distinct odour, it should be removed by extraction with one tenth its volume of carbon tetrachloride (analytical reagent grade).

Notes

1. The only interfering substance likely to be present in water is oxidised manganese. Its effect can be allowed for by developing the manganese colour in the "blank" as follows:—

Instead of using a "blank" cell, containing the sample only, in the left-hand compartment place 0.5 ml* of reagent 1 in a clean cell, add a crystal of potassium iodide and 1 drop* of a 0.5% solution of sodium arsenite ($NaAsO_2$). Make up to the 10 ml* mark with the sample, and mix. Add 0.5 ml* of reagent 2 and mix. Place this cell in the left-hand compartment as a blank: thus the colour due to manganese will have been developed equally in both fields, and cancels out.

2. All glassware used must be very thoroughly rinsed after making a test; this is particularly important, as only a trace of potassium iodide will cause a chloramine colour to develop.

The colour produced is reasonably stable at temperatures up to 80°F. Above this temperature, fading will occur and increase proportionately rapidly, but if the reading is taken immediately after mixing the reagents and sample, no significant error will be introduced.

3. If the solution becomes turbid on adding reagent 1 and 2, change the order to
    a reagent 2
    b sample
    c reagent 1

*these quantities refer to a 13.5 mm cell. For the 40 mm cell, double these figures.

COLORIMETRIC CHEMICAL ANALYTICAL METHODS

Chlorine Method 2
using acid *o*-tolidine

Principle of the method

For low concentrations of chlorine in potable and swimming waters, use is made of the reaction with *ortho*-tolidine, which gives a yellow colour in the presense of chlorine concentrations of the order of 5 parts per million and under. The same colour is given by nitrites in water, but if their presence is suspected a simple test is available. The yellow colour due to chlorine is destroyed by sodium thiosulphate (photographers' " hypo "), while that due to nitrites is not so destroyed. If, therefore, on the addition of a small crystal of " hypo " the yellow colour vanishes, this is confirmation that the colour was in fact due to chlorine. The test tube or cell must be carefully washed after each test, or subsequent tests would be spoilt by any residual " hypo."

Reagent required

o-tolidine	1 g
Hydrochloric acid (HCl)	100 ml
Redistilled water, zero chlorine demand, to produce	1000 ml

All chemicals should be of analytical reagent quality. (***o*-tolidine is considered carcinogenic, and must be handled with all due precautions.**)

The Standard Lovibond Comparator Discs

Comparator
3/2 A   0.1 to 1.0 parts per million chlorine (Cl) in steps of 0.1, omitting 0.9.
3/2 B   1.2 to 2.0   ,,   ,,   ,,   ,,   (5 standards only, 1.2, 1.4, 1.6, 1.8 and 2.0)
3/2 AB   0.15 to 2.0   ,,   ,,   ,,   ,,   (0.15, 0.25, 0.5, 0.75, 1.0, 1.25, 1.5, 1.75, 2.0)
KDD   0.1   0.3   0.6   0.9   1.1   parts per million chlorine. (5 standards only).
3/2 APC 1.0 to 5.0 parts per million chlorine in steps of 0.5. For use with 5 mm cells
3/2 APA 0.02 to 0.3   ,,   ,,   ,,   ,,   ⎱
3/2 APB 0.2 to 0.8   ,,   ,,   ,,   ,,   ⎰ For use only with 40 mm cells.

Nessleriser
NCA   0.01-0.09 parts per million chlorine
NCB   0.1 -0.5   ,,   ,,   ,,   ,,
NCAB   0.02-0.5   ,,   ,,   ,,   ,,

Technique

Fill both test vessels with the water to be tested and add the reagent to that in the right-hand compartment, so that the left-hand vessel contains a " blank " of the water behind the colour standard, to act as compensation for any inherent colour. Hold the comparator facing a standard source of white light such as the Lovibond White Light Cabinet or, failing this, north daylight, and rotate the disc until the colour of the liquid is matched by one of the glass standards. The value, expressed in parts per million, is then read from the indicator window.

The reagent is used in the proportion of 1 to 100. Mix the water and reagent and match immediately to obtain the figure for free chlorine. If the colour continues to develop (as indicated by a second reading at say 5 minutes) this indicates the presence of chloramines. In this case, take a reading after allowing to stand for 15-20 minutes, which indicates total residual chlorine. The difference between this and the original free chlorine reading indicates combined chlorine. Strongly alkaline water should be rendered acid, by the use of double the proper proportion of reagent, for full development of the colour.

COLORIMETRIC CHEMICAL ANALYTICAL METHODS — Chlorine Method 2

Discs 3/2 APA and 3/2 APB, must be used in conjunction with 40 mm cells. 20 ml of water are most conveniently employed, so 0.2 ml of reagent is required.

Discs 3/2A, 3/2B, 3/2AB and KDD Fill the test tubes to the 10 ml mark and add 0.1 ml of reagent.

Disc 3/2 APC must be used in conjunction with 5 mm cells. The cells hold 5 ml, so that it is more convenient to mix 10 ml of the water with 0.1 ml of reagent in another vessel, e.g. a test tube, and fill the right-hand cell with the mixture.

Discs NCA, NCB, and NCAB Use 0.5 ml reagent in the 50 ml sample in a Nessleriser cylinder.

Notes

1. It must be emphasized that the readings obtained by means of the Lovibond Nessleriser and disc are only accurate provided that Nessleriser glasses are used which conform to the specification employed when the discs were calibrated, namely that the 50 ml calibration mark shall fall at a height of 113 ± 3 mm measured internally.

2. For details of a Special Comparator for the limit testing of both chlorine and pH in water, apply to The Tintometer Ltd., Salisbury.

3. For high chlorine concentrations, up to 250 ppm, the potassium iodide test should be used.

Chlorine Method 3
using N,N-diethyl-*p*-phenylene diamine (Palin-DPD)

Principle of the method

Research in chlorine chemistry has resulted in the development of a very simple procedure for the differential determination of residual chlorine compounds in water. Depending upon the information required, it may be adapted to give total residual chlorine, or free and combined residual chlorine or complete separation into the free chlorine, mono-chloramine and dichloramine fractions. For the estimation of nitrogen trichloride, rarely required in practice, a simple supplementary procedure is provided. Differentiation, which with the new DPD indicator is remarkably clear-cut, into these various forms of residual chlorine is of the greatest importance in the control of modern chlorine processes of water treatment. With free chlorine the indicator gives a red colour. Subsequent addition of small amount of potassium iodide immediately causes monochloramine to produce a colour. Further addition of potassium iodide to provide a considerable excess evokes a rapid response from dichloramine. The colours produced require no stabilising agents and interference is suppressed by using EDTA as a chelating agent. With this method, correction of free chlorine readings for the presence of any nitrogen trichloride is unnecessary. When present, this compound is included with dichloramine, which in many of its properties it closely resembles.

Chlorine dioxide appears, to the extent of one-fifth of its total available chlorine content, with free available chlorine. A full response from chlorine dioxide, corresponding to its total available chlorine content, may be obtained if the sample is first acidified in the presence of iodide ion and subsequently brought back to an approximately neutral pH by addition of bicarbonate ion. Bromine, bromamine and iodine react with DPD indicator and appear with free available chlorine. DPD procedures for the determination of these halogens and related compounds have been developed, and are described in this book.

A novel feature of the Lovibond DPD methods lies in the use of compressed tablets which beside being far more convenient in use, permit of reagents being combined together to give procedures of exceptional simplicity.

In comparative studies by the Water Research Association (Tech. Papers Nos. 29:1963 47:1965 and 53:1966) it was concluded that the Palin-DPD procedure was the best method at present available for the determination of free chlorine and chloramines in water. This method is specified in British Standard 1427-:1962, Amendment 3:1968, and in German Standard Methods for Water Testing ("*Deutsche Einheits verfahren G4.6. Liefurung, Ansgabe 1971*").

Reagents required

Determination	DPD tablets
Total residual chlorine	No. 4 (or Nos. 1 & 3)
Free and combined chlorine	Nos. 1 & 3
Free chlorine, monochloramine and dichloramine (also nitrogen-trichloride).	Nos. 1, 2 & 3

Either Comparator tablets (for 10 ml sample volume) or Nessleriser tablets (for 50 ml sample volume) are required depending upon the instrument used.

The Standard Lovibond Discs

Disc	Range	Comparator — Chlorine ppm (mg/l) Steps
3/40A	0.1 to 1.0	0.1, 0.2, 0.3, 0.4, 0.5, 0.6, 0.7, 0.8, 1.0
3/40B	0.2 to 4.0	0.2, 0.4, 0.6, 1.0, 1.5, 2.0, 2.5, 3.0, 4.0
3/40C	0.25 to 3.0	0.25, 0.5, 1.0, 1.5, 2.0, 3.0
3/40D	0.5 to 6.0	0.5, 1.0, 2.0, 3.0, 4.0, 6.0
3/40E	0.02 to 0.3	0.02, 0.04, 0.06, 0.08, 0.10, 0.15, 0.2, 0.25, 0.3
3/40F	0.2 to 0.8	0.2, 0.25, 0.3, 0.35, 0.4, 0.5, 0.6, 0.7, 0.8
3/40H	1 to 10	1, 2, 3, 4, 5, 6, 7, 8, 9, 10
3/40K	0.5 to 6.0	0.5, 1.0, 1.5, 2.0, 2.5, 3.0, 4.0, 5.0, 6.0

COLORIMETRIC CHEMICAL ANALYTICAL METHODS — Chlorine Method 3

Disc 3/40C has three pH standards of values 7.0, 7.5 and 8.0, using diphenol purple.

Disc 3/40CZ has chlorine values 0.5, 1.0, 1.5, 2.0 and 4.0 with pH values of 7.0, 7.4, 7.6 and 8.0.

Disc 3/40D has the above three pH standards, using phenol red.

The discs in general are for use with 13.5 mm cells or test tubes except for—
3/40E and 3/40F — these require 40 mm cells.
3/40H — this requires a 5 mm cell and is a 10-hole disc.

Nessleriser – Chlorine ppm (mg/l).

NDPB — range 0.01 to 0.1 in steps of 0.01 omitting 0.09.
NDPC — range 0.02 to 0.2 in steps of 0.02 omitting 0.18.
NDP — range 0.05 to 0.5 in steps of 0.05 omitting 0.45.
NDPD — range 0.1 to 1.0 in steps of 0.1 omitting 0.9.

These discs are used with 50 ml tubes.

The chlorine values obtained by the use of these discs are identical with those obtained by the FAS titration method of Palin (*Standard Methods for the Examination of Water and Wastewater* 12th Edn., American Public Health Assoc. 1965). The master discs against which all reproductions are checked were tested and approved by Dr. Palin.

Technique

(a) Comparator, using discs 3/40 A, B, C, or D

For total residual chlorine

Place in the left-hand compartment, behind the colour standard of the disc, a 13.5 mm cell or test tube containing the sample only. Rinse a similar cell with the sample, and leave in it enough of the liquid just to cover the tablet when added. Drop into this prepared tube one No. 1 and one No. 3 tablet (or one No. 4 tablet, which is these two combined) and allow one or two minutes to disintegrate. Assist if necessary by crushing with a stirring rod. Then add the water sample up to the 10 ml mark, mix rapidly to dissolve the remains of the tablet, and place the cell in the right-hand compartment of the Comparator. Match after two minutes by holding the Comparator facing a standard source of white light, such as the Lovibond White Light Cabinet, or failing this, north daylight, and revolve the disc until the correct standard is found. The figure then shown in the indicator window represents parts per million of total residual chlorine present in the sample.

For differential estimation of free and combined Residual Chlorine

Prepare the tubes as above, one "blank" tube and one with just a few drops of the sample*, and to this tube add one No. 1 tablet only. After disintegration, make up to 10 ml, mix as before and match **at once.** This gives free residual chlorine. Then add one tablet No. 3, mix, and stand for two minutes. The colour then read off represents total residual chlorine. By deducting the first reading (free residual) the combined residual chlorine value is obtained.

For complete differentiation

Prepare the two tubes as above, and disintegrate one No. 1 tablet, make up to 10 ml and mix. Match **at once** and read off free residual chlorine as above (Reading No. 1).

Next add to the right-hand cell one tablet No. 2, mix vigorously to dissolve and match **at once** (Reading No. 2). From this deduct Reading No. 1 to obtain monochloramine value in parts per million.

Finally add one tablet No. 3, mix vigorously, stand for two minutes and match (Reading No. 3). From this deduct Reading No. 2 to obtain dichloramine value.

Nitrogen trichloride

Nitrogen trichloride, whose presence in the water would be indicated by its distinct odour, normally gives a colour, equivalent to half its available chlorine content, in the dichloramine fraction of the differential test. It may be caused to produce colour, corresponding to the remaining half, in the free chlorine fraction if the sample is first treated with potassium iodide. Using this as a basis, the following method for its estimation has been evolved.

* If monochloramine content is high, differentiation is improved by using one or two drops of distilled water instead of the sample for disintegrating the tablet, or the tablet may be crushed without any added water.

COLORIMETRIC CHEMICAL ANALYTICAL METHODS — Chlorine Method 3

Rinse the right-hand cell in the manner prescribed above. Add one DPD tablet No. 1 and allow to disintegrate, assisted if necessary by crushing. To 10 ml of the water sample in a separate tube add one DPD tablet No. 2 and mix vigorously to dissolve. Then add the 10 ml of sample so treated to the right-hand cell containing the remains of tablet No. 1 and mix rapidly to dissolve. With the left-hand cell containing the water sample only, evaluate the colour as before. From the reading thus obtained deduct the free chlorine reading (and the monochloramine, if any) and multiply by two. The result corresponds to nitrogen trichloride in parts per million chlorine. The next step is to subtract half of this result from the dichloramine reading as given by the normal differential procedure. The purpose of this is to correct the dichloramine reading, for that half of the nitrogen trichloride which appears with it. The total residual chlorine is then the sum of the various constituents. It is emphasized, however, that in practice it will usually be quite adequate to include any nitrogen trichloride with dichloramine and not to perform the further differentiation.

Using discs 3/40E and F – Measure 20 ml of sample. Add a few drops of sample to the 40 mm cell, then two DPD No. 1 Comparator-type tablets. Allow to disintegrate or crush and add remainder of sample. Mix to dissolve. For the chloramines follow with No. 2 and No. 3 tablets as in the standard procedure except that two of each are used instead of one.

If only simple free/combined differentiation is required omit the use of the No. 2 tablets.

If only total residual chlorine is required use two DPD No. 4 (Comparator-type) tablets per 20 ml sample or two each of No. 1 and No. 3 tablets.

Using disc 3/40H – The same general procedure is followed except that one No. 1 tablet (Comparator-type) is added to 4 ml of sample contained in a separate tube. A 13.5 mm moulded cell may conveniently be used for this purpose. After dissolving the tablet in the usual way the 5 mm cell is filled from this tube. Alternatively use one No. 1 tablet direct in the 5 mm cell with one or two drops of sample. Allow to disintegrate as before then fill cell with sample and dissolve.

Thereafter the subsequent tablets are added direct to the 5 mm cell, using one each of No. 2 and No. 3. Omit No. 2 as before if only free and combined chlorine are required.

When using DPD No. 4 (Comparator-type) tablets these are used in a similar manner to the DPD No. 1 tablets, that is to say one is used per 4 ml sample in a separate tube or one is used direct in the 5 mm cell.

(b) **Nessleriser**

The instructions given for the Comparator should be followed exactly except that 50 ml should be substituted for 10 ml and the special Nessleriser DPD tablets should be used.

Notes

1. When using excess potassium iodide as in the No. 3 tablet any copper present in the water sample would be able to produce a colour with the DPD indicator. This interference (up to approx. 10 ppm Cu) is prevented by the EDTA incorporated in the tablets. In the simple test for total residual chlorine, where potassium iodide is used from the start, very high concentrations of copper may give a colour until sufficient EDTA has dissolved for complete chelation. Such colours however, are of a transitory nature and disappear within the two minute period of the test. With the differential tests any copper present is, of course, chelated before the No. 3 tablet is added and cannot then produce colour.

2. Dissolved oxygen in water can produce a faint colour with the reagent if allowed to stand. The suppression of trace metal catalysis by the EDTA minimises this effect and there is no interference within the period of the test.

3. The only interfering substance likely to be present in water is oxidised manganese. Its effect can be allowed for by developing the manganese colour in the "blank" as follows:— To 10 ml of sample in a separate test tube add one DPD tablet No. 2 and 1 drop of 0.5% sodium arsenite ($NaAsO_2$) solution. Mix to dissolve. Rinse the Comparator cell or test tube as before and add 1 DPD tablet No. 1. Disintegrate and then add the 10 ml of arsenite-treated sample. Mix to dissolve remains of tablet and then place in left-hand side of Comparator. In this way the colour due to manganese will have been developed equally in both fields and thus cancels out. The same procedure may be used for the Nessleriser, substituting 50 ml of sample for the 10 ml. One drop of sodium arsenite is sufficient.

COLORIMETRIC CHEMICAL ANALYTICAL METHODS — Chlorine Method 3

4. All glassware used must be very thoroughly rinsed after making a test, since only a trace of potassium iodide will cause chloramine colour to develop. For the same reason handling the tablets, particularly DPD No. 1, should be avoided.

5. For maximum protection under the most extreme conditions all tablets are available individually packed in strips.

6. The quantity of indicator used in the tablets has been chosen to suit the range of chlorine concentrations covered by the discs. Samples containing higher concentrations **must** be diluted except when using disc 3/40H and the appropriate technique. Concentrations of chlorine above 8 ppm may entirely bleach the colour and give an apparently zero reading, but at this concentration the smell of chlorine would be very apparent. If there is any doubt about the need for dilution a simple check is to repeat the procedure using two DPD No. 1 (or No. 4 tablets as the case may be) instead of one. A very decided increase in colour would indicate dilution to be necessary, in which case the requisite amount of distilled water is added first to the reagent followed by the measured amount of sample.

7. For details of special portable sets for use with this test, apply to The Tintometer Limited, Salisbury.

8. It must be emphasised that the readings obtained by means of the Lovibond Nessleriser and disc are only accurate provided that Nessleriser glasses are used which conform to the specification employed when the discs were calibrated, namely that the 50 ml calibration mark shall fall at a height of 113 ± 3 mm measured internally.

9. Nos. 1 and 4 tablets disintegrate within one or two minutes when wetted. If it is desired to shorten the time taken for the tablets to disintegrate, they may be crushed with a clean glass or plastic rod. Tablets No. 2 and 3 dissolve quite readily by simple mixing.

Chlorine Method 4
using potassium iodide

Principle of the method
The method is based on the reaction of chlorine with potassium iodide in acid solution to liberate iodine. The colour of the iodine is matched against Lovibond permanent glass standards.

Reagents required
1. *Potassium iodide (KI)* 5 grain tablets (300 mg)
2. *Acetic acid* (CH_3COOH) 5%

The Standard Lovibond Comparator Discs 3/2 IOD, 3/2 ARP and 3/2 APH
3/2 APH. 2.0 to 10.0 parts per million chlorine (Cl) in steps of 1.0
3/2 ARP. 5.0 to 50.0   „   „   „   „   „ „ „ „ 5.0 omitting 45
3/2 IOD. 5.0 to 250   „   „   „   „   „ (5, 10, 25, 50, 75, 100, 150, 200, 250)
Disc 3/2 APH must be used with a 40 mm cell.

Technique
Fill both test vessels with the water to be tested and add the reagent to that in the right-hand compartment, so that the left-hand vessel contains a "blank" of the water behind the colour standard, to act as compensation for any inherent colour. Hold the comparator facing a standard source of white light such as the Lovibond White Light Cabinet or, failing this, north daylight, and rotate the disc until the colour of the liquid is matched by one of the glass standards. The value, expressed in parts per million, is then read from the indicator window.

Discs 3/2 IOD, 3/2 ARP. Fill the test tubes to the 10 ml mark, add 1 potassium iodide tablet to the right-hand tube, mix and match at once. Should the water be alkaline, add 0.5 ml of 5% acetic acid solution to render the water faintly acid. This is not critical, so long as the solution is acid to litmus paper. Nitrites could interfere if the pH is below 6.0, but the use of chlorine in such quantity as to give residuals falling within the range of these discs would oxidise any nitrites originally present, so that in practice such interference is unlikely to arise.

Disc 3/2 APH. To 50 ml sample add 1 ml of 5% acetic acid, or enough to render faintly acid to litmus paper, and 1 tablet of potassium iodide. Mix and transfer to a 40 mm cell. Match at once.

Chlorine Method 5

using the A.P.H.A. acid *o*-tolidine method

Principle of the method

In acid solution chlorine reacts with *o*-tolidine giving a yellow holoquinone colour. The intensity of this colour is proportional to the chlorine concentration, which is determined by comparing the colour with a series of Lovibond permanent glass standards.

This method measures both free and combined chlorine. If it is required to estimate the free and combined chlorine separately, an approximate differentiation may be obtained by using Hallinan's *o*-tolidine arsenite (OTA) method. This also provides a correction for interfering substances.

Where for any reason a more accurate estimation of the concentrations of free and combined chlorine is required, one of the alternative methods developed by Palin should be used (*Methods 1* and *3*).

Reagents required

1. *A.P.H.A. o-tolidine reagent*

o-tolidine	1 g*
Hydrochloric acid, (HCl)	150 ml
Redistilled water, zero chlorine demand, to produce	1000 ml

*Alternatively *o*-tolidine dihydrochloride may be used, in which case 1.35 g per litre of solution should be added.

_o_-tolidine is considered carcinogenic, and must be handled with all due precautions.

2. *Sodium arsenite solution*

Sodium arsenite, (NaAsO$_2$)	5 g
Redistilled water, zero chlorine demand, to produce	1000 ml

Caution. Sodium Arsenite is TOXIC. TAKE CARE TO AVOID INGESTION.
All chemicals should be of analytical reagent quality.

The Standard Lovibond Comparator Disc 3/47

This disc covers the range 0.1 to 1.0 ppm of chlorine in steps of 0.1. The colours of the 9 standards in this disc exactly match the modified Scott chromate-dichromate colour solutions as specified in the A.P.H.A. Standards. The calibration of the disc has been checked and approved by Dr. A. T. Palin.

Technique

Total residual chlorine

Fill a 10 ml comparator cell or test tube with the water to be tested. Place this filled cell in the left-hand compartment of the comparator to provide a blank and thus compensate for any inherent colour in the sample. To an identical cell or test tube add 0.5 ml of the *o*-tolidine solution (reagent 1), fill to the mark with sample, mix, and place in the right-hand compartment of the comparator.

Hold the comparator facing a standard source of white light, such as the Lovibond White Light Cabinet or, failing this, north daylight wherever possible, and rotate the disc until the colour of the liquid is matched by one of the Lovibond permanent glass standards. Matching should be carried out at the time of maximum colour development, but in any case no longer than 5 minutes after adding the reagent.

The concentration of chlorine in the sample, expressed in ppm is then read from the indicator window.

COLORIMETRIC CHEMICAL ANALYTICAL METHODS — Chlorine Method 5

OTA method for free and combined chlorine

To a 10 ml cell or test tube add 0.5 ml of *o*-tolidine solution (reagent 1), fill to the mark with sample and mix quickly. Immediately (within 5 seconds) add 0.5 ml of arsenite solution (reagent 2), mix quickly and match against the colour standards as before. Record this value as reading **A**.

To a second cell or test tube add 0.5 ml of arsenite solution (reagent 2), fill to the mark with sample and mix quickly. **Immediately** add 0.5 ml of *o*-tolidine solution (reagent 1), mix quickly and match against the colour standards as quickly as possible. Record this value as reading **B**. Match again in exactly 5 minutes. Record this value as reading **C**.

To a third cell or test tube add 0.5 ml of *o*-tolidine solution (reagent 1) and fill to the mark with sample. Mix quickly and match against the standards in exactly 5 minutes. Record this value as reading **D**. This represents the total amount of residual chlorine present plus the total amount of interfering colours.

The individual components can now be calculated as follows:—

Total residual chlorine = **D − C**
Free available chlorine = **A − B**
Combined available chlorine = Total minus Free
 = **B + D − (A + C)**.

Notes

1. If it is only necessary to ensure that the level of the residual chlorine falls within the recommended limits it is possible to use a simplified procedure. Details of a special comparator, designed for the limit testing of both chlorine and pH in water, may be obtained from Tintometer Ltd., Salisbury, England.

2. If for any reason it becomes necessary to determine higher concentrations of residual chlorine, this can be accomplished by diluting the sample with distilled water of zero chlorine demand. For example by diluting 1 ml of sample to 250 ml and then using the diluted sample in the test, the disc readings will be multiplied by 250 and then cover the range 25-250 ppm of chlorine.

To prepare zero chlorine demand water, add sufficient chlorine, to ammonia-free distilled water, to maintain a residual chlorine concentration of 0.5-1.0 ppm for 30 minutes. Then boil for 30 minutes and use within 48 hours of preparation.

The Determination of Chlorine Dioxide

and differential determination of chlorine dioxide and other forms of residual chlorine.

Palin[1] has introduced glycine as an improved reagent for these differential determinations. The glycine is now available in tablet form for both the Comparator and the Nessleriser. It is necessary to specify which type of tablet is required.

The glycine replaces the malonic acid as used in the original method for separating chlorine dioxide from free chlorine. This separation using glycine is based upon converting the free chlorine to monochloraminoacetic acid which no longer reacts in the first stage of the DPD test. The reaction is rapid and it is not necessary to wait the 2 minutes as previously specified with malonic acid.

Reagents required
DPD No. 1 tablets
DPD No. 3 tablets
DPD Glycine tablets
DPD Acidifying tablets } required only for Part 3
DPD Neutralising tablets

In each case order Comparator or Nessleriser tablets to conform to the disc in use.

Equipment required
Lovibond Comparator with two 13.5 mm moulded test tubes or Lovibond Nessleriser Mk. III with two 50 ml Nessler Cylinders.

The Standard Lovibond Discs
Comparator disc 3/40A covers the range 0.1 to 1.0 ppm Chlorine (Cl)
Comparator disc 3/40B covers the range 0.2 to 4.0 ppm Chlorine (Cl)
Nessleriser disc NDP covers the range 0.05 to 0.5 ppm Chlorine (Cl)
Nessleriser disc NDPC covers the range 0.02 to 0.2 ppm Chlorine (Cl)

The master discs against which all reproductions are checked, were tested and approved by Dr. A. T. Palin.

Procedure
Part 1 is for chlorine dioxide only.
Part 2 also provides for free and combined chlorine.
Part 3 provides for all the above plus any chlorite.

The presence of chlorite may result from incomplete reaction initially or from subsequent reduction of chlorine dioxide during the course of the treatment process.

Procedure – Part 1
(1) Rinse a 10 ml cell with sample leaving in one or two drops. Add to cell one DPD No. 1 tablet and allow to disintegrate or crush.
(2) Measure 10 ml of sample in a second cell and to this add one glycine tablet. Crush and dissolve with stirring rod.
Then add contents of this second cell to the prepared first cell.
(3) Mix and immediately match colour against disc. (Reading G).

Procedure – Part 2

(1) Rinse a 10 ml cell with sample leaving in one or two drops. Add to cell one DPD No. 1 tablet and allow to disintegrate or crush.

(2) To this cell add sample to mark, mix and immediately match colour against disc. (Reading A).

(3) Continue by adding to the same cell one DPD No. 3 tablet. Mix to dissolve and after approximately two minutes match colour against disc. (Reading C).

Procedure – Part 3

(This is required only if it is desired to test for the presence of chlorite).

(1) After taking Reading C as above add to the same colour mixture one Acidifying tablet. Mix to dissolve and stand for approximately two minutes.

(2) To this cell add one Neutralising tablet. Mix to dissolve and then match colour against disc. (Reading D).

Calculations

Results in ppm (mg/l) as Cl.

In the Absence of Chlorite

Chlorine Dioxide $= 5G$
Free Chlorine $= A - G$
Combined Chlorine $= C - A$
Total Available Chlorine $= C + 4G$

To express chlorine dioxide in terms of ClO_2 instead of as available chlorine multiply G by 1.9.

Reading D is required for the determination of chlorite. If D is greater than $C+4G$ then chlorite is present.

In the Presence of Chlorite

Chlorine Dioxide $= 5G$
Chlorite $= D - (C+4G)$
Free Chlorine $= A - G$
Combined Chlorine $= C - A$
Total Available Chlorine $= D$

Notes

1. It has been shown[2] that it is not possible to have both chlorine dioxide and dichloramine present together because they react with one another. Thus any combined chlorine found in the presence of chlorine dioxide may be assumed to be monochloramine.

2. In using the Lovibond Nessleriser with 50 ml sample volumes the corresponding Nessleriser type of tablets are required.

References

1. A. T. Palin, *Jour. Inst. Water Eng.*, 1974, **28,** 139.
2. C. B. Adams, J. M. Carter, J. W. Ogleby, *Proc. Soc. Water Treat. and Exam.*, 1966, **15,** 117.

The Determination of Chromium

Introduction
Chromium is widely used as an alloying element in metallurgy, for plating, for the tanning of leather and in the textile industry. Its determination is of interest in all of these applications and also in the control of industrial wastes and effluents. Chromate is added to cooling systems, in regions where the water supply is acidic, as a corrosion inhibitor. Its concentration in this application is usually controlled directly by the intensity of the chromate colour.

Methods of sample preparation
Air[1]
Draw the polluted air through an impinger type bubbler, containing 10 ml of normal caustic soda (NaOH) solution, at 100 litres an hour. Oxidise any Cr^{III} to Cr^V with bromine, remove excess bromine by boiling, and make the solution up to standard volume. See also this book "Chromic acid mist" page 485.

Beverages[2-4]
Beer[2,3] Evaporate 100 ml to dryness in a 200 ml beaker. Place the residue in an oven at 200°C, raise the temperature to 500°C, and ash the sample overnight. Cool the ash and moisten it with 2-3 ml of nitric acid. Re-evaporate and ash for 2 hours at 500°C. Repeat until no carbon remains. Treat the ash with 2 ml of dilute sulphuric acid and a few drops of water, filter and wash. Evaporate the filtrate and washings to a small volume on a water-bath and transfer to a 25 ml flask. Treat with 30% caustic soda until a slight opalescence appears, add 5 ml of sodium bromate ($NaBrO_3$) solution and heat for 1 hour on a water-bath. Cool, add 5 ml of 1.2% phenol (C_6H_5OH) solution to remove excess bromine and then add diphenylcarbazide(($C_6H_5NH.NH)_2CO$) reagent. Proceed as in *Method 2b*.

Whisky[4] Take a 10 ml sample and destroy the organic matter by digestion with sulphuric and nitric acids. When colourless neutralise and dilute to 25 ml with distilled water. Proceed as in *Method 2b*.

Biological materials[5,6]
Blood Digest 5 ml sample with 1 ml of sulphuric acid, 2 ml of perchloric acid ($HClO_4$) and 0.5 ml of nitric acid. When the digest darkens in colour, remove the heat and add a few drops of nitric acid, after the reaction subsides repeat the addition of nitric acid until there is no further reaction. Boil off excess perchloric acid, add 100 ml of water, boil, cool, and make up to volume.

Tissue Take 1 g of sample and proceed in the same manner as described for blood except add 2 ml of nitric acid immediately rather than gradually.

Chemicals[7]
Organic compounds Digest 1 g of sample with 1 ml of sulphuric acid, 5 ml of nitric acid and 1 ml of a 5% solution of sodium chloride. Determine iron, nickel and chromium on aliquots of this solution.

Foods and edible oils[8,9]
General[8] Ash the sample, fuse the residue with sodium carbonate (Na_2CO_3) and dissolve the melt in acid.

Preserves[9] Heat 1 g of sample, containing not more than 100 μg of chromium, in a platinum crucible at 550°C. Dissolve the residue in 5 ml of hot nitric acid, add 4 ml of 25% sulphuric acid, transfer the solution to a beaker with 50 ml of water, add 3% potassium permanganate ($KMnO_4$) solution one drop at a time, until a pink colour persists when the solution is swirled. Boil for 5 minutes and then set aside for 5 minutes. Neutralise the solution to litmus with 25% caustic potash (KOH) solution add 0.5 g of sodium peroxide (Na_2O_2), boil gently for 10 minutes, cool and make up to 100 ml.

Fuels and lubricants[10]
Solid fuel ash Fuse 1 g of sample with 6 g of sodium carbonate, extract the melt with hot water, add a few drops of alcohol (C_2H_5OH), to reduce any manganese which is present, filter and wash the precipitate ten times with 1% sodium carbonate solution. Combine the filtrate and washings and dilute to 100 ml.

Metals, alloys and plating baths[11-19]

Alloys[11,12] Oxidise the sample by fusion with sodium peroxide (Na_2O_2). Cool the melt and dissolve it in sulphuric acid. Heat the solution with ethanol (C_2H_5OH) until the acetaldehyde, formed during the reduction of chromate, has been decomposed. Dilute the acid solution and heat with a small excess of aqueous ammonia. Filter off the precipitated hydroxides. Dissolve the precipitate in sulphuric acid and treat with ammonium persulphate (($NH_4)_2S_2O_8$). Cool the solution and separate iron from chromium by passing through a column of KU-2 ion exchange resin[11].

Iron and steel[12-19] Weigh 1 g sample into a 400 ml flask and dissolve in 30 ml of 15% sulphuric acid. Oxidise the solution with the minimum amount of nitric acid and digest until red fumes no longer appear. Dilute to approximately 100 ml, heat to boiling and add a few drops of 1% potassium permanganate ($KMnO_4$) solution until the pink colour persists after boiling for 2 minutes. Cool slightly and add 25 ml of 25% caustic soda solution followed by 1 g of sodium peroxide (Na_2O_2). Reheat to boiling and boil for 3–4 minutes. Make up the solution, together with the precipitate, to exactly 500 ml with cold water and use an aliquot after filtration[14]. Alternatively[13], for high concentrations of chromium, dissolve the sample in a mixture of sulphuric and nitric acids, boil until no further red fumes are evolved and dilute the solution until the sulphuric acid concentration is about 20%. Measure the chromate colour using *Method 3*. Volke[17] recommends the following method for the rapid analysis of cast iron:— dissolve 0.2 g of sample in 50 ml of sulphuric acid. Add 1 g of ammonium persulphate ($(NH_4)_2S_2O_8$), 2 ml of 2% potassium permanganate solution and 50 ml of hot water. Heat to boiling, filter, cool and dilute the solution to 500 ml.

For the analysis of high purity iron[16,18,19], dissolve 0.1 g of sample in 10 ml of hydrochloric acid and 3 ml of nitric acid, heat almost to dryness and then dissolve the residue in 25 ml of 40% hydrochloric acid and shake the solution with 25 ml of isobutyl methyl ketone (i-$C_4H_9COCH_3$)— pentyl acetate ($CH_3COOC_5H_{11}$) (2:1) for 1 minute to remove iron. Evaporate the aqueous layer with 5 ml of nitric acid to dryness, dissolve the residue in 10 ml of 1.4% sulphuric acid, add 1 ml of 0.1% potassium permanganate ($KMnO_4$) solution and boil for 3 minutes. Shake the solution with 10 ml of 20% aqueous urea ($NH_2.CO.NH_2$) and sufficient sodium nitrite ($NaNO_2$) to produce a colourless solution and dilute to 25 ml.

Minerals, ores and rocks[11,12,20-22]

General[11,12,20] Fuse a 1 g sample with 8 g of sodium peroxide (Na_2O_2) in an iron crucible. Extract the melt with 100–150 ml of water and destroy the excess peroxide by boiling. Add 10–15 g of ammonium carbonate (($NH_4)_2CO_3$) and filter. Acidify the filtrate with acetic acid (CH_3COOH) or hydrochloric acid and add 10 ml of 0.5M barium chloride ($BaCl_2$) solution. Heat for 1 hour on a steam-bath. Filter through a Gooch crucible and wash the precipitate five times with ethanol-water (C_2H_5OH-H_2O,1:9). Dissolve the precipitate in 100 ml of hydrochloric acid (1:1) containing 2 g of potassium iodide (KI) and precipitate the chromium with aqueous ammonia (NH_4OH). Filter, dissolve the precipitate in sulphuric acid and determine chromium by *Method 2*.

Ilmenite[21] Grind sample to pass through 200 mesh and then fuse 1 g with 15 g of previously fused and cooled sodium bisulphate ($NaHSO_4$). Dissolve the melt in 50 ml of 2M sulphuric acid, add 30 ml of water, cool, filter and wash the residue with 0.5M sulphuric acid. Ignite the washed residue, treat with hydrofluoric (HF) and sulphuric acids, fuse any residue with 0.5 g of sodium bisulphate, leach the melt with 2M sulphuric acid add the leachings to the combined filtrate and washings and make up to 200 ml.

Titanomagnetite concentrate[22] Fuse 0.4 g of sample in an iron crucible with 7–8 g of sodium peroxide (Na_2O_2). Extract the melt with 100 ml of water, cool, dilute solution and residue to 250 ml, filter and determine chromium on an aliquot of this filtrate.

Miscellaneous[23-28]

Glass[23] Dissolve the sample in hydrofluoric acid (HF) solution in a platinum crucible. Add an equal volume of water and then sulphuric acid (1:1). Boil until the solution is clear. Add 0.1% silver nitrate ($AgNO_3$) solution, 0.05% manganese sulphate ($MnSO_4$) solution and saturated ammonium persulphate (($NH_4)_2S_2O_8$). Boil until no more oxygen is evolved and the determine chromium by *Method 2*.

Leather[24] Ash 2–5 g of sample, add 5 ml of sulphuric acid, 5 ml of nitric acid and 10 ml of 65% perchloric acid ($HClO_4$). Heat until the solution begins to turn orange in colour, continue until no further change occurs and then for a further 2 minutes. Cool rapidly, dilute to 200 ml and boil for 7–10 minutes to expel chlorine. Proceed by *Method 2*.

Maser crystals[25] Powder the sample by dropping red-hot portions into water. Grind in a steel or tungsten mortar, sift through silk or nylon and remove any contamination from the mortar by treatment with either aqua regia (1 volume of nitric acid plus 3 volumes of hydrochloric acid) if a steel mortar was used or a mixture of hydrofluoric and nitric acids if tungsten was used. Fuse the powdered sample with sodium carbonate (Na_2CO_3) and sodium tetraborate ($Na_2B_4O_7$), dissolve the cooled melt in $0.4N$ sulphuric acid, oxidise by boiling with potassium permanganate ($KMnO_4$) and determine chromium on this solution.

Tanning liquors[26] Evaporate the solution to dryness and treat the residue repeatedly with 1 ml of hydrogen peroxide (H_2O_2) and 1 ml of sulphuric acid until all the organic matter is destroyed. Dilute the hot acid solution to 10 ml and add 0.15 g of potassium sulphate (K_2SO_4). Cool and add 0.25–0.5 g of potassium thiocyanate (KSCN). Warm at 80°C for 1 hour. On cooling the solution should be violet in colour. If it is still green add more thiocyanate and warm again. Remove the iron by adsorption on a column of Wofatit P. Precipitate the chromium in the effluent with aqueous ammonia (NH_4OH), filter, dissolve the precipitate in sulphuric acid and proceed by *Method 2*.

Wood[27] Dry sawdust at 100–105°C to constant weight. Take 1 g of the dried sample in a 250 ml graduated flask, add 50 ml of $5N$ sulphuric acid and 10 ml of 100 volume hydrogen peroxide (H_2O_2) and heat on a water-bath at 70–75°C for 20 minutes with occasional swirling to keep the contents well mixed. Remove from the water-bath, add 150 ml of water, swirl, cool to room temperature, dilute to the mark and mix. Mix twice at 5 minute intervals and allow to settle for 10 minutes. Filter 10 ml off the supernatant liquid through a dry 7 cm Whatman No. 44 filter paper and discard the first 5 ml. Determine chromium on a 1 ml aliquot.

Wool and textiles[28] Either (a) wet ash the sample and then heat further in a Pyrex flask under reflux with 10 ml excess of perchloric acid until the colour changes to orange. Immerse flask in cold water and dilute the contents with 30 ml of water, boil the solution to expel chlorine and make up to 100 ml with M sulphuric acid and proceed by *Method 2 or 3*. Or (b) dry ash the sample, heat the ash on a steam-bath with 0.3 g of potassium chlorate ($KClO_4$) and 5 ml of nitric acid in a crucible covered with a watch-glass. When effervescence has ceased, evaporate to dryness, dissolve the residue in 5 ml of water and again evaporate. Dissolve the residue and make up to 100 ml with M sulphuric acid and proceed by *Method 2 or 3*.

Plants[29]

Ignite the sample and fuse the ash with potassium pyrosulphate ($K_2S_2O_7$). Dissolve the melt in hydrochloric acid, add 2 ml of 5% aluminium sulphate ($Al_2(SO_4)_3$) solution and coprecipitate chromium and aluminium. Remove iron and vanadium by extraction with 5% cupferron ($C_6H_5N(NO).ONH_4$) in carbon tetrachloride (CCl_4) from $1.5N$ sulphuric acid solution. Treat the aqueous layer with nitric acid to decompose the cupferron. Reprecipitate aluminium and chromium and filter. Dissolve the precipitate in 50 ml of $1.3N$ sulphuric acid. Oxidise with 30 mg of sodium bismuthate (Na_2BiO_2) and 3 drops of 10% silver nitrate ($AgNO_3$) solution. Determine chromium by *Method 2*.

Plastics and polymers[30]

Boil the sample for 20 minutes in 40 ml of hydrochloric acid potassium chloride (KCl) solution (3.7 ml of 10% hydrochloric acid+3.7 g of potassium chloride per litre) containing EDTA. Uncombined chromium passes into solution and is determined by *Method 2*.

Sewage and industrial wastes[31–33]

Industrial wastes[31] Take 100 ml of the effluent in a 250 ml Kjeldahl flask, add 0.1 g of sodium sulphite (Na_2SO_3) and 2 ml of sulphuric acid. Evaporate until white fumes are evolved. If necessary add nitric acid drop by drop until all the organic matter is oxidised. Boil off the red fumes. Add 10 ml of saturated ammonium oxalate (($COONH_4)_2.H_2O$) and again evaporate until white fumes are evolved. Cool. Dilute with 10 ml of water and transfer to a 25 ml calibrated flask and dilute to the mark. Mix well, take a 5 ml aliquot in a beaker, add 5 drops of phosphoric acid (H_3PO_4) and evaporate to white fumes. Cool, add 1 ml of 1% potassium permanganate ($KMnO_4$) solution, cover with a watch-glass and heat on a water-bath for 20 minutes. Neutralise to litmus with 15% caustic soda (NaOH) and then add 1 ml in excess. Add 2 ml of 3% hydrogen peroxide (H_2O_2) and simmer on a hot plate for 10 minutes. Cool. Dilute to 20 ml in a calibrated flask and filter. Take an accurate aliquot of the filtrate into a 25 ml volumetric flask, add 5 ml of 5% sulphuric acid, dilute to about 20 ml add diphenylcarbazide (($C_6H_5.NH.NH)_2CO$) solution, dilute to the mark and measure.

Sewage[32,33] To remove organic matter take a suitable volume of sample (to contain no more than 10 or 100μg of chromium, depending on whether a Nesslerizer or a Comparator is being used) in a beaker, add 5 ml of nitric acid and evaporate to small volume. Transfer the concentrated solution to a porcelain or silica basin and reduce the volume to about 15 ml on a water-bath. Add 2 ml of 60% perchloric acid ($HClO_4$) and evaporate to dryness. Remove the source of heat as soon as the sample becomes dry. Transfer to a furnace and ignite at just below 300°C for 40 minutes. Dissolve the residue in 2 ml of 5N hydrochloric acid plus a little distilled water. Warm gently. If the solution is clear, transfer to a 50 ml graduated flask and make up to the mark. If the solution is brownish then the process of addition of nitric and perchloric acids followed by evaporation to dryness, ignition and solution should be repeated until a colourless product is obtained. If a white precipitate or turbidity is obtained, filter through a porosity 4 sintered-glass filter and make filtrate up to 50 ml. Proceed using *Method 2*.

Soil[20,34,35]

Clays[34] Fuse the sample with sodium carbonate (Na_2CO_3), dissolve the melt in acid and make up to volume.

General[20,35] Calcine 1 g of soil for 2 hours at 450–500°C, fuse the residue with 3 g of potassium carbonate (K_2CO_3), dissolve the melt in 25 ml of hot water to which has been added 0.5 ml of 10% hydrochloric acid. Evaporate this solution to dryness. Add 5 ml of 10% hydrochloric acid and repeat the evaporation. Dry the residue at 120°C. Treat with 50 ml of water and 3 ml of 5N sulphuric acid. Filter off any insoluble material, wash the residue with hot water, combine the filtrate and the washings, neutralise to litmus with normal sulphuric acid and dilute to 100 ml.

Water[36,37]

Potable water[36] Use bromine-caustic soda solution (Br-NaOH) or potassium permanganate ($KMnO_4$) to oxidise Cr^{III} to Cr^V before the addition of diphenylcarbazide. Perchloric acid ($HClO_4$) is not recommended as the oxidising agent as it tends to give low results.

Sea water[37] Filter 2 litres of sample, acidify and then heat to 55°C with ferric chloride ($FeCl_3$) at pH 7.5 (borate buffer) and set aside overnight. Filter, dissolve the precipitate in hydrochloric acid and pass the solution through a column of DeAcidite FF ion exchange resin. Wash the column with hydrochloric acid and evaporate the percolate and washings with aqueous hydrogen peroxide (H_2O_2). Heat the residue with ceric sulphate ($Ce(SO_4)_2.4H_2O$) solution and sulphuric acid, remove the ceric salts by addition of sodium azide (NaN_3) and make the solution up to volume.

The individual tests

Method 1 was devised[14] for the determination of Cr in steel and uses a Lovibond Tintometer with calibration curves prepared from standard samples.

Method 2 is an adaptation of *Method 1* for Comparator or Nessleriser and was originally developed[32] for the determination of chromate and total chromium in sewage and sewage effluents, for which it is a standard procedure.

Method 3 uses the colour of the chromate ion directly and is suitable for the determination of chromate in the chromate treatment of industrial waters for corrosion inhibition.

References

1. W. Pilz, *Z. analyt. Chem.*, 1966, **219** (4), 350; *Analyt. Abs.*, 1967, **14**, 7161
2. A. R. Deschreider and R. Meaux, *Rev. Ferment.*, 1962, **17**, 73; *Analyt. Abs.*, 1963, **10**, 1584

3. A. R. Deschreider and R. Meaux, *Brass. et Malt.*, 1962, **12,** 387; *Analyt. Abs.*, 1964, **11,** 376
4. M. J. Pro and R. A. Nelson, *J. Assoc. off. agric. Chem.*, 1956 **39,**8 48; *Analyt. Abs.*, 1957, **4,** 703
5. V. D. Yablochkin, *Sudebno-Med. Ekspertiza*, 1963, **6,** 45; *Analyt. Abs.*, 1965, **12,** 1285
6. C. Y. Gooderson and F. J. Salt, *Lab. Pract.*, 1968, **17** (8), 921; *Analyt. Abs.*, 1969, **17,** 3617
7. D. Monnier, W. Haerdi and E. Martin, *Helv. Chim. Acta*, 1963, **46,** 1042; *Analyt. Abs.*, 1964, **11,** 223
8. A. F. Molodchenko and A. I. Lyubivaya, *Konserv. i. Ovoshchesush. Prom.*, 1963, (2), 36; *Analyt. Abs.*, 1964, **11,** 3338
9. J. Nosek, *Prum. Potravin,* 1968, **19** (5), 273; *Analyt. Abs.*, 1969, **17,** 1723
10. N. P. Fedorovskaya and L. V. Miesserova, *Trudy. Inst. goryuch. Iskop. Min. ugol'n. Prom. S.S.S.R.*, 1967, **23** (3), 3; *Analyt. Abs.*, 1969, **16,** 3044
11. B. Zagorchev, L. Bozadzhieva and E. Mitropolitska, *Compt. Rend. Acad. Bulg. Sci.,*1962, **15,** 483; Analyt. Abs., 1963, **10,** 2278
12. W. D. Nordling, *Chemist Analyst,* 1960, **49,** 79; *Analyt. Abs.*, 1961, **8,** 1543
13. V. N. Tolmachev and L. S. Prikhodko, *Izv. Vyssh. Ucheb. Zavednii. Khim. i Khim. Tekhnol.,* 1960, **3,** 985; *Analyt. Abs.,* 1962, **9,** 1917
14. B. Bagshawe, *J. Soc. Chem. Ind.,* 1938, **57,** 260
15. R. Goetz, *Giesserei,* 1965, **52** (1), 12; *Analyt. Abs.,* 1966, **13,** 2958
16. O. Kammori and A. Ono, *Japan Analyst,* 1965, **14** (12), 1137; *Analyt. Abs.*, 1967, **14,** 3201
17. G. Volke, *Giessereitechnik.*, 1968, **14** (6), 186; *Analyt. Abs.*, 1969, **17,** 1422
18. P. H. Scholes and D. V. Smith, *Metallurgia Manchr.*, 1963, **67,** 153; *Analyt. Abs.*, 1964, **11,** 1298
19. V. Kh. Aitova, *Uch. Zap. Permsk. Univ.*, 1963, **25,** 117; *Analyt. Abs.*, 1965, **12,** 645
20. A. de Sousa, *Chemist Analyst,* 1961, **50,** 9; *Analyt. Abs.*, 1961, **8,** 4154
21. E. S. Pilkington and P. R. Smith, *Analytica chim. Acta,* 1967, **39** (3), 321; *Analyt. Abs.* 1969, **16,** 650
22. V. L. Zolotavin and N. D. Fedorova, *Trudy. vses. nauchno. -issled. Inst. Standart. Obraztsov spektr.,* *Etalanov,* 1965, **2,** 92; *Analyt. Abs.*, 1967, **14,** 4711
23. N. A. Tananaev and L. I. Ganago, *Trudy. Uralsk. Politekhn. Inst.*, 1956, (57), 73; *Analyt. Abs.*, 1958, **5,** 84
24. Report of Chrome Liquors and Chrome Leather Anal. Cttee. Soc. Leather Trades Chem., *J. Soc. Leath. Tr. Chem.*, 1962, **46,** 385; *Analyt. Abs.*, 1963, **10,** 3779
25. R. C. Chirnside, H. J. Cluley, R. J. Powell and P. M. C. Proffitt, *Analyst,* 1963, **88,** 851
26. P. Spacu, E. Antonescu and I. Albescu, *Rev. Chim. Bucharest,* 1960, **11,** 230; Analyt. Abs., 1960, **7,** 5334
27. A. I. Williams, *Analyst,* 1970, **95** (7), 670
28. Shirley Inst., Test Leaflets Nos. Chem. 23 & 25, 1956; *Analyt. Abs.*, 1957, **4,** 192 & 194
29. Y. Yamamoto, *J. Chem. Soc. Japan pure Chem. Sect.*, 1960, **81,** 388; *Analyt. Abs.*, 1961. **8,** 2840
30. N. N. Pavlov, A. R. Kuznetsov and G. A. Arbuzov, *Izv. Vyssh. Ucheb. Zavedeni, Tekhnol. Legk. Prom.*, 1960, (3), 28; *Analyt. Abs.*, 1962, **9,** 789
31. Joint A.B.C.M.—S.A.C. Cttee. on Methods for the Anal. of Trade Effluents, *Analysti* 1956, 81, 607
32. Min. of Housing and Local Govt., *"Methods of Chemical Analysis as applied to Sewage and Sewage Effluents"*, H. M. Stationery Office, London, 1956
33. A. A. Christie, J. R. W. Kerr, G. Knowles and G. F. Lowden, *Analyst,* 1957, **82,** 336
34. F. Burriel-Marti, C. A. Herrero and F. J. V. Fuentis, *An. R. Soc. esp. Fis. Quim.*, B. 1967, **63** (7-8), 809; *Analyt. Abs.,* 1968, **15,** 6353
35. O. K. Dobrolyubskii and G. M. Viktorova, *Pochvovedenie,* 1969, (5), 126; *Analyt. Abs.*, 1970, **18,** 4369
36. K. Uesugi, *J. Waterwks. and Sewerage Assoc. Japan*, 1962, (329), 61; *Analyt. Abs.*, 1963, **10,** 4433
37. L. Chuecas and J. P. Riley, *Analytica chim. Acta,* 1966, **35** (2), 240; *Analyt. Abs.*, 1967, **14,** 5795

Chromium Method 1
using diphenylcarbazide

Principle of the method

The test is based on the colour reaction produced by diphenylcarbazide with dilute acidified chromate solutions.

The reagent also gives colour reactions with mercury, cadmium, magnesium, iron, copper, nickel, molybdenum and vanadium, but of these the first three are never found in steel, and iron, copper, and nickel are removed by the peroxidation treatment under the actual conditions obtaining in the test. Vanadium and molybdenum pass into the alkaline filtrate as soluble molybdates and vanadates together with chromium, and impart their characteristic colours on treatment with diphenylcarbazide. Thus the method is not applicable in the presence of vanadium and molybdenum.

Reagents required

1. *Diphenylcarbazide* Dissolve 1 g of diphenylcarbazide in 100 ml of 90% isopropyl alcohol and acidify with 3 drops of sulphuric acid (H_2SO_4 d 1.84). The solution so obtained darkens considerably on prolonged keeping, and should be freshly made every two or three days.

2. *Sulphuric acid* (H_2SO_4) 15% by volume.

Permanent standards

This test is carried out in the Lovibond Tintometer, and calibration curves are prepared by the operator to cover the appropriate range for which the test is required. The colour is matched by a combination of Lovibond red and blue glasses, but only the red value is plotted and considered significant for the purposes of this test. By choosing the appropriate fraction from the two prepared, and by varying the depth of cell employed, it is possible to cover a very wide range of concentrations. For example, using a 6″ depth, the author* quoted Lovibond values for chromium from 0.001 to 0.015 per cent and using a 1″ cell the range of chromium from 0.01 to 0.1 per cent. may be covered. The upper limit of applicability of the test is 0.2% chromium.

Technique

Take two aliquot parts of the sample solution, one of 50 ml and one of 100 ml. Transfer to separate 250 ml graduated flasks and treat each in turn with 50 ml of 15% H_2SO_4 (reagent 2) and 2 ml of diphenylcarbazide solution (reagent 1). Make up to the mark and mix thoroughly. Measure the solutions at an appropriate depth in a Lovibond Tintometer and compare with the colours of similarly treated standard solutions containing known amounts of chromium. With a little experience the colour of the two solutions will furnish a guide as to which of the sample solutions it is preferable to use and in what depth it should be viewed.

Note

The violet colour produced in the reaction fades slowly, but is sufficiently stable for determination in a reasonable time.

---

*B. Bagshawe *J. Soc. Chem. Ind.* 1938, 57, 260

COLORIMETRIC CHEMICAL ANALYTICAL METHODS

Chromium Method 2
using diphenylcarbazide

Principle of the method

Hexavalent chromium, in acid solution, reacts with diphenylcarbazide to give a red-violet colour the intensity of which is proportional to the concentration of hexavalent chromium present in the solution. This colour is compared with a series of Lovibond permanent glass standards. By analysing the sample for chromium before and after oxidation, the proportion of both chromate and of total chromium may be determined.

Reagents required

1. *Diphenylcarbazide solution* Dissolve 0.2 g diphenylcarbazide in 100 ml isopropyl alcohol. This solution must be kept as cool as possible and should be discarded immediately it becomes brown in colour.
2. *Sulphuric acid* (H_2SO_4) Cautiously add a volume of the concentrated acid to one volume of distilled water
3. *Silver nitrate* ($AgNO_3$) *solution*   5 g silver nitrate in 100 ml distilled water
4. *Sodium nitrite* ($NaNO_2$) *solution*   Dissolve 1 g sodium nitrite in 100 ml distilled water
5. *Ammonium persulphate* ((NH_4)$_2S_2O_8$)
6. *Phosphoric acid* (H_3PO_4 *sp. gr. 1.75*)

All chemicals used should be of analytical reagent quality.

The Standard Lovibond Discs, Comparator Disc 3/59 and Nessleriser Disc NOK

Disc 3/59 covers the range 10–100 µg of chromium (Cr) in the final 50 ml of solution in steps of 10 µg, i.e. 0.4 to 4.0 ppm on a 25 ml sample.

Disc NOK covers the range 2–10 µg of chromium (Cr) in 50 ml of solution, in steps of 1 µg, i.e. 0.08 to 0.4 ppm on a 25 ml sample.

To convert Cr to chromate in terms of CrO_4, multiply answer by 2.23.

Technique

a) Total chromium To 25 ml of the solution of the sample add 5 ml of dilute sulphuric acid (reagent 2). Remove any chloride ions which may be present by evaporating until white fumes appear. Cool and dilute to about 30 ml with distilled water. Bring solution nearly to the boiling point and then add 1 ml of silver nitrate solution (reagent 3), followed by about 1 g of ammonium persulphate (reagent 5). Boil for at least 10 minutes. If the solution at this stage has a pink tinge, due to the presence of manganese (see Note 2), add sodium nitrite solution (reagent 4) drop by drop until the colour just disappears. Care must be taken not to add too much nitrite solution as excess nitrite will reduce the chromate and lead to low results. Transfer the solution to a volumetric flask, add 1 ml phosphoric acid (reagent 6) and make up to 50 ml with distilled water. Mix, add 2 ml diphenylcarbazide solution (reagent 1) mix again and allow to stand for 5 minutes to ensure full colour development. Transfer to a Nessleriser tube and place in the right-hand compartment of the Nessleriser. Prepare a blank on the reagents by carrying out all the manipulations described above, including pre-treatment where this has been used, using distilled water in the place of the sample as starting material. Place a Nessleriser tube filled with this blank in the left-hand compartment of the Nessleriser and compare the colour of the sample with the Lovibond permanent glass standards in the disc using a standard source of white light such as the Lovibond White Light Cabinet or, failing this, north daylight.

If the chromium concentration is higher than the highest Nessleriser standard either repeat the colour measurement using a comparator and 13.5 mm tubes with the appropriate disc rather than the Nessleriser, or alternatively, repeat the determination using a smaller aliquot of the sample.

b) Chromium present as chromate Take an aliquot of the original sample solution before pre-treatment, filter if necessary through sintered glass, add 5 ml of sulphuric acid (reagent 2) and

COLORIMETRIC CHEMICAL ANALYTICAL METHODS — Chromium Method 2

dilute to 50 ml with distilled water. Add 1 ml phosphoric acid (reagent 6), mix, add 2 ml of diphenylcarbazide solution (reagent 1), mix again and allow to stand for 5 minutes. Prepare a blank using distilled water in the place of the sample. Compare the colour of the sample with the colours of the Lovibond permanent glass standards in the appropriate disc as above, using 50 ml tubes in the Nessleriser with disc NOK or 13.5 mm tubes with disc 3/59 in the Comparator.

Notes

1. Iron interferes with the determination of chromium by giving a yellow to brown coloration. If the ratio of iron to chromium does not exceed 100:1 the interference can be suppressed by the addition of phosphoric acid. If the proportion of iron is too high for suppression to be effective then the iron must be removed from the sample, by precipitation as ferric hydroxide, before the chromium is determined.

2. In the method for total chromium, manganese interferes if it is present in amounts greater than 0.2 ppm. The presence of manganese is indicated by the appearance of a pink colour during the chromium determination, and the interference can be removed by discharging the pink colour by titration with sodium nitrite.

3. The figures on the discs represent the concentration of chromium in the final solution. The chromium content of the original sample can be obtained from this by simple proportion.

4. The readings obtained by means of the Lovibond Nessleriser and disc are accurate only when the Nessleriser glasses used conform to the specification employed when the disc was calibrated; that is that the 50 ml calibration mark shall fall at a height of 113 ± 3 mm measured internally.

5. Correct results are obtained when using Diphenylcarbazide Laboratory Reagent (M.P. 164°C), but when using purified AnalaR grade reagent (M.P. 172–175°C) which produces a deeper colour, high results are found. If answers are multiplied by a factor of 0.8 when using this purified reagent, the answers will be correct.

If in doubt concerning the reagent, it is simple to carry out a test run using a known concentration of chromium.

6. Simplified instructions are available on request.

Chromium Method 3

Principle of the method

The yellow colour of the chromate ion is used as a measure of its concentration, where chromate corrosion inhibition treatment is being used for industrial waters. This colour is compared with a series of Lovibond permanent glass colour standards.

Reagents required

None

The Standard Lovibond Comparator Disc 4/35

This disc covers the range 100–600 parts per million of sodium chromate (Na_2CrO_4) in steps of 50 ppm.

Technique

Place 10 ml of the sample solution in a 13.5 mm cell or comparator tube and place this in the right-hand compartment of the comparator. Hold the comparator before a standard source of white light such as the Lovibond White Light Cabinet or, failing this, north daylight and match the colour of the solution against the standards in the disc.

The Determination of Cobalt

Introduction

Cobalt is used in a wide variety of industries, its applications varying from that as an alloying element in metallurgy to that as an accelerator in the oxidation of oils in the paint and varnish industries. Cobalt is also an essential trace element and its determination in soils, plants and animal foodstuffs, as well as in biological materials, is of importance.

Methods of sample preparation

Beverages[1-6]

Beer[1-3] Evaporate 50 ml of degassed sample with 1 ml of $18N$ sulphuric acid and ash the residue in a muffle furnace at 650°C. Moisten the carbon free ash with 1 ml of $6N$ hydrochloric acid, heat to dryness, add 10 ml of water, heat near the boiling point for 5 minutes, cool, filter and make the filtrate up to volume[2]. Alternatively[3] degas 50 ml of sample by boiling, cool and adjust the pH to between 5.0 and 5.5 with $4M$ sodium acetate (CH_3COONa). Add 2 ml of a 1% aqueous solution of nitroso-R salt ($C_{10}H_4(OH)(SO_3.Na)_2.NO$), boil for 1–2 minutes, cool rapidly and pass the solution through a 5 cm × 15 mm column of alumina (Al_2O_3) which has been washed with water. Elute the cobalt complex with M sulphuric acid and when the eluate is again colourless make the coloured portion up to 25 ml with M sulphuric acid and proceed by *Method 2*.

Spirits[4] Take a 20 ml sample and wet ash with sulphuric and nitric acids. Evaporate to dryness and ash the residue at 500°C. Dissolve the ash in nitric acid, evaporate to dryness, dissolve the residue in water with a little nitric and hydrochloric acids. Proceed by *Method 2*.

Wine[5,6] Ash a 500 ml sample, dissolve the ash in N hydrochloric acid, add 2 g of sodium acetate ($NaOOCCH_3$) to adjust the pH to approximately 7. Add 50% sodium citrate ($Na_3C_6H_5O_7$) solution drop by drop until the absence of colour with potassium thiocyanate (KSCN) shows that the iron is completely complexed. Proceed by *Method 2*.

Biological materials[7-13]

Blood[7-9] Evaporate 50–100 mg to dryness in a quartz vessel under an infra-red lamp. Ash the residue at 450–500°C. Dissolve the ash in 40 ml of warm $9M$ hydrochloric acid. Pass the solution through a column of Dowex 1–X8 ion exchange resin. Wash the column with $9M$ hydrochloric acid and elute the cobalt with 30 ml of $4M$ hydrochloric acid. Determine cobalt on the eluate. Alternatively[8] heat 10 ml of blood etc., or 5 g of tissue, with 5 ml of a 1:1 mixture of sulphuric and nitric acids, add 10 ml of perchloric acid ($HClO_4$) and continue heating until the solution is colourless. Evaporate to dryness. Dissolve the residue in 50 ml of $0.1N$ hydrochloric acid and adjust the pH to 2.5 with ammonia (NH_4OH). Extract copper by shaking with 10 ml of a 0.02% solution of dithizone in chloroform ($CHCl_3$). Add a drop of 30% hydrogen peroxide (H_2O_2) and extract the iron by shaking with 10 ml of a 1% solution of 8-hydroxyquinoline in chloroform. Add to the aqueous solution 3 ml of 10% ammonium tartrate (($CH(OH)COONH_4)_2$) solution, adjust to pH 5, add 5 ml of 2% sodium diethyldithiocarbamate and extract with three separate 10 ml portions of chloroform. Evaporate these extracts to dryness. To the residue add 0.2 ml of 10% sodium nitrate ($NaNO_3$) and heat with 0.25 ml of 1:1 nitric and sulphuric acids and 1 ml of perchloric acid under reflux until the solution is colourless. Evaporate to dryness. To the residue add acetate buffer and proceed by *Method 2*.

Tissue[8,10,11] Digest sample with hydrochloric and nitric acids and evaporate the solution to dryness. Dissolve the residue in 5 ml of hydrochloric acid and transfer to a column of Dowex-1 resin (H^+ form) with 4 ml of concentrated hydrochloric acid and then 4 ml of $4N$ hydrochloric acid. Elute the cobalt with 10 ml of $4N$ hydrochloric acid[10].

Alternatively digest sample in a Kjeldahl flask with sulphuric and nitric acids and remove copper by either passing hydrogen sulphide (H_2S) through the acidified digest or by making the digest very alkaline adding a fourfold excess of nitroso R salt, boiling to precipitate copper and determining cobalt by *Method 2*.

General warning[12] Strohal has warned that cobalt is partly volatilised during dry ashing even at temperatures as low as 110°C. He recommends wet oxidation using hydrogen peroxide[13].

Dairy products[14]

Milk Ash 100 ml of milk and dissolve the ash in dilute hydrochloric acid. Extract the copper into a 0.5% solution of dithizone in carbon tetrachloride (CCl_4) at pH 1.5. Buffer the aqueous phase to pH 9 and extract the cobalt with dithizone. Evaporate off the carbon tetrachloride and treat the residue with perchloric acid ($HClO_4$) to remove the dithizone. Evaporate to remove excess carbon tetrachloride. Dissolve the residue in a dilute mixture of hydrochloric and nitric acids and determine cobalt on this solution.

Food and Edible Oils[15,16]

Fish[15] Ash 50–300 g of sample, dissolve the ash in hydrochloric acid using 5 ml of acid for every 50 g of sample used. Evaporate to dryness, dissolve the residue in hot water, filter, wash the residue with hot water and combine the filtrate and washings. Cool and make up to volume.

General[16] Ash a sample containing 0.001 to 0.01 mg of cobalt in a silica dish at 500–550°C. Dissolve the ash in 10–15 ml of hydrochloric acid, evaporate to dryness and dissolve the residue in 5 ml of hydrochloric acid and 20 ml of water. Filter and adjust the volume of the filtrate to 50 ml. Add 10 ml of 1% α-nitroso-β-naphthol solution. Boil and allow to stand overnight. Filter, wash the precipitate with 10 ml of 5% acetic acid (CH_3COOH). Dissolve the precipitate with 2 ml of sulphuric acid and 2 ml of perchloric acid and boil until colourless. Heat until a yellow colour appears and continue heating until the solution becomes colourless once more. Dilute with 30 ml of water and extract with successive 5 ml portions of 0.15% dithizone in chloroform until no colour is formed in the extract. Add 10 ml of 25% ammonium citrate (($NH_4)_3C_6H_5O_7$) solution and add aqueous ammonia (NH_4OH) until the solution turns orange. Again extract with a 5 ml portion of dithizone solution and then with 5 ml of chloroform. Combine these extracts, evaporate to dryness, add 0.5 ml of sulphuric acid and 5 drops of perchloric acid. Heat until all the organic matter is destroyed, add 6 ml of water to the residue and one drop of phenolphthalein solution. Add caustic soda (NaOH) until the solution is alkaline and then hydrochloric acid until it is just acid. Proceed by *Method 2*.

Metals, alloys and plating baths[17–29]

Ferrous[17–20] Dissolve the sample in 8N hydrochloric acid and extract the solution with a 2:1 mixture of isobutyl methyl ketone (i-$C_4H_9COCH_3$) and phenylacetate ($CH_3COOC_6H_5$). Pass the aqueous phase through a column of Amberlite 1RA-400 anion exchange resin, elute the cobalt with 2.5N hydrochloric acid and make this eluate up to volume [17,18].

Alternatively[19,20] remove the iron, which interferes with *Method 2*, by precipitation with zinc oxide (ZnO).

Non-ferrous[21–29] *Aluminium and its alloys*[21]—dissolve 1 g of sample in 60 ml of 25% caustic soda (NaOH) and 1 ml of hydrogen peroxide (H_2O_2), boil for 10 minutes, dilute to 200 ml and filter while still hot. Dissolve the residue in hydrochloric acid to which a little bromine water has been added, transfer to a 100 ml flask, dilute to volume and determine cobalt on an aliquot by *Method 1*.

Beryllium[22]—dissolve the sample in acid, remove beryllium and other elements by extraction with acetyl acetone ($CH_3COCH_2COCH_3$) at pH 4.0. Determine cobalt on the aqueous layer by *Method 1*.

Copper alloys[23]—dissolve a 0.2 g sample in nitric acid, dilute the solution and electrolyse to remove copper. Evaporate the residual solution to dryness, dissolve the residue in 20 ml of hydrochloric acid and determine cobalt on this solution.

Kovar alloys and highly alloyed samples[24–26] Dissolve 0.2 g of the alloy[24] in 20–30 ml of aqua regia (1 part of nitric acid plus 3 parts of hydrochloric acid) and evaporate to dryness twice, adding 10 ml of hydrochloric acid after the first evaporation. Dissolve the residue in 10 ml of hydrochloric acid and pass the solution through a column containing 45 ml of a quaternary ammonium type anionite, such as Macroporous AV-17-8P, (Cl⁻ form) at 1 ml a minute. Wash the column with 90 ml of 8N hydrochloric acid, to remove nickel, followed by 30 ml of 6N hydrochloric acid, to remove manganese, and finally by 170 ml of 4N hydrochloric acid. Determine cobalt on this final eluate.

Highly alloyed metals should be dissolved in hydrochloric acid, with the addition of hydrofluoric acid if required, and evaporated to dryness. Take up the residue in $9M$ hydrochloric acid. Remove interfering metals either (i) by extraction with cupferron in chloroform followed by precipitation of cobalt, nickel and chromium by 2-nitroso-1-naphthol buffered with acetate to pH 5, dissolve the precipitate in $10N$ hydrochloric acid, extract the cobalt with benzene (C_6H_6) and determine cobalt on the extract[25]. Alternatively (ii) pass the acid solution through a column of Dowex-1 ion exchange resin, elute the cobalt with $4M$ hydrochloric acid and determine cobalt on the eluate[26].

Nickel[27,28] To determine traces of cobalt in nickel metal[27] dissolve 1 g of sample in hydrochloric acid, evaporate to dryness and redissolve the residue in 20 ml of $10.5N$ hydrochloric acid. Extract the cobalt from this solution by shaking with four successive 40 ml portions of purified tributyl phosphate (Bu_3PO_4) in white spirit. Combine the extracts, wash with four 40 ml portions of water and then evaporate with white spirit. Remove the organic matter by repeated evaporation with nitric acid and a few drops of hydrogen peroxide (H_2O_2). Dissolve the dry residue in water and determine cobalt on this solution by *Method 2*. Alternatively[28] dissolve 10 g of sample, containing 0.01–0.2 μg of cobalt in hydrochloric acid and make the solution $9N$ in this acid. Apply this solution to a 9×1.2 cm column of Amberlite CG-400 resin (100–200 mesh Cl^- form). Wash the column with 70 ml of $9N$ hydrochloric acid to elute the nickel and then elute the cobalt with 30 ml of $0.1N$ hydrochloric acid and determine by *Method 2*.

Tungsten and tungsten alloys[29] Dissolve the sample in a mixture of hydrofluoric and nitric acids and adjust the pH of the solution to approximately 6 with a buffer of ammonium tartrate $((CHOH.COONH_4)_2)$ and sodium borate ($Na_2B_4O_7$). Determine cobalt on an aliquot of this solution.

Minerals, ores and rocks[24, 30–32]

Dissolve 0.5 g of ore in 20–30 ml of aqua regia (1 part of nitric acid plus 3 parts of hydrochloric acid) and then proceed as described for Kovar alloys[24]. Or[30,32] dissolve the sample in hydrofluoric acid (HF) and perchloric acid ($HClO_4$). Remove the excess perchloric acid by evaporation. Dissolve the residue in hydrochloric acid and dilute with water. Add citrate buffer solution and extract with dithizone in carbon tetrachloride (CCl_4). Evaporate the extract and destroy the organic matter with sulphuric acid and hydrochloric acid. Evaporate, dissolve the residue in hydrochloric acid and determine cobalt on this solution. Alternatively[31] decompose the sample with a mixture of 10 ml of nitric acid and 20 ml of hydrochloric acid and add bromine if necessary to complete the solution process. Evaporate to dryness, add 25 ml of water and 1 ml of hydrochloric acid for every 50 ml of the final volume. Boil gently for a few minutes, cool and dilute to volume. The pH of this solution is 0.9 to 1.0 and must be re-adjusted to this value if further dilution proves to be necessary. Pipette 5 ml of the solution into a separating funnel containing a mixture of 5 ml of 20% sodium thiosulphate ($Na_2S_2O_3$), 3 ml of 8.3% trisodium phosphate (Na_3PO_4) and 10 ml of ammonium thiocyanate reagent (NH_4CNS) and proceed by *Method 1*.

Miscellaneous[33–38]

Animal-feeding salts[33] Mix 2–5 g of the finely ground sample with 5 ml of an acetate-thiourea buffer (pH 4) and proceed by *Method 1*.

Cement[34] Dissolve a 0.05 to 0.5 g sample in 10 ml of hydrochloric acid, add a few drops of nitric acid and evaporate to 3 ml. Add 10 ml of $9M$ hydrochloric acid, cool, centrifuge, and transfer the supernatant liquid to a Dowex 1-X8 column. Wash with $9M$ hydrochloric acid, elute the cobalt with $4M$ hydrochloric acid. To the eluate add 0.5 ml of 5% sodium chloride (NaCl) and evaporate to dryness. Dissolve the residue in 2 ml of water, add 2 ml of acetic acid — sodium acetate buffer (pH 6) and proceed by *Method 2*.

Cemented carbides[35] Dissolve sample in acid and pass the sample solution through a column of Diaion SA-100. Determine cobalt on the eluate.

Fertilizers[36] Digest the sample with a 5:1:2 mixture of nitric, sulphuric and 70% perchloric ($HClO_4$) acids, dilute the digest, centrifuge, decant the supernatant liquid, wash the deposit, re-centrifuge, add the supernatant liquid to the previous supernatant and make up to volume.

Thin magnetic films[37,38] Wash the film with alcohol (C_2H_5OH) and then dissolve in 2 or 3 drops of hot 1:1 nitric acid, dilute, boil to remove nitrogen oxides, cool and make up to volume.

Paper[39]

Heat 100 g of paper or wood pulp in a platinum dish at $575 \pm 25°C$ for 4 hours. Dissolve the ash in 10 ml of hydrochloric acid and evaporate to dryness on a water bath. Add 50 ml of water and 5 ml of perchloric acid ($HClO_4$), boil for 5 minutes and then add 2 ml of barium perchlorate ($Ba(ClO_4)_2$) solution to precipitate sulphates. Heat to the boiling point, filter, and make up cooled filtrate to 100 ml.

Plants[40-42]

General[40,41] Incinerate 150 g of sample at 540°C. Dissolve the ash in 15 ml of hydrochloric acid and evaporate on a water bath. Dissolve the residue in hot water and filter. To the filtrate add 2 drops of methyl orange solution and neutralise with 10% caustic soda (NaOH) solution. Add 5 ml of 30% sodium acetate ($NaOOCCH_3$) solution and boil. Filter off the precipitate, wash with 3 ml of sodium thiosulphate ($Na_2S_2O_3$), 5 ml of sodium dihydrogen phosphate (NaH_2PO_4) and water. Determine cobalt in the filtrate. Alternatively[41] extract the cobalt as its complex with 1-nitroso-2-naphthol ($C_{10}H_6(OH)NO$) in chloroform ($CHCl_3$) Wet ash organic phase, and determine cobalt on the diluted acid solution. Iron will interfere if the ratio of iron to cobalt concentrations exceeds 100:1. In this case extract the iron with isopropyl ether (($iC_3H_7)_2O$).

More[41] warns that hay is often contaminated with soil and that this has resulted in anomalously high cobalt contents being reported for hay.

Seaweed[42] Decompose the sample by digestion with hydrochloric acid. Treat the insoluble residue with hydrofluoric acid (HF), as this may contain up to 10% of the original cobalt content of the weed, and add this solution to the original solution. Add 1 ml of 3% hydrogen peroxide (H_2O_2) to ensure that all iron present is in the ferric state, when the phosphate present in the weed will mask it and prevent it from interfering, and extract the cobalt as its complex with 1-nitroso-2-naphthol. Wet ash to destroy the organic matter and proceed by *Method 2*.

Soil[43-46]

Fuse a 25 g sample with 1 g of potassium hydrogen sulphate ($KHSO_4$), allow to cool, add 5 ml of $0.5M$ hydrochloric acid and digest on a sand bath until the melt has disintegrated. Filter and determine cobalt on the filtrate. Alternatively[44] to a 20 g air-dried sample add 100 ml of N potassium nitrate (KNO_3) solution (pH 3) and sufficient nitric acid to neutralise any carbonate ($CaCO_3$) which is present. Shake for 1 hour, filter, and evaporate the filtrate to about 100 ml. Decolorise if necessary by boiling with 5–10 drops of 30% hydrogen peroxide (H_2O_2). Cool and make the solution up to 100 ml. Determine cobalt on a suitable aliquot.

If the soil contains an appreciable amount of copper, Veriginia[45] recommends that the copper be removed as its dithizone ($C_6H_5.N{:}N.CS.NH.NH.C_6H_5$) complex in carbon tetrachloride (CCl_4) solution and that cobalt be then determined on the aqueous phase.

For general soil survey analysis Vinogradova[46] considers that it is adequate to decompose 0.5—1 g of the finely ground sample with a mixture of sulphuric and hydrofluoric (HF) acids and to dilute the acid solution to 50 ml.

Water[47-50]

Natural waters[47] Evaporate 1 litre sample to dryness, moisten the residue with water and dissolve in 2 ml of nitric acid. Evaporate to dryness, add 10 ml of water, 0.5 ml of 1:1 nitric acid and 0.5 ml of 1:1 hydrochloric acid. Boil for 90 seconds, add nitroso-R salt and proceed by *Method 2*.

Sea water[48-50] Where the cobalt concentration is high enough for a direct determination[48] boil 1 litre of sample for 15 minutes with 20 ml of $6N$ hydrochloric acid, and add 5 ml of 20% ammonium citrate (($NH_4)_3C_6H_5O_7$) to mask iron, aluminium and calcium and use an aliquot of this solution for the determination. If it is necessary to concentrate the cobalt before it can be determined either [49] add sodium carbonate (Na_2CO_3) to 2 litres of sample, set aside for at least 7 days, filter off the precipitate, dissolve it in dilute hydrochloric acid and determine cobalt on this solution; or[50] co-precipitate the cobalt with iron and manganese by the addition of magnesium hydroxide ($Mg(OH)_2$), filter, dissolve the precipitate in acid and determine cobalt by *Method 2*.

Loss of cobalt during sample preparation

It has been demonstrated [12,51] that loss of cobalt can occur when ashing at temperatures even as low as 110°C, and the use of wet ashing techniques is therefore recommended wherever possible. Mays and co-workers[52] have also shown that the loss of cobalt during ignition of cobalt chloride ($CoCl_2$) deposited during the evaporation of hydrochloric acid solutions is much greater than the loss incurred when dry ashing in porcelain crucibles.

Separation from other ions

In addition to the separations mentioned in individual cases of sample preparation above, the following separation techniques have been reported:–

General[53] Pass the solution in a 4:1 mixture of acetone (($CH_3)_2CO$) and $3M$ hydrochloric acid through a column of Dowex 50–X8 resin (100–200 mesh, H^+ form). Wash the column with the same mixture of solvents and elute cobalt with a 9:1 mixture of acetone and $6M$ hydrochloric acid.

From iron and nickel[54-56] Dilute[54] 2 ml aliquot of solution in $6M$ hydrochloric acid to 20 ml with acetone (($CH_3)_2CO$), pass through an ion exchange column when iron is not absorbed and nickel and cobalt form green and blue bands respectively which are successively eluted with a 7:3 mixture of acetone and $2M$ hydrochloric acid.

Korkisch[55] recommends the use of Dowex 50 resin with elution of iron in a 4:1 mixture of tetrahydrofuran ($CH_2.(CH_2)_2.CH_2O$) and $3N$ hydrochloric acid, cobalt in a 9:1 mixture of tetrahydrofuran and $6N$ hydrochloric acid, and nickel in $6N$ hydrochloric acid.

Alternatively Lavrukhina[56], who also uses a solution in $6M$ hydrochloric acid and acetone, separates them on a column of Dowex 1–X8 or 2–X8 resin (AV–17 is not suitable) and elutes cobalt with $4N$ hydrochloric acid and nickel with $9-12N$ hydrochloric acid. Iron is not absorbed.

From nickel[57,58] Either[57] use a column of Zerolit 225 resin (H^+ form) with a solution containing nitroso-R salt and elute cobalt with $0.5N$ hydrochloric acid and nickel with $4N$ hydrochloric acid; or[58] add 2 g of malonic acid ($CH_2(COOH)_2$) to the sample solution, dilute to 100 ml, warm, add 2 g of sodium nitrite ($NaNO_2$) in 15 ml of water and cool, pass this solution through a column of Amberlite 1RA-400 (Cl^- form, 80–200 mesh) at 15 ml a minute, wash the column with 60 ml of 1% ammonium chloride (NH_4Cl) solution to remove the nickel, and then elute the cobalt with 80–100 ml of $2M$ ammonium chloride in dilute aqueous ammonia (NH_4OH) at 5 ml a minute.

From silver[59] Adjust the pH of the sample solution to within the range 1.5–4.0, treat this solution with sufficient E.D.T.A. (ethylenediaminetetra-acetic acid ($CH_2.N(CH_2.COOH)_2)_2$) to complex the cobalt and pass this through a column of Zerolit 225 resin. Elute the cobalt complex with water and the silver with 7% sodium nitrite ($NaNO_2$) solution.

The individual tests

Method 1 depends on the formation of the blue ammonium cobaltothiocyanate ion, originally described by Vogel[60]. This is extracted into amyl alcohol following the method of Powell[61].

Method 2 uses the reaction with nitroso-R salt (sodium-1-nitroso-2-hydroxy-naphthalene-3:5-disulphonate) as originally suggested by van Klooster[62]. The procedure described is essentially that due to MacNaught[63].

References

1. F. V. Harold and E. Szobolotzky, *J. Inst. Brewing*, 1963, **69**, 253; *Analyt. Abs.*, 1964, **11**, 1111
2. E. Segal and A. F. Lautenbach, *Proc. Am. Soc. Brew. Chem.*, 1964, 49; *Analyt. Abs.*, 1966, **13**, 3220
3. R. Parsons, *Proc. Am. Soc. Brew. Chem.*, 1964, 152; *Analyt. Abs.*, 1966, **13**, 3221
4. M. J. Pro and R. A. Nelson, *J. Assoc. off. agric. Chem.*, 1956, **39**, 848; *Analyt. Abs.*, 1957, **4**, 703
5. H. Eschnauer, *Z. Lebensmitt Untersuch.*, 1959, **110**, 196; *Analyt. Abs.*, 1960, **7**, 2470
6. T. Vondenhof and H. Beindorf, *Mschr. Brau.*, 1968, **21** (6), 156; *Analyt. Abs.*, 1969, **17**, 3070
7. W. Haerdi, J. Vogel, D. Monnier and P. E. Wenger, *Helv. Chim. Acta*, 1960, **43**, 869; *Analyt. Abs.*, 1961, **8**, 1666
8. F. A. Pohl and H. Demmel, *Analyt. Chim. Acta*, 1954, **10**, 554; *Analyt. Abs.*, 1954, **1**, 2183
9. E. D. Zhukovskaya and L. I. Idel'son, *Lab. Delo.*, 1963, **9** (4), 16; *Analyt. Abs.*, 1964, **11**, 2720
10. P. V. Kooi, B. Heagan and J. T. Lowman, *Proc. Soc. Exp. Biol. Med.*, 1963, **113**, 772; *Analyt. Abs.*, 1964, **11**, 4440
11. W. T. Binnerts, *Tijdschr. Diergeneesk.*, 1968, **93** (9), 558; *Analyt. Abs.*, 1969, **17**, 1548
12. P. Strohal, S. Lulic and O. Jelisavcic, *Analyst*, 1969, **94**, 678
13. J. L. Down and T. T. Gorsuch, *Analyst*, 1967, **92**, 398
14. F. Kiermeier and H. Winkelmann, *Z. LebensmittUntersuch.*, 1961, **115**, 309; *Analyt. Abs.*, 1962, **9**, 3435
15. B. Chojnicka and E. Szyszko, *Roczn. Zacl. Hig. Warsaw*, 1964, **15** 23; *Analyt. Abs.*, 1965, **12**, 1290
16. N. D. Sylvester and L. H. Lampitt, *J. Soc. Chem. Ind.*, 1940, **59**, 57
17. S. Hirano A. Mizucke, Y. Iida, and N. Kokubu, *Japan Analyst*, 1961, **10**, 326; *Analyt. Abs.*, 1963, **10**, 2298
18. H. Goto, Y. Kakita and M. Namiki, *J. Japan Inst. Metals Sendai*, 1961, **25**, 178; *Analyt. Abs.*, 1962, **9**, 1918
19. L. N. Krasil'nikova and K. N. Dolgorukova, *Sb. nauch. Trudy. vses. nauchno.-issled. gornometallurg. Inst. tsvet. Metall.*, 1965, (9), 30; *Analyt. Abs.*, 1967, **14**, 171
20. F. M. Tulyupa, Yu. I. Usatenko and V. A. Pavlichenko, *Zavodskaya. Lab.*, 1968, **34** (1), 14; *Analyt. Abs.*, 1969, **16**, 1878
21. H. Pohl, *Aluminium, Berlin*, 1959, **35**, 260; *Analyt. Abs.*, 1960, **7**, 542
22. J. O. Hibbits, A. F. Rosenburg and R. T. Williams, *Talanta*, 1960, **5**, 250; *Analyt. Abs.*, 1961, **8**, 2304
23. G. Lindley, *Metallurgia Manchr.*, 1960, **62**, 45; *Analyt. Abs.*, 1961, **8**, 2428
24. L. V. Kamaeva, V. V. Stepin and M. K. Makarov, *Trudy. vses. nauchno.-issled. Inst. standart Obraztsov*, 1967, **3**, 87; *Analyt. Abs.*, 1969, **16**, 3030
25. R. C. Rooney, *Metallurgia Manchr.*, 1960, **62**, 175; *Analyt. Abs.*, 1961, **8**, 2429
26. D. H. Wilkins and L. E. Hibbs, *Analyt. Chim. Acta*, 1959, **20**, 427; *Analyt. Abs.*, 1960, **7**, 541
27. V. T. Athavale, S. V. Gulavane and M. M. Tillu, *Analyt. Chim. Acta*, 1960, **23**, 487; *Analyt. Abs.*, 1961, **8**, 2430
28. Y. Iida, A. Mizuike and S. Hirano, *J. chem. Soc. Japan ind. Chem. Sect.*, 1964, **67** (12), 2042; *Analyt. Abs.*, 1966, **13**, 3583
29. G. Norwitz and H. Gordon, *Analyt. Chem.*, 1965, **37** (3), 417; *Analyt. Abs.*, 1966, **13**, 3550
30. R. E. Stanton, A. J. McDonald and I. Carmichael, *Analyst*, 1962, **87**, 134
31. R. S. Young and A. J. Hall, *Ind. Eng. Chem. Anal. Ed.*, 1946, **18**, 264
32. K. Kunz and E. Duczyminska, *Rudy i Metale Niezelazne*, 1963, **8**, 302; *Analyt. Abs.*, 1965, **12**, 2690
33. W. A. C. Campen and H. Dumoulin, *Chem. Weekbl.*, 1957, **53**, 398; *Analyt Abs.*, 1958, **5**, 290

34. W. Haerdi, J. Vogel and D. Monnier, *Analyt. Chim. Acta,* 1961, **24,** 365; *Analyt. Abs.,* 1961, **8,** 4658
35. K. Tada and H. Horigachi, *Japan Analyst,* 1963, **12,** 799; *Analyt. Abs.,* 1965, **12,** 595
36. J. F. Hodgson and V. A. Lazar, *J. Assoc. off. agric. Chem.,* 1965, **48** (2), 412; *Analyt. Abs.,* 1966, **13,** 4502
37. K. L. Cheng and B. L. Goydish, *Microchem. J.,* 1963, **7,** 166; *Analyt. Abs.,* 1964, **11,** 4291
38. A. S. Babenko and T. T. Volodchenko, *Zavodskaya Lab.,* 1967, **33** (9), 1059; *Analyt. Abs.,* 1968, **15,** 7282
39. O. Lagerstrom and H. Haglund, *Svensk. Pappn. Tidn.,* 1964, **67,** 4; *Analyt. Abs.,* 1965, **12,** 1829
40. E. Szyszko and B. Chojnicka, *Roczn. Zakl. Hig., Warsaw,* 1961, **12,** 125; *Analyt. Abs.,* 1962, **9,** 363
41. E. More, *Annls. agron.,* 1969, **20** (5), 527; *Analyt. Abs.,* 1970, **19,** 4424
42. T. Yamamoto, T. Fujita and M. Ishibashi, *J. chem. Soc. Japan, pure Chem. Sect.,* 1965, **86** (1), 49; *Analyt. Abs.,* 1967, **14,** 3381
43. R. E. Stanton and A. J. McDonald, *Trans. Inst. Min. Metall., London,* 1962, **71,** 511; *Analyt. Abs.,* 1963, **10,** 3480
44. A. N. Gyul'akhmedov, *Izv. Akad. Nauk. AzerbSSR, Ser. Biol. i Med. Nauk,* 1961, (7), 57; *Analyt. Abs.,* 1963, **10,** 1608
45. K. V. Verigina, *Mikroelementy v.S.S.S.R.,* 1963, (5), 50; *Analyt. Abs.,* 1965, **12,** 2024
46. E. N. Vinogradova, G. V. Prokhorova and T. N. Sevast'yanova, *Vest. mosk. gos. Univ. Ser. Khim.,* 1968, (5), 74; *Analyt. Abs.,* 1970, **18,** 1275
47. I. A. Mikhalyuk, *Lab. Delo.,* 1960, **6,** 26; *Analyt. Abs.,* 1961, **8,** 340
48. L. I. Rozhanskaya, *Trudy. sevastopol'. biol. Sta.,* 1964, **15,** 503; *Analyt. Abs.,* 1966, **13,** 5901
49. W. Forster and H. Zeitlin, *Analyt. chim. Acta,* 1966, **34** (2), 211; *Analyt. Abs.,* 1967, **14,** 3642
50. T. N. Krishnamoorthy and R. Viswanathan, *Ind. J. Chem.,* 1968, **6** (3), 169; *Analyt. Abs.,* 1969, **16,** 3280
51. G. R. Doshi, C. Sreekumaran, C. D. Mulay and B. Patel, *Current Sci.,* 1969, **38** (9), 206; *Analyt. Abs.,* 1970, **19,** 1474
52. D. L. Mays, W. V. Kessler, J. A. Chilko and E. D. Schall, *J. Assoc. off. agric. Chem.,* 1967, **50** (3), 735; *Analyt. Abs.,* 1968, **15,** 5705
53. J. Korkisch and H. Gross, *Sepn. Sci.,* 1967, **2** (2), 169; *Analyt. Abs.,* 1968, **15,** 2650
54. I. Hazan and J. Korkisch, *Analyt. chim. Acta,* 1965, **32** (1), 46; *Analyt. Abs.,* 1966, **13,** 2370
55. J. Korkisch, *Nature,* 1966, **210,** 626; *Analyt. Abs.,* 1967, **14,** 5881
56. A. K. Lavrukhina and N. K. Sazhina, *Zh. analit. Khim.,* 1969, **24** (6), 670; *Analyt. Abs.,* 1970, **19,** 4877
57. F. Burriel-Marti and C. Alvarez-Herrero, *Infcion. Quim. analit. pura- apl. Ind.,* 1966, **20** (1), 1; *Analyt. Abs.,* 1967, **14,** 2564
58. O. G. Balakrishnan Nambiar and P. R. Subbaraman, *Talanta,* 1967, **14** (7), 785; *Analyt. Abs.,* 1968, **15,** 6004
59. F. Burriel-Marti and C. Alverez-Herrero, *Revta Univ. ind. Santander* 1965, **7** (1–2), 17; *Analyt. Abs.,* 1966, **13,** 2224
60. H. W. Vogel, *Ber.,* 1879, **12,** 2314
61. A. D. Powell, *J. Soc. Chem. Ind.,* 1917, **36,** 273
62. H. S. van Klooster, *J. Amer. Chem. Soc.,* 1921, **43,** 746
63. K. J. MacNaught, *Analyst,* 1939, **64,** 23

Cobalt Method 1
using thiocyanate

Principle of the method

This test is based on the reaction between cobalt ions and concentrated thiocyanate solutions to give blue cobalt thiocyanate ions which are soluble in amyl alcohol. The colour of the amyl alcohol solution is measured by comparison with Lovibond permanent glass colour standards.

Reagents required

1. *Hydrochloric acid (HCl)* $5N$ 180 g/l
2. *Potassium permanganate* $(KMnO_4)$ $0.1N$ 3.16 g/l
3. *Ammonium thiocyanate* (NH_4CNS) $7.5M$ Aqueous solution 571 g/l
4. *Solvent mixture*   A mixture of equal volumes of amyl acetate and amyl alcohol
5. *Ammonium phosphate* $((NH_4)_2HPO_4)$ M Aqueous solution 132 g/l

All chemicals should be of analytical reagent quality.

The Standard Lovibond Comparator Disc 3/31

The disc covers the range 10 μg to 200 μg (0.01 to 0.2 mg) of cobalt (Co), with steps 10, 25, 50, 75, 100, 125, 150, 175, 200 μg.

This represents 1 to 20 ppm on a 10 ml sample.

Technique

Take 10 ml of the sample solution, add 1 ml of $5N$ hydrochloric acid (reagent 1) and 1 drop of $0.1N$ potassium permanganate (reagent 2); after mixing, add 5 ml of an approximately $7.5M$ aqueous solution of ammonium thiocyanate (reagent 3). Extract the mixture with 3 separate 3.3 ml portions of solvent mixture (reagent 4). After these 3 extractions, discard the aqueous layer and combine the solvent layers together. Shake with a mixture of 5 ml of an approximately M aqueous solution of ammonium phosphate (reagent 5), to prevent interference from any ferric iron present, and 5 ml of $7.5M$ aqueous solution of ammonium thiocyanate (reagent 3). After separation, reject the lower layer and wash the solvent layer once with a mixture consisting of 1 ml of M ammonium phosphate (reagent 5), 2 ml of $7.5M$ ammonium thiocyanate (reagent 3) and 1 ml of water. Draw off the lower layer. The colour of the upper layer of amyl acetate-amyl alcohol is then compared in a 13.5 mm test tube or cell with the standard disc in the comparator, held facing a standard source of white light, such as the Lovibond White Light Cabinet, or, failing this, north daylight.

The figures represent micrograms of Cobalt in the 10 ml of solution taken.

Notes

Nickel will not interfere with *Method 1*. Iron will interfere. If iron is present proceed as follows: Add 2 g malonic acid to a suitable volume of the solution, dilute to 100 ml with water. Warm. Add 2 g sodium nitrite in 15 ml of water. Cool, and pass through a column of Amberlite IRA 400 (chloride form 80–200 mesh) at the rate of 5 ml per minute, and wash the column with 60 ml of 1% ammonium chloride solution to remove nickel. Elute with 80–100 ml 2 molar ammonium chloride in dilute ammonium hydroxide at 5 ml per minute. Proceed with the test for cobalt on this eluate.

Cobalt Method 2
using nitroso-R salt

Principle of the method

The test is based on the reddish-orange colour produced by the reaction, under specified conditions, between cobalt and nitroso-R salt (sodium 1-nitroso-2-hydroxy-naphthalene-3:6-disulphonate). The intensity of this colour, which is proportional to the cobalt concentration, is measured by comparison with a series of Lovibond permanent glass colour standards.

Reagents required

1. *Sodium acetate* ($CH_3COONa.3H_2O$)
2. *Indicator solution*   A 0.2 per cent. w/v alcoholic solution of phenolphthalein
3. *Sodium hydroxide* ($NaOH$) 5N 200 g/l
4. *Sulphuric acid* (H_2SO_4) N 49 g/l
5. *Sulphuric acid*   Approximately $N/20$ 2.45 g/l
6. *Nitroso-R salt*   A 0.1 per cent. w/v aqueous solution of nitroso-R salt which should be freshly prepared each day.
7. *Nitric acid* (HNO_3)   Sp. gr. 1.42

All chemicals should be of analytical reagent quality

The Standard Lovibond Nessleriser Discs NTA and NTB

Disc NTA covers the range 1 μg to 9 μg (0.001 to 0.009 mg) of cobalt (Co) in steps of 1 μg.
Disc NTB covers the range 10 μg to 30 μg (0.01 to 0.03 mg) of cobalt (Co) in steps of 2.5 μg.

These ranges represent 0.2 to 1.8 ppm and 2 to 6 ppm respectively with a 5 ml sample.

Technique

To 5 ml of the solution under examination add 2 g of sodium acetate (reagent 1) and a few drops of indicator solution (reagent 2), warm the mixture on a boiling water-bath and render faintly alkaline with 5N sodium hydroxide (reagent 3); then add, first sulphuric acid approximately N (reagent 4) and finally sulphuric acid approximately $N/20$ (reagent 5) until the pink colour of the indicator almost disappears (approximately pH 8.3). Add 1 ml of the solution of nitroso-R salt (reagent 6), boil gently for one minute and, while still boiling, add slowly 2 ml of nitric acid (reagent 7). Continue the boiling for about a minute, then cool the mixture under the tap, transfer it to a Nessleriser glass, dilute with water to the 50 ml mark and place in the right-hand compartment of the Nessleriser. Conduct a control test starting with 5 ml of normal sulphuric acid (reagent 4) and employing the same reagents and procedure as above and transfer the final mixture to a Nessleriser glass, dilute with water to the 50 ml mark and place in the left-hand compartment of the instrument.

Stand the Nessleriser before a standard source of white light such as the Lovibond White Light Cabinet or, failing this, north daylight and compare the colour produced in the test solution with the colours in the standard disc, rotating the disc until a colour match is obtained.

The markings on the disc represent the actual amounts of cobalt (Co) producing the colours in the test. If a colour equivalent to 10 μg (0.01 mg) is produced in the test, then the quantity of solution taken for the test contains 10 μg (0.01 mg) of Co.

If the colour obtained is deeper than the standard colours, the test should be repeated on a smaller quantity of the solution.

Note

Lovibond Nessleriser discs are standardised on a depth of liquid of 113 $\pm$ 3 mm. The height, measured internally, of the 50 ml calibration mark on Nessleriser glasses used with the instrument must be within the same limits. Tests with Nessleriser glasses not conforming to this specification will give inaccurate results.

The Determination of Copper

Introduction

Traces of copper in a water supply often cause trouble in galvanized cisterns and hot-water tanks. Precautions to protect the galvanized steel should be taken if the copper content of the water approaches 0.1 ppm. The estimation of traces of copper is also important in the food industry, in biochemistry where copper is one of the essential trace elements in some systems, and in agricultural chemistry where copper is again an important trace element. In industrial chemistry copper is frequently determined as one of the major components of a mixture as well as at the trace level.

Methods of sample preparation

Beverages[1-11]

Beer[1-5] Take a 25 ml sample[1,2], add 1 drop of hexyl alcohol ($C_6H_{13}OH$) to prevent frothing, and 3 ml of 25% sulphuric acid. Place in a boiling water bath for 30 minutes, remove and cool to 25°C. Proceed as in *Method 1*. Alternatively[3] ash the sample, dissolve the ash in acid, add 2,2′-bipyridine and neutralise, to complex any iron, and proceed as in *Method 3*. Weyh[4] stresses that to avoid loss of copper by evaporation during ashing the temperature of the furnace should be kept between 560 and 580°C.

To minimise interference from the colouring matter in beer when using *Method 1* it is recommended[5] that 0.7 mg per litre of a standard copper solution be added to the sample before analysis as the interference has been shown to increase exponentially as the copper concentration falls below 0.6 mg per litre (ppm.).

Brandy[6] Evaporate 10 ml of sample to 3 ml, dilute to exactly 10 ml with water and use this solution.

Cider[7] Take a 10 ml sample, add 1.2 ml of $9N$ sulphuric acid and 10 ml of zinc dibenzyldithiocarbamate solution in carbon tetrachloride (CCl_4). Centrifuge for 10 minutes, pipette out the bottom layer, filter through cotton and measure.

Fruit cordial and must[8] Precipitate the copper by mixing a 5 g sample with 2 ml of methanol (CH_3OH) and 0.2% methanolic quercetin. After 2 hours collect the precipitate, ash, dissolve ash in hydrochloric acid and determine copper on this solution.

Mineral water[9] (i) Evaporate 100–500 ml sample, ash the residue, dissolve the ash in hydrochloric acid, evaporate and again dissolve the residue in hydrochloric acid. Alternatively (ii) acidify the sample to pH 1.5 with sulphuric acid and extract with dithizone in chloroform ($CHCl_3$). Apply the acid solution from (i) or (ii) to paper and develop the chromatogram with aqueous ammonia (NH_4OH) for 2 hours. Elute spots of $R_f=1.0$ with $2N$ sulphuric acid (2×10 ml) and estimate copper on the solution.

Wine[10,11] Pass 200 ml of sample through 10 ml of KU-2V ion exchange resin, wash the column with water, desorb the cations with 100 ml of 10% hydrochloric acid and wash the column with another 100 ml of water. Evaporate the eluate to dryness, dissolve the residue in 5 ml of nitric acid, transfer to a Kjeldahl flask and evaporate to dryness. Add 0.5 ml of sulphuric acid and then water a drop at a time. Add 5 ml of 25% aqueous ammonia (NH_4OH), filter off the ferric hydroxide and wash the precipitate with 1 ml of 25% aqueous ammonia. Add the washings to the filtrate, dilute to 25 ml and determine copper[10]. Alternatively[11] place 10 ml of sample in an ampoule, add 3 ml of $0.1N$ sulphuric acid, evaporate to dryness, heat the residue at 500–600°C for 15–20 minutes and cool. Add 1 ml of 1:1 sulphuric and nitric acids, place the ampoule in boiling water, add 3 ml of water and transfer the contents of the ampoule to a beaker using three 3 ml portions of water. Add 5 ml of 20% ammonium acetate (CH_3COONH_4) adjust the pH of the solution to 8.3–8.4 with aqueous ammonia and extract copper as its complex with diethyldithiocarbamate (($C_2H_5)_2N.CS.S.NH_2$).

Biological materials[12-21]

Blood[12-15] Wet ash a 5 ml sample with 4 ml of lead free nitric acid, 2 ml of perchloric acid ($HClO_4$) and 1 ml sulphuric acid. Heat until white fumes are evolved, dilute with 10 ml of water,

COLORIMETRIC CHEMICAL ANALYTICAL METHODS   Copper

boil, cool and dilute to about 30 ml. Proceed by *Method 1*[12]. Or[13], take 6 ml of sample, precipitate proteins with trichloracetic acid (Cl_3COOH), filter, dilute the filtrate to 11 ml, add 1 ml of 10% E.D.T.A. solution and 2 ml of aqueous ammonia (NH_4OH) and proceed by *Method 3*. Alternatively[14] wet ash 0.5–1.0 ml of sample by heating with 1 ml of nitric acid and then with 1 ml of 1:1 70% perchloric–70% nitric acids, dissolve the residue in 0.5 ml of sodium citrate dihydrate ($Na_3C_6H_5O_7.2H_2O$) and neutralise to phenolphthalein with N caustic soda (NaOH). Determine copper on this solution.

A simpler method[15] is to mix 2 ml of sample with 3 ml of $0.2N$ hydrochloric acid, add 50 mg of ascorbic acid (O.CO.C(OH):C(OH).CH.CH(OH).CH_2OH), cover with a layer of Parafilm and set aside. After 30 minutes add 1 ml of aqueous trichloracetic acid and after a further 15 minutes centrifuge for 10 mins at not less than 1500 rpm. Determine copper on an aliquot of the supernatant liquid.

Tissue[12,14,16,17] Take a 1–3 g sample and proceed as for blood[12,14]. Alternatively ash the sample, extract the ash with acetylacetone ($CH_3COCH_2COCH_3$) at controlled pH and proceed by *Method 3*[16].

Klewska[17] recommends that the tissue should be wet ashed with sulphuric and nitric acids. An aliquot of the resulting solution should then be diluted to 20 ml with water, 0.5 ml of sulphuric acid added and the solution cooled and then shaken vigorously for 3 minutes with zinc dibenzyldithiocarbamate solution. Proceed by *Method 1*.

Urine[18-21] Take a 30 ml sample, add a few drops of aqueous ammonia (NH_4OH) and 2 drops of 30% hydrogen peroxide (H_2O_2) and evaporate to half volume. Cool, repeat the addition of ammonia and peroxide and evaporate to 3 ml. Cool, add 2 ml of $6N$ hydrochloric acid, transfer to a centrifuge tube, add 3 ml of 20% trichloracetic acid (Cl_3CCOOH) dilute to 8 ml and centrifuge. To 6 ml of the supernatant liquid add 3 ml of 0.2% hydroxy ammonium chloride ($NH_2OH.HCl$) and neutralise exactly to phenolphthalein with aqueous ammonia (NH_4OH) or hydrochloric acid. Proceed by *Method 3*[18].

Alternatively[19] wet ash with sulphuric acid, cool, add 20 ml of water, 1 ml of 12% citric acid (C(OH)(COOH)($CH_2COOH)_2.H_2O$) make up to 50 ml and proceed by *Method 1*.

A simple test, suitable for field use, is described by Giorgio[20]. Mix 10 ml of sample with 1 ml of hydrochloric acid and set aside for 15 minutes, then shake with zinc dibenzyldithiocarbamate solution and proceed by *Method 1*.

Wilson and Klassen[21] advocate the use of nitric and perchloric ($HClO_4$) acids in addition to sulphuric acid for procedures based on wet ashing.

Chemicals[22, 23]

Phosphorus[22] Dissolve 1 g of sample in nitric acid, add perchloric acid ($HClO_4$) and evaporate until copious fumes are evolved. Add ammonium citrate (($NH_4)_3C_6H_5O_7$) and E.D.T.A. (($CH_2.N(CH_2COOH)_2)_2$) as masking agents and proceed by *Method 3*.

Sodium chloride[23] Treat a solution of the sample with ammonium citrate, E.D.T.A. and aqueous ammonia (NH_4OH) to bring pH to 9.0. Form the diethyldithiocarbamate complex of the copper and proceed by *Method 3*.

Dairy products[24-26]

Milk and milk powder Destroy organic matter by wet ashing with sulphuric acid, nitric acid and perchloric acid ($HClO_4$). Fat in butter and cream should first be removed by extraction with benzene (C_6H_6). Dilute the clear solution from ashing and adjust the pH to 4.4. Extract with diethyldithiocarbamate in carbon tetrachloride (CCl_4) and proceed by *Method 3*[24].

Alternatively precipitate fat and protein from an aqueous solution of the sample by heating with an equal volume of 10% trichloracetic acid (CCl_3COOH) on a boiling water bath for 15 minutes then filter. Determine copper in filtrate either by *Method 3*[25] or *Method 1*[26]. This is claimed[26] to be both faster and more accurate than techniques involving ashing.

Foods and edible oils[27-34]

Bread[27] Moisten 5 g of sample with 5 ml of 2% ethanolic magnesium acetate ($Mg(CH_3COO)_2$ in C_2H_5OH) solution, dry, and then ash at 500°C. This treatment prevents volatilization of the copper. Dissolve the ash in 5 ml of normal hydrochloric acid, filter and determine copper on the filtrate.

Fats[28-30] Boil 3 g of sample with 10 ml of $5N$ sulphuric acid for 5 minutes (for margarine samples which contain emulsifying agents use $10N$ acid). Transfer the solution to a separating funnel with 20 ml of chloroform ($CHCl_3$) and 5 ml of water. Separate the aqueous phase and determine copper on this solution[28]. Alternatively[29] ash the sample of fat and extract the ash with hydrochloric acid and determine copper on this acid extract by *Method 1*.

The method recommended by van Duin[30] is:—Weigh 25 g of sample into a large centrifuge tube, add 20 ml of carbon tetrachloride (CCl_4) and 15 ml of 15% trichloracetic acid (Cl_3CCOOH), heat for 30 minutes at 60°C, cool, shake for 30 seconds, centrifuge and set aside for a few hours. Determine copper on an aliquot of the supernatant liquid.

Fish[31] Clean the sample with distilled water, dry at 80°C for 48 hours and grind to a fine powder. Place 0.2 g of this powder in a size 0 crucible, char at 200°C then raise the temperature to 550°C and combust for 16 hours. Add 0.5 ml of 1:1 nitric acid, warm until solution is complete then dilute to volume.

General[32-34] Ash a 5–10 g sample of the air-dried product and periodically moisten during ashing with 3% hydrogen peroxide (H_2O_2) solution. Add 5 ml of 1:1 hydrochloric acid and 1 drop of peroxide. Evaporate to dryness on a water bath. Treat the residue with 5 ml of 10% hydrochloric acid, add water and filter. Determine copper on an aliquot of the filtrate[32]. Jones and Newman[33] recommend extracting the filtrate with neocuproin (2,9-dimethyl-1,1,10-phenanthroline) at pH 5–7 and then proceeding by *Method 1*. This takes advantage of the specificity of neocuproin and the sensitivity of zinc dibenzyldithiocarbamate. Alternatively[34] wet ash the sample with sulphuric and nitric acids, dilute the residue with water and treat with E.D.T.A. and ammonium citrate (($NH_4)_3C_6H_5O_7$) solution. Then add aqueous ammonia (NH_4OH) until the solution is blue or blue-green with thymol blue. Proceed by *Method 3*. Bismuth and tellurium if present can be removed by washing the carbon tetrachloride (CCl_4) layer, after the addition of diethyldithiocarbamate, with caustic soda (NaOH).

Fuels and lubricants[35,36]

Mineral oils Extract the copper from 10 ml of oil with three 10 ml portions of glacial acetic acid (CH_3COOH). Combine the acid extracts and adjust the pH with 1:1 aqueous ammonia (NH_4OH) until it is over 9. Add 1 g of citric acid ($C(OH)(COOH)(CH_2.COOH)_2.H_2O$) and proceed by *Method 3*. Compared with the use of cuproine this method has been found to be more accurate and more sensitive but slower[35].

Alternatively[36] dissolve the sample in carbon tetrachloride (CCl_4) and proceed by *Method 3*.

Metals, alloys and plating baths[37-49]

Aluminium[37] Dissolve the sample in 30 ml of 2:1 hydrochloric acid, add sulphuric acid until its concentration is about 5% and then dilute the solution to 100 ml.

Ferrous alloys and steel[38,39] Dissolve the sample by heating with 10 ml of a 3:3:14 mixture of sulphuric acid, phosphoric acid (H_3PO_4) and water. Destroy carbides with a few drops of 10% hydrogen peroxide (H_2O_2), cool and add 10 ml of water and boil. Adjust the pH to 4 and add 8 ml of 10% ammonium thiocyanate (NH_4SCN) and then add $2N$ ammonium fluoride (NH_4F) drop by drop until the solution is colourless. Proceed by either *Method 1* or *Method 3*.

Lead[40] Dissolve 0.1–0.2 g sample in 30–40 ml of 1:1 nitric acid, add 10 ml of 1:2 sulphuric acid, evaporate until copious white fumes are evolved, add 25 ml of water, filter and make filtrate up to 50 ml.

Mercury[41] Dissolve 5–10 g of the sample in 15–20 ml of nitric acid, evaporate to dryness, dissolve the residue in 30 ml of 1:3 hydrochloric acid. Make alkaline with aqueous ammonia (NH_4OH) and dilute to 100 ml.

Nickel[40] Dissolve 2 g of sample in 40 ml of a 3:1 mixture of hydrochloric and nitric acids (aqua regia) evaporate the solution until it is almost dry and then add 20 ml of hydrochloric acid. Continue the evaporation until it is almost dry again and then dissolve the residue in 20 ml of $8N$ hydrochloric acid. Pass this solution, at the rate of 1 ml a minute, through a column of Anionite TM which has previously been washed with 70 ml of $8N$ hydrochloric acid. Discard the washings and eluate. Then elute the copper with 120 ml of $0.5N$ hydrochloric acid, evaporate the eluate to a volume of 25–30 ml and make up to 50 ml with water.

Non-ferrous alloys[42-47] Dissolve the sample in nitric acid. Add sulphuric acid and evaporate to white fumes to remove excess nitric acid. Dilute. Separate copper by paper chromatography using butanol (C_4H_9OH)-hydrochloric acid-water (15:3:2). Elute the copper spot with hydrochloric acid and measure. Alternatively[43] prevent interference by treating the acid solution with a mixture of ammonium citrate $((NH_4)_3C_6H_5O_7)$ and E.D.T.A. at pH 8.8, and proceed as in *Method 3*.

For aluminium alloys which do not contain nickel[44] mix a solution of the alloy with sodium pyrophosphate $(Na_4P_2O_7)$ to complex iron, aluminium, magnesium, manganese, chromium and silicon and then convert the copper and zinc into complexes by adding aqueous ammonia (NH_4OH). Pass the solution through an 8 cm column of KU-2 resin (NH_4^+ form). Copper and zinc are adsorbed while the pyrophosphate complexes pass through. Discard the eluate. Then elute the zinc with 5% caustic soda (NaOH) solution at 4 ml a minute until the eluate gives no reaction with dithizone. Then elute the copper with normal hydrochloric acid until the eluate gives no reaction when tested with 10% potassium iodide (KI) and starch. Nickel interferes with this determination as it is eluted together with the copper.

To analyse tin and lead-tin alloys[45,46] dissolve 0.15–0.2 g of sample by heating with 20–25 ml of 1:2 nitric acid containing 4 g of tartaric acid $((CHOH.COOH)_2)$. When dissolution begins add about 11 drops of hydrochloric acid, heat to complete solution and drive off oxides of nitrogen. Cool, dilute to a known volume and determine copper on an aliquot of this solution.

For white metal alloys[47] dissolve 50 mg of sample in aqua regia (1 part nitric acid 3 parts hydrochloric acid) add sulphuric acid and heat until copious fumes are evolved, dilute with water and filter. Pass the filtrate through a column, 0.5 sq cm cross section, of Dowex 1-X8 resin (SCN^- form) which has been treated with 1.5M hydrochloric acid followed by M ammonium thiocyanate (NH_4SCN) then washed with water and dried at 50°C for 3 hours. Elute tin with 15 ml of a solution 0.5M in caustic soda (NaOH) and 0.5M in sodium chloride (NaCl) and then elute the copper with 50 ml of a solution 2M in ammonium chloride (NH_4Cl) and 2M in aqueous ammonia (NH_4OH).

Zinc[48] Dissolve 1 g of sample in 5 ml of hydrochloric acid add 30% aqueous hydrogen peroxide (H_2O_2) drop by drop, remove any excess peroxide by boiling for 2 minutes, cool, add 2 ml of 1% ascorbic acid $(O.CO.C(OH):C(OH).CH.CH(OH).CH_2OH)$ solution and dilute to 100 ml with water.

Zinc plating solutions[49] Boil 10 ml of electrolyte solution for 5 minutes with 1 ml of nitric acid, cool, add 1 ml of sulphuric acid, evaporate until salts begin to crystallise, cool and add water until the salts dissolve. Add 5 ml of ammonium citrate solution (210 ml of aqueous ammonia (NH_4OH), 15 ml of water and 200 g of citric acid $(C(OH)(COOH)(CH_2COOH)_2.H_2O)$, 4–5 drops of cresol red indicator solution and then 25% aqueous ammonia until the solution is yellow-pink in colour. Dilute to 50 ml and proceed by *Method 3*.

Minerals, ores and rocks[50–52]

Dissolve a 1 g sample in 20 ml of nitric acid:hydrochloric acid (3:1) partially evaporate and then add 20 ml of sulphuric acid:hydrochloric acid:water (5:2:3), evaporate to white fumes, and then add 40 ml of water. Boil, cool and filter. Wash the residue with 2% sulphuric acid. Add 1 g of tartaric acid $((CH(OH)COOH)_2)$ to the filtrate. Neutralise with aqueous ammonia (NH_4OH) add 15 ml of 80% acetic acid (CH_3COOH), boil and add 10 ml of a 5% aqueous solution of potassium ethyl xanthate $(CH_3.CH_2.O.CSSK)$. Filter off the precipitate, wash with hot water, acidify with acetic acid and dissolve in sulphuric acid and nitric acid. Evaporate the solution to dryness, dissolve the residue in hot water and estimate copper on the solution[50].

Alternatively[51] dissolve 20 g of powdered sample by heating with hydrochloric and hydrofluoric (HF) acids. Fuse the insoluble residue with potassium pyrosulphate $(K_2S_2O_7)$ and dissolve the melt in 2N sulphuric acid. Combine the two solutions, add an excess of freshly prepared zinc sulphide (ZnS) and then add aqueous ammonia (NH_4OH) until the pH of the solution is approximately 5. Collect the precipitate and heat it with 2N sulphuric acid. Dissolve the insoluble residue in nitric acid, evaporate the combined solutions until fumes are evolved and then dilute to 50 ml with water.

To determine copper in silumin[52] dissolve 0.2 g of sample in 25 ml of 1:3 hydrochloric acid plus 15 ml of 65% nitric acid. Evaporate to dryness and heat residue gently at 100–110°C. Evaporate twice more with 30% hydrochloric acid, dissolve the residue in 30% hydrochloric acid and add 100 ml of water. Filter off silica. Treat the silica with a mixture of hydrofluoric and sulphuric acids, fuse the residue with sodium carbonate (Na_2CO_3), dissolve the melt in 1:1 hydrochloric acid and filter. Combine filtrate with the previous filtrate, evaporate the solution to dryness, dissolve the residue in $8N$ hydrochloric acid and pass the solution through a column containing 5 g of anionite AV-17 which has previously been washed with $8N$ hydrochloric acid. Wash the column with a further 200 ml of $8N$ hydrochloric acid and discard this eluate. Elute copper together with iron with 100 ml of $2.5N$ hydrochloric acid. To an aliquot of this eluate add 5 ml of 20% potassium citrate ($K_3C_6H_5O_7.H_2O$) solution and 0.5 ml of a 1% solution of dimethylglyoxime ($CH_3.C(:NOH).C(:NOH).CH_3$) in 5% caustic soda (NaOH) and 1 drop of phenol red indicator solution. Adjust the pH to 8.9 with aqueous ammonia (NH_4OH) and dilute to 25 ml. Proceed by *Method 3*.

Miscellaneous[9, 53-60]

Drugs and pharmaceutical preparations[9,53,54] Ignite a 5 g sample at 450°C. Dissolve the ash in hydrochloric acid:nitric acid (3:1) and evaporate to dryness. Dissolve the residue in hydrochloric acid and chromatograph as described for mineral waters under beverages above[9].

Alternatively[53,54] wet ash the sample with 10 ml of sulphuric acid and complete the decomposition by adding perchloric acid ($HClO_4$) and hydrogen peroxide (H_2O_2). Evaporate the solution until dense fumes are evolved, take up the residue in aqueous acetic acid (CH_3COOH) and extract the copper with a solution of dithizone ($C_6H_5.N:N.CS.NH.NH.C_6H_5$) in carbon tetrachloride (CCl_4) at pH 2.7. Wet ash the organic layer with sulphuric acid, take up the residue in water and dilute to a known volume.

Dyes[55] Dissolve 1-2 g of the sample in sulphuric acid+nitric acid, heat until fumes appear, add a little water, centrifuge, wash the residue and make up the clear solution to a known volume. To an aliquot add 1 ml of phosphoric acid (H_3PO_4), 5 ml of water and 4 ml of reagent in carbon tetrachloride (CCl_4). Proceed by *Method 1*.

Gluten[56] Ash 4-5 g of sample, treat the ash with nitric acid and re-ignite at 450°C. Dissolve the residue in 2 ml of hydrochloric acid and dilute the solution to 50 ml.

Organic matter[57] Destroy the organic matter by wet ashing, dilute the residue with water and filter the solution. To an aliquot of the filtrate add 10 ml of a solution containing 20 g of ammonium citrate (($NH_4)_3C_6H_5O_7$) and 5 g of E.D.T.A. in 100 ml of water. Add $6N$ of aqueous ammonia (NH_4OH) solution until the mixture is blue or blue-green when tested with thymol blue indicator. Proceed by *Method 3*.

Phosphors and semi-conductors[58] Dissolve 2 g of sample by heating with 4-5 ml of nitric acid and 5-10 ml of water, boil to expel oxides of nitrogen and add aqueous ammonia (NH_4OH) until the precipitate which initially forms re-dissolves. Cool and proceed by *Method 3*.

Photographic gelatin[59] Dissolve the sample in water. Pass the solution through a column of Dowex 50W cation exchange resin and elute metals with $3N$ hydrochloric acid. Extract copper into a 0.005% dithizone solution in carbon tetrachloride (CCl_4) from acid medium at pH 2.1-2.2. Discard the aqueous layer, evaporate the solvent, destroy organic matter with sulphuric acid and nitric acid, evaporate until dense fumes appear, take up the residue in water and dilute to a known volume.

Wood[60] Dry a sample of sawdust at 100-105°C to constant weight. Take 1 g of the dried sample in a 250 ml graduated flask, add 50 ml of $5N$ sulphuric acid and 10 ml of 100 volumes hydrogen peroxide (H_2O_2) and heat on a water bath at 70-75°C for 20 minutes, occasionally swirl the flask to mix the contents. Remove the flask from the water bath, add 100 ml of water, swirl, cool to room temperature make up to volume and mix. Mix twice more at 5 minute intervals and then allow to settle for 10 minutes. Filter 10 ml of the supernatant liquid through a dry 7 cm Whatman No. 44 filter paper, discard the first 5 ml of filtrate and determine copper on 1 ml aliquot of the remaining filtrate.

Paper[61]

Ash a 10 g sample, add 5 ml of $6M$ hydrochloric acid and evaporate to dryness on a steam bath. Dissolve the residue in water, reheat any undissolved material with hydrochloric acid and repeat. Add 10 ml of 5% E.D.T.A. solution and 5 drops of phenolphthalein. Neutralise with aqueous ammonia (NH_4OH) to a faint pink colour and cool. Extract with reagent in carbon tetrachloride (CCl_4) and proceed by *Method 1 or 3*.

Plants[62-67]

General[62-65] Ash the sample, treat the ash with hydrochloric acid, followed by sodium citrate ($Na_3C_6H_5O_7.2H_2O$) to complex any iron. Extract the copper with an organic solution of the reagent as in *Method 1* or *3*[62]. Alternatively[63-65] take 0.5–0.6 g of fresh, or 0.2 g of air-dried, sample, wet ash with sulphuric acid, and potassium sulphate (K_2SO_4) neutralise the solution with aqueous ammonia (NH_4OH) and add 1 g of potassium dihydrogen phosphate (KH_2PO_4). Extract the solution with the reagent in carbon tetrachloride (CCl_4) and proceed by *Method 1* or *3*.

Leaves[66] Dry ash 1 g of sample at 450°C. Digest the ash with 2 ml of a 3:1 mixture of nitric and perchloric ($HClO_4$) acids then boil for 10 minutes with 10 ml of $0.5N$ hydrochloric acid and 1 ml of 0.5% sodium nitrite ($NaNO_2$) solution. Dilute to 20 ml and set aside overnight for silica to separate. Filter and determine copper on an aliquot of the filtrate.

Rape seed[67] Ignite the sample at 500°C. Extract the ash twice with 1:3 hydrochloric acid and dilute combined extracts to known volume. Proceed by *Method 1*.

Plastics and polymers[55,68]

Latex[91] Take sufficient sample to yield about 5 g of dried solids, and dry this in a suitable dish. Wrap 5 g of the dried sample in an ashless filter paper, place in a silica crucible, transfer to a muffle furnace at 550°C and ash for 2–3 hours. Cool, add 15 ml of 1:2 nitric acid, digest on a steam-bath for 30–60 minutes and then wash the contents of the crucible into a small flask. Dilute the liquid to about 25 ml and filter. To the filtrate add 5 ml of 50% citric acid ($C(OH)(COOH)(CH_2COOH)_2.H_2O$) and then add aqueous ammonia (NH_4OH) until the solution is strongly alkaline. Make the solution up to a known volume and determine the copper on an aliquot using *Method 3*.

Rubber Either[55] proceed as described under "*Miscellaneous—dyes*" above or[68] ash the sample in a silica crucible, extract the ash with hydrochloric acid and dilute the extract to a known volume. It is important to use a silica crucible as it has been shown[68] that the use of porcelain crucibles for ashing leads to anomalously high results.

Soil[69-73]

To 1–2 g of ignited soil add 20–30 ml of hydrochloric acid:nitric acid (3:1). Evaporate, add 2 ml of sulphuric acid and heat until fumes appear. Cool, add 40–60 ml of water, boil, filter and wash the residue with warm dilute sulphuric acid (4 ml/l). Dilute the filtrate and washings to a known volume and proceed[69]. Alternatively[70] take 20 g of air-dried soil, add 100 ml of N potassium nitrate (KNO_3) at pH 3 and add sufficient nitric acid to neutralise any carbonate ($CaCO_3$). Shake for 1 hour, filter and evaporate the filtrate to dryness. De-colourise the residue if necessary by heating with hydrogen peroxide (H_2O_2), cool and make up to 100 ml. Take a 10 ml aliquot, add 5 ml of 5% ammonium nitrate (NH_4NO_3) solution and 1–2 drops of an ethanolic phenolphthalein solution and 5 ml of 10% E.D.T.A. solution. Shake, if no pink colour appears add aqueous ammonia (NH_4OH) until pink, and then extract with the reagent in carbon tetrachloride (CCl_4) solution (*Method 1* or *3*).

Verigina[71] recommends decomposing the sample with sulphuric and hydrofluoric acids (HF), dissolving the residue in 10 ml of hydrochloric acid, filtering and extracting copper from the filtrate with dithizone. Geering[72] merely extracts the soil at pH 1 and then also extracts the copper with dithizone. Busev[73] endorses the technique described by Nemodruk[69].

Water[69, 74-81]

No pre-treatment is necessary unless the sample has to be concentrated. *Method 1* is recommended for the examination of boiler feed water[75], *Method 2* is the standard method[74] for industrial waters and *Method 4* has been developed[79] especially for the determination of traces of copper in domestic water supplies.

If concentration is required either[80] add 10 ml of Dowex 50W-X2 resin (200–400 mesh, H^+ form) to 1 litre of sample and stir for 10–15 minutes, suck off liquid through a sintered glass filter, wash resin with water, elute copper with $5M$ calcium chloride ($CaCl_2$) and determine by *Method 3*; or[81] make sample $0.1N$ in hydrochloric acid and extract copper into a solution of dithizone in carbon tetrachloride (CCl_4). Re-extract into $6N$ hydrochloric acid and proceed by *Method 3*.

Separation from other ions

From cadmium[82] Pass solution through a column of Dowex 50-X8 resin, elute cadmium with water and copper with $4N$ hydrochloric acid.

From silver[83] Pass solution through a column of Lewatit MN resin saturated with sodium arsenate ($Na_2HAsO_4.7H_2O$) and elute copper with $2N$ hydrochloric acid.

Suppression of interferences

Metals other than copper produce coloured dithiocarbamates and precautions have to be taken to avoid interferences. The use of citric acid in *Method 3* removes interference from ferric iron, but ferrous iron in quantities exceeding 1 mg interferes and must be oxidised to ferric. Alkali metals and small amounts of calcium, aluminium, zinc, nitrates, sulphates and phosphates do not interfere. Reduce chromium with sulphurous acid (H_2SO_3) until no colour is produced when the reaction mixture is spotted onto an external indicator consisting of a solution of diphenylamine in sulphuric acid. After reduction heat the solution to boiling with 2 g of citric acid, allow to cool and then make alkaline with an excess of aqueous ammonia. This procedure enables traces of copper to be determined in the presence of 0.2 g of chromium. In the presence of manganese add 2 g of citric acid and 5 ml of 5% sulphur dioxide (SO_2) solution followed by excess aqueous ammonia. Cobalt and nickel can be suppressed[84] by adding 10 ml of 0.02N E.D.T.A. and 2.5 ml of a 0.1% solution of magnesium to the test solution before the addition of citric acid (*Method 3*). Alternatively[85] add 1 ml of 0.5% dimethylglyoxime to the sample solution before adding the aqueous ammonia. Centrifuge and decant the supernatant liquid from the precipitated nickel. The cobalt remains in the aqueous layer on extraction. In the presence of bismuth[86] remove nickel and cobalt with dimethylglyoxime, develop the colour due to copper+bismuth and extract. Add 1 ml of 0.5% potassium cyanide (KCN) solution for each 0.01% of copper+bismuth present. This completely removes the copper colour leaving the bismuth blank which is deducted from the total to give the copper concentration.

The individual tests

Method 1 uses zinc dibenzyldithiocarbamate, as suggested by Martens and Githens[87]. It is the method recommended[75] for the examination of boiler feed water.

Method 2 uses dithio-oxamide (rubeanic acid) which, it is claimed[88], is less affected by the presence of chromium and manganese.

Method 3 is the Haddock and Evers[89] modification of the Callan and Henderson[90] diethyldithiocarbamate procedure. This is the standard test for copper in industrial waters[74].

Method 4 also uses diethyldithiocarbamate and is the test especially developed by the British Non-Ferrous Metals Research Association[79] for the determination of copper in domestic water supplies. A special kit is available from Tintometer Ltd. for carrying out this test in the field.

Method 5 uses the direct measurement of the colour of copper sulphate ($CuSO_4.5H_2O$) as a means of controlling its concentration in copper-plating baths.

References

1. H. Trommsdorff, *Mschr. Brauerie*, 1962, **15**, 78; *Analyt. Abs.*, 1963, **10**, 366
2. I. Stone, R. Ettinger and C. Gantz, *Analyt. Chem.*, 1953, **25**, 893
3. I. Stone, *Ind. Eng. Chem. Analyt.*, 1942, **14**, 479
4. H. Weyh, W. Hagen and Un Hua Pek, *Brauwissenschaft*, 1968, **21** (12), 472; *Analyt. Abs.*, 1970, **18**, 2784
5. H. Trommsdorff, *Mschr. Brauerei*, 1963, **16** (6), 101; *Analyt. Abs.*, 1964, **11**, 4550
6. V. Chioffi and G. Osti, *Boll. Lab. chim. prov.*, 1965, **16** (6), 633; *Analyt. Abs.*, 1967, **14**, 1708
7. C. F. Timberlake, *Chem. & Ind.*, 1954, **47**, 1442
8. H. E. Jungnickel and W. Klinger, *Z. Lebensmitt-Untersuch*, 1963, **119**, 399; *Analyt. Abs.*, 1964, **11**, 374
9. A. Smoczkiewiczowa and T. Smoczkiewicz-Szczepansk, *Acta Polon. Pharm.*, 1962, **19**, 65; *Analyt. Abs.*, 1962, **9**, 3476
10. G. I. Beridze and G. D. Gudzhedzhiani, *Vinodelie i Vinogradarstvo S.S.S.R.*, 1962, (7), 5–7; *Analyt. Abs.*, 1964, **11**, 375
11. L. Laporta, *Boll. Lab. chim. prov.*, 1964, **15**, 315; *Analyt. Abs.*, 1965, **12**, 5467
12. N. A. Brown and R. G. Hemingway, *Res. Vet. Sci.*, 1962, **3**, 345; *Analyt. Abs.*, 1963, **10**, 4786
13. F. J. Gareia Canturri and M. P. Sanchez de Rivera, *An. Acad. Farm. Madrid*, 1964, **30**, 51; *Analyt. Abs.*, 1965, **12**, 4041
14. R. E. Stoner and W. Dasler, *Clin. Chem.*, 1964, **10** (9), 845; *Analyt. Abs.*, 1966, **13**, 268

15. J. V. Joossens and J. H. Claes, *Revue belge Path. Med. exp.*, 1965, **31**, 254; *Analyt. Abs.*, 1966, **13**, 6355
16. P. C. van Erkelens, *Analyt. chim. Acta*, 1961, **25**, 129; *Analyt. Abs.*, 1962, **9**, 1155
17. A. Klewska and M. Strycharska, *Chemia analit.*, 1967, **12** (6), 1325; *Analyt. Abs.*, 1969, **16**, 798
18. V. Fiserova-Bergerova, *Pracovni Lekarstvi*, 1958, **10**, 25; *Analyt. Abs.*, 1960, **7**, 1478
19. D. Kilshaw, *J. Med. Lab. Technol.*, 1963, **20**, 295; *Analyt. Abs.*, 1964, **11**, 1837
20. A. J. Giorgio, G. E. Cartwright and M. M. Wintrobe, *Amer. J. Clin. Path.*, 1964, **41**, 22; *Analyt. Abs.*, 1965, **12**, 1852
21. J. F. Wilson and W. H. Klassen, *Clinica chim. Acta*, 1966, **13** (6), 766; *Analyt. Abs.*, 1967, **14**, 6281
22. G. Norwitz, J. Cohen and M. E. Everett, *Analyt. Chem.*, 1964, **36**, 142
23. B. Riva, *Ann. Chim. Roma*, 1963, **53**, 1898; *Analyt. Abs.*, 1965, **12**, 1416
24. R. Alifax and M. Bejambes, *Lait*, 1958, **38**, 146; *Analyt. Abs.*, 1959, **6**, 2329
25. C. C. J. Olling, *Neth. Milk & Dairy J.*, 1963, **17**, 295; *Analyt. Abs.*, 1964, **11**, 1487
26. A. C. Smith. *J. Dairy Sci.*, 1967, **50** (5), 664; *Analyt. Abs.*, 1968, **15**, 5018
27. P. Leman and J. Saerens, *Ann. Falsif. Expert. Chim.*, 1963, **56**, 257; *Analyt. Abs.* 1965, **12**, 336
28. R. Gaigl, *Z. Lebensmitt Untersuch.*, 1963, **119**, 506; *Analyt. Abs.*, 1964, **11**, 1917
29. M. Karvanek, *Sb. Vys. Sk. Chem-Technol., Odd. Fak. Potrav. Technol.*, 1961, **5** (1), 203; *Analyt. Abs.*, 1964, **11**, 2388
30. H. van Duin and C. Brons, *Neth. Milk & Dairy J.*, 1963, **17**, 323; *Analyt. Abs.*, 1964; **11**, 3352
31. G. A. Knauer, *Analyst*, 1970, **95**, 476
32. V. S. Gryuner and L. A. Evmenova, *Vopr. Pitaniya*, 1961, **20**, 66; *Analyt. Abs.*. 1963; **10**, 797
33. P. D. Jones and E. J. Newman, *Analyst*, 1962, **87**, 637
34. I.U.P.A.C., *Ann. Falsif. Exp. Chim.*, 1962, **55**, 38; *Analyt. Abs.*, 1962, **9**, 3896
35. J. M. Howard and H. O. Spauschus, *Analyt. Chem.*, 1963, **35**, 1016
36. T. P. Labuza and M. Karel, *J. Food Sci.*, 1967, **32** (5), 572; *Analyt. Abs.*, 1968, **15**, 6281
37. M. I. Abramov, *Uch. Zap. Azerb. Univ. Ser. Khim. Nauk.*, 1963, (3), 31; *Analyt. Abs.* 1965, **12**, 1654
38. Sh. T. Talipov, K. G. Nigai and E. L. Abramova, *Zavodskaya Lab.*, 1963, **29**, 804; *Analyt. Abs.*, 1964, **11**, 3534
39. O. P. Bhargava, *Talanta*, 1969, **16** (6), 743; *Analyt. Abs.*, 1970, **19**, 1273
40. V. N. Podchainova and T. G. Malkina, *Trudy vses. nauchno-issled. Inst. standardt. Obraztsov*, 1967, **3**, 100; *Analyt. Abs.*, 1969, **16**, 1144
41. B. M. Lipshits, A. M. Andreichuk and G. S. Agafonova, *Zavodskaya Lab.*, 1964, **30** (9), 1075; *Analyt. Abs.*, 1966, **13**, 62
42. K. N. Munshi and A. K. Dey, *J. prakt. chem.*, 1962, **18**, 233; *Analyt. Abs.*, 1963, **10**, 2625
43. G. de Angelis and M. Gerardi, *Ric. Sci. R. C., A*, 1961, **1**, 67; *Analyt. Abs.*, 1962, **9**, 5091
44. A. N. Kharin and N. N. Soroka, *Zhur. Prikl, Khim.*, 1964, **37**, 672; *Analyt. Abs.*, 1965, **12**, 3244
45. V. P. Zhivopistsev and E. A. Selezneva, *Uch. Zap. Permsk. Univ.*, 1963, **25** (2), 84; *Analyt. Abs.*, 1965, **12**, 1087
46. S. Onuki, K. Watanuki and Y. Yoshino, *Japan Analyst*, 1965, **14** (4), 339; *Analyt. Abs.*, 1966, **13**, 6803
47. K. Kawabuchi, M. Kaya and Y. Ouchi, *Japan Analyst*, 1966, **15** (6), 543; *Analyt. Abs.*, 1967, **14**, 5978
48. J. Balcarek, *Hutn. Listy*, 1968, **23** (3), 208; *Analyt. Abs.*, 1969, **17**, 62
49. G. S. Semikozov, E. G. Kruglova and I. P. Kalinkin, *Izv. Vyssh. Ucheb. Zavedenii Khim. i. Khim. Tekhnol.*, 1964, **7**, 194; *Analyt. Abs.*, 1965, **12**, 4446
50. M. Szelag and M. Kozlicka, *Chem. Anal. Warsaw*, 1962, **7**, 815; *Analyt. Abs.*, 1963, **10**, 1711
51. N. N. Lapin and A I. Vovk, *Izv. Vyssh. Ucheb. Zavedenii Khim. i. Khim. Tekhnol.*, 1966, **9** (1), 27; *Analyt. Abs.*, 1967, **14**, 3860

52. O. V. Morozova, Z. E. Mel'chekova and V. V. Stepin, *Trudy vses. nauchno-issled. Inst. standardt. Obraztsov Spektr. Etalonov,* 1964, **1,** 88; *Analyt. Abs.,* 1966, **13,** 3496
53. J. Hollos, *Gyogyszereszet,* 1963, **7,** 204; *Analyt. Abs.,* 1964, **11,** 5684
54. J. Hollos and I. Horvath-Gabair, *Pharmazie,* 1965, **20** (4), 207; *Analyt. Abs.,* 1966, **13,** 4367
55. J. Kocher, *Chim. Analyt.,* 1962, **44,** 161; *Analyt. Abs.,* 1962, **9,** 4151
56. K. Roszek-Masiak, *Chemia. analit.,* 1964, **9** (6), 1125; *Analyt. Abs.,* 1966, **13,** 1988
57. Metallic Impurities in Organic Matter Sub. Cttee., Analyt. Methods Cttee. Soc. Anal. Chem., *Analyst,* 1963, **88,** 253
58. I. P. Kalinkin and V. B. Aleskovskii, *Izv. Vyssh. Ucheb. Zavedinii Khim. i. Khim. Tekhnol,* 1963, **6,** 553; *Analyt. Abs.,* 1965, **12,** 1708
59. J. Saulnier, *Sci. ind. Photogr.,* 1964, **35,** 1; *Analyt. Abs.,* 1965, **12,** 4037
60. A. I. Williams, *Analyst,* 1970, **95** (7), 670
61. Scand. Pulp Paper and Board Testing Cttee., *Norsk. Skogind.,* 1962, **16,** 391; *Analyt. Abs.,* 1963, **10,** 1875
62. C. Huffman and D. L. Skinner, *U.S. Geol. Survey Proff. Paper,* 1961, **424-B, 331;** *Analyt. Abs.,* 1962, **9,** 1586
63. L. N. Lapin and M. A. Rish, *Trudy Tadzh. Uchit. Inst.,* 1957, **4,** 71; *Analyt. Abs.,* 1959, **6,** 998
64. K. R. Middleton, *Analyst,* 1965, **90,** 234
65. E. R. Page, *Analyst,* 1965, **90,** 435
66. E. G. Bradfield, *J. Sci. Food Agric.,* 1964, **15,** 469; *Analyt. Abs.,* 1965, **12,** 5957
67. M. Karvanek, *Prumysl. Potravin,* 1964, **15,** 282; *Analyt. Abs.,* 1965, **12,** 5472
68. L. de Roo, J. F. M. Tertoolen and C. Buijze, *Analyt. chim. Acta,* 1963, **29,** 82; *Analyt. Abs.,* 1964, **11,** 3533
69. A. A. Nemodruk and V. V. Stasyuchenko, *Zhur. Analyt. Khim.,* 1961, **16,** 407; *Analyt Abs.,* 1962, **9,** 1245
70. A. Gyulakhmedov, *Izv. Akad. Nauk. AzerbSSR. Ser. Biol. i Med. Nauk.,* 1961, (7), 57; *Analyt. Abs.,* 1963, **10,** 1608
71. K. V. Verigina, *Mikroelementy v. S.S.S.R.,* 1963, (5), 50; *Analyt. Abs.,* 1965, **12,** 2024
72. H. R. Geering and J. F. Hodgson, *J. Assoc. offic. agric. Chem.,* 1966, **49** (5), 1057; *Analyt. Abs.,* 1968, **15,** 1117
73. A. I. Busev and A. A. Nemodruk, *Pochvovedenie,* 1967, (1), 110; *Analyt. Abs.,* 1968, **15,** 2323
74. Brit. Stand. Inst., "Methods of Testing Water Used in Industry", B.S. 2690, London, 1964
75. U.K.A.E.A. Report PG 434 (W) 1963; *Analyt. Abs.,* 1963, **10,** 3939
76. A. L. Wilson, *Analyst,* 1962, **87,** 884
77. D. C. Abbott and J. R. Harris, *Analyst,* 1962, **87,** 497
78. T. K. Kolesnikova, *Gidrokhim Materialy,* 1961, **32,** 165; *Analyt. Abs.,* 1962, **9,** 2942
79. Brit. Non-Ferrous Met. Res. Assoc. Development Report D 59, 1960
80. M. Pavlik, *Chemicky Prum.,* 1965, **15** (6), 365; *Analyt. Abs.,* 1966, **13,** 5895
81. T. V. Gurkina and A. M. Igoshin, *Zh. analit. Khim.,* 1965, **20** (7), 778; *Analyt. Abs.,* 1967, **14,** 1730
82. F. Burriel-Marti and C. Alvarez-Herrero, *An. R. Soc. esp. Fis. Quim., B,* 1966, **62** (1), 45; *Analyt. Abs.,* 1967, **14,** 2436
83. W. Kielczewski, *Chem. Analyt. Warsaw,* 1963, **8,** 691; *Analyt. Abs.,* 1965, **12,** 40
84. A. Jewsbury, *Analyst,* 1953, **78,** 363
85. L. I. Butler and H. O. Allen, *J. Assoc. off. agric. Chem.,* 1942, **25,** 567
86. P. L. Wilmot and F. J. Raymond, *Analyst,* 1950, **75,** 24
87. R. I. Martens and R. E. Githens, *Analyt. Chem.,* 1952, **24,** 991
88. N. L. Allport and G. H. Skrimshire, *Quart. J. Pharm.,* 1932, **5,** 461
89. L. A. Haddock and N. Evers, *Analyst,* 1932, **57,** 495
90. T. Callan and J. A. R. Henderson, *Analyst,* 1929, **54,** 650
91. Brit. Standards Inst., "Methods of Testing Rubber Latex, Part 2: Chemical and Physical Tests," B.S. 1672: Part 2: 1954: Amendment No. 1: 1959, London, 1959

Copper Method 1
using zinc dibenzyldithiocarbamate

Principle of the method

Zinc dibenzyldithiocarbamate combines with copper forming a mixture of zinc and copper dithiocarbamates. On making this mixture, in carbon tetrachloride, acid by means of dilute sulphuric acid, the zinc is transferred to the aqueous phase while the copper dithiocarbamate remains in the methyl chloroform. The following metals interfere; bismuth, nickel and cobalt, which form yellow complexes with the reagent; mercury (II), silver and antimony hinder the extraction of the copper complex.

Reagents required

1. *Zinc dibenzyldithiocarbamate* Dissolve 0.05 g of zinc dibenzyldithiocarbamate $((((C_6H_5.CH_2)_2\ N.CS.S)_2Zn)$ in 100 ml methyl chloroform (1,1,1-trichloroethane) analytical reagent grade.
2. *Sulphuric acid, copper-free.* 10% by volume in distilled water.

The Standard Lovibond Comparator Disc 3/39

The disc covers the range 2.5, 5, 10, 15, 20, 25, 30, 40, 50 μg (0.0025 to 0.05 mg) of copper (Cu). This corresponds to 1 to 20 ppm of copper if a 2.5 ml sample is used.

Technique

Take a suitable volume of the sample, which contains enough copper to bring it within the range of the disc, and make it approximately Normal in respect to sulphuric acid, i.e. adjust the sulphuric acid content to approximately 5% w/v by adding acid or water as necessary. Add 10 ml of reagent 1 and shake vigorously in a separating funnel for 30 seconds. Separate the yellow methyl chloroform layer and match in a 13.5 mm cell in the Lovibond Comparator against the standard colour disc, using a standard source of white light such as the Lovibond White Light Cabinet or, failing this, north daylight.

The answer represents micrograms of copper in the amount taken.

A blank test must be performed on the amount of sulphuric acid used to acidify the sample, and any copper found deducted from the above answer.

Note

Simplified instructions for this test are available.

Copper Method 2
using dithio-oxamide (rubeanic acid)

Principle of the method

The test is based on the colours produced by the action of dithio-oxamide (rubeanic acid) on copper. Dithio-oxamide also yields colours with nickel and cobalt but in solutions containing free acetic acid the sensitivity of the reactions with those metals is greatly depressed although the delicate response to copper persists. Colours or precipitates are also produced with platinum, palladium and silver, while iron, if present in more than a faint trace, gives rise to a brown tint and interferes with the determination of copper. Traces of mercury and most other heavy metals tend to prevent the full development of the colour due to copper, but small amounts of aluminium, bismuth, chromium, calcium, manganese and the alkali metals do not interfere. As the colour is due to the formation of a compound in a colloidal state, the solution to be tested should be as free as possible from electroytes.

Reagents required

1. *Acetic acid* (CH_3COOH)   Approx. $5N$ (300 g/l)
2. *Ammonium acetate*   A 20% w/v aqueous solution of ammonium acetate (NH_4OOCCH_3 analytical reagent grade)
3. *Gum acacia*   A 1% w/v aqueous solution of gum acacia prepared by dissolving 10 grammes of gum acacia in 1 litre of water, boiling to destroy oxidases, filtering and preserving with thymol
4. *Dithio-oxamide*   A 0.2% w/v solution if dithio-oxamide in 90% ethyl alcohol

The Standard Lovibond Nessleriser Discs NDA, NDB

Disc NDA covers the range 2.5 µg to 40 µg (0.0025 to 0.04 mg) of copper (Cu) in steps of 5 µg.

Disc NDB covers the range 30 µg to 100 µg (0.03 to 0.1 mg) of copper (Cu) in steps of 10 µg.

This equals 0.125 to 2 and 1.5 to 5 ppm on a 20 ml sample.

Technique

To a suitable quantity of the solution under examination (which should be adjusted to approximately neutral, pH 7.0) contained in a Nessleriser glass, add 1 ml of $5N$ acetic acid (reagent 1), 5 ml of ammonium acetate solution (reagent 2) and 1 ml of gum acacia solution (reagent 3); dilute with distilled water to 50 ml and add 0.5 ml of dithio-oxamide solution (reagent 4). Mix and allow to stand for five minutes, for the colour to develop, and then place in the right-hand compartment of the Nessleriser. In another Nessleriser glass place 1 ml of reagent 1, 5 ml of reagent 2, and 1 ml of reagent 3; dilute with distilled water to 50 ml and add 0.5 ml of reagent 4. Mix and place in the left-hand compartment of the Nessleriser. Stand the Nessleriser before a standard source of white light such as the Lovibond White Light Cabinet or, failing this, north daylight, and compare the colour produced in the test solution with the colours in the standard disc, rotating the disc until a colour match is obtained.

The markings on the disc represent the actual amount of copper (Cu) producing the colours in the test. Should the colour in the test solution be deeper than the standard colour glasses, a fresh test should be conducted using a smaller quantity of the sample under examination The standardization of the glasses has been carried out at temperatures between 20° and 25°C and it is recommended that the same condition should be observed when using the disc.

Notes

1. Distilled water prepared in metal stills usually contains traces of copper and gives a distinct colour in this test. Distilled water (analytical reagent grade) or water re-distilled from glass should be used throughout the test.

2. It must be emphasized that the readings obtained by means of the Lovibond Nessleriser and disc are only accurate provided that Nessleriser glasses are used which conform to the specification employed when the discs were being calibrated, namely that the 50 ml calibration mark shall fall at a height of 113 mm plus or minus 3 mm measured internally.

COLORIMETRIC CHEMICAL ANALYTICAL METHODS

Copper Method 3
using sodium diethyldithiocarbamate

Principle of the method

Sodium diethyldithiocarbamate forms a yellow complex when added to alkaline solutions containing traces of copper. This complex is soluble in organic solvents and extraction with methyl chloroform separates the copper complex from complexes with some other ions which would otherwise interfere. The intensity of the yellow colour of the extract is proportional to the copper concentration and is measured by comparison with Lovibond permanent glass colour standards.

Reagents required

1. *Citric acid*   A 20 per cent w/v aqueous solution of citric acid ($C_6H_8O_7.H_2O$)
2. *Aqueous ammonia* (NH_4OH)   Solution of ammonia containing 10 per cent w/v NH_3
3. *Sodium diethyldithiocarbamate*   A 0.1 per cent aqueous solution
4. *Methyl chloroform* (1,1,1-trichloroethane) $CH_3.CCl_3$
5. *Sodium sulphate anhydrous* (Na_2SO_4)

All chemicals should be of analytical reagent quality.

The Standard Lovibond Comparator Disc 3/5

The disc covers the range 2.5, 5, 10, 15, 20, 25, 30, 40, 50 μg (0.0025 to 0.050 mg) of copper (Cu), which equals 0.25 to 5.0 parts per million if a 10 ml sample is taken.

Technique

Transfer 10 ml of the solution under examination (which should be adjusted to approximately neutral, pH 7.0) to a small separating funnel, add 10 ml of the citric acid solution (reagent 1) and 6 ml of reagent 2; mix and add 10 ml of the sodium diethyldithiocarbamate solution, (reagent 3), followed immediately by 2.5 ml of methyl chloroform (reagent 4). Shake vigorously, allow to separate and run the lower layer into a dry 10 ml measure. Repeat the extraction with three more portions, each of 2.5 ml, of methyl chloroform and transfer them to the same 10 ml measure. If necessary add more methyl chloroform to produce 10 ml, add about 0.5 g of anhydrous sodium sulphate, shake and pour off the clear liquid into one of the comparator test tubes and place in the right-hand compartment of the comparator. At the same time carry out a blank determination using if possible 10 ml of the water which was employed as a solvent for the sample under examination, and the same quantities of all the reagents, and place the tube containing the carbon tetrachloride extract of this in the left-hand compartment of the comparator. This is to compensate for the presence of any copper already present in this water. Hold the comparator facing a standard source of white light such as the Lovibond White Light Cabinet or, failing this, north daylight, and compare the colour produced in the test solution with the colours in the standard disc, rotating the latter until a colour match is obtained. The amount of copper present in the test solution is then shown in the indicator recess in the right-hand bottom corner of the comparator.

Note

If the colour produced in the test is deeper than the standard colours the test should be repeated using a smaller quantity of the original liquid under examination, previously diluted to 10 ml with distilled water. Extremely minute traces of copper may be determined by evaporating an appropriate volume of the liquid under examination to 10 ml and then applying the method as described above. Alternatively, use *Method 4*.

Copper Method 4
using sodium diethyldithiocarbamate

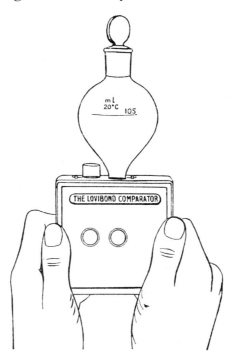

Principle of the method

Sodium diethyldithiocarbamate forms yellow copper diethyldithiocarbamate when added to alkaline solutions containing traces of copper. This coloured compound is soluble in organic solvents and extraction with methyl chloroform separates the copper diethyldithiocarbamate from the diethyldithiocarbamates of other metals, such as iron, which would otherwise interfere with the determination. The intensity of the yellow colour of the methyl chloroform extract is proportional to the copper concentration, and is measured by comparison with Lovibond permanent glass colour standards.

Reagents required

1. *Mixed reagent* The following parts, by weight, of reagents are ground together until the mixture is homogeneous, and the mixture is then stored in a brown bottle:—

    15 parts ammonium chloride (NH_4Cl)

    15 parts sodium carbonate (Na_2CO_3)

    30 parts ammonium citrate (($NH_4)_3C_6H_5O_7$ (copper free))

    0.4 parts sodium diethyldithiocarbamate

    This reagent is available from Tintometer Ltd.

2. *Methyl chloroform* (1,1,1-trichloroethane) $CH_3.CCl_3$

    All chemicals should be of analytical reagent quality.

The Standard Lovibond Comparator Disc 3/109

This disc covers the range 0.0125, 0.025, 0.05, 0.075, 0.10, 0.125, 0.15, 0.20, 0.25 mg/l (parts per million) of Copper (Cu) based on a 100 ml sample.

COLORIMETRC CHEMICAL ANALYTICAL METHODS   Copper Method 4

Technique

Into both the special extractor (See diagram) and the square test tube measure 5 ml of methyl chloroform. Measure into the extractor 100 ml of the water to be tested. Add 2-3 g (1 heaped capful) of the mixed reagent (reagent 1) to both the test tube and to the extractor using the funnel provided. Shake the extractor, to dissolve the reagent, and allow to stand for 5 minutes. To the test tube add pure copper-free water (preferably demineralised) to the 10 ml mark, fit the cap and shake to dissolve the reagent, and allow to stand for 5 minutes.

After standing for the required time, shake both the extractor and the test tube for 3 minutes to extract the yellow colour into the methyl chloroform layer (Note 1). Place the test tube in the left-hand compartment of the Lovibond Comparator, and the extractor tube in the right-hand compartment. Match the colour in the extractor tube against the permanent glass standards in the disc, using a standard source of white light such as the Lovibond White Light Cabinet, or failing this, north daylight.

Notes

1. "Bubbles" in the methyl chloroform layer may be cleared by gently rotating the tubes in their holders, or by warming the tube in water at 50° to 60°C for a few moments.

2. A special field kit containing:
      A Lovibond Comparator
      Disc (code 3/109) containing 9 permanent glass colour standards
      Glass extractor (100 ml capacity) to fit comparator
      Calibrated square test tube fitted with cap and 'O' ring
      Glass bottle of methyl chloroform
      Polythene wash bottle for distilled water
      Bottle of "mixed reagent" in powder form
      Polythene funnel for pouring powder into tube and extractor
may be obtained from Tintometer Ltd., Salisbury, England.

3. This method was devised by the British Non-Ferrous Metals Research Association.

Copper Method 5

Principle of the method

No reagent is used, the colour of the copper sulphate electro-plating solution itself being an adequate indication when viewed in a suitable apparatus.

The Standard Lovibond Comparator Disc 3/16

This disc covers the range 125 g/l to 325 g/l of copper sulphate in acid solution (H_2SO_4, 50 g/l), in nine steps of 25 g. The amount of copper metal may be calculated from the factor given.

In order to enhance the difference between the varying depths of the same blue colour, a special screen is interposed in front of the solution. This has the effect of making the colours range from a purple to a blue grey, and in consequence small variations in colour are readily distinguishable.

A 5 mm Lovibond Comparator cell must be used in conjunction with this disc.

Technique

The solution to be tested must be filtered bright before use, as a cloudiness due to suspended matter will render matching difficult.

Fill a 5 mm cell with the solution, and place it in the right-hand compartment, so that it comes behind the centre of the disc. Hold the comparator facing a standard source of light such as the Lovibond White Light Cabinet or, failing this, north daylight, and compare the colour of the solution with the colours in the standard disc, revolving the latter until a match is obtained. The value in grams per litre of copper sulphate is then read from the indicator recess. Intermediate values may be estimated.

To convert to ounces per gallon, divide the answer by $6\frac{1}{4}$. To convert to grams of copper metal per litre, divide the disc answer by 4 (or, more exactly, multiply by 0.254).

The Determination of Fluoride

Introduction

The fluoridation of public water supplies requires careful control of the fluoride concentration. The optium concentration for the reduction of the incidence of dental caries is 1 ppm of fluoride. This level is the sum of natural and added fluoride. A concentration of 5 ppm or more of fluoride is dangerous as it may produce mottling of the teeth and, in extreme cases, fluorosis of the bony structure. On the other hand waters containing little or no fluoride are associated with bad teeth formation and dental decay, especially in young children[1].

This need for the accurate control of the fluoride content of water supplies has produced a demand for a simple and reliable method of determining the fluoride concentration. Among the many methods which have been suggested, colorimetric methods based on the reaction between fluoride and a zirconium-dye lake[2] are believed, at the present time, to be the most satisfactory.

Methods of sample preparation

Air[3,4]

The field test for the determination of hydrogen fluoride in air[3] is unsuitable for use in atmospheres also polluted with fluoride-containing dusts, on account of the aluminium or phosphate which are almost certainly also present in the dust and which would interfere with the method used. For testing such atmospheres[4] collect the fluoride, vapour and dust, on an alkali impregnated filter paper. Release the fluoride by treatment with perchloric acid in a micro diffusion chamber and trap it in an alkali coating on the lid of the diffusion vessel. Transfer to a volumetric flask and proceed either by *Method 1* or *2*, or by the "Total Fluoride" method described in Section F of this book.

Chemicals[5]

Organic compounds Burn the sample in an oxy-hydrogen flame and pass the combustion products over a bed of quartz heated to 1000°C. Absorb the decomposition products in boric acid (H_3BO_3) solution and make this solution up to known volume.

Minerals, ores and rocks[6–8]

For the indirect determination of fluoride as fluorosilicic acid (H_2SiF_2) heat 5 mg of finely powdered sample with syrupy phosphoric acid as described in *Method 3*[6].

For the direct determination as fluoride fuse 200 mg – 1 g of sample with sodium carbonate (Na_2CO_3) and then either[3] digest the melt with 10 ml of $2N$ sulphuric acid, transfer to a steam-distillation apparatus with 45 ml of $18N$ sulphuric acid, steam distil at 145–150°C, pass distillate through a column of Deacidite FF ion exchange resin (OH form) until 300–400 ml of distillate have passed, and then elute the fluoride with 25 ml of $0.1M$ sodium acetate (CH_3COONa); or[8] digest the melt with ammonium carbonate (($NH_4)_2CO_3$), set aside for 12 hours then make just acid and dilute to 500 ml. Proceed using *Method 2* or *3*.

Miscellaneous[9]

Wood Cut the sample into sections less than 0.3 mm thick. Extract these sections with 1 ml of a solution of $2N$ caustic soda (NaOH) and 1 drop of 50 volumes hydrogen peroxide (H_2O_2) for 5 minutes at 130°C. Determine fluoride on an aliquot of this extract.

Water[10–18,22–25]

If the water sample is free from interfering substances no preliminary treatment is necessary. In *Method 1*[10] free chlorine, sulphates (500 ppm), orthophosphates (2 ppm), metaphosphates (0.5 ppm), aluminium (0.1 ppm), iron (5 ppm), ammonium ion (5 ppm), bicarbonate (220 ppm), chloride (1000 ppm), nitrate (200 ppm), calcium (200 ppm) and magnesium (200 ppm) all give erroneous results. Chloride can be removed by treatment with 0.1 ml of sodium arsenite solution. If impurity levels in excess of those listed are present in the sample then a Willard-Winter[11,12] steam distillation of the sample from perchloric acid ($HClO_4$) must be carried out. This is reported[10] to lead to results which may be up to 8% low.

However *Method 2*, which is a novel development of the zirconium-alizarin reaction[13-18], cancels out all interference with the exception of that from aluminium or high phosphate concentration. The aluminium may be removed by a simple pre-treatment of the sample with an ion exchange resin[15] or a supplementary procedure is available to compensate for this interference.

Methods of concentration

If it is necessary to measure fluoride concentrations below the lowest concentration for which a standard is provided in the disc either take[19] a 250 ml sample add 20 ml of a magnesium hydroxide ($Mg(OH)_2$) suspension, heat for 5 minutes at 90–100°C and decant; dissolve the precipitate in $0.4M$ nitric acid at 30–40°C and dilute to 50 ml with twice-distilled water; or alternatively[20] pass the sample through a column of Dowex 2-X8 resin (Cl^- form) and elute fluoride with $0.005M$ beryllium acetate (($CH_3COO)_2Be$) in $0.1M$ acetic acid (CH_3COOH).

The individual tests

Method 1 is based on the bleaching effect of fluoride on a lake formed by zirconium and alizarin Red S, and was developed by Cooke et al[10] from the original test due to Megragian and Maier[21], specifically for use in the field. The temperature-colour development time must be carefully controlled. It is specified in B.S.1427[22].

Method 2 is Palin's[13-18] development of the zirconium-alizarin test. This method, which uses tablet reagents, overcomes all interference except those from aluminium and high phosphate concentrations[26] and requires no precise control of time or temperature. A simple supplementary procedure is available for the removal of aluminium interference. An alternative procedure for measuring the amount of added fluoride is described and this procedure is also free from interference.

Method 3 is the indirect method developed by Shapiro[6] in which the fluoride in the sample is converted into the volatile fluorosilicic acid (H_2SiF_6) which is separated by distillation, dissolved in hydrochloric acid and the silicon content of the solution measured by one of the standard methods for silicon. This method is appropriate to samples other than water.

References

1. "*Manual of British Water Engineering Practice*", The Institute of Water Engineers, London, 3rd. Edn., 1961
2. R. D. Scott, *J. Amer. Water Wks. Assoc.*, 1941, **33**, 2018
3. B. S. Marshall and R. Wood, *Analyst*, 1968, **93**, 821
4. B. S. Marshall and R. Wood, *Analyst*, 1969, **94**, 493
5. D. E. Willis and W. T. Cave, *Analyt. Chem.*, 1964, **36**, 1821
6. L. Shapiro, *Prof. Pap. U.S. geol. Surv.*, 575D, 1967, D233-D235; *Analyt. Abs.*, 1969, **16**, 186
7. W. H. Evans and G. A. Sergeant, *Analyst*, 1967, **92**, 690
8. A. Hall and J. N. Walsh, *Analyt. chim. Acta*, 1969, **45** (2), 341; *Analyt. Abs.*, 1970, **19**, 151
9. A. I. Williams, *Analyst*, 1969, **94**, 300
10. J. R. Cooke, E. J. Dixon and R. Sawyer, *Proc. Soc. Water Treatment and Exam.*, 1965, **14**, 145
11. H. H. Willard and O. B. Winter, *Ind. Eng. Chem. Anal. Ed.*, 1933, **5**, 7
12. "*Standard Methods for the Exam. of Water Sewage and Industrial Wastes*", Amer. Pub. Health Assoc., New York, 10th Edn., 1955
13. A. T. Palin, *J. Inst. Water Engineers*, 1963, **17**, 347; *Analyt. Abs.*, 1965, **12**, 1460
14. A. T. Palin, *J. Inst. Water Engineers.*, 1964, **18**, 47; *Analyt. Abs.*, 1965, **12**, 2014
15. A. T. Palin, *J. Inst. Water Engineers*, 1966, **20**, 135
16. A. T. Palin, British Patent No. 1,006,908
17. A. T. Palin, British Patent No. 1,053,184; *Analyt. Abs.*, 1967, **14**, 6478

18. A. T. Palin, *J. Amer. Water Wks. Assoc.*, 1967, **59** (2), 255; *Analyt. Abs.*, 1968, **15**, 2997
19. M. Y. Shapiro and V. G. Kolesnikova, *Zhur. analit, Khim.*, 1963, **18**, 507; *Analyt. Abs.*, 1964, **11**, 1982
20. F. S. Kelso, J. M. Matthews and H. P. Kramer, *Analyt. Chem.*, 1964, **36**, 577
21. S. Megregian and F. J. Maier, *J. Amer. Water Wks. Assoc.*, 1952, **44**, 239
22. British Standards Inst. "*Routine control methods of testing water used in industry*". B.S.1427: 1962, Amendment 1: 1966
23. G. I. Carver, *Water Treatment and Exam.*, 1968, **17**, 216
24. K. C. G. Stroud, *Water Treatment and Exam.*, 1968, **17**, 226
25. H. D. Thornton, *Water Treatment and Exam.*, 1968, **17**, 240
26. S. J. Patterson, N. G. Bunton and N. T. Crosby, *Proc. Soc. Water Test. Eng.*, 1969, **18**, 182

COLORIMETRIC CHEMICAL ANALYTICAL METHODS

Fluoride Method 1

Principle of the method

Zirconyl chloride and sodium alizarin sulphonate react, in acid solution, to form a brilliant reddish-violet lake. This lake is destroyed by fluoride ions and the pale yellow colour of the alizarin reappears. Varying concentrations of fluoride produce a range of colours from red to yellow. These colours are matched against Lovibond permanent glass colour standards. The purity of both the alizarin and the zirconyl chloride is critical, and methods are therefore given for the preparation of purified samples of these two compounds.

Reagents required

1. *Recrystallised alizarin red S* Heat 2 g of alizarin red S to boiling with about 25 ml of 80% v/v ethyl alcohol (C_2H_5OH) in a flask fitted with a short air-condenser. Add more alcohol as necessary until no more solid dissolves, and then filter the solution, while still hot, through a medium grade filter paper. Allow the solution to cool. Filter off the yellow-brown needles which separate, using suction. Recrystallise a second time using the same procedure. Dry at about 105°C for about an hour and store the purified material in a desiccator.

2. *Pure zirconyl chloride* Dissolve 2 g of zirconyl chloride ($ZrOCl_2.8H_2O$) in about 6 ml of water, and filter if there is any undissolved material. Add 20 ml of hydrochloric acid (HCl) while stirring. Filter off the precipitated zirconyl chloride and remove as much as possible of the HCl by suction. Dry the precipitate to constant weight over saturated aqueous sodium hydroxide (NaOH) in an evacuated desiccator at a pressure of 15-20 mm of mercury. A second purification stage may be neceessary if a precipitate occurs in reagent 4 on standing. A 0.900% solution of zirconyl chloride in de-ionised water should have a specific gravity of 1.00480-1.00505 at 20°C.

3. *Acid zirconium alizarin red S reagent*

i) Dissolve 0.045 g of the freshly dried zirconyl chloride (reagent 2) in a few ml of water in a 100 ml measuring flask.

ii) Dissolve 0.035g of recrystalised alizarin red S (reagent 1) in a few ml of water.

iii) Dilute 35.0 ml of sulphuric acid (H_2SO_4) to 250 ml with water.

iv) Add the alizarin red S solution to the zirconyl chloride solution and mix. Add 50 ml of the diluted H_2SO_4, make up to the 100 ml mark with distilled water and mix. Stand for at least one hour and store in the dark.

4. *Sodium arsenite solution* Dissolve 1.0 g of $NaAsO_2$ in 100 ml of distilled water.

All the chemicals used should be of analytical reagent quality.

As an alternative to reagent 3, Palin A-Z tablets may be used (Note 1).

The Standard Lovibond Nessleriser Disc NOO

This disc contains 9 standards corresponding to 0.5, 0.6, 0.7, 0.8, 0.9, 1.0, 1.1, 1.2 and 1.3 ppm of free fluoride (F^-). This method will not measure fluoride complexed with aluminium (Note 5).

The master disc against which all reproductions are checked was tested and approved by Dr. J. R. Cooke (Note 2).

Technique

Place 50 ml of the sample in a Nessler cylinder (Note 3) add about 0.1 ml of sodium arsenite solution and mix. Add 1 ml of freshly shaken reagent 3 and immediately mix. Place the sample tube in a bath at 20°C for 20 minutes. After this time remove the tube from the bath, wipe off any water from the outside of the tube and place the tube in the right-hand compartment of the Nessleriser. Place the Nessleriser before a standard source of white light, such as the Lovibond White Light Cabinet or, failing this, north daylight, and compare the colour of the sample with the colours in the disc.

COLORIMETRIC CHEMICAL ANALYTICAL METHODS — Fluoride Method 1

Notes

1. Instead of making up reagent 3 it may be more convenient to use Palin A-Z tablets, obtainable from Tintometer Ltd. These tablets contain an acid alizarin-zirconium mixture. Alongside the sample cylinder prepare a second cylinder containing 1 ppm F. in distilled water, for which purpose standard fluoride tablets are available. Add 1 A-Z tablet to each cylinder, crush to dissolve, and match when both readings fall within the disc range. Add to or subtract from the sample reading, as appropriate, the difference between the standard cylinder reading and the figure 1.0 (A. T. Palin, *Water Treatment and Exam*. 1965, 14, 189). The distilled water used should be at or about the same temperature as the sample.

2. The procedure using the liquid reagent was developed by Dr. J. R. Cooke and his colleagues at the Laboratory of the Government Chemist, and permission to quote from his original paper is gratefully acknowledged.

3. Nessler tubes should be used which correspond to B.S.S. 612:1952, i.e. tubes in which the height of the 50 ml calibration mark falls at 113 ± 3 mm measured internally.

4. A reaction time of 20 minutes at $20 \pm 1°C$ is considered to be the best compromise between the time and temperature requirements of this reaction, and the disc was standardised under these conditions. However, if the temperature is held at $20°C$ in a thermostatted bath then the time can vary from 13 to 32 minutes without introducing an error of more than ± 0.1 ppm. Alternatively a colour development time of exactly 20 minutes will give results of the same accuracy for sample temperatures from 14 to $33°C$. The need for close adherence to time and temperature is obviated by the use of Palin tablets (see reference above).

5. The following concentrations of impurities will induce errors of up to 0.1 ppm:—sulphate 500 ppm, orthophosphate 2 ppm, chloride 1000 ppm, bicarbonate 220 ppm, nitrate 200 ppm, calcium 200 ppm, magnesium 200 ppm, aluminium 0.1 ppm, iron 5 ppm, ammonium ions 5 ppm, and metaphosphate 0.5 ppm. Of these only aluminium, and to a smaller extent bicarbonate, are likely to cause significant interference. If impurities in excess of these levels are known to be present then the sample must be steam distilled from perchloric acid ($HClO_4$), or *Method 2* used instead. Samples subjected to steam distillation may give results up to 8% low.

Fluoride Method 2
using acid alizarin-zirconium (Palin A-Z)

Principle of the method

Zirconyl chloride and sodium alizarin sulphonate react, in acid solution, to form a brilliant reddish-violet lake. This colour is destroyed by fluoride ions to give the pale yellow colour of the alizarin sulphonate. Thus varying amounts of fluoride ion produce a range of colours from red to yellow and the concentration of fluoride can be measured by comparing the colour produced with a range of Lovibond permanent glass colour standards.

In the Palin version of the method a novel procedure is introduced in order to cancel out interference. This makes use of a prepared blank in which the fluoride has been rendered inactive by complexing it with an excess of an aluminium salt. Elimination of any remaining interference caused by the possible presence of residual alum in waters which have undergone treatment with aluminium sulphate, or similar types of coagulant, requires a simple preliminary treatment with an ion-exchange resin. The only remaining interference is that due to phosphates. Normally there is insufficient natural phosphate in waters to interfere significantly. Where waters are treated with hexametaphosphate amounts up to 2 ppm cause no interference.

Compensation is also provided for variations in reagent quality, temperature, and time of colour development.

Reagents required

To simplify the use of this test in the field all the necessary reagents have been prepared in tablet form. The following tablets are required:—

Standard Excess-AL procedure
  i) A-Z tablets (acid alizarin-zirconium mixture).
  ii) Excess-AL tablets (containing aluminium ammonium sulphate).

Supplementary procedure
  iii) Resin tablets (for the elimination of residual aluminium interference).

The Standard Lovibond Nessleriser Discs NOM and NOT

Disc NOM is a wide range disc for general use in determining natural and added fluoride. It has nine standards which cover the range 0-1.6 ppm of fluoride (as F), in steps of 0.2.

Disc NOT is a narrow range disc for use in the control of water fluoridation. It has nine standards with increments of 0.1 ppm, covering 0.2 to 0.5 and 1.1 to 1.5 inclusive. Intermediate values are found "by difference" as explained below.

The master discs against which all reproductions are checked were tested and approved by Dr. A. T. Palin.

Technique

Standard Excess-AL procedure Place 50 ml of sample in a Nessler cylinder and add one "Excess-AL" tablet. Crush the tablet with the flattened end of a stout glass rod and stir until the powder is completely dissolved. Any turbidity which develops at this stage can be ignored. Place a further 50 ml of sample in another Nessler cylinder. To each cylinder add one "A-Z" tablet. Crush and stir rapidly until all the powder is dissolved. If the same rod is used for both, make sure that it is thoroughly rinsed before being transferred to the second cylinder. Place a further 50 ml of sample in a third Nessler cylinder and place this in the left-hand compartment of the Nessleriser. Allow the two treated samples to stand until the developed colours are both within the range of the standards on the disc (15-60 minutes). Place the cylinders in turn in the right-hand compartment of the Nessleriser and match the colour against the standards in the disc using a standard source of white light such as the Lovibond White Light Cabinet or, failing this, north daylight. The difference between the fluoride readings from these two cylinders gives the fluoride content of the sample in parts per million.

Allowing the samples to stand for too long before matching will result in the "Excess-AL" colour being deeper than the lowest standard. Too short a development time on the other hand will result in the colour of the sample in the second cylinder being lighter than the highest standard. In using the narrow range disc NOT, a shorter time period, say 15–30 minutes, is desirable.

Supplementary resin treatment Add about 150 ml of sample to a stoppered bottle of 200 to 250 ml capacity, followed by one resin tablet. Replace the stopper and shake. After disintegration of the tablet, which is rapid, continue shaking for about half a minute. Filter through a Whatman No. 1 filter paper discarding the first 10 to 15 ml of filtrate. Then collect two separate 50 ml portions in Nessler cylinders. This treatment will eliminate interference from "residual alum".

Thereafter follow the normal procedure by adding one "Excess-AL" tablet to the first cylinder, dissolving and then adding one "A-Z" tablet to each cylinder. If the original sample is turbid a portion of this should be filtered direct to provide 50 ml for the cylinder placed in the left-hand compartment of the Nessleriser.

Notes

1. To determine *Total Fluoride*, as normally required for fluoridation control at waterworks and in distribution systems, follow the Excess-AL technique.

This result includes both natural and artificial fluoride. If residual alum interference is expected apply the preliminary resin treatment.

2. To determine *Free Fluoride* in circumstances where some of the fluoride may be complexed with natural aluminium or "residual alum" use the Excess-AL method without preliminary resin treatment. The difference between this result and that for total fluoride will correspond to complexed fluoride.

3. To check the *Fluoride Dose* at treatment works determine total fluoride before and after fluoridation. The difference represents artificially added fluoride.

4. In the case of water samples which are turbid or discoloured by the presence of iron, it may be found desirable for correct colour matching to acidify the 50 ml of sample placed in the left-hand compartment of the Nessleriser by the addition of 2-3 ml of dilute sulphuric acid.

5. The readings obtained by means of this disc and a Nessleriser are accurate only provided that Nessler cylinders are used which conform to the specification of those used in the original calibration i.e. that the 50 ml calibration mark shall fall at a height of 113 ± 3 mm from the bottom of the cylinder when measured internally.

Fluoride Method 3
using fluorosilicic acid

Principle of the method
The fluoride is converted into the volatile fluorosilicic acid (H_2SiF_6) which is distilled into hydrochloric acid. The silicon content of this solution, which is proportional to the fluoride content, is determined by one of the standard methods for the determination of silica.

Reagents required
1. *Phosphoric acid* (H_3PO_4) Syrupy
2. *Hydrochloric acid* (*HCl*) 0.4% Dissolve 12.5 ml of hydrochloric acid (sp. gr. 1.16) in water and dilute to 1 litre.
3. Reagents for method chosen for determination of silicon.

The Standard Lovibond Comparator Disc 3/13 and Nessleriser Discs NN, and NV
Disc 3/13 covers the range 2.5 to 25.0 ppm of silica (SiO_2).
Disc NN covers the range 0.05 to 1.0 mg of silica (SiO_2).
Disc NV covers the range 0.2 to 1.0 ppm of silica (SiO_2).

To convert the disc readings to the equivalent fluoride concentrations multiply by 1.9. For many purposes doubling the reading will be sufficiently accurate.

Technique
Finely powder the sample and transfer 5 mg of the powder to the bottom of a borosilicate-glass test tube, by means of a long-stemmed funnel, so that no sample is left on the sides of the tube. Add 3 drops of syrupy phosphoric acid (reagent 1) and place 3 drops in an empty test tube to act as a blank. Fit into each tube a rubber stopper carrying a short piece of thick-walled glass capillary tubing and wrap a 1 inch strip of wet filter paper around the upper part of each tube immediately below the stopper. Heat the bottom of each tube over a low Meker burner flame holding the tube at 30° to the horizontal. In about 1 minute a thick coating of silica will form as a result of the attack of the phosphoric acid on the test tube. Increase the flame temperature and heat the lower third of the tube strongly for 30 seconds. Set aside to cool for between 5 and 40 minutes. During this time prepare two 150 ml beakers, one for each tube, by adding 50 ml of 0.4% hydrochloric acid (reagent 2) to each. Take a piece of glass tubing 18 inches long and of 3 mm external diameter and bend this into a U-tube to fit into a 100 ml measuring cylinder, with one arm $5\frac{1}{2}$ inches long. Transfer the acid from one of the beakers to the 100 ml graduated cylinder, remove the filter paper and stopper from one of the test tubes, invert the test tube and insert the short arm of the U-tube to the bottom of the inverted tube. Using the long arm of the U-shape as a handle lower the test tube to the bottom of the cylinder. When the acid has risen two thirds up the test tube, prevent any further rise by placing a finger tip on the open end of the U-tube, to prevent the sample being wetted. Raise and lower the test tube 5 times to well wash its neck then remove it and discard it. Transfer the acid from the measuring cylinder back to the beaker. Repeat for the other test tube washing with acid from a different beaker. Determine the silica content of these solutions by Silicon *Method 1* or *Method 2,* and subtract the silica content of the blank from that of the sample.

The Determination of Hydrazine

Introduction

The use of hydrazine for the treatment of boiler feed water has been recommended by a number of workers[1,2,3]. Its main advantage in this application is claimed[1] to be its superior performance, as compared with sodium sulphite, in the removal of dissolved oxygen.

Several methods have been suggested for the determination of hydrazine, including oxidation with permanganate[4], potentiometric titration with iodine chloride[5] and colorimetric estimation using 2-nitro-1,3-indanedione[6].

Methods of sample preparation

Air[7]

Draw the air at not more than 5 litres a minute through 10 ml of 8% v/v sulphuric acid in an impinger type bubbler and proceed by *Method 2*.

Water[8–11]

No sample preparation is normally required.

The individual tests

Method 1 is the standard[9,10] *p*-dimethylaminobenzaldehyde test which is based on an A.S.T.M. test[8] and has been specially developed for the examination of boiler feed waters. The analytical conditions for this test have recently been critically examined by Gamble[12] who claims that the best conditions for colour development are given by a solution which is $0.2M$ in hydrochloric acid, $11.4~\mu M$ in hydrazine (N_2H_4) and $0.04M$ in para-dimethylaminobenzaldehyde ((CH_3)$_2NC_6H_4CHO$).

Method 2 is a modification of *Method 1* which uses a single stable reagent of para-dimethylaminobenzaldehyde dissolved in normal sulphuric acid.

References

1. E. R. Woodward, *Power,* 1956, **100,** 80
2. E. R. Woodward, *Petroleum Refiner,* 1956, **35,** 208
3. G. Genin, *Chaleur & ind.,* 1957, **38,** 57
4. M. Issa and R. M. Issa, *Analyt. chim. Acta,* 1956, **14,** 578
5. J. Cihalik and K. Terebova, *Chem. Listy,* 1956, **50,** 1768
6. G. Vanags and M. Mackanova, *Zhur. Anal. Khim.,* 1957, **12,** 149
7. W. Pilz and E. Stelzol, *Z. analyt. Chem.,* 1966, **219** (5), 416; *Analyt. Abs.,* 1967, **14,** 7161
8. A.S.T.M. Designation, D1385, 57T
9. Brit. Standards Inst., "*Routine Control Methods of Testing Waters used in Industry*", B.S.1427, London, 1962
10. Brit. Standards Inst., "*Methods of Testing Water used in Industry*", B.S.2690, London, Part 2, 1965
11. Brit. Standards Inst., "*Treatment of Water for Marine Boilers*", B.S.1170, London, 1968
12. D. S. Gamble, *Can. J. Chem.,* 1968, **46** (8), 1365; *Analyt. Abs.,* 1969, **17,** 1398

Hydrazine Method 1
using p-dimethylaminobenzaldehyde

Principle of the method

p-Dimethylaminobenzaldehyde reacts specifically with hydrazine giving a yellow colour. The intensity of this colour, which is proportional to the concentration of hydrazine, is estimated by comparison with Lovibond permanent glass standards.

Reagents required

1. *p-Dimethylaminobenzaldehyde reagent*

p-dimethylaminobenzaldhyde (See Note 1)	4.0 g
Methyl alcohol (CH_3OH)	200 ml
Hydrochloric acid (HCl 36%)	20 ml

2. *Hydrochloric acid, 5N* Approx. 600 ml conc. acid per litre

The Standard Lovibond Nessleriser Disc NOH

The disc covers the range 0 to 10 microgrammes of hydrazine (N_2H_4) in steps 0, 0.5, 1, 2, 3, 4, 6, 8, 10 µg. On a 10 ml sample this is equivalent to 0–1.0 ppm, or on a 25 ml sample this equals 0 to 0.4 ppm.

Technique

Place a 10 or 25 ml fresh sample of the boiler feed water (Note 2) in a 50 ml Nessleriser cylinder (Note 3). Add 1 ml of $5N$ hydrochloric acid and 10 ml of the p-dimethylaminobenzaldehyde reagent and mix thoroughly. Adjust the volume to 50 ml with distilled water, mix and allow to stand for 10 minutes. Place the cylinder in the right-hand compartment of the Nessleriser. Fill an identical cylinder with 10 ml of unreacted reagent and 40 ml of distilled water, and place in the left-hand compartment. Match the colour of the sample with that of the permanent glass standards in the disc, using a standard source of white light, such as the Lovibond White Light Cabinet or, failing this, north daylight.

Notes

1. It is essential to use a specially purified grade of p-dimethylaminobenzaldehyde for this determination, as the yellow colour of solutions prepared from ordinary grades of this material prevents correct readings being obtained. The colour of reagent 1, in a filled Nessleriser cylinder should not exceed that of the zero standard in the disc. A prepared and standardised reagent is available from BDH Chemicals Ltd.

2. The sample must not be collected at a temperature exceeding 21°C (70°F), and an efficient cooling coil should be fitted at the sampling point if necessary, to ensure that this temperature is not exceeded.

3. The readings obtained by means of the Lovibond Nessleriser and disc are only accurate provided that the Nessleriser glasses which are used conform to the specification employed when the disc was calibrated, that is that the 50 ml calibration mark shall fall at a height of 113 ± 3 mm measured internally.

Hydrazine Method 2
using *p*-dimethylaminobenzaldehyde

Principle of the method
This modification of *Method 1* uses a single stable reagent of *p*-dimethylaminobenzalhyde dissolved in normal sulphuric acid. This reagent reacts specifically with hydrazine to produce a yellow colour. The intensity of this colour, which is proportional to the hydrazine concentration, is measured by comparison with a series of Lovibond permanent glass colour standards.

Reagent required
A 1% solution of *p*-dimethylaminobenzaldehyde in normal sulphuric acid. If stored in the dark at a temperature not exceeding 25°C this reagent is stable for at least two years (Note 1). It is essential to use a specially purified grade of *p*-dimethylaminobenzaldehyde for this test.

The Standard Lovibond Comparator Disc 3/85
This disc contains standards corresponding to 0, 0·1, 0·2, 0·3, 0·4, 0·5, 0·6, 0·8, and 1·0 ppm of hydrazine (N_2H_4).

Technique
Place 5 ml of water sample (Note 2) in a comparator tube, add 5 ml of the reagent and mix by shaking the tube. Place the tube in the right-hand compartment of the comparator and place the comparator in front of a standard source of white light, such as the Lovibond White Light Cabinet or, failing this, north daylight. After 2 minutes match the colour of the sample solution with the colours in the disc and read off the hydrazine concentration from the indicator window in the comparator.

Notes
1. To check the reagent during storage carry out the test as described above using tap water as the sample. After shaking, the colour of the solution should not be darker than the zero standard in the disc.
2. The sample must not be tested at a temperature exceeding 21°C. An efficient cooling coil should be fitted at the sampling point or the sample must be cooled immediately under running water. Turbid samples must be filtered before the test is carried out.

The Determination of Hydrogen Peroxide

Principle of the method

The test is based on the reaction between hydrogen peroxide and a titanium (1V) salt in acid solution to produce a yellow pertitanic acid complex. The intensity of the yellow colour, which is proportional to the concentration of hydrogen peroxide, is measured by comparison with Lovibond permanent glass colour standards.

Reagent required

Potassium titanium oxalate reagent. Dissolve 5.0 g of Potassium titanium oxalate ($K_2TiO(C_2O_4)_2.2H_2O$) in 100 ml of 10% (by volume) sulphuric acid. CAUTION:— the reagent solution is toxic and corrosive.

All chemicals used in the preparation of reagents should be of analytical reagent quality.

The Standard Lovibond Comparator Disc 296473

The disc covers the range 3, 5, 10, 15, 20, 30, 40, 50, 70 ppm. H_2O_2.

Technique

Fill two 13.5 mm Comparator test tubes to the 10 ml mark with the sample under test. To one of the tubes add 1.0 ml of the reagent by means of a dropping pipette or similar means, a mouth pipette must not be used. The contents of the tube should be mixed thoroughly and the tube then placed in the right hand compartment of the Comparator. The other, untreated tube, should be placed in the left hand compartment. After two minutes place the Comparator before a source of light such as the Lovibond white light cabinet, or, failing this, north daylight and match the colour of the sample solution with the standards in the disc.

Notes

1. Higher concentrations of Hydrogen Peroxide may be measured by diluting a known volume of sample with demineralised or distilled water in the test tube, proceeding as above and then multiplying the disc reading by the dilution factor.

2. Acknowledgement is made to Laporte Industries Limited for assistance in developing this disc.

The Determination of Hydrogen Sulphide

Introduction

Hydrogen sulphide is a very poisonous gas which is encountered in many branches of industry. It is also found during the decomposition of organic matter containing sulphur. Typical industries in which it is encountered include chemical, artificial silk, and dyeing works, coke oven plants and gas works, grease and petroleum refining works etc. Hydrogen sulphide is also encountered in sewage works and during the cleaning of sewers, septic tanks and cesspools.

Despite its strong characteristic smell, hydrogen sulphide is an insidious poison. The smell may not be recognised in concentrations at which it is highly dangerous. Furthermore, workers rapidly become accustomed to the smell, as all students of chemistry know, and increases in the concentration may pass un-noticed. Hydrogen sulphide can cause sudden unconsciousness without premonitory symptoms and the effects of accidents arising in this manner are often as serious as the direct effects of the poison.

It is reported[1,2] that concentrations as low as 50 ppm can produce irritation to the eyes, nose and respiratory tract and exposures to concentrations of 200 ppm, even for short periods, will result in serious toxic effects. Concentrations of 1000 ppm or higher are believed to be rapidly fatal. If it is found that the atmosphere contains a concentration of 20 ppm or more, this indicates that working conditions are unsatisfactory.

Methods of sample preparation

Air[2-4]

If *Method 2*[2] is being used no sample preparation is required. Otherwise[3] absorb the hydrogen sulphide (H_2S) from one cubic foot of air in a Dreschel gas wash bottle fitted with a sintered glass (G1 or G2 porosity) distribution tube, containing 10 ml of a 1% solution of cadmium acetate (($CH_3COO)_2Cd.2H_2O$) rendered slightly acidic with acetic acid (CH_3COOH), and proceed by *Method 1*.

For the determination of hydrogen sulphide in the presence of sulphur dioxide (SO_2) it is recommended[4] that the sulphur dioxide be scrubbed out of the sample stream with a tube containing sodium acetate (CH_3COONa).

Beverages[5]

Wine Take 50 ml of wine and place this in a flask fitted with an inlet tube and air condenser. Heat to 50–60°C and aspirate 5 litres of air through the wine. Estimate hydrogen sulphide in the effluent by *Method 2*. It is claimed[5] that as little as 0.01 mg of hydrogen sulphide per litre of wine can be detected by this method, and this is far less than the lowest concentration which can be detected in the wine by taste.

Dairy products[6]

Cheese Take 10 g of freshly grated cheese and place this in a flask fitted with inlet and exit tubes, the inlet tube ending 4 mm above the bottom of the flask. Add 20 ml of distilled water and 4 ml of 12.5% sulphuric acid. Immerse the flask in a glycerol bath and boil at 105–108°C for 45 minutes while passing nitrogen through the flask at 150–200 ml per minute. Either determine hydrogen sulphide directly on the nitrogen stream by *Method 2*, or collect the hydrogen sulphide as described for air above, and proceed by *Method 1*.

Miscellaneous[2,3]

Town gas No sample preparation is required for *Method 2*[2]. Alternatively[3] absorb the hydrogen sulphide as described for air above and proceed by *Method 1*.

Sewage and industrial wastes[7-9]

The same precautions should be taken as are described for water.

Water[10]

Care must be taken to avoid loss of hydrogen sulphide during and after sampling, either by loss of free hydrogen sulphide to the atmosphere or by oxidation. Loss of hydrogen sulphide can be avoided by the addition of a little zinc acetate (($CH_3COO)_2Zn.2H_2O$) to the water immediately on sampling and then proceed by *Method 1*. Alternatively[4] release the hydrogen sulphide from the sample or from the precipitated sulphide by boiling with dilute sulphuric acid and either trap the hydrogen sulphide as before for estimation by *Method 1* or determine the evolved hydrogen sulphide directly by *Method 2*.

The individual tests

Method 1 uses the methylene blue colour reaction originally described by Lindsay[10] and subsequently investigated and modified by Fogo and Popowsky[11]. Methyl and ethyl mercaptans are not absorbed by the cadmium acetate solution and therefore do not interfere. Interference by organic matter, alkalinity, nitrites and nitrates is also avoided in this method[12].

Method 2 is the standard method described by the Department of Employment[2] for the analysis of industrial atmospheres. It uses the formation of a stain of lead sulphide by passing the hydrogen sulphide through a paper impregnated with lead acetate.

References

1. Ministry of Labour, *"Toxic Substances in Factory Atmospheres"*, H.M. Stationery Office, London, 1960
2. Department of Employment, *"Methods for the Detection of Toxic Substances in Air, Booklet No. 1, 'Hydrogen Sulphide',"* H.M. Stationery Office, London, 1969
3. A. E. Sands, M. A. Grifins, H. W. Wainwright and M. W. Wilson, *U.S. Bureau of Mines Rept. Invest.*, No. 4547, 1949, 19
4. A. A. Pavlenko and T. A. Kuz'mina, *Trudy glav. geofiz. Obs. A.I. Voeikova*, 1968, (234), 188; *Analyt. Abs.*, 1970, **18**, 1291
5. T. Standenmeyer, *Z. Lebensmitt Untersuch*, 1961, **115**, 16; *Analyt. Abs.*, 1962, **9**, 874
6. S. Poznanski, *Roczn. Technol. Chem. Zywnosci*, 1962, **9**, 17; *Analyt. Abs.*, 1963, **10**, 2479
7. Min. of Housing and Local Govt., *"Methods of Chemical Analysis as applied to Sewage and Sewage Effluents"*, H.M. Stationery Office, London, 1956
8. R. Pomeroy, *Sewage Wks. J.*, 1936, **8**, 572
9. R. Pomeroy, *Sewage Wks. J.*, 1941, **13**, 502
10. W. G. Lindsay, *Columbia School of Mines Quart.*, 1902, **23**, 24
11. J. F. Fogo and M. Popowsky, *Analyt. Chem.*, 1949, **21**, 732
12. F. H. Heath and F. A. Lee, *J. Amer. Chem. Soc.*, 1923, **45**, 1643

Hydrogen Sulphide Method 1
(H_2S in solution)
using dimethyl-*p*-phenylene diamine dihydrochloride

Principle of the method

Hydrogen sulphide reacts with dimethyl-*p*-phenylene diamine dihydrochloride, in the presence of ferric chloride in a hydrochloric acid medium, to produce methylene blue. The intensity of the methylene blue colour, which is proportional to the hydrogen sulphide concentration, is measured by comparison with a series of Lovibond permanent glass colour standards.

Reagents required

1. *Dimethyl-p-phenylene diamine solution* 0.5 g of dimethyl-*p*-phenylene diamine dihydrochloride in 100 ml of 1:1 hydrochloric acid. This solution should be practically colourless. Decolourise with charcoal if necessary. It is advisable to check each batch of this reagent against known hydrogen sulphide standards and the disc.
2. *Ferric chloride solution* 54 g of ferric chloride ($FeCl_3.6H_2O$) in 1 litre of concentrated hydrochloric acid. Dilute 1 volume with 4 volumes of distilled water immediately before use.

The Standard Lovibond Comparator Disc 6/14

This disc covers the range 0.2–1.6 ppm of hydrogen sulphide (H_2S) in steps of 0.2 ppm when 1 cu ft of atmosphere is sampled as in the test on page 503. For solutions, the corresponding microgram amounts in the volume of solution used for the analysis are given below.

This disc is designed to be used with 40 mm cells.

Technique

Take 10 ml of the sample solution in a 50 ml volumetric flask or Nessler cylinder, add 4 ml of reagent 1 followed by 1 ml of diluted reagent 2. Shake to ensure that any precipitated sulphide is dissolved, and make up to 50 ml. Transfer to a water-bath at 40°C and maintain at this temperature for 10 minutes to develop the colour. Cool and transfer to a 40 mm comparator cell. Place this cell in the right-hand compartment of the comparator.

At the same time as reagents 1 and 2 are added to the sample solution, prepare a blank from 4 ml of reagent 1 plus 1 ml of reagent 2 and dilute to 50 ml. Place this blank in the water-bath at the same time as the test solution, and after 10 minutes cool, transfer to a 40 mm comparator cell and place this cell in the left-hand compartment of the comparator. Hold the comparator facing a standard source of white light, such as the Lovibond White Light Cabinet or, failing this, north daylight, and compare the colour of the sample with the standards in the disc. Read off the answer from the indicator window in ppm of a 1 cu ft sample and convert this to micrograms of hydrogen sulphide by means of the Table below. In a 1 ml sample of solution containing hydrogen sulphide these figures correspond to parts per million.

Note

Disc reading ppm of H_2S on 1 cu ft sample	Equivalent micrograms H_2S
0.2	8.7
0.4	17.3
0.6	26.0
0.8	34.7
1.0	43.4
1.2	52.0
1.4	60.6
1.6	69.4

Hydrogen Sulphide Method 2
using lead acetate

Principle of the method

The concentration of hydrogen sulphide is measured by drawing 125 ml of the air through a test paper impregnated with lead acetate. The complete absorption of the hydrogen sulphide is ensured by the incorporation of glycerol into the test paper.

A suitable rubber bulb aspirator and test paper holder for use with this test may be obtained from Siebe Gorman Limited, Davis Road, Chessington, Surrey.

Reagents required

1. Lead acetate $(CH_3COO)_2 Pb.3H_2O)$ Analytical reagent quality.
2. Acetic acid (CH_3COOH) Glacial
3. Glycerol $(CH_2(OH)CH(OH)CH_2OH)$ Pure

Preparation of test papers

Dissolve 10 g of lead acetate in 90 ml of distilled water. Add 5 ml of glacial acetic acid and 10 ml of glycerol to the solution and mix well. This solution should be freshly prepared each time a batch of test papers is made.

Place the solution in a 100 ml measuring cylinder. Immerse strips of Whatman No. 3 MM chromatographic paper 20 mm wide and 100 mm long, vertically in this solution for one minute. Allow the excess liquid to drain off, suspend the papers vertically and allow to dry as completely as possible at room temperature in an atmosphere free from hydrogen sulphide. When dry, cut off the top and bottom 25 mm of the prepared papers and store the remaining 50 mm in stoppered wide-necked, dark glass bottles. Use the test papers within 14 days of preparation.

The Standard Lovibond Comparator Disc 6/38

This disc, which replaces Disc 6/2, contains four standards corresponding to 5, 10, 20 and 40 parts per million of hydrogen sulphide in a 125 ml sample. This disc should be used in conjunction with the Lovibond Comparator and a Test Paper Viewing Stand.

Technique

Fit the test paper centrally between the gaskets of the test paper holder and screw up the holder until it is finger-tight. Then attach the holder to the aspirator by means of a piece of rubber tubing, and draw air steadily through the paper at a rate of 125 ml per minute i.e., one inflation of the aspirator. When the aspirator has apparently fully inflated allow a further 15 seconds to elapse for the full capacity to be utilised.

Remove the test paper and place this in the right-hand aperture of the Test Paper Viewing Stand. Place the comparator and stand before a source of white light, such as the Lovibond White Light Cabinet or, failing this, north daylight and compare the colour of the stain with the Standards in the disc.

The stain should be only on that side of the test paper exposed to the gas entering the holder. If the stain is visible on both sides of the paper this indicates that the hydrogen sulphide has not been fully absorbed by the paper. Fresh test papers should be prepared and the test repeated with these.

COLORIMETRIC CHEMICAL ANALYTICAL METHODS

The Determination of Iodine
using diethyl-*p*-phenylene diamine (Palin DPD)

Introduction

Iodine is occasionally used in place of chlorine for the treatment of water. As the standard Palin DPD test for chlorine also reacts to iodine, this test can be used to determine residual iodine in treated waters, provided that chlorine is not also present. The use of the tabletted reagent makes this test especially easy and convenient.

Principle of the method

In the presence of free iodine diethyl-*p*-phenylene diamine gives a red colour. The intensity of this colour, which is proportional to the iodine concentration, is measured by comparison with a series of Lovibond permanent glass colour standards.

Reagent required

DPD tablet No. 1

The Standard Lovibond Comparator Discs 3/77A and 3/77B

Disc 3/77A covers the range 0.4 to 3.6 ppm of iodine.
Disc 3/77B covers the range 0.7 to 14.0 ppm of iodine.

Technique

Place in the left-hand compartment of the comparator a 13.5 mm comparator cell containing the sample only. Rinse an identical cell with the sample and leave enough liquid in the tube just to cover the tablet when added. Drop into this prepared tube one No. 1 DPD tablet and disintegrate. Add the water sample up to the 10 ml mark, mix rapidly and place the cell in the right-hand compartment of the comparator. Place the comparator before a standard source of white light, such as the Lovibond White Light Cabinet or, failing this, north daylight. Match the colour of the sample against the standard colours in the disc.

The Determination of Iron

Introduction

The determination of iron is of interest in most laboratories as it occurs, either as a major constituent or at the trace level, in so many industrial and natural products. Iron is frequently found in natural waters and, for most purposes, it is considered that its concentration in a water supply should not exceed 0.5 ppm. The presence of even this amount of iron can be a considerable nuisance in some industrial applications such as laundries and textile processing plants. Standard methods for the control of the iron content of water supplies have been specified[1,2].

Methods of sample preparation

Beverages[3-10]

Beer[3,4] Evaporate a 250 ml sample to dryness, ash the residue and dissolve the ash in 100 ml of 1:1 hydrochloric acid. To avoid loss of iron by evaporation the ashing temperature should be maintained at 560–580°C[4].

Fermentation broths[5] Digest a sufficient volume of sample to contain 0.02–0.15 mg of iron for 2 hours with 1 ml of 72% perchloric acid ($HClO_4$) and 1 ml of sulphuric acid in a Kjeldahl flask. Cool, add 1 ml of freshly prepared 20% sodium dithionite ($Na_2S_2O_4.H_2O$) solution and boil for 1 hour. Cool, slowly add 1–2 ml of water and then add 1 drop of 1% ethanolic *para*nitrophenol ($HOC_6H_4NO_2$ dissolved in C_2H_5OH) add 19% aqueous ammonia solution (NH_4OH) until the colour changes and then acidify to pH 2 with $10N$ sulphuric acid.

Wine[6-13] Wet[7,8] or dry[6] ash a 10–15 ml sample in a platinum dish. Dissolve the residue in acid, filter if necessary, dilute to a suitable volume and determine iron on an aliquot.

Alternatively[9] boil a 20 ml sample with 20 ml of 110 volume hydrogen peroxide (H_2O_2) on a sand bath until the volume of the solution is reduced to 2–5 ml. Add 3–4 ml of aqueous ammonia (NH_4OH) a drop at a time to the cooled concentrate until the solution becomes opalescent and smells of ammonia. Make up to 100 ml with normal hydrochloric acid and determine iron on an aliquot.

If it is required to determine both total iron and ferrous iron[10] then take 2–5 ml of sample, add 5 ml of acetate buffer (pH 4) and 2 ml of a 10% solution of hydroxylammonium chloride ($HONH_3Cl$). Heat on a boiling water bath for a few minutes and then allow to cool. Proceed by *Method 4*. To determine ferrous iron proceed by *Method 4* on the sample without any pretreatment and omit the addition of sodium pyrosulphite (*Method 4* reagent 1).

For a rapid field test using only simple equipment the use of *Method 4* without any sample preparation is recommended[11-13]

Biological materials[14-32]

Blood, plasma and serum[14-23] Planas[14] has examined the effect of the anticoagulant, used when taking the blood sample, on the subsequent determination of iron. He concludes that heparin is the best anticoagulant and that the next best is sodium citrate ($Na_3C_6H_5O_7.2H_2O$). To 5 ml of blood[15] add 20 ml of water and filter. To 2 ml of the filtrate in a 50 ml flask add 5 ml of sodium hypochlorite (NaOCl) solution and stand for 2 minutes to decolourise. Add 25 ml of water and 10 ml of 1:1 sulphuric acid. Mix. Add water almost to the mark, add 1 drop of octan-2-ol to prevent foaming, dilute to the mark, mix, and filter. Take a 10 ml aliquot of the filtrate and proceed by *Method 2*. Ceriotti[16] and Forman[17,18] prefer to use *Method 4* for the final determination.

For plasma[19,20] take a 2 ml sample, add 1 ml of $0.3N$ hydrochloric acid and heat on a water bath at 56°C for 20 minutes. Add 2 ml of 20% trichloracetic acid (CCl_3COOH) mix and centrifuge for 10 minutes. Remove the clear supernatant liquid and determine iron by *Method 2*. Alternatively[21,22] take 2 ml of serum, add 2 drops of redistilled thioglycollic acid ($HSCH_2COOH$), 1 ml of $2N$ hydrochloric acid and 0.5 ml of 40% trichloracetic acid. Stir with a glass rod for 45 seconds and then centrifuge for 10 minutes. Separate the supernatant liquid and then use an aliquot to determine iron by *Method 1*. For micro samples and for samples stored by drying on filter paper[23] take 0.1 ml of sample, hydrolyse with 3 ml of ethanolic potash (KOH in C_2H_5OH) at 70°C for 5 hours. Evaporate the aqueous phase to dryness, add 2 ml of nitric acid, 0.5 ml of sulphuric acid and 0.5 ml of perchloric acid. Heat to fumes, dissolve the residue in hydrochloric acid and determine iron on this solution.

Faeces[24,25] To 10 g of homogenised 24-hour faeces add 35 ml of $4N$ hydrochloric acid, boil for several minutes, cool, filter and make up to 50 ml by washing filter with $4N$ hydrochloric acid and adding the washings to the filtrate.

Tissue[26-28] Wet ash the sample. Evaporate the acid solution to dryness and ash the residue at 300°C. Dissolve the ash in 2 ml of $2N$ sulphuric acid, dilute to a suitable volume and determine iron on this solution.

Urine[25,29-32] Evaporate a 50 ml sample to dryness with 3 successive portions of nitric acid (5 ml, 1 ml, 1 ml). Dissolve the residue in 1 ml of sulphuric acid, decolourise with 1 ml of 30% hydrogen peroxide (H_2O_2) evaporate to fumes and cool. Dissolve the residue in water and determine iron on this solution[29-32].

The non-haem iron in urine may be determined directly on the supernatant liquid after precipitation of proteins with trichloracetic acid (Cl_3CCOOH) and separation of the precipitate by centrifugalization[25]. Barry[32] prefers the use of *Method 4* for the determination of iron on the solution from wet ashing.

Chemicals[33-36]

Catalysts[33] Dissolve the sample in hydrochloric acid.

Niobium compounds[34] Fuse the sample with either potassium bisulphate ($KHSO_4$) or caustic potash (KOH), extract the melt with $0.5N$ sulphuric acid and proceed by *Method 4*.

Organic compounds[35] Ignite 1 g of sample with 1 ml of sulphuric acid, 5 ml of nitric acid and 1 ml of a 5% solution of sodium chloride (NaCl), filter and dilute filtrate to a known volume.

Phosphorus[36] Dissolve 0.5 g of the sample in nitric acid and evaporate the solution until 1 ml remains. Add 75 ml of water, evaporate until volume remaining is about 35 ml, add 10 ml of a 10% solution of hydroxylammonium chloride ($HONH_3Cl$) set aside at room temperature for 15 minutes and determine iron on an aliquot.

Dairy products[37,38]

Milk and milk products Remove the fat by extraction with benzene (C_6H_6), decompose the organic matter by wet ashing, dissolve the residue in water and determine iron on this solution[37].

Alternatively[38] dilute an aliquot of the sample, containing 100–500 μg of iron, to 100 ml with water, then heat to boiling on a water bath. Add 50 ml of a 20% solution of trichloracetic acid (Cl_3CCOOH) and 25 ml of 24% hydrochloric acid, swirl continuously and keep on the water bath for 20 minutes with continuous mixing. Filter, and determine iron on the filtrate.

Food and edible oils[25, 39-45]

Fish[39] Clean the sample with distilled water and dry at 80°C for 48 hours. Grind. Place 0.2 g of the powder in a size 0 porcelain crucible, char at 200°C, raise the temperature to 550°C and combust for 16 hours. Add 0.5 ml of 1:1 nitric acid and when solution is complete dilute to a known volume.

Flour[40] Warm 5 g of sample with 20 ml of 30% aqueous hydrogen peroxide (H_2O_2) and 5 ml of sulphuric acid until solution is complete. Cool and dilute to 25 ml.

General[25,41] Either wet ash with nitric and perchloric ($HClO_4$) acids[25] or[41] ignite a 5–10 g sample. Evaporate the ash twice with 5 ml portions of 10% hydrochloric acid. Dissolve the dry residue in water and make up to 50 ml. Filter 5 or 10 ml of this solution into a flask, add 50 ml of water, 1 ml of hydrochloric acid and 3–5 drops of 30% hydrogen peroxide (H_2O_2). Make up to a convenient volume and determine iron on an aliquot using *Method 2*.

Molasses[42] Take a 5 g sample, dissolve it in 20 ml of water and pass the solution at 2 ml a minute through a column of a cation exchange resin. Wash the column with water and elute the iron with 95 ml of $2N$ hydrochloric acid. Determine iron in the eluate using *Method 2*.

Vegetable fats and oils[43-45] Ash the fat and extract the ash with hydrochloric acid. Shake the acid extract with a solution of lead diethyldithiocarbamate ((($C_2H_5)_2N.CS.S)_2Pb$) in chloroform ($CHCl_3$). Reject the chloroform layer and determine iron on the aqueous layer[43]. For vegetable oils either[44] carbonize the sample by heating under an infra-red heater and then heating in a muffle furnace at a temperature programmed to increase from 200 to 450°C in 12 hours or[45] wet ash the sample with 5 ml of mixed acids (perchloric acid:sulphuric acid:nitric acid, 2:1:1) for 30 minutes keeping the temperature below 150°C. Raise the temperature to 250°C and finally ignite the residue at 800°C. Dissolve the ash in 0.5 ml nitric acid, heat to dryness in a platinum crucible, evaporate with 0.5 ml of 10% sodium carbonate (Na_2CO_3) solution and then fuse the residue. Dissolve the melt in water and dilute to 25 ml. Determine iron on this solution.

Fuels and lubricants[46–48,119]

Coal ash[46] Grind the sample to pass through 250 mesh. Mix 0.5 g of the fine powder with 2 g of lithium metaborate ($LiBO_2$) in a platinum crucible and heat in a muffle furnace at 900°C, until the melt is clear. If the sample is expected to contain more than 15% of iron add 30–50 mg of ammonium vanadate (NH_4VO_3) to act as a flux. Cool the melt and transfer it to a beaker containing 8 ml of nitric acid and 150 ml of water. Stir vigorously until it has dissolved completely, add 2.5 g tartaric acid (($CHOH.COOH)_2$) and make up to 250 ml.

Mineral oils[47,48,119] To 0.2–20 g of oil add 2 ml of saturated salt (NaCl) solution, 20 ml of nitric acid and 15 ml of perchloric acid ($HClO_4$). Heat until the exothermic reaction starts and then stop heating. Reheat at intervals until the volume remaining is about 6 ml. Add 10 ml of water, boil for 1 minute, cool and dilute to 25 ml. Determine iron on an aliquot[47]. Alternatively[48] dissolve 0.5–2.0 g of the sample in 1 ml of light petroleum, transfer to a 50 ml separating funnel, add 5 ml of light petroleum and 1 ml of 10% hydroxylammonium chloride ($HONH_3Cl$). Add 5 ml acetate buffer (8.3 g of anhydrous sodium acetate and 1 ml of anhydrous acetic acid in 100 ml) separate the layers and determine iron on the aqueous phase by *Method 3*. See also I.P. Standard Methods[119].

Metals, alloys and plating baths[49–59]

Aluminium[49] Dissolve the sample in 30 ml of 2:1 hydrochloric acid, add sulphuric acid until its concentrations is about 5% and then dilute the solution to 100 ml. Determine iron on an aliquot.

Aluminium alloys[50] Dissolve a 1 g sample in 25 ml of 1:1 hydrochloric acid and up to 3 ml of nitric acid. Filter. Wash the residue with dilute hydrochloric acid and shake the filtrate and the washings for 30 seconds with 20 ml of isobutyl methyl ketone (($CH_3)_2CHCH_2COCH_3$). Wash the organic layer with 2 separate 20 ml portions of 1:1 hydrochloric acid and then with 2×20 ml of water. Combine the aqueous layers and determine iron on an aliquot.

Bismuth[51,52] Dissolve the sample in hydrochloric acid, add sodium thiocyanate (NaSCN) and extract the iron complex into isobutanol (($CH_3)_2CHCH_2OH$). Alternatively[52] isolate the iron by passing the solution through a column of Dowex-1 ion exchange resin, elute vanadium and manganese with $9N$ hydrochloric acid, then cobalt with $4N$ hydrochloric acid and finally elute the iron with $0.5N$ hydrochloric acid.

Cobalt[53] Dissolve 3 g of the sample in a mixture of hydrofluoric (HF) and nitric acids, evaporate the solution three times with hydrochloric acid and then dilute with hydrochloric acid to 25 ml. Extract the iron into two 5 ml portions of 2-chloroethyl ether (($ClCH_2CH_2)_2O$) and re-extract into 10 ml of $0.6N$ hydrochloric acid.

Copper[53] Dissolve 5 g of the sample in hydrochloric acid add a few drops of nitric acid, partly evaporate and then dilute to 50 ml with hydrochloric acid. Extract the iron with two 7 ml portions of 2-chloroethyl ether. Wash the extract free from copper with hydrochloric acid and then re-extract into 5 ml of $0.6N$ hydrochloric acid.

Magnesium alloys[54] Take 5 g of cleaned sample and dissolve it in hydrochloric acid. Evaporate the solution to remove any excess acid and then re-dissolve the residue in 100 ml of water. Treat a 20 ml aliquot with hydroxylammonium chloride ($HONH_3Cl$) and sodium acetate (CH_3COONa) solution and adjust the pH to 3.0 ± 0.5 and determine the iron by *Method 3* or *Method 5*.

Nickel[55,56] Either[55] dissolve the sample in hydrochloric acid, add sodium thiocyanate (NaSCN) extract the thiocyanate complex into isobutyl methyl ketone (($CH_3)_2CHCH_2COCH_3$) or[56] dissolve 0.5 g of the sample in 40 ml of 1:1 nitric acid, evaporate the solution to dryness, dissolve the residue in 50 ml of water, pass this solution through a column of KU-2 resin (Na^+ form), wash the column with 300 ml of $0.1M$ glycine (NH_2CH_2COOH) at pH 10, and then elute the iron with 70 ml of $2.5N$ hydrochloric acid.

Plating baths[57] To 5 ml of the electrolyte solution add 1 ml of 1:3 sulphuric acid and 4–5 drops of 3% hydrogen peroxide (H_2O_2). Heat the solution to boiling and add 5–7 ml of aqueous ammonia (NH_4OH) saturated with ammonium chloride (NH_4Cl). Filter, wash the precipitate with 10% ammonium chloride solution and then with hot water. Dissolve the precipitate in hot 1:20 sulphuric acid and determine iron by *Method 2*.

Silver alloys[8] Dissolve 1 g of sample in 5 ml of 1:1 nitric acid, dilute with water and add 1:1 aqueous ammonia (NH_4OH) drop by drop, while heating the solution, until the solution turns blue in colour. After 20 minutes collect the precipitate, wash it with dilute aqueous ammonia and dissolve it in 20 ml of hot 1:1 hydrochloric acid. Determine iron on an aliquot of this solution.

Tungsten and tungsten alloys[59] Dissolve the sample in a mixture of hydrofluoric (HF) and nitric acids, adjust the pH to about 6 with ammonium tartrate (($CHOH.COONH_4$)$_2$)-sodium borate ($Na_2B_4O_7$) buffer solution and determine iron on an aliquot of this solution.

Minerals, ores and rocks[46,60-68]

Chrome ores[60] Fuse a 0.5-1.0 g sample with a mixture of sodium peroxide (Na_2O_2) and hydroxide (NaOH). Dissolve the melt in 20 ml of water. To the solution add 60 ml 1:1 sulphuric acid, make up to 100 ml and determine iron on an aliquot.

Dolomite[61] Fuse the sample with sodium carbonate (Na_2CO_3) dissolve the melt in hydrochloric acid and make up to known volume with water.

Ilmenite ore[62] Decompose 0.2-0.3 g of the sample with sulphuric acid and hydrofluoric acid (HF), heat the residue to remove silicon and excess fluoride and then fuse it with potassium pyrosulphate ($K_2S_2O_7$). Dissolve the melt in dilute sulphuric acid and dilute to a known volume.

Rocks[46,63-66] Treat as described above for coal ash[46]. Or[63] dissolve a 20 g sample in hydrochloric acid, extract the iron into isobutyl methyl ketone ((CH_3)$_2CHCH_2COCH_3$), back extract the iron with hydrochloric acid containing hydroxylammonium chloride ($HONH_3Cl$) and determine.

To determine ferrous iron[64] mix 60-150 mg of sample with six times its weight of 'Rowledge Mixture' (sodium fluoride (NaF)-boric acid (H_3BO_3) 21:31), fuse at 1000°C for 1 minute in a platinum crucible under helium or argon. Cool, dissolve the melt in 30 ml of 1:2 sulphuric acid and 25 ml of saturated boric acid (H_3BO_3) under carbon dioxide and estimate ferrous iron on this solution using *Method 4*. Care should be taken not to exceed the 1 minute fusion time as this will result in ferric iron being reduced to ferrous. Similarly it is claimed[65] that it is impossible to obtain an accurate figure for ferrous iron in the presence of organic matter as this results in the reduction of ferric iron to ferrous. Conversely[66] atmospheric oxidation of ferrous iron to ferric may result if the sample is ground for more than 30 seconds. For accurate determination of ferrous iron the sample should be ground carefully by hand preferably in a non-oxidising medium such as acetone.

Silumin[67] Dissolve 0.2 g of sample in 25 ml of 1:3 hydrochloric acid and 15 ml of 65% nitric acid. Evaporate to dryness, heat the residue gently at 100-110°C, evaporate twice more adding 30% hydrochloric acid each time, dissolve the final residue in the minimum amount of 30% hydrochloric acid and add 100 ml of water. Filter off the residual silica. Save the filtrate and treat the precipitate with a mixture of sulphuric and hydrofluoric (HF) acids, fuse the residue with 1 g of sodium carbonate (Na_2CO_3), dissolve the melt in 1:1 hydrochloric acid and filter. Add this filtrate to the previous filtrate, evaporate to dryness, dissolve the residue in $8N$ hydrochloric acid and pass the solution through a column of 5 g of anionite AV-17 which has previously been washed with $8N$ hydrochloric acid. After passing the sample solution, wash the column with 200 ml of $8N$ hydrochloric acid. Desorb the iron, together with any copper, with 100 ml of $2.5N$ hydrochloric acid.

Sulphide ores[68] Fuse 0.2 g of the finely pulverised sample with 2-3 g of potassium pyrosulphate ($K_2S_2O_7$). Dissolve the cooled melt in water, acidify with hydrochloric acid and determine iron on an aliquot.

Miscellaneous[69-80]

Cellulosic fibres[69] *The standard TAPPI method*. Ash a 10 g sample in a silica or platinum dish, dissolve the residue in 10 ml of 1:1 hydrochloric acid and reduce the volume to 4-5 ml. Add sufficient 10% ammonium persulphate ((NH_4)$_2S_2O_8$) solution to make the final solution about $0.02M$ in this reagent and then add potassium thiocyanate (KSCN) reagent. Dilute to 50 ml and compare the colour with a standard, using a blank (*Method 2*).

Cement[70,71] Dissolve a 0.25 g sample by boiling with 10 ml of hydrochloric acid containing 0.5 g of ammonium chloride (NH_4Cl). Filter and determine iron on the filtrate[70]. Alternatively[71] fuse the sample with sodium carbonate (Na_2CO_3) and caustic soda (NaOH) and dissolve the melt in dilute hydrochloric acid. Filter and save the filtrate. Dissolve the insoluble residue in hydrofluoric acid (HF), dilute, filter, add this filtrate to the original filtrate and dilute to a known volume.

Ceramics[72] Decompose a sufficient quantity of the sample to contain 1–5 mg of iron, under an atmosphere of carbon dioxide, by means of a mixture of sulphuric and hydrofluoric (HF) acids in a closed polythene vessel on a water bath. Mask the fluoride ion with boric acid (HBO_3), shake the mixture with $6N$ hydrochloric acid and isopropyl ether $((CH_3)_2CHOCH(CH_3)_2)$. Separate the phases and determine iron on the organic layer.

Drugs[73] Wet oxidise the sample with a mixture of sulphuric and perchloric ($HClO_4$) acids and hydrogen peroxide (H_2O_2) and make up to volume.

Fertilizers[74] Grind the sample, in an agate mortar, to pass through a No. 100 sieve, place 0.5 g (limestone) or 0.2 g (fertilizer) of the ground sample in a 75 ml nickel crucible. Remove any organic matter which may be present by heating at 900°C for 15 minutes without the lid. Cool. Mix 0.3 g of potassium nitrate (KNO_3) with the sample and then add 1.5 g of caustic soda (NaOH) pellets. Cover the crucible with its nickel lid and heat to a dull red heat for 5 minutes, over a gas flame, not in a muffle furnace. Remove the crucible from the flame and swirl the melt around the sides of the crucible. Cool, add about 50 ml of water and warm to disintegrate the fused cake. Transfer the contents of the crucible to a 150 ml beaker containing 15 ml of $5N$ perchloric acid ($HClO_4$). Scrub the crucible and lid with a rubber tipped glass rod and wash any residue into the beaker. Transfer to a 100 ml volumetric flask and dilute to volume.

Thin magnetic films[75] Dissolve the sample in a few drops of nitric acid and dilute to 10 ml with $0.05N$ perchloric acid ($HClO_4$).

Maser crystals[76] Powder the sample by dropping molten pieces into water and then grind in a steel or tungsten mortar. Sift through silk or nylon and remove any contamination from the mortar by washing with hydrofluoric (HF) and nitric acids (for tungsten mortar) or with aqua regia (for steel mortar). Sinter the powdered solid with sodium carbonate (Na_2CO_3) and sodium borate ($Na_2B_4O_7$) dissolve the sinter in hydrochloric acid, make the solution up to a known volume, and determine iron on an aliquot.

Photographic gelatin[77] Dissolve the gelatin in water. Pass the solution through a column of Dowex 50W cation exchange resin and elute the adsorbed metals with $3N$ hydrochloric acid. Add ammonium thiocyanate (NH_4SCN) and then extract the iron into isobutyl methyl ketone $((CH_3)_2CHCH_2COCH_3)$.

Slags[78,79] Dissolve a 0.5 g sample in hydrochloric acid+nitric acid with the addition of hydrofluoric acid (HF) if required. Dilute the solution to 250 ml and determine iron on an aliquot.

Slime[80] Fuse 0.2 g of dried slime with 2 g of sodium carbonate (Na_2CO_3), extract the melt with 50 ml of hot water, rapidly heat the solution to the boiling point and filter. Dilute the filtrate to 200 ml.

Paper[81]

Take a strip 2.3–8.5 cm wide and 10–20 cm long, cut into small pieces, place in a platinum dish, add 25 ml of 10% sodium carbonate (Na_2CO_3) and then ignite. Treat the ash with 6 ml of 1:1 nitric acid and make up to 50 ml with normal nitric acid.

Plants[82–89]

Chop up the sample, to pass through 0.4–0.7 mm mesh, dry at 105°C for 3 hours, wet ash up to 0.5 g of the dried material, evaporate until white fumes are evolved, dissolve the residue in water and make up to a known volume.

Alternatively[87–89] dry ash 1 g of plant material at 450°C. Digest the ash with 2 ml of a 3:1 mixture of nitric and perchloric ($HClO_4$) acids and then boil for 10 minutes with 10 ml of $0.5N$ hydrochloric acid and 1 ml of 0.5% sodium nitrite ($NaNO_2$) solution. Dilute to 20 ml, set aside overnight for silica to separate, filter and determine iron on an aliquot of the filtrate.

Plastics and polymers[90–92, 118]

Polypropylene[90] Heat 1 g of the sample in a silica flask until black and viscous, then cool and add 5 ml of 30% hydrogen peroxide (H_2O_2) and reheat. Repeat until the liquid is colourless. Remove excess peroxide by boiling and make up to 100 ml with water.

Rubber[91,118] Heat a 2–5 g sample in a porcelain crucible taking care to avoid flaming, and then ash the residue at 660–700°C. Treat the ash with 40–50 ml of 1:1 hydrochloric acid in 10 ml portions. Filter and make the filtrate up to 200 ml. Precipitate the iron with aqueous ammonia (NH_4OH). Filter, re-dissolve the precipitate in hydrochloric acid and determine iron on this solution. See alternatively British Standard 1672[118].

Silicone polymers[92] Decompose the polymer by heating with sulphuric acid and destroy the organic matter by adding 1 drop of nitric acid. Remove insoluble silica by filtration and determine iron on the filtrate.

Sewage and industrial wastes[93]

Place 500 ml of the sample in a beaker, add 5 ml of nitric acid and evaporate to a small volume. Transfer the concentrate to a porcelain or silica basin and further reduce the volume to about 15 ml on a water bath. Add 2 ml of 60% perchloric acid ($HClO_4$) and gently evaporate to dryness. Ignite at just below 300°C for 40 minutes. Dissolve the residue by adding 2 ml of $5N$ hydrochloric acid and a little water and warm. If the residue is not colourless repeat the addition of nitric and perchloric acids and evaporate to dryness until a colourless residue is obtained. Dilute the clear solution to 100 ml and use a suitable aliquot to determine iron by *Method 1*.

Soil[65,87,88,95–97]

Use the alternative method for plants given above[87]. Or[95] place 0.5 g of soil in a centrifuge tube with 10 ml of citrate buffer solution at pH 4.75, add 0.05 g of sodium hyposulphite ($Na_2S_2O_4$) stopper the tube and shake in a water bath at 50°C until the soil has bleached. Continue shaking for an extra 10 minutes. Filter and determine iron on an aliquot of the filtrate. Other extractants which have been suggested for the determination of iron in soil are water[97] and molar potassium chloride (KCl)[88].

The warning given by Pruden and Bloomfield[65] regarding the effect of organic matter in minerals on the determination of ferrous iron, applies even more strongly to soils. Because of the inevitable presence of organic matter the determination of ferrous iron is of little or no value.

Water[1,2,98–106]

In many instances no sample preparation is required. If concentration is necessary evaporate 100 ml to dryness. If much organic matter is present ignite the residue carefully, cool, and moisten the ash with 5 ml of 1:1 hydrochloric acid. Warm for 2–3 minutes. Add 5–10 ml of hot water, warm for a few minutes longer and wash into a volumetric flask. Add a few drops of $0.2N$ potassium permanganate ($KMnO_4$) until the colour persists for at least 5 minutes. Cool and dilute to 20 ml and determine iron on a suitable aliquot[98]. Alternatively evaporate a suitable volume of sample with 2–3 ml of nitric acid until the residual volume is about 50 ml. Add permanganate if necessary to destroy organic matter. Add aqueous ammonia (NH_4OH) to the hot solution and continue heating until the smell of ammonia is barely perceptible. Filter and wash the precipitate, with water containing a little ammonia, at 70–80°C. Dissolve the precipitate in 5 ml of hydrochloric acid and transfer to a volumetric flask. Dilute to volume and determine iron on an aliquot.

For sea water[99] measure a 100 ml sample into a 500 ml flask, add 6 ml of sulphuric acid and evaporate to white fumes. While still warm add 85 ml of distilled water and cover the flask with a watch glass. Heat on a steam bath until the residue has dissolved. Add a 0.63% solution of potassium permanganate drop by drop until the colour persists. Decolourise with 1 ml of bromine water and boil to remove excess bromine. Cool, transfer the solution to a 100 ml volumetric flask, dilute to volume and determine iron on an aliquot using *Method 2*. Alternatively[100] adjust the pH of the sea water to be between 2 and 9 and extract the iron, as its 1:10 phenanthroline complex, into propylene carbonate ($CH_3.\underline{CH.O.CO.O.CH_2}$).

For process water in the petroleum industry[101], filter a 5–10 ml sample, evaporate to dryness, ignite the residue gently, to destroy any organic matter, dissolve the residue in 5 ml of $6N$ hydrochloric acid and make up to 10 ml.

For rain water the use of *Method 4* is recommended. If concentration is necessary the iron-bathophenanthroline complex will further complex with perchlorate ions (ClO_4^-) and this new complex is extractable into nitrobenzene ($C_6H_5NO_2$) from solutions of pH 4–8[102].

Guenther[103] recommends that when using *Method 2*, hydrogen peroxide (H_2O_2) should be used in place of potassium permanganate ($KMnO_4$) as he claims that this accelerates colour development and prevents interference from nitrites.

If it is necessary to determine the different forms of iron present in water, then the following procedures may be followed. For "reactive iron"[104] collect the sample in a vessel containing enough hydrochloric acid to make the final acid concentration $0.1N$. Treat 200 ml of the sample with 2 ml of hydroxylammonium chloride ($HONH_3Cl$) and adjust the pH using aqueous ammonia (NH_4OH) to the yellow end-point of m-cresol purple indicator. Add acetate buffer solution (pH 4) and proceed by *Method 4*. For "non-reactive iron"[105] take a 200 ml sample, adjust it to be normal in respect to its hydrochloric acid concentration and then boil in a vessel covered with asbestos paper until the final volume is 5–10 ml. Cool, dilute to 200 ml and then proceed as for "reactive iron". To determine "total iron"[106] collect the sample in a polythene bottle containing 4 ml of mercapto acetic acid (thioglycollic acid, $CH_2(SH).COOH$), cover the open neck of the bottle with an inverted beaker and heat the bottle for 1 hour at 80°C. Cool, mix 200 ml of the treated sample with 10 ml of acetate buffer (pH 4) and proceed as for "reactive iron".

Separation from other ions[107–115]

From aluminium[107] Milligram amounts can be separated by reducing ferric to ferrous iron, complexing the ferrous iron with 2,2'-bipyridyl, passing the solution at pH 5.5 through a column of Lewatit MN ion exchange resin saturated with neutral sodium arsenate ($Na_2HAsO_4.7H_2O$) solution. The reduced iron complex passes through the column and is quantitatively removed with a 1% solution of ammonium acetate (CH_3COONH_4). The adsorbed aluminium can be subsequently removed with $0.5N$ hydrochloric acid.

From cadmium, uranium and zinc[108] Apply the solution of the mixed ions to a 10 × 0.8 cm column of zirconium phosphate (NH_4^+ form) and elute cadmium with $2.5N$ ammonium chloride, zinc with $5N$ ammonium chloride, uranium with $0.5N$ hydrochloric acid and iron with $4N$ hydrochloric acid.

From cobalt, copper, manganese and zinc[109] Pass a solution of the metal tartrates through a column of Dowex 2-X8 resin (tartrate form) elute cobalt, copper, manganese and zinc with $2.12 \times 10^{-3}M$ tartaric acid (($CHOH.COOH)_2$) at pH 4.5 at a rate of 3–4 drops a minute, then elute iron with molar hydrochloric acid.

From cobalt and nickel[110-112] Dilute a 2 ml aliquot of the solution in $6M$ hydrochloric acid to 20 ml with acetone (CH_3COCH_3). Pass this solution through an ion exchange column when iron is not absorbed but nickel and cobalt form green and blue bands respectively which can be successively eluted with a 7:3 mixture of acetone and $2M$ hydrochloric acid[110]. Or[111] pass the hydrochloric acid solution through a column of Dowex 50 resin and elute the iron with a 4:1 mixture of tetrahydrofuran ($CH_2.(CH_2)_2.CH_2O$) and $3N$ hydrochloric acid. Lavrukhina[112] recommends Dowex 1-X8 or 2-X8 resins (AV 17 is not suitable) for use with the hydrochloric acid—acetone mixture.

Generally from other metals[113-115] Iron is quantitatively extracted from $2M$ hydrochloric acid solution by trioctylamine[113]. Alternatively[114] iron can be adsorbed from hydrochloric acid solution onto a column of Fluoropak saturated with 2-octanone. It is eluted with 2-propanol and it is claimed that 1000–5000 mole ratios of other ions do not interfere. Pietsch[115], claims that iron can be separated from all other metals, apart from vanadium, by extraction from hydrochloric acid concentrations greater than $8N$ into a 2% solution of triphenyl arsine oxide $((C_6H_5)_3AsO)$ in chloroform ($CHCl_3$). The recovery is stated to be better than 95%.

The individual tests

Method 1 is the thioglycollic acid test as originally suggested by Lyons[116], citric acid being used to suppress interference by other ions. This is the standard method for the determination of iron in water[1,2], sewage[94], rubber latex[118], and mineral oils[119].

Method 2 uses the thiocyanate method originally proposed by Ossian[117]. This method is still widely used and an extensive literature on it has been published. It is the basis of the official method for the determination of iron in cellulosic fibres[69].

Method 3 uses o-phenanthroline and may be used to measure ferrous, ferric and total iron.

Method 4 is the method using bathophenanthroline which was developed[11-13] to provide a rapid and simple method, for the determination of iron in wine, with a minimum of sample preparation. It has also been adopted[2] for the determination of iron in water.

Method 5 also uses *o*-phenanthroline and is a modification of *Method 3* developed for use with lower iron concentrations.

References

1. Brit. Standards Inst., *"Routine Control Methods of Testing Water Used in Industry"*, B.S, 1427, London, 1962
2. Brit. Standards Inst., *"Methods of Testing Water Used in Industry"*, B.S. 2690, Part 1, London, 1964
3. T. Vondenhof and H. Beindorf, *Mschr.Brau.*, 1968, **21** (6), 156; *Analyt. Abs.*, 1969, **17**, 3070
4. H. Weyh, W. Hagen and Un Hua Pek, *Brauwissenschaft*, 1968, **21** (12), 472; *Analyt. Abs.*, 1970, **18**, 2784
5. L. Carta de Angeli and F. Dentice di Accadia, *R.C. Inst. Sup. Sanit.*, 1962, **25**, 966; *Analyt. Abs.*, 1964, **11**, 1848
6. H. Konrad, *Nahrung*, 1960, **4**, 365; *Analyt. Abs.*, 1961, **8**, 325
7. H. Eschnauer, *Z. LebensmittUntersuch*, 1959, **109**, 474; *Analyt. Abs.*, 1960, **7**, 1931
8. I. G. Mokhnachev and L. G. Serdyuk, *Vinodelie i Vinogradarstvo S.S.S.R.*, 1964, **24** (1), 22; *Analyt. Abs.*, 1965, **12**, 2506
9. L. Laporta, *Boll. Lab. Chim. Provinciali*, 1964, **15** (3), 239; *Analyt.Abs.*, 1965, **12**, 4859
10. O. Colagrande and A. Del Re, *Industrie agrarie*, 1969, **7** (5), 206; *Analyt. Abs.*, 1970, **19**, 1751
11. H. Tanner, *Schweiz.Z.Obst.u. Weinbau*, 1967, **103** (76) Jahrgang, 180
12. W. M. Banick and G. F. Smith, *Anal. Chim. Acta*, 1957, **16**, 464
13. L. Nykaenen, *Z. Lebensm.Unters.Forsch.*, 1961, **114**, 181
14. J. Planas, *Rev. Esp. Fisiol,.* 1963, **19**, 73; *Analyt. Abs.*, 1964, **11**, 4439
15. H. V. Connerly and A. R. Briggs, *Clin.Chem.*, 1962, **8**, 151; *Analyt. Abs.*, 1962, **9**, 4411
16. G. Ceriotti and L. Spandrio, *Biochim.Appl.*, 1964, **11**, 104; *Analyt. Abs.*, 1965, **12**, 4694
17. D. T. Forman, *Tech. Bull. Registry Med. Technologists*, 1964, **34** (6), 93; *Analyt. Abs.*, 1965, **12**, 5293
18. D. T. Forman, *Amer. J. Clin. Path.*, 1964, **42**, 103; *Analyt. Abs.*, 1965, **12**, 5972
19. E. Scala, A. Castaldo and G. Ruiz, *Boll. Soc. Ital. Biol. Sper.*, 1960, **36**, 1048; *Analyt. Abs.*, 1961, **8**, 1665
20. E. A. Efimova, *Lab. Delo.*, 1963, **9** (5), 19; *Analyt. Abs.*, 1964, **11**, 2719
21. D'A. Kok and F. Wild, *J. Clin. Path.*, 1960, **13**, 241; *Analyt. Abs.*, 1961, **8**, 2553
22. I. Dezso and T. Fulop, *Mikrochim. Acta*, 1961, (1), 154; *Analyt. Abs.*, 1961, **8**, 3399
23. M. Mancini, *Biochim. Appl.*, 1959, **6**, 404; *Analyt. Abs.*, 1960, **7**, 3448
24. H. J. Ybema, B. Leijnse and W. F. Wiltink, *Clinica chim.Acta*, 1965, **11** (2), 178; *Analyt.Abs.*, 1966, **13**, 5011
25. J. F. Goodwin and B. Murphy, *Clin.Chem.*, 1966, **12** (2), 58; *Analyt.Abs.*, 1967, **14**, 3378
26. V. Kauppinen, *Suomen.Kem.B.*, 1961, **34**, 27; *Analyt.Abs.*, 1961, **8**, 3856
27. G. Ceriotti and L. Spandrio, *Clin. Chim. Acta.*, 1961, **6**, 233; *Analyt.Abs.*, 1961, **8**, 5133
28. K. Beyermann and K. Cretius, *Clin. Chim.Acta.*, 1961, **6**, 111; *Analyt.Abs.*, 1961, **8**, 5114
29. C. Vecchione, S. Fati and P. Piccoli, *Biochim.Appl.*, 1959, **6**, 257; *Analyt.Abs.*, 1960, **7**, 2361
30. P. Collins and H. Diehl, *Analyt.Chem.*, 1959, **31**, 1692
31. E. S. N. Littlejohn and D. N. Raine, *Clinica chim.Acta*, 1966, **14** (6), 793; *Analyt.Abs.*, 1968, **15**, 1507
32. M. Barry, *J.clin.Path.*, 1968, **21** (2), 166; *Analyt.Abs.*, 1969, **16**, 2600
33. A. S. Kuznetsova, *Zavodskaya* Lab., 1968, **34** (4), 410; *Analyt.Abs.*, 1969, **17**, 817
34. H. Klimezyk, Z. Kubas, K. Pytel and L. Suski, *Hutnik.Katowice*, 1968, **35** (10), 462; *Analyt.Abs.*, 1970, **18**, 1613

35. D. Monnier, W. Haerdi and E. Martin, *Helv.chim.Acta,* 1963, **46,** 1042; *Analyt.Abs.,* 1964, **11,** 223
36. G. Norwitz, J. Cohen and M. E. Everett, *Analyt.Chem.,* 1964, **36,** 142
37. R. Alifax and M. Bejambes, *Lait,* 1958, **38,** 146; *Analyt.Abs.,* 1959, **6,** 2329
38. P. Boersma, *Neth.Milk and Dairy J.,* 1963, **17,** 289; *Analyt.Abs.,* 1964, **11,** 1488
39. G. A. Knauer, *Analyst,* 1970, **95,** 476
40. K. Lauber and H. Aebi, *Mitt.Geb.Lebensmittelunters.u.Hyg.,* 1966, **57** (5), 363; *Analyt. Abs.,* 1967, **14,** 7840
41. N. M. Rusin and V. F. Rabinovitch, *Inform.Byull.Mosk.Nauch.Inst.Sanit. i Gigieny,* 1957, (10–11), 66; *Analyt.Abs.,* 1959, **6,** 341
42. S. Zagrodski and Z. Olszenko-Piontkowa, *Gaz.Cukr.,* 1956, **58,** 281; *Analyt.Abs.,* 1957, **4,** 3754
43. M. Karvanek, *Sb.Vys.Sk.Chem.-Technol., Odd.Fak.Potrav.Technol.,* 1961, **5** (1), 203; *Analyt.Abs.,* 1964, **11,** 2388
44. P. G. Pifferi and A. Zamorani, *Industrii agrarie,* 1968, **6** (4), 213; *Analyt.Abs.,* 1969, **17,** 1778
45. T. Takeuchi and T. Tanaka, *J.Chem.Soc.Japan, Ind.Chem.Sect.,* 1961, **64,** 305; *Analyt. Abs.,* 1963, **10,** 1590
46. P. L. Boar and L. K. Ingram, *Analyst,* 1970, **95,** 124
47. J. S. Forrester and J. L. Jones, *Analyt.Chem.,* 1960, **32,** 1443
48. T. P. Labuza and M. Karel, *J.Food Sci.,* 1967, **32** (5), 572; *Analyt.Abs.,* 1968, **15,** 6281
49. M. I. Abramov, *Uch.Zap.Azerb.Univ.Ser.Khim.Nauk.,* 1963, (3), 31; *Analyt.Abs.,* 1965, **12,** 1654
50. N. Tajima and M. Kuroke, *Japan Analyst,* 1960, **9,** 798; *Analyt.Abs.,* 1962, **9,** 3632
51. G. Signorelli, *Ann.Chim.Rome,* 1960, **50,** 1057; *Analyt.Abs.,* 1961, **8,** 2376
52. D. H. Wilkins and L. E. Hibbs, *Analyt.Chim.Acta,* 1959, **20,** 427; *Analyt.Abs.,* 1960, **7,** 541
53. P. I. Artyukhin, E. N. Gilbert, V. A. Pronin and V. M. Moralev, *Zavodskaya Lab.,* 1967, **33** (8), 926; *Analyt.Abs.,* 1968, **15,** 6657
54. U. K. Atomic Energy Authority, Report PG453(S) 1963; *Analyt.Abs.,* 1964, **11,** 148
55. H. Goto, Y. Yakita and M. Namiki, *J.Japan Inst.Metals, Sendai* 1961, **25,** 181; *Analyt. Abs.,* 1962, **9,** 1918
56. R. N. Golovatyi and V. V. Oshchapovskii, *Ukr.Khim.Zhur.,* 1963, **29,** 187; *Analyt. Abs.,* 1964, **11,** 1293
57. G. Loshkareva, *Trudy Sverdlovsk Gorn.Inst.,* 1960, (3), 95; *Analyt.Abs.,* 1962, **9,** 687
58. Z. Skorko-Trybula and J. Chwastowska, *Chem.Anal.Warsaw,* 1963, **8,** 859,; *Analyt. Abs.,* 1965, **12,** 1091
59. G. Norwitz and H. Gordon, *Analyt.Chem.,* 1965, **37** (3), 417; *Analyt.Abs.,* 1966, **13,** 3550
60. H. Grubitsch, *Radex Rdsch.,* 1959, (1), 460; *Analyt.Abs.,* 1960, **7,** 2118
61. A. K. Kutarkina and R. G. Syrova, *Trudy ural'.nauchno-issled.khim.Inst.,* 1964, (11), 19; *Analyt.Abs.,* 1966, **13,** 1264
62. G. E. Lunina and E. G. Romanenko, *Zavodskaya Lab.,* 1968, **34** (5), 538; *Analyt.Abs.,* 1969, **17,** 1418
63. I. Morimoto and S. Tanaka, *Japan Analyst,* 1962, **11,** 861; *Analyt.Abs.,* 1964, **11,** 1738
64. Yu. N. Novikova, *Zh.analit.Khim,* 1968, **23** (7), 1057; *Analyt.Abs.,* 1970, **18,** 929
65. G. Pruden and C. Bloomfield, *Analyst.* 1969, **94,** 688
66. J. G. Fitton and R. C. O. Gill, *Geochim.cosmochim.Acta,* 1970, **34** (4), 518; *Analyt. Abs.,* 1970, **19,** 2239
67. O. V. Morozova, Z. E. Mel'chekova and V. V. Stepin, *Trudy vses.nauchno-issled.Inst. Standartn.Obraztosov Spektr.Etalonov,* 1964, **1,** 88; *Analyt.Abs.,* 1966, **13,** 3496
68. V. Skrivanek and P. Klein, *Z.analyt.Chem.,* 1963, **192,** 386; *Analyt.Abs.,* 1963, **10,** 4049
69. *Tech.Assoc.Pulp and Paper Ind.,* 1959, **42,** 147A; *Analyt.Abs.,* 1960, **7,** 1094
70. N. S. Frumina, *Zavodskaya.Lab.,* 1959, **25,** 148; *Analyt.Abs.,* 1959, 6, 4776
71. K. Kawagaki, M. Saito and K. Hirokawa, *J.chem.Soc.Japan, ind.Chem.Sect.,* 1965, **68** (3), 465; *Analyt.Abs.,* 1967, **14,** 5434
72. Y. Abe and A. Naruse, *Japan Analyst,* 1964, **13** (5), 417; *Analyt.Abs.,* 1966, **13,** 3596

73. J. Hollos and I. Hovarth-Gabai, *Pharmazie,* 1965, **20** (4), 207; *Analyt.Abs.,* 1966, **13,** 4367
74. P. Chichilo, *J.Assoc.off.agric.Chem.,* 1964, **47,** 1019
75. K. L. Cheng and B. L. Goydish, *Microchem.J.,* 1963, **7,** 166; *Analyt.Abs.,* 1964, **11,** 4291
76. R. C. Chirnside, H. J. Cluley, R. J. Powell and P. M. C. Proffitt, *Analyst,* 1963, **88,** 851
77. J. Saulnier, *Sci.Ind.Photogr.,* 1964, **35,** 1; *Analyt.Abs.,* 1965, **12,** 4037
78. Y. Endo and H. Takagi, *Japan Analyst,* 1960, **9,** 998; *Analyt.Abs.,* 1962, **9,** 3737
79. S. Wakamatsum, *J.Iron St.Inst.Japan,* 1959, **45,** 717; *Analyt.Abs.,* 1960, **7,** 2744
80. T. I. Badeeva, L. A. Molot, N. S. Frumina and K. G. Petrikova, *Uch.Zap.Saratovsk. Univ.,* 1962, **75,** 100; *Analyt.Abs.,* 1964, **11,** 875
81. T. Matsuo, S. Komatsu and K. Tozawa, *Japan Analyst,* 1962, **11,** 1194; *Analyt.Abs.,* 1964, **11,** 1822
82. H. E. Jansen, M. J. M. Luykx, B. D. Hartong, W. J. Klopper and C. G. T. P. Schneller, *Int.Tijdschr.Brouw Mout.* 1958–59, **18,** 158; *Analyt.Abs.,* 1960, **7,** 1928
83. E. J. Rubins and G. R. Hagstrom, *J.Agric.Food Chem.,* 1959, **7,** 722; *Analyt.Abs.,* 1960, **7,** 2868
84. C. Quarmby and H. M. Grimshaw, *Analyst,* 1967, **92,** 305
85. D. E. Quinsland and D. C. Jones, *Talanta,* 1969, **16** (2), 282; *Analyt.Abs.,* 1970, **18,** 4203
86. K. R. Middleton, *Analyst,* 1964, **89,** 421
87. W. J. Wark, *Analyt.Chem.,* 1954, **26,** 203
88. J. Paul, *Mikrochim.Acta,* 1966, (6), 1075; *Analyt.Abs.,* 1968, **15,** 494
89. E. G. Bradfield, *J.Sci.Food Agric.,* 1964, **15,** 469; *Analyt.Abs.,* 1965, **12,** 5957
90. K. Novak and V. Mika, *Chem.Prumysl.,* 1963, **13,** 360; *Analyt.Abs.,* 1964, **11,** 4414
91. N. M. Vital'skaya and N. F. Pantaeva, *Kauchuk i Rezina,* 1962, (6), 53; *Analyt.Abs.,* 1963, **10,** 2811
92. S. Fujiwara and H. Narasaki, *Analyt.Chem.,* 1964, **36,** 206
93. A. A. Christie, J. R. W. Kerr, G. Knowles and G. F. Lowden, *Analyst,* 1957, **82,** 336
94. Min. of Housing and Local Govt., "*Methods of Chemical Analysis as Applied to Sewage and Sewage Effluents*", H.M. Stationery Office, London, 1956
95. D. E. Coffin, *Canad.J.Soil Sci.,* 1963, **43,** 7; *Analyt.Abs.,* 1963, **10,** 5423
96. P. Fontana and G. Rossi, *Ann.Chim.Roma,* 1959, **49,** 310; *Analyt.Abs.,* 1960, **7,** 784
97. M. L. Tsap, *Vestn.Sel'sk.-Khoz.Nauk.,* 1963, (2), 122; *Analyt.Abs.,* 1964, **11,** 1133
98. Amer. Pub. Health Assoc., "*Standard Methods of Water Analysis*", 5th Edit., New York, 1923
99. T. G. Thompson, R. W. Bremmer and I. M. Jamieson, *Ind.Eng.Chem.Analyt.Ed.,* 1932, **4,** 288
100. B. G. Stephens and H. A. Suddeth, *Analyt.Chem.,* 1967, **39** (12), 1478; *Analyt.Abs.,* 1969, **16,** 144
101. I. L. Bagbanly and T. R. Mirzoeva, *Azerb.neft.Khoz.,* 1967, (3), 43; *Analyt.Abs.,* 1968, **15,** 5696
102. F. P. Gorbenko, L. Ya. Zakarova, I. A. Shevchuk and F. V. Lapshin, *Gidrokhim. Mater.,* 1965, **40,** 194; *Analyt.Abs.,* 1967, **14,** 1733
103. G. Guenther, *Fortschr.Wasserchem.Iher.Grenzgeb.,* 1967, **5,** 305; *Chem.Abs.,* 1968, **68,** 16018
104. A. L. Wilson, *Analyst,* 1964, **89,** 389
105. A. L. Wilson, *Analyst,* 1964, **89,** 402
106. A. L. Wilson, *Analyst,* 1964, **89,** 442
107. J. Gera, *Chem.Anal.Warsaw,* 1964, **9,** 541; *Analyt.Abs.,* 1965, **12,** 4542
108. I. Gal and N. Peric, *Mikrochim.ichnoanalyt.Acta.* 1965, (2), 251; *Analyt.Abs.,* 1966, **13,** 2838
109. G. F. Pitstick, T. R. Sweet and G. P. Morie, *Analyt.Chem.,* 1963, **35,** 995
110. I Hazan and J. Korkisch, *Analytica chim. Acta,* 1965, **32** (1), 46; *Analyt.Abs.,* 1966, **13,** 2370
111. J. Korkisch, *Nature,* 1966, **210,** 626; *Analyt.Abs.,* 1967, **14,** 5881
112. A. K. Lavrukhina and N. K. Sazhina, *Zhur.analit.Khim.,* 1969, **24** (6), 670; *Analyt.Abs.,* 1970, **19,** 4877
113. B. E. McClellan and V. M. Benson, *Analyt.Chem.,* 1964, **36,** 1985

114. M. A. Wade and S. S. Yamamura, *Analyt.Chem.*, 1964, **36,** 1861
115. R. Pietsch, *Mikrochim.Acta,* 1967, (4), 708; *Analyt.Abs.,* 1968, **15,** 5893
116. E. Lyons, *J.Amer.Chem.Soc.,* 1927, **49,** 1916
117. H. Ossian, *Pharm.Centralb.,* 1837, **13,** 205
118. Brit. Standards Inst. "*Methods of testing rubber latex*", B.S.1672: Part 2 :1954 London
119. Institute of Petroleum (U.K.) "I.P. Standards", 31st Edition, 1972 (London) Test 120/48

Iron Method 1

using thioglycollic acid

Principle of the method

The method adopted is that of matching the colour produced by the action of thioglycollic acid on iron, both ferric and ferrous, in the presence of ammonia. The most satisfactory results are obtained in the presence of alkali citrate, and the Lovibond colour glasses have been standardised on the colours produced in solutions containing ammonium citrate.

Reagents required

1. *Citric acid* Solution of citric acid ($C_6H_8O_7.H_2O$) containing 20 per cent w/v.
2. *Thioglycollic acid* ($HSCH_2COOH$)
3. *Aqueous ammonia* Solution of ammonia containing 10 per cent. w/w NH_3.

All chemicals should be of analytical reagent quality.

The Standard Lovibond Comparator Discs 3/6 and APFE

Disc 3/6 covers the range 10 µg to 100 µg of iron (Fe) in steps of 10 µg, which equals 2.0 to 20.0 parts per million if a 5 ml sample is taken.

Disc APFE covers the range 0.5-5.0 ppm in steps of 0.5 on a 10 ml sample, and must be used with 40 mm cells.

The Standard Lovibond Nesseriser Discs NEA, NEB and NEAB

Disc NEA covers the range 2 µg to 18 µg (0.002 to 0.018 mg) of iron (Fe) in steps of 2 µg.
Disc NEB covers the range 20 µg to 60 µg (0.02 to 0.06 mg) of iron (Fe) in steps of 5 µg.
Disc NEAB covers the range 5 µg to 60 µg (0.005 to 0.06 mg) of iron (Fe), with steps 5, 10, 15, 20, 25, 30, 40, 50, 60 µg.

This covers the range 0.04 to 1.2 ppm with a 50 ml sample.

Technique for disc 3/6

To a suitable volume of the solution to be tested for iron, which should be neutral or slightly acid, add 0.5 ml of citric acid (reagent 1); mix and add 1 drop of thioglycollic acid (reagent 2). Again mix, add 1 ml of aqueous ammonia (reagent 3), or sufficient to render distinctly alkaline and dilute with distilled water to 10 ml. Place this solution in one of the comparator test tubes in the right-hand compartment of the comparator. In the left-hand compartment place the other test tube containing a "blank" test composed of 0.5 ml of reagent 1, 1 drop of reagent 2, 1 ml of reagent 3 and sufficient distilled water to produce 10 ml. Hold the comparator facing a standard source of white light such as the Lovibond White Light Cabinet or, failing this, north daylight, and compare the colour produced in the test solution with the colours in the standard disc, rotating the latter until a colour match is obtained. The amount of iron present in the test solution is then indicated in the right-hand corner of the comparator.

The markings on the disc represent the actual amounts of iron producing the colour in the test; thus, if a colour equivalent to 50 µg is produced when 5 ml of the original solution is taken for the test, the amount of iron present in the original solution is 10 parts per million.

Technique for disc APFE

To 10 ml of the solution to be tested for iron, which should be neutral or slightly acid, add 0.5 ml of citric acid (reagent 1); mix, and add 1 drop thioglycollic acid (reagent 2). Again mix, add 1 ml ammonia (reagent 3). Place in a 40 mm comparator cell in the right-hand compartment of the comparator. In the left-hand compartment place a similar cell containing a "blank" made as above but using distilled water instead of sample. Match after 5 minutes; the answer read from the disc is in parts per million.

Technique for discs NEA etc.

Fill one of the Nesseriser glasses to the 50 ml mark with distilled water; to this add 2 ml of citric acid (reagent 1) and 0.1 ml of thioglycollic acid (reagent 2) and render alkaline with ammonia (about 2 ml of reagent 3). Mix and place in the left-hand compartment of the Nessleriser. Fill the other Nessleriser glass to the 50 ml mark with the solution under examination, add 2 ml of reagent 1 and 0.1 ml of reagent 2, mix and render alkaline with ammonia (about 2 ml of reagent 3). After mixing, allow to stand for five minutes and then place in the right-hand compartment of the instrument. Stand the Nessleriser before a standard source of white light and compare the colour produced in the test solution with the colours in the standard disc, rotating the disc until a colour match is obtained.

The markings on the discs represent the actual amounts of iron (Fe) producing the colour in the test. Thus, if a colour equivalent to 10 μg is produced in the test, the amount of iron present in the 50 ml of solution under test is 0.01 mg and the solution therefore contains 0.2 part of iron per million.

Notes

1. Lovibond Nessleriser discs are standardized on a depth of liquid of 113 $\pm$ 3 mm. The height measured internally, of the 50 ml calibration mark on Nessleriser glasses used with the instrument must be within the same limits. Tests with Nessleriser glasses not conforming to this specification will give inaccurate results.

2. This method is used in British Standard 1672 Part 2 : 1954 "*Methods of Testing Rubber Latex,*" and in Institute of Petroleum method I.P.120/48.

Iron Method 2
using thiocyanate

Principle of the method

The colour standards are designed to match the red colours produced by the well-known reaction between ferric iron and ammonium thiocyanate in acid solution. In order to avoid the variation of colour which occurs in the presence of other metallic radicals, the coloured compound is extracted from aqueous solution by means of a mixed solvent consisting of equal volumes of amyl alcohol and amyl acetate.

Reagents required

1. *Ammonium thiocyanate* (NH_4SCN)   Solution, approximately 7.5M, containing 570 g of ammonium thiocyanate per litre
2. *Dilute hydrochloric acid* (HCl)   Approximately 5N (about 600 g conc. acid/litre)
3. *Potassium permanganate* ($KMnO_4$)   Solution 0.1N (3 g/l)
4. *Solvent mixture*   A mixture of equal volumes of amyl acetate and amyl alcohol

All chemicals should be of analytical reagent quality

The Standard Lovibond Comparator Disc 3/11

The disc covers the range 0.002, 0.004, 0.006, 0.008, 0.01, 0.0125, 0.015, 0.02, 0.025 mg of iron (Fe). This equals 0.4 to 5.0 parts per million if a 5 ml sample is used.

Technique

Place 5 ml of the solution under test into a small stoppered separating funnel, acidify with 1 ml of dilute hydrochloric acid (reagent 2), add 2 drops of potassium permanganate solution (reagent 3) and mix: then add 5 ml of ammonium thiocyanate solution (reagent 1) and 10 ml of the mixed solvent (reagent 4). Shake vigorously and allow to separate. Reject the lower (aqueous) layer and transfer the upper layer to one of the 10 ml comparator test tubes, and place in the right-hand compartment of the comparator. Carry out a blank test, using 5 ml of distilled water with the quantities of the reagents stated above, and place the resulting solution in the left-hand compartment of the comparator, so that it comes behind the glass colour standards. Hold the comparator facing a standard source of white light such as the Lovibond White Light Cabinet or, failing this, north daylight, and compare the colour produced in the test solution with the colours in the standard disc, rotating the latter until a colour match is obtained. If the colour produced is deeper than the standard colours, the solution may be diluted with more of the organic mixed solvent until the colour is within the range of the disc, and the necessary calculation made. Alternatively, the test may be carried out on a smaller quantity of the original solution after dilution with distilled water to 5 ml.

The markings on the disc represent the actual amounts of iron producing the colour in the test. Thus, if a colour equivalent to 0.02 mg is produced when 5 ml of the original solution is taken for the test, the amount of iron present in the original solution is 0.4 mg per 100 ml or 4 parts per million.

Iron Method 3
using *o*-phenanthroline

Principle of the method

This determination is based on the orange-red complex $((C_{12}H_8N_2)_3Fe)^{++}$ formed by *o*-phenanthroline with ferrous iron. Ferric iron is reduced to ferrous iron with hydroxylamine hydrochloride for the determination of total iron. The colour has been shown to be independent of pH within the range 2–9. The intensity of this colour is proportional to the amount of iron present and is determined by comparison with Lovibond permanent glass standards.

Reagents required

1. *Hydroxylamine hydrochloride* ($NH_2.OH.HCl$) 10% aqueous solution (See Note 2)
2. *Buffer solution* Mix equal volumes of $3N$ HCl (about 360 g conc. acid/l) and $5N$ (385 g/l) ammonium acetate (CH_3COONH_4).
3. *o-phenanthroline* 0.5% solution in 50/50 alcohol/water mixture

All chemicals should be of analytical reagent quality.

The Standard Lovibond Comparator Disc 3/54

This disc covers the range 10, 20, 30, 40, 50, 60, 80, 120, 140 micrograms of iron (Fe). The range equals 2 to 28 ppm if a 5 ml sample is taken.

Technique

Transfer 5 ml of the sample solution to a standard comparator tube. Add 1 ml of distilled water, 1 ml of hydroxylamine solution (reagent 1), mix well and allow to stand for 15 minutes. Then add 1 ml of buffer solution (reagent 2), 0.2 ml of *o*-phenanthroline solution (reagent 3) and make up to the 10 ml mark with distilled water. Mix well, using a thin glass rod bent at one end as a stirrer, and then place the tube in the right hand compartment of the Lovibond Comparator. Place an identical tube filled with distilled water in the left-hand compartment. After 20 minutes match the colour of the sample with the permanent glass standards in the disc, using a standard source of white light such as the Lovibond White Light Cabinet or, failing this, north daylight. The figure obtained represents micrograms of iron present in the volume of sample taken.

If the colour is darker than the highest standard, repeat the test using a small aliquot of the sample and increasing the initial addition of distilled water to maintain the volume of solution at 6 ml before the addition of the hydroxylamine solution. Calculate the iron concentration proportionately.

Notes

1. This modification of the original test was developed by A. G. Pollard, and W. Y. Magar, Imperial College, University of London.

2. It has been found that, instead of adding 1 ml of a 10% solution of hydroxylamine hydrochloride, it is possible to add roughly 0.1 g of the solid without any adverse effect on the determination. The same technique **cannot** be used with reagent 3, the *o*-phenanthroline **must** be used in the form of the specified solution, which is stable for at least 6 months if stored in a dark bottle with a ground glass stopper.

3. This method estimates total iron. Ferrous iron may be determined, by omitting the reduction with hydroxylamine, and ferric iron may then be obtained by difference.

Iron Method 4
using batho-phenanthroline

Principle of the method

Batho-phenanthroline (4,7-diphenyl-1,10-phenanthroline) reacts with ferrous iron to form a red complex. The intensity of the colour of this complex, which is proportional to the amount of ferrous iron present, is measured by comparison with a series of Lovibond permanent glass colour standards. To determine total iron the ferric iron is reduced to ferrous by means of sodium pyrosulphite.

Reagents required

1. *Sodium pyrosulphite*   5% aqueous solution of $Na_2S_2O_5$.
2. *Buffer solution*   Mix 6 volumes of a saturated aqueous solution of potassium chloride (KCl) with 6 volumes of pH 4 buffer solution (50 ml of $0.1M$ potassium hydrogen phthalate ($C_6H_4COOH.COOK$) plus 0.1 ml of $0.1M$ sodium hydroxide (NaOH), or a reliable commercial buffer of this pH, such as Merck, B.D.H., etc.) and 1 volume of N NaOH.
3. *Batho-phenanthroline* $0.001M$ Dissolve 0.0334 g of 4,7-diphenyl-1,10-phenanthroline in warm absolute alcohol (C_2H_5OH) in a 100 ml flask and make up to the mark with cold absolute alcohol.
4. *Carbon tetrachloride* (CCl_4).
5. *Anhydrous sodium sulphate* (Na_2SO_4).

All chemicals used in the preparation of reagents should be of analytical reagent quality.

The Standard Lovibond Comparator 3/66

This disc contains standards corresponding to 1, 2, 3, 4, 5, 6, 7, 8, and 9 mg of iron (Fe) per litre.

Technique

To 1 ml of sample add, and shake after each addition, 3 ml of sodium pyrosulphite solution (reagent 1), 6.5 ml of buffer solution (reagent 2) and 5 ml of batho-phenanthroline solution (reagent 3). Pour the resulting solution into a 25 ml separating funnel (Note 2). Place the funnel for 5 minutes in hot water, and then cool immediately. Add 7 ml of carbon tetrachloride (reagent 4) and shake vigorously. Place the separating funnel in a hand centrifuge (Note 2) and centrifuge for 10 revolutions. Transfer the red (carbon tetrachloride) layer to a standard comparator test tube or 13.5 mm comparator cell and place this tube in the right-hand compartment of the comparator. Prepare a reagent blank, simultaneously with the preparation of the sample, by using 1 ml of distilled water instead of sample and subjecting this to the same treatment as the sample. Place an identical tube filled with the reagent blank in the left-hand compartment of the comparator. Place the comparator before a standard source of white light, such as the Lovibond White Light Cabinet or, failing this, north daylight, and match the colour of the sample solution against the standards in the disc (Notes 3-5). Read off the corresponding iron concentration in mg per litre from the indicator window in the comparator.

COLORIMETRIC CHEMICAL ANALYTICAL METHODS — Iron Method 4

Notes

1. Acknowledgement is made to Mr. H. Tanner of the Swiss Federal Horticultural Experimental Station, Wädenswil, Switzerland for making this test available.

2. A separating funnel which will fit into the bucket of a hand centrifuge must be used for this determination.

3. It is important that the colours be matched immediately after the test tubes are inserted into the comparator.

4. If the red colour of the sample solution is more intense than the 9 mg/litre standard, repeat the test, dilute the final solution with an equal volume of water and re-match. Then double the result read from the disc.

5. If the carbon tetrachloride solution is turbid as the result of the entrainment of a small amount of water, add a small amount of sodium sulphate (reagent 5) to the solution, stopper the tube with a clean rubber stopper, invert the tube several times and then match the colour immediately the sodium sulphate has settled.

Iron Method 5
using *o*-phenanthroline

Principle of the method

Ferrous iron forms a coloured complex with *o*-phenanthroline. To determine total iron, ferric iron is reduced with hydroxylamine hydrochloride and the resulting ferrous iron is complexed with *o*-phenanthroline. The colour of the complex is measured by comparison with a series of Lovibond permanent glass colour standards.

Reagents required

1. *Buffer solution* Dissolve 40 g of ammonium acetate (CH_3COONH_4) and 50 ml of glacial acetic acid (CH_3COOH) in distilled water and make up to 100 ml.
2. *Hydroxylamine hydrochloride solution* Dissolve 20 g of hydroxylamine hydrochloride in 100 ml of distilled water.
3. *o-Phenanthroline solution* Dissolve 0.5 g of 1,10-phenanthroline hydrochloride in 100 ml of distilled water.

All chemicals used for the preparation of these reagents should be of analytical reagent quality.

The Standard Lovibond Comparator Disc 3/88

This disc contains standards corresponding to 0.05, 0.1, 0.15, 0.2, 0.25, 0.3, 0.5, 0.75 and 1.0 ppm of iron (Fe) and is calibrated for use with 40 mm cells.

Technique

Take 20 ml of sample (Note 1) and in succession add 2 ml of buffer solution (reagent 1), 1 ml of hydroxylamine hydrochloride solution (reagent 2) and 2 ml of *o*-phenanthroline solution (reagent 3). Mix thoroughly. At the same time prepare a blank solution by using 20 ml of distilled water and the same quantities of the reagents. Allow the sample and blank solutions to stand for 15 minutes to ensure that reduction of any ferric iron is complete and transfer them to two identical 40 mm comparator cells. Place the cell containing the sample solution in the right-hand compartment of the comparator and that containing the blank in the left-hand compartment. Place the comparator before a standard source of white light, such as the Lovibond White Light Cabinet or, failing this, north daylight, and compare the colour of the sample with the standard colours in the disc.

If only ferrous iron is to be determined replace the 1 ml of hydroxylamine hydrochloride solution with 1 ml of distilled water. In that case the colour of the solution can be matched immediately after adding the reagents.

Notes

1. It is essential to ensure that all the iron in the sample is completely dissolved. It is therefore recommended that 1 litre of sample be taken and that this be immediately treated with 5 ml of 20% "iron free" hydrochloric acid. Use 20 ml of this treated sample for the test.
2. Silver, aluminium, copper, magnesium, manganese, nickel, lead, titanium, zinc and phosphate ions, in the quantities likely to be encountered in water samples, will not interfere with this determination. Interference from cobalt ions will occur if these are present in higher concentration than the iron.

The Determination of Lead

Introduction

The toxic effects of lead are cumulative and the necessity of reducing lead contamination of foods to a minimum is now widely recognised. Statutory limits range from 0.2 ppm for "ready to drink" beverages to 20 ppm for food colouring materials. The majority of foods are required to conform to a limit of 2 ppm. The dust of industrial atmospheres is often heavily contaminated with lead; it has been found in tea packed in lead-lined chests; and traces may remain, especially on fruit, from the use of some agricultural sprays.

Lead contamination of water, beer and cider has occassionally arisen as the result of the use of lead pipes and lead-glazed tanks. The World Health Organisation[1] has set a limit of 0.1 mg/l (ppm) on the lead content of piped water supplies in Europe (in some parts of America the limit is 0.05 ppm). They also point out that the lead content of water may exceed this limit after prolonged standing in lead pipes but stress that "in no instance should the lead content exceed 0.3 mg/l after 16 hours' contact with the pipes. Lead is also used as a stabilizer in some plastic pipes and the possibility of its being leached out must be borne in mind."

Lead is also an important element metallurgically, and the measurement of its concentration in alloys, ores, minerals, soils and industrial wastes is of importance.

Methods of sample preparation

In the preparation of samples for the determination of lead (Pb) it is especially important that all reagents used should be "Pb-free".

Air[2]

Trap the air-borne lead by drawing the air sample through a filter paper, as described later in detail in Section F; disintegrate the paper and dissolve the lead by shaking the paper, in a stoppered tube, with nitric acid and hydrogen peroxide (H_2O_2). Determine the lead content of this solution by *Method 1*. Alternatively, see method in Section F.

Beverages[3-6]

Mineral waters[3] Heat a 50 ml sample in a 800 ml Kjeldahl flask, add 0.5 g of iodine pentoxide (I_2O_5) and boil for 10 minutes. Remove the flame, add 25 ml of nitric acid and evaporate to small volume. Repeat with a further 25 ml of nitric acid. Wash the residue into a silica basin and evaporate to dryness. Ignite at 480°C for 30 minutes. Cool, moisten the residue with nitric acid, evaporate to dryness and re-ignite. Cool, wash the ash into a separating funnel with 2×25 ml of 1% nitric acid and 5 ml of ammonium acetate (CH_3COONH_4) solution. Extract the mixture with re-distilled chloroform ($CHCl_3$) to remove free iodine. Discard the chloroform layer and determine lead on the aqueous layer by *Method 1*.

Spirits[4] Place a 10 ml sample in a silica basin, add 5 ml of "ash acid" solution (25 g of anhydrous sodium sulphate (Na_2SO_4) in 500 ml of 10% sulphuric acid) and evaporate to half volume on a water bath. Add 3 ml of nitric acid and a little water and evaporate to dryness. Char the residue by gentle heating and then ash at 450°C for 15 minutes. Cool, moisten the residue with nitric acid, dry and re-ash and repeat until a white ash is obtained. Wash the ash into a separating funnel with 25 ml of hot 1% nitric acid, 5 ml of hot 20% ammonium acetate (CH_3COONH_4) solution and finally 25 ml of cold 1% nitric acid and determine lead by *Method 1*. Recovery of lead is claimed to be complete for dry wines and spirits and 88% for sweet wines.

Wines[3-6] Either of the methods outlined above may also be used for wines[3,4]. Alternatively[5] place a 20 ml sample in a flask, add 2 ml of sulphuric acid and distill off the alcohol. Add nitric acid drop by drop to oxidise the boiling liquid but not to produce an excess of acid. Dilute the residue and determine lead by *Method 1*. A quicker method[6] is to pass 50 ml of wine through a column of 30 g of Dowex-50 ion-exchange resin which has been washed with $4N$ hydrochloric acid. Elute the lead with $4N$ hydrochloric acid and proceed by *Method 1*.

Biological materials[3,7-16]

Blood[7,8] Take 1 ml of whole blood, add 10 ml of 5% trichloracetic acid (CCl_3COOH), set aside for 1 hour, mixing occasionally, and then centrifuge. Decant the supernatant liquid into a separating funnel, extract the precipitate with a further 10 ml of trichloracetic acid solution and again centrifuge. Decant the supernatant liquid into the same separating funnel and determine lead by *Method 1*[7]. Alternatively[8] take 3 ml of whole blood in a 50 ml glass stoppered tube, add 5 ml of nitric acid, 3 ml of 60% perchloric acid and 0.5 ml of sulphuric acid, set aside for 30 minutes and then heat on a sand bath at 200 ± 5°C until only sulphuric acid remains and then continue heating until no further fumes are evolved. Cool, add 3 ml of 50% ammonium citrate ((NH_4)$_3C_6H_5O_7$) solution, 5 ml of concentrated aqueous ammonia (NH_4OH sp.gr. 0.880) solution and proceed by *Method 1*.

Faeces[9] Places 1 g of dried faeces in a 500 ml beaker. Carefully cover with pure fuming nitric acid and warm gently until the vigorous reaction has ceased. Evaporate the solution, with the addition of more nitric acid if necessary, until it consists of a clear brown liquid containing a little insoluble matter and a little fat. Cool, dilute to 100 ml and filter. Determine lead by *Method 1* on an aliquot of this filtrate.

Tissue[10,11] Ash the sample, dissolve the ash in hydrochloric acid and extract an aliquot of this solution with a solution of sodium diethyldithiocarbamate in diethyl ether ($C_2H_5OC_2H_5$). Take the ether layer, remove the ether by evaporation on a water bath and destroy the organic matter by heating with sulphuric acid and 100 volume hydrogen peroxide (H_2O_2). Evaporate to white fumes and dissolve the residue in hydrochloric acid. Repeat the extraction process in the presence of sodium citrate ($Na_3C_6H_5O_7 \cdot 5H_2O$) and potassium cyanide (KCN) at pH 7.5—8.0. Remove the ether and destroy the organic matter with sulphuric acid and hydrogen peroxide (H_2O_2) as before. Dilute the digest with water and determine lead by *Method 1*[10]. Alternatively[11], digest the sample with 4 ml of water, 1 ml of sulphuric acid, 2 ml of nitric acid and 3 ml of perchloric acid ($HClO_4$) and add 0.5–1.0 ml of perchloric acid as required until the solution is colourless. Cool and dilute to a known volume.

Urine[3,8,12-16] Use the method described above for mineral waters[3]. Alternatively[12,13] adjust a 10 ml sample to approximately pH 4.5 by the addition of dilute acetic acid (CH_3COOH) solution or dilute aqueous ammonia (NH_4OH) using bromo-cresol green as the indicator. Place in a centrifuge tube and add 1 ml of saturated ammonium oxalate (($COONH_4$)$_2 \cdot H_2O$) solution and 0.5 ml of calcium chloride ($CaCl_2$) solution. Stand for 20 minutes and then centrifuge. Decant the supernatant fluid and allow the tube to drain onto a filter paper for 2—3 minutes. Digest the precipitate with 0.5 ml of perchloric acid ($HClO_4$) until the solution is colourless. Determine lead by *Method 1*[9]. Machata[14] recommends that before the determination of lead any thallium, bismuth and zinc should be removed by extraction of the sample solution with dithizone at pH 2.

Forman and Garvin[15] recommend that lead should be separated from the urine by ion exchange in the following manner. Adjust the pH of the urine to 3.5 with nitric acid, filter and pass 50 ml of the filtrate through a 30 × 1.2 cm tube containing a 7 cm depth of a slurry of Dowex A-1 resin (Na^+ form) at 5 ml a minute and discard the percolate. Desorb the lead with 18 ml of $2N$ hydrochloric acid, at 4 ml a minute, followed by 10 ml of water; make these combined percolates up to 50 ml and determine lead on an aliquot. Vinter[8] recommends wet ashing whereas Browett[16] prefers evaporation followed by dry ashing and solution of the ash in dilute nitric acid.

Chemicals[17-21]

Calcium phosphate[17,18] Dissolve 1 g of sample in 10–15 ml of $2N$ hydrochloric acid and proceed by *Method 2*[17]. Alternatively[18] dissolve the sample in $2N$ sulphuric acid, heat until white fumes are evolved, add 15 ml of a 1:2 ethanol (C_2H_5OH) water mixture and set aside overnight. Filter, wash the residue with a few ml of a water, ethanol, sulphuric acid (20:10:1) mixture. Dissolve the precipitated lead sulphate in hot 10% ammonium acetate (CH_3COONH_4) and proceed by *Method 2*.

General[19] Aqueous solutions of chemicals which do not contain metals giving coloured sulphides are determined directly by *Method 2*. Otherwise[19] dissolve the sample in 20 ml of 57% perchloric acid ($HClO_4$), evaporate until copious white fumes appear, dissolve the residue in a little water, add iron alum ($NH_4Fe(SO_4)_2 \cdot 12H_2O$), precipitate hydroxides with aqueous ammonia (NH_4OH), collect the precipitate, wash it with hot water, dissolve it in 30 ml of $4N$ hydrochloric acid, dilute the solution with an equal volume of water and pass it through a column of anionite AN-31 (Cl^- form). Wash the column with $2N$ hydrochloric acid and desorb lead with $0.02N$ hydrochloric acid.

Stibophen (Fouadin, sodium-antimony-bispyrocatechol-3,5-sodium disulphonate)[20]. Heat a 1 g sample with 10 ml of nitric acid and 5 ml of sulphuric acid until white fumes are evolved, cool, dilute with 15 ml of water, add 1 g of tartaric acid ($C_4H_6O_6$), cool, neutralise by the addition of 20% caustic soda (NaOH) and add 10 ml in excess and proceed by *Method 2*.

Zinc and its compounds[21] Dissolve 1 g of sample in 33% (w/w) acetic acid (CH_3COOH), dilute the solution to 100 ml with water and determine lead on a suitable aliquot using *Method 2*.

Foods and edible oils[22-32]

Wet oxidise a 5 g sample with 5 ml of sulphuric acid and a sufficiency of nitric acid. When oxidation is complete, cool, dilute the residue with water and re-heat to white fumes. Cool, dilute with 20 ml of water, add 2 g of citric acid ($C_6H_8O_7$) boil for a few minutes and cool. Neutralise the solution to litmus by the addition of aqueous ammonia (NH_4OH sp.gr. 0.880) and add 0.5 ml in excess. Add 1 ml of 10% potassium cyanide (KCN) solution and extract with a 0.1% solution of dithizone (diphenylthiocarbazone) in chloroform ($CHCl_3$) (3 × 10 ml). Wash each extract in turn with 10 ml of water. Combine the extracts and remove chloroform by evaporation. Add 0.7 ml of sulphuric acid heat to white fumes and then add nitric acid drop by drop until all organic matter is destroyed. Add 5 ml of water and re-heat to white fumes. Add 15 ml of 1:2 ethanol (C_2H_5OH) and stand overnight. Filter through a 5 cm Whatman No. 44 filter paper which has been previously washed with hot 5N hydrochloric acid and then with hot water just before use. Wash the residue 3 times with a few ml of a mixture of 20 ml of water, 10 ml of absolute ethanol and 1 ml of sulphuric acid. Dissolve the precipitated lead sulphate by boiling 10 ml of 10% ammonium acetate (CH_3COONH_4) in the flask in which the precipitation was carried out and pass the hot solution through the filter. Repeat the operation with the same 10 ml of solution and finally wash the filter 3 times with 5 ml of hot water containing ammonium acetate. Proceed by *Method 2*[22-25]. If the sample contains much calcium phosphate (e.g. bone material, baking powder etc.) the calcium sulphate which precipitates after the wet oxidation may contain lead. In such cases transfer the acid mixture, suitably diluted with water, to a 100 ml centrifuge-tube and wash out the digestion flask with ammonium acetate solution. Separate the precipitate by centrifuging. Decant the supernatant liquid back into the digestion flask, wash the residue with hot water (2 × 20 ml) and add the washings to the flask. Add 4 g of sodium carbonate (Na_2CO_3) and 90 ml of hot water to the residue in the centrifuge-tube and immerse the tube in boiling water for 4 hours and stir frequently. Wash down with water, cool, centrifuge and add the supernatant liquid to the previous supernantant. This solution now contains all the "soluble lead". Treat with 4 g of citric acid, render alkaline with aqueous ammonia and extract with dithizone as described above[26]. Alternatively Lockwood[27] recommends charring the sample with calcium nitrate ($Ca(NO_3)_2$), extracting the ash with hydrochloric acid and extracting the lead at pH 7 with a carbon tetrachloride (CCl_4) solution of diethylammonium diethyldithiocarbamate, evaporating the solvent from the organic layer, destroying the residual organic matter with sulphuric acid + perchloric acid, diluting the solution with water and determining lead on this solution by either *Method 1 or 2*. Elridge and Garratt[28] recommend the ignition of a 3–4 g sample in a Mahler-Cook bomb followed by the determination of lead on the solution so formed. The official method of the Society for Analytical Chemistry[29] is to wet ash the sample with sulphuric acid and perchloric acid, or dry ash with magnesium nitrate ($Mg(NO_3)_2$) at not more than 500°C and dissolve the ash in hydrochloric acid. Make the solution just alkaline with aqueous ammonia and then acid with 10 ml of 5N hydrochloric acid. Add sodium iodide (NaI), remove excess with sodium metabisulphate ($Na_2S_2O_5$) and extract the liquid with a solution of diethylammonium diethyldithiocarbamate in chloroform. Add dilute sulphuric acid to the organic layer, remove the solvent by evaporation, heat the residue with perchloric acid, dissolve in hydrochloric acid and proceed by *Method 1*. Neumann[30] suggests that dithizone should be used for the separation of lead and recommends suppression of interfering ions by the use of masking agents[31]. Johnson and Polhill[32] use an aliquot of the ash solution which contains less than 40 μg of lead and apply this to a column of the chloride form of Amberlite IRA–400 ion exchange resin. The column is then washed with 25 ml of ammonium chloride (NH_4Cl) solution and the lead eluted with 0.01N hydrochloric acid and determined by *Method 1*.

Fuels and lubricants[33,34,69]

Weigh out a 45 g sample and transfer it into a 250 ml separating funnel. Add 10 ml of di-isobutylene (($CH_3)_2C=CHC(CH_3)_3$) and mix. Add bromine solution slowly until a marked excess of bromine persists for more than 4 minutes. Add 30 ml of chloroform ($CHCl_3$) and 100 ml of 1% nitric acid, shake for 1 minute and discard the organic layer. Wash the acid solution with a further 25 ml of chloroform and again discard the organic layer. Estimate lead on the aqueous layer by *Method 1*[33]. Alternatively[34] dilute 25 ml of the sample with 25 ml of a lead-free heavy distillate and shake at room temperature with 50 ml of M aqueous iodine chloride (ICl). Separate the phases, wash the upper, hydrocarbon, phase with water and combine the aqueous phase and the washings. Evaporate to low volume and oxidise, by repeated evaporation with 5 ml portions of nitric acid, until the residue is white. Cool, dissolve the residue in water and make the solution up to a known volume.

Aviation turbine fuels and light petroleum distillates[69] Place 5 ml of sample, at a temperature of $60 \pm 5°F$, in a separating funnel, add 10 ± 0.2 ml of M iodine monochloride (ICl) solution, and shake the mixture for 3 minutes. Allow the layers to separate and then run the lower, aqueous, layer into a flask. Wash the organic layer remaining in the funnel with 3×10 ml of distilled water and add the washings to the liquid in the flask. Boil this aqueous solution on a hot plate for 1 minute, cool, and then add saturated sodium sulphite ($Na_2SO_3.7H_2O$) solution drop by drop until the solution is colourless. Make alkaline and proceed by *Method 1*.

Metals alloys and plating baths[35-42]

Alloys[35,36] Dissolve 1 g of sample in 50 ml of a 1:1 mixture of hydrochloric and nitric acids. Evaporate the solution to dryness and dissolve the residue in 1:1 nitric acid. Dilute this solution to 200 ml and pass through a column of pumice powder impregnated with cadmium sulphide (6–7 g pumice, impregnated with 5% cadmium acetate—$(CH_3COO)_2Cd.2H_2O$—solution, exposed to hydrogen sulphide and then washed with water). Lead is retained on this column while other metals pass through. Wash the column twice with water and then elute the lead with 5 ml of nitric acid and proceed by *Method 1*[35]. For silver alloys[36] dissolve the sample in 5 ml of 1:1 nitric acid add 75 ml of water and 2 ml of 2% aluminium nitrate ($Al(NO_3)_3$) solution. Heat nearly to boiling and add aqueous ammonia drop by drop until the solution becomes dark blue. Set aside for 20 minutes then filter, wash the precipitate with dilute aqueous ammonia and then dissolve it in 20 ml of 1:1 nitric acid. Make up this solution to 50 ml with water.

Bismuth[37] Dissolve 1–2 g of the metal in 10–15 ml of 50% nitric acid, evaporate to dryness, cool, add 10–15 ml of hydrogen bromide (HBr), evaporate to dryness and then heat the residue at $280 \pm 50°C$. Repeat this treatment until no yellow bismuth bromide ($BiBr_3$) remains. Dissolve the final residue in a 1:3 mixture of hydrogen bromide and water, add 1.5 ml of a 0.1 ppm solution of zinc, to facilitate extraction, adjust the pH to 3.5 with aqueous ammonia and proceed by *Method 1*.

General[38-42] Dissolve up to 1 g of the sample in acid with the addition of either hydrogen bromide[38] or nitric acid[39] to ensure oxidation, dilute the solution to a suitable volume and estimate lead on an aliquot by *Method 1*.

High purity metals[42] Dissolve up to 2 g of the sample in 20 ml of 1:1 nitric acid, add 20 ml of perchloric acid and heat until white fumes are evolved. Dissolve the residue in 150 ml of water, add 2 ml of 7% potassium chromate (K_2CrO_4) solution, make ammoniacal and add 2 ml of 4% barium chloride ($BaCl_2$) solution and filter. Dissolve the precipitate in 10 ml of 1:1 perchloric acid, heat to white fumes and dissolve the residue in 1 ml of hydrochloric acid and heat to dryness. Dissolve the residue in dilute hydrochloric acid and determine lead by *Method 1*.

Minerals, ores and rocks[43,44]

Take sufficient sample to contain about 15 μg of lead. Dissolve it in 10 ml of aqua regia (1 part nitric acid, 3 parts hydrochloric acid) and 5 ml of hydrogen fluoride (HF) in a Teflon beaker on a steam bath. Convert the salts to chlorides by repeated evaporation with hydrochloric acid. Dissolve the chlorides in 25 ml of N hydrochloric acid, cool to 5–10°C and transfer to a 125 ml separating funnel together with 0.5 g of cupferron and 20 ml of cold chloroform ($CHCl_3$). Shake for 2 minutes and separate the lower layer. Increase the acidity of the aqueous phase to $1.5N$ by adding 3 ml of $6N$ hydrochloric acid. Add 10 ml of a 1% solution of diethylammonium diethyldithiocarbamate in chloroform and destroy the residual organic matter with 3 ml of nitric acid and 0.5 ml of perchloric acid. Evaporate to dryness. Dissolve the residue in 1 ml of 1:1 nitric acid and 5 ml of water and determine lead by *Method 1*[13]. Maynes and McBride[44] describe individual treatments for Lyndochite, Feldspar, Apatite and Sphene but use the above extraction procedure after ensuring the solution of the sample.

Miscellaneous[45-49]

Drugs[45,46] Wet oxidise the sample with sulphuric and perchloric acids and hydrogen peroxide (H_2O_2). Evaporate the solution to the point at which copious white fumes appear, take up the residue in aqueous acetic acid (CH_3COOH), extract copper with dithizone in carbon tetrachloride (CCl_4) at pH 2.7, then adjust the pH to 4.9 with ammonium acetate buffer solution and extract with dithizone in chloroform. Adjust the pH to 8.5 and proceed by *Method 1*.

Essential oils[17] Remove the volatile components by boiling the sample with dilute hydrochloric acid and then extract organic matter with carbon tetrachloride (CCl_4). Determine lead on the aqueous layer with *Method 1*.

Paint[18] Ash sufficient of the sample to contain 0.1–0.3 mg of lead, extract the ash with 2N nitric acid and dilute to a known volume.

Photographic gelatin[19] Dissolve the gelatin in water, pass this solution through a column of Dowex 50-W cation exchange resin and elute metals with 3N hydrochloric acid. Proceed by *Method 1*.

Plants[50,51]

Dry the sample at 80°C overnight, then take 1–2 g of the dried sample and wet ash with sulphuric, nitric and perchloric ($HClO_4$) acids until the residue is colourless. Dissolve the residue in water, make this solution up to a known volume and proceed by *Method 1*.

Sewage and industrial wastes[5,53]

Place a 500 ml sample in a beaker, add 5 ml of nitric acid and evaporate to a small volume. Transfer to a porcelain or silica basin and further reduce the volume to about 15 ml on a water bath. Add 2 ml of 60% perchloric acid ($HClO_4$) and gently evaporate to dryness. Ignite at below 300°C for about 40 minutes. Dissolve the residue in 2 ml of 5N hydrochloric acid and a little warm water. If the solution is coloured, add 2 ml of 60% perchloric acid and 2 ml of nitric acid and repeat the process of evaporation followed by ashing and solution of the ash until a clear solution is obtained. If a white turbidity is observed filter through a sintered glass filter. Add 50 ml of hot 40% ammonium acetate (CH_3COONH_4) solution to the basin in which the acid treatment was carried out and pass this solution through the same filter. Add this second filtrate to the first and determine lead by *Method 1*[52]. Alternatively[53] evaporate a sample containing not more than 100 μg of lead with nitric acid and destroy the organic matter either by ashing at low temperature or by wet ashing. Dissolve the residue in dilute nitric acid and extract with dithizone solution in the presence of potassium cyanide (KCN) and aqueous ammonia (NH_4OH). Re-extract the lead with dilute nitric acid and then proceed by *Method 1*.

Soil[54,55]

Soil extracts which have been prepared by any of the standard extraction procedures, examples of which have been cited earlier in the sample preparation sections for aluminium and ammonia determinations, can be analysed by the second procedure described below for water samples[54].

To determine lead in soil, disintegrate a 5 g sample of the soil with perchloric acid ($HClO_4$) or with aqua regia (3 parts of hydrochloric acid to 1 part of nitric acid) followed by perchloric acid. Evaporate until white fumes are evolved and then dissolve the residue in dilute nitric acid. Filter and then adjust the pH of the filtrate to 9.0 with aqueous ammonia (NH_4OH) and proceed by *Method 1*.

Water[54,56–60]

Measure 250 ml of sample into a 500 ml beaker, add 2 drops of methyl orange indicator, just acidify with nitric acid and add 4 ml in excess. Reduce the total volume to 10–15 ml and transfer the solution to a small glass dish using the minimum amount of wash water. Evaporate the solution to dryness on a steam bath. Wash down the sides of the dish with 2 ml of nitric acid and again evaporate to complete dryness. Ignite the residue in a muffle furnace for 20 minutes at 490–500°C. Cool, add 3 ml of nitric acid and 15 ml of water. Heat for 5 minutes on a steam bath. Filter, wash the filter paper with small amounts of hot 20% nitric acid and cool. Add 2 ml of 10% hydroxylamine hydrochloride ($NH_2OH.HCl$) solution and determine lead by *Method 1*[56]. Alternatively[54,60] to a 500 ml sample add 10 ml of a 5N calcium chloride ($CaCl_2$) solution, 2 ml of a 5% sodium diethyl-dithiocarbamate solution and 10 ml per litre of a saturated sodium citrate ($Na_3C_6H_5O_7.5H_2O$) solution. Adjust the pH to 8.5–9.0 add 15 ml of carbon tetrachloride (CCl_4) and shake for 5 minutes. Transfer the carbon tetrachloride layer to a separating funnel and extract with 0.1N hydrochloric acid (4 × 5 ml) to remove zinc. Wash the carbon tetrachloride layer with metal-free water and shake with 3N hydrochloric acid (2 × 5 ml). Combine the acid extracts and determine lead by *Method 1*. It is recommended[57] that for low lead concentrations the lead be concentrated on an ion exchange column. Elute the column with 15% hydrochloric acid and determine lead on the eluate. The application of *Method 2* to the determination of lead in water has been studied by Reith[59] who claims that samples which are colourless, which do not contain a high proportion of dissolved solids or traces of other heavy metals may be examined directly for lead by this method. However in the presence of such contamination one of the methods described above is to be preferred.

Separation from other ions[61-63]

From mercury, silver and thallium[61,62] Silver, lead and mercury are successively eluted from Dowex 50W-X8 resin by $0.25M$, $0.5M$ and $4M$ ammonium acetate (CH_3COONH_4) respectively[61], or may be separated by paper chromatography using acidic solvents containing chloroform, methanol, etc.[62].

From magnesium or calcium[63] Lead may be separated, from an aqueous solution of pH 7, by extraction into diphenylarsinic acid solution in chloroform.

General[64-66] Using Dowex 1-X8 resin (100-200 mesh, NO^-_3 form) and a 9:1 mixture of tetrahydrofuran ($CH_2.(CH_2)_2.CH_2O$) and $5N$ nitric acid as solvent and eluant, lead is retained on the resin and can be desorbed with a 4:1 mixture of tetrahydrofuran and $2.5N$ nitric acid[64]. Or, using a 19:1 mixture of methanol (CH_3OH) and $5N$ nitric acid as the solvent, and the same resin, after washing the column with a 9:1 mixture of methanol and $6N$ hydrochloric acid, the lead can be eluted with normal hydrochloric acid[65]. Fritz[66] recommends elution with $0.6M$ hydrogen bromide (HBr).

The individual tests

Method 1 is the popular dithizone procedure originally proposed by Fischer[67]. It is the standard method recommended for the determination of lead in water[56], sewage[52], effluents[53], food[29], fuel oil[69] and tin and tin alloys[11]. It should be used wherever it is necessary to determine lead in the presence of other metals.

Method 2 is the sodium sulphide method. This is claimed[68] to be the best method for lead in the absence of interfering metals. It is the method recommended for the determination of lead in food-colouring materials[22] and in pharmaceutical chemicals[20].

References

1. World Health Organisation, *"European Standards for Drinking Water"*, W.H.O., Copenhagen, 1970, p. 25
2. B. E. Dixon and P. Metson, *Analyst,* 1960, **85**, 122
3. N. Greenblau and J. P. van der Westhuyzen, *J. Sci. Food Agric.*, 1956, **7**. 186; *Analyt. Abs.,* 1957, **4**, 273
4. J. P. van der Westhuyzen, *S. Afr. J. Agric. Sci.*, 1959, **2**, 183; *Analyt. Abs.*, 1960, **7**, 1932
5. J. Bonastre, *Chim. Anal.*, 1957, **39**, 104; *Analyt. Abs.*, 1957, **4**, 3131
6. R. A. Edge and N. Penny, *J. Sci. Food Agric.*, 1958, **9**, 401; *Analyt. Abs.*, 1959, **6**, 1930
7. E. Berman, *Amer. J. Clin. Path.*, 1961, **36**, 549; *Analyt. Abs.*, 1962, **9**, 2411
8. P. Vinter, *J. med. Lab. Techno.*, 1964, **21** (4), 281; *Analyt. Abs.*, 1966, **13**, 3683
9. R. A. Kehoe, F. Thamann and J. Cholak, *J. Amer. Med. Assoc.*, 1935, **104**, 90
10. S. L. Tompsett, *Analyst,* 1956, **81**, 330
11. A. Dyfverman, *Ark. Kemi*, 1967, **27** (1), 79; *Analyt. Abs.*, 1968, **15**, 6147
12. C. C. Lucas and J. R. Ross, *J. Biol. Chem.*, 1935, **111**, 285
13. U.K.A.E.A. Report IGO-AM/W-169, 1958; *Analyt. Abs.*, 1958, **5**, 3429
14. G. Machata and H. Neuniger, *Wien. Med. Wochschr.*, 1960, **110** (2), 39; *Analyt. Abs.*, 1960, **7**, 4402
15. D. T. Forman and J. E. Garvin, *Clin. Chem.*, 1965, **11** (1), 1; *Analyt. Abs.*, 1966, **13**, 2508
16. E. V. Browett and R. Moss, *Analyst,* 1965, **90**, 715
17. J. R. Nicholls, *Analyst,* 1931, **56**, 594

18. D. W. Kent-Jones and C. W. Herd, *Analyst,* 1933, **58,** 152
19. V. V. Stepin, V. I. Ponosov, G. N. Emasheva and N. A. Zobnina, *Trudy vses. nauchno-issled. Inst. standart. Obraztsov,* 1968, **4,** 100; *Analyt. Abs.,* 1970, **18,** 900
20. British Pharmacopaeia, 1958
21. T. T. Cocking, *Chem. and Drugg.,* 1906, **79,** 507
22. Second Report to the Analytical Methods Cttee. of the Soc. Pub. Analysts and other Anal. Chemists, of the sub-cttee. on the detn. of As, Pb and other poisonous metals in food-colouring matls., *Analyst,* 1935, **60,** 541
23. N. L. Allport and G. H. Skrimshire, *Analyst,* 1932, **57,** 440
24. G. W. Monier-Williams, Reports on Public Health and Medical Subjects No. 88 (1938), *"Lead in Food"*
25. A. G. Francis, C. O. Harvey and J. L. Buchan, *Analyst,* 1929, **54,** 725
26. G. Roche-Lynch, R. H. Slater and T. G. Osler, *Analyst,* 1934, **59,** 787
27. H. C. Lockwood, *Analyst,* 1954, **79,** 143
28. D. A. Eldridge and D. C. Garratt, *Analyst,* 1954, **79,** 146.
29. Report of the Pb Panel of the Metallic Imp. in Foodstuffs sub-cttee. of the Anal. Methods Cttee. of the Soc. Anal. Chem., *Analyst,* 1954, **79,** 397
30. F. Neumann, *Z. analyt. Chem.,* 1957, **155,** 346; *Analyt. Abs.,* 1957, **4,** 3112
31. D. Abson and A. G. Lipscomb, *Analyst,* 1957, **82,** 152
32. E. I. Johnson and R. D. A. Polhill, *Analyst,* 1957, **82,** 938
33. M. E. Griffing, A. Rozek, L. J. Snyder and S. R. Henderson, *Analyt. Chem.,* 1957, **29,** 190
34. R. Moss and K. Campbell, *J. Inst. Petrol.,* 1967, **53,** 89; *Analyt. Abs.,* 1968, **15,** 2723
35. W. Stolper, *Neue Hutte,* 1963, **8,** 424; *Analyt. Abs.,* 1964, **11,** 3684
36. Z. Skorko-Trybula and J. Chivastowska, *Chem. Anal. Warsaw,* 1963, **8,** 859; *Analyt. Abs.,* 1965, **12,** 1091
37. K. Nishimura, A. Tsuchibuchi and T. Aoyama, *Japan Analyst,* 1964, **13** (3), 220; *Analyt. Abs.,* 1966, **13,** 2309
38. K. Ota and S. Mori, *Japan Analyst,* 1956, **5,** 442; *Analyt. Abs.,* 1957, **4,** 1486
39. R. Socolovschi, *Rev. Chim. Bucharest,* 1960, **11,** 348; *Analyt. Abs.,* 1961, **8,** 93
40. S. Wakamatsu, *Japan Analyst,* 1956, **5,** 509; *Analyt. Abs.,* 1957, **4,** 2203
41. Brit. Stand. Inst., *"Methods for the Sampling and Analysis of Tin and Tin Alloys, Pt. 5, Lead in Ingot Tin and Tin-Antimony Solders"*, B.S. 3338 Part 5, London, 1961
42. T. Yanagihara, N. Matano and A. Kawase, *Japan Analyst,* 1959, **8,** 576; *Analyt. Abs.,* 1960, **7,** 4195
43. J. J. Warr and F. Cuttitta, *U.S. Geol. Survey Profess. Paper No. 400-B* 1960; *Analyt. Abs.,* 1961, **8,** 2347
44. A. D. Maynes and W. A. E. McBride, *Analyt. Chem.,* 1957, **29,** 1259
45. J. Hollos, *Gyogyszereszet,* 1963, **7,** 204; *Analyt. Abs.,* 1964, **11,** 5684
46. J. Hollos and I. Horvath-Gabai, *Pharmazie,* 1965, **20** (4), 207; *Analyt. Abs.,* 1966, **13,** 4367
47. Brit. Standards Inst., BS 2073 Addendum No. 1 (1964), *"Methods of testing essential oils"*, No. 15, *"Determination of lead"*, London, 1964; *Analyt. Abs.,* 1965, **12,** 2891
48. E. Hoffmann, *Z. analyt. Chem.,* 1965, **208** (6), 423; *Analyt. Abs.,* 1966, **13,** 3671
49. J. Saulnier, *Sci. ind. Photogr.,* 1964, **35,** 1; *Analyt. Abs.,* 1965, **12,** 4037
50. K. Riebartsch and G. Gottschalk, *Z. analyt. Chem.,* 1965, **214** (3), 179; *Analyt. Abs.,* 1967, **14,** 851
51. Z. Kerin, *Mikrochim. Acta,* 1968, (5), 927; *Analyt. Abs.,* 1970, **18,** 596
52. Min. of Housing and Local Govt., *"Methods of Chemical Analysis as Applied to Sewage and Sewage Effluents"*, H.M. Stationery Office, London, 1956

53. Joint A.B.M.-S.A.C. Cttee. on Methods for the Anal. of Trade Effluents, *Analyst,* 1956, **81,** 607
54. A. D. Miller and R. I. Libina, *Zh. analyt. Chim.,* 1958, **13,** 664; *Analyt. Abs.,* 1959, **6,** 2792
55. A. Duca and D. Stanescu, *Stud. Cercet. Chim. Cluj.,* 1957, **8,** 75; *Analyt. Abs.,* 1958, **5,** 1717
56. Brit. Stand. Inst., *"Methods of Testing Water Used in Industry"*, B.S. 2690, London, 1956
57. V. B. Aleskovskii, R. I. Libina and A. D. Miller, *Trudy Leningr. Tekhnol. Inst. Lensoveta,* 1958, (48), 5; *Analyt. Abs.,* 1960, **7,** 1698
58. D. C. Abbott and J. R. Harris, *Analyst,* 1962, **87,** 387
59. J. F. Reith and J. de Beus, *Chem. Weekblad,* 1935, **32,** 205
60. T. V. Gurkina and A. M. Igoshin, *Zh. analit. Khim.,* 1965, **20** (7), 778; *Analyt. Abs.,* 1967, **14,** 1730
61. A. K. De and S. K. Majumdar, *Talanta,* 1963, **10,** 201; *Analyt. Abs.,* 1964, **11,** 24
62. R. P. Bhatnagar and K. D. Sharma, *Analyt. chim. Acta.,* 1964, **30,** 310; *Analyt. Abs.,* 1965, **12,** 3190
63. R. Pietsch and G. Nagl, *Z. analyt. Chem.,* 1964, **203,** 253; *Analyt. Abs.,* 1965, **12,** 5098
64. J. Korkisch and F. Feik, *Analyt. Chem.,* 1964, **36,** 1793
65. S. S. Ahluwalia and J. Korkisch, *Z. analyt. Chem.,* 1965, **208** (6), 414: *Analyt. Abs.,* 1966, **13,** 3521
66. J. S. Fritz and R. G. Greene, *Analyt. Chem.,* 1963, **35,** 811
67. H. Fischer and G. Leopoldi, *Wiss. Veroffentlich. Siemens-Konzern,* 1933, **12,** 44; *Angew. Chem.,* 1934, **47,** 90
68. N. L. Allport and J. E. Brocksopp, *"Colorimetric Analysis, Vol. II"*, Chapman and Hall, London, 1963
69. Inst. Petroleum, *"IP Standards for Petroleum and its Products, Part I Section* 2, *p.*1014, *IP 224/68 Tentative"*, London, 31st Edn., 1972

COLORIMETRIC CHEMICAL ANALYTICAL METHODS

Lead Method 1
using dithizone (diphenylthiocarbazone)

Principle of the method

The lead is extracted from alkaline solution by dithizone in chloroform solution. The excess dithizone is removed by treatment with sodium cyanide and the intensity of the red colour of the lead dithizonate is measured by comparison with Lovibond permanent glass colour standards. Hydroxylamine hydrochloride is added to prevent oxidation of the dithizone by any ferric iron present.

Reagents required, all analytical reagent grade

1. *Aqueous ammonia* (NH_4OH) *solution* .880
2. *Citric acid* ($C_6H_8O_7.H_2O$) Solution of 10 g in 100 ml water
3. *Dithizone solution* 4 mg dithizone (diphenylthiocarbazone) in 100 ml chloroform This must be freshly made.
4. *Sodium cyanide* ($NaCN$) Solution of 5 g in 100 ml water
5. *Chloroform* ($CHCl_3$)
6. *Bromo-thymol blue* 0.04% indicator solution
7. *Hydroxylamine hydrochloride* ($HONH_3Cl$) Solution of 20 g in 100 ml water

The Standard Lovibond Comparator Disc 5/17

The disc includes standards for the range 0.002 to 0.02 mg of lead (Pb), in steps of 0.002. This covers the range 0.2 to 2.0 ppm if 10 ml sample is taken, or 0.08 to 0.8 ppm on a 25 ml sample (see Note 1).

Technique

To 10 ml of the sample solution add 1 ml dilute citric acid (reagent 2), 0.1 ml of Bromothymol blue indicator (reagent 6) and aqueous ammonia (reagent 1) drop by drop until the mixture is just blue. Transfer to a 25 ml separating funnel containing 1 ml of the dithizone solution (reagent 3) 1 ml of sodium cyanide solution (reagent 4), and 1 ml of hydroxylamine hydrochloride solution (reagent 7), and shake until the red colour in the chloroform reaches a maximum. The chloroform layer should then be run off and the extraction repeated with a further 1 ml of reagent 3. If the second chloroform extract shows an appreciable pink colour, a third or even a foutrh extraction with 1 ml portions of reagent 3 must be made. The combined chloroform extracts are then placed in a separating funnel and shaken with 10 ml of water, 0.1 ml of reagent 4 and 0.1 ml of reagent 1, when the green colour due to excess of reagent 3 goes into the aqueous layer. Transfer the chloroform layer to a graduated vessel, make up to 5 ml with chloroform, and place in a test tube or 13.5 mm cell in the right-hand compartment of the Comparator. Parallel with the above work, a "blank" test is carried out, using 10 ml of distilled water in pace of sample: this blank is placed in the left-hand compartment of the Comparator.

Hold the Comparator facing a standard source of white light such as the Lovibond White Light Cabinet or, failing this, north daylight, and rotate the disc until the colour of the solution is matched by one of the Lovibond permanent glass standards. The value, expressed as mg of lead (Pb) in the original volume of sample taken, is read from the indicator at the bottom right-hand corner. If the colour is deeper than the darkest standard, the test must be repeated using a smaller quantity of the original sample.

Note

1. In the case of a 25 ml sample, a 50 ml separating funnel is required and the volume of citric acid (reagent 2), used, is increased to 2 ml.
2. All glassware used for this test must be scrupulously clean; it should be boiled with nitric acid and thoroughly rinsed with distilled water before use.

Lead Method 2
using sodium sulphide

Principle of the method

This method is based on the colours produced by the addition of sodium sulphide to ammoniacal solutions containing lead. These colours are measured by comparison with Lovibond permanent glass colour standards. The conditions adopted for the standardisation of the colours are based upon the recommendations in the Second Report to the Analytical Methods Committee of the Society of Public Analysts and Other Analytical Chemists of the Sub-Committee on the Determination of Arsenic, Lead and other Poisonous Metals in Food-colouring Materials.

Reagents required, (all of analytical reagent grade)

1. *Ammonium acetate* (CH_3COONH_4) A 10% w/v aqueous solution

2. *Ammonium citrate* A 10% w/v aqueous solution prepared by dissolving 8.75 g of citric acid ($C_6H_8O_7.H_2O$) in water, neutralising with ammonia and diluting with water to 100 ml.

3. *Aqueous ammonia* (NH_4OH) A solution containing 10% w/w NH_3

4. *Potassium cyanide* (KCN) A 10% w/v aqueous solution

5. *Sodium sulphide* (Na_2S) A 10% w/v aqueous solution

The Standard Lovibond Nessleriser Disc NF

The disc covers the range from 10 µg to 100 µg (0.01 to 0.1 mg) of lead (Pb), in steps of 10 µg. This covers the range 0.5 to 5 ppm if a 20 ml sample is taken.

Technique

To a suitable quantity of the solution under examination (which should be approximately neutral and not exceed 25 ml in volume) contained in a Nessleriser glass, add 10 ml of ammonium acetate solution (reagent 1), 5 ml of ammonium citrate solution (reagent 2), 5 ml of aqueous ammonia (reagent 3), and 1 ml of potassium cyanide solution (reagent 4) and dilute to 50 ml with distilled water; mix and then add 2 drops of sodium sulphide solution (reagent 5). Again mix and place in the right-hand compartment of the Nessleriser. In another Nessleriser glass place 10 ml of reagent 1, 5 ml of reagent 2, 5 ml of reagent 3 and 1 ml of reagent 4 and dilute to 50 ml with distilled water; mix and add 2 drops of reagent 5. Again mix and place in the left-hand compartment of the instrument. Stand the Nessleriser before a standard source of white light such as the Lovibond White Light Cabinet or, failing this, north daylight, and compare the colour produced in the test solution with the colours in the standard disc, rotating the disc until a colour match is obtained.

The markings on the disc represent the actual amounts of lead producing the colour in the test. Thus, if a colour equivalent to 30 µg is produced in the test, the amount of lead present in the quantity of solution taken for the test is 0.03 mg of Pb.

Note

It must be emphasized that the readings obtained by means of the Lovibond Nessleriser and disc are only accurate provided that Nessleriser glasses are used which conform to the specification employed when the discs were being calibrated, namely that the 50 ml calibration mark shall fall at a height of 113 mm plus or minus 3 mm measured internally.

The Determination of Magnesium

Introduction

Magnesium is one of the elements essential to plant growth and it is present in all plant and animal tissues. It is also an important alloying element in metallurgy, and its salts are responsible, in part, for "hardness" in water.

The determination of magnesium is thus of interest in agricultural, biochemical and metallurgical laboratories, as well as in the control of water supplies.

Methods of sample preparation

Biological materials[1-21]

Blood[1-14] Take 0.3 ml of sample, add 1.2 ml of a 10% solution of trichloracetic acid (CCl_3COOH) and centrifuge. Pipette 0.75 ml of the clear supernatant liquid, neutralise with 10% caustic soda (NaOH) and dilute to 3 ml. Proceed by *Method 1*[1,10,13]. Andraesen[3] suggests the addition of 0.5 ml of a 0.1% solution of polyvinyl alcohol to stabilise the colour. This suggestion is also supported by Lewis[5], Schmid[7], Martinek[11] and Masson[14]. Schmid[7] also recommends the addition of calcium chloride ($CaCl_2$) solution to intensify the colour. However Baron and Bell[2] claim that *Method 1* is accurate, reproducible and independent of the calcium content. Alternatively[4,6,8] treat the serum with trichloracetic acid (CCl_3COOH) and centrifuge as before. Remove calcium from the supernatant liquid by precipitation with ammonium oxalate (($COONH_4)_2.H_2O$). Filter. Precipitate the magnesium as magnesium ammonium phosphate ($MgNH_4PO_4$) and proceed by *Method 2*.

Anast[9] reports that with the blood of patients receiving intravenous calcium gluconate the normal Titan Yellow procedure (*Method 1*) yields erroneously low results, and suggests that the sample should be wet ashed before magnesium is determined. Burcar et al[12] also recommend wet ashing as a general method of sample preparation for blood.

Tissue[1,15-20] Ash 25 mg of sample in a platinum crucible at 600°C. Dissolve the ash in 1 drop of 5% sulphuric acid, dilute to 3 ml and proceed by *Method 1*[15,16,19]. In the presence of zinc or a large excess of calcium, such as in bone, remove zinc with dithizone at pH 5, render the solution alkaline and chelate the calcium with E.D.T.A. (ethylene-diamine-tetra-acetic-acid) and then determine magnesium directly using *Method 1*[17]. Hofer[18] and Conradi[20] both recommend the use of ion exchange resins to separate magnesium from the acid solution of the ash, but operate the resins in different ways. Hofer[18] adds the solution to Dowex 50–X8 (H^+ form) resin, where the magnesium is absorbed. After washing the column with normal hydrochloric acid he desorbs the magnesium with $2N$ hydrochloric acid. Conradi[20] on the contrary uses a strongly basic column of resin, purely to absorb phosphate ions, and determines magnesium on an aliquot of the eluate.

Urine[3,7,21] Dilute 0.3 ml of sample to 3 ml with water and proceed by *Method 1*. Neither the natural colour nor the phosphates present interfere with the test[21].

Foods and edible oils[1,22]

Ash 0.1 g sample in a platinum dish at below red heat. Dissolve the ash in 1 drop of 5% sulphuric acid and dilute to 3 ml. Proceed by *Method 1*[1]. Alternatively[22] heat the sample with sulphuric acid and perchloric acid ($HClO_4$) until a clear digest is obtained. Pass the digest through an ion exchange column to remove interfering anions, and determine magnesium on the eluate by *Method 1*.

Fuels and lubricants[23]

Coal ash Grind the sample until it passes through a 250 mesh sieve. Mix 0.5 g of the ground sample with 2 g of lithium metaborate ($LiBO_2$) in a platinum crucible and heat in a muffle furnace at 900°C until the melt is clear. If the sample contains more than 15% of iron, add 30–50 mg of ammonium metavanadate (NH_4VO_3) to act as a flux. Cool the melt and transfer it to a beaker containing 8 ml of nitric acid and 150 ml of water. Stir vigorously until dissolution is complete. Add 2.5 g of tartaric acid (($CHOH.COOH)_2$) and make up to 250 ml.

Metals, alloys and plating baths[24-29]

Aluminium[24] Dissolve the sample in 20% caustic soda (NaOH) add 10 ml of 10% sodium carbonate (Na_2CO_3) solution and 1 ml of 3% hydrogen peroxide (H_2O_2) and boil. Cool, filter, wash the precipitate with a solution of 3% sodium carbonate and 5% caustic soda and then dissolve it in 20 ml of 1:3 hydrochloric acid plus 2 ml of 3% hydrogen peroxide. Remove interfering anions as pyridine-thiocyanate complexes by adding 20 ml of 20% potassium thiocyanate (KCNS) solution and excess of 20% pyridine. Stand for 30 minutes, filter, wash the precipitate with pyridine potassium thiocyanate solution. Dissolve the precipitate in hydrochloric acid, make up to a known volume and determine magnesium on an aliquot by *Method 1*.

Cast iron[25] Dissolve the sample in acid, dilute with water, remove iron by complexing with 1-nitroso-2-naphthol and extract the complex with chloroform ($CHCl_3$). Determine magnesium on the aqueous phase by *Method 1*.

General[26] Dissolve the sample in hydrochloric acid, extract the interfering metals with ether ($C_2H_5OC_2H_5$) and then by formation and extraction of their cupferron and diethyldithiocarbamate complexes into chloroform ($CHCl_3$). Add diethylamine (($C_2H_5)_2NH$) and 8 hydroxy quinoline to the aqueous phase and extract into chloroform. Re-extract the magnesium into $2N$ hydrochloric acid and determine by *Method 1*.

Lead and lead alloys[27] Dissolve a 0.1–2.0 g sample in nitric acid. Evaporate the solution with hydrochloric acid, filter off the precipitated lead chloride ($PbCl_2$). Remove heavy metals from the filtrate with diethyldithiocarbamate and determine magnesium on the residual solution by *Method 1*.

Nickel and nickel alloys[28,29] Dissolve a 2–10 g sample in aqua regia (hydrochloric acid:nitric acid 3:1), allow the solution to evaporate at room temperature, moisten the residue with hydrochloric acid, add water and dissolve by heating. Filter off silica, neutralise the solution with sodium carbonate (Na_2CO_3), heat to drive off carbon dioxide, add potassium thiocyanate (KSCN) and pyridine and proceed as for aluminium above[28]. Alternatively[29] dissolve the sample in a mixture of hydrochloric acid and hydrogen peroxide (H_2O_2), remove excess peroxide by adding freshly precipitated nickel hydroxide ($Ni(OH)_2$) and make the solution up to a known volume.

Minerals, ores and rocks[23,24,30-34]

Bauxite[24] Either fuse the sample with twice its weight of sodium carbonate (Na_2CO_3), leach out the melt with hydrochloric acid, separate silica by dehydration and excess aluminium with caustic soda (NaOH) and sodium carbonate, or fuse the sample with caustic soda and sodium carbonate and leach with water. Proceed as for aluminium above[28].

Calcium carbide or lime[30] Ershov claims that the interference arising from large amounts of calcium in Method 1 can be suppressed by the addition of sucrose and starch. Dissolve 0.5–1.0 g of the sample in acid, filter and dilute to 200 ml. Treat 5–10 ml with a calcium salt to bring the calcium concentration to about 100 mg of calcium oxide, add 5 ml of 1% triethaneolamine ($N(C_2H_4OH)_3$) to mask iron, 5 ml of 0.5% starch solution, normal caustic soda until the solution is neutral, 50 ml of 20% sucrose and then dilute to 100 ml with water. Proceed by *Method 1*.

Dolomite[31] Dissolve 0.2 g of sample by first moistening with water and then refluxing with 3 ml of 1:1 hydrochloric acid until the evolution of carbon dioxide ceases, then boil for 3–5 minutes. Cool, neutralize to methyl orange with 2.5% aqueous ammonia (NH_4OH), filter and make up to 100 ml with water.

General[23,32] Either use the procedure described for use with coal ash[23] above or [32] pre-ignite an empty graphite crucible for 30 minutes at 950°C and cool. Take care not to disturb the powdery inside surface. Mix 0.2 g of rock powder with 1.4 g of lithium metaborate ($LiBO_2$) in a porcelain crucible and then transfer the mixture to the pre-ignited graphite crucible. Fuse in a muffle furnace at 900°C for exactly 15 minutes. Pour the melt into 100 ml of 4% nitric acid contained in a plastic beaker with a flat bottom, and stir with a Teflon coated stirring rod. Wash the solution into a 250 ml flask and dilute to volume with 4% nitric acid.

Silicates[33,34] Dissolve the sample in acid, using hydrogen fluoride (HF) if necessary, and precipitate the magnesium as magnesium ammonium phosphate ($MgNH_4PO_4$). Proceed by *Method 2*. Alternatively[34], digest 1 g of sample overnight with hydrogen fluoride add 5 ml of 50% perchloric acid ($HClO_4$) and evaporate to dryness. Repeat the evaporation three times with further 5 ml portions of perchloric acid, adding 2 ml of 1:1 sulphuric acid to the first of these three portions. Heat until fuming ceases. Dissolve the residue in 5–10 ml of perchloric acid and dilute to 200 ml. Dilute a 20 ml aliquot of this solution to 40 ml, adjust the pH to 2 with 8% caustic soda solution, add 2 g of sodium succinate (($CH_2.COONa)_2.6H_2O$) and a Whatman precipitation accelerator and boil for 2–3 minutes. Filter, wash the precipitate with 0.05% sodium succinate solution, dissolve the precipitate in 1 ml of perchloric acid, re-precipitate, filter and wash. Dilute the combined filtrates to 200 ml and proceed by *Method 1*.

Miscellaneous[35-39]

Cement[35] Transfer 1 g of sample into a 200 ml platinum or porcelain evaporating dish, disperse it with 15 ml of water, add 10 ml of hydrochloric acid, heat and stir until decomposition is complete. Evaporate on a steam bath until nearly dry, cool, add 8 ml of hydrochloric acid and then, after 2 minutes, add 50 ml of hot water. Stir, cover, digest on the steam bath for 5 minutes then filter, through a Whatman No. 40 paper, into a 500 ml flask. Transfer the filter paper and precipitate to a platinum crucible, heat gently to first dry and then char the paper without igniting it and then ash at 1100°C for 30 minutes. Add 0.1 g of lithium chloride (LiCl) to the residue, fuse, cool, dissolve the melt in hydrochloric acid, filter and add this filtrate to the original filtrate. Make the combined filtrates up to 500 ml and determine magnesium by *Method 1*.

Dusts and bonded deposits[36] Fuse the sample with solid caustic soda (NaOH), cool and extract the melt with a mixture of water, sulphuric acid and hydrogen peroxide (H_2O_2) at 60°C for 1 hour. Remove silica by dehydration with sulphuric acid, filter, extract titanium and vanadium into a chloroform ($CHCl_3$) solution of cupferron from acid solution of pH less than 1.0. Separate the organic phase and determine magnesium on an aliquot of the aqueous phase.

Pickling brine[37] Boil 25 ml of the sample with 1 ml of 0.5% copper chloride ($CuCl_2$) solution and 1 ml of 20% caustic soda (NaOH). Wash the precipitate with $0.01N$ caustic soda and dissolve it in 5 ml of warm $0.3N$ hydrochloric acid. Determine magnesium on this solution by *Method 1*.

Plant nutrient solutions[38] may be estimated directly after dilution, if necessary, to bring the magnesium concentration within the range 1–10 ppm.

Slag[39] Finely grind 0.2 g of sample, moisten with 20 mg of water, add 10 ml of hydrochloric acid and heat until dissolution is complete. Boil the solution with 1–2 drops of nitric acid and 20 ml of water, precipitate the sesquioxides with 20% aqueous ammonia (NH_4OH) and 10–20 ml of a 10% solution of ammonium persulphate (($NH_4)_2S_2O_8$). Filter and wash the precipitate. Boil the combined filtrate and washings for 10–12 minutes, cool and dilute to 250 ml. Determine magnesium on an aliquot by *Method 1*.

Plants[40-45]

Ash the sample, treat the ash with hydrochloric acid and ash again. Dissolve the residue in hydrochloric acid. Precipitate calcium as the oxalate in the presence of lactic acid, at pH 5, heat on a water bath and filter. Treat the filtrate with 8-hydroxyquinoline at pH 10. Filter. Remove heavy metals from the precipitate by washing with ethanol-chloroform-ammonium hydroxide ($EtOH–CHCl_3–NH_4OH$). Dissolve the residue in hydrochloric acid and determine magnesium by *Method 1*[40]. Alternatively[41,42] wet ash 1 g of sample in nitric-perchloric acids ($HNO_3–HClO_4$), heat to fumes, cool, dilute, boil, filter and adjust the volume of the filtrate to 100 ml. Take an aliquot, adjust the pH to 3–4 with aqueous ammonia (NH_4OH) and apply to a column of Deacidite-E ion exchange resin (Cl^- form). Elute the column with water until 100 ml is collected and determine magnesium on this eluate by *Method 1*. Or[43] dissolve the ash of a 0.1 g sample in 5 ml of $0.5N$ hydrochloric acid and dilute to 10 ml. Add 2 ml of a solution containing 5 g of 1,2-di-(2-aminoethoxy) ethane N,N,N′,N′–tetra-acetic acid dissolved in the minimum amount of 10% caustic soda (NaOH). Adjust the pH to about 7 with dilute hydrochloric acid, add 0.054 g of aluminium chloride ($AlCl_3.6H_2O$) dissolved in a small amount of water, then add 20 ml of triethanolamine ($N(C_2H_4OH)_3$) and dilute the solution to 200 ml. Proceed by *Method 1*.

The interference which may arise due to the presence of manganese in some plants may be prevented by the use of triethanolamine together with excess potassium ferricyanide ($K_3Fe(CN)_6$) and 50% caustic soda solution[45]. The brown colour intitially formed disappears in 10–15 minutes.

Soil[46-48]

Available magnesium[46,47] Extract the soil by any standard procedure. To the extract add 5% ammonium citrate (($NH_4)_3C_6H_5O_7$) solution at pH 7 and pass the solution through a column of Dowex-50 ion exchange resin (K^+ form). Elute magnesium from the column with 0.025M ammonium citrate solution at pH 7.5 and proceed by *Method 1*[46.] Alternatively [47] shake 5 g of air dried soil with 50 ml of 0.025N calcium chloride ($CaCl_2$) solution for 2 hours, filter and determine magnesium on an aliquot of this solution.

Total magnesium[48] Take an aliquot of the hydrochloric acid extract. Neutralise with 4N caustic soda (NaOH) until a slight precipitate appears. Carefully add 1:9 hydrochloric acid drop by drop until this precipitate disappears. Precipitate iron and aluminium with a slight excess of 1:1 caustic soda. If there is a considerable precipitate re-dissolve it and re-precipitate. Combine the filtrates and washings and dilute to a known volume and determine magnesium on an aliquot.

Water[49-59]

No sample preparation is required unless interfering elements are known to be present (see *Method 1* Note 2). Magnesium can be separated from interfering elements by precipitation as the oxalate[50], followed by solution of the precipitate in acid. See also separation by ion exchange below.

Separation from other ions

From alkali metals[51] Extract the magnesium with 8-hydroxyquinoline in chloroform ($CHCl_3$), remove solvent by evaporation, ignite, dissolve the residue in hydrochloric acid and proceed by *Method 1*.

From aluminium[52] Absorb the sample solution on a 14 × 1 cm column of Dowex 50–X8 (H^+ form), and elute magnesium with 200 ml of 0.7N hydrochloric acid.

From barium, calcium and strontium[53-57] Pass the test solution through a column of ion exchange resin and desorb magnesium with the appropriate solvent. Fritz[53,55] recommends absorption onto Amberlyst XN-1002 resin from 90% isopropanol (($CH_3)_2CHOH$) containing 0.5M nitric acid followed by elution of the absorbed magnesium with a 1:9 mixture of 0.5M nitric acid and either 90% isopropanol[53], or 95% ethanol (C_2H_5OH)[55]. Nelson[54] recommends a mixture of 6.4M perchloric ($HClO_4$) and 2.6M hydrochloric acids for elution. Tanaka[56] recommends 0.7M hydrochloric acid, while Winowski[57] recommends 0.018% ammonium oxalate (($COONH_4)_2H_2O$).

Alternatively[57] these ions may be separated by chromatography on paper which has been treated with 0.01M to 0.1M bis-(2-ethyl-hexyl) hydrogen phosphate in cyclohexane with 0.2mM to 10M hydrochloric acid as the developing agent.

General[59,60] Extract Zeo-Karb 225 (2ONC) ion exchange resin with hot 6N hydrochloric acid until it is iron free, wash to remove excess acid and pack into short polythene tubes to form columns 8 cm × 0.7 cm. Wash each column with 25 ml of 15% potassium chloride (KCl) solution to convert the resin to the K^+ form and then with water to remove Cl^-. Plasma or serum should be diluted with 2 or 3 volumes of normal saline and applied to the column and the column then washed with saline; urine, tissue extracts and ash solutions need no dilution and the column can be washed with water. Elute the column with 15% potassium chloride at a flow rate of 6–10 ml per hour. Magnesium is recovered from the 3rd to 8th mls inclusive[59]. Alternatively[60] remove interfering ions by extracting the magnesium with 8–hydroxyquinoline in chloroform ($CHCl_3$) at pH 7.2.

Heavy metals[61] Precipitate calcium and magnesium with ammonium hydrogen phosphate $((NH_4)_2HPO_4)$, in the presence of 2 mg of any tervalent lanthanum compound to act as a carrier and of potassium cyanide (KCN) to mask any remaining traces of heavy metals. Filter, dissolve the precipitate in hydrochloric acid, pass through a column of Dowex 50-X8 resin (H^+ form) and desorb magnesium with $0.7 N$ hydrochloric acid.

The individual tests

Method 1 Titan Yellow is absorbed from solution by magnesium hydroxide [62-65]. The factors affecting this adsorption have been studied by Bradfield[66] who claims that the best protective colloid for use is polyvinyl alcohol at a concentration of 0.01% and that the optimum caustic soda concentration is $0.6 - 0.8 N$.

Method 2 This is an indirect method which uses the precipitation of magnesium as magnesium ammonium phosphate ($MgNH_4PO_4$). The precipitate is dissolved in acid and the concentration of phosphorus in the solution, which is proportional to the magnesium concentration, is determined by one of the standard methods for phosphorus (see "The determination of phosphate" *Methods 1–8*).

A comparison of the relative merits of these two methods has been carried out by Heaton[67] and Butler[68]. Heaton concludes that *Method 1* is quicker, subject to the removal of interfering anions, and that while it is satisfactory for serum and urine, it is less satisfactory for blood-cells, fæces and food. *Method 2* is more reliable and can be applied to a much wider range of materials, but it takes much longer. Butler concludes that *Method 2* is the more accurate. He further claims that the method of sample preparation used had no consistent effect on the results.

References

1. G. P. Sanders, *J. Biol. Chem.*, 1931, **90**, 747
2. D. N. Baron and J. L. Bell, *J. Clin. Pathol.*, 1957, **10**, 280; *Analyt. Abs.*, 1958, **5**, 2311
3. E. Andreasen, *Scand. J. Clin. Lab. Invest.*, 1957, **9**, 138; *Analyt. Abs.*, 1958, **5**, 1589
4. P. Janella, *Biochem. Appl.*, 1959, **6**, 187; *Analyt. Abs.*, 1960, **7**, 2357
5. W. H. P. Lewis, *J. Med. Lab. Technol.*, 1960, **17**, 32; *Analyt. Abs.*, 1960, **7**, 3862
6. J. K. Aikawa and E. L. Rhoades, *Amer. J. Clin. Path.*, 1958, **31**, 314; *Analyt. Abs.*, 1960, **7**, 3863
7. E. Schmid and M. Seibold, *Klin. Wochschr.*, 1960, **38**, 947
8. P. D. Spare, *Amer. J. Clin. Path.*, 1962, **37**, 232; *Analyt. Abs.*, 1962, **9**, 4330
9. C. S. Anast, *Clin. Chem.*, 1963, **9**, 544; *Analyt. Abs.*, 1964, **11**, 2253
10. H. H. Sky-Peck, *Clin. Chem.*, 1964, **10**, 391; *Analyt. Abs.*, 1965, **12**, 4681
11. R. G. Martinek and R. E. Berry, *Clinica chim. Acta*, 1964, **10** (4), 365; *Analyt. Abs.*, 1966, **13**, 789
12. P. J. Burcar, A. J. Boyle and R. E. Mosher, *Clin. Chem.*, 1964, **10** (11), 1028; *Analyt. Abs.*, 1966, **13**, 1383
13. H. Haury, *Arzneimittel-Forsch*, 1965, **15** (5), 579; *Analyt. Abs.*, 1966, **13**, 4996
14. M. Masson, *Path. Biol. Paris*, 1966, **14** (23-24), 1202; *Analyt. Abs.*, 1968, **15**, 1502
15. J. Carles, *Bull. Soc. Chim. Biol.*, 1957, **39**, 445; *Analyt. Abs.*, 1957, **4**, 3393
16. D. Glick, E. F. Freier and M. J. Ochs, *J. Biol. Chem.*, 1957, **226**, 77; *Analyt. Abs.*, 1958, **5**, 180
17. H. G. McCann, *Analyt. Chem.*, 1959, **31**, 2091
18. M. Hofer, *Experientia*, 1963, **19**, (7), 367; *Analyt. Abs.*, 1964, **11**, 3230
19. G. Peres and G. Zwingelstein, *Bull. Soc. Sci. Vet. Lyon*, 1962, **64**, 347; *Analyt. Abs.*, 1964, **11**, 1380
20. G. Conradi, *Chemist Analyst*, 1967, **56** (4), 87; *Analyt. Abs.*, 1969, **16**, 284
21. J. Becka, *Biochem. Z.*, 1931, **233**, 118
22. R. Santini and J. M. de Jesus, *Amer. J. Clin. Path.*, 1959, **31**, 181; *Analyt. Abs.*, 1960, **7**, 639
23. P. L. Boar and L. K. Ingram, *Analyst*, 1970, **95**, 124
24. A. S. Andreev, A. Marshikova and G. V. Telyatnikov, *Trudy. Leningr. Politekh. Inst.*, 1959, (201), 51; *Analyt. Abs.*, 1960, **7**, 3161
25. A. K. Babko and N. V. Lutokhina, *Ukr. Khim. Zhur.*, 1959, **25**, 226; *Analyt. Abs.*, 1960, **7**, 3259

26. I. Nachev and D. Filipov, *Compt. Rend. Acad. bulg. Sci.*, 1969, **22** (7), 767; *Analyt. Abs.* 1970, **19**, 2090
27. S. D. Gurev and N. F. Saraeva, *Zavodskaya. Lab.*, 1959, **25**, 795; *Analyt. Abs.*, 1960 **7**, 1675
28. A. S. Andreev, A. N. Novikov and F. Cherny, *Trudy. Leningr. Politekh. Inst.*, 1959, (201), 46; *Analyt. Abs.*, 1960, **7**, 3269
29. F. H. Harrison, *Metallurgia Manchr.*, 1964, **70** (11), 251; *Analyt. Abs.*, 1966, **13**, 1288
30. V. A. Ershov and E. A. Kachanova, *Zavodskaya Lab.*, 1967, **33** (1), 28; *Analyt. Abs.*, 1968, **15**, 1878
31. A. I. Cherkesov and Yu V. Pushinov, *Zavodskaya Lab.*, 1964, **30** (9), 1053; *Analyt. Abs.*, 1966, **13**, 49
32. J. C. Van Loon and C. M. Parissis, *Analyst*, 1969, **94**, 1057
33. W. Bodenheimer and M. Gaon, *Bull. Res. Council Israel, A*, 1958, **7**, 117; *Analyt. Abs.*, 1959, **6**, 2497
34. W. H. Evans, *Analyst*, 1968, **93**, 306
35. A. Nestoridis, *Analyst*, 1970, **95**, 51
36. J. Grant, *J. appl. Chem.* 1964, **14** (12), 525; *Analyt. Abs.*, 1966, **13**, 1782
37. M. Suzuka, H. Kondo and S. Hirano, *Japan Analyst*, 1958, **7**, 47; *Analyt. Abs.*, 1958, **5**, 3627
38. G. S. Fawcett and R. H. Stoughton, "*The Chemical Testing of Plant Nutrient Solutions*" The Tintometer Ltd., Salisbury, England, 1944
39. Z. I. Kuzmina, *Zavodskaya Lab.*, 1964, **30** (10), 1215; *Analyt. Abs.*, 1966, **13**, 653
40. J. Dumas, *Fruits d'Outre Mer*, 1958, **13**, 161; *Analyt. Abs.*, 1959, **6**, 653
41. C. Bould, E. G. Bradfield and G. M. Clarke, *J. Sci. Food Agric.*, 1960, **11**, 229; *Analyt. Abs.*, 1961, **8**, 343
42. E. G. Bradfield, *Analyst*, 1960, **85**, 666
43. E. G. Bradfield, *Analyst*, 1961, **86**, 269
44. W. Plant, J. O. Jones and D. J. D. Nicholas, "*The Diagnosis of Mineral Deficiencies in Crops by Means of Chemical Tissue Tests*", Annual Report of the Long Ashton Research Station, Bristol University, 1944
45. E. M. Chenery, *Analyst*, 1964, **89**, 365
46. S. K. Tobia and N. E. Milad, *J. Agric. Food Chem.*, 1958, **6**, 358; *Analyt. Abs.*, 1959, **6**, 763
47. R. J. Hall, G. A. Gray and L. R. Flynn, *Analyst*, 1966, **91**, 102
48. J. L. Steenkamp, *J. S. African Chem. Inst.*, 1930, **13**, 64
49. Brit. Standards Inst., "*Methods of Testing Water Used in Industry*", B.S. 2690, London, 1964
50. A. D. Hirschfelder, E. R. Series and V. G. Haury, *J. Biol. Chem.*, 1934, **104**, 635
51. W. I. Stephen and A. M. Weston, *Mikrochim. Ichnoanal. Acta.* 1964, (2–4), 179; *Analyt. Abs.*, 1965, **12**, 3754
52. M. Tanaka, *J. Chem. Soc., Japan, Pure Chem. Sect.*, 1963, **84**, 582; *Analyt. Abs.*, 1965, **12**, 5045
53. J. S. Fritz and H. Waki, *Analyt. Chem.*, 1963, **35**, 1079
54. F. Nelson, J. H. Holloway and K. A. Kraus, *J. Chromatog.*, 1963, **11**, 258; *Analyt. Abs.*, 1964, **11**, 3008
55. J. S. Fritz, H. Waki and B. B. Garralda, *Analyt. Chem.*, 1964, **36**, 900
56. M. Tanaka, *J. Chem. Soc. Japan, Pure Chem. Sect.*, 1964, **85**, 117; *Analyt. Abs.*, 1965, **12**, 5670
57. Z. Winowski, *Chemia analit.*, 1967, **12**, (6), 1271; *Analyt. Abs.*, 1969, **16**, 579
58. E. Cerrai and G. Ghersini, *J. Chromatog.*, 1964, **15** (2), 236; *Analyt. Abs.*, 1965, **12**, 5669
59. D. E. Stevenson and A. A. Wilson, *Clin. Chim. Acta*, 1961, **6**, 298; *Analyt. Abs.*, 1961, **8**, 5118
60. E. Asmus and W. Klank, *Z. analyt. Chem.*, 1964, **206** (2), 88; *Analyt. Abs.*, 1966, **13**, 565
61. Z. Marezenko, *Chim. anal.*, 1964, **46** (6), 286; *Analyt. Abs.*, 1965, **12**, 5653
62. I. M. Kolthoff, *Biochem. Z.*, 1922, **128**, 344
63. I. M. Kolthoff, *Chen. Weekblad*, 1927, **24**, 254

64. H. D. Barnes, *J. S. African Chem. Inst.*, 1918, **11,** 67
65. J. Becka, *Biochem. Z.,* 1931, **233,** 118
66. E. G. Bradfield, *Anal. Chim. Acta.* 1962, **27,** 262; *Analyt. Abs.,* 1963, **10,** 1716
67. F. W. Heaton, *J. Clin. Path.,* 1960, **13,** 358; *Analyt. Abs.,* 1961, **8,** 3390
68. E. J. Butler, D. H. S. Forbes, C. S. Munro and J. C. Russell, *Analyt. chim. Acta.,* 1964, **30** (6), 524; *Analyt. Abs.,* 1965, **12,** 5279

Magnesium Method 1
using Titan Yellow

Principle of the method

Titan yellow is absorbed from solution by magnesium hydroxide. It gives an orange coloured solution with 1 ppm of magnesium, a red colour with 5 ppm and a red flocculant precipitate with over 10 ppm. Solutions to be tested must be diluted with distilled water until the magnesium concentration lies within this range. The colours are measured by comparison with Lovibond permanent glass colour standards. The reagent is also standardised by comparison with a separate Lovibond Standard.

Reagents required

1. *Titan Yellow reagent* This is prepared **freshly** from Titan Yellow solution (0.1 per cent in 20 per cent alcohol) by diluting with distilled water and standardising as described below. The stock solution should be kept in the dark. The diluted reagent deteriorates rapidly, and should always be freshly prepared for use.

2. Hydroxylamine hydrochloride ($NH_2OH \cdot HCl$) (analytical reagent grade)   0.08 g
   Sucrose ($C_{12}H_{22}O_{11}$) (analytical reagent grade) .. ..   5.00 g
   Distilled water to make   .. ..   100 ml

3. *8N Sodium hydroxide solution* (32 g NaOH in 100 ml). This should be kept in small polythene bottles, well closed to prevent absorption of carbon dioxide from the atmosphere. It deteriorates on keeping, and should not be used when more than one month old, unless checked by titrating against a standard acid.

All chemicals should be of analytical reagent quality.

Preparation of Titan Yellow Reagent:

Slide the screen marked "Titan Yellow" into the recess in the centre boss of the Comparator. (An arrow will be seen on the screen moulding to indicate the direction in which the screen should be inserted.) Fit the disc in the Comparator and rotate it until "Titan Yellow" appears in the indicator window of the Comparator. The neutral grey standard in the disc now appears in front of the left hand cell compartment. Place a test tube of distilled water in the left-hand compartment. In the right-hand compartment place a test tube containing 0.1 ml (accurately measured) of Titan Yellow solution. Add distilled water from a graduated burette, until, after thorough mixing, the colour of the solution, viewed through the violet glass of the screen, matches the colour of the grey glass. Note carefully the volume of water used, to which mentally add 0.1 ml (i.e. the volume of Titan Yellow solution used). This gives the final total volume required of the Titan Yellow reagent. Now prepare the reagent by diluting 0.2 ml of the strong solution with sufficient distilled water to produce a volume equal to the final volume noted above. The reagent will thus be double strength to the solution prepared against the screen.

The Standard Lovibond Comparator Disc 3/28 (formerly APMG)

The disc contains colour standards for 0, 1, 2, 3, 4, 5, 7 and 10 parts per million of magnesium and one grey standard labelled "Titan Yellow" for use with the screen marked "Titan Yellow" in standardising the Titan Yellow solution. There is also a second dulling screen, to be used when actually matching the unknown solution.

Technique

In a clean test tube or 13.5 mm cell place:

suitably diluted sample solution	3.0 ml
hydroxylamine hydrochloride solution	0.5 ml and mix well
Add Titan Yellow reagent, diluted as explained above ..	0.5 ml
and then 8N sodium hydroxide solution	0.5 ml

Mix, and allow to stand for exactly five minutes from the last addition, then match in the comparator.

While waiting, place the screen marked "Magnesium" in the centre boss instead of the Titan Yellow screen. In the left-hand compartment place a test tube of distilled water, and in the right-hand one the test solution plus reagents.

Compare the colour of the test solution with the colours on the disc, rotating the latter until a colour match is obtained. If the colour is near to that marked 10 on the disc. the test should be repeated, using a more dilute nutrient solution, as there is little change in colour for concentrations above 10 ppm of magnesium. The markings on the disc indicate parts of magnesium per million in the diluted solution; therefore the result indicated must be multiplied by the dilution figure of the original solution.

Notes

1. Avoid using a mouth pipette for the sodium hydroxide solution. One with a rubber bulb is preferable. Where a reagent bottle with a pipette incorporated in the stopper is used for the sodium hydroxide, the solution as it gets low in the bottle leaves a white deposit of carbonate in the pipette where it is not immersed. This will detract for the accuracy of the pipette, and should be removed by washing with a 1 in 10 dilution of concentrated hydrochloric acid, followed by a thorough washing with distilled water.

2. The colours on the disc have been calibrated on the assumption that calcium is present in concentrations of not more than 500 ppm. Iron, manganese, aluminium, silica and proteins interfere if present in high concentrations. Phosphate in excess of 100 ppm reduces the colour intensity somewhat.

COLORIMETRIC CHEMICAL ANALYTICAL METHODS

Magnesium Method 2
using di-ammonium hydrogen phosphate

Principle of the method
The magnesium is separated as magnesium ammonium phosphate. This is filtered off, dissolved in acid, and the concentration of phosphate in the resulting solution, which is proportional to the concentration of magnesium, is determined by any of the standard phosphate methods.

Reagents required
1. *Diammonium hydrogen phosphate* Dissolve 17.4 g of dipotassium hydrogen phosphate (K_2HPO_4) and 100 g of ammonium chloride (NH_4Cl) in 900 ml of distilled water. Add 50 ml of ammonium hydroxide (NH_4OH) and dilute to 1 litre. 1 ml of this solution will precipitate 2.4 mg of magnesium
2. *Ammonium hydroxide* 10% solution of NH_4OH
3. *Ammoniacal alcohol* To 90 ml of absolute alcohol (C_2H_5OH) add 10 ml of 10% NH_4OH
4. *Hydrochloric acid* 0.1 N HCl

These reagents should be prepared from analytical reagent quality chemicals. In addition, the reagents specified for the chosen method of phosphate determination will also be required. These reagents are listed in the appropriate phosphate method.

The Standard Lovibond Phosphate Discs
The following discs are available for the determination of phosphate:—

Comparator Disc	Range		Nessleriser Disc	Range		Phosphate Method
3/7	2– 22 ppm	P_2O_5	NMB	0.2–2.0 ppm	P_2O_5	1
3/4	1– 20 ,,	,,	NMC	0.2–2.4 ,,	,,	2
5/14	20–160 ,,	P				3
3/12	0– 80 ,,	PO_4				4
3/38	10–100 ,,	,,				5
3/51	2– 80 ,,	,,				6
3/60	0–100 ,,	,,				7
3/70	0–100 ,,	,,	NOP	0–100 ppm	PO_4	8
			NOX	1–10 ppm	PO_4	8

Technique
Take sufficient of the sample solution to contain approximately 0.05 mg of magnesium (0.005 mg if Nessleriser is used), place in a 25 ml centrifuge tube and add, drop by drop, 1 ml of reagent 1 followed by 2 ml of reagent 2. Thoroughly scratch the insides of the tube with a glass rod to initiate precipitation and allow to stand overnight. Centrifuge, wash the precipitate twice with reagent 2 and the once with reagent 3. Dry the precipitate at 70°C, dissolve in 10 ml of reagent 4, make up to 25 ml with distilled water and determine phosphate by one of the standard methods. Convert the phosphate reading from the disc used to the corresponding magnesium concentration by using the appropriate correlation factor.

Correlation factors
1 ppm P = 0.784 ppm Mg
1 ppm PO_4 = 0.256 ppm Mg
1 ppm P_2O_5 = 0.342 ppm Mg

The Determination of Manganese

Introduction

Manganese (Mn) is an important alloying element in metallurgy. It is also an essential element in plant and animal growth. The determination of manganese is of importance in a wide variety of industries as the applications below illustrate.

Methods of sample preparation

Beverages[1]

Whisky Evaporate 25 ml to dryness, burn off organic matter and take the residue to fuming with sulphuric and phosphoric acids. Dilute and proceed by *Method 3*.

Biological materials[2-10]

Blood, serum or cerebro-spinal fluid[2-4] Serum may be treated directly by *Method 1*[2], but blood or cerebro-spinal fluid must be mineralised. Take a sample containing 1–5 μg of manganese (3–5 g of blood, or cerebro-spinal fluid) and evaporate to dryness in a porcelain or silica crucible. Heat to red heat, cool, add 3 ml of nitric acid and 5 ml of water. Evaporate to dryness and repeat the addition and evaporation three times. To the final residue add 3 ml of nitric acid 10 ml of water and proceed by *Method 2*[3,4].

Bone[5] Ash the sample and dissolve the residue with 2 g of citric acid ($C(OH)(COOH)(CH_2COOH)_2.H_2O$) for each gram of ash and the minimum amount of hydrochloric acid necessary to achieve solution. Adjust the pH to 5.2 with dilute aqueous ammonia (NH_4OH) add 0.4 g of sodium diethyldithiocarbamate (($C_2H_5)_2N.CS.SNa.3H_2O$), adjust the pH to 5.3 with normal aqueous ammonia, dilute to 200 ml with water and extract with 25 ml of chloroform ($CHCl_3$). Readjust the pH of the aqueous phase to 5.3 with $3N$ hydrochloric acid, add a further 0.4 g of sodium diethyldithiocarbamate and again extract with 25 ml of chloroform. Combine the extracts, evaporate to dryness, decompose the residue with nitric, sulphuric and perchloric ($HClO_4$) acids and determine manganese on this solution.

General[6-8] Ash the sample at not more than 600°C, dissolve the ash in acid and precipitate manganese either as the diethyldithiocarbamate complex [6,8] or as its hydroxide[7]. Either extract the diethyldithiocarbamate complex, on the original acid solution, with benzene and determine manganese by *Method 1* or *4* on the aqueous layer[6], or extract the precipitated complex into chloroform ($CHCl_3$) at pH 8–9 in the presence of potassium cyanide (KCN) and sodium citrate ($Na_3C_6H_5O_7.2H_2O$), then re-extract the manganese into $2.5N$ hydrochloric acid, make up to volume and proceed by *Method 3*[8]. Dissolve the precipitated hydroxides, in the alternative separation procedure[7], in nitric acid and proceed by *Method 2* or *3*.

Hair[9] Wet ash the sample with a 10:1:1 mixture of nitric, sulphuric and perchloric ($HClO_4$) acids and determine manganese on this solution by *Method 1* or *4*.

Tissue[10] Digest the sample with either perchloric ($HClO_4$) or trichloracetic (Cl_3CCOOH) acid and determine manganese by *Method 1* or *4*.

Urine[3,4] Mineralise 10 ml of sample by the method described for blood.

It has been reported[11] that manganese is partly volatilized during dry ashing even at temperatures as low as 110°C and it is recommended that wet oxidation, preferably using hydrogen peroxide (H_2O_2)[12] be used rather than dry ashing.

Dairy products[13,14]

Cheese[13] Ash a 10 g sample at 600°C, with the addition of a few drops of hydrogen peroxide (H_2O_2) as necessary, until no black particles remain. Dissolve the ash in 8 ml of 10% hydrochloric acid, filter and dilute filtrate to a known volume with water.

Milk[13,14] Evaporate 100 ml to dryness, heat residue at 220°C for 6 hours and then ash and proceed as for cheese[13]. Alternatively[14] take 100 ml of sample, remove the calcium as phosphate in the presence of trisodium-N-hydroxyethyl-ethylenediamine-triacetate, and determine manganese by *Method 3*.

Foods and edible oils[15-19]

Fish[15] Clean the sample with distilled water and dry at 80°C for 48 hours. Grind, place 0.2 g in a size 0 crucible, char at 200°C, raise temperature to 550°C and combust for 16 hours. Add 0.5 ml of 1:1 nitric acid, warm until solution is complete and then dilute to a known volume.

General[16,17] Ash sample with 5% ethanolic magnesium acetate (($CH_3.COO)_2$ $Mg.4H_2O$ dissolved in C_2H_5OH) and then heat to 910°C until the ash is white. Cool, add 5 ml of a 1:1:1 mixture of water, 85% phosphoric acid (H_3PO_4) and 65% nitric acid, heat for 2–3 minutes, add a further 3–4 ml of water, filter and wash residue 3–4 times with hot water. Combine washings and filtrate and make up to a known volume[16]. Alternatively[17] dissolve the ash in 50 ml of 30% nitric acid, add 2 ml of syrupy phosphoric acid (H_3PO_4), heat to boiling point and proceed by *Method 2*.

Vegetable oils[18] Carbonize a 60 g sample under an infra red lamp and then heat in a muffle furnace programmed to increase its temperature from 200°C to 450°C in 12 hours. Heat the residue with 5 ml of $0.5N$ hydrochloric acid on a boiling water bath for about 1 hour, filter and dilute the filtrate to be $0.1N$ in hydrochloric acid. Make up to a known volume with $0.1N$ hydrochloric acid.

The formaldoxime (*Methods 1* and *4*) and periodate (*Method 3*) methods have been compared for food analysis[19]. It is claimed that the formaldoxime methods are marginally faster and more accurate than periodate, but that they are limited in their application by their lack of sensitivity and by the necessity to remove iron from the sample solution. Both methods are free of bias, but from a practical viewpoint the periodate method is preferred.

Fuels and lubricants[20-22]

Petroleum products Take 50 ml of sample, extract with 3.8 ml of bromine (Br) and then with sulphuric acid:nitric acid:water (1:1:2) and determine manganese on the aqueous extract[20]. Alternatively[21] decompose manganese compounds with a carbon tetrachloride (CCl_4) solution of bromine and extract with phosphoric acid (H_3PO_4).

Kyriakopoulos[22] recommends the extraction of manganese and lead compounds with potassium chlorate ($KClO_3$) and nitric acid, the removal of the lead by heating to fumes with sulphuric acid, followed by filtration and the dilution of the filtrate to a known volume.

Metals, alloys and plating baths[23-36]

Aluminium and aluminium alloys[23,24] Dissolve the sample in caustic soda (NaOH) solution, followed by nitric and sulphuric acids if required. Determine manganese on this solution by *Method 3*[23]. Alternatively[24] dissolve a 5 g sample in 80 ml of 20% caustic soda solution, dilute to 300 ml, add 3% aqueous hydrogen peroxide (H_2O_2) and boil for 5 minutes. Add 15 ml of a ferric nitrate ($Fe(NO_3)_3$) solution containing 1 mg of iron per ml and boil to decompose any remaining peroxide. Filter, wash the precipitate with 1% caustic soda solution containing 0.03% of hydrogen peroxide, dissolve the precipitate in 15 ml of a 6:10:7:37 mixture of sulphuric, nitric and phosphoric acids and water, plus 1 ml of 3% hydrogen peroxide, concentrate the solution by evaporation until the volume is less than 40 ml and proceed by *Method 2*.

Chromium[25,26] Dissolve a 4 g sample in 20 ml of perchloric acid ($HClO_4$) heat the solution to 110°C and then add 1:1 hydrochloric acid dropwise and distil off the chromoxy chloride (CrO_2Cl_2). When the solution becomes green stop adding hydrochloric acid and heat to oxidise chromium. When the volume has been reduced to 5 ml, cool, add 10 ml of perchloric acid and distil until colourless. Evaporate the residue nearly to dryness and then dissolve it in water and make up to 25 ml. Proceed either by *Method 1*[25] or by *Method 3*[26].

Kovar alloys and ores[27] Dissolve 0.2 g of alloy, or 0.5 g of ore, in 20–30 ml of a 1:3 mixture of nitric and hydrochloric acids, evaporate to dryness twice adding 10 ml of hydrochloric acid after the first evaporation, dissolve the residue in 10 ml of hydrochloric acid and pass the solution through a column of 45 ml of macroporous AV-17-8P anionite resin (Cl^- form) at 1 ml a minute. Wash the column with 90 ml of $8N$ hydrochloric acid to remove nickel then desorb manganese with 30 ml of $6N$ hydrochloric acid.

Magnesium alloys[28] Dissolve the sample in acid and adjust the pH of the solution to 3–4 with 10% caustic soda (NaOH). Add 5 ml of acetate buffer (pH 4–5.5) and 4 ml of 2% sodium diethyldithiocarbamate solution and extract the manganese with 10 ml portions of carbon tetrachloride (CCl_4). Heat the combined extracts with 10 ml of 1:1 sulphuric acid to remove carbon tetrachloride and then evaporate to fuming with 3–5 ml of nitric acid to remove organic matter. Dissolve the residue in water and determine manganese on this solution by *Method 3*.

Magnetic alloys[29] Dissolve the sample in hydrochloric acid and nitric acid, evaporate to dryness and dissolve the residue in $9N$ hydrochloric acid. Pass this solution through a column of Dowex-1 ion exchange resin, elute the manganese with $9N$ hydrochloric acid and determine the manganese on the eluate by *Method 1, 3* or *4*.

Nickel[30,31] Dissolve the sample in acid and precipitate the hydroxides of iron and manganese on a column of Lewatite M II resin which has previously been washed with aqueous ammonia (NH_4OH). Elute the manganese with sulphuric acid and determine by *Method 2*[30]. Alternatively[31] determine manganese directly on the original acid solution by *Method 3*.

Steel[32-36] Dissolve the sample in 50 ml of nitric acid, boil to remove oxides of nitrogen and proceed by *Method 2*[32,33]. Alternatively[34-36] dissolve 2.5–100 mg of sample in a 5:1 mixture of perchloric ($HClO_4$) and phosphoric (H_3PO_4) acids, add nitric and hydrofluoric (HF) acids, heat to white fumes of perchloric acid, dilute to a known volume with water and proceed by *Method 3*.

Minerals, ores and rocks[27,37-40]

Cobalt ores[27] Proceed as described for Kovar alloys above.

Minerals and ores[37,38] Dissolve the sample in hydrochloric acid and nitric acid. Evaporate repeatedly with hydrochloric acid and filter off any undissolved residue. Treat this residue with sulphuric acid and hydrogen fluoride (HF), evaporate to fumes and dissolve the residue in hydrochloric acid. Evaporate an aliquot of the combined solutions to dryness on a sand bath. To the residue add 3 g of citric acid ($C_6H_8O_7.H_2O$) and dilute to 100 ml. Adjust the pH to 3.3–4 and dilute to 150 ml. Pour this solution through a column of Amberlite IR-120 resin at a rate of 2–3 ml per minute and wash with 3×13 ml and 1×50 ml of water. Elute manganese with 15 ml of acetate buffer at pH 4.6 and determine manganese on the eluate[37]. Alternatively[38] the ore can be brought into solution by treatment with hydrochloric acid followed by fusion of the insoluble residue with potassium hydrogen sulphate ($KHSO_4$) and any final residue with fusion mixture ($Na_2CO_3+K_2CO_3$).

Rock[39,40] Fuse 0.5 g of sample with sodium carbonate (Na_2CO_3), dissolve the melt in sulphuric acid and precipitate silica by evaporation to white fumes, filter, treat the precipitate with sulphuric and hydrofluoric (HF) acids to remove silica, fuse any residue with sodium carbonate and dissolve the melt in sulphuric acid. Add this solution to the original filtrate, neutralise with aqueous ammonia (NH_4OH) and precipitate the sesqui oxides with hexamine (($CH_2)_6N_4$). Heat the mixture to boiling, filter, wash the precipitate with hot 1% hexamine solution three to four times and dilute the cooled filtrate to 250 ml. Proceed on an aliquot by *Method 1* or *4*[39].

Alternatively[40] pre-ignite an empty graphite crucible for 30 minutes at 950°C and cool, taking care not to disturb the powdery inside surface. Mix 0.2 g of powdered rock with 1.4 g of lithium metaborate ($LiBO_2$) in a porcelain crucible and then transfer the mixture to the preignited graphite crucible. Fuse the mixture at 900°C in a muffle furnace for exactly 15 minutes, pour the melt into 100 ml of a 1:24 nitric acid water mixture contained in a plastic beaker with a flat bottom. Stir with a Teflon coated stirring rod until dissolved and then wash the solution into a 250 ml flask and dilute to volume with 1:24 nitric acid.

Miscellaneous[41-48]

Animal feeding stuffs[41] Ignite 5 g of sample for 2 hours at 550°C. Cool, add 10 ml of $3N$ hydrochloric acid, evaporate to dryness, on a boiling water bath, add a further 10 ml of the acid and re-evaporate. Dissolve the residue in a further 10 ml of the acid, dilute with 10 ml of hot water, filter and dilute the filtrate with water to 100 ml. Evaporate an aliquot of this solution to dryness under an infra-red lamp, dissolve the residue in 5 ml of $10N$ hydrochloric acid and pass the solution through Wofatit SBW resin. Elute interfering elements with 3 ml of

10.5N hydrochloric acid and then elute manganese with 10 ml of 8N hydrochloric acid and proceed by *Method 1 or 4*.

Cement[42] Transfer a 1 g sample into a 200 ml platinum or porcelain evaporating dish and disperse it with 15 ml of water, add 10 ml of hydrochloric acid and heat and stir until decomposition is complete. Evaporate nearly to dryness on a steam bath, cool, add 8 ml of hydrochloric acid and then, two minutes later, add 50 ml of hot water. Stir, cover, digest for 5 minutes and then filter through a Whatman No. 40 filter paper, into a 500 ml flask. Wash the precipitate ten times with hot dilute (1+99) hydrochloric acid and then five times with hot water. Add the washings to the filtrate in the flask. Transfer the precipitate with the filter paper to a platinum crucible, ash, moisten the ash with a few drops of water, add 6 drops of perchloric acid ($HClO_4$) and about 15 ml of hydrofluoric acid (HF). Evaporate on a low heat hot-plate and then ignite at 1100°C for 5 minutes. Add 0.1 g of lithium chloride (LiCl) to the ash and fuse. Cool, dissolve the melt in hydrochloric acid and add this solution to the filtrate and washings in the flask. Dilute to the mark with water and determine manganese on an aliquot.

Fertilizers and liming materials[43] Grind the sample in an agate mortar until it will pass a No. 100 sieve. Place 0.5 g (limestone or 0.2 g of silicate) in a 75 ml nickel crucible, remove any organic matter which may be present by heating at 900°C for 15 minutes without the lid. Cool, mix the sample with 0.3 g of potassium nitrate (KNO_3), add 1.5 g of caustic soda (NaOH) pellets, cover the crucible with a nickel lid and heat at a dull red heat over a gas flame (not in a furnace) for 5 minutes. Remove from the flame and swirl the melt around the sides of the crucible. Cool, add about 50 ml of water and warm to disintegrate the fused cake. Transfer the contents of the crucible to a 150 ml beaker containing 15 ml of 5N perchloric acid. Scrub the crucible and lid with a rubber tipped glass rod and wash any residue into the beaker. Transfer to a 100 ml volumetric flask and dilute to the mark.

Glass[44] Dissolve the sample in aqueous hydrofluoric acid (HF), remove excess hydrofluoric acid by boiling and make the solution up to known volume. Determine manganese on an aliquot by *Method 2 or 3*.

Textiles[45-47] Remove the organic matter by dry ashing at low temperature or by wet oxidation. Remove iron, titanium, zirconium, molybdate and vanadate by forming the cupferron complex and extracting this with dichloromethane (CH_2Cl_2). Mask cobalt, copper and nickel by cyanide. Remove borate interference by using a large excess of the formaldoxime reagent. Remove interference by chromium by previous precipitation with ferric hydroxide ($Fe(OH)_3$) as a collector, cobalt by leaving the alkaline solution to stand for 30 minutes and copper by extracting iron, adding sodium diethyldithiocarbamate and extracting the brown copper complex with dichloromethane. Proceed by *Method 1 or 4*[45,46].

Alternatively[47] dissolve the sample in perchloric acid ($HClO_4$) and proceed by *Method 2 or 3*.

Wood pulp[48] Ash about 20 g of pulp (10 g if the manganese concentration is over 0.5%). To the residue add 3 drops of 5% sodium sulphite (Na_2SO_3) solution. Dissolve the mixture in up to 5 ml of 1.5M nitric acid. Evaporate to dryness on a steam bath and add a few drops of 1.5M nitric acid and transfer to a 25 ml flask with water. Place the flask on a steam bath and add 1 ml of sodium periodate-phosphoric acid solution (50 g $NaIO_4$ and 200 ml of 85% H_3PO_4 per litre) and heat for 5 minutes. Dilute with water and adjust the final volume to 25 ml. Proceed by *Method 3*.

Plants[49-54]

General[49-51] Digest 1 g of dried plant material with nitric acid until free from solids and fume the digest with perchloric acid ($HClO_4$) until most of the acid is removed. Dilute, filter if necessary, and take an aliquot of the filtrate containing 10–50 µg of manganese and mix with a solution of N-hydroxy-ethylethylenediamine triacetic acid (HEEDTA), neutralise the free perchloric acid with caustic soda (NaOH) and proceed by *Method 1*[49]. Alternatively[50] ash 0.1 g of plant material at 550°C, dissolve the residue in dilute hydrochloric acid and evaporate to dryness. Ignite gently, dissolve the ash in 7N hydrochloric acid, extract with di-isopropyl ether (iPr_2O), reject the ether extract and evaporate the aqueous layer to dryness. Dissolve the residue in dilute hydrochloric acid, boil and filter. Determine manganese on the filtrate by *Method 3*.

Dobritskaya[51] recommends the following method:—Take 3 g of air-dried sample and ignite for 2–3 hours at 500°C. Add 5 ml of 1:4 nitric acid (for a carbonate ash) or 5 ml of 10% sodium nitrate ($NaNO_3$) solution (for a silicate ash), evaporate to dryness and then re-ignite for 30 minutes. Cool, moisten the residue with water, dissolve in 20 ml of 20% sulphuric acid, boil for 10 minutes, filter, and collect the filtrate. Wash the precipitate with hot dilute sulphuric acid and add the washings to the filtrate. Dry the filter paper with the precipitate, ignite for 15–20 minutes, cool, add 5 ml of hydrofluoric acid (HF), evaporate to dryness, dissolve the residue by warming with 5 ml of 20% sulphuric acid and add this solution to the collected filtrate and washings. Evaporate until the volume is 25–30 ml, add 1 ml of syrupy phosphoric acid (H_3PO_4) and 1 ml of 1% silver nitrate ($AgNO_3$) solution, heat on a water bath for 10 minutes, filter off the precipitated silver chloride (AgCl) and proceed by *Method 2*.

Leaves[52] Dry ash 1 g of plant material at 450°C, digest the ash with 2 ml of a 3:1 mixture of nitric and perchloric ($HClO_4$) acids, and then boil for 10 minutes with 10 ml of 0.5 N hydrochloric acid and 1 ml of 0.5% sodium nitrite ($NaNO_2$) solution. Dilute to 20 ml, set aside overnight for the silica to separate, filter and determine manganese on an aliquot.

Rape seed[53] Ignite the sample at 500°C. Extract the ash twice with 1:3 hydrochloric acid and dilute the solution to a known volume. Extract an aliquot with 10 ml of lead diethyldithiocarbamate solution in chloroform ($CHCl_3$) to remove copper. To the aqueous phase add 2 ml of 1% cupferron solution and 10 ml of chloroform to extract iron. To the remaining aqueous phase add 2 ml of 5% tiron solution, 2 drops of bromocresol green solution and then normal aqueous ammonia (NH_4OH) drop by drop until a blue-green colour appears. Precipitate nickel with 1 ml of 0.8% nioxime solution and extract the complex into benzene, manganese is retained in the aqueous phase, in which it may be determined by *Method 1* or *4*.

Seaweed[54] Evaporate 0.25 g of ash with hydrochloric acid and nitric acid and extract with 20 ml of 3.6 N sulphuric acid. Filter and determine manganese on the filtrate by *Method 2*.

Sewage and industrial wastes[55-58]

Place 500 ml of sample in a beaker, add 5 ml of nitric acid and evaporate to small volume. Transfer to a porcelain or silica basin and reduce the volume to about 15 ml on a water bath. Add 2 ml of 60% perchloric acid ($HClO_4$) and evaporate to dryness. Ignite at 300°C for 40 minutes. Dissolve the residue by adding 2 ml of 5 N hydrochloric acid and a little distilled water and heating gently. Repeat oxidation, evaporation, ignition and solution processes until a colourless solution is obtained. Make up the solution to known volume and take an aliquot containing not more than 75 µg of manganese. To this aliquot add 5 ml of 1:1 sulphuric acid and remove chloride by evaporation to fuming, adjust the volume of the solution to 25 ml and proceed by *Method 3*[55,56]. Alternatively[57] take two 20 ml samples and place them in Erlenmeyer (conical) flasks. To each add 40 ml of nitric acid and 15 ml of sulphuric acid, evaporate to white fumes and keep at that temperature for about 5 minutes. Repeat until the residue is white. Dilute to about 50 ml with water and filter or, preferably, centrifuge. Develop the colour in one solution by *Method 3*. Reduce the manganese in the other solution by adding 3 drops of hydrochloric acid and use this as a blank.

After comparing several alternative methods Sikorowska[58] recommends those based on oxidation of manganese to permanganate. Evaporate the sample in a silica basin then ignite the residue and dissolve the ash in 1 ml of 1:1 nitric acid. Dilute the solution with 10 ml of water, filter and wash the residue with water. Add the washings to the filtrate and then dilute to a known volume. Proceed by *Method 2* or *3*.

Soil[37,38,51,59-63]

Pour 500 ml of N hydrochloric acid on 100 g of dry soil and stand for 24 hours. Filter through ash-free filter paper, evaporate the filtrate to dryness in a porcelain dish, dissolve the residue in 10–15 ml of 1:1 nitric acid and transfer to a porcelain crucible. Evaporate to dryness and heat at dull red for 30 minutes. Cool, add 10 ml of nitric acid and repeat the evaporation and ignition. Treat the residue with 10 ml of hydrochloric acid, transfer the solution to a column of Anionite AB-17 resin (Cl-form) and elute the manganese with 50 ml of 8 N hydrochloric acid. Determine manganese on this elute by *Method 1* or *Method 3*[60]. Alternatively[37] "total manganese" may be determined by extracting the soil with nitric acid, "assimilable manganese" by extraction with aqueous ammonium acetate ($CH_3COO.NH_4$) solution, and "reducible manganese" by extraction with ammonium acetate and quinol.

Other workers[38,51,59,61-63] recommend decomposing the sample either by fusion with sodium carbonate (Na_2CO_3) or by heating with hydrofluoric (HF) and sulphuric acids.

Water[50,58,64-69]

To up to 10 litres of sample add 8% caustic soda (NaOH) solution up to pH 12. This co-precipitates manganese with magnesium hydroxide ($Mg(OH)_2$). If the water sample is low in magnesium add 0.5 ml of 30% magnesium sulphate ($MgSO_4$) solution. Decant the supernatant liquid, filter and wash the precipitate with 4% caustic soda, dissolve in 10 ml of 20% phosphoric acid (H_3PO_4), wash with water and make the volume up to 30 ml. Proceed by *Method 3*[64,68]. Goto[65] suggests using *Method 1* on the untreated sample while Yuen[50] prefers *Method 3*. Alternatively[66] take 100 ml of sample, add 10 ml of 1:3 nitric acid and 1 ml of sulphuric acid. Evaporate until most of the sulphuric acid has been driven off as white fumes but do not take to dryness. Cool, take up with 50 ml of water and 20 ml of 1:3 nitric acid (oxide free). Dilute to 100 ml and determine manganese. The American Public Health Association recommend[67] a double oxidation with sodium bismuthate, but if all organic matter is destroyed as described above, *Method 2* may be applied directly to the resulting solution.

As in the case of sewage, Sikorowska[58] considers that methods based on oxidation of the manganese to permanganate are the most useful. As a result of an investigation of methods of analysing lake waters Delfino[69] reaches the same conclusion.

Separation from other ions

In addition to the separation methods mentioned in specific applications above the following more general methods may be used:-
(i) *from iron* Precipitate iron as benzoate in a boiling solution buffered with hexamine at pH 3-5 in the presence of a small amount of ammonium salts[70].
(ii) Extract manganese as tetraphenyl arsonium permanganate into nitrobenzene. This is claimed[71] to be highly specific with $87 \pm 5\%$ recovery.
(iii) Form a complex with 1-(2-pyridylazo)-2-naphthol (PAN) and extract with ethyl ether ($C_2H_5OC_2H_5$) at pH 9-10[72].
(iv) *from nickel, cobalt and zinc*[73]. Pass the solution in $5N$ hydrochloric acid in 60% methanol (CH_3OH) through a 10 × 1.2 cm column of Amberlite CG 400 resin (Cl^- form), and elute the manganese with $10N$ hydrochloric acid.

The individual tests

Method 1 This is the formaldoxime hydrochloride test discovered by Bach[74]. A recent study[75] has shown that the relative error of this test is less than 3%. If a precipitate is formed on the addition of alkali the modified procedure due to Sideris[76] should be used.

Method 2 This uses sodium bismuthate to oxidise manganese to permanganate, as originally advised by Reddrop and Ramage[77]. According to Davidson and Capen[78] this method yields slightly low results when applied to plant materials.

Method 3 This uses periodate oxidation of manganese to permanganate followed by permanganate oxidation of the leuco base to the brilliant green carbinol[63]. It is more sensitive than the previous two methods.

Method 4 This is a modification[79] of *Method 1* in which the interference from ferrous ions is prevented by complexing them with E.D.T.A. and hydroxylammonium chloride.

For food analysis it is considered [19] that *Methods 1 and 4* are marginally faster and more accurate than *Method 3* but the latter is more adaptable and is therefore generally to be preferred.

Method 3 is the official method for the determination of manganese in aluminium and aluminium alloys[23].

Methods 2 and 3 are preferred[58-69] for the analysis of sewage and water.

References

1. M. J. Pro and R. A. Nelson, *J.Soc.off.agric.Chem.*, 1956, **39**, 848; *Analyt.Abs.*, 1957, **4**, 703
2. H. Holesovska-Kozakova, *Chem.Listy.*, 1960, **54**, 967; *Analyt.Abs.*, 1961, **8**, 1144
3. U. S. Durgakeri and R. A. Bellare, *J.Sci.Ind.Res.India, C*, 1961, **20**, 314; *Analyt.Abs.*, 1962, **9**, 2412
4. A. A. Fernandez, C. Sobel and S. L. Jacobs, *Analyt.Chem.*, 1963, **35**, 1721; *Analyt.Abs.*, 1964, **11**, 4438
5. W. B. Healy, *Analyt.Chim.Acta*, 1966, **34** (2), 238; *Analyt.Abs.*, 1967, **14**, 3376
6. M. Karvanek and J. Karvankova, *Prumysl.Potravin*, 1963, **14**, 271; *Analyt.Abs.*, 1964, **11**, 5128
7. R. A. D'yachenko and N. P. D'yanchenko, *Lab.Delo*, 1964, **10** (10), 593; *Analyt.Abs.*, 1966, **13**, 810
8. F. Dittel, *Z.analyt.Chem.*, 1967, **228** (6), 412; *Analyt.Abs.*, 1968, **15**, 5464
9. C. H. Henkens and L. J. Mebius, *Tijdschr.Diergeneesk.*, 1965, **90** (9), 640; *Analyt.Abs.*, 1967, **14**, 3377
10. W. Barthy, B. M. Notton and W. C. Werkheiser, *Biochem.J.*, 1957, **67**, 291; *Analyt.Abs.*, 1958, **5**, 1593
11. P. Strohal, S. Lulic and O. Jelisavcic, *Analyst*, 1969, **94**, 678
12. J. L. Down and T. T. Gorsuch, *Analyst*, 1967, **92**, 398
13. H. Hanni, *Schweiz.Milchztg.*, 1967, **93**, (34), *Wiss.Beil No. 113*, 934; *Analyt.Abs.*, 1969, **16**, 375
14. W. B. Healy, *J.Agric.Food Chem.*, 1958, **6**, 606; *Analyt.Abs.*, 1959, **6**, 2330
15. G. A. Knauer, *Analyst*, 1970, **95**, 476
16. J. B. Firbas, G. H. Vaccari and L. R. Silvar, *Boln.Soc.quim.Peru*, 1965, **31** (2), 57; *Analyt.Abs.*, 1966, **13**, 6511
17. S. M. Zucas and A. A. R. Arruda, *Revta Fac.Farm.Bioquim.Univ.S. Paulo*, 1968, **6** (1), 23; *Analyt.Abs.*, 1969, **17**, 3012
18. P. G. Pifferi and A. Zamorani, *Industrie agrarie*, 1968, **6** (4), 213; *Analyt.Abs.*, 1969, **17**, 1778
19. H. Mlodecki, J. Chmielnicka, B. Pawlowska and A. Piotrowska, *Chemia analit.*, 1965, **10** (6), 1267; *Analyt.Abs.*, 1967, **14**, 1671
20. M. Pedinelli, *Chim. e Ind.*, 1959, **41**, 1180; *Analyt.Abs.*, 1960, **7**, 3838
21. E. D. Steinke, R. A. Jones and M. Brandt, *Analyt.Chem.*, 1961, **33**, 101
22. G. B. Kyriakopoulos, *J.Inst.Petrol.*, 1968, **54**, 369; *Analyt.Abs.*, 1970, **18**, 2544
23. Brit.Standards Inst., "*Methods for the Analysis of Aluminium and Aluminium Alloys, 10 Manganese (absorptiometric method)*", B.S. 1728, London, 1957
24. N. Sudo and S. Inoue, *Japan Analyst*, 1969, **18** (6), 717: *Analyst.Abs.*, 1970, **19**, 3746
25. J. Chwastowska, *Chem.Anal.Warsaw*, 1962, **7**, 731; *Analyt.Abs.*, 1963, **10**, 1782
26. A. A. Fedorov, F. A. Ozerskaya, R. D. Malinina, Z. M. Sokolova and F. V. Linkova, *Sb.Trud.Tsentr.Nauch-Issled.Inst.Chern.Metallurg.*, 1960, (19), 7; *Analyt.Abs.*, 1962, **9**, 2284
27. L. V. Kamaeva, V. V. Stepin and M. K. Makarov, *Trudy vses.nauchno.–issled.Inst.standart.Obraztsov*, 1967, **3**, 87; *Analyst.Abs.*, 1969, **16**, 3030
28. V. N. Tikhonov and A. P. Nikitina, *Zavodskaya Lab.*, 1962, **28**, 662; *Analyt.Abs.*, 1963, **10**, 52
29. D. H. Wilkins and L. E. Hibbs, *Analyt.Chim.Acta*, 1959, **20**, 427; *Analyt.Abs.*, 1960, **7**, 541
30. W. Kemula, K. Brajter, S. Cieslik and A. Lipinska-Kostrowcka, *Chem.Anal.Warsaw*, 1960, **5**, 229; *Analyt.Abs.*, 1960, **7**, 5255
31. *Proc.Amer.Soc.Test.Mat.*, 1952, **52**, 160; *Analyt.Abs.*, 1954, **1**, 2100
32. A. A. Blair, *J.Amer.Chem.Soc.*, 1904, **26**, 793
33. W. Blum, *J.Amer.Chem.Soc.*, 1912, **34**, 1395
34. S. Meyer and O. G. Koch, *Mikrochim.Ichnoanal.Acta*, 1964, (2–4), 216; *Analyt.Abs.*, 1965, **12**, 3890
35. P. H. Scholes and C. Thulborne, *Analyst*, 1964, **89**, 466
36. R. Goetz, *Giesserei*, 1965, **52** (1), 12; *Analyt.Abs.*, 1966, **15**, 2958

37. P. Povondra and Z. Sulcek, *Coll.Czech.Chem.Commun.*, 1959, **24**, 2398; *Analyt.Abs.*, 1960, **7**, 1744
38. D. N. Grindley, E. H. W. Burden and A. H. Zaki, *Analyst*, 1954, **79**, 95
39. N. I. Vetrova, *Trudy vses.nauchno.–issled.geol.Inst.*, 1964, **117**, 21; *Analyt.Abs.*, 1965, **12**, 4535
40. J. C. Van Loon and C. M. Parissis, *Analyst*, 1969, **94**, 1057
41. J. Tusl, *Chemicke Listy*, 1967, **61** (11), 1500; *Analyt.Abs.*, 1969, **16**, 987
42. A. Nestoridis, *Analyst*, 1970, **95**, 51
43. P. Chichilo, *J.Assoc.off.agric.Chem.*, 1964, **47**, 1019; *Analyt.Abs.*, 1965, **12**, 6778
44. T. S. Hermann, *Analyt.chim.Acta.*, 1964, **31** (3), 284; *Analyt.Abs.*, 1966, **13**, 150
45. R. H. McKinlay, *Fibres.*, 1954, **15** (2), 39; *Analyt.Abs.*, 1954, **1**, 2726
46. A. G. Hamlen, *J.Text.Inst.Trans.*, 1956, **47**, T445; *Analyt.Abs.*, 1957, **4**, 1282
47. L. Benisek, *Textil-Praxis*, 1964, **19** (1), 16; *Analyt.Abs.*, 1965, **12**, 1834
48. Scand. Pulp Paper and Board Testing Cttee., *Svensk.Papp-Tidn.*, 1962, **65**, 821; *Norsk Skogind.*, 1962, **16**, 447; *Analyt.Abs.*, 1963, **10**, 1500
49. E. G. Bradfield, *Analyst*, 1957, **82**, 254
50. S. H. Yuen, *Analyst*, 1958, **83**, 350
51. Yu. I. Dobritskaya, *Mikroelementy v. S.S.S.R.*, 1963, (5), 55; *Analyt.Abs.*, 1965, **12**, 1288
52. E. G. Bradfield, *J.Sci.Food Agric.*, 1964, **15**, 469; *Analyt.Abs.*, 1965, **12**, 5957
53. M. Karnakev, *Prumysl.Potravin*, 1964, **15** (2), 282; *Analyt.Abs.*, 1965, **12**, 5472
54. M. Ishibashi and T. Yamato, *J.Chem.Soc. Japan, Pure Chem.Sect.*, 1958, **79**, 1184; *Analyt.Abs.*, 1959, **6**, 2598
55. Min. of Housing and Local Govt:, *"Methods of Chemical Analysis as applied to Sewage Effluents"*, H.M. Stationery Office London, 1956
56. A. A. Christie, J. R. W. Kerr, G. Knowles and G. F. Lowden, *Analyst*, 1957, **82**, 336
57. E. J. Serfass and R. F. Muraca, *Plating*, 1955, **42**, 147
58. C. Sikorowska and W. Dozanska, *Roczn.panst.Zakl.Hig.*, 1966, **17** (1), 79; *Analyt.Abs.*, 1967, **14**, 3636
59. C. Bighi and G. Trabanelli, *Ann.Chim.Roma*, 1954, **44**, 371; *Analyt.Abs.*, 1954, **1**, 3143
60. I. K. Tsitovich, *Zhur.Analit.Khim.*, 1962, **17**, 621, *Analyt.Abs.*, 1963, **10**, 3481
61. A. Duca and D. Stanescu, *Stud.Cercet.Chim.Cluj.*, 1957, **8**, 75; *Analyt.Abs.*, 1958, **5**, 1717
62. W. Y. Magar, Ph.D. Thesis, London Univ., 1961
63. A. de Sousa, *Rev.fac.cienc.Lisbon*, 1954, 2a Ser. B3, 177
64. E. Hluchan and J. Mayer, *Chem.Listy.*, 1953, **47**, 846; *Analyt.Abs.*, 1954, **1**, 2834
65. K. Goto, T. Komatsu and T. Furukawa, *Analyt.Chim.Acta.*, 1962, **27**, 331; *Analyt.Abs.*, 1963, **10**, 2935
66. W. D. Collins and M. D. Foster, *Ind.Eng.Chem.*, 1924, **16**, 586
67. Amer.Pub.Health Assoc., *"Standard Methods of Water Analysis"*, 6th Editn., New York, 1925, pp 50–1
68. V. V. Mokievskaya, *Trudy Inst.Okeanol.*, 1965, **79**, 3; *Analyt.Abs.*, 1966, **13**, 3244
69. J. J. Delfino and G. F. Lee, *Envir.Sci.Technol.*, 1969, **3** (8), 761; *Analyt.Abs.*, 1970, **19**, 3476
70. R. Ripan and I. Pirvu, *Stud.Cercet.Chim.Cluj.*, 1959, **10**, 129; *Analyt.Abs.*, 1960, **7**, 1749
71. J. M. Matuszek and T. T. Sugihara, *Analyt.Chem.*, 1961, **33**, 35
72. S. Shibata, *Analyt.Chim.Acta.*, 1961, **25**, 348; *Analyt.Abs.*, 1962, **9**, 1806
73. L. Mazza, R. Frache, A. Dadone and R. Ravasio, *Gazz.chim.ital.*, 1967, **97**, 1551; *Analyt.Abs.*, 1968, **15**, 5224
74. A. Bach, *Compt.Rend.*, 1899, **128**, 363
75. A. Okac and M. Bartusek, *Z.analyt.Chem.*, 1960, **178**, 198; *Analyt.Abs.*, 1961, **8**, 2883
76. C. P. Sideris, *Ind.Eng.Chem.Analyt.Ed.*, 1937, **9**, 445
77. J. Reddrop and H. Ramage, *J. Chem.Soc.*, 1895, **67**, 268
78. J. Davidson and R. G. Capen, *J.Assoc.off.agr.Chem.*, 1931, **14**, 547
79. Deutsche Einheitsverfahren, E2

Manganese Method 1
using formaldoxime hydrochloride

Principle of the method

The procedure consists in the treatment of the manganese solution with formaldoxime hydrochloride, followed by the addition of $4N$ sodium hydroxide, until a wine-red coloration appears. The interference of iron is obviated by its extraction with ether after rendering the solution $6N$ with respect to hydrochloric acid, as is ensured by the following technique.

Reagents required

1. *Formaldoxime hydrochloride* $((CH_2 NOH)_3.HCl)$ A 5 per cent. aqueous solution
2. $4N$ *Sodium hydroxide* $(NaOH)$ 160 g/l
3. *Hydrochloric acid* (HCl) Concentrated
4. *Ether* $(C_2H_5OC_2H_5)$

The Standard Lovibond Comparator Disc 3/21

The disc covers the range from 5 μg to 60 μg (0.005 to 0.06 mg) of manganese (Mn), in steps 5, 10, 15, 20, 25, 30, 40, 50, 60 μg. This covers the range 1 to 12 ppm with a 5 ml sample.

Technique

To an appropriate proportion of the solution to be tested add 5 ml of concentrated hydrochloric acid (reagent 3), and dilute to 10 ml with water. Place in a separator, add 5 ml of ether (reagent 4) and shake. Transfer the aqueous layer to another separator, add another 5 ml of ether, and shake again. Transfer the aqueous layer to a small (50 ml) flask, and warm on a steam-bath to remove the dissolved solvent. Dilute the cooled solution to 50 ml with water. To 10 ml of this solution add 1 ml of formaldoxime hydrochloride (reagent 1), and sufficient $4N$ sodium hydroxide (reagent 2) to produce the colour. Dilute to 15 ml with water, mix, and transfer to a 13.5 mm cell or test tube and place in the right-hand compartment of the Comparator. In the left-hand compartment, place a similar cell of distilled water. With the Comparator facing a standard source of white light (Note 3) rotate the disc until a match is obtained. The figure shown indicates the amount of manganese present in the volume of solution actually taken for the test, and should be corrected by multiplying by the appropriate factor in order to obtain the value for the original solution.

Notes

1. Traces of copper produce a blue-violet colour with the reagent, but interference from this metal is obviated by adding a few drops of a 10% solution of potassium cyanide. The test is not influenced by the presence of relatively large concentrations of mercury, lead, bismuth, silver, cadmium, tin, antimony, arsenic, aluminium, zinc, or the alkaline earths. Molybdate, tungstate, and vanadate do not interfere; if beryllium is present the solution under test should be rendered alkaline with ammonium carbonate. Chromium, cobalt, and nickel produce colorations with the reagent, and should be removed during sample preparation.

2. The formaldoxime method is not applicable to solutions which give precipitates when they are made alkaline.

3. The recommended standard source is the Lovibond White Light Cabinet. In the absence of a standard source, north daylight should be used.

The Standard Lovibond Nesseriser Disc NZZ

This disc covers the range 0.025, 0.05, 0.075, 0.10, 0.15, 0.20, 0.30, 0.40, and 0.50 ppm of manganese (Mn) based on a 40 ml sample.

COLORIMETRIC CHEMICAL ANALYTICAL METHODS Manganese Method 1

Technique

To 40 ml of the sample add 5 ml of concentrated hydrochloric acid (reagent 3). Place in a separating funnel, add 5 ml of ether (reagent 4) and shake. Transfer the aqueous layer to another separator, add another 5 ml of ether and shake again. Transfer the aqueous layer to a beaker and warm on a steam bath to remove the volatile solvent, cool and transfer to a Nessler tube and adjust the temperature of the solution to $20 \pm 2°C$. Add 1 ml of formaldoxime hydrochloride solution (reagent 1) and sufficient sodium hydroxide (reagent 2) to produce the colour. Make the volume up to the 50 ml mark with distilled water. Mix thoroughly and then place the Nessler tube in the right hand compartment of the Nessleriser. Place an identical tube filled with distilled water in the left hand compartment and place the Nessleriser before a standard source of white light, such as the Lovibond White Light Cabinet, or, failing this, north daylight. Five minutes after the addition of the sodium hydroxide, match the colour in the Nessler tube with the standards in the disc.

Periodically a blank determination should be carried out on the reagents using distilled water instead of the sample and omitting the ether extraction stage. It should not be necessary to carry out a blank with each test unless the periodic check gives a significant blank.

Note

The results obtained by the use of the Nessleriser and disc are accurate only if the Nessleriser tubes are used which conform to the specification used when the discs were calibrated, that is that the 50 ml mark falls at a height of 113 ± 3 mm from the bottom of the tube measured internally.

Manganese Method 2
using sodium bismuthate

Principle of the method

This test is based on the permanganate colour produced by the oxidation of manganese by sodium bismuthate in nitric acid solution. To provide a blank solution for comparison when an inherently coloured solution is examined, the colour due to permanganate is discharged in a quantity of the test solution by hydrogen peroxide.

Reagents required (all of analytical reagent grade)

1. *Sodium bismuthate* ($NaBiO_3$)
2. *Nitric acid* (HNO_3) concentrated
3. *Hydrogen peroxide* (H_2O_2) 20 volumes

The Standard Lovibond Nessleriser Disc NG and Lovibond Comparator Disc APMN

Disc NG covers the range from 10 μg to 90 μg (0.01 to 0.09 mg) of manganese (Mn) in steps of 10 μg. This equals 20 to 180 ppm if 1 gram sample is used.

Disc APMN is used with the Lovibond 1000 Comparator Large Cells Attachment and a 40 mm cell. It covers the range 1-5 ppm of manganese in steps of 0.5 ppm.

Technique

(a) *Nessleriser*

Boil 1 g (or a suitable quantity) of the substance under test, which should be free from chloride, with 2 g of sodium bismuthate (reagent 1), 15 ml of nitric acid (reagent 2) and 35 ml of water until the reaction is **just** complete and a clear solution results. Cool rapidly under the tap, dilute with distilled water to 100 ml and transfer to two 50 ml Nessleriser glasses. Place one of these in the right-hand compartment of the Nessleriser; to the contents of the other add 2 drops of hydrogen peroxide (reagent 3), mix and place it in the left-hand compartment of the Nessleriser. Stand the Nessleriser before a standard source of white light (Note 2) and compare the colour produced in the test solution with the colours in the standard disc, rotating the disc until a colour match is obtained.

The markings on the disc represent the amount of manganese producing the colour in the test. Thus, if a colour equivalent to 50 μg is produced in the test carried out under the above conditions, using 1 g of the substance being tested, the amount of manganese in 0.5 g of the substance is 0.05 mg, which is equivalent to 100 parts per million.

(b) *Comparator*

Take 20 ml of the test solution and place them in a 500 ml Pyrex flask. Evaporate to about 3 ml, add 10 ml of nitric acid (reagent 2) and boil. Cool, add 1 g of sodium bismuthate (reagent 1) and wash down any solid from the sides of the flask with the minimum amount of distilled water. Boil until the characteristic permanganate colour appears. Cool quickly under a tap and transfer the solution to a 20 ml volumetric flask; wash out the Pyrex flask with up to 5 ml distilled water, and add the washings to the volumetric flask. Make up to 20 ml with distilled water. Place 10 ml in each of two 4 cm cells and insert these into the comparator. To the cell in the left-hand compartment, add 2 drops of hydrogen peroxide (reagent 3), and wait until the permanganate colour in this cell has disappeared. Match the colour of the right-hand cell against the standards in the disc using a standard source of white light (Note 2). The figure shown in the indicator window of the Comparator is the Mn concentration of the original solution in ppm. If the colour of the sample solution is outside the range of the standards in the disc then use either more or less of the sample as required, but bring the final volume up to 20 ml as above and calculate the final concentration proportionately.

Notes

1. Lovibond Nessleriser discs are standardized on a depth of liquid of 113 ± 3 mm. The height, measured internally, of the 50 ml calibration mark on Nessleriser glasses used with the instrument must be within the same limits. Tests with Nessleriser glasses not conforming to this specification will give inaccurate results.

2. The recommended standard source is the Lovibond White Light Cabinet. In the absence of a standard source north daylight should be used.

Manganese Method 3
using leucomalachite green

Principle of the method
Manganese ions are oxidised to permanganate by potassium periodate. The permanganate ions oxidise the leuco base to the brilliant green carbinol. The intensity of the green colour, which is proportional to the manganese concentration, is measured by comparison with Lovibond permanent glass standards.

Reagents required
1. *Buffer solution* Dissolve 50 g of sodium acetate trihydrate ($CH_3COONa.3H_2O$) in 300 ml of water. Add 400 ml glacial acetic acid (CH_3COOH) and make up to 1 litre with water
2. *Potassium periodate* (KIO_4) 0.2% solution in water
3. *Leuco malachite green solution* Dissolve 0.1 g of leuco malachite green in 100 ml of water containing 1.25 ml of concentrated hydrochloric acid (HCl). Heat to 80°C to help solution, taking care not to allow the temperature to exceed 80°. Cool and make up to 250 ml with water

All chemicals used in the preparation of these reagents should be of analytical reagent quality and distilled water should be used where water is required.

The Standard Lovibond Comparator Disc 3/55
The disc covers the range 0.03 to 0.3 µg of manganese (Mn), in steps of 0.03 µg. This covers the range 0.3 to 3 ppm if 0.1 ml sample is used.

Technique
Pipette 0.1 ml of the sample solution into a 10 ml comparator tube, add 6 ml of distilled water followed successively by 1 ml of buffer solution (reagent 1) and 1 ml of potassium periodate solution (reagent 2). Mix thoroughly, after each addition, by stirring the solution with a stirrer made from 4 mm glass rod. Stand for 20 minutes, add 0.5 ml of leuco malachite green solution (reagent 3) and dilute to the 10 ml mark with distilled water. Mix thoroughly by stirring and place the tube in the right-hand compartment of the Comparator. Place an identical tube filled with distilled water in the left-hand compartment and match the colour of the test solution with the Lovibond permanent glass standards in the disc, using a standard source of white light (Note 4). The figures obtained from the disc represent micrograms of manganese in the 10 ml of solution in the tube, i.e. in 0.1 ml of the sample solution. If the colour of the test solution is darker than that of the highest standard in the disc then dilute the original solution, repeat the test, and multiply the final result by the dilution factor.

Notes
1. Owing to the extreme sensitivity of this test it is essential that great care be taken to prevent contamination of the glass-ware used with traces of manganese. All glassware should therefore be rinsed in standard chromic-sulphuric acid cleaning mixture immediately prior to the test and then thoroughly rinsed in distilled water.
2. The test should be carried out, if possible, at a temperature between 20°C and 30°C. At temperatures below 20°C colour development is slow. At temperatures above 30°C colour development is rapid and erratic. If ambient temperatures are above 30°C the comparator tube and its contents should be cooled during the period of colour development.
3. The colour is stable for two hours. After this time the colour gradually fades.
4. The recommended standard source is the Lovibond White Light Cabinet. In the absence of a standard source, north daylight should be used.

Manganese Method 4
using formaldoxime

Principle of the method

In alkaline solution manganese ions react with formaldoxime to produce an orange-red complex which is stable for several hours.

The intensity of this orange-red colour, which is proportional to the manganese concentration, is measured by comparison with Lovibond permanent glass colour standards. Ferrous ions also form a coloured complex with formaldoxime, which would interfere with the determination of manganese. In this method this interference is prevented by forming a colourless complex of the ferrous ions with E.D.T.A. and hydroxylammonium chloride, additional ferrous ions being added to the sample solution to ensure efficient reaction with E.D.T.A.

Reagents required
1. *Formaldoxime solution* Dissolve 40 g of hydroxylammonium chloride ($HONH_3Cl$) and 8 g of paraformaldehyde (($HCHO$)n) in distilled water and make up to 1 litre
2. *Ferrous ammonium sulphate solution* Dissolve 140 mg of ferrous ammonium sulphate (($NH_4)_2Fe(SO_4)_2 6H_2O$) in distilled water. Carefully add 1 ml of concentrated sulphuric acid (H_2SO_4, density 1.84 g/ml) and make up to 1 litre
3. *Ammonia solution* Dilute 750 ml of ammonia solution (d = 0.91 g/ml, equivalent to 540 ml of 0.880 ammonia) to 1 litre with distilled water
4. *EDTA solution* Dissolve 37.224 g of ethylenediaminetetra-acetic acid disodium salt in distilled water and dilute to 1 litre
5. *Hydroxylammonium chloride solution* Dissolve 100 g of hydroxylammonium chloride ($HONH_3Cl$) in distilled water and make up to 1 litre

All chemicals used in the preparation of these reagents should be of analytical reagent quality.

The Standard Lovibond Nessleriser Disc CCA

This disc contains standards corresponding to 0.05, 0.1, 0.15, 0.2, 0.25, 0.3, 0.5, 0.75, and 1.0 ppm of manganese (Mn) based on the use of 100 ml of sample and 250 mm Nessleriser tubes (Note 1).

Technique

Take 100 ml of sample (Note 2) and add successively 10 ml each of formaldoxime, ferrous ammonium sulphate and ammonia (reagents 1–3). After about 5 minutes add 10 ml each of EDTA and hydroxylammonium chloride (reagents 4 and 5). Shake continuously during the whole of this process of addition and mixing.

Prepare a blank in the same way using 100 ml of distilled water in place of the sample.

Allow both the sample and blank to stand for one hour then transfer each solution into a 250 mm Nessleriser tube, up to the 250 mm mark. Place the tube containing the sample in the right-hand compartment of the Nessleriser and that containing the blank in the left-hand compartment. Place the Nessleriser before a standard source of white light such as the Lovibond White Light Cabinet, or, failing this, north daylight, and match the colour of the sample with the standards in the disc. The readings show ppm (mg/l) of manganese.

Notes
1. The readings obtained with this disc will be accurate only if Nessleriser tubes, identical to those used to calibrate the discs are used. These tubes have a calibrated mark 250 mm from the base measured internally and may be obtained from The Tintometer Limited, code reference DB420, and are used in conjunction with a Lovibond 1000 Comparator with Nessler attachment.
2. Organic substances in the concentrations found in surface waters will not interfere with this method. The presence of calcium and manganese ions in combined concentrations greater than 300 ppm would lead to high results. In this case dilute the sample solution with

distilled water to bring the calcium magnesium concentration below 300 ppm (30°DH—German degrees of hardness) and multiply the answer found by the appropriate factor for the dilution.

Phosphate ions in concentrations greater than 5 ppm will result in low results if only calcium is present, but will not interfere in the presence of magnesium.

Cobalt and nickel ions will interfere in concentrations of 1 ppm or above.

Turbid samples must be filtered before determination.

The Determination of Nickel

Introduction

Nickel (Ni) is widely used industrially in plating processes, as a catalyst, as an alloying constituent, and as the pure metal. It also occurs as a trace constituent in most plant and animal tissues.

Methods of sample preparation

Air[1]

Draw the air at 0.5 litres per minute through an absorbant tube filled with filter paper, to collect solid particles and then through 3 ml of 1.5% chloramine B ($C_6H_5SO_2NNaCl$) in ethanol (C_2H_5OH) treated with 0.1 ml of 1:1 hydrochloric acid to absorb nickel carbonyl ($Ni(CO)_4$). Dilute the solution with 2 ml of water and 0.2 ml of 10% caustic soda (NaOH) solution. Determine nickel on this solution by *Method 2*. Treat the paper from the absorbant tube with 2—4 ml of 3% hydrochloric acid and determine nickel on this solution also by *Method 2*.

Beverages[2]

Evaporate 250 ml of the sample to dryness, ash the residue and dissolve the ash in 100 ml of 1:1 hydrochloric acid. Determine nickel on an aliquot part of this solution using *Method 2*.

Biological materials[3-6]

General[3,4] Evaporate or dry the sample in a porcelain dish and ash, add 10 ml of hydrochloric acid and evaporate to dryness. Repeat 2 or 3 times. Extract the residue with 10 ml of water containing 2 ml of 4N hydrochloric acid. If the sample contains iron in excess, precipitate this with aqueous ammonia (NH_4OH) and filter. Re-dissolve the precipitate in 10 ml of dilute hydrochloric acid as above. Re-precipitate and filter. Repeat twice more. If little or no iron is present sufficient must be added to combine with any phosphate (PO_4^{3-}). If phosphate is in excess, neutralise the cold solution with 1:1 aqueous ammonia, add 8—10 ml of 10% ammonium acetate (CH_3COONH_4) solution and enough ferric chloride ($FeCl_3$) to tint the liquid yellow-red. Boil and filter, wash the precipitate and precipitate any excess iron with aqueous ammonia as before. Evaporate the united filtrates to dryness, dissolve the residue in water and add 6 ml of 4N hydrochloric acid. Pass hydrogen sulphide (H_2S) into the hot solution for 30 minutes. Allow the sulphides to settle, filter and wash with hydrogen sulphide solution. Evaporate the solution to dryness, re-dissolve the residue in water and add 10% caustic soda (NaOH) solution until no further ammonia (NH_3) is evolved. Nickel is precipitated as the hydroxide ($Ni(OH)_2$). Convert to the sesquioxide (Ni_2O_3) by the addition of 1—2 ml of bromine water. Filter, wash the precipitate and dissolve in hydrochloric acid. Evaporate this solution to dryness and re-dissolve in 1:100 hydrochloric acid. Proceed by *Method 2*[3].

Alternatively[4] dissolve the ashed sample in 5 ml of 1:4 hydrochloric acid, filter and dilute the filtrate to 25 ml with water. To a 5 ml aliquot add 8 ml of water and extract any copper with a 0.01% solution of dithizone in carbon tetrachloride (CCl_4) until the colour of the dithizone solution remains unchanged. Make the aqueous fraction of the solution ammoniacal with aqueous ammonia and extract the nickel with chloroform ($CHCl_3$) and then re-extract with 0.5N hydrochloric acid. Adjust the volume of the aqueous layer to 10 ml and proceed by *Method 2*.

Serum[5] Freeze dry a sample in a Kjeldahl flask and digest the dry residue with nitric acid and hydrogen peroxide (H_2O_2). Add dimethylglyoxime ($CH_3.C(:NOH).C(:NOH).CH_3$) solution and extract nickel into chloroform ($CHCl_3$). Wash the chloroform phase with aqueous ammonia (NH_4OH), re-extract the nickel into 0.5N hydrochloric acid and proceed by *Method 2*.

Urine[6] Wet ash 100 ml of sample with nitric acid and hydrogen peroxide (H_2O_2) and evaporate to dryness. Dissolve the residue in 8 ml of water and 1 ml of normal hydrochloric acid and then adjust the acidity to pH 4 by the addition of normal caustic soda (NaOH). Add 0.5 ml of a 0.5% solution of sodium diethyldithiocarbamate, 0.4 ml of cadmium sulphate ($CdSO_4$) and 0.3 ml of 0.4N sodium sulphide (Na_2S) solution. Set aside for 20 minutes and then centrifuge at 3000 r.p.m. for 15 minutes. Decant the supernatant liquid, dissolve the precipitate in 5 drops of aqua regia ($HNO_3 + 3HCl$) and evaporate to dryness. Dissolve the residue in 1 ml of water, add 1 ml of a 20% solution of potassium sodium tartrate (COOK.CHOH.CHOH.COONa.$4H_2O$) adjust the acidity to pH 4 and add 2—5 mg of ammonium sulphate (($NH_4)_2SO_4$) and 2 ml of 25% aqueous ammonia (NH_4OH). Centrifuge for 10 minutes and estimate nickel in the supernatant liquid using *Method 2*.

Chemicals[7]

Organic compounds Ignite 1 g of sample with 1 ml of sulphuric acid, 5 ml of nitric acid and 1 ml of 5% sodium chloride (NaCl) solution. Dilute the solution to a known volume and determine nickel on an aliquot by *Method 2*.

Foods and edible oils[8-15]

General[8] Ash the sample at 500°C, dissolve the ash in 1:3 hydrochloric acid and filter. Make the filtrate up to a known volume and determine nickel on an aliquot using *Method 2*.

Fats[9-12] Ash the sample and extract the ash with hydrochloric acid. Shake the acid extract with a solution of lead diethyldithiocarbamate in chloroform. Separate the aqueous phase, add 1% solution of cupferron ($C_6H_5.N(NO).ONH_4$), extract the precipitated iron complex into chloroform. Determine nickel on the aqueous phase by *Method 2*[9]. Alternatively[11,12] melt 25 g of sample, together with a little kaolin, dry at 120°C, ignite and then calcine the residue at 600°C for 1 hour, cool, add 2 ml of 30% hydrogen peroxide (H_2O_2) evaporate to dryness, add 2 ml of hydrochloric acid, evaporate, dissolve the residue in 2 ml of 0.1N hydrochloric acid and wash the crucible with 1 ml of 0.1N hydrochloric acid. Add these washings to the solution, filter, make up the filtrate to a known volume and determine nickel on an aliquot by *Method 2*.

Filajdic[10] has compared several methods for the determination of nickel in fats and concludes that the method developed by Liberman for the analysis of edible oils[13], which is described below, gives the most accurate results.

Oils[13-14] Take a 50 g sample in a porcelain basin, add 1 g of magnesium nitrate ($Mg(NO_3)_2$) and place a small cone of filter paper in the oil to act as a wick. Ignite and allow the oil to burn away. Ignite the residue at 550—600°C, cool, add 1—2 ml of 50% nitric acid and warm. Decant into a 50 ml graduated flask and repeat the washing with 50% nitric acid several times to dissolve all the nickel, adding all the washings to the 50 ml flask. Neutralise with 50% aqueous ammonia (NH_4OH), filter and wash the precipitate with 1% aqueous ammonia. Transfer the filtrate and washings to the 50 ml flask and make up to volume with 1% aqueous ammonia. Proceed by *Method 2*[13].

Pifferi[14], after ashing the sample, considers that extraction of the ash with 0.5N hydrochloric acid gives an adequate recovery.

Wheat[15] Wet oxidise the sample with sulphuric acid + nitric acid until a white residue remains on heating to fuming. Add water and reheat to white fumes. Rinse the residue into a beaker, reduce the volume by evaporation to about 20 ml, cool, add 20% sodium tartrate (($CH(OH)COONa)_2.H_2O$) and heat to 80°C. Add dimethylglyoxime solution and adjust the pH to 8—9 with aqueous ammonia (NH_4OH). Stand overnight, filter and treat the precipitate with 1 ml of aqueous bromine solution. Wash the precipitate with water and repeat the bromine treatment and washing. Boil the combined filtrates and washings to remove bromine, cool and proceed by *Method 2*.

Fuels and lubricants[16]

Digest the sample with sulphuric and nitric acids with the addition of 30% aqueous hydrogen peroxide (H_2O_2) as required until the solution is clear and then dilute to a known volume and determine nickel on an aliquot by *Method 2*.

Metals, alloys and plating baths[17-36]

Aluminium and its alloys[17,18] Dissolve a 0.5 g sample in 10 ml of hydrochloric acid followed by 10 ml of nitric acid and 15 ml of perchloric acid ($HClO_4$). Heat to fumes, cool, add 25 ml of water, warm and filter. Wash the residue with warm water and add the washings to the filtrate. To this filtrate add 75 ml of aqueous ammonia (NH_4OH), 100 ml of 20% ammoniacal ammonium citrate (($NH_4)_3C_6H_5O_7$) solution, a further 20 ml of aqueous ammonia and then water to bring the total volume to 200 ml. To 100 ml of this solution add 10 ml of 1:1 nitric acid and electrolyse at 5 amps with stirring. After 5 minutes, add 1 ml of 50% sulphuric acid,

1 ml of 10% sulphamic acid (NH_2SO_3H) solution and 2 drops of 1% hydrochloric acid. When all the copper is deposited, filter, dilute the filtrate to 250 ml and determine nickel on an aliquot by *Method 2*[17]. Alternatively[13] dissolve the sample in caustic soda (NaOH) solution, filter and discard the filtrate. Dissolve the residue in hydrochloric acid and determine nickel by *Method 2*.

Cobalt and its compounds[19, 20] Dissolve a sample containing about 1 g of cobalt in acid and add this solution to 200 ml of a reagent containing 20 g of disodium hydrogen phosphate (Na_2HPO_4), 25 g of ammonium chloride (NH_4Cl), 2.5 g of hydroxylamine hydrochloride ($NH_2OH.HCl$) and 150 ml of saturated aqueous ammonia (0.880 NH_4OH) per litre. Boil the mixture, cool and filter. Determine the nickel in the filtrate by *Method 2*[19]. Alternatively[20] dissolve 0.2 g of alloy, or 0.5 g of ore, in 20—30 ml of aqua regia ($HNO_3 + 3HCl$), evaporate to dryness twice with successive 10 ml portions of hydrochloric acid and dissolve the residue in 10 ml of hydrochloric acid and pass the solution through a column containing 45 ml of a quaternary ammonium amonite, such as Macroporous AV–17–8P, (Cl^- form) at 1 ml a minute. Wash the column with 90 ml of 8N hydrochloric acid to remove the nickel, make these washings up to 100 ml and determine nickel on an aliquot by *Method 2*.

Copper alloys[21-25] Dissolve the sample in nitric acid and electrodeposit copper from a slightly acid solution. Determine nickel on the residual solution by *Method 2*[21]. Alternatively the interference of copper may be prevented by masking with E.D.T.A.[22]; by extracting the nickel dimethylglyoxime complex with chloroform ($CHCl_3$) from a solution of pH 6.5[23]; or by mixing 1—2 ml of the sample solution with 5—10 ml of 2N picolinic acid (N:CH.CH:CH.CH:C COOH) and 10 ml of water, adjusting pH to 3, adding 5 ml of dioxan ($CH_2.CH_2.O.CH_2.CH_2.O$) and applying the mixture to a column containing about 1.5 g of Amberlite IRA–401 previously washed with 0.05M picolinic acid in 70% alcohol (C_2H_5OH), eluting the copper with 0.05M picolinic acid in 40 ml of 30% dioxan in water and then eluting the nickel with 2N hydrochloric acid[24].

Copper alloys which also contain tin are dissolved in 10—15 ml of hydrochloric acid to which 1—2 ml of 30% aqueous hydrogen peroxide (H_2O_2) has been added, the solution is then evaporated to dryness, the residue is dissolved in 10 ml of 8N hydrochloric acid and the solution is passed through a column of amonite EDE—10P (Cl^- form), the nickel is eluted with 100 ml of 8N hydrochloric acid[25].

Iron and steel[26-30] Dissolve a 1 g sample in 50 ml of nitric acid and treat the solution with 4 g of potassium chlorate ($KClO_3$). Filter and dilute the filtrate to known volume. Treat an aliquot with 15 ml of 50% citric acid ($C_6H_6O_7$) solution, 5 ml of aqueous bromine solution and 5 ml of aqueous ammonia (NH_4OH). Proceed by *Method 2*. Ferrosilicon and silico-manganese alloys (0.5 g) should be dissolved in 10 ml of nitric acid, 10 ml of hydrofluoric acid and 10 ml of 1:1 sulphuric acid and then treated as for steel solutions [26-30]. Goetz[28] and Volke[30] both recommend the addition of ammonium persulphate (($NH_4)_2S_2O_8$) to the acid to assist dissolution.

High purity manganese and cobalt[31] Separate the nickel by paper chromatography using Whatman No. 1 paper and isopropanol (($CH_3)_2CHOH$)—ethylacetate ($CH_3COOC_2H_5$)—hydrochloric acid solvent system. Add sodium pyrophosphate ($Na_4P_2O_7$) to prevent interference by iron and citric ($C(OH)(COOH)(CH_2COOH)_2$) and tartaric (($CHOH.COOH)_2$) acids, or E.D.T.A., to prevent interference from heavy metals.

Metal-ceramic alloys[32] Take a 0.5 g sample, add 5—8 ml of hot 1:1 nitric acid and then 25—30 ml of hot water. Heat on a hot water bath until no further nitrous fumes are evolved. Filter, wash the residue with hot 0.5% nitric acid. Heat the residue with 40 ml of 9% oxalic acid (($COOH)_2$) solution at 80°C for 30 minutes. Add 8—10 ml of nitric acid and heat until the alloy is completely dissolved. If any tungstic oxide (WO_3) precipitates add 1 ml of 3% gelatine solution, to coagulate the precipitate, 3 ml of nitric acid and dilute to 50—80 ml with hot water. Filter into the flask containing the original filtrate. Use an aliquot of the combined filtrates to determine nickel by *Method 2*.

High purity molybdenum[33] Dissolve 1 g of sample in aqua regia ($HNO_3 + 3HCl$), boil to remove excess acid, cool, make just alkaline with 50% aqueous ammonia (NH_4OH) and make up to 100 ml. Determine nickel on an aliquot by *Method 2*.

Nickel alloys[13,34] Dissolve about 40 mg of alloy in 4 ml of 50% nitric acid, heating if necessary. Cool and make up to about 50 ml with water and neutralise with 50% aqueous ammonia. Filter and wash the precipitate with 1% aqueous ammonia. Transfer the filtrate and washings to a 250 ml flask, make up to the mark with 1% aqueous ammonia and proceed by *Method 2*[13]. Alternatively[34] separate the nickel from other metals by passing the acide solution through a column of Dowex-1 anion exchange resin. Elute aluminium, nickel and titanium with $7M$ hydrochloric acid. Pass the eluate through two columns in series, one containing an anionic resin and the other a cationic resin. Elute with $0.8M$ hydrofluoric acid $+0.06M$ hydrochloric acid. Aluminium then passes through, nickel is retained on the cationic resin and titanium on the anionic. Elute the nickel with $4M$ hydrochloric acide, neutralise with aqueous ammonia and proceed by *Method 2*.

Tantalum[35] Dissolve the sample in a mixture of hydrofluoric and nitric acids, remove most of the hydrofluoric acid by gentle heating and keep the nickel in solution by the addition of citric acid. Complex any remaining hydrofluoric acid with boric acid (H_3BO_3) and dilute the solution to a known volume.

Tungsten and its alloys[33,36] Proceed as for molybdenum[33] but use a mixture of nitric and hydrofluoric acids for the dissolution of the sample. Alternatively[36] use ammonium tartrate – sodium borate buffer solution to adjust the pH to about 6 before proceeding by *Method 2*.

Minerals, ores and rocks[20,37,38]

Proceed as described above for cobalt[20]. Alternatively[37] decompose 0.5 g of the sample in 20 ml of hydrochloric acid and 5 – 10 ml of nitric acid (for oxide ores) or 20 ml of nitric acid and 10 – 20 ml of hydrochloric acid. Add 10 ml 1:1 sulphuric acid, heat until white fumes are evolved, cool, dilute with water to about 50 ml, filter, wash precipitate with hot water and make up filtrate and washing to 250 ml. To a 25 ml. aliquot add 2 ml of 50% citric acid ($C_6H_6O_7$) neutralise with aqueous ammonia and add 1 ml in excess. Add dimethylglyoxime solution and extract with $5+5+3$ ml of chloroform ($CHCl_3$). Re-extract the combined chloroform extracts with $10+5$ ml of $0.5N$ hydrochloric acid and determine nickel on this solution by *Method 2*.

If it is advantageous to concentrate the nickel after acid solution of the sample, co-precipitate its complex with furil α-dioxime with 2,4-dinitroaniline solution in acetone. Ignite the precipitate, which is free from iron, aluminium, titanium, calcium, magnesium, manganese, copper and other elements that usually accompany nickel, dissolve the residue in hydrochloric acid and make up to a known volume[38].

Miscellaneous[39-41]

General[39] Dissolve 0.25 – 0.5 g of sample in aqua regia (HCl:HNO_3–3:1) and proceed by: *Method 2*.

Maser crystals[40] Finely grind 0.02 g of sample and fuse at 400 – 450°C with caustic potash (KOH) in a crucible of high purity zirconium. Extract the melt with dilute sulphuric acid, make up to known volume and determine nickel on an aliquot by *Method 2*.

Thin magnetic films[41] Dissolve the sample in a few drops of nitric acid and then dilute to 10 ml with $0.05N$ perchloric acid ($HClO_4$). Proceed by *Method 2*.

Plants[42]

Rape seed. Ignite the sample at 500°C, extract the ash twice with 1:3 hydrochloric acid and dilute the extract to a known volume. Proceed by *Method 2*.

Sewage and industrials wastes[43,44]

Wet oxidise a sample with nitric and perchloric acids, evaporate the solution to dryness and ignite the residue at 300°C. Dissolve the ash in hydrochloric acid, filter if neccesary and proceed by *Method 2*.

Soil[45-48]

Fuse a 0.1 g sample of soil with potassium hydrogen sulphate ($KHSO_4$), treat the melt with 5 ml of N hydrochloric acid for 30 minutes, filter and dilute the filtrate to 10 ml. Separate nickel by forming the dimethylglyozime complex in alkaline citrate solution, extracting this with benzene-amyl alcohol (C_6H_6–$C_5H_{11}OH$, 4:1) and re-extracting the nickel into 15% caustic potash (KOH) and determine nickel on the aqueous phase by *Method 2*[45]. Alternatively pour 500 ml of N hydrochloric acid onto 100 g of dry soil and stand for 24 hours. Filter the extract through ash-free filter paper, evaporate the filtrate to dryness in a porcelain dish, dissolve the dry residue in 10 – 15 ml of 1:1 nitric acid and transfer to a porcelain crucible Evaporate to dryness and ignite the residue for 30 minutes at dull red heat. Cool, add 10 ml of nitric acid and repeat the process of evaporation and ignition. Treat the residue with 10 ml of hydrochloric acid, transfer the solution to a column of anionite AB-17 resin (Cl$^-$ form). Elute the nickel with 50 ml of 12N hydrochloric acid and apply *Method 2* to the eluate.

Vinogradova[48] recommends that for total nickel the soil be decomposed with a mixture of sulphuric and hydrofluoric acids.

Water[49-52]

Boiler feed water[49] Collect 200 ml of sample in sufficient hydrochloric acid to give a final acid concentration of 0.1N, heat to the boiling point, cool and treat with aqueous hydrogen peroxide (H_2O_2) and potassium sodium tartrate (COOK.CHOH.CHOH.COONa.4H_2O), make the solution alkaline to phenolphthalein and then just acidify, add furil α-aldoxime ((C(:NOH). C:CH.CH:CH.O)$_2$) and aqueous ammonia (NH_4OH) and extract into chloroform ($CHCl_3$). Take the chloroform layer, evaporate to dryness, destroy organic matter by heating with sulphuric acid and hydrogen peroxide and dilute the final solution to a known volume.

General[50] Concentrate the nickel by the "sinking particle method" by adding fine (100 – 150 μ) cation exchange resin particles to the sample and allowing to settle. These particles gradually sink and so traverse the whole volume of the solution removing the nickel in the process. Filter off the resin, elute the nickel and determine by *Method 2*.

Sea water[51,52] Critical evaluation has shown that when concentrating nickel by precipitation with a 5% solution of sodium carbonate (Na_2CO_3) the precipitate must be aged for 7 days before collection and analysis[51]. Alternatively[52] add dimethylglyoxime to a 750 ml sample adjust the *pH* to lie between 9 and 10 and extract the nickel complex into chloroform. Re-extract into normal hydrochloric acid, oxidise with aqueous bromine and proceed by *Method 2*.

Separation from other ions[53-58]

From cobalt[54,57] Either treat the solution with nitroso R salt and then adsorb on a 20×2.5 cm column of Zerolit 225 resin (H$^+$ form), eluting the cobalt with 0.5N hydrochloric acid and then eluting the nickel with 4N hydrochloric acid[53], or, alternatively[54], add malonic acid to the sample solution then add 2 g of sodium nitrite ($NaNO_2$) in 15 ml of water, pass the solution through a column of Amberlite IRA-400 resin (Cl$^-$ form) at 5 ml a minute, wash the column with 60 ml of 1% ammonium chloride solution and determine nickel on this solution.

From cobalt and iron[53,55,58] Pass the sample solution through a column of Dowex 50 resin and elute the iron with a 4:1 mixture of tetrahydrofuran and 3N hydrochloric acid; then elute the cobalt with a 9:1 mixture of tetrahydrofuran and 6N hydrochloric acid and finally elute the nickel with 6N hydrochloric acid[55]. Alternatively[53,58] complex cobalt and nickel with a 6N hydrochloric acid – acetone mixture, adsorb the complexes onto Dowex 1-X8 or 2-X8 resin and elute cobalt with 4N hydrochloric acid and nickel with 9N – 12N hydrochloric acid. Iron does not form a complex and is not adsorbed.

From cobalt, manganese and zinc[56] Make the sample solution 5M in hydrochloric acid and 60% in methyl alcohol (CH_3OH), pass this solution thorugh a 10×1.2 cm column of Amberlite CG400 (Cl$^-$ form) when the other ions are adsorbed but nickel passes through.

The individual tests

Method 1 was specifically developed for the determination of nickel in plating solutions, and depends on the blue colour formed by nickel and aqueous ammonia (NH_4OH) in the presence of ammonium salts.

Method 2 is the popular dimethylglyoxime method and is the standard method for the determination of nickel in sewage[44], and magnesium and its alloys[59]. This method has been examined by Gregorowicz et al[60] who established that 1 ml of a 0.1% ethanolic solution of dimethylglyoxime was the optimum for 0.01 – 0.1 mg of nickel in 50 ml of solution (0.2 – 2.0 ppm): Yamasaki and Matsumoto[61] showed that there are two complexes formed between nickel and dimethylglyoxime with molar ratios of 1:2 and 1:3 respectively. The latter is the more stable and its formation is favoured by the use of excess dimethylglyoxime and by high *pH*. Excess bromine is to be avoided. Iron (<230×nickel), chromium (<3×), manganese (<2×), copper (<2×) and cobalt (<2.5×) do not interfere.

References

1. A. A. Belyakov, *Zavodskaya. Lab.*, 1960, **26**, 158; *Analyt. Abs.*, 1960, **7**, 4510
2. T. Vondenhof and H. Beindorf, *Mschr. Brau*, 1968, **21** (6), 156; *Analyt. Abs.*, 1969, **17**, 3070
3. H. W. Armit and A. Harden, *Proc. Roy. Soc. (London)*, 1906, **77B**, 420
4. S. D. Taktakishvili, *Lab. Delo*, 1964, **10** (3), 153; *Analyt. Abs.*, 1965, **12**, 3416
5. F. W. Sanderman, *Clin. Chem.*, 1967, **13** (2), 115; *Analyt. Abs.*, 1968, **15**, 2783
6. L. E. Gorn and A. D. Miller, *Lab. Delo*, 1966, **12** (3), 163; *Analyt. Abs.*, 1967, **14**, 4147
7. D. Monnier, W. Haerdi and E. Martin, *Helv. chim. Acta*, 1963, **46**, 1042; *Analyt. Abs.*, 1964, **11**, 223
8. M. Karvanek, *Sb. vys. Sk. chem-technol. Praze, Potrav. Technol.*, 1964, **8** (1), 13; *Analyt. Abs.*, 1966, **13**, 1506
9. M. Karvanek, *Sb.vys. Sk. chem-technol., Odd. Fak., Potrav. Technol.*, 1961, **5** (1), 203; *Analyt. Abs.*, 1964, **11**, 2388
10. M. Filajdic, D. Vilicic and N. Ekart, *Kemija Ind.*, 1964, **13** (9), 673; *Analyt. Abs.*, 1966, **13**, 399
11. A. Rudnicki and H. Niewiadomski, *Chemia analit.*, 1968, **13** (4), 755; *Analyt. Abs.*, 1969, **17**, 3803
12. K. Medrzycka and H. Niewiadomski, *Chemia analit.*, 1969, **14** (4), 771; *Analyt. Abs.*, 1970, **19**, 2603
13. A. Liberman, *Analyst*, 1955, **80**, 595
14. P. G. Pifferi and A. Zamorani, *Industrie agrarie*, 1968, **6** (4), 213; *Analyt. Abs.*, 1969, **17**, 1778
15. I. Hoffman, *Analyst*, 1962, **87**, 650
16. I. Buzas, *Magy. Kem. Lap.*, 1968, **23** (12), 716; *Analyt. Abs.*, 1970 **18**, 2537
17. C. Goldberg, *Foundry*, 1956, **84**, 118; *Analyt. Abs.*, 1958, **5**, 56
18. Brit. Standards Inst., B.S. 1728, Part 15, 1966 – "*Methods for the analysis of aluminium and aluminium alloys. Part 15, Nickel (photometric method)*", London, 1966
19. E. M. Goldstein, *Chemist Analyst*, 1954, **43** (2), 42; *Analyt. Abs.*, 1954, **1**, 2424
20. L. V. Kamaeva, V. V. Stepin and M. K. Makarov, *Trudy vses. nauchno-issled. Inst. standart. Obraztsov*, 1967, **3**, 87; *Analyt. Abs.*, 1969, **16**, 3030
21. S. Barabas and W. C. Cooper, *Metallurgia, Manchr.*, 1957, **56**, 101; *Analyt. Abs.*, 1958, **5**, 790
22. I. Kalinichenko and A. A. Kinyazeva, *Izv. Vyssh. Ucheb. Zavedenii, Khim. i Khim. Tekhnol.*, 1960, **3**, 418; *Analyt. Abs.*, 1961, **8**, 4665
23. C. M. Dozniel, *Chim. Anal.*, 1962, **44**, 436; *Analyt. Abs.*, 1963, **10**, 2604
24. W. Kemula and K. Brajter, *Chemia analit.*, 1966, **11** (2), 373; *Analyt. Abs.*, 1967, **14**, 3817
25. M. N. Kruglova, L. S. Studentskaya and V. V. Stepin, *Trudy vses. nauchno-issled. Inst. standart-Obraztsov.*, 1968, **4**, 146; *Analyt. Abs.*, 1970, **18**, 779
26. T. Imai and S. Nagumo, *J. Chem. Soc. Japan, Ind. Chem. Sect.*, 1957, **60**, 544; *Analyt. Abs.*, 1958, **5**, 1528
27. M. Simek, *Hutn. Listy*, 1964, **19** (9), 663; *Analyt. Abs.*, 1966, **13**, 173
28. R. Goetz, *Giesserei.*, 1965, **52** (1), 12; *Analyt. Abs.*, 1966, **13**, 2958
29. A. T. Pilipenko and N. N. Maslie, *Ukr. Khim. Zh.*, 1968, **34** (2), 174; *Analyt. Abs.*, 1969, **16**, 3025
30. G. Volke, *Giessereitechnik.*, 1968, **14** (6), 186; *Analyt. Abs.*, 1969, **17**, 1422
31. M. Steffek, *Chem. Listy*, 1963, **57** (9), 972; *Analyt. Abs.*, 1965, **12**, 129
32. V. A. Obolonchik and K. D. Modylevskaya, *Zavodskaya, Lab.*, 1957, **23**, 912; *Analyt. Abs.*, 1958, **5**, 2099

33. R. Puschel and E. Lassner, *Mikrochim. ichnoanalyt. Acta,* 1965, (1), 17; *Analyt. Abs.,* 1966, **13,** 2336
34. L. E. Hibbs and D. H. Wilkins, *Talanta,* 1959, **2,** 16; *Analyt. Abs.,* 1959, **6,** 4672
35. R. D. Gardner, C. H. Ward and W. H. Ashley, *Rep. atom. Energy Commn. U.S.L.A.* 3152, 1964; *Analyt. Abs.,* 1966, **13,** 124
36. G. Norwitz and H. Gordon, *Analyt. Chem.,* 1965, **37** (3), 417; *Analyt. Abs.,* 1966, **13,** 3550
37. K. Isono, *Japan Analyst,* 1957, **6,** 557; *Analyt. Abs.,* 1958, **5,** 2653
38. K. S. Pakhomova, L. P. Volkova and V. V. Gorshkov, *Zh.analit.Khim.,* 1964, **19** (9), 1085; *Analyt. Abs.,* 1966, **13,** 175
39. L. Doubek, *Hutn.List.,* 1957, **12,** 430; *Analyt. Abs.,* 1958, **5,** 112
40. E. M. Dodson, *Analyt.Chem.,* 1962, **34,** 966
41. K. L. Cheng and B. L. Goydish, *Microchem.J.,* 1963, **7,** 166; *Analyt. Abs.,* 1964, **11,** 4291
42. M. Karnakev, *Prumysl.Potravin,* 1964, **15** (6), 282; *Analyt. Abs.,* 1965, **12,** 5472
43. A. A. Christie, J. R. W. Kerr, G. Knowles and G. F. Lowden, *Analyst,* 1957, **82,** 336
44. Min. of Housing and Local Govt., "*Methods of Chemical Analysis as applied to Sewage and Sewage Effluents,*" H.M. Stationery Office, London, 1956
45. A. N. Chowdhury and B. Das Sarma, *Analyt.Chem.,* 1960, **32,** 820
46. I. K. Tsitovich, *Zhur.analit.Khim.,* 1962, **17,** 621; *Analyt. Abs.,* 1963, **10,** 3481
47. N. Mizuno and K. Hayashi, *Japan Analyst,* 1967, **16** (1), 38; *Analyt. Abs.,* 1968, **15,** 7012
48. E. N. Vinogradova, G. V. Prokhorova and T. N. Sevast'yanova, *Vest.mosk.gos, Univ.Ser.Khim.,* 1968, (5), 74; *Analyt. Abs.,* 1970, **18,** 1275
49. A. L. Wilson, *Analyst,* 1968, **93,** 83
50. V. B. Aleskovskii, A. D. Miller and E. A. Sergeev, *Trudy Komiss. Anal.Khim.Akad. Nauk, SSSR.,* 1958, **8,** 217; *Analyt. Abs.,* 1959, **6,** 2369
51. W. Forster and H. Zeitlin, *Analyt.chim.Acta,* 1966, **35** (1), 42; *Analyt. Abs.,* 1967, **14,** 5075
52. E. Kentner, D. B. Armitage and H. Zeitlin, *Analyt.chim.Acta,* 1969, **45** (2), 343; *Analyt. Abs.,* 1970, **19,** 823
53. I. Hazan and J. Korkisch, *Analyt.chim.Acta,* 1965, **32** (1), 46; *Analyt. Abs.,* 1966, **13,** 2370
54. F. Burriel-Marti and C. Alvarez-Herrero, *Infcion Quim.analit.pura apl.Ind.,* 1966, **20** (1), 1; *Analyt. Abs.,* 1967, **14,** 2564
55. J. Korkisch, *Nature,* 1966, **210,** 626; *Analyt. Abs.,* 1967, **14,** 5881
56. L. Mazza, R. Frache, A. Dadone and R. Ravasio, *Gazz.chim.ital.,* 1967, **97,** 1551; *Analyt. Abs.,* 1968, **15,** 5224
57. O. G. Balakrishnan Nambiar and P. R. Subbaraman, *Talanta,* 1967, **14** (7), 785; *Analyt. Abs.,* 1968, **15,** 6004
58. A. K. Lavrukhina and N. K. Sazhina, *Zh.analit.Khim.,* 1969, **24** (6), 670; *Analyt. Abs.,* 1970, **19,** 4877
59. Brit. Standards Inst., B.S. 3907, Part 7, London, 1969
60. Z. Gregorowicz, S. Grochowski and J. Kubala, *Chem.Anal. Warsaw,* 1957, **2,** 1322; *Analyt. Abs.,* 1958, **5,** 1230
61. K. Yamasaki and C. Matsumoto, *J.Chem.Soc. Japan, Pure Chem.Sect.,* 1957, **78,** 833; *Analyt. Abs.,* 1958, **5,** 1231

COLORIMETRIC CHEMICAL ANALYTICAL METHODS

Nickel Method 1
using ammonium citrate

Principle of the method

In the presence of ammonium salts, ammonium hydroxide forms a blue complex with nickel ions.

Reagent required

Dissolve 10 g of citric acid ($C_6H_8O_7.H_2O$) in 60 ml of 10% w/w aqueous ammonia (NH_4OH) and dilute to 100 ml with distilled water.

The Standard Lovibond Comparator Disc 3/14

The glasses in the disc are numbered 1 to 9, each colour corresponding to the amounts of nickel shown in the table.

Technique

Measure carefully the specified quantity (see table below) of the plating solution into one of the test-tubes, add 5 ml of the reagent, fill the test-tube to the 10 ml mark with water, mix well and place in the right-hand compartment of the comparator. In the left-hand compartment place a test-tube full of water. Hold the comparator facing a standard source of white light, such as the Lovibond White Light Cabinet or, failing this, north daylight, and compare the colour produced in the solution with the colours in the disc, rotating the latter until a colour match is obtained. If the colour produced in the solution is weaker than the No. 1 standard glass or stronger than the No. 9 standard glass the test should be repeated using a larger or smaller quantity of the plating solution (see table).

Note

By carrying out the test on different quantities of the plating solutions, the standard disc may be used to determine the strengths of nickel ammonium sulphate solutions (nickel double salt) or of the much stronger nickel sulphate solutions (nickel single salt) normally used in electroplating. The accompanying table shows the quantities of the plating solutions which should be taken for the test according to the type of solution in use. Thus, if the plating solution contains nickel ammonium sulphate and 4 ml be taken for the test, the range of the disc extends from 4 ounces to 12 ounces per gallon. If 3 ml be taken for the test, the range of the disc is from 5⅓ ounces to 16 ounces per gallon. For the much stronger nickel sulphate solutions, correspondingly smaller quantities must be taken. Thus, the disc will cover the range from 16 to 48 ounces per gallon of crystallised nickel sulphate when 0.7 ml of the plating solution is taken for the test.

If it is desired to know the strength of the solution in terms of metal nickel, then the quantities in the third section of the table should be taken. By the use of this table all calculations are avoided, and the content of the solutions may be obtained directly from a single observation.

Table.

Amount of the Plating Solution taken for test		No. 1	No. 2	No. 3	No. 4	No. 5	No. 6	No. 7	No. 8	No. 9
					Colour Standards					
3 ml	Nickel Ammonium Sulphate	$5\frac{1}{3}$	$6\frac{2}{3}$	8	$9\frac{1}{3}$	$10\frac{2}{3}$	12	$13\frac{1}{3}$	$14\frac{2}{3}$	16
4 ml		4	5	6	7	8	9	10	11	12
0.7 ml	Nickel Sulphate	16	20	24	28	32	36	40	44	48
1.4 ml		8	10	12	14	16	18	20	22	24
2.1 ml		$5\frac{1}{3}$	$6\frac{2}{3}$	8	$9\frac{1}{3}$	$10\frac{2}{3}$	12	$13\frac{1}{3}$	$14\frac{2}{3}$	16
2.8 ml		4	5	6	7	8	9	10	11	12
0.6 ml	Metal Nickel	4	5	6	7	8	9	10	11	12
1.2 ml		2	$2\frac{1}{2}$	3	$3\frac{1}{2}$	4	$4\frac{1}{2}$	5	$5\frac{1}{2}$	6
1.8 ml		$1\frac{1}{3}$	$1\frac{2}{3}$	2	$2\frac{1}{3}$	$2\frac{2}{3}$	3	$3\frac{1}{3}$	$3\frac{2}{3}$	4
2.4 ml		1	$1\frac{1}{4}$	$1\frac{1}{2}$	$1\frac{3}{4}$	2	$2\frac{1}{4}$	$2\frac{1}{2}$	$2\frac{3}{4}$	3

The figures in the chart represent ounces per gallon. To convert to grams per litre multiply by $6\frac{1}{4}$.

Nickel Method 2
using dimethyl-glyoxime

Principle of the method

Nickel is often estimated gravimetrically by precipitation of the nickelous complex formed with dimethyl glyoxime. In the presence of oxidising agents however this divalent complex is converted into a soluble red tetravalent complex and this modified procedure serves as a very sensitive method for the estimation of traces of nickel.

Interference from small amounts of iron is prevented by the use of ammonium citrate, Appreciable quantities of other metals, including cobalt, copper and manganese, interfere and special procedures are necessary in their presence.

Reagents required

1. *Ammonium citrate* $((NH_4)_3C_6H_5O_7)$ Solution 10% w/v
2. *Bromine water* Saturated
3. *Ammonium hydroxide* (NH_4OH) 10% w/w ammonia
4. *Dimethyl-glyoxime* 1% solution in Industrial Methylated Spirits

The Standard Lovibond Comparator Disc 3/36

The colours in the disc correspond to 1, 2, 3, 4, 5, 6, 7, 8 and 10 ppm of nickel (Ni) using 10 ml of the solution under test.

Technique

The sample solution should be rendered neutral with ammonia.
To 10 ml of this solution add, in the following order:—

    5 ml ammonium citrate solution (reagent 1)
    2 ml bromine water (reagent 2)
    2 ml ammonia solution (reagent 3)
    1 ml dimethyl glyoxime solution (reagent 4)

Mix thoroughly after each addition. Transfer to a Lovibond test tube or 13.5 mm comparator cell and place in the right-hand compartment of the Lovibond Comparator. In the left-hand compartment place a blank of the reagents which has been prepared in an exactly similar manner.

Hold the comparator facing a standard source of white light, such as the Lovibond White Light Cabinet or, failing this, north daylight, and compare the colour of the solution with the colours in the standard disc, rotating the latter until the nearest match is obtained.

The figure read in the indicator window represents parts per million of nickel in 10 ml of the neutral solution.

Notes

1. Interference can be caused either by the precipitation of a metal as the hydroxide or by complex formation between the interfering ion and the reagent.

In the former case separation can be effected by double precipitation or in certain cases by a special technique such as electrolysis (see Introduction).

When complex formation takes place, advantage can be taken of the fact that nickelous dimethyl glyoxime is soluble in chloroform whilst complexes formed with the majority of other metals are insoluble. Copper dimethyl glyoxime is soluble in chloroform, but unlike the nickel complex this can be decomposed with dilute ammonia. The nickel can then be back-extracted from the chloroform by treatment with dilute hydrochloric acid. This procedure is suitable for the estimation of nickel in cobalt salts.

2. It should be noted that in the presence of oxidising agents (e.g. manganese) a trace of hydroxylamine sulphate may be necessary to prevent the formation of the tetravalent nickel complex which is **insoluble** in chloroform.

The Determination of Nitrate

Introduction

The determination of the nitrate ion (NO_3^-) is of importance in agricultural chemistry for the testing of soils, fertilizers and animal feedstuffs; in the food industry for the control of pickling solutions and the examination of meat products; in the examination of both potable and boiler-feed waters and in the control of industrial effluents.

Methods of sample preparation

Biological materials[1-5]

Blood, plasma or serum[1,2]   Take 1 ml of serum, add 8 ml of distilled water and precipitate the protein with 1 ml of 5% mercuric chloride ($HgCl_2$) solution containing 1.5 ppm of copper. Shake for 10 minutes and then centrifuge. Determine the nitrate on the supernatant liquid[1]. Alternatively[2] precipitate protein with tungstic acid (H_2WO_4) and then precipitate chlorides and some of the remaining organic matter with a mixture of copper sulphate ($CuSO_4.5H_2O$) and silver sulphate (Ag_2SO_4) solutions. Precipitate excess copper and silver with calcium hydroxide ($Ca(OH)_2$) and magnesium carbonate ($MgCO_3$) and filter. Determine nitrate on the filtrate using *Method 2*. Methods 1 and 5 are stated[2] to be unsatisfactory for this application.

Stomach contents[3]   Boil sample with water, filter, percolate the filtrate through a column of alumina and determine nitrate on the percolate.

Tissues[4]   Suspend 10 g of cominuted sample in 40 ml of water, add 2 ml of $5N$ sulphuric acid to eliminate nitrite and pass a current of nitrogen through the liquid for 10 minutes. Precipitate proteins by adding 9 ml of 20% tungstophosphoric acid ($H_3PO_4.12WO_3.xH_2O$), dilute to 100 ml and filter. Determine nitrate on the filtrate by *Method 2*.

Urine[2,5]   Either proceed by the second method described above for blood[2] or[5], dilute 1 ml of sample to 10 ml with a saturated aqueous solution of silver sulphate (Ag_2SO_4), immerse the flask in warm water for 10 minutes, filter and determine nitrate on this filtrate by *Method 2*. This is used as a test for exposure to diethylene glycol dinitrate.

Chemicals[6]

Propellants   Extract 0.5 g of sample overnight with acetone (CH_3COCH_3) at room temperature. Filter through asbestos and wash the residue with acetone. Dissolve the residue in water and determine nitrate on this solution.

Dairy products[2,7]

Milk and milk products   Either proceed by the second method described above for blood[2] or[7], dilute 40 ml of liquid milk, or 4 g of dried milk, to 150 ml, warm to 50°C, clarify by the addition of 10 ml of 12% zinc sulphate ($ZnSO_4$) solution and 10 ml of $0.5N$ caustic soda (NaOH), maintain at 50°C for 10 minutes, cool, dilute to 200 ml, filter and determine nitrate by *Method 4*.

Foods and edible oils[8-13]

Baby foods[8]   Extract by shaking sample with aqueous ammonia (NH_4OH)—hydrochloric acid buffer solution at pH 9.6, alumina cream (prepared from aluminium sulphate ($Al_2(SO_4)_3$), potassium sulphate (K_2SO_4) and aqueous ammonia (NH_4OH)) and water. Filter and determine nitrate on the filtrate.

Brines and pickle solution[9]   Brines may be examined directly by *Method 2*, but Landmann[9] recommends that the sample be diluted 1:10 with water before analysis.

Dry cure mix[9]   Dissolve 1 g of the mixture in 100 ml of distilled water and dilute 10 ml of this solution to 100 ml. Proceed by *Method 3*.

Flour meal[10]   Mix a sample, containing not more than 3 mg of nitrate with 20 ml of water. Dilute to 50 ml and measure 20 ml of this dilution into a 50 ml flask. Remove nitrite by bubbling carbon dioxide through the mixture for 10 minutes, after the addition of 1 ml of methanol (CH_3OH) and 1 ml of $5N$ sulphuric acid. Add 7–8 ml of $2N$ caustic soda (NaOH) solution (only 2–3 ml are required in absence of the nitrite treatment) and precipitate protein with 10 ml of 20% trichloracetic acid (Cl_3CCOOH). Filter and determine nitrate on the filtrate.

Meat and meat products[9,11-13] Transfer 1 g of the minced sample of meat product to a 100 ml volumetric flask containing 20 ml of water, immerse the flask in boiling water for 15 minutes and periodically agitate the contents. Allow the mixture to cool, render just acid with dilute sulphuric acid, using bromocresol green as an indicator, and oxidise any nitrite by adding $0.2N$ potassium permanganate ($KMnO_4$) solution drop by drop until a faint pink colour persists for 1 minute. Add successively, shaking after each addition, sufficient of a saturated aqueous solution of silver sulphate (Ag_2SO_4) to precipitate any chloride which is present, 5 ml of 25% basic lead acetate (($CH_3COO)_2Pb.Pb(OH)_2$) and 5 ml of alumina cream. Dilute with water to 100 ml, shake and filter. Mix an appropriate proportion of the filtrate with 3 times its volume of 85% sulphuric acid, adjust the temperature of the mixture to 35°C, add 1 ml of a 1% solution of 2,4–xylen–1–ol in glacial acetic acid (CH_3COOH) and proceed as in *Method 2*[9,11,12].

Stoya[13] recommends that the nitrate be reduced to nitrite by mixing 10 ml of the deproteinated and clarified sample solution with 2 ml of 25% aqueous ammonia (NH_4OH), 4 ml of 10% cadmium sulphate solution and 0.5 g of zinc powder. The solution is then filtered and the nitrate concentration determined by the Griess Ilosvay method (Nitrite *Methods 1 and 2*).

Metals, alloys and plating baths[14]

Chromium plating baths Take 1 ml sample, precipitate chromate with 5 ml of $2N$ aqueous ammonia (NH_4OH) and 4 ml of 15% barium chloride ($BaCl_2$) solution. Centrifuge, decant the supernatant liquid and determine nitrate.

Minerals, ores and rocks[15]

Dissolve the sample in sulphuric acid, make the solution alkaline with caustic soda (NaOH) and distil out any ammonia which is present. Reduce nitrate to ammonia using Devarda's alloy (45% aluminium, 50% copper, 5% zinc) and caustic soda. Distil out the ammonia and determine by Nessler reagent (*Method 5* or Ammonia *Method 1*).

Miscellaneous[3,16,17]

Animal feeding stuffs[3] Boil sample with water, filter, percolate the filtrate through a column of alumina and determine nitrate on the filtrate.

Fertilizers[16,17] Weigh out sufficient sample to contain not more than 0.06 g of nitrate nitrogen and place this in a 650 ml Kjeldahl flask. Add 25 ml of water and 50 ml of normal chromous solution (chromous chloride—$CrCl_2$—stored under nitrogen), swirl to mix and then set aside for at least a minute. Add sulphuric acid (30 ml for 30 minutes digestion, 35 ml for 60 minutes and 40 ml for 120 minutes), 0.7 g of mercuric oxide (HgO), 15 g of potassium sulphate (K_2SO_4) and 2–3 g of anti-bumping material, and then swirl to oxidise any residual chromous solution. Digest for the appropriate period, cool, add 175 ml of distilled water, distil off the ammonia and determine by *Method 5* (or Ammonia *Method 1*).[16] Alternatively[17] use a powdered Raney catalyst, containing equal amounts of aluminium and nickel, in the sulphuric acid solution of the sample to effect reduction of nitrate to ammonia. As the nickel provides a catalytic surface for the reduction, relatively little of the alloy is required and the resulting salts do not interfere with the Kjeldahl process.

Plants[18-21]

Macerate the sample with water, filter and determine nitrate on the filtrate [18-20]. Alternatively[21] shake 125 mg of dried and powdered sample with 50 mg of Dowex 50W–X8 resin (H^+ form) and 25 ml of water for 15 minutes. Filter, transfer the filtrate to a column of Dowex 1–X8 resin (Cl^- form), discard the percolate, wash the column with successive portions of water, $0.01N$ hydrochloric acid, and a further portion of water and then elute the nitrate with two 5 ml portions of normal sodium chloride (NaCl) followed by 9 ml and 5 ml portions of water, make up to 25 ml and determine nitrate on an aliquot.

Sewage and industrial wastes[22-26]

Organic matter, nitrite and chloride ions interfere with the determination of nitrate and must be removed from the sample. Organic matter can be removed in the following way. Take 50 ml sample and add a few drops of caustic soda (NaOH) solution to bring the *p*H above 8.5. Shake with 0.5 g of active carbon, filter, discard the first portion of the filtrate.

The active carbon used should be tested on solutions of known nitrate content to ensure that it does not absorb nitrate. Nitrite may be removed by the addition of 3 ml of a 0.1% solution of sodium azide (NaN_3) to the sample before the initial evaporation in *Method 1*. Alternatively, Bock et al[22] recommend that the sample is shaken with a solution of triphenyl tin ($SnPh_3$) in benzene (C_6H_6). This extracts 97% of the nitrite ions and less than 0.2% of the nitrate ions. Chloride may be removed by adding 1 ml of glacial acetic acid (CH_3COOH) to 10 ml of the sample, followed by 0.1 g of silver sulphate (Ag_2SO_4 nitrate free) and after shaking, filter through Whatman No. 32 filter paper. Any colour may be removed in the following manner. To a suitable volume of sample add 2 ml of a 0.5% suspension of chalk ($CaCO_3$) in water and evaporate to a volume of 10 ml. Add 1 ml of 30% hydrogen peroxide (H_2O_2). Cover the vessel with a watch-glass and digest on a steam-bath for 2 hours. Remove the watchglass and evaporate to dryness. Heat the dry residue for an additional 30 minutes to decompose peroxides. Proceed as in *Method 1*. N.B. This treatment will oxidise any nitrite to nitrate and give a result equal to the combined concentrations of these two ions.

Edwards et al[23] claim that reduction of nitrate to nitrite by means of zinc and estimation of nitrite using the Ilosvay procedure (Nitrite *Method 1*) is a more satisfactory technique for use with industrial waters and wastes.

Jenkins and Medsker report[24] that *Method 3* is unsuitable for use with raw sewage.

The official method[25] is *Method 1*, but Ormerod[26] prefers a modification of *Method 2*, which he claims is particularly useful for coloured or turbid effluents. The modified procedure treats 2 ml of sample, containing nor more than 20 μg of nitrate, with 0.5 ml of a 0.1% solution of sulphamic acid (NH_2SO_3H) and then with 2 ml of a 2% solution of mercuric sulphate ($HgSO_4$) in 20% sulphuric acid (to prevent interference by chloride ions). Then add 15 ml of 85% sulphuric acid and continue as described in *Method 2*.

Soils[19-21,27-30]

Shake 50 g of dry pulverised soil (or 25 g of peat) with 250 ml of water containing 5 ml of N copper sulphate ($CuSO_4.5H_2O$) solution for 10 minutes. Add 0.4 g of calcium hydroxide ($Ca(OH)_2$) and 1 g of magnesium carbonate ($MgCO_3$), shake for 5 minutes to precipitate copper and iron and filter through a coarse filter paper. Discard the first 20 ml of the filtrate. Re-filter if necessary through the same paper and determine nitrate on the filtrate using *Method 1 or 4*[27].

Some workers[19,20] recommend the reduction of nitrate to ammonia with Devarda's alloy and determination of the ammonia by Nesslerization.

Alternative solutions which have been recommended for the extraction of nitrate from soil are 0.2% aqueous calcium sulphate[21]; 0.02N copper sulphate containing 0.01% sulphamic acid (NH_2SO_3H) and made 0.1N in sulphuric acid[28]; 5 g of silver sulphate (Ag_2SO_4), 71 g of sodium sulphate (Na_2SO_4), 8.3 g of copper sulphate ($CuSO_4.5H_2O$) and 5.6 ml of sulphuric acid per litre[29]; and 2N potassium chloride (KCl)[30].

Water[22-24,31-39]

Interference by organic matter, nitrites and chlorides can be prevented by the methods already described above for the analysis of sewage.

For salt water *Method 1 or 4* is recommended by Montgomery and Dymock[33], whereas Jenkins and Medser[24] prefer *Method 3*. Fadda and Alamanni[32] claim that *Method 2* is preferable to *Methods 1, 3 and 4* for general use. Harrison[35] recommends reduction of nitrate and nitrite to ammonia by boiling with chromous chloride ($CrCl_2$) solution (*Method 5*), removing the ammonia by steam distillation and determining ammonia by Nessler reagent (see ammonia methods). Nitrite is determined separately by the Ilosvay method (Nitrite *Method 1*) and deducted from the total to give the nitrate. Several other authors recommend the reduction of nitrate to nitrite as the best means of determining nitrate in sea water. Wood[36] advocates treating the sample with E.D.T.A. and then passing it through a column of "copperised cadmium" (cadmium filings shaken with copper sulphate solution and then kept in an inert atmosphere). Reduction of nitrate to nitrite is claimed to be quantitative. Isaeva[37] recommends adding 2 ml of ammoniacal buffer solution (250 g of NH_4Cl and 100 ml of 6% NH_4OH in

1 litre) to 100 ml of sample and then passing 70—75 ml of this solution through a Jones reductor, containing cadmium amalgam, at 2 ml a minute. Discard the first 20 ml of percolate and determine nitrite on an aliquot of the remainder. Matsunaga[38] carries out the reduction at pH 3 using zinc powder in the presence of ammonium chloride (NH_4Cl).

Fadrus[39] has examined the influence of acidity and temperature on the use of *Method 3* for the determination of nitrate in water. He claims that in the presence of nitrite pre-treatment of the sample with urea allows up to 200 mg of nitrite nitrogen per litre of water to be tolerated.

The individual tests

Method 1 is the phenol disulphonic acid method of Sprengel[40] and is one of the standard methods for sewage and effluents[25].

Method 2 is the 2,4–xylen–1–ol method suggested by Blom and Treschow[41] and is the standard method for water[31].

Method 3 is the Haase[42] brucine method and is widely used for the examination of boiler and river water.

Method 4 is a modification of *Method 1* which has been especially developed[43] for soil analysis in the field. It is less susceptible to nitrite interference than *Method 1*.

Method 5 uses the reduction of nitrate to ammonia followed by estimation of ammonia with Nessler reagent. It was specifically developed[35] to provide a simple test for water control in the brewing industry.

References

1. R. H. Diven, W. J. Pistor, R. E. Reed, R. J. Trautman and E. R. Watts, *Amer.J.Vet. Res.*, 1962, **23**, 497; *Analyt.Abs.*, 1963, **10**, 4292
2. T. Greweling, K. L. Davison and C. J. Morris, *J.Ag.Food Chem.*, 1964, **12** (2), 139; *Analyt.Abs.*, 1965, **12**, 3410
3. E. Mehnert, *Arch.exp.Vet.Med.*, 1967, **21** (5), 1191; *Analyt.Abs.*, 1969, **16**, 401
4. J. Davidek, S. Klein and A. Zackova, *Z.Lebensmitt.Untersuch.*, 1963, **119**, 342; *Analyt.Abs.*, 1964, **11**, 1484
5. V. Vasak, *Pracovni Lek.*, 1965, **17** (2), 47; *Analyt.Abs.*, 1966, **13**, 5004
6. G. Norwitz, *Analyt.Chem.*, 1962, **34**, 227
7. P. B. Manning, S. T. Coulter and R. Jenness, *J.Dairy Sci.*, 1968, **51** (11), 1725; *Analyt.Abs.*, 1970, **18**, 1987
8. L. Kamm, G. G. McKeown and D. M. Smith, *J.Assoc.off.agric.Chem.*, 1965, **48** (5), 892; *Analyt.Abs.*, 1967, **14**, 1014
9. W. A. Landmann, M. Saeed, Katherine Pik and D. M. Doty, *J.Assoc.off.agric.Chem.*, 1960, **43** (3), 531
10. S. Klein, *Prumysl.Potravin.*, 1958, **9**, 110; *Analyt.Abs.*, 1958, **5**, 3893
11. W. C. McVey, *J.Assoc.off.agric.Chem.*, 1935, **18**, 459
12. R. Grau and A. Mirna, *Z.analyt.Chem.*, 1957, **158**, 182; *Analyt.Abs.*, 1958, **5**, 2390
13. W. Stoya, *Dt.Lebensmitt.Rdsch.*, 1969, **65** (5), 144; *Analyt.Abs.*, 1970, **19**, 1724
14. I. R. Klyachko and I. S. Schipkova, *Zavodskaya Lab.*, 1961, **27**, 145; *Analyt.Abs.*, 1961, **8**, 3762
15. U.K.A.E.A. Report PG 225 (S), 1962
16. E. L. Nelson, *J.Assoc.off.agric.Chem.*, 1960, **43** (3), 468
17. J. A. Brabson and W. G. Burch, *J.Assoc.off.agric.Chem.*, 1964, **47** (6), 1035; *Analyt.Abs.*, 1966, **13**, 2065
18. M. P. Morris and A. Gonzalez-Mas, *J.Agric. Food Chem.*, 1958, **6**, 456; *Analyt.Abs.*, 1959, **6**, 1558
19. R. P. Murphy, *J.Sci. Food Agric.*, 1957, **8**, 231; *Analyt.Abs.*, 1957, **4**, 3489
20. R. F. Milton and W. D. Duffield, *Lab Practice*, 1954, **3**, 318; *Analyt.Abs.*, 1954, **1**, 2738

21. A. S. Baker, *J.Agric. Food Chem.*, 1967, **15** (5), 802; *Analyt.Abs.*, 1968, **15**, 7633
22. R. Bock, H-T. Hiederauer and K. Behrends, *Z.analyt.Chem.*, 1962, **190**, 33; *Analyt.Abs.*, 1963, **10**, 1055
23. G. P. Edwards, J. R. Pfafflin, L. H. Schwartz and P. M. Lauren, *J.Wat.Pollut.Control Fed.*, 1962, **34**, 1112; *Analyt.Abs.*, 1963, **10**, 2048
24. D. Jenkins and L. L. Medsker, *Analyt.Chem.*, 1964, **36**, 610
25. Min. of Housing and Local Govt., "*Methods of Chemical Analysis as applied to Sewage and Sewage Effluents*", H.M. Stationery Office, London, 1956
26. N. S. Ormerod, *J.Proc.Inst.Sewage Purif.*, 1966, (5), 471; *Analyt.Abs.*, 1967, **14**, 5798
27. H. J. Harper, *Ind.Eng.Chem.*, 1924, **16**, 180
28. A. L. Clarke and A. C. Jennings, *J.Agric.Food Chem.*, 1965, **13** (2), 174; *Analyt.Abs.*, 1966, **13**, 5911
29. R. H. Laby and T. C. Morton, *Nature*, 1966, **210**, 298; *Analyt.Abs.*, 1967, **14**, 5080
30. J. Keay and P. M. A. Menage, *Analyst*, 1970, **95**, 379
31. Brit. Standards Inst., "*Methods of Testing Water Used in Industry*" B.S. 2690, London, 1964
32. M. Fadda and U. Alamanni, *Igiene Mod.*, 1961, **54**, 232; *Analyt.Abs.*, 1962, **9**, 5473
33. H. A. C. Montgomery and J. F. Dymock, *Analyst*, 1962, **87**, 374
34. A. M. Hartley and R. I. Asai, *J.Amer.Wat.Wks.Assoc.*, 1960, **52**, 255; *Analyt.Abs.*, 1960, **7**, 4516
35. G. A. F. Harrison, *Talanta*, 1962, **9**, 533
36. E. D. Wood, F. A. J. Armstrong and F. A. Richards, *J.mar.biol.Assoc. U.K.*, 1967, **47**, 23; *Analyt.Abs.*, 1968, **15**, 7628
37. A. B. Isaeva and A. N. Bogoyavlenskii, *Okeanologiya*, 1968, **8** (3), 539; *Analyt.Abs.*, 1969, **17**, 1210
38. K. Matsunaga and M. Nishimura, *Analyt.chim.Acta*, 1969, **45** (2), 350; *Analyt.Abs.*, 1970, **19**, 822
39. H. Fadrus and J. Maly, *Z.analyt.Chem.*, 1964, **202** (3), 164; *Analyt.Abs.*, 1965, **12**, 4259
40. H. Sprengel, *Pogg.Annalen*, 1864, **121**, 188
41. J. Blom and C. Treschow, *Z.Pflan.Dungung Bodenk*, 1929, **13a**, 159
42. L. W. Haase, *Chem.Ztg.*, 1926, **50**, 372
43. J. E. Eastoe and A. G. Pollard, *J.Sci. Food Agric.*, 1950, **1**, 266

Nitrate Method 1
using phenol-2:4-disulphonic acid

Principle of the method

The test is based on the colour produced by the nitration of phenol-2:4-disulphonic acid.

Reagents required

1. *Phenol-disulphonic acid solution*   Approximately 25% w/v in concentrated sulphuric acid, made as follows:—

   Phenol (C_6H_5OH) (analytical reagent grade) .. .. 25 g
   Concentrated sulphuric acid (H_2SO_4 nitrogen free) .. ..158 ml

   Dissolve and add
   Fuming sulphuric acid, containing about 20% sulphur trioxide .. .. .. 67 ml

   Heat the mixture on a boiling water-bath for two hours.

2. *Ammonia solution* (NH_4OH)   Analytical reagent grade containing 10% w/v ammonia.

The Standard Lovibond Discs, Comparator Disc 3/17 and Nessleriser Disc NHP

Disc 3/17 cover the range 5, 10, 20, 30, 40, 50, 60, 80, 100 µg of nitrate nitrogen (N). This equals 1.0 to 20.0 parts per million if a 5 ml sample is taken.

Disc NHP covers the range 2.5, 5, 7.5, 10, 12.5, 15, 20, 25, 30 µg of nitrate nitrogen. This equals 0.5 to 6 ppm on a 5 ml sample.

To convert to nitrate ion (NO_3^-) multiply answer by 4.428.

Technique

Place a suitable quantity of the solution under examination in a small porcelain dish and evaporate to dryness on a boiling water-bath.

To the cooled residue add 1 ml of the phenol-disulphonic acid solution (reagent 1), taking care that the reagent makes contact with the whole of the solid material derived from the sample, and allow to stand for 10 minutes. Then add 10 ml of water, cool the mixture, add 10 ml of ammonia solution (reagent 2), again cool, and dilute with water to 25 ml. At the same time, prepare a "blank" solution containing the same quantities of the reagents, omitting the sample under test. Fill one of the Comparator test tubes with the test solution and place in the right-hand compartment of the Comparator. Fill the other test tube with the "blank" solution and place in the left-hand compartment of the Comparator.

Hold the Comparator facing a standard source of white light such as the Lovibond White Light Cabinet or, failing this, north daylight, and compare the colour produced in the test solution with the colours in the standard disc, rotating the latter until a match is obtained. The matching of the colours should be carried out at 20°—25°C. The markings on the disc represent the nitrate nitrogen in the amount of sample taken for the test (not the amount in the comparator test tube). Thus, if a colour equivalent to 30 µg is produced in a test carried out on 5 ml of a sample of water, then the amount of nitrogen present as nitrate is 0.03 mg, which corresponds to 6 parts per million of water.

For use with the Lovibond Nessleriser the procedure is identical except that the final volume is made up to 50 ml and the measurements are carried out in standard Nessleriser tubes.

Notes

1. Nitrites in excess of 1 part per million will interfere.

2. In examining water supplies it will generally be necessary to remove chlorides, since interference from this source may be expected if the chloride content exceeds 2 parts per million. Chlorides may be removed by adding 1 ml of glacial acetic acid to 10 ml of the sample, followed by 0.1 gram of solid silver sulphate (nitrate free) and, after shaking, filtering through a Whatman No. 32 paper. The test for nitrate may then be applied to 5 ml or other suitable quantity of the filtrate. The acetic acid is employed in order to prevent the adsorption of nitrate ions by the silver chloride precipitate.

Alternatively, a greater tolerance to the presence of chlorides is obtainable by a modification of this technique, described by Eastoe and Pollard (*Method 4*).

Nitrate Method 2
using 2:4-xylen-1-ol

Principle of the method

The test depends upon the formation of 5-nitro-2:4-xylen-1-ol which is volatile in steam and can therefore be distilled into dilute sodium hydroxide with which it forms a highly coloured salt.

Reagents required

1. *Sodium hydroxide (NaOH)* Solution $2N$
2. *Sulphuric acid (H_2SO_4)* 85% w/w (nitrogen free)
3. *2:4-Xylen-1-ol solution* 1% w/v in glacial acetic acid

The Standard Lovibond Nessleriser Disc NH

The disc covers the range 2, 4, 6, 8, 10, 12, 15, 18, 25 µg (0.002 to 0.025 mg) of nitrate nitrogen (N), equivalent to 1.0 to 12.5 ppm on a 2 ml sample.

To convert to nitrate ion (NO_3) multiply answer by 4.428.

Technique

Add 2 ml of the sample to 4 ml of distilled water and 15 ml of 85% w/w sulphuric acid (reagent 2). Adjust the temperature of the liquid to 35°C add 1 ml of the xylenol solution (reagent 3) and maintain the mixture at 35°C for half an hour; then dilute with 100 ml of distilled water and transfer to an ordinary distillation apparatus. Distil the mixture and collect 40 ml in a Nessleriser glass containing 10 ml of $2N$ sodium hydroxide (reagent 1). Adjust the temperature of the distillate to 20°C, and place the Nessleriser glass in the right-hand compartment of the Nessleriser. Fill another glass to the 50 ml mark with water and place in the left-hand compartment of the instrument. Stand the Nessleriser before a standard source of white light such as the Lovibond White Light Cabinet or, failing this, north daylight, and compare the colour produced in the test solution with the colours in the standard disc, rotating the disc until a match is obtained.

The markings on the disc represent the actual amounts of nitrate nitrogen producing the colour in the test. Thus, if a colour equivalent to 8 µg is produced, the amount of nitrogen present as nitrate in the quantity of solution taken for the test is 0.008 mg.

Note

It is important that the temperature of distillates should be adjusted to 20°C, which is the temperature adopted for the standardisation of the nitrate disc, since the colour intensity of the nitro-compound in alkaline solution increases by 0.62% per 1°C.

Nitrate Method 3
using brucine

Principle of the method

A solution of brucine, in concentrated suphuric acid, is nitrated by nitrates to give a deep red colour, fading to reddish-yellow.

Reagents required

1. *Brucine*   Dissolve 5 g in 90 ml glacial acetic acid (CH_3COOH) and 10 ml water
2. *Sulphuric acid* (H_2SO_4)   Concentrated, nitrogen free

The Standard Lovibond Comparator Disc 3/32

The disc covers the range 1 to 9 parts per million nitrogen (N) as nitrate, in steps of 1 ppm.

To convert to nitrate ion (NO_3) multiply answer by 4.428.

Technique

To 1 ml of the sample add 0.1 ml of brucine solution (reagent 1) and 2 ml of sulphuric acid (reagent 2) slowly from a dropping pipette while the tube is cooled in ice. Shake, holding well away from the face and taking care not to spill any of the liquid on the hands, and allow to stand for 7 minutes.

Place in a Lovibond 13.5 mm cell or test tube in the right-hand compartment of the Comparator, and use a blank of 1 ml distilled water, plus the reagents in the above proportions, in a similar cell in the left-hand Compartment. Hold the comparator facing a standard source of white light such as the Lovibond White Light Cabinet or, failing this, north daylight, and match against the colour standards in the disc. Place a wedge of cotton wool in the right-hand compartment, so that the test tube is raised and the small amount of solution covers the window, or a special clip may be obtained from The Tintometer Ltd. to raise the tube.

The readings show parts per million of nitrogen in the 1 ml sample taken. To convert to nitrate ion (NO_3^-) multiply answer by 4.428.

Notes

1. Organic substances in the quantities found in natural waters are not likely to interfere.
2. Chlorides in greater quantity than 1000 parts per million interfere, and must be removed by silver sulphate (see *Method 1*).
3. Nitrite interferes in proportion to its molecular weight, and should therefore be estimated separately and due allowance made. In most industrial waters the nitrite ion is present as less than 1 ppm, and in this concentration may be ignored.
4. Ammonia below 65 ppm does not interfere.

Nitrate Method 4
using phenoldisulphonic acid

Principle of the method

Phenol-2,4-disulphonic acid is nitrated by nitrates to 6-nitrophenol-2,4-disulphonic acid, which gives rise to an intense yellow colour on the addition of alkali. The intensity of this colour, which is proportional to the nitrate concentration, is determined by comparison with Lovibond permanent glass standards.

Reagents required

1. *20% (w/w) phenoldisulphonic acid*
   Phenol (C_6H_5OH) (analytical reagent grade)   25 g
   Sulphuric acid (H_2SO_4) (analytical reagent grade)   100 ml
   Fuming sulphuric acid, 20% (w/w) SO_3 67 ml
   Dissolve the phenol in the sulphuric acid, slowly stir in the fuming sulphuric acid, and heat the mixture in a water bath at 100°C for 2 hours (Note 1).

2. *Diluted phenoldisulphonic acid* Slowly add 100 ml of reagent 1 to 40 ml of distilled water. Keep the mixture cool and well stirred by swirling the flask under running water. This diluted reagent is supersaturated with respect to solid phenoldisulphonic acid at ordinary temperatures. Should crystals separate, they may be redissolved readily by placing the bottle containing the reagent in warm water. For this reason it is preferable to dilute reagent 1 as required.

3. *Diluting solution* Dissolve 5 g of ammonium citrate (($NH_4)_3C_6H_5O_7$) in 125 ml of 0.880 ammonia solution (NH_4OH) and dilute to 1 litre with distilled water.

The Standard Lovibond Comparator Disc 3/56

The disc covers the range 10, 20, 40, 60, 80, 100, 120, 160, 200 micrograms of nitrate nitrogen (N), equivalent to 10–200 ppm on a 1 ml sample.

To convert to nitrate ion (NO_3^-) multiply answer by 4.428.

Technique

(a) *Laboratory method* Pipette 1 ml of the sample solution into the bottom of an ordinary thin-walled test tube, and add 1.5 ml of reagent 2 down the side of the tube from a small burette. Thoroughly mix the contents of the tube, by gently twirling the tube between the fingers, and then place the tube in a vigorously boiling water bath. Take care that the part of the tube containing the reaction mixture is immersed in the water.

After 3 minutes remove the tube from the water bath, and dilute almost to the 25 ml mark with diluting solution (reagent 3). Mix well, cool to 20°C in a water bath, dilute to the mark, mix again and transfer to a 13.5 mm comparator tube or rectangular comparator cell. Place in the right-hand compartment of the Comparator and match against the permanent Lovibond glass standards in the disc, using a standard source of white light such as the Lovibond White Light Cabinet or, failing this, north daylight.

(b) *Rapid method* To 1 ml of the sample solution in a dry test tube, graduated at 25 ml, add 1 ml reagent 2, running the reagent straight into the extract. Shake gently, set the tube aside for 2 minutes, dilute to the mark with reagent 3 and match against the disc as in (a) above.

Notes

1. 20% (w/w) phenoldisulphonic acid reagent (reagent 1) may be obtained ready made if preferred.

2. It has been shown that varying the time of heating or the temperature of the final solution both affect the intensity of the colour produced. The details of the technique given must therefore be followed exactly.

3. This modification of Chamot and Pratt's original method does not differentiate between nitrate and nitrite, both ions producing the same colour. If it is required to estimate nitrate in the presence of nitrite, then one of the methods described in the Introduction should be used.

4. The following are the maximum amounts, in micrograms, of ions which can be tolerated without causing interference in the determination of nitrate.

Br^-	2	Na^+	6,000	Fe^{2+}	15
I^-	10	K^+	4,000	SO_3^{2-}	80
NH_4^+	2,000	Fe^{3+}	200	Cu^{2+}	500

Nitrate Method 5
using Nessler's reagent

Principle of the method

Nitrate ions are reduced to ammonium ions by boiling with chromous chloride solution. Ammonia is released from the ammonium ions, by making the solution alkaline, and removed by steam distillation. The distillate is reacted with Nessler's reagent producing a yellow colour, the intensity of which is proportional to the ammonia concentration, and is measured by comparison with Lovibond permanent glass standards.

This method does not differentiate between nitrate and nitrite ions. If significant concentrations of nitrite ions are present in the samples, the nitrite concentration must be determined separately and allowance made for it in the calculation of the nitrate concentration.

Reagents required

All chemicals used should be of analytical reagent quality. Water used in preparing reagents should be distilled or deionised, and should be free of ammonia.

1. *Chromous chloride solution* This reagent is prepared by the reduction of potassium chromate using nitrogen-free zinc in the following manner:—Wash granulated zinc by boiling for a few minutes with hydrochloric acid diluted with an equal volume of water. (This operation must be carried out in an efficient fume cupboard). Rinse thoroughly with water. Place 2 g of potassium dichromate ($K_2Cr_2O_7$), 10 ml of concentrated hydrochloric acid (HCl), 5 ml of water and a large excess (about 10 g) of the purified zinc, in a 125 ml conical flask. Boil for 1 to 2 minutes until the colour of the solution becomes pale blue. Cool immediately, and cover the surface of the liquid with a layer of petroleum ether (b.p. 100—120°C) to inhibit oxidation. This reagent must be prepared immediately before use.

2. *Caustic soda solution* Dissolve 200 g of sodium hydroxide (NaOH) in water and dilute to 1 litre in a 2 litre Pyrex flask. Heat to boiling, add an excess of chromous chloride solution (10 to 20 ml), boil the solution for about 30 minutes, filter through a Whatman No. 54 filter paper, cool and adjust the final volume to 1 litre.

3. *Nessler's reagent* (See Note 1). Dissolve 35 g of potassium iodide (KI) and 12.5 g of mercuric chloride ($HgCl_2$) in 800 ml of water. Slowly add a cold saturated solution of mercuric chloride until, after repeated shaking, a slight red precipitate remains. Add 120 g of caustic soda; shake until dissolved and finally add a little more of the saturated solution of mercuric chloride. Make up to 1 litre with water. Allow to stand, with occasional shaking for several days. Use the clear supernatant liquid for the tests.

The Standard Lovibond Nessleriser Discs, NAA, NAB, NAC and NAD

Disc NAA covers the range 1 μg to 10 μg of ammonia (NH_3).

Disc NAB covers the range 10 μg to 26 μg of ammonia.

Disc NAC covers the range 28 μg to 60 μg of ammonia.

Disc NAD covers the range 60 μg to 100 μg of ammonia.

The conversion of ammonia concentration to the equivalent nitrate concentration is given in Note 5.

Technique

Pipette 50 ml of the water sample into a 1 litre flask, and add 200 ml of ammonia-free distilled water. Connect the flask to a distillation apparatus fitted with a dropping-funnel (See figure).

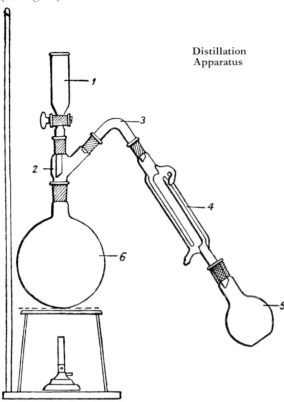

Distillation Apparatus

1. Dropping funnel   Cat. No. D. 3/12
2. Multiple adaptor   Cat. No. M.A. 2/3
3. Bend   Cat. No. S.H. 1/22
4. Double surface condenser. Cat. No. C. 5/12
5. 500 ml Pyrex, flat bottom flask, medium neck B. 24
6. 1,000 ml Pyrex, round bottom flask, short neck B.24
   Also required:—   B. 24 stopper   B. 19 stopper
Catalogue numbers refer to:—
        Quickfit & Quartz Ltd., Mill Street, Stone, Staffs.

Turn off the water supply to the condenser and heat the solution until steam emerges from the end of the condenser. Turn on the condenser cooling water and collect the distillate. Continue heating until 50 ml of distillate have been collected. Test this preliminary distillate for ammonia, in the manner described below for the final distillate, and then discard it. If the preliminary distillate gives a positive reaction with the Nessler reagent, collect a further 50 ml and test this.

When the preliminary distillate has been shown to be free of ammonia, pipette 1 ml of chromous chloride solution (reagent 1) into the flask and boil for about 15 seconds to ensure thorough mixing. Add 2 ml of caustic soda solution (reagent 2) and continue heating until two successive 50 ml portions have been collected in Nessler tubes (see Note 2). To each tube add 2 ml of Nessler's reagent (reagent 3 and Note 1) and mix. Place one of the tubes in the right-hand compartment of the Nessleriser. Prepare a blank on the reagents by carrying out the whole of the above procedure but omitting the 50 ml of sample and substituting an additional 50 ml of distilled water, i.e. 250 ml in all. By using replicate distillation equipment the blank and samples can be treated simultaneously. For the blank collect only the first 50 ml of the final distillate, add Nessler's reagent as above, and place in the left-hand compartment of the Nessleriser. Compare the colour of the sample with the permanent Lovibond glass standards in the disc, using a standard source of white light such as the Lovibond White Light Cabinet or, failing this, north daylight. Note the measurement and repeat with the second fraction of the sample distillate. Add these two disc readings together.

This combined figure gives the total content of reducible nitrogen (nitrate plus nitrite) in the sample, expressed as micrograms of ammonia. This can be converted to micrograms of nitrate (NO_3^-) by multiplying by 3.65 or by use of the Table (see Note 5).

The presence of nitrite in the sample may be detected by any suitable qualitative test, of which the Tofranil test is probably the simplest. If nitrite is shown to be present, its concentration can be determined by the standard Nessleriser Nitrite test and due allowance made for its contribution to the final answer.

The concentration of ammonia in the original sample can be obtained by adding 2 ml of Nessler's reagent directly to a fresh 50 ml of the sample in a Nessler tube and measuring, using the same discs. In this case the disc reading should be quoted directly and not referred

COLORIMETRIC CHEMICAL ANALYTICAL METHODS — Nitrate Method 5

to the conversion Table. Neither is there any need to correct the final answer for the initial ammonia content as the 'free' ammonia is removed by distillation before the production of ammonia by reduction of nitrates and nitrites.

Notes

1. The published formulae for Nessler's reagent vary considerably. When using the Lovibond permanent glass standards for the determination of ammonia it is important that the reagent employed should correspond to that used for preparing the original standard colours. The formula used for that purpose is given above (reagent 3) and this must be followed exactly.

2. Readings obtained by the use of the Lovibond Nessleriser and discs are accurate only provided that the Nessler tubes used conform to the specification employed when the discs were calibrated, namely that the 50 ml calibration mark shall fall at a height of 113 ± 3 mm when measured internally.

3. If it is found that the colour which develops in the first 50 ml of the final distillate falls outside the range of the discs, then the test should be repeated, using either less or more sample as appropriate, and making up to 250 ml total volume with distilled water as before. The final result is then related to the volume of sample used. If the result is divided by the sample volume then the figure obtained will be in micrograms per ml, i.e. in parts per million (w/v).

4. A possible source of interference is albuminoid matter in the water, but the conditions of this test have not proved sufficiently severe to cause decomposition of the albumin to ammonia in any of the waters yet examined.

5. **Conversion Table**

Disc	Reading ($\mu g\ NH_3$)	Equivalent nitrate concentration $\mu g\ NO_3^-$	ppm NO_3^-
NAA	1	4	0.08
	2	7	0.14
	3	11	0.22
	4	15	0.30
	5	18	0.36
	6	22	0.44
	7	26	0.52
	8	29	0.58
NAB	10	36	0.72
	12	44	0.88
	14	51	1.02
	16	58	1.16
	18	66	1.32
	20	73	1.46
	22	80	1.60
	24	87	1.74
	26	95	1.90
NAC	28	102	2.04
	32	117	2.34
	36	131	2.62
	40	146	2.92
	44	160	3.20
	48	175	3.50
	52	190	3.80
	56	204	4.08
NAD	60	219	4.38
	65	237	4.74
	70	255	5.10
	75	273	5.46
	80	292	5.84
	85	310	6.20
	90	328	6.56
	95	346	6.92
	100	365	7.30

The Determination of Nitrite

Introduction

The presence of nitrite in a raw water supply usually denotes bacterial activity and is often taken as indicative of contamination by sewage. Similarly the presence of nitrite in meat pickling brines is indicative of the bacterial reduction of nitrate. The permitted content of nitrite in uncooked pickled meat is limited to 200 ppm calculated as sodium nitrite[1]. The determination of nitrite is thus important in the control of water supplies, of sewage effluents and of pickled meats and pickling brines.

Nitrogen dioxide (NO_2) is the most toxic of the various nitrogen oxides. It may be evolved during the use of explosives, in welding operations, in chemical processes and in vehicle exhaust fumes. The maximum permitted concentration in air in industrial atmospheres is 5 ppm. Measurement of the nitrogen dioxide concentration in air is conveniently carried out by absorbing the gas in a suitable liquid and determining the nitrite so formed.

Methods of sample preparation

Air

Sample the air at up to 0.5 litres per minute into a freshly prepared mixture of the sulphanilic acid ($NH_2C_6H_4SO_3H$) and 1-naphthylamine-7-sulphonic acid ($C_{10}H_6(NH_2).SO_3H$) reagents and proceed by *Method 1*. See also Section F of this book.

Biological materials[2,3]

Serum[2] Precipitate the protein by adding 1 ml of 5% mercuric chloride solution to 1 ml of sample and 8 ml of water. Pipette off 1 ml of the supernatant fluid and proceed by *Method 1*.

Tissue[3] Suspend 10 g of cominuted sample in 40 ml of water, precipitate proteins by adding 9 ml of 20% tungstophosphoric acid ($H_3PO_4.12WO_3 + xH_2O$) dilute to 100 ml and filter. To 5 ml of the filtrate add 0.01N potassium permanganate and proceed by *Nitrate Method 2*.

Dairy products[4]

Dilute 40 ml of liquid milk or 4 g of dried milk to 150 ml, warm to 50°C, clarify by the addition of 10 ml of 12% zinc sulphate ($ZnSO_4$) solution and 10 ml of 0.5N caustic soda (NaOH). Maintain the temperature of the solution at 50°C for 10 minutes then cool, dilute to 200 ml, filter and determine nitrite on an aliquot.

Foods and edible oils[5-8]

Brines may be analysed directly by *Method 2*.

General[5] Extract 1 g of the minced sample with successive quantities of hot, nitrite free, distilled water and collect the extracts in a 500 ml flask. Add sufficient hot water to bring the volume up to about 300 ml and heat on a steam bath for 2 hours. Add 5 ml of a saturated solution of mercuric chloride ($HgCl_2$), cool and dilute to 500 ml. Filter and take a suitable aliquot in a Nessleriser glass. Dilute to 50 ml and proceed by *Method 2*.

Meat and meat products[6-8] Suspend the meat sample in a hot 30% solution of zinc sulphate ($ZnSO_4$), cool, filter and determine by *Method 2*[6]. Alternatively[7] homogenize 10 g of sample with three successive 30 ml portions of water, filter, dilute to 100 ml, add 1 drop of glacial acetic acid (CH_3COOH), boil to precipitate protein, filter and make up to volume. For pork products[8] finely grind 2—5 g of the sample with water, transfer to a 100 ml flask, dilute to volume, shake for 15 minutes, filter and determine nitrite on an aliquot.

Miscellaneous[9]

Animal feeds Boil the sample with water, filter, percolate the filtrate through a column of alumina and determine nitrite on an aliquot of the percolate.

Plants[10-12]

Extract the ground plant material with $0.1N$ hydrochloric acid. Neutralise and dilute to known volume. Place 1 ml of this solution in a 12 ml centrifuge tube, add 9 ml of 20% acetic acid (CH_3COOH) solution. Add 0.3 to 0.5 g of a powder mixture, containing 75 g of citric acid ($C_6H_8O_7.H_2O$), 4 g of sulphanilic acid ($NH_2.C_6H_4.SO_3H$), 2 g of 1-naphthylamine-7-sulphonic acid ($C_{10}H_6(NH_2).SO_3H$) and 100 g of barium sulphate ($BaSO_4$), stopper the tube and shake for 50—60 seconds. Centrifuge until the supernatant liquid is clear and match the colour of the supernatant layer[10,11]. Alternatively[12] free the extract from proteins (cf 6 above), buffer at pH 10 with borate and place an aliquot in a semi-micro Kjeldahl distillation apparatus. Add 40% caustic soda (NaOH) solution and remove ammonia and amide nitrogen by steam distillation at 100°C. Add 20% ferrous sulphate ($FeSO_4$) solution and continue distillation to remove nitrite nitrogen. Proceed by *Method 3*.

Sewage and industrial wastes[13-15]

Take 0.1 to 10 ml of clarified sewage, dilute to 25 ml and neutralise to pH 6.5—7.5. Dilute to 50 ml and proceed by *Method 1*[13,14].

Metal finishing effluents[15]. Reduce the nitrite in the sample to ammonia (NH_3) by means of aluminium foil in alkaline solution for 8 hours at not less than 20°C. Add aqueous sodium sulphide (Na_2S) solution and proceed by *Method 3*.

Soil[10-12]

Proceed as described for plants above.

Water[13,16,17]

If the sample is coloured, or if suspended solids are present, add 2 ml of aluminium hydroxide ($Al(OH)_3$) suspension to 100 ml of sample. Stir thoroughly and allow to settle. Filter and discard the first 25 ml of the filtrate. Neutralise the filtrate, or the original sample if clear and colourless, to pH 6.5—7.5 and proceed by *Method 1*.

The individual tests

Method 1 is essentially the classical Griess[18] diazotisation of sulphanilic acid and coupling with 1-naphthylamine as modified by Ilosvay[19], and by the British Standards Institution[16]. 'The conditions for optimum results have been checked by Barnes and Folkard[20]. This is the standard method for the determination of nitrite in water[17] and sewage[13,14].

It has recently been shown [21,16] that 1-naphthylamine may be carcinogenic. This reagent has therefore been replaced by Cleve's acid (1-naphthylamine-7-sulphonic acid). N. T. Crosby[22] has demonstrated that this replacement of the reagent has no effect on the colour standards and the old discs may therefore be used with the new reagent.

Method 2 is an adaptation of *Method 1* which was especially developed for the determination of nitrite in meat and pickling brines. This method is calibrated directly in terms of sodium nitrite ($NaNO_2$). The remarks about *Method 1* apply equally to *Method 2*.

Method 3 uses the reduction of nitrite to ammonia under controlled conditions and the determination of the ammonia by Nessler reagent, as described earlier in this book.

Method 4 is the comparator method using Cleve's acid[16,22].

References

1. Statutory Instrument 1971 No. 882 Preservatives in Food (Amendment)
2. R. H. Diven, W. J. Pistor, R. E. Reed, R. J. Trautman and R. E. Watts, *Amer.J. Vet.Res.,* 1962, **23,** 497; *Analyt.Abs.,* 1963, **10,** 4292
3. J. Davidek, S. Klein and A. Zackova, *Z. Lebensmitt.Untersuch;* 1963, **119,** 342; *Analyt.Abs.,* 1964, **11,** 1484
4. P. B. Manning, S. T. Coulter and R. Jenness, *J. Dairy Sci.,* 1968, **51** (11), 1725; *Analyt.Abs.,* 1970, **18,** 1987
5. M. Z. Barakat and I. Sadek, *Food Technol.,* 1964, **18** (2), 120; *Analyt.Abs.,* 1965, **12,** 3033

6. R. Grau and A. Mirna, *Z. analyt.Chem.*, 1957, **158**, 182; *Analyt.Abs.*, 1958, **5**, 2390
7. J. Davidek and T. Beniak, *Prum.Potravin*, 1965, **16** (8), 425; *Analyt.Abs.*, 1966, **13**, 7111
8. R. Truhaut and Nguyen Phu Lich, *Annls.Falsif.Expert.Chim.*, 1966, **59**, 401; *Analyt. Abs.*, 1968, **15**, 444
9. E. Mehnert, *Arch.exp.vet.Med.*, 1967, **21** (5), 1191; *Analyt.Abs.*, 1969, **16**, 401
10. J. T. Woolley, G. F. Hicks and R. H. Hageman *J.Agric.Food Chem.*, 1960, **8**, 481; *Analyt.Abs.*, 1961, **8**, 2667
11. J. L. Nelson, L. T. Kurtz and R. H. Brag, *Analyt.Chem.*, 1954, **26**, 1081.
12. J. E. Varner, W. A. Bulen, S. Vanecko and R. C. Burrell; *Analyt.Chem.*, 1953, **25**, 1528
13. Amer. Public Health Assoc., *"Standard Methods for the Examination of Water, Sewage and Industrial Wastes"*, 10th Ed., New York, 1955
14. Min. of Housing and Local Govt., *"Chemical Analysis as applied to Sewage and Sewage Effluents"*, H.M. Stationery Office, London, 1956
15. E. J. Serfass and R. F. Muraca, *Plating*, 1956, **43**, 233; *Analyt.Abs.*, 1957, **4**, 875
16. Brit. Standards Inst., *"Routine Control Methods of Testing Water Used in Industry"*, B.S. 1427, London, 1962. Amendment 3:1968
17. Brit. Standards Inst., *"Methods of Testing Water Used in Industry"*, B.S. 2690, London, 1964
18. P. Griess, *Ber.*, 1879, **12**, 426
19. M. L. Ilosvay, *Bull.Soc.Chim.*, 1889, **49**, 388
20. H. Barnes and A. R. Folkard, *Analyst*, 1951, **76**, 599
21. Chester Beatty Res. Inst., *"Precautions for Laboratory Workers who handle Aromatic Amines"*, London, 1966
22. N. T. Crosby, *Proc.Soc.Wat.Treat. and Exam.*, 1967, **16** (1), 51

Nitrite Method 1
using Cleve's Acid

Principle of the method

The well known Griess test, as modified by Ilosvay, depends on the diazotisation of sulphanilic acid by nitrous acid. The compound thus formed is then coupled with 1-naphthylamine-7-sulphonic acid to produce a red azo dye, the colour of which is measured by comparison with a series of Lovibond permanent glass colour standards.

Reagents required

Various formulae for the reagent have been suggested and experiment has shown that the colours produced by these vary slightly; it is therefore important to employ a reagent corresponding with that used for standardising the colours. The following formula has been adopted.
1. *Sulphanilic acid* Dissolve 0.5 g of sulphanilic acid ($NH_2.C_6H_4.SO_3H$) in 30 ml of glacial acetic acid (CH_3COOH) and add 120 ml of distilled water. Filter.
2. *Naphthylamine sulphonic acid* Dissolve 0.2 g of Cleve's acid (1-naphthylamine-7-sulphonic acid ($C_{10}H_6(NH_2)SO_3H$)) in 30 ml of glacial acetic acid, add 120 ml of distilled water and filter. Store in dark.

Should either of these reagent solutions become coloured on storage, shake with a small quantity of zinc dust and refilter.

All chemicals used in the preparation of these reagents should be of analytical reagent quality.

The Standard Lovibond Nessleriser Disc NJ

The disc covers the range 0.05 μg to 1 μg of nitrogen (N) present as nitrite, in the following steps:—.05, 0.1, 0.2, 0.3, 0.4, 0.5, 0.6, 0.8, 1.0 μg equal to 0.001 to 0.02 ppm on a 50 ml sample.

To convert to nitrite ion (NO_2) multiply the answer by 3.286.

Technique

Fill one of the Nessleriser glasses to the 50 ml mark with distilled water and place in the left-hand compartment of the Nessleriser. Fill the other Nessleriser glass to the mark with the solution under examination, add 2 ml of reagent 1 and an equal quantity of reagent 2, mix, allow to stand for 30 ± 5 minutes, and then place in the right-hand compartment of the instrument. Stand the Nessleriser before a standard source of white light such as the Lovibond White Light Cabinet or, failing this, north daylight and compare the colour produced in the test solution with the colours in the standard disc, rotating the disc until a colour match is obtained. Should the colour in the test solution be deeper than the standard colour glasses, a fresh test should be carried out using a smaller quantity of the water under examination and diluting to 50 ml with distilled water before adding the reagent. The sample when tested must be at a temperature of 20°C.

The markings on the disc represent the actual amounts of nitrogen (N) present as nitrite producing the colours in the test. Thus, if on adding the reagent to 50 ml of water, a colour equivalent to 1 μg is produced, the amount of nitrogen present as nitrite in the water will be 0.02 part per million, as 1 μg per ml is equivalent to 1 ppm.

Note

It must be emphasized that readings obtained by the Lovibond Nessleriser and discs are only accurate provided that Nessleriser glasses are used which conform to the specification employed when the discs were calibrated, namely that the 50 ml calibration mark shall fall at a height of 113 mm plus or minus 3 mm measured internally.

COLORIMETRIC CHEMICAL ANALYTICAL METHODS

Nitrite Method 2
using Cleve's Acid

Principle of the method

Sulphanilic acid is first diazotised with sodium nitrite and then coupled with Cleve's acid (1-naphthylamine-7-sulphonic acid) to produce a red azo dye. The intensity of the red colour produced under the conditions of the test is proportional to the concentration of sodium nitrite in the sample solution, and is measured by comparison with a series of Lovibond permanent glass colour standards.

Reagents required

1. *Sodium bromide (NaBr)* A 10% solution in distilled water.
2. *Sulphanilic acid solution* Dissolve 0.5 g of sulphanilic acid ($NH_2.C_6H_4.SO_3H$) in 30 ml of glacial acetic acid (CH_3COOH d.1.05) and then dilute to 150 ml with distilled water.
3. *Cleve's acid solution* Dissolve 0.2 g of Cleve's acid (1-naphthylamine-7-sulphonic acid, $C_{10}H_6(NH_2).SO_3H$) in 120 ml of warm water, cool, filter, and dilute to 150 ml with glacial acetic acid.

All chemicals used in the preparation of these reagents should be of analytical reagent quality.

The Standard Lovibond Comparator Discs 3/93A and 3/93B

Disc 3/93A covers the range 200-1000 ppm (0.02-0.1%) of sodium nitrite ($NaNO_2$) in steps of 100 ppm (0.01%).

Disc 3/93B covers the range 1000-2000 ppm (0.1-0.2%) of sodium nitrite in steps 0.1, 0.11, 0.12, 0.14, 0.15, 0.16, 0.18, 0.2%.

These discs must be used in conjunction with the 25 mm cells.

Technique

Dilute 10 ml of the sample solution to 1 litre with distilled water. Place 2 ml of this diluted solution in a 50 ml Nessler tube. Add 1 ml of sodium bromide solution (reagent 1), 5 ml of sulphanilic acid solution (reagent 2) and 5 ml of Cleve's acid solution (reagent 3). Mix, make up to 50 ml with distilled water and mix again. Prepare a blank of 1 ml of reagent 1, 5 ml of reagent 2, and 5 ml of reagent 3, mix and make up to 50 ml with distilled water. Set aside the test and blank solutions for 30 minutes at 20°C, for the colour to develop. At the end of this time fill a 25 mm cell with the test solution and place this in the right-hand compartment of a Lovibond Comparator. Fill an identical cell with the blank solution and place this in the left-hand compartment. Place the comparator before a standard source of white light, such as the Lovibond White Light Cabinet or, failing this, north daylight, and compare the colour of the test solution with the standards in the disc.

Reference

Statutory Instrument 1971 No. 882. The Preservatives in Food (Amendment) Regulation 1971.

Ham, Bacon and Pickled Meat. Maximum permitted nitrite concentration, 200 ppm.

Nitrite Method 3
using ferrous sulphate and Nessler reagent

Principle of the method

Free ammonia is removed from the sample solution by gentle heating at pH 10, amide nitrogen is removed by adding sodium hydroxide and then steam distilling at 100°C. The nitrites present in the sample are then reduced to ammonia by means of ferrous sulphate and the ammonia is distilled and then determined by means of Nessler's reagent and the standard ammonia discs.

Reagents required

1. *Borate buffer* A saturated solution of sodium tetraborate ($Na_2B_4O_7.10H_2O$) adjusted to pH 10 with $0.1M$ sodium hydroxide (NaOH) solution
2. *Boric acid* 4% solution of H_3BO_3
3. *Sodium hydroxide* 40% solution of NaOH
4. *Ferrous sulphate* 20% solution of $FeSO_4.7H_2O$

These reagents should be prepared from chemicals of analytical reagent quality.

In addition Nessler's reagent will be required. This should be prepared in accordance with the instructions accompanying the particular ammonia disc which is being used. This is most important, as the recipes for Nessler's reagent vary somewhat and it is imperative to use the one which was used for the calibration of the disc.

The Standard Lovibond Ammonia Discs

The following discs are available for the determination of ammonia:—

Nessleriser Disc	Ammonia Method	Range
NAA	1	0.02–0.2 ppm NH_3
NAB	1	0.2 –0.52 ,, ,,
NAC	1	0.56–1.2 ,, ,,
NAD	1	1.2 –2.0 ,, ,,

Technique

Place the sample, in 1–20 ml of water, in the distillation flask of a suitable steam-distillation apparatus (such as a micro-Kjeldahl distillation equipment). Wash the sample into the flask with 5 ml of borate buffer (reagent 1) and distil off any free ammonia by heating on a water-bath at 50–55°C for 5 minutes. Allow a slow stream of air to pass through the equipment, to prevent bumping, and collect the effluent in boric acid solution (reagent 2) for the estimation of free ammonia if required. Change the receiver and add 15 ml of caustic soda (reagent 3) to the distillation flask and steam distil at 100°C for 5 minutes. This removes any ammonia formed from amide nitrogen, and this may also be collected and determined if required. Again change the receiver and add 5 ml of ferrous sulphate solution (reagent 4) to the contents of the distillation flask. Continue the steam distillation for a further 10 minutes, and determine the ammonia collected in the receiver by the standard method. This is ammonia which was originally present in the sample as nitrite.

Convert the reading given on the disc for ammonia into the equivalent amount of nitrite by **multiplying by 2.7**, or to convert ammonia readings into nitrogen (N) present as nitrite multiply by 0.82.

Note

The results obtained by the use of the Nessleriser and disc are accurate only if Nessler tubes are used which conform to the specification used when the disc was calibrated, that is that the 50 ml calibration mark falls at a height of 113 ± 3 mm from the bottom of the tube, measured internally.

Nitrite (4)
using Cleve's acid

Introduction

In view of the statement in the report of the Chester Beatty Research Institute *"Precautions for Laboratory Workers who handle aromatic amines"*, 1966, that the Griess-Ilosvay's reagent 1-naphthylamine is carcinogenic, the British Standards Institution Committees concerned with B.S. 1427 and 2690 have substituted a new reagent for the determination of nitrites. This reagent, Cleve's acid (1,7-naphthylamine sulphonic acid) is claimed to be non-carcinogenic and recommended (N. T. Crosby, *Proc. Water Treat.* 1967. 16. Part 1.51) in place of Griess-Ilosvay reagent.

Principle of the method

Nitrites, in acid solution, will diazotise sulphanilic acid. The resulting diazo compound couples with 1,7-naphthylamine sulphonic acid to form a purplish-pink azo dye. The intensity of the colour of this dye, which is proportional to the nitrite concentration, is measured by comparison with a series of Lovibond permanent glass colour standards.

Reagents required

1. *Sulphanilic acid solution* Dissolve 0.5 g of sulphanilic acid ($NH_2.C_6H_4.SO_3H$) in a mixture of 120 ml of water and 30 ml of glacial acetic acid (CH_3COOH). Filter the solution and store it in the dark.
2. *1,7-naphthylamine sulphonic acid solution* Add 120 ml of water to 0.2 g of 1,7-naphthylamine sulphonic acid ("Cleve's acid") and warm. Filter, cool, and then add 30 ml glacial acetic acid. Store in the dark.
3. *Sodium bromide solution* Dissolve 10 g of sodium bromide (NaBr) in 100 ml of distilled water.

Should reagents 1 or 2 become coloured on storage, shake with a small quantity of zinc dust and refilter.

All chemicals used should be of analytical reagent grade quality.

The Standard Lovibond Comparator Disc 3/83

This disc covers the range 1 to 15 µg of nitrogen (N) present as nitrite (NO_2^-) in steps of 1, 2, 3, 5, 7, 9, 11, 13, 15, µg.

Technique

Measure a volume of the sample (up to 40 ml) which will contain not more than 15 µg of nitrogen, present as nitrite, into a 50 ml volumetric flask and make up to 40 ml with de-ionised water. Adjust the temperature of the sample to 20-25°C and maintain it at a temperature not higher than 25°C throughout the test. Add 1 ml of sodium bromide solution (reagent 3) and mix. Add 2 ml of sulphanilic acid solution (reagent 1), mix, add 2 ml of 1,7-naphthylamine sulphonic acid solution (reagent 2), mix, dilute to the mark with water, and mix again. Measure 40 ml of distilled water into a second 50 ml volumetric flask and treat this in exactly the same manner as the sample. This will provide a blank. After 45 minutes transfer 10 ml of the contents of the sample flask into a standard comparator tube or 13.5 mm cell and place this in the right-hand compartment of the Comparator. Fill an identical tube, or cell, with the blank solution and place this in the left-hand compartment. Place the comparator in front of a standard source of white light, such as the Lovibond White Light Cabinet or, failing this, north daylight, and match the colour of the sample with the standards in the disc. Read off the corresponding amount of nitrite nitrogen from the indicator window in the comparator and calculate the corresponding concentration by means of the formula:—

$$\text{Nitrogen Nitrite concentration in ppm} = \frac{\text{Disc reading (in µg N)}}{\text{Volume of sample taken (in ml)}}$$

To convert to nitrite ion (NO_2^-) multiply the answer by 3.286.

Notes

1. 1,7-naphthylamine sulphonic acid may be obtained from B.D.H. Chemicals Ltd., Poole, Dorset.

2. Ammonia, urea, organic amines, strong reducing agents, strong oxidising agents and complexing agents are known to interfere with this test. The following concentrations of other ions do not interfere when present in the test solution in not more than the following amounts, see British Standard 2690, Part 7, 1968

Chloride (Cl)	5000 μg
Sulphate (SO$_4$)	5000 μg
Calcium (Ca)	2500 μg
Copper (Cu)	1000 μg
Magnesium (Mg)	1000 μg
Iron (Ferrous)	1000 μg
Iron (Ferric)	500 μg
Metaphosphate (as P$_2$O$_5$)	500 μg

The Determination of Nitrogen

Introduction

Many methods have been published for the determination of total nitrogen in a wide variety of materials, ranging from animal feeding stuffs to urine. Without exception these methods are based on the reduction of all nitrogen compounds to ammonia, followed by the determination of the ammonia with Nessler's reagent. This was mentioned in the earlier section on the determination of ammonia, but it was considered to be worthwhile to collect the methods of sample preparation for the determination of total nitrogen into a separate section.

Methods of sample preparation

Biological materials[1-5]

General[1-3] Wet ash the sample with sulphuric acid containing potassium sulphate (K_2SO_4) and a catalyst, usually mercury, in a Kjeldahl flask. Remove the ammonia formed, by distillation, after neutralising the acid with sodium bicarbonate ($NaHCO_3$) and adding 1 g of potassium bromide (KBr). Determine the ammonia by Ammonia *Method 1*. Fleck[3] has reviewed available methods for the determination of nitrogen in biological materials and has concluded that the Kjeldahl wet ashing procedure is the best. He considers that the use of a mercury catalyst is essential, as this gives a digestion temperature between 370 and 410°C, and claims that a suitable mixture for the reduction of 0.3–2 mg of nitrogen is 1.5 ml of $36N$ sulphuric acid, 1.2 g of potassium sulphate and 50 mg of mercuric oxide (HgO).

Serum[4] To 0.1 ml of serum add 0.4 ml of a 10% solution of trichloroacetic acid (CCl_3COOH) and filter or centrifuge. To 0.2 ml of the filtrate, or of the supernatant fluid, add 0.3 ml of 1:1 perchloric acid ($HClO_4$), digest for 12 minutes at 200–250°C, cool and add Nessler's reagent directly to this solution.

Urine[5] Heat sufficient sample to contain 5–15 µg of nitrogen with 0.75 ml of 70% perchloric acid for 15 minutes. Dilute the solution to a known volume and determine nitrogen on an aliquot, using Ammonia *Method 1*.

Chemicals[5]

Organic chemicals Use the method described above for urine.

Dairy products[5]

Milk Use the method described above for urine.

Foods and edible oils[5-7]

Eggs[5] Use the method described above for urine.

Meat extracts[5] Use the method described above for urine.

Sugar[6] Digest 1 g of sample with 1 g of a mixture of potassium sulphate (K_2SO_4), mercuric sulphate ($HgSO_4$) and selenium in the proportions 32:5:1 and 5 ml of sulphuric acid in a 50 ml Kjeldahl flask, for 20 minutes. Neutralise, distil off the ammonia and determine by Ammonia *Method 1*.

Vegetable oils[7] Digest 0.1–0.2 g of oil in a 100 ml Kjeldahl flask for two hours with a mixture containing selenium, sulphuric acid, copper sulphate ($CuSO_4.5H_2O$) and potassium sulphate. Cool, dilute, neutralise to methyl red with $10N$ alkali, filter off any copper hydroxide ($Cu(OH)_2$) which precipitates and make up to a known volume. Determine nitrogen on an aliquot, using Ammonia *Method 1*.

Fuels and lubricants[8]

Coal Treat 100–150 mg of the finely ground sample with 4 ml of a solution containing 32.6 g of chromic oxide (Cr_2O_3), 5 g of potassium persulphate ($K_2S_2O_8$) and 0.25 g of cobaltic oxide (Co_2O_3) in 100 ml of water, and heat under reflux with 10 ml of sulphuric acid at 200°C for 50–60 minutes. Treat the solution with excess alkali, distil off the ammonia and determine by Ammonia *Method 1*.

Metals, alloys and plating baths [9]

Cast iron Dissolve 0.5 g of sample in 12 ml of water, 3 ml of sulphuric acid and 8–10 drops of hydrofluoric acid. Heat, then add several drops of 10% barium chloride ($BaCl_2$) solution and centrifuge for 4–5 minutes. Pour the supernatant liquid into a flask. Heat the precipitate with 2 ml of sulphuric acid until fumes are evolved. After 90 minutes at the boiling point, transfer this solution to the flask containing the original solution using 10–12 ml of water to wash out the tube. To the contents of the flask add 20 ml of 50% caustic soda solution (NaOH), steam distil and collect the distillate in 1 ml of $0.01N$ sulphuric acid. Proceed by Ammonia *Method 1*.

Miscellaneous [9–13]

Feeding stuffs [9] Digest 0.1–1.0 g of finely powdered sample in 20 ml of sulphuric acid to which has been added 10–12 g of a mixture of 100 parts of potassium sulphate and 1 part of selenium. Continue the digestion for 30 minutes after the solution has cleared. Dilute the cooled digest to 200 ml and filter if necessary. Take 1 ml of this filtrate, dilute to 20–30 ml, add Nessler's reagent, make up to 50 ml and measure the colour.

Fertilizers [10–13] To not more than 0.6 of sample, containing not more than 42 mg of nitrate, add 1.7 g of Raney catalyst powder, 3 drops of tributyl citrate and 150 ml of a solution containing 10.6% potassium sulphate in 20% sulphuric acid. Stand for 10 minutes, then boil for 10 minutes, add mercuric oxide and then follow the usual Kjeldahl procedure [12]. Burch [10] claims that the best composition for the Raney catalyst is 10% cobalt, 40% nickel and 50% aluminium. The addition of cobalt to the usual 50:50 mixture of nickel and aluminium speeds up the reduction of nitrates, which in $8N$ sulphuric acid are reduced to ammonia in 10 minutes.

Alternatively [11] transfer a sample containing not more than 60 mg of nitrogen to a 500–800 ml Kjeldahl flask, add 1.2 g of chromium powder (100 mesh) and then 35 ml of water. Set aside for 10 minutes, but swirl the flask occasionally to keep the contents mixed. After the 10 minutes have elapsed add 7 ml of hydrochloric acid and set aside until a visible reaction occurs. This usually takes place within 1–5 minutes. Then heat the flask to the boiling point and boil for not more than 5 minutes. Cool, add 22 g of potassium sulphate, 1 g of mercuric oxide and 1.5 g of alundum, to act as an anti-bumping agent. Add 25 ml of sulphuric acid and proceed by the normal Kjeldahl procedure.

Morris et al [13] have compared available methods for the determination of nitrogen in fertilizers and conclude that the Kjeldahl method is more precise than the Dumas method.

Plants

Ekpet [14] recommends that 3% of salicylic acid ($C_6H_4(OH)COOH$) be added to the sulphuric acid in the normal Kjeldahl digestion. After this addition the mixture is cooled and 10 minutes later 1.5 g of sodium thiosulphate ($Na_2S_2O_3.5H_2O$) is added and then the digestion is continued in the normal way. It is claimed that the modified technique gives better recoveries of added nitrate than either the normal Kjeldahl or the sulphuric acid-hydrogen peroxide digestions.

Soils [15,16]

To 1–2 g of soil in a 100 ml pyrex flask add 2.5–5 ml of water and, after 1–2 minutes, 5–10 ml of sulphuric acid, stir well and then set aside for 30 minutes. Add 4–6 drops of perchloric acid ($HClO_4$), close the flask with a small funnel and boil for 5–7 minutes, until the sulphuric acid condenses on the sides of the flask above the reaction mixture. Stop boiling, add 2 drops of perchloric acid and then continue boiling. Repeat this procedure until the liquid above the soil is colourless, then cool, dilute to 100 ml, filter and determine nitrogen on an aliquot by Ammonia *Method 1* [15].

To determine "dispersible ammoniacal nitrogen" [16] weigh 1.5 g of sample into a stoppered cylinder, add 15 ml of water and 1 drop of an inoculating suspension obtained from a previous soil sample having a high ammonia content. Incubate at 36°C for 1 week, then add 1 drop of $5N$ sulphuric acid, shake, filter and treat a 5 ml aliquot with 0.5 ml of 20% sodium silicate ($Na_2SiO_3.5H_2O$) solution in $0.1N$ caustic soda (NaOH) and Nessler reagent. Shake for 2 hours, centrifuge and measure colour of supernatant fluid.

See also the determination of "Potentially Available Nitrogen".

References

1. M. Ashraf, M. Illahi, M. K. Bhatty and R. A. Shah, *Pakistan J. sc. ind. Res.*, 1963, **6**, 17; *Analyt. Abs.*, 1964, **11**, 1384.
2. L. T. Mann, *Analyt. Chem.*, 1963, **35**, 2179.
3. A. Fleck and H. N. Munro, *Clinica chim. Acta*, 1965, **11** (1), 2; *Analyt. Abs.*, 1966, **13**, 2510
4. L. Wuensch, *Z. analyt. Chem.*, 1969, **248** (1–2), 29; *Analyt. Abs.*, 1970, **19**, 5007
5. D. S. Galanos and V. M. Kapoulas, *Analyt. chim. Acta*, 1966, **34** (3), 360; *Analyt. Abs.*, 1967, **14**, 4026
6. A. Kagaya, *Proc. Res. Soc. Japan's Sugar Refineries' Technologists*, 1963, **12**, 35; *Analyt. Abs.*, 1964, **11**, 3342
7. M. K. Govind Rao and S. Raghavendar Rao, *Indian J. Technol.*, 1963, **1** (12), 464; *Analyt. Abs.*, 1965, **12**, 1999
8. L. L. Okhapkina, A. P. Bykova and G. A. Evstratova, *Zavodskaya Lab.*, 1965, **31** (3), 277; *Analyt. Abs.*, 1966, **13**, 3647
9. P. C. Williams, *Analyst*, 1964, **89**, 276
10. W. G. Burch and J. A. Brabson, *J. Assoc. off. agric. Chem.*, 1965, **48** (6), 1111
11. C. W. Gehrke, J. P. Ussary, C. H. Perrin, P. R. Rexroad and W. L. Spangler, *J. Assoc. off. agric. Chem.*, 1967, **50**, 965
12. J. A. Brabson and T. C. Woodis, *J. Assoc. off. analyt. Chem.*, 1969, **52** (1), 23; *Analyt. Abs.*, 1969, **18**, 3466
13. G. F. Morris, R. B. Carson and W. T. Jopkiewicz, *J. Assoc. off. analyt. Chem.*, 1969, **52**, 943
14. D. M. Ekpete and A. H. Cornfield, *Analyst*, 1964, **89**, 670
15. A. A. Meshcheryakov, *Pochevovdenie*, 1963, (5), 96; *Analyt. Abs.*, 1964, **11**, 4005
16. M. Pavageau, *Revta Quim. ind. Riode J.*, 1969, **38** (443), 11; *Analyt. Abs.*, 1970, **19**, 768

The Determination of Potentially Available Nitrogen (P A N)
using Nessler's reagent

Introduction

In the absence of a reliable method of measuring the nitrogen-supplying power of soils, nitrogen fertiliser requirements have to be arrived at by indirect and often erroneous means. This test was developed[1] to enable an accurate estimate to be made of the amount of nitrogen potentially available in the soil.

Most of the nitrogen present in soil is combined in organic substances which are not readily susceptible to bacterial decomposition. However the nitrogen which is contained in proteins may be released by bacterial action. The protein composition of soils has been shown to be relatively constant[2,3]. It has therefore been assumed that a chemical determination of the readily hydrolysable soil protein will provide a measure of the nitrogen available for plant growth. This assumption has been checked by measuring the organic matter,[4] total nitrogen,[5] rate of nitrification by incubation for 30 days,[5] and, using this technique, the potentially available nitrogen (P A N). It was shown[1] that all these measurements were closely related. Subsequent tests with plants confirmed that the P A N tests gave a measure of the nitrogen supplying power of soils that is as reliable as the more time-consuming determinations of organic matter or total nitrogen content.

Principle of the method

The protein in the soil is hydrolysed by treatment with sulphuric acid. The ammonia thus liberated is extracted with water and reacted with Nessler's reagent. The intensity of the yellow colour thus formed, which is proportional to the concentration of ammonia and thus to the concentration of hydrolysable protein, is measured by comparison with a series of Lovibond permanent colour standards.

Reagents required

1. *Sulphuric acid solution* Dilute 2.0 ml of concentrated sulphuric acid to 1000 ml with distilled water, by adding the acid slowly to the water.
2. *Nessler's reagent* The published formulae for Nessler's reagent vary considerably. It is important that when using Lovibond discs the reagent employed should correspond to that used for standardising the colours of the discs. The following formula must therefore be used:—

   Dissolve 35 g of potassium iodide (KI) and 12.5 g of mercuric chloride ($HgCl_2$) in 800 ml of distilled water in a 1 litre volumetric flask. Add a cold saturated solution of mercuric chloride until a slight red precipitate remains despite repeated shaking. Then add 120 g of sodium hydroxide (NaOH). Shake the flask until the sodium hydroxide has dissolved, add a little more of the saturated solution of mercuric chloride and make the total volume up to 1 litre with distilled water. Allow the solution to stand for several days, shaking the flask occasionally, and use the clear supernatant liquid as the test reagent.
3. *Sodium phosphate ($Na_3PO_4.12H_2O$) solution* Dissolve 5 g of sodium phosphate in 100 ml of distilled water.
4. *Sodium chloride (NaCl) solution* Dissolve 0.5 g of sodium chloride in 100 ml of distilled water.

All chemicals used in the preparation of reagents should be of analytical reagent quality.

The Standard Lovibond Discs NAB, NAC and NAD

Disc NAB covers the range 10–26 µg of ammonia in steps of 2 µg. This range corresponds to 20–52 ppm of P A N calculated as ammonia (Note 1).

Disc NAC covers the range 28–60 µg of ammonia in steps of 4 µg. This range corresponds to 56–120 ppm of P A N calculated as ammonia (Note 1).

Disc NAD covers the range of 60–100 µg of ammonia in steps of 5 µg. This range corresponds to 120–200 ppm of P A N calculated as ammonia (Note 1).

Technique

Pass the air-dried soil sample through a 1.0 mm screen. Place 0.5 g of screened soil in a 100 ml beaker, add 2 ml of sulphuric acid solution (reagent 1) and heat on a steam bath until dry. Allow the beaker to remain on the steam bath for at least 15 minutes after the soil appears to be dry. Add 50 ml of distilled water and 1 drop of sodium phosphate solution (reagent 3). Stir thoroughly and then filter, into a Nessler cylinder (Note 2), through a Whatman No. 2 filter paper which has been washed with sodium chloride solution (reagent 4) immediately prior to filtering the soil suspension. Add 2 ml of Nessler's reagent (reagent 2) to the filtrate shake well to mix, and place the cylinder in the right-hand compartment of the Nessleriser. Fill an identical Nessler cylinder with distilled water and place this in the left-hand compartment. Place the Nessleriser before a uniform source of white light, such as the Lovibond White Light Cabinet or, failing this, north daylight, and match the colour of the test solution with the standards in the disc.

For the interpretation of the results see Note 3.

Notes

1. As the standards are calibrated in micrograms of ammonia it is convenient to express the P A N content of the soil in ppm of ammonia. If for any reason the P A N content is required in terms of nitrogen rather than ammonia, then multiply the ammonia figure by 0.8.

2. It is important that Nessler cylinders are used which conform to the specification of those which were employed for the calibration of the discs, namely with the 50 ml calibration mark at a height of 113 $\pm$3 mm measured internally.

3. It has been found[1] that the P A N value for soil falls within the limits 30–180 ppm when calculated in terms of ammonia. Crops grown on soils having P A N values of 50 ppm or less showed symptoms of nitrogen deficiency. The crop yield varied linearly with the P A N content and addition of nitrogen, from a mixture of sodium nitrate and ammonium sulphate, at a level calculated to increase the P A N figure by 60 ppm, produced increases in yield varying from 200% when the original P A N level was 39 ppm to 27% when the original P A N level was 115 ppm.

References

1. E. R. Purvis and M. W. M. Leo, *J. Agric. Food Chem.*, 1961, **9**, 15
2. J. M. Bremner, *Biochem. J.*, 1950, **47**, 538
3. D. I. Davidson, F. J. Sowden and H. J. Atkinson, *Soil Sci.*, 1951, **71**, 347
4. E. R. Purvis and G. E. Higson, *Ind. Eng. Chem. Anal. Edn.*, 1939, **11**, 19
5. A. L. Prince, *Soil Sci.*, 1945, **59**, 47

The Determination of "α-amino" Nitrogen
using copper acetate

Introduction

This method has been developed for the estimation of α-amino nitrogen in the sugar beet. The term "α-amino nitrogen" refers to the nitrogen in those substances not eliminated during the normal purification processes in the beet sugar factory. These nitrogenous constituents can have a pronounced influence on the ultimate extraction efficiency, and the measure of their prevalence in the beet itself must therefore be of considerable value. Of the total harmful nitrogen, that present in the form of amino acids is regarded as the most objectionable.

Principle of the method

This method, due to Stanek and Pavlas, makes use of the blue colour produced by amino acids in the presence of copper acetate.

Reagent required

Dissolve 10 g of copper nitrate ($Cu(NO_3)_2.3H_2O$) in 700 ml distilled water, and add 250 g sodium acetate ($CH_3COONa.3H_2O$). Prepare solution in the cold, from chemicals of analytical reagent quality, then make up to 1,000 ml and filter.

The Standard Lovibond Comparator Disc 3/27

The disc covers the range 10–90 mg of Nitrogen (in steps of 10 mg) in 100 g of fresh sugar beet, and is designed for use with a 1″ cell. The colour standards were matched against solutions prepared from mono sodium glutamate (see Note 4).

Technique

To 2 ml of the copper reagent add 20 ml of freshly filtered clarified solution of half-normal strength, i.e. 13 g beet in 100 ml solution, prepared by the customary digestion method from beet brei or cossettes, and which contains no added acetic acid. Place the well-shaken coloured solution in the 1″ cell in the right-hand compartment of the comparator, so that it comes behind the centre of the disc. Rotate the disc until the nearest colour match is obtained. The figure shown in the aperture at the lower right-hand side of the comparator face gives the value for the sample under test in milligrammes noxious nitrogen per 100 g beet, or parts per hundred thousand. Readings should be taken using a standard light source (Note 3). As the filtrates tested are perfectly clear and colourless, no second cell is necessarily required in the left-hand side of the comparator, but for greater ease in reading, distilled water may be used as a blank.

Notes

1. The standard colours for calibrating this disc have been prepared by using a solution of mono-sodium glutamate of such strength that 1 ml made up to 100 ml with distilled water gives, on addition of 10 ml of the special copper reagent, a blue colour equivalent to 10 mg of α-amino nitrogen per 100 g of beet.
2. This disc was prepared with the co-operation of the Central Laboratory of the British Sugar Corporation, Ltd., and their help is gratefully acknowledged.
3. A suitable standard source of white light is available in the Lovibond White Light Cabinet.
4. If asparagine is substituted for sodium glutamate, the same results are obtained up to 50 ppm. Beyond this, the curve obtained from asparagine flattens more quickly.

The reading multiplied by 0.0929 = millimoles of amino acid per litre of test solution.

In the case of asparagine, the reading multiplied by 12.26 = mg asparagine per litre of test solution, because only the α-amino nitrogen and not the amide nitrogen in asparagine reacts under the test conditions.

Reference

Stanek & Pavlas: *Z. Zuckerind cechoslovak.*, Rep. **59**, 129 (1934/35); **60**, 46 (1935/36)

The Determination of Dissolved Oxygen

Introduction

The presence of dissolved oxygen in boiler feed-water is a serious cause of corrosion. This is believed[1,2] to be caused by the electrolysis which arises when metal is in contact with water containing zones of different dissolved oxygen content. These can result from imperfect mixing of the liquid, or from the presence of scale, paint, dirt, or even a gas bubble, on one portion of the metal surface. It is recommended[3] that the dissolved oxygen content of the feed-water to low pressure (less than 250 p.s.i.) boilers should not exceed 0.2 ml per litre and preferably should be below 0.1 ml per litre. For high pressure boilers (over 250 p.s.i.) the concentration should be not greater than 0.02 and preferably should be below 0.01 ml per litre.

The dissolved oxygen content is also of great importance in the control of effluents[4] and in the control of fermentation processes.

Dissolved oxygen is essential for the respiration of fish and other aquatic animals and is also necessary in a river or estuary to prevent the onset of anaerobic conditions and the development of objectionable smells. The amount present in unpolluted river water saturated with oxygen at 15°C is about 10 ppm, but in sea water the value is about 8 ppm at this temperature. The determination of dissolved oxygen is also useful for controlling the efficiency of an activated sludge sewage treatment plant. Dissolved oxygen is the most important index of the quality of an estuary water; it is advisable to have at least 50% of saturation (e.g. 5 ppm at 15°C) if fisheries are to thrive, and at least 30% of saturation (i.e. 3 ppm at 15°C) to avoid nuisance by smell. When the dissolved oxygen in an estuary falls as low as 5–10% of saturation on account of pollution, any nitrate present is first reduced by bacterial action to nitrogen, after which there is a further drop in dissolved oxygen to zero. The estuary is then liable, especially in warm weather, to become anaerobic and to develop bad smells of hydrogen sulphide.

There are several ways of expressing the dissolved oxygen content of a water sample, namely:—

1. parts by weight of oxygen per 1,000,000 parts by volume of sample (i.e. mg per litre, or ppm).

2. ml of oxygen (0°C, 760 mm pressure) per litre of sample.

3. as a percentage of saturation. In order to calculate this figure, which is particularly useful in estuary surveys, the saturation values for water in equilibrium with air at various temperatures must be obtained from a Solubility Table.

The following Table gives figures for the solubility of oxygen in fresh water and salt water at various temperatures when the water is in equilibrium with air under a pressure of 760 mm of mercury:—

Solubility of oxygen in equilibrium with air at 760 mm pressure.*

TEMPERATURE °C	FRESH WATER (ppm of dissolved oxygen)	SALINE WATER (Correction (ppm) to be subtracted for each 1 gram of total salts per kilogram of saline water)
0	14.63	0.0925
1	14.23	0.0890
2	13.84	0.0857
3	13.46	0.0827
4	13.11	0.0798
5	12.77	0.0771
6	12.45	0.0745
7	12.13	0.0720
8	11.84	0.0697
9	11.55	0.0675
10	11.28	0.0653
11	11.02	0.0633
12	10.77	0.0614
13	10.53	0.0595
14	10.29	0.0577
15	10.07	0.0559
16	9.86	0.0543
17	9.65	0.0527
18	9.46	0.0511
19	9.27	0.0496
20	9.08	0.0481
21	8.91	0.0467
22	8.74	0.0453
23	8.57	0.0440
24	8.42	0.0427
25	8.26	0.0415
26	8.12	0.0404
27	7.97	0.0393
28	7.84	0.0382
29	7.70	0.0372
30	7.57	0.0362

*Reproduced by courtesy of the Editor of the Journal of Applied Chemistry.

The figures in this Table are taken, by permission, from the article by H. A. C. Montgomery, N. S. Thom and A. Cockburn. (*J. Appl. Chem.* 1964. 14. 280.)

The following formulæ show the relation between the various ways of expressing dissolved oxygen:

(a) ml of dissolved oxygen per litre $\times$ 1.43
= ppm of dissolved oxygen

(b) ppm of dissolved oxygen $\times$ 0.7
= ml of dissolved oxygen per litre

(c) dissolved oxygen, expressed as % of saturation
$$\frac{\text{ppm of dissolved oxygen found in sample} \times 100}{\text{ppm of dissolved oxygen at temperature of sample}}$$
(from above Table)

It is possible for the % saturation figure to exceed 100% sometimes. From the above Table it will be seen that fresh water saturated with air (20% of oxygen, 79% of nitrogen) contains 9.08 ppm of dissolved oxygen at 20°C, but water in equilibrium with *pure* oxygen at 20°C contains about 40 ppm of dissolved oxygen, i.e. more than four times as much. Figures exceeding 100% of saturation are obtained in relatively unpolluted rivers when plant life (algae, etc.) under the influence of bright sunshine, converts dissolved carbon dioxide into oxygen and utilises the carbon for the synthesis of sugars, etc., a very important natural phenomenon known as "photo-synthesis". It follows that the dissolved oxygen figure for such a river will be higher during the day than at night and higher on bright summer days than on dull winter ones. In polluted rivers, where there is little or no photo-synthesis, a higher % saturation figure is generally obtained in winter, when biochemical reactions using up dissolved oxygen proceed at a much slower rate than at the higher temperatures of summer.

Methods of sample preparation

Beverages[5]

Beer and wort Method 1 is recommended[5] for the analysis of this type of sample.

Sewage and industrial wastes[4,6-10]

For polythionate, thiosulphate and sulphate wastes, preliminary oxidation using alkaline hypochlorite and the subsequent removal of excess chlorine by sulphurous acid has been suggested[6]. The sulphurous acid treatment is also specified for wastes containing residual chlorine. Suspended matter such as raw sewage, sludge or river mud should be removed as follows[7]. Collect a sample of about 1 litre in a glass stoppered bottle with the usual precautions against aeration. Add 10 ml of a 10% solution of potassium aluminium sulphate ($KAl(SO_4)_2 \cdot 12H_2O$) and follow this with 1-2 ml of 0.880 aqueous ammonia (NH_4OH). Re-stopper the bottle and rotate for one minute. Allow the floc to settle for 10 minutes, syphon the clear supernatant liquid into the dissolved oxygen bottle and estimate. For activated sludge and other samples containing readily oxidisable organic matter a deactivating liquid must be added to the sample to stop deoxygenation. Ruchoft and Placak[8] recommend a mixture prepared by dissolving 32 g of sulphamic acid ($NH_2SO_2 \cdot OH$) in 475 ml of distilled water and adding a solution of 50 g of copper sulphate ($CuSO_4 \cdot 5H_2O$) in 500 ml of distilled water and 25 ml of glacial acetic acid (CH_3COOH). Heat must not be used to dissolve the sulphamic acid, nor should the mixed reagent be exposed to heat at any time. For each 100 ml of sample measure 1 ml of this reagent into the empty sampling bottle, then fill the bottle to overflowing and insert the stopper. Thoroughly mix the contents, allow the sludge to settle and then syphon the top liquid into a dissolved oxygen bottle and use the sodium azide modification of Winkler's method (*Method 2* or *3*).

Spent sulphite liquor present in industrial wastes can give low results for dissolved oxygen when the Winkler method and its modifications (*Methods 2* and *3*) are used for the determination. This is caused by the sulphite consuming part of the liberated iodine. Baker[9] suggests that the extent of this consumption can be measured by adding 5 ml of 0.025N iodine solution and 1.4 ml of sulphuric acid to the volume of water used for the dissolved oxygen determination (200 ml for *Method 2*, 1,000 ml for *Method 3*) and titrate with 0.025N sodium thiosulphate ($Na_2S_2O_3$) solution using starch as an indicator. Subtract this titre from 5.0 to obtain the iodine demand and add the oxygen equivalent of this to the dissolved oxygen figure obtained. Alternatively Felicetta[10] recommends the titration of the sulphite with standard permanganate ($KMnO_4$) solution to obtain the same figure for oxygen correction.

Water[5,11-23,35]

The important aspect of sampling is the complete exclusion of atmospheric oxygen. Many methods of doing this have been proposed[5,11-17,20], and approved methods are described in the individual tests. The use of an ion exchange resin to remove interfering cations is recommended[18,19]. Attention is also drawn[16] to the dangers of sampling through rubber, silicone rubber or polystyrene tubing, and the use of nylon tubing is recommended for temperatures up to 90°C. Owing to their permeability to oxygen, polyethylene, polypropylene, furfuraldehyde polymer (Tygon) and polytetrafluoroethylene tubing have been reported[20] to be unsuitable for sampling water for this test.

For heavily polluted waters such as swamps[21], preliminary clarification enables the standard methods to be applied. Sea water may be examined by normal techniques but the time of agitation of sample with reagents must be limited to one minute[22]. A micro method based on *Method 1* has been described by St. John[23]. This uses a gas-tight syringe as both sampler and cell.

The individual tests

Method 1 is the indigo-carmine method[5,12,24-26,35] originated by Buchoff et al[24].

Method 2 is the Alsterberg[27] modification of the Winkler method[28-30] and is one of the standard tests for sewage[4] and for industrial waters[5,12]. Disc 3/3 used in this method may also be used for the determination of Permanganate Value, (see page 601).

Method 3 is a more sensitive modification of the Alsterberg method[27] devised by Arnott and MacPheat[31].

Method 4 was devised[3,32,33] as a simple control method for boiler feed-water and is a modification of the earlier test devised by McCrumb and Kenny[34].

References

1. U. R. Evans and T. P. Hoar, *Trans. Faraday Soc.*, 1934, **30**, 424
2. U. R. Evans, *"Metallic Corrosion, Passivity and Corrosion"*, Edward Arnold & Co., London, 1937
3. W. Francis, *"Boiler House and Power Station Chemistry"*, Edward Arnold & Co., London, 1940
4. Min. of Housing and Local Govt., *"Methods of Chemical Analysis as Applied to Sewage and Sewage Effluents"*, H.M. Stationery Office, London, 1956
5. P. Jenkinson and J. Compton, *Proc. Amer. Soc. Brew. Chem.*, 1960, 73; *Analyt. Abs.*, 1961, **8**, 5267
6. Amer. Pub. Health Assoc., *"Standard Methods for the Examination of Water and Sewage"*, 9th edit., New York, 1946, pp 132-134
7. Ibid., p 136
8. C. C. Ruchhoft and O. R. Placak, *Sewage Wks. J.*, 1942, **14**, 638
9. M. C. Baker, *T.A.P.P.I.*, 1965, **48** (4); 81A, *Analyt. Abs.*, 1966, **13**, 4495
10. V. F. Felicetta and D. R. Kendall, *T.A.P.P.I.*, 1965, **48** (6), 362
11. Brit. Standards Inst., *"Methods of Testing Water Used in Industry"*, B.S. 2690, London, Part 2, 1965
12. Brit. Standards Inst., *"Routine Control Methods of Testing Water Used in Industry"*, B.S. 1427, London, 1962, and amendments
13. H. A. C. Montgomery and A. Cockburn, *Analyst*, 1964, **89**, 679
14. J. Bargh, *Chem. and Ind.*, 1959 (42), 1307
15. E. C. Potter and G. E. Everitt, *J. Appl. Chem.*, 1959, **9**, 642
16. E. C. Potter and G. E. Everitt, *J. Appl. Chem.*, 1960, **10**, 48
17. E. C. Potter and J. F. White, *J. Appl. Chem.*, 1957, **7**, 317
18. E. C. Potter and G. E. Everitt, *J. Appl. Chem.*, 1959, **9**, 645
19. E. C. Potter and J. F. White, *J. Appl. Chem.*, 1957, **7**, 459
20. G. E. Everitt, E. C. Potter and R. G. Thompson, *J. Appl. Chem.*, 1965, **15** (8), 398; *Analyt. Abs.*, 1967, **14**, 423
21. L. C. Beadle, *J. Expl. Biol.*, 1958, **35**, 556; *Analyt. Abs.*, 1959, **6**, 4603
22. K. Grasshoff, *Kieler Meeresforsch*, 1962, **18**, 42; *Analyt. Abs.*, 1963, **10**, 4436
23. P. A. St. John, T. D. Winefordner and W. S. Silver, *Analyt. chim. Acta*, 1964, **30** (1), 49; *Analyt. Abs.*, 1965, **12**, 2017
24. L. S. Buchoff, N. M. Ingber and J. H. Brady, *Analyt. Chem.*, 1955, **27**, 1401
25. G. P. Alcock and K. B. Coates, *Chem. and Ind.*, 1958, 554
26. P. Harsch, *Nature* (Lond.), 1952, **169**, 792
27. G. Alsterberg, *Biochem. Z.*, 1925, **159**, 36
28. L. W. Winkler, *Ber.*, 1888, **21**, 2843
29. S. Rideal and W. T. Burgess, *Analyst*, 1909, **34**, 193
30. N. L. Allport and J. E. Brocksopp, *"Colorimetric Analysis"*, 2nd edit., Vol. II, Chapman & Hall, London, 1963
31. J. Arnott and J. MacPheat, *Engineering*, 1953, **176**, 103
32. J. Haslam and G. Moses, *Chem. and Ind.*, 1938, 344
33. H. J. Meyer and C. Brack, *Water Pollut. Abstr.*, 1952, **25**, 130
34. F. R. McCrumb and W. R. Kenny, *J. Amer. Waterworks Assoc.*, 1929, **21**, 400
35. Brit. Standards Inst. *"Treatment of Water for Marine Boilers"*, B.S. 1170, London, 1968

Oxygen Method 1
using indigo-carmine

Principle of the method

Reduced indigo-carmine is a bright yellow-green colour. On oxidation it changes to orange, to red, to purple, to blue, and finally to blue-green in the completely oxidised form. The indigo-carmine is reduced by glucose in the presence of potassium hydroxide. Glycerol is used to provide air-stability and to sharpen the colours.

Reagents required

1. *Indigo-carmine stock solution:*

Indigo-carmine—reagent grade	0.018 g
Glucose—reagent grade	0.20 g
Distilled water	to 5 ml
Glycerol—reagent grade	75 ml

    This reagent is stable for 90 days if stored in a refrigerator, and for 6 weeks at ambient temperature if stored in the dark. A combined tablet is available containing the required amounts of indigo-carmine and glucose (Note 1).

2. *Potassium hydroxide (KOH)* Dissolve 37.5 g of potassium hydroxide in 62.5 ml of distilled water.

3. *Leuco reagent* In a small bottle mix 8 ml of reagent 1 and 2 ml of reagent 2. After thorough mixing allow the reagent to stand until the initial dark red colour changes to lemon yellow (approximately 10 minutes). The leuco reagent should be freshly prepared on the day on which it is to be used.

The Standard Lovibond Nessleriser Disc NOE

The disc covers the range 0 to 0.120 ml/l oxygen (O_2) in steps 0, 0.005, 0.01, 0.015, 0.03, 0.055, 0.08, 0.1, 0.12 ml/l. The disc must be used in conjunction with the special glass reaction vessels described. To convert ml/l to parts per million, multiply by 1.43.

Technique

The sample must be taken and the reagent added in the absence of atmospheric oxygen. For this purpose a special glass reaction vessel has been developed for use with the Lovibond Nessleriser (see Figure). Sampling may be carried out by the "Submerged Bottle Method" which is probably the most reliable method to use. However, with experience, a satisfactory sample can be obtained by slowly lowering the reaction vessel from the end of a submerged and rigid vertical sampling tube through which the water is flowing at a rate of about a litre per minute at room temperature.

With a dropping pipette the vial attached to the inner surface of the special reaction vessel is filled with leuco reagent (reagent 3). Any entrapped air bubbles are allowed to rise, and disperse, so that the vial is completely full and free of air. It is then sealed by means of a glass ball which may be placed in position by sliding the ball down a glass tube placed over the end of the vial. The sample is allowed to flow through the reaction vessel for at least 10 minutes, to remove any reagent on the outside of the vial. The reaction vessel is then removed from the sampling line and the stopper inserted, under water, to prevent the trapping of any air.

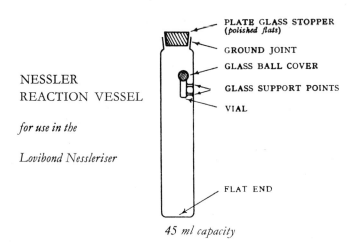

NESSLER REACTION VESSEL

for use in the

Lovibond Nessleriser

45 ml capacity

The reaction vessel is now removed from the water; inverted to allow the ball to fall off the vial, and the contents thoroughly mixed by shaking. Five minutes after mixing, the reaction vessel is placed in the right-hand compartment of the Nessleriser and a similar modified Nessleriser tube filled with the sample without reagents is placed in the left-hand compartment to serve as a blank. The colour developed in the sample is then compared with the colours of the permanent glass standards, using a standard source of white light, such as the Lovibond White Light Cabinet. In the absence of a standard source use north daylight wherever possible.

Notes

1. The preparation of the indigo-carmine reagent can be simplified by the use of special indigo-carmine-glucose tablets prepared by B.D.H. Chemicals Ltd., Poole, Dorset. Place one indigo-carmine-glucose tablet in a dry 100 ml measuring cylinder and add 5 ml of distilled water from a pipette. Swirl the cylinder until the tablet has completely dissolved. This needs a few minutes and, as the solution is deep blue, careful observation is necessary to ensure that no particles of the tablet remain undissolved.

To the measuring cylinder add analytical reagent grade glycerol to give a total volume of 80 ml. Mix the contents thoroughly with a long glass rod. This stock solution will keep for one month if kept cool and in the dark.

2. No interference has been found from nickel, cupric or zinc ions in concentrations of 1 ppm nor from ferric ion at 3 ppm. Sulphite and hydrazine do not interfere at the low concentrations normally found in boiler feed water. Of the metals tested, only ferrous ion has been found to interfere. This ion, or intolerable concentrations of other ions, can be removed by a mixed bed ion-exchange demineraliser in series with the sampling line.

3. The temperature of the sample must not exceed 70°F (21°C). Higher temperatures invalidate the test.

4. The adaptation of this test to the Nessleriser is due to the "Alfloc" Water Treatment Service of I.C.I. Ltd., to whom acknowledgment is made.

5. Conforms to A.S.T.M. D888–66.

Oxygen Method 2
using Alsterberg's method

Principle of the method

In the Winkler method for the determination of dissolved oxygen in water, the sample contained in a completely filled bottle is treated with a solution of manganese chloride and a solution containing potassium hydroxide and potassium iodide. Manganese hydroxide is precipitated and this absorbs the dissolved oxygen present in the sample, forming the higher oxides of manganese. On subsequently acidifying the mixture with sulphuric acid, the higher oxides of manganese react with the potassium iodide present, liberating iodine in an amount equivalent to that of the dissolved oxygen originally present. Instead of titrating the liberated iodine as in the original Winkler method, the amount present may be determined by measuring the depth of the yellow colour produced. This can be carried out readily by comparing the colour with a series of Lovibond glasses standardised on the colours produced by known quantities of iodine liberated in the course of the determination of dissolved oxygen by the Winkler method. In Alsterberg's modification of this method nitrites, which interfere, are decomposed by the use of sodium azide.

Reagents required

    Ampoules (a) white glass containing manganese chloride ($MnCl_2$) 0.2 g, distilled water (air free) to 2 ml

                     (b) amber glass containing potassium iodide (KI) 0.3 g, potassium hydroxide 1.0 g, sodium azide (NaN_3) 0.020 g, distilled water (air free) to 2 ml

    Concentrated sulphuric acid

    Compressed tablets of potassium iodide, 5 grain ($=0.324$ g)

Standard Lovibond Discs, Comparator Disc 3/3, and Nessleriser Discs NKA and NKB

Disc 3/3 covers the range from 4 to 12 parts of dissolved oxygen per million by weight, in steps of 1, which equals 2.7 to 8.3 ml per litre or 4 to 12 mg per litre. This disc may also be used for the determination of the Permanganate Value of water, see page 601.

Disc NKA: for the determination of dissolved oxygen in boiler feed water and covering the range from 0.05 to 1.0 ml of dissolved oxygen per litre in steps of 0.1 from 0.1 upwards. This equals 0.07 to 1.4 mg per litre (ppm).

Disc NKB: for the determination of dissolved oxygen in river water and other natural waters and covering the range from 0.4 to 1.6 part of dissolved oxygen per million by conducting the test as set out below, the range may be extended to 16 parts per million or mg per litre.

In order to increase the range of utility of the disc the yellow solution obtained in the test, may, if necessary, be suitably diluted so as to bring the colour within the range of the standard glasses. Simple dilution with water, however, is not satisfactory, as this also reduces the concentration of the potassium iodide, and the yellow colour of the iodine-potassium iodide compound is destroyed and replaced by the much paler brown colour of iodine. It is therefore essential to add potassium iodide when diluting the solution. The potassium iodide can be added conveniently in the form of a compressed tablet.

Technique

A wide necked glass bottle which contains about 250 ml when full, having a ground-in glass stopper, is filled to the brim with the water to be examined. The temperature of the sample is noted. Make a file mark near each end of one of the white ampoules and one of the amber ampoules. Break off one tip of the white ampoule, place the open end of the ampoule under the surface of the water in the bottle and then break off the other tip, thus allowing the contents to flow into the bottle. Transfer the contents of the amber ampoule to the bottle in the same manner, insert the stopper in the bottle so that no bubble of air is trapped, invert the bottle several times in order to mix the contents thoroughly and allow to stand for five minutes for the precipitate to settle. Remove the stopper and by means of a pipette add rapidly 2 ml of concentrated sulphuric acid, the tip of the pipette being just below the surface of the water: again stopper the bottle and mix the contents thoroughly by inverting the bottle several times.

Comparator

One 13.5 mm test tube is filled with an untreated sample of the water and the other tube with the treated solution under examination. The former is placed in the left-hand compartment of the comparator and the latter in the right-hand compartment.

The comparator is held about 18 inches from the eyes facing a standard source of white light such as the Lovibond White Light Cabinet, or, failing this, north daylight. The disc is revolved until the two apertures show the same colour, and the amount of dissolved oxygen in parts per hundred thousand is read at the indicator recess near the right-hand bottom corner of the comparator.

Nessleriser

With the brown solution thus obtained fill one of the Nessleriser glasses to the 50 ml mark and place it in the right-hand compartment of the Nessleriser. In the left-hand compartment place 50 ml of the untreated water to which has been added a tablet of potassium iodide and 0.5 ml of analytical reagent grade sulphuric acid. (In the case of a water which has been proved to remain unaltered in colour on the addition of potassium iodide and sulphuric acid, it will be sufficient if the Nessleriser glass in the left-hand compartment is filled with the untreated water.) Stand the Nessleriser before a standard source of white light and compare the colour in the test solution with the colours in the standard disc, rotating the disc until a colour match is obtained.

Notes

1. Should the colour of the treated water be deeper than that of the standard glasses, an aliquot of the treated water should be diluted with the untreated water, and after a tablet of potassium iodide and 0.1 ml of concentrated sulphuric acid have been added, the colour of the resulting solution should be compared with the standard glasses.

When it has been necessary to dilute the treated water before matching the colour, the figures indicated should be multiplied by the corresponding dilution figure.

2. Many natural waters contain iron, which in the ferric state will liberate iodine from an acidified solution of potassium iodide. To allow for any yellow colour due to iodine from this source, it is advisable to add a tablet of potassium iodide and 0.1 ml of concentrated sulphuric acid to the untreated water under test and to use this solution in the tube in the left-hand compartment instead of an untreated blank.

3. By dividing the oxygen content, ascertained by this method, by the concentration of dissolved oxygen necessary to saturate the water at the temperature of sampling (these values can be obtained from published tables or page 292 previous), and multiplying by 100, the percentage saturation is obtained.

4. If a more sensitive test is required, the modification of the Winkler method, described as *Method 3* should be used.

5. The preparation of the necessary discs was suggested by members of the staff of the Ministry of Agriculture and Fisheries Research Station at Alresford, who assisted in the standardization.

Oxygen Method 3
using a modification of Alsterberg's method

Principle of the method

The method is essentially that due to Winkler, but it has been so modified that the iodine liberated is measured colorimetrically and not titrated as in the original procedure, and interference by nitrite ions is prevented by the use of sodium azide.

Water contained in a completely filled bottle is treated with a solution of manganese chloride and a solution containing potassium hydroxide and potassium iodide. Manganese hydroxide is precipitated and this absorbs the dissolved oxygen present in the sample, forming the higher oxides of manganese. On subsequently acidifying the mixture with sulphuric acid, these react with the potassium iodide present, liberating iodine in an amount equivalent to that of the dissolved oxygen originally present. The iodine is extracted by shaking with carbon tetrachloride and the colour matched in the Nessleriser with a disc containing a series of yellow glasses, each of which has been carefully matched to correspond to a definite concentration of iodine. The oxygen content of the water is indicated by the figure on the disc corresponding to the standard colour which matches that of the test solution.

Boiler water often contains copper and ferric iron and, as these will liberate iodine from potassium iodide, it is essential to carry out a blank determination. In an acid solution, potassium iodide is unaffected by the presence of dissolved oxygen; thus, by adding the reagents in the reverse order, the iodine liberated is equivalent solely to the interfering ions.

Reagents required

1. White glass ampoules each containing 0.2 g of manganous chloride ($MnCl_2$) in sufficient air-free water to produce 2 ml.
2. Amber glass ampoules each containing 0.3 g of potassium iodide (KI) 1 g of potassium hydroxide (KOH) and 0.020 g of sodium azide (NaN_3) in sufficient air-free water to produce 2 ml.
3. Sulphuric acid (analytical reagent grade)
4. Carbon tetrachloride (analytical reagent grade)

The Standard Lovibond Nessleriser Discs NYA and NYB

The following two Nessleriser discs are available:—

Disc NYA covers the range 0.001, 0.003, 0.005, 0.007, 0.009, 0.012, 0.015, 0.018, 0.020 ml dissolved oxygen per litre (0.0014—0.028 parts per million).

Disc NYB covers the range 0.025, 0.03, 0.035, 0.04, 0.045, 0.05, 0.06, 0.08, 0.1 ml dissolved oxygen per litre (0.035—0.14 parts per million).

Technique

The water used in the test should be cooled by an efficient coil to below 65°F and preferably to a temperature less than that of the surrounding air. This ensures that the air is not subsequently drawn in by contraction of the sample and also facilitates the easy withdrawal of the stopper.

Fill a bottle, of about 1,110 ml capacity and fitted with a ground-in glass stopper, with the water to be tested. In order to avoid contamination from atmospheric oxygen it is necessary to fill the bottle by means of a tube passing to the bottom and to allow the water to overflow so as to displace the air in the bottle and also the water which has come in contact with the air; the tube should be withdrawn whilst the water is still running. Make a file mark near each end of one of the white ampoules and one of the amber ampoules. Break off one tip of the white ampoule, place the open end of the ampoule under the surface of the water in the bottle and then break off the other tip, thus allowing the contents to flow into the bottle. Transfer the contents of the amber ampoule to the bottle in the same manner, insert the

COLORIMETRIC CHEMICAL ANALYTICAL METHODS — Oxygen Method 3

stopper in the bottle so that no bubble of air is trapped, invert the bottle several times in order to mix the contents thoroughly and allow to stand for five minutes. Remove the stopper and by means of a pipette add rapidly 2 ml of sulphuric acid; again stopper the bottle and mix the contents thoroughly by inverting the bottle several times. Transfer 1,000 ml to a separating funnel and extract by shaking twice with 20 ml portions of carbon tetrachloride and, finally, with a further 10 ml.

Pour the combined extracts into a Nessleriser glass and place it in the right-hand compartment of the Nessleriser.

A blank is carried out by treating a further 1,100 ml of the water under test in a similar manner except that the reagents are added in the **reverse** order. The combined extracts are placed in the left-hand compartment of the Nessleriser.

Stand the Nessleriser before a standard source of white light such as the Lovibond White Light Cabinet or, failing this, north daylight, and compare the colour in the test solution with the colours in the standard disc, rotating the disc until a colour match is obtained.

The figures on both Disc NYA and Disc NYB indicate millilitres of dissolved oxygen per litre.

Notes

1. It is essential that the water be completely free from suspended matter and also from sulphites. Prior to and during a de-aeration test, the sulphite supply should be cut off. When there is any doubt about the purity of the water it is advisable to verify that it does not absorb iodine. In order to test for absorption, a measured amount of iodine solution is added to a litre of acidified water, extracted with 50 ml of carbon tetrachloride and matched on the disc. For example, if 8.00 ml of N/1000 iodine were added initially, then the colour of the carbon tetrachloride extract should correspond to 0.045 ml per litre of oxygen.

2. For the estimation of higher concentrations of dissolved oxygen in water, the original Alsterberg's method is suitable, see *Method 2*.

3. It must be emphasised that readings obtained by the Lovibond Nessleriser and disc are only accurate provided that Nessleriser glasses are used which conform to the specification employed when the discs were calibrated, namely that the 50 ml mark shall fall at a height of 113 mm plus or minus 3 mm measured internally.

Oxygen Method 4
using acid *ortho*-tolidine

Principle of the method

Manganous hydroxide is generated *in situ* from manganous chloride and potassium hydroxide. It is oxidised to manganic compounds by the dissolved oxygen and these react with dilute hydrochloric acid to produce manganous chloride and chlorine. The free chlorine is then determined by the yellow colour produced on reaction with acid *ortho*-tolidine. The intensity of this colour, which is proportional to the chlorine concentration and thus to the dissolved oxygen concentration, is measured by comparison with a series of Lovibond permanent glass colour standards.

Reagents required

All chemicals used in the preparation of these reagent solutions must be of analytical reagent quality.
1. *Oxygen-free distilled water* The dissolved oxygen is best removed from distilled water by sweeping out with a stream of nitrogen, from which oxygen has been removed by scrubbing with alkaline pyrogallol[4]. The stream of nitrogen should be passed through the water for 24 hours and the deoxygenated water should be stored under nitrogen.
2. *Manganous chloride solution* 40 g of manganous chloride crystals ($MnCl_2.4H_2O$) are dissolved in, and the solution made up to 100 ml with, oxygen-free distilled water.
3. *Potassium hydroxide solution* 70 g of potassium hydroxide (KOH) are dissolved in, and the solution made up to 100 ml with, oxygen-free distilled water.
4. *Ortho-tolidine solution* 1 g of pure *ortho*-tolidine (m. pt. 129.2°C) is dissolved in 100 ml of concentrated hydrochloric acid and the solution is diluted to 1 litre with oxygen-free distilled water. **o-Tolidine is regarded as carcinogenic, and must be handled with all due care.**

The Standard Lovibond Nessleriser Discs NOWA and NOWB

Disc NOWA covers the range 0.002—0.014 ml of oxygen per litre of water.
Disc NOWB covers the range 0.016—0.079 ml of oxygen per litre of water.
To convert the disc readings to ml of oxygen per litre see Table.

Technique

Use air-tight connections in all apparatus used for sampling and testing. When it is necessary to sample hot water, introduce a water-cooled coil into the sampling line, as shown in Figure 1.

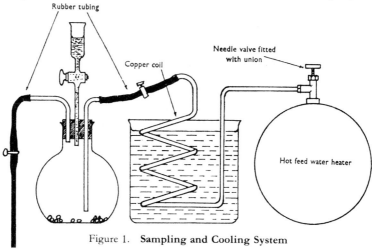

Figure 1. **Sampling and Cooling System**

COLORIMETRIC CHEMICAL ANALYTICAL METHODS — Oxygen Method 4

Take the sample into a 500 ml Winkler flask containing a few glass beads, as shown in Figure 2. Allow the water to flow for at least 10 minutes, before taking the actual sample, to displace all traces of air. Take care that air bubbles do not form around the stopper of the Winkler flask during sampling.

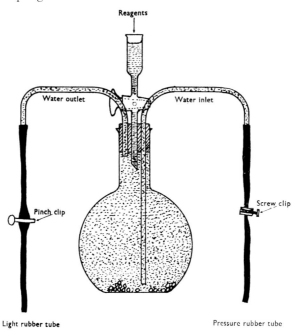

Figure 2. **Winkler Flask.**

After the sample has been taken add 1 ml of manganous chloride solution (reagent 2) to the Winkler flask, by means of the graduated funnel, and then add 1 ml of the potassium hydroxide solution (reagent 3). If the reagents refuse to flow into the Winkler flask, momentarily open the pinch-cock on the inlet rubber tubing to relieve the internal pressure. Close the pinch-cock before air can be drawn in after the reagents. Mix by momentarily inverting the flask.

Allow to stand for 10 minutes and then add 2 ml of concentrated hydrochloric acid. When the solution becomes clear transfer 50 ml to a Nessler cylinder (Note 1), add 0.5 ml of the *ortho*-tolidine solution (reagent 4) and allow to stand for 5 minutes.

Place the sample cylinder in the right-hand compartment of the Nessleriser. Place an identical cylinder filled with untreated water in the left-hand compartment and match the colour of the sample against the permanent glass colour standards in the disc using a standard source of white light (Note 2). The readings obtained from the disc are converted into ml of oxygen per litre of water by means of the following table.

Conversion of Disc Readings to ml O_2/litre. To convert to ppm, multiply ml/l by **1.43**

Disc reading	ml O_2/litre	Disc reading	ml O_2/litre
1	0.002	10	0.016
2	0.003	11	0.024
3	0.005	12	0.032
4	0.006	13	0.040
5	0.008	14	0.047
6	0.010	15	0.055
7	0.011	16	0.063
8	0.013	17	0.071
9	0.014	18	0.079

Notes

1. The readings obtained with the Nessleriser and discs are accurate only when cylinders are used which are identical with those used to calibrate the disc, i.e. cylinders for which the 50 ml calibration mark falls at 113 ± 3 mm from the bottom, measured internally.

2. In the laboratory the Lovibond White Light Cabinet should be used. In the field, or in laboratories not equipped with a standard illuminant, north daylight should be used wherever possible.

The Determination of Ozone

Introduction

The ozonisation of water is a well established process of sterilization. The advantages of this process include the removal of unpleasant tastes and odours and, in the absence of excessive amounts of iron or manganese, a considerable reduction of colour. Ozonisation is also used for the purification of water in swimming baths and for the production of high quality water at breweries and mineral water factories.

Ozone is also one of the components of the atmosphere which is of interest in studies of atmospheric pollution. It is a toxic gas which can be generated by industrial operations, such as arc welding, which are often carried out in enclosed spaces. A maximum permissible concentration of 0.1 ppm has been laid down[1] for ozone in industrial atmospheres.

Methods of sample preparation

Air[2-6]

Pass air at 5 litres a minute through a bubbler containing a solution of 1 g of potassium iodide (KI) and 0.4 g of caustic soda (NaOH) in 100 ml of water. Add 2 drops of 3% hydrogen peroxide (H_2O_2) solution and evaporate until the total volume is less than 20 ml. Add 1:5 acetic acid (CH_3COOH) until the pH is 3.8 and then dilute to 25 ml and determine the ozone using *Method 2*.

Water

No sample preparation is normally required. However the ozone residuals are very unstable and the determination must be carried out immediately the sample is taken. Stability is improved at low temperatures and low pH values. The samples should be collected as quickly as possible and without aeration.

The individual tests

Method 1 This method uses the reaction between ozone and diethyl *p*-phenylene diamine in the presence of potassium iodide to produce a stable red colour. It is based on an earlier method[7] using dimethyl *p*-phenylene diamine and is presented in an improved form as a modification of the well known Palin D.P.D. method[8-12] for the determination of residual chlorine in water. The values obtained by this method are identical with those obtained by the FAS titration method[13,14] as applied to the determination of ozone, and with the results obtained by the iodometric method. The tablets used for this method contain E.D.T.A. to complex any copper ions which may be present. These could otherwise produce a colour with the D.P.D. in the presence of potassium iodide.

Method 2 This method uses the reaction between ozone and potassium iodide, in acid solution, to release a stoichiometric quantity of iodine. This method is simple and is claimed[15] to give consistent and accurate results.

References

1. Department of Employment, *Safety Health and Welfare New Series No. 8*, "Dust and Fumes in Factory Atmospheres, 1968", H. M. Stationery Office, London, 1968
2. D. H. Byers and B. E. Saltzman, *Amer. ind. Hyg. Assoc. J.,* 1958, **19,** 251
3. A.S.T.M. Method D1609-60
4. G. A. Hunold and W. Pietrulla, *Z. analyt. Chem.,* 1961, **178,** 271
5. W. Leiche, "*Analysis of Air Pollutants*", Ann Arbor, 1970, p. 131
6. See also Section F, page 540
7. A. T. Palin, *Water and Water Engineering,* 1953, **57,** 277
8. A. T. Palin, *Analyst,* 1945, **70,** 203
9. A. T. Palin, *J. Amer. Water Wks. Assoc.,* 1957, **49,** 783
10. A. T. Palin, *Baths Service,* 1958, **17,** 21
11. A. T. Palin, *Water and Water Engineering,* 1958, **62,** 30
12. Chlorine *Method 3*, page 141
13. A. T. Palin, *Proc. Soc. Water Treatment and Exam.,* 1957, **6,** 133
14. Amer. Pub. Health Assoc., "*Standard Methods for the Examination of Water and Wastewater*", 11th Edn., 1960
15. A. Elphick, personal communication, Dec. 1965

COLORIMETRIC CHEMICAL ANALYTICAL METHODS

Ozone Method 1
using diethyl-p-phenylene diamine (Palin DPD)

Principle of the method

Ozone reacts instantaneously with DPD in the presence of potassium iodide to give a stable red colour. The intensity of this colour, which is proportional to the ozone concentration, is measured by comparison with a series of Lovibond permanent glass colour standards. Oxidising agents, such as halogens, must be absent.

Reagents required

Comparator DPD Comparator tablets No. 4 (or Nos. 1 and 3 together)

Nessleriser DPD Nessleriser tablets Nos. 1 and 3 or No. 4.

The Standard Lovibond Discs, Comparator 3/67 and Nessleriser NOR

Disc 3/67   covers the range 0.1 to 1.0 ppm (mg/1) of ozone (O_3) in steps of 0.1.
Disc 3/67S covers the range 0.05 to 0.45 ppm (mg/1) of ozone (O_3) in steps of 0.05.
Disc NOR   covers the range 0.01 to 0.30 ppm of ozone (O_3), in the following steps 0.01, 0.02, 0.04, 0.06, 0.1, 0.15, 0.2, 0.25 and 0.3 ppm.

Technique

Comparator Fill a 13.5 mm cell or comparator tube with the sample and place this in the left-hand compartment of the comparator. Wet an identical cell with the sample and leave just enough liquid to cover the tablets when added. Drop into this prepared cell one No. 1 and one No. 3 tablet (or one No. 4 tablet which is equivalent to these two combined) and stand until the tablets have disintegrated, or crush. Then add the sample up to the 10 ml mark, mix until the remains of the tablets have dissolved, and then place the cell in the right-hand compartment of the comparator. Match the colour immediately (Note 1) with the standards in the disc using a standard source of white light such as the Lovibond White Light Cabinet or, failing this, north daylight. When the correct match is found the figure then showing in the indicator window represents parts per million of ozone in the sample.

Nessleriser The instructions given for the comparator test should be followed with the following amendments:—

Substitute 50 ml Nessleriser tubes for 10 ml test tubes, and 50 ml quantities for 10 ml, and use Nessleriser DPD tablets.

Notes

1. In the presence of potassium iodide, copper would produce a colour with the DPD indicator. Interference by up to 10 ppm of copper is prevented by the EDTA incorporated in the tablets. There may be a transitory colour until sufficient EDTA has dissolved to completely chelate the copper. Therefore if the presence of copper is suspected allow the tubes to stand for two minutes, after placing in the comparator, before matching.

2. Dissolved oxygen in the water can produce a faint colour with the reagent if the mixture is allowed to stand. The suppression of trace metal catalysis by the EDTA minimises this effect and there is no interference within the period of the test.

3. The only interfering substance, other than halogens, which may possibly be present in water is oxidised manganese. The effect of this can be allowed for by developing the manganese colour in the "blank" tube as follows:—

To 10 ml of the sample in a separate tube add 1 drop of sodium arsenite ($NaAsO_2$) solution and mix thoroughly. Rinse the comparator cell with the sample as before, add the DPD tablets and proceed as for the sample but instead use the 10 ml of the prepared solution. Finally place this tube in the left-hand compartment of the comparator. In this way any colour due to the manganese will develop equally in both tubes and will cancel out during the colour matching. The same procedure is followed with the Nessleriser, using 1 drop of the arsenite solution to 50 ml sample, and Nessleriser size tablets.

4. To prevent accidental contamination, handling of the DPD tablets should be avoided. By shaking the tablet into a bottle top, it is a simple matter to use the top to convey the tablet to the comparator cell.

5. This test was primarily developed for determining the concentration of ozone dissolved in water. However, ozone in the atmosphere can be determined by bubbling a suitable volume of air through a neutral potassium iodide solution, and then proceeding with the test on this solution as described. The result obtained represents the concentration in the solution, from which the concentration in the air may be calculated by reference to the volume of air sampled.

6. The results obtained by the use of the Nessleriser and disc are accurate only provided that Nessleriser tubes are used which conform to the specification of the tubes used in the calibration of the disc. That is that the 50 ml mark falls at a height of 113 ± 3 mm measured internally.

COLORIMETRIC CHEMICAL ANALYTICAL METHODS

Ozone Method 2
using potassium iodide

Principle of the method

In acid solution ozone reacts with potassium iodide to release a stoichiometric quantity of iodine. The intensity of the colour of the iodine solution is proportional to the amount of iodine present and therefore is proportional to the ozone concentration. The intensity of this colour is determined by comparison with a series of Lovibond permanent glass colour standards.

Reagents required

1. *Sulphuric acid*   10% solution of H_2SO_4
2. *Potassium iodide*   10% solution of KI

The chemicals used should be of analytical reagent quality and the solutions should be made up in good quality distilled water.

The Standard Lovibond Nessleriser Disc NOS

This disc contains colour standards corresponding to 0.2, 0.4, 0.6, 0.8, 1.0, 1.5, 2.0, and 2.5 ppm (mg/l) of ozone (O_3).

Technique

Carry out the determination **immediately** the sample is taken, as ozone residuals are very unstable. The stability is much improved at low temperatures and low pH values. Collect the sample as quickly as possible and without producing aeration.

Measure 100 ml of the sample and add this to a freshly prepared mixture of 5 ml of reagent 1 and 5 ml of reagent 2. Mix well, transfer 50 ml of the mixture to a Nessleriser tube and place this tube in the right-hand compartment of the Nessleriser. Place the Nessleriser before a standard source of white light such as the Lovibond White Light Cabinet or, failing this, north daylight, and rotate the disc until one of the standards matches the developed colour of the sample. The figure in the indicator window then corresponds to the concentration of ozone in the sample in parts per million.

Prepare a blank by placing a further portion of the ozonated sample in a gas-washing bottle and passing a vigorous stream of pure air through the sample for several minutes. This treatment removes the ozone from the sample but will leave behind any other oxidising agents. Take 100 ml of this de-ozonated sample and proceed as before. If the water is known to be free of nitrites, iron, manganese or similar substances then the untreated water may be used as the blank. Determine the apparent ozone concentration of the blank and subtract this from the result obtained with the ozonated sample to give the true ozone concentration.

The Determination of Phosphorus

Introduction

The determination of phosphorus is a widely used analytical control test. Phosphates are added to boiler feed-water to prevent scaling and are an essential constituent of fertilizers, of the body fluids, of soil and of plants, and are also found in many industrial products from steels to detergents.

The following are a selection of recommended procedures. These determine only orthophosphate. Pyrophosphate is converted to orthophosphate, by boiling with 10% of sulphuric acid for two minutes, metaphosphate requires 20 minutes boiling for conversion. Total phosphorus concentration may thus be found, and the various forms by difference.

Methods of sample preparation

Air[1-3]

Particulates[1,2] Trap the particles on membrane filters, transfer to a 150 ml conical flask, add 20 ml of a nitric: perchloric acid ($HClO_4$) mixture, heat to fuming, cool and add 3 ml of water. Repeat evaporation, wash into 25 ml volumetric flask, make up to mark and estimate using molybdovanadate technique (*Methods 5 or 7*)[1,2].

Vapours[3] It is reported that membrane filters have a large absorptive capacity for the vapours of organophosphorus compounds, such as insecticides, and may be used for vapour, as well as particulate, sampling. After sampling the filter is treated in the same manner as when used for sampling particulates.

Beverages[4,5]

Wines Pipette a 5 ml sample into a 100 ml Pyrex beaker and evaporate to dryness on a steam-bath. Add 15 ml of nitric acid, some anti-bumping material, cover with a watch-glass and heat gently until the residue dissolves. Boil gently for 10–15 minutes, cool, add 4 ml of 70% perchloric acid ($HClO_4$) and boil gently until the solution fumes copiously and becomes colourless or nearly colourless. Remove the watch-glass as soon as the solution begins to fume, but do not evaporate to dryness. (If the solution is brown add 2 ml of nitric acid and boil again.) Cool slightly, add about 25 ml of water, boil for a few minutes and then transfer the solution to a 100 ml volumetric flask. Rinse the beaker with water and add these rinsings to the flask bringing the total volume up to 50–60 ml. Add 20 ml of molybdovanadate reagent, dilute to 100 ml and measure the colour after 15 minutes[4]. Alternatively[5] digest 1 ml of wine with 5 ml of $8N$ sulphuric acid, on a sand-bath, until white fumes are evolved. Add potassium perchlorate ($KClO_4$) a little at a time until the mixture is colourless. Cool, dilute to 50 ml with water and proceed by *Method 6*.

Biological materials[6-22]

Blood, plasma and serum[6,15] Use *Method 3*[6,8]. Alternatively[9,11] add 0.1 ml of blood or serum to 2 ml of 10% trichloroacetic acid ($Cl_3C.COOH$), shake, stand for 10 minutes and centrifuge. Use 1 ml of the supernatant liquid and proceed by *Method 6*. Other workers[12,13] recommend the use of perchloric acid ($HClO_4$) for deproteinization of the sample and extraction of the ammonium molybdate complex into an organic solvent. Typically[13] add 2 ml of 10% perchloric acid to 0.02 ml of sample, then add 0.5 ml of ammonium molybdate reagent, extract with 2 ml of butyl acetate ($CH_3COO(CH_2)_3CH_3$) and determine phosphorus on the organic layer.

Protein may also be removed by coagulation[14]. Place 0.25 ml of serum in a phial and swirl to distribute the serum evenly over the bottom. Immerse the bottom of the phial in boiling water for two minutes and then allow to cool. If the coagulum is not well formed take another 0.25 ml of serum, add 1 drop of 22% bovine serum-albumin solution and treat as before. Add 6 ml of $0.3N$ sulphuric acid down the side of the phial to avoid disrupting the clot, and incubate for either 60 minutes at 37°C or for 90 minutes at 25°C. Determine phosphorus on the supernatant liquid by *Method 2*.

Opinions differ as to the best method to use for the determination of phosphorus in blood. Fiske[6], Kay[7] and King[8] all recommend the use of amino-naphthol sulphonic acid (*Method 3*) as the reducing agent, which they claim to be especially useful in biochemical applications. Hall[12] recommends stannous chloride (*Method 1*), Alimova[15] claims that amidol is more sensitive than ascorbic acid (*Method 6*), while London[14] recommends metol (*Method 2*).

Lipids[16] Evaporate the sample solution, containing more than 1 μg of phosphorus, to dryness, heat with 1 ml of nitric acid and then with 1 ml of 70% perchloric acid ($HClO_4$) for each 0.2 g of lipid. Evaporate to remove excess perchloric acid, dilute and make up to a known volume.

Tissue[10,17-21] Wet ash the sample. To the extract in a silica tube add 0.3 to 0.7 ml of 70% perchloric acid ($HClO_4$). Heat at 200°C for 40 minutes. Transfer to a 10 ml flask, add 2 ml of the ammonium molybdate reagent, mix well, add 4 ml of acetone (($CH_3)_2CO$), make up to the mark and estimate[17]. Alternatively[10] mineralise about 2 g of sample with sulphuric and nitric acids, dilute to 50 ml and treat a 2.5 ml aliquot as described above for blood.

Urine[6,7,8,11,12,15,18-22] Use *Method 3*. Cook[19] claims that mannitol interferes if present at concentrations greater than 0.5 g/100 ml and recommends that its concentration should be reduced below this critical level by dilution with water or trichloracetic acid ($Cl_3C.COOH$). Schriever[21] discusses the effects of reducing agents in urine and claims that the interference can be removed by warming 10 ml of the urine with 2 ml sulphuric acid and 30 ml of N potassium permanganate until the colour fades, then neutralise, dilute to 200 ml and use a suitable aliquot. Iwanaga[20] recommends the removal of interfering substances by applying the urine, at pH 5.4, to paper, developing the chromatogram with methanol, 85% formic acid and water (CH_3OH : $HCOOH$: H_2O, 16:3:1) removing the phosphorus spot at R_f 0.70 and extracting with $8N$ aqueous ammonia (NH_4OH) or ashing and determining phosphorus on the extract.

Chemicals[22-27]

Organic compounds Wet oxidise the sample in a mixture of sulphuric and perchloric ($HClO_4$) acids and determine phosphorus by *Method 5 or 7*[22,23,27]. Saliman[24] has suggested an alternative digestion reagent consisting of 50 ml of re-distilled hydriodic acid (HI), 0.6 g of calcium hydroxide ($Ca(OH)_2$), 50 ml of water and 500 g of phenol (C_6H_5OH) made up to 1 litre with acetic acid (CH_3COOH). The recovery of phosphorus as phosphate is claimed to be better than 95%. Maruyama[25] recommends that one should fuse the sample with 1 g of a 2:1 mixture of sodium carbonate (Na_2CO_3) and potassium nitrate (KNO_3) in a platinum crucible for about 20 minutes, dissolve the cooled melt in 5 ml of nitric acid, dilute to 2000 ml and determine phosphorus on a 2 ml aliquot. Instead of fusion Yu[26] suggests one should combust about 4 mg of sample in an oxygen flask, absorb the combustion products in 12 ml of $0.4N$ sulphuric acid, boil the solution for 5 minutes with 50 mg of ammonium persulphate (($NH_4)_2S_2O_8$) and make up to a known volume.

Dairy products[22,28,29]

Milk Heat enough sample to contain 5–15 μg of phosphorus with 0.75 ml of 70% perchloric acid ($HClO_4$) for 15 minutes. Dilute the solution to a known volume and determine phosphorus on an aliquot[22]. The method recommended by the Federation Internationale de Laiterie[28] is to dry ash the sample at 500–550°C. A modified ashing procedure is that suggested by Bedessem[29]. Weigh 5 g of well mixed sample into a platinum dish. Add 1 ml of a saturated solution of magnesium nitrate ($Mg(NO_3)_2.6H_2O$) in ethanol (C_2H_5OH). Evaporate to dryness on a boiling water bath then transfer to a hot plate and increase the temperature until no further charring occurs. Ash for 6 hours at 525°C, mix ash with water and wash the sides of the dish with 2 ml of 1:1 sulphuric acid. Filter through a Whatman No. 40 filter paper into a 1 litre volumetric flask. Thoroughly wash the dish, the ash and the filter paper and add washings to the filtrate. Dilute to volume and determine phosphorus on an aliquot.

Foods and edible oils[22,30-33]

Eggs[22] Use first method for milk described above.

Fruit and fruit products[30] Ash sample, dissolve the ash in 10 ml of 1:3 hydrochloric acid and evaporate the solution to dryness on a steam bath. Dissolve the residue in 10 ml of 1:9 hydrochloric acid on the steam bath and transfer the solution to a 100 ml volumetric flask. Cool, dilute to volume, mix thoroughly and proceed by *Method 5 or 7*. This has been recommended as the official method for fruit and fruit products after comparison with gravimetric, titrimetric and other photometric methods.

Meat and meat extracts[22,31] Use the first method for milk described above[22]. Alternatively[31] dry 2.5 g of sample at 125°C for 30 minutes, ash at 550°C and then dissolve the cooled ash, in 25 ml of 1:4 nitric acid, by heating on a sand bath for 30 minutes. Cool and make the solution up to known volume.

Vegetable oils[32] Mix 6 g of the sample with 1 g of zinc oxide (ZnO) and ignite at 500°C. Dissolve the ash in sulphuric acid and make up to a known volume.

Wheat products[33] Extract 0.5 g of the sample with 50 ml of $0.5N$ sulphuric acid. Filter the extract and determine phosphorus on an aliquot part of the filtrate.

Metals, alloys and plating baths[34-46]

Aluminium-silicon alloys[34] Dissolve 1 g of the sample in hydrochloric acid under a stream of nitrogen and collect the phosphine which is evolved in a freshly prepared solution of sodium hypobromite (NaOBr). Dissolve any residual sample by adding a mixture of nitric, sulphuric and hydrofluoric (HF) acids. Evaporate this solution until copious fumes are evolved and then dilute to a known volume. Estimate phosphorus both in an aliquot of this solution and in an aliquot of the hypobromite solution and combine the results.

Chromium[35] Dissolve sufficient sample to contain between 100 and 500 µg of phosphorus in 3–5 ml of nitric acid and 10–15 ml of perchloric acid ($HClO_4$), by heating in an open beaker. Add hydrochloric acid until no further fumes of chromyl chloride (CrO_2Cl_2) are evolved and then distill off residual chromyl chloride until all the chromium has been removed. Heat the residual solution until fumes of perchloric acid appear then add 20 ml of water, heat to the boiling point, filter, wash the residue three or four times with small portions of hot 1% hydrochloric acid and then dilute the combined filtrate and wash liquors to 50 ml. To a 20 ml aliquot of this solution add 4 ml of ferric nitrate ($Fe(NO_3)_3$) solution (dissolve 180 g of ferric nitrate ($Fe(NO_3)_3.9H_2O$) in 500 ml of water, add 5 ml of nitric acid, filter and make up the filtrate to 1 litre), add 1:1 aqueous ammonia (NH_4OH) until hydroxides start to precipitate and then add hydrochloric acid until the precipitate re-dissolves. Add 10 ml of 20% hydroxylammonium chloride ($HONH_3Cl$) solution or 12 ml of sodium sulphite (Na_2SO_3) solution, heat the solution to the boiling point, cool and determine phosphorus on this solution. Use a second 20 ml aliquot as a blank.

Copper alloys[36-38] Take 0.2–2 g of sample depending on the phosphorus content and place in 150 ml beaker. Add 15 ml of mixed nitric: hydrochloric: sulphuric acids (8:3:4) and heat until the sample has dissolved. Add 1 ml of 3% hydrogen peroxide (H_2O_2) and heat for 3 minutes. Cool and transfer to a 50 ml volumetric flask. Make up to volume and use an aliquot for the determination using *Method 5 or 7*[36,37]. Alternatively[38] dissolve 100 g of filings by heating with 20 ml of 3% sodium fluoride (NaF) solution and 20 ml of 1:2 nitric acid. Expel nitrous gases by boiling, oxidise any residual phosphite to phosphate with 2 ml of 5% potassium permanganate ($KMnO_4$) solution, reduce any excess permanganate with a few drops of 3% hydrogen peroxide (H_2O_2) solution. To the resulting solution add 25 ml of $0.25M$ E.D.T.A. (($CH_2.N(CH_2COOH)_2)_2$) solution, digest for a short time, cool, make up to a known volume and proceed by vanadomolybdate (*Method 5 or 7*).

Ferrovanadium[39,40] Dissolve the sample by heating with 10 ml of nitric acid then evaporate twice after adding hydrochloric acid and dissolve the residue in water. Dilute to 120 ml with hot water, reduce iron and vanadium by the addition of 15 ml of 10% hydroxylammonium chloride ($HONH_3Cl$) solution, adjust the concentration of hydrochloric acid to be $0.3N$ and then pass the solution first through a filter of paper pulp and then through a column containing 25 g of cationite resin KU-2. Wash the filter and the column with 50 ml of $0.2N$ hydrochloric acid and determine phosphorus on the percolate.

Iron and steel[39,41-46] Dissolve 0.5 g of the sample in 25 ml of gently boiling nitric acid, add 30 ml of water, heat to boiling and filter off any residual carbon. To the filtrate add 0.1 g of potassium permanganate ($KMnO_4$) and boil for 2–3 minutes. Add hydrogen peroxide (H_2O_2) drop-wise to the boiling solution until the colour of the permanganate disappears and boil for a further 2–3 minutes. Cool and determine phosphorus using the vanadomolybdate method (*Method 5 or 7*)[41-43].

For alloy steels Pakalns[44] recommends that the sample be dissolved in a dilute mixture of nitric and hydrochloric acids to which 0.5 ml of 40% hydrofluoric acid has been added and then evaporate to fumes after adding perchloric acid. For samples containing more than 18% of tungsten dissolve in 10 ml of water, 7 ml of 15M nitric acid and 0.5 ml of 40% hydrofluoric acid and then evaporate to fumes with perchloric acid. Determine by vanadomolybdate (*Method 5 or 7*).

For the rapid determination of phosphorus in cast iron[45,46], dissolve 0.2 g of the sample in 50 ml of sulphuric acid saturated with ammonium persulphate (($NH_4)_2S_2O_8$), add 2 ml of 2% potassium permanganate ($KMnO_4$) solution and 50 ml of hot water. Heat, filter, cool and dilute to 500 ml. Determine phosphorus on an aliquot of this solution.

Minerals, ores and rocks[47-51]

Bauxite[47] Extract phosphorus from acid solution as the molybdophosphate in iso-butanol and reduce with stannous chloride (*Method 1*).

Chromium ores[48] Boil the sample with 1:1 nitric acid, neutralise, extract the phosphomolybdate complex with ether ($C_2H_5OC_2H_5$) and reduce with stannous chloride (*Method 1*).

Iron ores[49] Dissolve the sample in a mixture of nitric and perchloric acids, convert the phosphorus to vanadomolybdophosphate and extract into isobutyl methyl ketone (($CH_3)_2$CHCH$_2$COCH$_3$) after the addition of citric acid ($C(OH)(COOH)(CH_2COOH)_2 \cdot H_2O$) to mask arsenic and iron.

Ores generally[47,50] Dissolve 1 g of finely divided sample in a mixture of nitric: hydrochloric acids (1:3) and evaporate to dryness. Add 0.5 ml of hydrochloric acid and again evaporate. Repeat to render the silicic acid completely insoluble. To the residue add 1–2 ml of hydrochloric acid, dilute, filter and make up to 100 ml. To remove interfering elements pass 20 ml of this solution through a column containing 10 g of phenolsulphonic resin in the H^+ form, wash with water to give a total volume of 100 ml and determine phosphorus on this solution.

Rocks[51] Weigh 0.1 g of the finely powdered sample into a test tube, add 3 ml of 1:1 nitric acid, heat to the boiling point and boil gently for 30 minutes. If the solution is then colourless, dilute to 20 ml, shake and leave insoluble material to settle. If the solution is coloured, treat with a few drops of potassium permanganate ($KMnO_4$) solution until the pink colour persists, destroy excess permanganate with a few drops of 1% sodium sulphite (Na_2SO_3) solution, boil for 2–3 minutes, cool and dilute to 20 ml. Determine phosphorus on a 0.1–2.0 ml aliquot using vanadomolybdate (*Methods 5* or *7*).

Miscellaneous[52-66]

Agricultural feeds and mineral supplements[52-55] Weigh 0.5–2.0 g of the sample, depending on the expected phosphorus content, into a 150 ml beaker and ash at 600°C. Cool, add 40 ml of 1+3 hydrochloric acid and a few drops of nitric acid. Bring to the boil on a hot plate then cool and make up to a final volume of 200 ml with distilled water. Filter and determine phosphorus on an aliquot part of the filtrate using vanadomolybdate (*Methods 5* or *7*)[52]. This method was claimed by the authors[52] to be more rapid than, and as accurate as, the official Association of Official Agricultural Chemists standard method which uses molybdenum blue for the determination. This claim has been substantiated by collaborative tests comparing the two methods[53]. Stuffins[54] recommended Kjeldahl digestion, but Rash[55] after comparing various methods of sample preparation recommended dry ashing, claiming that the recovery of phosphorus using wet ashing varied from 51% to 96%.

Cement[56] Fuse the sample with sodium carbonate (Na_2CO_3) and caustic soda (NaOH) and dissolve the melt in dilute hydrochloric acid. Filter, make up to volume and determine phosphorus on an aliquot part of this solution.

Detergents[57] Shen and Dyroff claim that by measuring the molybdenum blue colour 40, 70 and 100 minutes after the addition of reagents and then comparing the results with those from known mixtures of phosphate and silicate it is possible to determine both radicals in detergent preparations.

Dust[58] Ignite 75–300 mg of dust at 700°C for 3 hours. Heat the residue for 10 minutes on a water bath with 10–15 ml of 40% hydrofluoric acid (HF) and then evaporate three times with successive 5 ml portions of 6N hydrochloric acid. Dissolve the residue in 5 ml of 6N hydrochloric acid and pass the solution through a column of Lewatit KSN resin. Elute the column with 150 ml of 0.15N hydrochloric acid, dilute to 250 ml and use an aliquot.

Fertilizers and liming materials [59-61] Kowalski[59], having examined all the available methods recommends vanadomolybdate (*Methods 5* and *7*) as most suitable for use with fertilizers. Lasiewicz[60] recommends that, according to the material being examined and to the nature of the information which is required, the sample should be dissolved in sulphuric acid+nitric acid, or water, or 2% citric acid, or ammonium citrate solution. Alternatively[61] grind the sample in an agate mortar until it will pass through a No. 100 sieve. Place 0.5 g of a limestone sample, or 0.2 g of a silicate, in a 75 ml nickel crucible. Remove any organic matter by heating at 900°C for 15 minutes without the lid. Cool. Mix 0.3 g of potassium nitrate (KNO_3), with the ignited sample add 1.5 g of caustic soda (NaOH) pellets, cover the crucible and heat for 5 minutes to a dull red heat over a gas flame (*not* in a furnace). Remove the flame and swirl the melt around the sides of the crucible. Cool, add about 50 ml of water and warm to disintegrate the fused cake. Transfer the solution and cake to a 150 ml beaker containing 15 ml of $5N$ perchloric acid. Scrub the crucible and lid with a rubber-tipped glass rod and wash any residue into the beaker. Transfer the solution to a 100 ml volumetric flask and dilute to volume.

Hair[62] Wash hair cuttings with water and then with light petroleum and dry. Ash for 12 hours at 500°C, treat the ash with normal hydrochloric acid and water, filter and dilute the filtrate to a known volume.

Ion-exchange resins[63] Fuse 0.2 g of sample in a porcelain crucible with potassium carbonate (K_2CO_3). Dissolve the melt in hydrochloric acid and determine phosphorus on this solution.

Leather[64] Oxidise the sample with a mixture of nitric, sulphuric and perchloric ($HClO_4$) acids make this solution up to a known volume and determine phosphorus as vanadomolybdate (*Methods 5* or *7*).

Potato starch[65] Digest 5 g of the starch with 50 ml of 1:1 nitric acid and 25 ml of perchloric acid ($HClO_4$). Dilute the final digest solution to 200 ml with water and determine phosphorus by *Method 1*.

Textiles[66] Extract soluble phosphate by boiling a 20 g sample with water. Remove arsenic by treatment with hydrobromic acid (HBr) and dilute the solution to 500 ml. Determine phosphorus on a 10 ml aliquot using another 10 ml aliquot as a blank.

Plants[67,68]

Digest 0.5 g of dried plant material with 2 ml of perchloric acid ($HClO_4$) and 10 ml of nitric acid for about 6 hours on a sand bath. When the residue is colourless, cool, add $0.5N$ perchloric acid, filter and dilute the filtrate to 100 ml[67]. To prevent interference from arsenic residues from crop spraying, treat the extract with a reducing mixture of $5N$ sulphuric acid, 10% sodium hyposulphite ($Na_2S_2O_4$) and 1% sodium thiosulphate ($Na_2S_2O_3$) in the proportion 1:2:2, before adding mixed molybdate reagent. Allow the mixture to stand for at least 40 minutes before measuring the colour[68].

Plastics and polymers[26,69]

Either[26] combust about 4 mg of the sample in an oxygen flask and absorb the combustion products in about 12 ml of $0.4N$ sulphuric acid, boil the solution for 5 minutes with 50 mg of ammonium persulphate (($NH_4)_2S_2O_8$) and make up to a known volume; or[69] wet ash using sulphuric and nitric acids and hydrogen peroxide (H_2O_2).

Sewage and industrial wastes [70-72]

If the sample is coloured or turbid, clarify by adding 1 ml of sulphuric acid followed by 1 ml of 10% barium chloride ($BaCl_2$). Allow the precipitate of barium sulphate ($BaSO_4$) to settle, and centrifuge. Alternatively to the neutral sample add silica gel (about 5% by volume) shake vigorously and filter through three thicknesses of filter paper. Repeat the filtration until the solution is colourless. For the determination of soluble inorganic phosphate in sewage *Method 1* is the standard method[70].

For the determination of total phosphorus digest 5-10 ml of sample with 0.3-0.5 ml sulphuric acid. Boil until blackening or discoloration appears. Keep simmering so that the sulphuric acid condenses on the sides of the tube. When the liquid is colourless cool, dilute to 20 ml with distilled water and transfer to a 50 ml flask. Add 1 drop of phenolphthalein indicator solution and then N caustic soda (NaOH) until the solution is just pink. Make up to the mark and estimate phosphorus in an aliquot using *Method 1*.

The use of ascorbic acid instead of stannous chloride as the reducing agent has been suggested by Fogg and Wilkinson[71] who claim that this produces a more stable molybdenum blue colour. See *Method 6*.

An alternative method for sewage, due to Namili[72] removes the necessity for preliminary clarification. To 50 ml of sample add 1 ml sulphuric acid (1:50) and 15–20 ml of butanol (C_4H_9OH). Shake for 1 minute. To the aqueous layer add 5 ml of 2% ammonium molybdate ((NH_4)$_2MoO_4$) in 8.5N sulphuric acid and 0.25 ml of 2% stannous chloride ($SnCl_2$) in 1.2N hydrochloric acid. Extract with 10 ml of butanol for several minutes, separate and measure the colour of the organic layer.

Soil[68,73-84]

Available phosphorus[73-78] Dry the soil and grind it to pass through a 36 mesh sieve. Wash some DeAcidite FF-510 resin (larger than 30 mesh) with acetone (CH_3COCH_3) and dry it in air. Mix 2 ml of the ground and dried soil sample with 2.9 g of resin in a volume of distilled water dependent on the expected phosphate content (e.g. 100 ml for agricultural soil, 1000 ml for greenhouse soil) and shake the mixture by tumbling for 16 hours at 20°C. After this time pour the mixture onto a stretched Terylene net of approximately 30 mesh to separate the resin from the soil and the water. Wash the adhering soil particles from the resin with deionised water. Transfer the resin to a glass tube 12 mm in diameter and recover the phosphate by leaching with normal sodium sulphate (Na_2SO_4) solution[73].

A number of other extraction methods have also been recommended:–
(a) Extract the soil with 30 times its volume of acetate buffer[74] (Morgan's reagent).
(b) Shake a 5 g sample of air-dried and sieved (2 mm) soil for 4 hours at $20 \pm 1°C$ with 100 ml of a solution of 0.1N ammonium lactate ($NH_4.C_3H_5O_3$) and 0.4N acetic acid(CH_3COOH) at pH 3.7, and filter[75].
(c) Extract the soil with calcium lactate ($Ca(C_3H_5O_3)_2$)[76].
(d) Boil 20 g of soil for 5 minutes with 70 ml of hydrochloric acid and then digest on a water bath for 48 hours. Add water, filter and dilute to 250 ml. Treat 15 ml of this extract with 0.5 ml of 20% sodium permanganate ($NaMnO_4$) solution and heat on a sand bath for 15 minutes. Cool, dilute to about 30 ml, add 6 ml of 10% potassium ferrocyanide ($K_4Fe(CN)_6.3H_2O$) to precipitate iron. Filter. Add 5 ml of 10% manganous sulphate ($MnSO_4.4H_2O$) and shake. After one hour add 1:1 aqueous ammonia (NH_4OH) until the blue colour turns purple. Add 3.5 ml of 2N sulphuric acid and transfer to a 100 ml flask. Dilute to volume, filter and use aliquot of the filtrate.
(e) Shake 5 g of soil with 50 ml of 1% citric acid ($C(OH)(COOH)(CH_2.COOH)_2.H_2O$) solution and heat to vigorous boiling. Shake the suspension three times during heating and filter. To 2–10 ml of the filtrate, depending on the phosphorus content, add 2 ml of 30% sulphuric acid and 20 ml of 0.5N potassium permanganate ($KMnO_4$), shake for 2–3 minutes, then heat to the boiling point and boil for 5 minutes. Add 5 ml of 10% glucose ($O.(CH.OH)_4.CH.CH_2OH.H_2O$) solution and continue heating until the brown colour of precipitated manganese dioxide (MnO_2) disappears. Cool, dilute to a known volume and proceed by *Method 1*.

Total phosphorus[79-81] To 1–2 g of soil in a 100 ml Pyrex flask add 2.5–5 ml of water and, after a few few minutes, 5–10 ml of sulphuric acid, stir well and set aside for 30 minutes. Then add 4–6 drops of perchloric acid ($HClO_4$), close the flask with a small conical funnel and boil for 5–7 minutes until sulphuric acid condenses on the sides of the flask above the reaction mixture. Stop boiling, add a further 2 drops of perchloric acid and again boil. Repeat this step until the solution above the soil is colourless. Cool, dilute to 100 ml and filter. Take a 10 ml aliquot of the filtrate add 5 ml of 12% potassium ferrocyanide ($K_4Fe(CN)_6.3H_2O$) solution and 5 ml of 10% manganous sulphate ($MnSO_4.4H_2O$) solution, set aside for 10 minutes and then titrate with 10% aqueous ammonia (NH_4OH) until the colour of the solution changes to purple-lilac. Add 5 ml of 1.4N sulphuric acid, dilute to exactly 100 ml, filter and determine phosphorus on an aliquot part of the filtrate[79].

Sherrell[80] has compared the relative merits of three procedures—decomposition by a mixture of nitric and hydrofluoric acids, digestion with perchloric acid, and fusion with sodium carbonate (Na_2CO_3). He found the mixed acid decomposition to be satisfactory. Digestion with perchloric acid did not completely extract the phosphorus. Fusion was satisfactory provided that the melt was then extracted with 9N sulphuric acid and *not* with water alone. He also found that

Method 5 or 7 was the most satisfactory procedure to use after mixed acid decomposition and Method 6 after fusion. Syers[81] claims that fusion is better than mixed acid decomposition.

Lake cores[82] Extract the sample with dilute hydrochloric or sulphuric acid (pH 1.1) and determine phosphorus by Method 5 or 7.

General[66,83,84] Alexander[83] having compared the two methods recommends Method 6 as preferable to Method 1, whereas Guiot[84] finds these two methods equally suitable. The procedure detailed earlier for the suppression of arsenic interference in the analysis of plants[68] is equally applicable to soils.

Water[85-93,108]

For concentrations of 1–80 ppm of phosphorus the standard method[85,86] is Method 5, while for 0–1 ppm of phosphorus it is Method 1. In natural waters which also contain arsenic (As), the molybdenum blue test gives high results. This can be overcome[87] by taking 50 ml of sample, treating it with 7.5 ml of $8N$ nitric acid and the molybdate reagent, extracting the solution with 10 ml of a mixture of chloroform ($CHCl_3$) and butanol (C_4H_9OH) (3:1) and after 15 minutes shaking the organic layer with the stannous chloride solution. Murphy[88] proposed a modification of Method 6 in which the ammonium molybdate and ascorbic acid reagents are combined in a single reagent. He also showed that in the presence of 1 ppm of antimony, which he added to the test solution as potassium antimony tartrate ($KSbOC_4H_4O_6$) solution, colour development is speeded up.

To analyse diesel engine cooling water take a 5 ml sample, add 1 or 2 drops of hydrochloric acid, a few drops of a saturated solution of sodium sulphite (Na_2SO_3) and 5 ml of the ammonium molybdate reagent (*Method 1*). Stir with a tin rod until colour development is complete and then compare the colour with standards[89].

Jones[90] has investigated five methods for the determination of phosphorus in sea water and recommends the Murphy[88] modification of Method 6.

In cases where organically bound phosphorus is present in the water a comparison of the standard acid digestion method and the persulphate oxidation[96] showed that the latter was more efficient[91].

If there is a need to determine a concentration of phosphorus below the minimum level detectable by the normal tests Kar[92,93] recommends concentration of the phosphate by precipitation with a carrier of barium sulphate ($BaSO_4$) in the following manner:– To 980 ml of sample, which should not contain more than 15 μg of arsenic, add 10 ml of $0.18M$ barium chloride ($BaCl_2$), adjust the pH to be between 7.2 and 8.0 by the addition of $0.5N$ aqueous ammonia (NH_4OH) solution and add 1.5 g of barium sulphate. Stir, set aside for 15 minutes, stir again and then set aside overnight. Mix the supernatant liquid with a further 1.5 g of barium sulphate, set aside for at least 10 hours and then syphon off, and discard, the supernatant liquid. Combine the two suspensions, centrifuge and wash the residues with water at pH 8.5. Leach phosphorus and arsenic from the carrier with three successive 7 ml portions of $0.15N$ hydrochloric acid, and separate the phosphorus from the arsenic by extraction from this hydrochloric acid solution into amyl alcohol ($C_5H_{11}OH$) and then re-extraction into an ammoniacal buffer solution at pH 8.5[92]. This precipitation can also be carried out using strontium sulphate ($SrSO_4$), in the presence of lead chloride ($PbCl_2$), as the carrier[93]. Using these techniques it is claimed that as little as 0.0025 ppm of phosphorus may be determined.

Separation from other ions[94-98]

From iron[94] Pass the sample solution at pH 5 through a column (12 cm × 1 sq cm) of Wofatit KPS 200 resin, at a rate of 4–5 ml a minute. Wash the column with an equal volume of water then elute phosphorus with 30 ml of $0.5N$ hydrochloric acid.

From silicon[95-98] It is claimed[95] that interference by silicon in the determination of phosphorus in organophosphorus compounds, can be eliminated by carrying out the digestive oxidation with hot alkaline ammonium persulphate (($NH_4)_2S_2O_8$)[96] in polypropylene test tubes rather than in glass. Alternatively[97] extract the phosphorus as phosphomolybdic acid at pH 1.1 into a 1:1 mixture of benzene (C_6H_6) and isobutanol (($CH_3)_2CH.CH_2.OH$). Re-extract the phosphorus into 10% aqueous ammonia (NH_4OH). It is claimed that this procedure will separate 1 part of phosphorus from 10,000 parts of silicon.

Macdonald[98] preferentially precipitates phosphomolybdate by slightly acidifying the test solution with nitric acid, treating with ammonium molybdate reagent at 50–55°C and setting aside at this temperature for 30 minutes. Then allow to cool for a further 30 minutes, filter and wash the precipitate with ammonium nitrate (NH_4NO_3) solution. Determine phosphorus on the precipitate and silicon on the filtrate.

Interferences in the determination of phosphorus by the molybdenum blue reaction[99-101]

Hydrogen peroxide[99] When hydrogen peroxide (H_2O_2) is used in wet ashing of the sample any residual peroxide will result in low phosphorus results. The addition of a few drops of potassium permanganate ($KMnO_4$) solution to the digest has been found to destroy excess peroxide without adversely affecting the determination of phosphorus.

Arsenic[100] Phosphorus can be determined in the presence to up to 300 times as much arsenic provided the arsenic is reduced to the trivalent form. If the final 10 ml of sample solution contains 40–400 ppm of arsenic the preferred reducing agent is sodium metabisulphite ($Na_2S_2O_5$), which also prevents interference by iron. When the arsenic concentration is 0.5–5 ppm sodium thiosulphate ($Na_2S_2O_3.5H_2O$) is satisfactory. Thiourea ($NH_2.CS.NH_2$) sodium sulphite (Na_2SO_3) and sodium hydrogen sulphite ($NaHSO_3$) are not satisfactory reducing agents for arsenic.

pH and time[101] The *p*H must be below 0.7 to prevent direct reduction of molybdenum, and for the most rapid results the acid concentration should be between 0.3N and 0.5N. The colour should be measured within two minutes of adding the reducing agent (*Methods 3* and *6*).

The individual tests

Method 1 is the standard Deniges method[102] using stannous chloride as the reducing agent, as modified by Homan and Pollard[103]. Henriksen[104] recommended that the stannous chloride should be dissolved in glycerol to increase its stability. This method is standard for phosphoric acid in cane sugar.

Method 2 is Tschopp's[105] modification of Deniges method, using metol as the reducing agent. This is claimed to reduce interference from silicates.

Method 3 is King's modification[8] of the Fiske and Subbarow procedure[6] using aminonaphthol sulphonic acid as the reducing agent. This method is especially useful in biochemical applications.

Method 4 uses hydroquinone as the reducing agent.

Method 5 is the simplified vanadomolybdate method which was originally developed for estimating phosphorus in iron and steel[42,43] and is now recommended[30] as the best method for fruit.

Method 6 uses ascorbic acid as the reducing agent which is claimed[71] to increase colour stability and to reduce interference from other ions including silicates.

Method 7 is a single reagent version of the vanadomolybdate method and is especially suitable for the control of the phosphate treatment of boiler-feed water and cooling water.

Method 8 is an adaptation of *Method 7* which uses the reagent in tablet form and was specifically developed for use in the field.

Bolotov[106] claims that stannous chloride (*Method 1*) is the most sensitive reducing agent, whereas Shafran[107] claims that of stannous chloride (*Method 1*), hydrazine, and ascorbic acid (*Method 6*), ascorbic acid is the best.

References

1. N. A. Talvitie and E. Perez, *Analyt. Chem.*, 1962, **34**, 866
2. D. Zach, *Chem. Tech. Berl.*, 1969, **21** (11), 711; *Analyt. Abs.*, 1970, **19**, 5246
3. J. W. Miles, L. E. Fetzer and G. W. Pearce, *Envir. Sci. Technol.*, 1970, **4** (5), 420
4. R. A. Nelson, *J. Assoc. offic. agric. Chem.*, 1962, **45**, 624
5. L. Laporta, *Boll. Lab. Chim. Provinciali*, 1964, **15** (2), 123; *Analyt. Abs.*, 1965, **12**, 3545
6. C. H. Fiske and Y. Subbarow, *J. Biol. Chem.*, 1932, **99**, 375
7. H. D. Kay, *J. Biol. Chem.*, 1932, **99**, 85
8. E. J. King, *Biochem. J.*, 1932, **26**, 292
9. E. Baginski and B. Zak, *Clin. Chim. Acta.* 1960, **5**, 834; *Analyt. Abs.*, 1961, **8**, 5128
10. M. Y. Shapiro and V. I. Muzychenko, *Lab. Delo*, 1963, **9**, 27; *Analyt. Abs.*, 1964, **11**, 1387

11. A. Negrin, *Clinica chim. Acta,* 1964, **10** (3), 262; *Analyt. Abs.,* 1966, **13,** 276
12. R. J. Hall, *J. Med. Lab. Technol.,* 1963, **20,** 97; *Analyt. Abs.,* 1964, **11,** 680
13. T. Hayami, *Japan Analyst,* 1962, **11,** 822; *Analyt. Abs.,* 1964, **11,** 1846
14. M. London and J. H. Marymont, *Clin. Chem.,* 1964, **10** (5), 417; *Analyt. Abs.,* 1965, **12,** 5284
15. M. M. Alimova, *Lab. Delo,* 1964, (6), 346; *Analyt. Abs.,* 1965, **12,** 5285
16. B. C. Black and E. G. Hammond, *J. Amer. oil Chem. Soc.,* 1965, **42** (11), 1002; *Analyt. Abs.,* 1967, **14,** 1573
17. A. A. Hirata and D. Appleman, *Analyt. Chem.,* 1959, **31,** 2097
18. G. Stearns and E. Warweg, *Amer. J. Diseases of Children,* 1935, **49,** 79
19. B. S. Cook and D. H. Simmonds, *J. Lab. Clin. Med.,* 1962, **60,** 160; *Analyt. Abs.,* 1963, **10,** 1120
20. T. Iwanaga, *J. Chem. Soc. Japan, pure Chem. Sect.,* 1960, **81,** 984; *Analyt. Abs.,* 1960, **7,** 2873
21. K. Schriever, *Arch. Pharm. Berlin,* 1959, **292,** 555; *Analyt. Abs.,* 1960, **7,** 2873
22. D. S. Galanos and V. M. Kapoulas, *Analytica chim. Acta,* 1966, **34** (3), 360; *Analyt. Abs.,* 1967, **14,** 4026
23. T. Salvage and Jean P. Dixon, *Analyst,* 1965, **90,** 24
24. P. M. Saliman, *Analyt. Chem.,* 1964, **36,** 112
25. M. Maruyama and K. Hasegawa, *Ann. Rep. Tekamine Lab.,* 1961, **13,** 173; *Analyt. Abs.,* 1964, **11,** 186
26. H. Y. Yu and I. H. Sha, *Chem. Bull. Peking,* 1965, (9), 557; *Analyt. Abs.,* 1967, **14,** 191
27. L. I. Diuguid and N. C. Johnson, *Microchem. J.,* 1968, **13** (4), 616; *Analyt. Abs.,* 1970, **18,** 2472
28. J. Pien, *Annls. Falsif. Expert. chim.,* 1966, **59,** 431; *Analyt. Abs.,* 1968, **15,** 448
29. R. V. Bedessem. P. Alioto and R. H. Moubry, *J. Assoc. offic. analyt. Chem.,* 1969, **52,** 917
30. B. Estrin and F. E. Boland, *J. Assoc. off. analyt. Chem.,* 1970, **53,** 575
31. M. Okamoto, J. W. Shafir and C. L. Ettinger, *J. Assoc. offic. analyt. Chem.,* 1969, **52** (3), 634; *Analyt. Abs.,* 1970, **19,** 700
32. P. G. Pifferi and A. Zamorani, *Industrie agrarie,* 1968, **6** (4), 213; *Analyt. Abs.,* 1969, **17,** 1778
33. J. Pomeranz, *Chemist Analyst,* 1954, **43,** 37; *Analyt. Abs.,* 1954, **1,** 2522
34. G. Matelli and V. Vicentini, *Allumin. nuova Metall.,* 1969, **38** (12), 627; *Analyt. Abs.,* 1970, **19,** 4771
35. A. A. Fedorov and G. P. Sokolova, *Sb. Trud. Tsentr. nauchno-issled. Inst. Chern. Metallurg.,* 1963, (31), 175; *Analyt. Abs.,* 1964, **11,** 1686
36. Amer. Soc. Materials, Philadelphia, Pa., *"1950 Book of Methods for Chemical Analysis of Metals",* Philadelphia, 1950
37. H. K. Lutwak, *Analyst,* 1953, **78,** 661
38. W. Weber, *Metall.,* 1964, **18,** 1188; *Analyt. Abs.,* 1965, **12,** 3211
39. K. Vigh, J. Inczedy and L. Erdey, *Magyar Kem. Foly.,* 1963, **69** (2), 73; *Analyt. Abs.,* 1965, **12,** 679
40. V. V. Stepin and I. A. Onorina, *Trudy vses. nauchno-issled. Inst. standart. Obraztsov.,* 1968, **4,** 136; *Analyt. Abs.,* 1970, **18,** 943
41. G. Lindly, *Analyt. chim. Acta,* 1961, **25,** 334; *Analyt. Abs.,* 1963, **10,** 1914
42. R. Schroder, *Stahl u. Eisen,* 1918, **38,** 316
43. G. Misson, *Chem. Zeit,* 1922, **32,** 633; *Ann. chim. anal. chim. appl.,* 1922, **4,** 267
44. P. Pakalns, *Analyt. chim. Acta,* 1970, **49,** 511
45. R. Goetz, *Giesserie.,* 1965, **52** (1), 12; *Analyt. Abs.,* 1966, **13,** 2958
46. G. Volke, *Giesserietechnik,* 1968, **14** (6), 186; *Analyt. Abs.,* 1969, **17,** 1422

47. L. Erdy and V. Fleps, *Acta Chim. Acad. Sci. Hung.*, 1957, **11**, 195; *Analyt. Abs.*, 1958, **5**, 1168
48. A. A. Fedorov, G. P. Sokolova and G. P. Volkova, *Sb. Trudy. tsent. nauchno-issled. Inst. chern. Metall.*, 1966, (49), 73; *Analyt. Abs.*, 1967, **14**, 6766
49. Brit. Standards Inst., B.S. 4158, Part 2, 1970, London, 1970
50. A. Lewandowski and H. Witkowski, *Prace Kom. Mat. Przyr. Pozan. Tow. Przyr. Nauk.*, 1959, **7**, 3; *Analyt. Abs.*, 1961, **8**, 1479
51. P. D. Malhotra and J. K. Sacher, *Amer. Miner.*, 1969, **54**, (1–2), 313; *Analyt. Abs.*, 1970, **18**, 3900
52. P. F. Parks and Dorothy E. Dunn, *J. Assoc. offic. agric. Chem.*, 1963, **46** (5), 836; *Analyt. Abs.*, 1965, **12**, 395
53. M. Heckman, *J. Assoc. offic. agric. Chem.*, 1964, **47** (3), 509; *Analyt. Abs.*, 1966, **12**, 5519
54. C. B. Stuffins, *Analyst*, 1967, **92**, 107
55. A. E. Rash, *J. Assoc. offic. analyt. Chem.*, 1968, **51** (4), 771; *Analyt. Abs.*, 1969, **17**, 3090
56. K. Kawagaki, M. Saito and K. Hirokawa, *J. Chem. Soc. Japan, ind. Chem. Sect.*, 1965, **68** (3), 465; *Analyt. Abs.*, 1967, **14**, 5434
57. C. Y. Shen and D. R. Dyroff, *Analyt. Chem.*, 1962, **34**, 1367
58. J. Fitzek and H. Stegemann, *Beitr. Silikose-Forsch*, 1957, (47), 41; *Analyt. Abs.*, 1958, **5**, 2515
59. W. Kowalski and M. Szwanenfeld, *Przem. Chem.*, 1955, **11**, 698; *Analyt. Abs.*, 1957, **4**, 727
60. K. Lasiewicz and H. Zawadzka, *Chem. Anal. Warsaw*, 1956, **1**, 53; *Analyt. Abs.*, 1957, **4**, 2006
61. P. Chichilo, *J. Assoc. offic. agric. Chem.*, 1964, **47**, 1019; *Analyt. Abs.*, 1965, **12**, 6778
62. S. H. Kamel, *J. vet. Sci., U.A.R.*, 1965, **2** (2), 103; *Analyt. Abs.*, 1967, **14**, 1538
63. F. M. Shemyakin and E. N. Zelenina, *Zavodskaya Lab.*, 1969, **35** (6), 657; *Analyt. Abs.*, 1970, **19**, 1463
64. Internat. Union of Leather Chemists' Societies, Leather Analysis Commission, *J. Soc. Leath. Trades Chem.*, 1969, **53** (10), 389; *Analyt. Abs.*, 1970, **19**, 4084
65. B. Mica, *Prum. Potravin.*, 1964, **15** (10), 526; *Analyt. Abs.*, 1966, **13**, 960
66. J. T. McAloren and G. F. Reynolds, *Talanta*, 1963, **10**, 145; *Analyt. Abs.*, 1964, **11**, 233
67. L. Duval, *Ann. Agron., Paris*, 1962, **13**, 469; *Analyt. Abs.*, 1964, **11**, 1386
68. J. C. van Schouwenberg and I. Walinga, *Analyt. chim. Acta*, 1967, **37** (2), 271; *Analyt. Abs.*, 1968, **15**, 3010
69. N. V. Mikhailov, T. T. Strashnova and G. M. Terekhova, *Khim. Volokna*, 1963, (4), 66; *Analy. Abs.*, 1965, **12**, 4016
70. Min. of Housing and Local Govt., "*Methods of Chemical Analysis as Applied to Sewage and Effluents*", H. M. Stationery Office, London, 1956
71. D. N. Fogg and N. T. Wilkinson, *Analyst*, 1958, **83**, 406
72. H. Namiki, *Japan Analyst*, 1961, **10**, 945; *Analyt. Abs.*, 1963, **10**, 4437
73. I. J. Cooke and J. Hislop, *Soil Sci.*, 1963, **96** (5), 308; *Analyt. Abs.*, 1965, **12**, 2542
74. D. M. Kheifets, *Pochvovedenie*, 1962, (5), 114; *Analyt. Abs.*, 1963, **10**, 3955
75. H. Riehm, *Agrochima*, 1958, **3**, 49; *Analyt. Abs.*, 1961, **8**, 2193
76. E. Rauterberg and H. Ossenberg-Neuhaus, *Pft. Ernahr. Dung.*, 1958, **82**, 46; *Analyt. Abs.*, 1959, **6**, 1561
77. R. G. Warren and A. J. Pugh, *J. Agric. Sci.*, 1930, **20**, 532
78. A. E. Beridze, *Pochvovedenie*, 1964, (11), 101; *Analyt. Abs.*, 1966, **13**, 441
79. A. M. Meshcheryakov, *Pochvovedenie*, 1963, (5), 96; *Analyt. Abs.*, 1964, **11**, 4005
80. C. G. Sherrell and W. M. H. Saunders, *N. Z. J. agric. Res.*, 1966, **9** (4), 972; *Analyt. Abs.*, 1968, **15**, 3649
81. J. K. Syers, J. D. H. Williams and T. W. Walker, *N. Z. J. agric. Res.*, 1968, **11** (4), 757; *Analyt. Abs.*, 1970, **18**, 2031

82. D. A. Wentz and G. F. Lee, *Envir. Sci. Technol.,* 1969, **3** (8), 750; *Analyt. Abs.,* 1970, **19,** 3431
83. T. G. Alexander and J. A. Robertson, *Can. J. Soil Sci.,* 1968, **48** (2), 217; *Analyt. Abs.,* 1969, **17,** 3093
84. J. Guiot, *Bull. Rech. agron. Gembloux,* 1969, **4** (1), 94; *Analyt. Abs.,* 1970, **19,** 4428
85. Brit. Standards Inst., *"Routine Control Methods of Testing Water Used in Industry",* B.S. 1427, London, 1962 and amendments
86. Brit. Standards Inst., *"Methods of Testing Water Used in Industry",* B.S. 2690, London, 1966. Part 3
87. M. Sakanoue, *J. Chem. Soc. Japan pure Chem. Sect.,* 1960, **81,** 242; *Analyt. Abs.,* 1961, **8,** 3050
88. J. Murphy and J. P. Riley, *Analyt. chim. Acta,* 1962, **27,** 31
89. A. K. Usova, O. F. Kutumova and B. S. Beinisovich, *Dokl. Acad. Nauk. Uz. S.S.R.,* 1963, (10), 18; *Analyt. Abs.,* 1965, **12,** 1140
90. P. G. W. Jones, *J. mar. biol. Assoc. U.K.,* 1966, **46,** 19; *Analyt. Abs.,* 1968, **15,** 487
91. D. E. Sanning, *Wat. Sewage Works.,* 1967, **114** (4), 131; *Analyt. Abs.,* 1968, **15,** 4326
92. K. R. Kar and Gurbir Singh, *Mikrochim. Acta,* 1968, (3), 560; *Analyt. Abs.,* 1969, **17,** 1400
93. K. R. Kar, M. M. Bhutani and Gurbir Singh, *Mikrochim. Acta,* 1968, (6), 1198; *Analyt. Abs.,* 1970, **18,** 1594
94. M. Kozhukharov and N. Gudev, *Zh. analit. Khim.,* 1963, **18,** 280; *Analyt. Abs.,* 1964, **11,** 972
95. R. B. Lew and F. Jakob, *Talanta,* 1963, **10,** 322; *Analyt. Abs.,* 1964, **11,** 629
96. J. Kolmerton and J. Epstein, *Analyt. Chem.,* 1958, **30,** 1536
97. A. K. Lavrukhina and A. I. Zaitseva, *Trudy. Kom. analit. Khim.,* 1968, **16,** 210; *Analyt. Abs.,* 1969, **16,** 2429
98. A. M. G. Macdonald and F. H. Van der Voort, *Analyst,* 1968, **93,** 65
99. J. G. Mortimer and D. N. Raine, *Analyt. Biochem.,* 1964, **9** (4), 492; *Analyt. Abs.,* 1966, **13,** 1888
100. L. Duval, *Chim. analyt.,* 1967, **49** (6), 307; *Analyt. Abs.,* 1968, **15,** 5300
101. S. R. Crouch and H. V. Malmstadt, *Analyt. Chem.,* 1967, **39** (10), 1084; *Analyt. Abs.,* 1968, **15,** 6597
102. G. Deniges, *Compt. Rend.,* 1920, **171,** 802
103. W. M. Holman and A. G. Pollard, *J. Soc. Chem. Ind.,* 1937, **56,** 339T
104. A. Henriksen, *Analyst,* 1963, **88,** 898
105. E. & E. Tschopp, *Helv. Chim. Acta.,* 1923, **15,** 793
106. M. P. Bolotov and P. V. Karetnikov, *Lab. Delo.,* 1965, **11** (1), 30; *Analyt. Abs.,* 1966, **13,** 2513
107. I. G. Shafran, M. V. Pavlova, S. A. Titova and L. D. Yakusheva, *Trudy. vses. nauchno. -issled. Inst. Khim. React.,* 1966, (28), 56; *Analyt. Abs.,* 1967, **14,** 6767
108. Brit. Standards Inst. *"Treatment of water for marine boilers"* B.S. 1170 London 1968

Phosphate Method 1
using ammonium molybdate and stannous chloride (Deniges' method)

Principle of the method

The blue colour produced is obtained by the reduction of phosphomolybdic acid with stannous chloride solution. As the depth of colour is dependent on the proportion of added reagents, temperature, and time of reaction, it is essential that strict attention be paid to all details in order to obtain correct results.

Reagents required

1. *Ammonium molybdate solution* Dissolve 10 g of ammonium molybdate $((NH_4)_6Mo_7O_{24}.4H_2O)$ in 100 ml of distilled water. Dissolve 150 ml of sulphuric acid in 150 ml of distilled water, adding the acid cautiously to the water and keeping the solution continuously stirred and cooled. When the diluted acid is cold add the ammonium molybdate solution to it.

This reagent is stable for long periods if stored in a boro-silicate glass or polythene bottle and protected from the action of light.

2. *Stannous chloride solution* Dissolve 2.5 g stannous chloride $(SnCl_2.2H_2O)$ in 100 ml glycerol.

This reagent is stable for several weeks.

All chemicals used in the preparation of these reagents should be of analytical reagent quality.

The Standard Lovibond Discs, Comparator Disc 3/7 and Nessleriser Disc NMB

Disc 3/7 covers the range 20, 40, 60, 80, 100, 130, 160, 190, 220 μg (0.02 to 0.22 mg) of phosphate, calculated as P_2O_5. This equals 2.0 to 22.0 parts per million if a 10 ml sample is used.

Nessleriser Disc NMB covers the range 2, 4, 6, 8, 10, 12, 14, 17, 20 μg (0.002 to 0.02 mg) of phosphate calculated as P_2O_5.

The markings on the disc represent the actual amounts of phosphate, calculated as phosphoric oxide (P_2O_5), producing the colour in the test. Thus if a colour equivalent to 10 μg is produced in the test, the amount of P_2O_5 present in the volume of solution taken for the test is 0.01 mg. On a 10 ml sample, this means that the range is 0.2 to 2.0 ppm.

To convert P_2O_5 to PO_4, multiply answer by 1.34.

Technique

Dilute a suitable volume of the solution under test with distilled water to 45 ml; transfer to a flask immersed in a water-bath at 25°C. When the solution has reached the temperature of the bath, add 1 ml of ammonium molybdate (reagent 1) and mix thoroughly. Then add 0.15 ml of stannous chloride (reagent 2), mix, and dilute to 50 ml with water at 25°C and allow to stand for five minutes.

Comparator

Transfer 10 ml to a test tube and place in the right-hand compartment in the comparator. Fill a test tube with the untreated solution under test, diluted with distilled water in the same proportions, and place in the left-hand compartment, to provide a background for the glass colour standards. Hold the comparator facing a standard source of white light such as the Lovibond White Light Cabinet or, failing this, north daylight, and compare the colour produced in the test solution with the colours in the standard disc, rotating the latter until a colour match is obtained. If the colour produced is deeper than the standard colours, the test should be repeated, using a smaller volume of the solution under examination.

COLORIMETRIC CHEMICAL ANALYTICAL METHODS   Phosphate Method 1

Nessleriser

For use with Lovibond Nessleriser place the whole 50 ml in a Nessleriser glass in the right-hand compartment of the Nessleriser and in the left-hand compartment place a Nessleriser glass containing the same volume of the solution under test diluted with distilled water to 50 ml. Stand the Nessleriser before a standard source of white light and compare the colour produced with the colours in the standard disc, rotating the disc until a colour match is obtained. Should the colour in the test solution be deeper than the standard colour glasses, a fresh test should be carried out using a smaller quantity of the solution under examination.

Notes

1. The colour is produced only in the presence of ortho-phosphates; meta and pyro-phosphates must therefore be completely hydrolysed before testing. Free mineral acids and alkalis depress the colour and must therefore be neutralised. Certain organic acids, such as citric, oxalic, and tartaric (but not acetic acid), inhibit the development of the blue colour if present in appreciable amounts. If a substantial amount of organic matter is present it must be removed, to avoid possibilities of interference. Ferric iron if exceeding 1 part per million in the final test solution, should be reduced to the ferrous state, for which purpose a Jones reductor is recommended. Arsenate must be reduced to arsenite, by means of hydrogen sulphide in acid solution.

2. It must be emphasised that readings obtained with the Lovibond Nessleriser and disc are only accurate provided that Nessleriser glasses are used which conform to the specification employed when the discs were calibrated, namely that the 50 ml mark shall fall at a height of 113 ± 3 mm measured internally.

Phosphate Method 2
using ammonium molybdate and metol (Tschopp's method)

Principle of the method

The test is based on the colours produced by the reduction of phosphomolybdic acid with p-methylamino-phenol sulphate (metol). Numerous variations of this test have been proposed from time to time and many different reducing agents have been recommended. The technique here adopted is a modification of that advocated by Ernst and Emilio Tschopp which experience has shown to be most reliable for use in conjunction with permanent standards since, under the conditions of the test, uniform colours are produced, and the test is rendered relatively insensitive to silica.

Reagents required

1. *Ammonium molybdate solution* Slowly add 150 ml of sulphuric acid to 100 ml of distilled water, keeping the mixture continuously stirred and cooled during the addition of the acid. Dissolve 10 g of ammonium molybdate $((NH_4)_6Mo_7O_{24}.4H_2O)$ in 100 ml of distilled water and add this solution slowly to the diluted sulphuric acid.
   This reagent is stable if stored in a boro-silicate glass bottle and protected from the action of light.
2. *Metol solution* Dissolve 40 g of sodium metabisulphite $(Na_2S_2O_5)$ 1 g of sodium sulphite $(Na_2SO_3.7H_2O)$ and 0.2 g of metol (p-methylaminophenol sulphate $(CH_3.NH.C_6H_4.OH)_2.H_2SO_4$) in 200 ml of distilled water. Stable for several months.

All chemicals used in the preparation of these reagents should be of analytical reagent quality.

The Standard Lovibond Discs, Comparator Disc 3/4 and Nessleriser Disc NMC

Disc 3/4 covers the range from 5, 10, 20, 30, 40, 50, 60, 80, 100 μg (0.005 to 0.1 mg) of phosphate calculated as phosphorus pentoxide (P_2O_5). This value multiplied by 0.4367 gives the value in terms of phosphorus (P). The range is equivalent to 1 to 20 ppm of P_2O_5 for a 5 ml sample.

Disc NMC covers the range 5, 10, 15, 20, 25, 30, 40, 50, 60 μg (0.005 to 0.06 mg) of phosphate calculated as phosphorus pentoxide (P_2O_5). This is equivalent to 0.2 to 2.4 ppm of P_2O_5 for a 25 ml sample.

To convert P_2O_5 to PO_4, multiply answer by 1.34.

Technique

(a) *Comparator*

To 5 ml of the solution under examination, contained in one of the graduated test-tubes, add 1 ml of reagent No. 1 and heat the tube in a boiling water-bath for 15 minutes. Then add 1 ml of reagent No. 2 and continue the heating in a boiling water-bath for a further 15 minutes. Cool the solution, dilute it with distilled water to 10 ml and place it in the right-hand compartment of the Comparator. At the same time and in the same manner prepare a blank solution from 5 ml of distilled water and place in the left-hand compartment of the Comparator. Hold the Comparator facing a standard source of white light such as the Lovibond White Light Cabinet or, failing this, north daylight, and compare the colour produced in the test solution with the colours in the standard disc, rotating the latter until a colour match is obtained.

(b) *Nessleriser*

To 25 ml of the solution under examination, or a suitable quantity diluted to 25 ml, contained in one of the Nessleriser glasses, add 5 ml of reagent No. 1 and immerse the Nessleriser glass in a boiling water-bath for 15 minutes; then add 5 ml of reagent No. 2 and continue the heating for a further 15 minutes. Cool the solution, dilute it with distilled water to 50 ml and place it in the right-hand compartment of the Nessleriser. At the same time and in the same manner prepare a blank solution from 25 ml of distilled water and place it in the left-hand compartment of the Nessleriser. Stand the Nessleriser before a standard source of white light and compare the colour produced in the test solution with the colours in the standard disc, rotating the disc until a colour match is obtained.

COLORIMETRIC CHEMICAL ANALYTICAL METHODS — Phosphate Method 2

If the colour produced in the test is deeper than the standard colours, the original solution should be suitably diluted and the test carried out on the diluted solution, the result being corrected accordingly.

Notes

1. Nitrates may interfere, if present to the extent of more than 1 mg NO_3^-, using the comparator, or more than 3 mg using the Nessleriser.

2. The test is five thousand times less sensitive to silica than to phosphate.

3. Iron can be tolerated in concentrations up to one thousand times the amount of phosphate.

4. It is essential that the solutions to be examined for phosphate should be neutral, since uniformity of colour production is dependent on the final concentration of free acid in the solution. If the solution contains a buffer salt such as sodium acetate, a quantity of sulphuric acid equivalent to the acetate must be added before carrying out the test.

5. Arsenites react with phosphates to give a similar colour and should be removed by precipitation with hydrogen sulphide followed by boiling and filtration. Sugars, and also lactates, citrates, tartrates, oxalates and other organic salts, depress the intensity of colour produced by phosphates, and these compounds, if present, should be removed.

6. It must be emphasised that the readings obtained with the Lovibond Nessleriser and disc are only accurate provided that Nessleriser glasses are used which conform to the specification employed when the discs were calibrated, namely that the 50 ml mark shall fall at a height of 113 ± 3 mm measured internally.

Phosphorus Method 3
using 1-amino-2-naphthol-4-sulphonic acid

Principle of the method

The method is based on King's modification of the Fiske and Subbarrow procedure. The phosphate is combined with molybdic acid, and reduced with amino-naphthol-sulphonic acid to give a blue colour, the intensity of which is proportional to the amount of phosphate ion.

Reagents required

1. *Perchloric acid* 60% solution of $HClO_4$
2. *Ammonium molybdate* 5% solution of $(NH_4)_6Mo_7O_{24}.4H_2O$
3. *Aminonaphthol-sulphonic acid solution* Dissolve 0.2 g of 1-amino-2-naphthol-4-sulphonic acid $(NH_2.C_{10}H_5(OH).SO_3H)$, 12 g of sodium metabisulphite $(Na_2S_2O_5)$ and 2.4 g of sodium sulphite $(Na_2SO_3.7H_2O)$ in 100 ml of distilled water.

This reagent is stable for up to 2 weeks if stored in a dark bottle.

All chemicals used in the preparation of these reagents should be of analytical reagent quality.

The Standard Lovibond Comparator Disc 5/14

The disc covers the range 0.02, 0.03, 0.04, 0.05, 0.06, 0.08, 0.1, 0.12, 0.16 mg of phosphorus (as P), present in the amount of sample taken.

With the 5 ml sample suggested, this represents 4 to 32 ppm P, or 12 to 96 ppm PO_4.

Technique

Place 5 ml of sample in a 10 ml test-tube, and add 0.7 ml of perchloric acid (reagent 1) 0.7 ml of ammonium molybdate (reagent 2), and 0.5 ml of amino-naphthol-sulphonic acid solution, (reagent 3) and finally distilled water to the 10 ml mark. Mix, and leave for 10 minutes for the colour to develop. At the end of this time, place the solution in a comparator test tube (or 13.5 mm cell) in the right-hand compartment of the Comparator, and a "blank" of distilled water in the left-hand compartment so that it comes behind the colour standards in the disc. Hold the Comparator facing a standard source of white light such as the Lovibond White Light Cabinet or, failing this, north daylight, and rotate the disc until the colour of the liquid is matched by one of the glass standards. The value, expressed as mg of phosphorus in the 5 ml of test solution, is then read off from the indicator window at the bottom right-hand corner of the Comparator. Multiplied by 200, this result gives ppm of phosphorus (P).

To convert P to PO_4, multiply the answer by 3. To convert PO_4 to P_2O_5 multiply by 0.75.

Phosphate Method 4
using hydroquinone and ammonium molybdate

Principle of the method

The test consists of adding ammonium molybdate to an acid solution of the phosphate. The ammonium phosphomolybdate formed is then reduced to a lower state of oxidation resulting in a blue-coloured compound said to have the composition $(MoO_2.4MoO_3)_2.H_3PO_4$. The reducing agent used in this method is hydroquinone.

Reagents required

1. *Sulphuric acid solution* To about 500 ml of distilled water slowly add 65 ml of sulphuric acid and dilute the mixture to exactly 1 litre.

2. *Acid—molybdate solution* Dissolve **without heating** 8.8 g of ammonium molybdate $((NH_4)_6Mo_7O_{24}.4H_2O,)$ in about 100 ml of distilled water. To about 300 ml of distilled water add 38 ml of sulphuric acid. Add the diluted acid to the ammonium molybdate solution and dilute the mixture to exactly 500 ml.

3. *Hydroquinone solution* Dissolve **completely** 5 g of hydroquinone and then add 0.3 ml of sulphuric acid, in 500 ml of distilled water.
   This solution slowly darkens in colour, but will keep for about 3 weeks in an amber glass bottle in the dark.

4. *Carbonate sulphite* Dissolve in 500 ml of distilled water 130 g of anhydrous potassium carbonate (K_2CO_3) and 24 g of sodium sulphite $(Na_2SO_3.7H_2O)$. This is stable up to 6 months.

All chemicals used in the preparation of these reagents should be of analytical reagent quality.

The Standard Lovibond Comparator Disc 3/12

The disc covers the range 0—80 mg per litre (=parts per million) of phosphate in steps of 10 mg, calculated as PO_4, based on a 5 ml sample. The equivalent in terms of phosphorus (P), is 0 to 26.1 parts per million.

Technique

Filter the sample brilliantly clear, so that only soluble phosphate is determined. To 5 ml of this filtrate, in a Nessler tube, add the reagents in the following order. During and after each addition, swirl the sample to mix thoroughly:—

        2 ml of molybdate solution (reagent 2)
        1 ml of hydroquinone solution (reagent 3)

The solution should preferably be kept at a temperature of $25°$ C $\pm\ 2°$. Allow to stand for 5 minutes, to develop the green phosphate colour. During this time, measure 2 ml of the carbonate-sulphite solution (reagent 4) into a separate Nessler tube. At the end of the 5 minutes quickly pour the solution under test into the 2 ml of carbonate-sulphite solution, pouring backwards and forwards a few times to ensure thorough mixing. Transfer to a 13.5 mm test tube and place in the right-hand aperture of the Comparator. A "blank" sample is prepared at the same time and in the same manner, substituting 2 ml of the sulphuric acid solution (reagent 1) for the 2 ml of molybdate solution. This is placed in a test tube in the left-hand aperture of the Comparator, and compensates for any inherent colour in the boiler water or any slight colour of the hydroquinone solution.

Hold the Comparator facing a standard source of white light such as the Lovibond White Light Cabinet or, failing this, north daylight, and compare the colour produced in the test solution with the colours in the standard disc, rotating the latter until a colour match is obtained. If the colour produced is deeper than the disc colours, the test should be repeated using a smaller volume of the boiler filtrate and adding distilled water to give a total volume of 5 ml. The degree of dilution must be allowed for when calculating the final answer.

The blue colour of the solution fades, so comparison should be made immediately the colour has been developed. Bubbles in the solution which cling to the side of the tube render matching difficult, and should be removed by gently tapping the tube.

The figures shown at the indicator window represent mg per litre (=parts per million) of PO_4 in the boiler water, when 5 ml of the sample is taken for the test.

Phosphate Method 5

using a simplified vanadate/molybdate method

Principle of the method

The method is based on the reaction of phosphates with molybdates in the presence of vanadate to form yellow phosphovanadomolybdate.

Reagents required

1. *Ammonium vanadate solution* Dissolve 2.5 g of ammonium vanadate (NH_4VO_3) in distilled water, add 20 ml of nitric acid and make up to 1 litre with distilled water.
2. *Ammonium molybdate solution* Dissolve 5 g of ammonium molybdate (($NH_4)_6Mo_7O_{24} \cdot 4H_2O$) in 100 ml of distilled water.

All chemicals used in the preparation of these reagents should be of analytical reagent quality.

The Standard Lovibond Comparator Disc 3/38

The disc covers the range of 10 to 100 parts per million (mg per litre) of phosphate in steps of 10 ppm, calculated as PO_4, based on a 5 ml sample.

Technique

Place 5 ml of the solution to be tested in a Lovibond Comparator test tube, add 1 ml of reagent 1 and 1 ml of reagent 2. Make up to the 10 ml mark with distilled water mix, and place in the right-hand compartment of the Comparator. In the left-hand compartment, behind the glass colour standards, place a "blank" of 1 ml of reagent 1 made up to 10 ml with the untreated sample. Match after exactly 15 minutes, holding the Comparator facing a standard source of white light such as the Lovibond White Light Cabinet or, failing this, north daylight. The figure read from the indicator window represents parts per million of phosphate.

Note

The colour produced differs slightly when testing solutions of neutral reactions as compared with solutions which are acid, but it is always possible to identify the appropriate step in the scale.

Phosphate Method 6

using ascorbic acid and ammonium molybdate

Principle of the method

The method depends, as do many previous methods, on the reduction of molybdophosphate to molybdenum blue. The reduction is carried out by ascorbic acid, at the boiling point of the solution, and the colour so formed is compared with permanent Lovibond glass standards.

Reagents required

1. *Ammonium molybdate solution* Dissolve 10 g of ammonium molybdate $((NH_4)_6Mo_7O_{24}.4H_2O)$ in 100 ml of distilled water. Carefully add 150 ml of sulphuric acid to 150 ml of water, stirring and cooling the mixture throughout the addition. To this cooled diluted acid slowly add the ammonium molybdate solution. This reagent is stable if stored in a boro-silicate glass bottle and protected from the action of light.

2. *Ascorbic acid* $(C_6H_8O_6)$

3. *Normal sulphuric acid* Carefully add 49 g of sulphuric acid to 500 ml of distilled water while cooling. When cold dilute to 1 litre with distilled water.

All chemicals used in the preparation of these reagents should be of analytical reagent quality.

The Standard Lovibond Comparator Disc 3/51

This disc covers the range 10, 25, 50, 75, 100, 150, 200, 300, 400 μg of phosphate (calculated as PO_4). This is equivalent to 2—80 ppm of PO_4 based on a 20 ml sample.

Technique

Measure 20 ml of the sample (Note 2) into a 100 ml calibrated flask. Add sufficient N sulphuric acid (reagent 3) to neutralise the alkalinity, and dilute to 100 ml with distilled water. Measure 25 ml of this solution into a 100 ml beaker, add 15 ml of distilled water, and 4 ml of ammonium molybdate solution (reagent 1) and mix. Add 0.1 g of ascorbic acid, cover with a watch glass, heat to the boiling point, and boil gently for 1 minute only. Cool the solution quickly, transfer to a 50 ml flask, and dilute to 50 ml with distilled water.

Carry out a blank test on the reagents with distilled water in the place of the sample. Transfer a portion of the test solution to a standard test tube and place in the right-hand compartment of the Lovibond Comparator. Transfer a portion of the blank solution to an identical test tube and place this in the left-hand compartment. Compare the colour of the test solution with the colours of the permanent glass standards in the disc using a standard source of white light such as the Lovibond White Light Cabinet or, failing this, north daylight.

Notes

1. Up to 1,000 ppm of iron, chloride equivalent to 10% of sodium chloride, nitrate equivalent to 2,500 ppm of sodium nitrate and sulphate and perchlorate equivalent to at least 25% of the equivalent sodium salt are claimed to have no effect on this determination.

   The presence of 2,500 ppm of soluble silica can also be tolerated.

2. For the determination of phosphate present in boiler-feed water as hexametaphosphate, it is recommended that the hexametaphoshphate be hydrolysed by neutralisation of the sample, addition of 1 ml of hydrochloric acid (sp gr 1.18) and evaporation of the solution to dryness.

Phosphate Method 7

using vanadomolybdate single reagent

Principle of the method

In the presence of vanadates, phosphates react with molybdates to form yellow phosphovanadomolybdate. The intensity of this yellow colour, which is proportional to the amount of phosphate present, is determined by comparison with a series of Lovibond permanent glass standards. If the boiler water is coloured by the presence of organic matter, this is removed before the test by treatment with a suitable oxidising agent.

Reagents required

1. *Vanadomolybdate reagent* Solution A: 20 g of ammonium molybdate tetrahydrate $((NH_4)_6Mo_7O_{24}.4H_2O)$ dissolved in 250 ml of distilled water.
   Solution B: 1 g of ammonium metavanadate (NH_4VO_3) dissolved in 40 ml of nitric acid (sp. gr. 1.42) and 200 ml of distilled water.
   Mix solutions A and B, add 100 ml nitric acid (sp. gr. 1.42) and dilute to 1 litre with water.

2. *Oxidising mixture* Intimately mix, by grinding in a mortar, 100 g of potassium persulphate $(K_2S_2O_8)$ and 60 g of sodium carbonate (Na_2CO_3), and sieve through a 40 mesh sieve.

The Standard Lovibond Comparator Disc 3/60

This disc covers the range 10 to 100 ppm of phosphate, calculated as PO_4, in 9 steps of 10, omitting 90.

Technique

Filter the water through a No. 42 Whatman filter paper into a 50 ml stoppered measuring cylinder. Use the first few ml of filtrate to rinse out the cylinder. Filter between 20 and 25 ml of water, and then add an equal volume of reagent 1. Stopper the cylinder and mix the contents. Rinse one of the comparator 13.5 mm tubes or cells with this solution, then fill the cell and place it in the right-hand compartment of the comparator. In the left-hand compartment place a blank prepared by mixing equal volumes of reagent 1 and distilled water. Allow the colour to develop for at least three minutes and then match the colour of the sample tube against that of the standards in the disc, using a standard source of white light such as the Lovibond White Light Cabinet or, failing this, north daylight. If the colour is deeper than the 100 ppm colour standard the solution may be diluted with distilled water and due allowance made for the degree of dilution.

If the original water sample is coloured, measure 50 ml into a 150 ml beaker, add 1-2 g of reagent 2, boil until colourless, cool and make up to 50 ml with distilled water. Filter through No. 42 Whatman filter paper and proceed as above.

Note

This test was developed in collaboration with the Water Treatment Section, Technical Service Department of Albright and Wilson (Mfg.) Ltd.

Phosphate Method 8

using Palin vanadomolybdate tablets

Introduction

The vanadomolybdate method has been simplified by the use of the reagent in tablet form.

Principle of the method

In the presence of vanadates, phosphates react with molybdates to form yellow phosphovanadomolybdate. The intensity of the yellow colour, which is proportional to the amount of phosphate present, is determined by comparison with a series of Lovibond permanent glass colour standards.

Reagent required

Palin Phosphate tablet, either Comparator or Nessleriser type, according to which instrument is being used.

The Standard Lovibond Discs, Nessleriser Discs NOP and NOX, and Comparator Disc 3/70

Discs NOP and 3/70 cover the range 0 to 100 ppm of phosphate, calculated as PO_4, in 9 steps of 10 ppm omitting 70 and 90.

Disc NOX covers the range 1 to 10 ppm PO_4 in steps of 1 ppm, omitting 9 ppm.

The master discs, against which all reproductions are checked, were tested and approved by Dr. A. T. Palin.

Technique

Nessleriser discs

NOP. Filter a sample of cold boiler water through a No. 42 Whatman filter paper into a 100 ml measuring cylinder. Place 50 ml of this filtrate in a Nessler tube and insert this in the left-hand compartment of the Nessleriser to serve as a blank. Fill a second 50 ml Nessler tube with filtrate and add one Phophate test tablet. Crush the tablet with the flattened end of a glass rod and continue mixing and crushing until the tablet is completely dissolved. After approximately 10 minutes place this Nessler tube in the right-hand compartment of the Nessleriser and match against the standards in the disc, using a standard source of White Light such as the Lovibond White Light Cabinet. Failing this, north daylight should be used.

NOX. Place 50 ml distilled water in one Nessler tube and 50 ml filtered sample in the other. To both tubes, add a Phosphate test tablet, crush and proceed as above, the tube containing distilled water plus tablet being used as a blank in the left-hand compartment.

Comparator disc 3/70

Exactly the same technique as for disc NOP is used, but take 10 ml in a calibrated comparator test tube instead of 50 ml in a Nessler tube, and use tablets labelled Comparator instead of those labelled Nessleriser.

COLORIMETRIC CHEMICAL ANALYTICAL METHODS — Phosphate Method 8

Notes

1. Appreciable amounts of natural colour in the water sample may have to be allowed for; a colour of 60° Hazen for instance will increase the apparent phosphate reading by about 1 ppm.

To correct for natural colour place 50 ml untreated distilled water in a Nessler tube in the left-hand compartment and 50 ml of untreated sample (filtered if necessary) in a Nessler tube in the right-hand compartment of the Nessleriser. Estimate the equivalent reading using the Phosphate disc and subtract this figure from the subsequent phosphate test result.

2. Testing boiler water. The acid reaction at which the vanadomolybdate test is carried out may itself modify the "natural" colour of the sample. Thus in cases where boiler water samples remain appreciably coloured after the preliminary filtration, it is desirable to acidify the "blank" so as to allow for any such change when matching the colours.

This may be conveniently carried out by adding to the blank tube one "Phosphate Blank-Acidifying Tablet" using either Comparator or Nessleriser type as required. Crush and dissolve the tablet in the portion of sample contained in the blank tube and place this prepared blank in the left hand compartment of the Comparator or Nessleriser.

Continue the test in the normal way by adding one Palin Vanadomolybdate tablet to the appropriate volume of sample, dissolving as before, standing approximately 10 minutes, then placing in the right-hand compartment.

3. It must be emphasised that the readings obtained with the Lovibond Nessleriser are only accurate provided that Nessleriser glasses are used which conform to the specification employed when the discs were calibrated, namely that the 50 ml mark shall fall at a height of 113 ± 3 mm measured internally.

The Determination of Potassium

Introduction

Potassium occurs widely as a trace element in most materials. It is an essential element to plant and animal growth and methods have been developed for its analysis in a wide variety of materials, including biological tissues and fluids, plants, soils, food, fertilisers, ores and pharmaceuticals.

Methods of sample preparation

Biological materials[1]

Ash the sample, dissolve the ash in warm $2M$ hydrochloric acid, remove phosphate by passing the solution through a column of a strongly basic anion-exchange resin and make the percolate up to a known volume.

Foods and edible oils[2]

Fish Wet ashing (see General Introduction p. vii) gives more precise results than dry ashing as potassium losses occur at temperatures of 550°C and above.

Fuels and lubricants[3]

Coal ash Grind sample to pass through a 250 mesh sieve and then mix 0.5 g of this finely ground ash with 2 g of lithium metaborate ($LiBO_2$) in a platinum crucible. Heat the mixture in a muffle furnace at 900°C until the melt is clear. If the sample is likely to contain more than 15% of iron add 30–50 mg of ammonium vanadate (NH_4VO_3) to act as a flux. Cool the melt and transfer it to a beaker containing 8 ml of nitric acid and 150 ml of water. Stir vigorously until dissolution is complete. Add 2.5 g of tartaric acid ($(CHOH.COOH)_2$) to mask the iron and make the volume up to 250 ml.

Minerals, ores and rocks[3,4]

Either use the method described above for the analysis of coal ash[3] or[4], pre-ignite an empty graphite crucible for 30 minutes at 950°C and cool, taking care not to disturb the powdery inside surface. Mix 0.2 g of the powdered sample with 1.4 g of lithium metaborate ($LiBO_2$) in a porcelain crucible and transfer the mixture to the prepared graphite crucible. Fuse at 900°C in a muffle furnace for exactly 15 minutes. Pour the melt into 100 ml of 1:24 nitric acid: water contained in a plastic beaker with a flat bottom, and stir with a Teflon-coated stirring rod. Wash the solution into a 250 ml flask and dilute to volume with 1:24 nitric acid.

Miscellaneous[5,6]

Cement[5] Transfer 1 g of sample into a 200 ml platinum or porcelain evaporating dish, disperse with 15 ml of distilled water, add 10 ml of hydrochloric acid, heat and stir until dissolution is complete. Evaporate nearly to dryness on a steam bath, add 8 ml of hydrochloric acid, wait for 2 minutes and then add 50 ml of hot water. Stir, cover the dish with a clock-glass and then digest for 5 minutes. Filter through a Whatman No. 40 paper into a 500 ml volumetric flask. Wash the residue on the filter paper ten times with hot dilute (1+99) hydrochloric acid, five times with hot water and add the washings to the solution already in the flask. Transfer the filter paper and residue to a platinum crucible, warm gently to dry and then ignite, without firing, the filter paper and then ignite at 1100°C for 30 minutes. Moisten the ash with several drops of water followed by 6 drops of perchloric acid ($HClO_4$) and about 15 ml of hydrofluoric acid. Evaporate on a low-heat hot-plate and, when dry, ignite at 1100°C for 5 minutes. Fuse the ash with 0.1 g of lithium chloride ($LiCl.H_2O$), cool, dissolve the melt in hydrochloric acid, add this solution to that already in the flask and make the volume up to 500 ml.

Potato starch[6] Digest 5 g of starch with 50 ml of 1:1 nitric acid and 25 ml of perchloric acid. Dilute the solution to 200 ml with water.

Plants[7]

Wet ash the sample (General Introduction p. ix) and mix 1 ml of the solution with 9 ml of pH 4.8 buffer (dissolve 120 g of sodium acetate ($CH_3COONa.3H_2O$) in 700 ml of 35% formaldehyde (HCHO) solution, add 28 ml of anhydrous acetic acid (CH_3COOH) and dilute to 1 litre with water) and 0.5 ml of 0.15% gum acacia solution. Set aside for 15 minutes and then proceed by *Method 1*.

Separation from other ions[8,9]

From lithium and sodium[8] Apply the sample solution in ethanol (C_2H_5OH)—water—hydrochloric acid to a column of Dowex 50W-X8 resin (H^+ form). Wash the column with 60% ethanol and elute lithium with $0.6N$ hydrochloric acid in 60% ethanol. Wash the column with 40% ethanol and then elute sodium with $0.4N$ hydrochloric acid in 40% ethanol. Finally wash the column with 20% ethanol and elute potassium with $0.6N$ hydrochloric acid in 20% ethanol.

General[9] A solution of α-hexyl-(2,4-dinitro-N-picryl-1-naphthylamine) in nitrobenzene can be used to separate potassium from aqueous solution. Several ions, especially sodium, interfere but repeated extraction is claimed to give good results. A molar ratio of the reagent to potassium of at least 2:1 is required for complete extraction.

The individual tests

Method 1 This method was developed[10] to enable potassium to be determined in the presence of a large excess of sodium. The potassium is precipitated as its complex with sodium cobaltinitrite and the cobalt concentration in the precipitate is determined by use of the standard method for cobalt, *Method 2*, (p. 167).

References

1. G. Conradi, *Chemist Analyst,* 1967, **56** (4), 87; *Analyt. Abs.,* 1969, **16,** 284
2. M. H. Thompson, *J. Assoc. offic. agric. Chem.,* 1964, **47** (4), 701; *Analyt. Abs.,* 1965, **12,** 6717
3. P. L. Boar and L. K. Ingram, *Analyst,* 1970, **95,** 124
4. J. C. Van Loon and C. M. Parissis, *Analyst,* 1969, **94,** 1057
5. A. Nestoridis, *Analyst,* 1970, **95,** 51
6. B. Mica, *Prum. Potravin.,* 1964, **15** (10), 526; *Analyt. Abs.,* 1966, **13,** 960
7. S. Fortini and S. Panella, *Ann. Staz. Chim.-agr. Roma,* 1964, III, No 227; *Analyt. Abs.,* 1965, **12,** 5956
8. V. Nevoral, *Z. analyt. Chem.,* 1963, **195,** 332; *Analyt. Abs.,* 1964, **11,** 2990
9. T. Iwachido, S. Ukai and K. Toei, *Bull. Chem. Soc. Japan,* 1967, **40** (3), 694; *Analyt. Abs.,* 1968, **15,** 3801
10. C. P. Sideris, *Ind. Eng. Chem. Anal. Ed.,* 1937, **9,** 145

Potassium Method 1
using sodium cobaltinitrite

Principle of the method

The potassium is precipitated as sodium potassium cobaltinitrite ($K_2NaCo(NO_2)_6$) and the cobalt content of the precipitate is determined in the standard manner using nitroso-R salt. The potassium concentration is then calculated from the measured cobalt concentration.

Reagents required

1. *N Sodium hydroxide* ($NaOH$)
2. *N Hydrochloric acid* (HCl)
3. *Sodium cobaltinitrite reagent* Dilute a 12.5% solution of sodium cobaltinitrite ($Na_3Co(NO_2)_6$) with an equal volume of 95% ethyl alcohol (C_2H_5OH).
4. *Acetone* (CH_3COCH_3)
5. *2N Sulphuric acid* (H_2SO_4)

All the chemicals used should be of analytical reagent quality.

In addition the reagents for the determination of cobalt by *Method 2* will also be required (see p. 167).

The Standard Lovibond Nessleriser Discs NTA and NTB

Disc NTA covers the range 1–9 μg of cobalt (Co) equivalent to 1.3–12 μg of potassium (K). Disc NTB covers the range 10–30 μg of cobalt equivalent to 13–40 μg of potassium.

Technique

Take 5 ml of the sample solution, which should contain up to 1 mg of potassium. Make the solution slightly alkaline with sodium hydroxide (reagent 1) and boil for 15 minutes to remove any ammonia which might be present. Filter and wash any precipitate with hot water. Adjust the *p*H of the filtrate and washings to 4.5 with hydrochloric acid (reagent 2), transfer to a 15 ml centrifuge tube and evaporate to a volume of not more than 0.5 ml. Add 5 ml of reagent 3 and place the tube and contents in a refrigerator for at least one hour. Centrifuge, remove supernatant liquid and wash the precipitate first with 10 ml of water and then with 5 ml portions of acetone (reagent 4) until the washings are colourless. Dissolve the precipitate in 5 ml of 2N sulphuric acid (reagent 5) warming if necessary, and adjust the cobalt concentration of this solution, to lie within the range of the disc being used, by dilution if required. Proceed by Cobalt *Method 2*.

Notes

1. Providing that the ratio of sodium to potassium concentration is greater than 22:1 then the composition of the potassium sodium cobaltinitrite may be taken as $K_2NaCo(NO_2)_6$. The factor for converting the weight of cobalt found from the disc reading into the equivalent weight of potassium is 1.327. For most purposes it is accurate enough if the cobalt figure is multiplied by 4/3.

2. The results obtained by the use of the Nessleriser and disc are accurate only if Nessler tubes are used which conform to the specification used when the disc was calibrated, that is that the height of the 50 ml calibration mark shall fall at 113 ±3 mm from the bottom of the tube, measured internally.

The Determination of Silicon

Introduction

Silicon is, next to oxygen, the most abundant of the elements. In combination with oxygen as silica (SiO_2) it amounts to roughly one quarter of the earth's crust. Silica is found as an impurity in most materials and its determination is of importance in a wide range of industries.

Methods of sample preparation

Air[1]

To determine organosilicon compounds in the air, sample the contaminated atmosphere into sulphuric acid, add nitric acid, evaporate the acid solution until copious white fumes are evolved, adding additional nitric acid if necessary to produce a colourless residue, take up the residue in water and determine silicon by *Method 2*.

Biological materials[2-9]

Blood and serum[2,3] For total silicon in serum[2] ash the sample, treat the ash with hydrofluoric acid and remove the volatile silicon tetrafluoride (SiF_4) by boiling. Trap this silicon tetrafluoride in molybdate solution and proceed by *Method 1*. Alternatively[3] treat the sample of blood or serum with ion exchange resin, to remove any reducing substances which may be present, and determine silicon on the eluate. This method is also recommended for the determination of silicon in urine.

Tissues[4-7] Wet ash the sample in a platinum dish, add molybdate to the strongly acid solution and heat at 50°C. Filter off the precipitated phosphate, make slightly alkaline and proceed by *Method 1*[4]. If this method is not sufficiently sensitive, reduce the silicomolybdate to molybdenum blue and proceed by *Method 2*[5,6]. Alternatively[7] char the sample and then heat until a white ash is left, disperse the ash in hydrochloric acid, precipitate interfering ions (Fe, Mg and Mn) as hydroxides with caustic soda (NaOH) solution, filter and make the filtrate up to a known volume.

Urine[3,8,9] Treat the sample with an ion exchange resin and determine silicon on the eluate as recommended for blood[3]. Other methods for urine remove the interfering phosphates either by precipitating most of the phosphate with calcium chloride ($CaCl_2$) in aqueous ammonia (NH_4OH) at pH 7, filtering, treating the filtrate with molybdate and hydrochloric acid, extracting the phosphomolybdate with ethyl acetate ($CH_3COOC_2H_5$) and discarding the organic layer[8], or by taking 2 ml of sample, shaking with 2 ml of ammonium chloride (NH_4Cl), then adding molybdate and shaking for 3 minutes, extracting the phosphomolybdate with 20 ml of ethyl acetate and determining silicon on the aqueous phase[9].

Chemicals[10-12]

Gallium phosphide[10] Grind the sample to pass through a 150 mesh sieve. Take 0.1 g of the finely ground sample, which should not contain more than 40 μg of soluble silicon, and dissolve this in 3 ml of a 4:1 mixture of hydrochloric and nitric acids. Destroy any excess nitric acid by the addition of formic acid and then add a 10% aqueous solution of ammonium tartrate (($CHOH.COONH_4)_2$). Pass the solution through a column of AG3-X resin (20–50 mesh, Cl^- form) and determine silicon on the eluate.

Organosilicon compounds[11,12] Fuse the sample with magnesium, decompose the magnesium silicide with sulphuric acid and absorb the gaseous silicon hydrides in bromine water to form silicic acid. Proceed by either *Method 1* or *Method 2*[11]. Alternatively[12] take 0.2 g of sample in a platinum crucible, add 3 ml of sulphuric acid, cool, add 0.2 g of ammonium nitrate (NH_4NO_3) and shake gently and then warm. Repeat the addition of ammonium nitrate with warming until a colourless solution is obtained. Make this solution up to a known volume and determine silicon on an aliquot.

Foods and edible oils[13]

Wet ash 0.1–0.5 g of dried sample in a platinum dish, wash down the sides of the dish with a little water, evaporate to dryness and ignite gently. Fuse the residue with 1–2 g of anhydrous sodium carbonate (Na_2CO_3) until a clear melt is obtained. Cool, dissolve the melt in water, filter, neutralise and adjust the pH to 4.5–5.0 with $6N$ sulphuric acid. Use a suitable aliquot for the determination of silicon by either *Method 1* or *Method 2*.

Fuels and lubricants[14]

Coal ash Grind the sample until it passes through a 250 mesh sieve. Mix 0.5 g of the finely ground sample with 2 g of lithium metaborate ($LiBO_2$) in a platinum crucible and heat the mixture in a muffle furnace at 900°C until the melt is clear. If the sample contains more than 15% of iron, add 30–50 mg of ammonium metavanadate (NH_4VO_3) to act as a flux. Cool the melt and transfer it to a beaker containing 8 ml of nitric acid and 150 ml of water. Stir vigorously until dissolution is complete, add 2.5 g of tartaric acid (($CHOH.COOH)_2$), to mask the iron, and make the volume up to 250 ml.

Metals, alloys and plating baths[15-33]

Aluminium[15] Dissolve 0.08–1.0 g of sample, depending on the expected silicon content, in 7 ml of hydrochloric acid, 3 ml of water and 2 ml of nitric acid. Add 3 drops of a 1% aqueous solution of mercuric chloride ($HgCl_2$) to accelerate dissolution. Evaporate off excess acid and adjust the pH to lie between 1.1 and 1.3. Add 1 ml of $0.3N$ sulphuric acid, 1 ml of 6.25% aqueous ammonium molybdate (($NH_4)_6Mo_7O_{24}.4H_2O$), set aside for 10 minutes and then add 2 ml of normal sulphuric acid and 2 ml of a 10% solution of oxalic acid (($COOH)_2.2H_2O$) mixing thoroughly after each addition. Extract the silicon complex into 4 ml of pentanol ($CH_3(CH_2)_4OH$), wash the organic layer with two 1 ml portions of normal sulphuric acid and proceed by *Method 2*.

Beryllium and its alloys[16,17] Take 0.5 g of sample and dissolve it in 50% hydrochloric acid. Treat this solution with a 5% solution of boric acid (H_3BO_3), filter off any insoluble matter and fuse this with sodium borate ($Na_2B_4O_7$), dissolve the melt in dilute hydrochloric acid and add this to the main solution. Determine silicon on an aliquot of this combined solution.

Copper and its alloys[16,18-20] Dissolve 10 g of sample in 80 ml of 1:1 nitric acid in a polythene beaker, add 5 ml of 1:99 hydrofluoric acid (HF), heat on a water bath for 15 minutes and make up to 500 ml. To 50 ml of this solution add 10 ml of saturated boric acid (H_3BO_3) solution, adjust the pH to 1.2 with aqueous ammonia (NH_4OH) and proceed by *Method 1*[18]. Alternatively[16] dissolve 0.2 g of sample in 20 ml of 1:6 sulphuric acid and 5 ml of 30% hydrogen peroxide (H_2O_2), boil to decompose any excess of peroxide and make up to 250 ml. Determine silicon on a 50 ml aliquot. Nikitina[19] recommends solution in nitric acid in the presence of ammonium persulphate (($NH_4)_2S_2O_8$) to promote the formation of the active α-SiO_2. For the analysis of high purity copper, and brass, Donaldson[20] has developed the following method:—To 1 g of sample add 5 ml of water, 4 ml of 1:1 sulphuric acid and 2 ml of nitric acid. Heat on a steam bath for 1 hour, cool to room temperature, add 2 ml of 1:3 hydrofluoric acid and then heat on a water bath at 60–70°C for 30 minutes. Cool, add 20 ml of 5% boric acid solution, set aside for 15 minutes and then add 5 ml of 8% ammonium molybdate (($NH_4)_6Mo_7O_{24}.4H_2O$) solution. Neutralise the excess acid with silica-free aqueous ammonia, adjust the pH to 2.0 ± 0.2, dilute to about 50 ml and then heat on a steam bath for 30 minutes, cool, and dilute to exactly 100 ml. Take a 10 ml aliquot, add 5 ml of 40% sulphuric acid and 10 ml of pentanol ($C_5H_{11}OH$-n-amyl alcohol), shake the mixture for 2 minutes, separate the layers and discard the aqueous layer. Wash the organic layer with 10 ml of 8% sulphuric acid, discard the aqueous layer and determine silicon on the organic layer.

Ferro-molybdenum[21] Dissolve 0.2–0.3 g of sample by heating with 15–20 ml of 1:3 nitric acid. Cool, neutralise to litmus with 8% caustic soda (NaOH) solution, add 25 ml of 20% ammonium chloride (NH_4Cl) solution and boil for 1–2 minutes. Filter, wash the precipitate 10–12 times with 2% ammonium chloride solution, then ash the paper and precipitate. Fuse the residue with 2 g of sodium carbonate (Na_2CO_3) at 900–1,000°C, extract the melt with 50 ml of hot water, add 2 ml of 1:1 nitric acid and dilute to 100 ml. Determine silicon on an aliquot.

Iron, steel and ferrous alloys[22-32] Dissolve 0.2 g of the sample in acid and oxidise with nitric acid[22-24], potassium nitrate (KNO_3)[25], potassium permanganate ($KMnO_4$)[26], hydrogen peroxide (H_2O_2)[27] or ammonium persulphate (($NH_4)_2S_2O_8$)[23]. Boil, cool, make up to known volume and determine silicon on an aliquot. Alternatively[29] fuse 0.1 g of sample with 3 g of sodium peroxide (Na_2O_2), extract the melt with hot water and then with 15 ml of hydrochloric acid without heating. Filter this solution, dilute to 250 ml and determine silicon on an aliquot.

Janousek[30] uses a solvent consisting of 20 g of ammonium fluoride (NH_4F) in 1 litre of $4N$ nitric acid, which he stores in polythene bottles. To 0.1 g of sample in a platinum crucible add 10 ml of this solvent and immerse the crucible, to half its height, in water at 90°C. After 10 minutes add 10 ml of saturated boric acid (HBO_3) solution, cool and dilute to 250 ml.

Goetz[31] uses a sulphuric-nitric acid mixture to which he adds ammonium persulphate. For a rapid analysis of cast iron Volke[32] recommends the following:— Dissolve 0.2 g of sample in 50 ml of sulphuric acid containing ammonium persulphate, add 2 ml of 2% potassium permanganate solution and 50 ml of hot water. Heat, filter, cool and dilute the filtrate to 500 ml.

Nickel[33] Warm 0.5 g of sample with 9 ml of $4N$ nitric acid until dissolved, cool, dilute to 25 ml, extract with 10 ml of butanol (C_4H_9OH), discard the organic layer and determine silicon on the aqueous layer.

Minerals, Ores and Rocks[34-44]

Fuse the sample (0.15–0.25 g) using a suitable fusion mixture. Dissolve the melt in water, acidify and determine silicon on an aliquot of this solution. If necessary separate silicon from phosphorus by butanol (C_4H_9OH) extraction[34]. Recommended fusion mixtures are a mixture of sodium and potassium carbonates ($Na_2CO_3+K_2CO_3$)[35-37], caustic soda (NaOH) pellets[38-40], sodium peroxide (Na_2O_2)[41] and lithium metaborate ($LiBO_2$)[42,43]. Alternatively[44] dissolve the sample in hydrofluoric and sulphuric acids, pass dry, carbon dioxide-free air through the solution and into acidified ammonium molybdate solution.

Miscellaneous[45,57]

Cement[45,46] Fuse the sample with sodium carbonate (Na_2CO_3) and caustic soda (NaOH) and dissolve the melt in dilute hydrochloric acid. Filter, make the filtrate up to a known volume and determine 'soluble' silicon. Dissolve the insoluble residue in hydrofluoric acid (HF), dilute to a known volume and determine 'insoluble" silicon[45]. Alternatively[46] transfer 1 g of sample into a platinum or porcelain evaporating dish, disperse with 15 ml of water, add 10 ml of hydrochloric acid, heat and stir until decomposition is complete. Evaporate nearly to dryness on a steam bath, cool, add 8 ml of hydrochloric acid and then, after 2 minutes, add 50 ml of hot water. Stir, cover the dish, digest for 5 minutes, then filter through a Whatman No. 40 paper into a 500 ml volumetric flask. Wash the residue ten times with hot, dilute (1+99) hydrochloric acid and five times with hot water. Add the washings to the filtrate in the flask. Transfer the filter paper and residue to a platinum crucible, dry, ignite without firing the paper, and then ash at 1100°C for 30 minutes, cool, add 0.1 g of lithium chloride (LiCl), fuse, cool, extract melt with hydrochloric acid, add the solution to the filtrate and washings in the flask, make up to volume and determine silicon on an aliquot.

Detergents[47] Dissolve 1 g of sample in warm water and make up to 100 ml. Determine silicon on an aliquot by *Method 2*.

Dust and bonded deposits[48] Fuse the sample with caustic soda (NaOH) and extract the melt with water, sulphuric acid and hydrogen peroxide (H_2O_2) at 60°C for 1 hour. Determine silicon on an aliquot of this solution.

Fertilizers and liming materials[49,50] Grind the sample, to pass through a No. 100 sieve, and dry at 105°C. Place 0.5 g of the dry powder in a nickel crucible and add 1.5 g of caustic soda (NaOH) pellets, cover with a nickel lid and heat to dull red heat for about 5 minutes. Remove from the heat, swirl melt around the sides of the crucible and cool. To the cooled residue add about 50 ml of water and warm. Transfer to a beaker containing 400 ml of water and 20 ml of 1:1 hydrochloric acid, wash the crucible and add washings to the liquid in the beaker, then transfer the contents of the beaker to a 1 litre volumetric flask and dilute to the mark with water. Determine silicon on an aliquot part of this liquid by *Method 2*.

General[51-53] Dissolve 0.25 g of sample in 25 ml of acid mixture (72 ml of sulphuric acid and 91 ml of nitric acid per litre), heat until no further brown fumes are evolved, cool and dilute to 250 ml. Determine silicon on an aliquot[51]. Alternatively[52], decompose the sample with perchloric acid ($HClO_4$) plus hydrochloric acid or nitric acid and finally heat until fuming. Add hydrofluoric acid (HF) and distil off the silicon tetrafluoride (SiF_4) from a platinum still into molybdate solution. Ehrlich and Keil[53] also recommend a distillation method.

Leather[54] Oxidise the sample in a Parr or Wurzschmitt bomb under alkaline conditions and determine silicon by *Method 1*.

Slags[55,56] Fuse 0.5 g of the sample with sodium peroxide (Na_2O_2) in an iron crucible at 800—850°C. Cool, leach the melt with water, neutralise the solution with hydrochloric acid and dilute to 500 ml. Determine silicon on an aliquot by *Method 1*[55] or *Method 2*[56].

Welding fluxes[23,57] Fuse 0.1 g of sample with 4 g of potassium carbonate (K_2CO_3) and 25 g of borax ($Na_2B_4O_7.10H_2O$) at 950—1000°C for 20 minutes in a covered platinum crucible. Dissolve the melt in 350 ml of water, 7 ml of sulphuric acid and 6 ml of a 7% solution of thiourea ($CS(NH_2)_2$). Dilute the solution to 500 ml and determine silicon on an aliquot.

Plants[34,58]

Use the butanol (C_4H_9OH)—chloroform ($CHCl_3$) technique for extracting phosphomolybdate, as described for water, and proceed as described[34]. Alternatively mill the dried sample and then wet ash with sulphuric acid. Evaporate the solution to dryness, fuse the ignited residue with sodium sulphite (Na_2SO_3) and extract the melt with dilute hydrochloric acid. Neutralise the solution, make up to standard volume and determine the silicon on an aliquot[58].

Soil[34,59,60]

Either use the butanol (C_4H_9OH) extraction technique to remove interfering substances from the soil extract[34] or shake 2.5 g of soil for 30 minutes with 2×100 ml of Tamm's reagent (63 g of oxalic acid (($COOH)_2$) and 124 g of ammonium oxalate (($COONH_4)_2$), in 1 litre of solution of pH 3.3) and filter; to 5—15ml of the filtrate add 1:1 sulphuric acid (0.8 ml for 5 ml of solution or 1.3 ml for 15 ml of solution) and saturated potassium permanganate ($KMnO_4$) solution in an amount 0.1—0.2 ml less than is required to destroy the oxalate, heat and add $0.1N$ potassium permanganate until a faint pink colour persists, cool and proceed by *Method 1* or *Method 2*[59]. Alternatively[60] finely crush the soil, extract first with $8N$ hydrochloric acid and then with $0.5N$ sodium carbonate (Na_2CO_3) solution. Combine the extracts, add $7N$ sodium carbonate to give a solution containing about 1% hydrochloric acid and having a pH of approximately 1. Repeat the extraction procedure eight times on each sample.

Water[34,61-74]

The standard procedures[61,62] use *Method 1* for the relatively high levels of silicon (1—25 ppm) such as are found in natural waters, and *Method 2* for the lower concentrations (0.2—1 ppm) which are required in boiler-feed waters. Phosphates interfere and can be removed by the use of tartaric acid (($C_2H_3O_3)_2$) or sodium citrate ($Na_3C_6H_5O_7.2H_2O$)[61,62]. Alternatively treat 50 ml of the sample with 25 ml of Sorensen's borate buffer, pH 10, (6 parts of borate solution, containing 12.40 g of boric acid (H_3BO_3) and 100 ml of N caustic soda (NaOH) per litre, plus 4 parts of $0.1N$ caustic soda)[70-72] and 1 ml of $2N$ calcium chloride ($CaCl_2$) solution. Mix, stand for 2 hours and filter. Determine silicon on 50 ml of the filtrate using *Method 1*, and multiply the result by 1.5 to allow for the dilution with buffer. As Sorensen's buffer invariably contains silica derived from the bottle, a blank determination should be carried out and the necessary correction made. For low silica contents the silica may be concentrated either by adding 7.5 g of N hydrofluoric acid (HF) to a litre sample of water, passing the solution down a column of a strongly basic ion exchange resin (eg Permutit ES) and recovering the silica by washing the column with a saturated aqueous solution of boric acid (H_3BO_3)[64] or alternatively by neutralising the silica solution with 10% caustic potash (KOH) solution and adding 0.5 ml in excess and then adding the alkaline solution to the ammonium molybdate reagent together with 10 ml of N hydrochloric acid, 8 ml of butanol (C_4H_9OH) and making up the total volume to 100 ml. Shake for 5—10 minutes, add 5 ml of hydrochloric acid and extract phosphomolybdate with butanol—chloroform (C_4H_9OH : $CHCl_3$—1:4). Extract the aqueous layer with butanol (2×15 ml) and determine silicon on the alcoholic extract[34]. The intensity of the silicomolybdate colour is reduced by high salt concentrations, the effect of sulphate being greater than that of chloride. To determine silicon in waters of high salt concentration, such as sea water[67-69], either dilute the sample sufficiently for the salt effect to become negligible; calibrate in solutions of the same salt concentration[68]; or measure the apparent silica concentration in the sample (A) and in the sample diluted with an equal volume of water (B) and then calculate the true content (S) by

$$S = A + 2(2B - A)$$

Morrison[73] recommends evaporating not more than 10 ml of the sample to dryness, fusing the residue with 0.5±0.002 g of sodium carbonate (Na_2CO_3), dissolving the melt in water, transferring the solution to a polythene bottle containing 10 ml of 2.8% v/v sulphuric acid, rinse out crucible and add rinsings to bottle contents and continue rinsing until a total volume of 60±5 ml is obtained. Dilute to a known volume and determine silicon by *Method 1*.

Webber[74] recommends the following method:— Shake 1 litre of sample with a mixed anionic-cationic exchange resin, collect the resin on a Millipore filter supported on a porous polythene disc. Place the disc and the resin in a platinum crucible, add sodium carbonate solution, dry, heat to volatilise resins and fuse. Warm the melt with aqueous hydrogen peroxide (H_2O_2), dissolve in water, add this solution to 4 ml of $2N$ sulphuric acid and dilute to 100 ml.

Separation from other ions[75-77]

From arsenic[75] Place the sample in a platinum dish and three times add excess hydrochloric acid and evaporate to dryness. The final residue is free from arsenic.

From phosphorus[76,77] Extract the phosphorus as phosphomolybdic acid at pH 1.1 into a 1:1 benzene (C_6H_6)-isobutanol ((CH_3)$_2$CHCH$_2$OH) mixture. Re-extract phosphorus into 10% aqueous ammonia (NH_4OH)[76]. Alternatively[77] slightly acidify the test solution with nitric acid, treat at 50—55°C with ammonium molybdate reagent, maintain at this temperature for 30 minutes then cool for 30 minutes, filter, wash the precipitate with ammonium nitrate (NH_4NO_3) solution, combine the filtrate and washings and determine silicon on this solution.

The individual tests

Method 1 is the standard silicomolybdate test. Phosphates interfere and may be removed by the use of borate buffer as described under *Water* above. Alternatively[78] interference may be prevented by the addition of 2 ml of M citric acid ($C_6H_8O_7.H_2O$) to the test solution. It has been suggested that boiling the acid solution of silicomolybdic acid promotes the formation of the active α–form and leads to more consistent results[79,80].

Method 2 is the more sensitive molybdenum blue method which was developed[65] for the determination of low levels of silicon in boiler waters. Morrison and Wilson[82] have studied this method and recommend optimum reduction conditions.

Note

Colorimetric methods only determine crystalloid, or soluble, silica. For total silica gravimetric methods must be used except in those instances where special methods such as hydrofluoric (HF) solution have been used to ensure that all the silica is in the soluble form.

References

1. F. D. Krivoruchko, *Zavodskaya Lab.*, 1963, **29**, 927; *Analyt. Abs.*, 1964, **11**, 4564
2. P. Aumonier and R. Quilichini, *Bull.Soc.Pharm. Bordeaux*, 1962, **101**, 41; *Analyt.Abs.*, 1963, **10**, 3785
3. S-C. Wang, S-C Lin and L-W Tsai, *Acta Biochim.Sinica*, 1959, **2**, 63; *Analyt.Abs.*, 1960, **7**, 3870
4. M. Antonielli, *R.C.Acad.Lincei*, 1956, **20**, 813; *Analyt. Abs.*, 1957, **4**, 2701
5. J. Tuma, *Mikrochim.Acta*, 1962, 513; *Analyt. Abs.*, 1962, **9**, 3360
6. H. Baumann, *Hoppe-Seyl.Z.*, 1960, **319**, 38; *Analyt.Abs.*, 1961, **8**, 229
7. Z. E. Estes and R. M. Faust, *Analyt.Biochem.*, 1965, **13** (3), 518; *Analyt. Abs.*, 1967, **14**, 2093
8. J. Paul, *Biochem. J.*, 1960, **77**, 202; *Analyt.Abs.*, 1961, **8**, 1139
9. J. Paul and W. F. R. Pover, *Analyt.chim.Acta*, 1960, **22**, 185; *Analyt.Abs.*, 1960, **7**, 4187
10. C. L. Luke, *Analyt.Chem.*, 1964, **36** (10), 2036
11. J. Jenik and M. Jurecek, *Coll.Czech.Chem.Commun.*, 1961, **26**, 967; *Analyt.Abs.*, 1961, **8**, 4685
12. S. Kohama, *Bull.chem.Soc.Japan*, 1963, **36**, 830; *Analyt.Abs.*, 1964, **11**, 4365
13. D. F. Boltz, *"Colorimetric Determination of Non-metals"*, Interscience Publishers, London, 1958, p.72
14. P. L. Boar and L. K. Ingram, *Analyst*, 1970, **95**, 124
15. A. Golkowska, *Chemia analit.*, 1965, **10**, 749; *Analyt. Abs.*, 1967, **14**, 606
16. K. Kida, M. Abe, S. Nishigaki and T. Kusaka, *Japan Analyst*, 1961, **10**, 358; *Analyt. Abs.*, 1963, **10**, 2203

17. U.K. At.Energ.Authority, A.E.R.E.—AM39, 1959; *Analyt.Abs.*, 1960, **7**, 33
18. M. I. Miyamoto, *Japan Analyst,* 1961, **10**, 433; *Analyt.Abs.,* 1963, **10**, 2166
19. E. I. Nikitina, *Zavodskaya Lab.,* 1958, **24**, 398; *Analyt.Abs.,* 1959, **6**, 462
20. E. M. Donaldson and W. R. Inman, *Tech.Bull.Mines Brch.Can.,* T.B.77, 1965; *Analyt. Abs.,* 1967, **14**, 1304
21. V. I. Kurbatova and E. V. Silaeva, *Trudy.vses.nauchno.-issled.Inst.Standart.Obraztsov. Spektr.Etalanov.,* 1964, **1**, 79; *Analyt.Abs.,* 1966, **13**, 3573
22. M. Pangrac. *Hutn.Listy.,* 1960, **15**, 807; *Analyt.Abs.,* 1961, **8**, 2422
23. V. F. Mal'tsev and L. P. Luk'yanenko, *Trudy.Nauk.Tekh.Ob.Chernoi Metallurg.Ukr. Resp. Pravl.,* 1956, **4**, 111; *Analyt.Abs.,* 1958, **5**, 4127
24. M. del C. C. Ferrer and R. de la Cierva, *Inst.Hiero Acero,* 1961, **14**, 392; *Analyt.Abs.,* 1961, **8**, 4647
25. S. Meyer and O. G. Koch, *Mikrochim.Acta,* 1961, 82; *Analyt.Abs.,* 1961, **8**, 3286
26. W. S. Sobers, *Foundry,* 1957, **85**, 234; *Analyt.Abs.,* 1959, **6**, 2166
27. K. Narita, *J.Chem.Soc. Japan. pure Chem.Sect.,* 1957, **78**, 1367; *Analyt.Abs.,* 1958, **5**, 2994
28. W. F. Saunders and C. H. Cramer, *Analyt.Chem.,* 1957, **29**, 1139
29. E. Y. Shmulevich, *Zavodskaya.Lab.,* 1962, **28**, 811; *Analyt.Abs.,* 1963, **10**, 575
30. I. Janousek and D. Cechova, *Hutn. Listy.,* 1964, **19** (2), 128; *Analyt.Abs.,* 1965, **12**, 2802
31. R. Goetz, *Giesserei,* 1965, **52** (1), 12; *Analyt.Abs.,* 1966, **13**, 2958
32. G. Voke, *Giessereitechnik,* 1968, **14** (6), 186; *Analyt.Abs.,* 1969, **17**, 1422
33. F. H. Harrison, *Metallurgia Manchr.,* 1962, **66**, 300; *Analyt.Abs.,* 1963, **10**, 4181
34. M. Lheureux and J. Cornil, *Ind.Chim.Belge,* 1959, **24**, 634; *Analyt.Abs.,* 1960, **7**, 1695
35. G. Picasso, *Metallurg.Ital.,* 1962, **54**, 394; *Analyt.Abs.,* 1963, **10**, 1379
36. P. G. Jeffrey and A. D. Wilson, *Analyst,* 1960, **85**, 478
37. I. I. Tokarev and A. I. Novachok, *Trudy.Nauch.-Tekh.Ob.Chernoi Metallurg.Ukr. Resp.Pravl.,* 1956, **4**, 115; *Analyt.Abs.,* 1958, **5**, 4037
38. T. W. Bloxam, *Analyst,* 1961, **86**, 420
39. J. P. Riley and H. P. Williams, *Mikrochim Acta.,* 1959, 804; *Analyt.Abs.,* 1960, **7**, 2247
40. T. Sato and A. Ikegami, *Japan Analyst,* 1961, **10**, 433; *Analyt.Abs.,* 1958, **5**, 2905
41. C. B. Belcher and L. B. Skelton, *Analyt.chim.Acta,* 1960, **22**, 567; *Analyt.Abs.,* 1961, **8**, 69
42. J. C. Van Loon and C. M. Parissis, *Analyst,* 1969, **94**, 1057
43. P. L. Boar and L. K. Ingram, *Analyst,* 1970, **95**, 124
44. C. Hozdic, *Analyt.Chem.,* 1966, **38** (11), 1626; *Analyt.Abs.,* 1968, **15**, 194
45. K. Kawagaki, M. Saito and K. Hirokawa, *J.chem.Soc., Japan, ind. Chem.Sect.,* 1965, **68** (3), 465; *Analyt.Abs.,* 1967, **14**, 5434
46. A. Nestoridis, *Analyst,* 1970, **95**, 51
47. A. L. Olsen, E. A. Gee, V. McLendon and D. D. Blue, *Ind.Eng.Chem.* (Anal.Ed.,), 1944, **16**, 462
48. J. Grant, *J.appl.Chem.Lond.,* 1964, **14** (12), 525; *Analyt.Abs.,* 1966, **13**, 1782
49. P. Chichilo, *J.Assoc.offic.agric.Chem.,* 1963, **46** (4), 603
50. P. Chichilo, *J.Assoc.offic.agric.Chem.,* 1964, **47**, 1019
51. Z. Konecny, *Hutn.List.,* 1959, **14**, 903; *Analyt.Abs.,* 1960, **7**, 2660
52. B. D. Holt, *Analyt.Chem.,* 1960. **32**, 124
53. P. Ehrlich and T. Keil, *Z.anal.Chem.,* 1959, **166**, 254; *Analyt.Abs.,* 1959, **6**, 4706
54. Int. Union of Leather Chemists' Societies, Leather Analysis Commission, J.Soc. Leath.Trades Chem., 1969, **53** (10), 389; *Analyt.Abs.,* 1970, **19**, 4084
55. G. L. Povolotskaya, *Metallurg. i Khim.Prom.Kasakh.Nauch-Tekh.Sb.,* 1961, **4** (14), 87; *Analyt.Abs.,* 1963, **10**, 1004
56. S. Y. Fainberg, A. A. Blyakhman and S. M. Stankova, *Zavodskaya.Lab.,* 1957, **23**, 647; *Analyt.Abs.,* 1958, **5**, 32
57. V. F. Mal'tsev and L. P. Luk'yanenko, *Zavodskaya.Lab.,* 1958, **24**, 537; *Analyt.Abs.,* 1959, **6**, 876
58. R. J. Volk and R. L. Weintraub, *Analyt.Chem.,* 1958, **30**, 1011
59. N. I. Belyaeva, *Pochvovedenie,* 1962, (7), 104; *Analyt.Abs.,* 1963, **10**, 3954
60. B. Dabin, J. C. Brion, P. Pelloux, J. Rivoalen and F. Robin, *Cah.ORSTOM. Pedol.,* 1968, **6** (2), 225; *Analyt.Abs.,* 1970, **18**, 599

61. Brit.Standards Inst., *"Routine Control Methods of Testing Water used in Industry"*, B.S. 1427, London, 1962 and amendments
62. Brit.Standards Inst., *"Methods of Testing Water used in Industry"*, B.S. 2690, Part 3, London, 1966
63. B. Visintin and S. Monteriolo, *Ann.Chim.Roma,* 1961, **51**, 266; *Analyt.Abs.,* 1961, **8**, 5284
64. R. Wickbold, *Z.analyt.Chem.,* 1959, **171**, 81; *Analyt.Abs.,* 1960, **7**, 2662
65. W. E. Bunting, *Ind.Eng.Chem. (Anal.Ed.)*, 1944, **16**, 612
66. C. Milani, *Chim. e Ind.,* 1956, **38**, 587; *Analyt.Abs.,* 1957, **4**, 1047
67. G. S. Brien, *Analyt.Chem.,* 1958, **30**, 1549
68. A. B. Isaeva, *Trudy.Inst.Okeanol.Acad.Nauk.,* 1958, **26**, 234; *Analyt.Abs.,* 1959, **6**, 4188
69. I. Iwasaki and T. Taruntani, *Bull.chem.Soc.Japan,* 1959, **32** (1), 32; *Analyt.Abs.,* 1959, **6**, 3795
70. S. P. L. Sorensen, *Biochem Z.,* 1909, **21**, 131; 1910, **22**, 352
71. S. P. L. Sorensen, *Ergebn.Physiol.,* 1912, **12**, 393
72. *"Internat.Critical Tables",* 1926, Vol. 1, p 82
73. I. R. Morrison and A. L. Wilson, *Analyst,* 1963, **88**, 446
74. H. M. Webber and A. L. Wilson, *Analyst,* 1969, **94**, 110
75. S. A. Kiss and R. B. Doszpod, *Magy.Kem.Lap.,* 1968, **23** (9), 530; *Analyt.Abs.,* 1970, **18**, 115
76. A. K. Lavrukhina and A. I. Zaitseva, *Trudy.Kom.analit.Khim.,* 1968, **16**, 210; *Analyt. Abs.,* 1969, **16**, 2429
77. A. M. G. Macdonald and F. H. Van der Voort, *Analyst,* 1968, **93**, 65
78. S. Misumi and T. Tarutani, *Japan Analyst,* 1961, **10**, 1113; *Analyt.Abs.,* 1963, **10**, 4100
79. W. Kemula and S. Rosolowski, *Chem.Anal.Warsaw,* 1960, **5**, 419; *Analyt.Abs.,* 196,1 **8**, 1444
80. L. H. Andersson, *Acta Chem.Scand.,* 1958, **12**, 495; *Analyt.Abs.,* 1958, **5**, 3662
81. I. R. Morrison and A. L. Wilson, *Analyst,* 1963, **88**, 88

COLORIMETRIC CHEMICAL ANALYTICAL METHODS

Silica Method 1
using ammonium molybdate

Principle of the method

The colour standards are designed to match the colours produced by the addition of ammonium molybdate and sulphuric acid to solutions containing silicon. In this test the concentration of the sulphuric acid present in the mixture is important, and the presence of either too little or too much free acid results in diminution of the intensity of the yellow colour due to silicon. It is essential, therefore, when carrying out the test to adhere strictly to the conditions described below, under which the colour glasses have been standardised. In particular, the test **must** be carried out within the stated temperature range.

Reagent required

1. A 10 per cent w/v aqueous solution of ammonium molybdate $((NH_4)_6Mo_7O_{24}.4H_2O)$ (analytical reagent grade).
2. $2N$ sulphuric acid (H_2SO_4).
3. Mix 1 volume of the ammonium molybdate solution with 2 volumes of the sulphuric acid.

The Standard Lovibond Discs, Comparator Discs 3/13 and Nessleriser Disc NN and NN Special

Disc 3/13 covers the range from 2.5 to 25.0 (in steps of 2.5 up to 20) parts per million of silicon, calculated as silica (SiO_2), and is designed for use with a 40 mm cell.

Disc NN covers the range .05, .1, .2, .3, .4, .5, .6, .8, 1.0 mg of silicon calculated as SiO_2. Using a 50 ml sample, this equals 1 to 20 ppm.

Disc NN Special contains only two standards, 2 and 3 ppm of SiO_2. This disc has been prepared for use in the control of "jet-boost" water for aircraft.

Technique

(a) *Comparator*

Fill one of the 4 cm glass cells with distilled water and place in the left-hand compartment of the comparator. If the solution under test is not colourless, the cell should be filled with the solution instead of with distilled water. In a suitable flask place 25 ml of the solution under examination, at 25°—35C*, add 3 ml of the reagent, mix, allow to stand for 10 minutes and then pour into the other 4 cm cell in the right-hand compartment of the instrument. Hold the comparator facing a standard source of white light such as the Lovibond White Light Cabinet or, failing this, north daylight and compare the colour produced with the colours in the standard disc, rotating the latter until a colour match is obtained. Should the colour in the test solution be deeper than the standard colour glasses, a fresh test should be carried out using a smaller quantity of the solution under examination and diluting to 25 ml with distilled water before adding the reagents, due correction being made to the answer obtained. The figures on the disc represent parts per million of silica when 25 ml of the solution are taken for the test.

(b) *Nessleriser*

Fill one of the Nessleriser glasses to the 50 ml mark with distilled water and place in the left-hand compartment of the Nessleriser. If the solution under test is not colourless, the Nessleriser glass should be filled with the solution instead of with distilled water. Fill the other Nessleriser glass to the mark with the solution under examination at 25°—35°C*, add 6 ml of the reagent, mix, allow to stand for ten minutes and then place in the right-hand compartment of the instrument. Stand the Nessleriser before a standard source of white light as described for the Comparator above—and compare the colour produced with the colours in the standard disc, rotating the disc until a colour match is obtained. Should the colour in the test solution be deeper than the standard colour glasses, a fresh test should be carried out using a smaller quantity of the solution under examination and diluting to 50 ml with distilled water before adding the reagents.

*It is important that the test should be carried out at the temperature stated.

COLORIMETRIC CHEMICAL ANALYTICAL METHODS — Silica Method 1

Notes

1. Most colourless salts, even when present in relatively large quantities, are without influence upon the colour produced in the test, provided the concentration of free acid is not unduly disturbed. Phosphates, however, must be absent, since they respond to the test and yield a yellow colour similar to that produced by silica.

British Standard 1427:1962 gives a method to prevent interference from phosphate in the concentrations usually present in water.

This uses trisodium citrate ($Na_3C_6H_5O_7.2H_2O$) 30% w/v in water. After adding the ammonium molybdate reagent and standing for 10 minutes as in the standard method, add to both the blank and the test 1 ml citrate in the case of the comparator test and 2 ml in the case of the Nessleriser test. Mix and match.

2. It must be emphasized that the readings obtained by means of the Lovibond Nessleriser and disc are only accurate provided that Nessleriser glasses are used which conform to the specification employed when the discs were being calibrated, namely that the 50 ml calibration mark shall fall at a height of 113 mm plus or minus 3 mm measured internally.

Silica Method 2
using 1-amino-2-naphthol-4-sulphonic acid

Principle of the method

The silicic acid reacts with ammonium molybdate in acid solution to form silicomolybdate, and this is reduced with 1-amino-2-naphthol-4-sulphonic acid to produce molybdenum blue. Tartaric acid is added to suppress interference by phosphates. The intensity of the blue colour, which is proportional to the concentration of silica, is measured by comparison with Lovibond permanent glass colour standards.

Reagents required

1. *Sulphuric acid-molybdate reagent* To 75 g of ammonium molybdate $((NH_4)_6Mo_7O_{24}.4H_2O)$ dissolved in silica-free distilled water, add 322 ml of $10N$ sulphuric acid and make up to 1 litre.

2. *Tartaric acid reagent* Dissolve 10 g of tartaric acid $(C_4H_6O_6)$ in 100 ml of silica-free distilled water.

3. *1-Amino-2-naphthol-4-sulphonic acid reagent* (a) Dissolve 90 g of sodium metabisulphite $(Na_2S_2O_5)$ in 800 ml of silica-free distilled water.
   (b) Dissolve 14 g of sodium sulphite hydrated $(Na_2SO_3.7H_2O)$ in approximately 100 ml of silica-free distilled water.
   To solution (b) add 1.5 g of 1-amino-2-naphthol-4-sulphonic acid $(NH_2C_{10}H_4(OH)(SO_3H))$ mix until dissolved and add to solution (a). Make up total volume to 1 litre.

All chemicals used in the preparation of reagents should be of analytical reagent quality.

The Standard Lovibond Nessleriser Disc NV

The disc covers the range 0.2–1.0 ppm silica (SiO_2), in steps of 0.1 ppm.

Technique

To a 50 ml of sample, at a temperature between 20° and 30°C in a Nessleriser glass, add 2 ml of sulphuric acid-molybdate (reagent 1). After five minutes, add 4 ml of tartaric acid (reagent 2) and 1 ml of 1-amino-2-naphthol-4-sulphonic acid (reagent 3). Mix and allow to stand for twenty minutes, making quite certain that the temperature does not fall below 20°C. Place this Nessleriser glass in the right-hand compartment of the Nessleriser, and a similar glass containing a blank of the sample in the left-hand compartment. Match the colour of the solution against the colours in the disc using a standard source of white light such as the Lovibond White Light Cabinet or, failing this, north daylight.

Notes

1. It is important that the test should be carried out at between 20° and 30°C, since it has been found that at temperatures under 20°C the reaction does not proceed to completion and low results are obtained.

2. It must be emphasized that the readings obtained by means of the Lovibond Nessleriser and disc are only accurate provided that Nessleriser glasses are used which conform to the specification employed when the discs were calibrated, namely that the 50 ml calibration mark shall fall at a height of 113 ± 3 mm measured internally.

The Determination of Sodium

Introduction

Sodium occurs widely as a contaminant in other materials, and the majority of sodium salts are readily soluble. For this reason especial care must be taken during sodium estimations not to introduce sodium into the sample solution from either the reagents which are being used or from contaminated glassware. For this reason all bottles used for the storage of reagents should be of borosilicate glass.

The determination of sodium is of importance in the analysis of biological fluids and tissues and also in the analysis of soil extracts.

Methods of sample preparation

Biological materials[1-4]

Biological fluids[1,2] Precipitate the protein with trichloro-acetic acid (CCl_3COOH) and filter. Make up the filtrate to known volume and determine sodium on an aliquot.

Tissues[3,4] Extract the sodium either by digestion with trichloro-acetic acid[3] or by boiling the tissue sample with acidified water under reflux.[4] Filter, make the filtrate up to known volume and determine sodium on an aliquot.

Foods and edible oils[5]

Fish It has been shown that wet ashing (General Introduction p.vii) leads to results which are at least as precise as those obtained by dry ashing at 550°C.

Fuels and lubricating oils[6,7]

Coal ash[6] Mix 0.5 g of the finely ground (through 250 mesh) sample with 2 g of lithium metaborate ($LiBO_2$) in a platinum crucible, and heat the mixture in a muffle furnace at 900°C until the melt is clear. If the sample contains more than 15% of iron add 30-50 mg of ammonium metavanadate (NH_3VO_4) to act as a flux. Cool the melt and transfer it to a beaker containing 8 ml of nitric acid and 150 ml of water. Stir vigorously until dissolution is complete, add 2.5 g of tartaric acid (($CHOH.COOH)_2$), to complex the iron, and make the volume up to 250 ml.

Fuel oil[7] Place 1 g of sample in a silica crucible in a calorimetric bomb containing 1 g of water. Assemble the bomb, increase the pressure to 25 atmospheres with oxygen and fire. After 30 minutes wash out the inside of the bomb with distilled water, make the washings up to a known volume and determine sodium on an aliquot part of this solution.

Minerals, ores and rocks[6,8]

Either use the method described above for the analysis of coal ash[6] or[8], pre-ignite a graphite crucible for 30 minutes at 905°C and cool. Take care not to disturb the powdery inside surface of the crucible. Mix 0.2 g of the powdered sample with 1.4 g of lithium metaborate ($LiBO_2$) in a porcelain crucible and then transfer this mixture to the prepared graphite crucible and heat in the muffle furnace at 900°C for exactly 15 minutes. Pour the melt into 100 ml of a 1:24 solution of nitric acid in water, contained in a plastic beaker with a flat bottom, and stir with a Teflon-covered glass rod. Wash the solution into a 250 ml volumetric flask and make up to volume with more 1:24 nitric acid.

Miscellaneous[9]

Cement Fuse the sample with a mixture of sodium carbonate (Na_2CO_3) and caustic soda (NaOH), dissolve the melt in dilute hydrochloric acid, filter and make up to a known volume.

Soil[10]

Soil extracts Evaporate a sample of the extract to dryness and ignite gently. Grind the residue and extract by boiling with 10 ml of water for 5 minutes. Filter and wash well on the filter with hot water. Dilute the filtrate to 25 ml and determine sodium on an aliquot.

Separation from other ions[11]

From lithium and potassium Apply the sample, dissolved in a mixture of ethanol (C_2H_5OH), water and hydrochloric acid, to a column of Dowex 50W-X8 resin (H^+ form). Wash the column with 60% ethanol and then elute the lithium with 0.6N hydrochloric acid in 60% ethanol. Again wash the column, this time with 40% ethanol and then elute the sodium with 0.4N hydrochloric acid in 40% ethanol.

The individual test

Method 1 is the Shawarbi and Pollard[10] modification of Woelfel's method[2] in which sodium is precipitated as sodium manganese uranyl acetate, and the manganese in the precipitate is determined by oxidation to permanganate. Calcium, magnesium, iron, aluminium, manganese and phosphate do not interfere. Potassium does interfere giving high results if present in concentrations in excess of three times the sodium concentration.

References

1. R. F. L. Maruna, *Clin. Chim. Acta.,* 1957, **2,** 581; *Analyt. Abs.,* 1958, **5,** 1588
2. W. C. Woelfel, *J. Biol. Chem.,* 1938, **125,** 219
3. M. S. Moumb and J. V. Evans, *Analyst,* 1957, **82,** 522
4. G. Conradi, *Chemist Analyst,* 1967, **56** (4), 87; *Analyt. Abs.,* 1969, **16,** 284
5. M. H. Thompson, *J. Assoc. offic. agric. Chem.,* 1964, **47** (4), 701; *Analyt. Abs.,* 1965, **12** 6717
6. P. L. Boar and L. K. Ingram, *Analyst,* 1970, **95,** 124
7. J. E. Rayner, *J. Inst. Fuel,* 1964, **37,** 30; *Analyt. Abs.,* 1965, **12,** 2289
8. J. C. Van Loon and C. M. Parissis, *Analyst,* 1969, **94,** 1057
9. K. Kawagaki, M. Saito and K. Hirokawa, *J. chem. Soc. Japan, ind. Chem. Sect.,* 1965, **68** (3), 465; *Analyt. Abs.,* 1967, **14,** 5434
10. M. Y. Shawarbi and A. G. Pollard, *J. Soc. Chem. Ind.,* 1943, **62,** 71
11. V. Nevoral, *Z. analyt. Chem.,* 1963, **195,** 332; *Analyt. Abs.,* 1964, **11,** 2990

Sodium Method 1

using manganous uranyl acetate

Principle of the method

Sodium is precipated as sodium manganese uranyl acetate, the precipitate is dissolved in $2N$ sulphuric acid and the manganese is oxidised to permanganate with potassium periodate. The permanganate colour is then compared with Lovibond permanent glass standards.

Reagents required

1. *Manganous uranyl acetate stock solution*
   Uranyl acetate ($UO_2.(CH_3COO)_2.2H_2O$) .. 160 g
   Manganous acetate (($CH_3COO)_2Mn.4H_2O$) 490 g
   30% acetic acid ($CH_3.COOH$) .. .. 138 ml
   Distilled water .. .. 1500 ml

   After the chemicals have dissolved the solution is made up to 2 litres, allowed to stand for 24 hours, filtered and stored in a dark bottle.

2. *Dilute alcoholic mangahous uranyl acetate solution*
   Reagent 1 .. .. 180 ml
   95% alcohol (EtOH) .. .. .. 60 ml

   Mix, stand for 4 hours, filter through No. 42 Whatman filter-paper. This solution is stable for 4 weeks if stored in a dark bottle.

3. *Zinc uranyl acetate stock solution*
   Uranyl acetate .. .. 160 g
   Zinc acetate (($CH_3COO)_2 Zn.2H_2O$) .. 440 g
   30% acetic acid .. .. 138 ml
   Distilled water .. .. 1500 ml

   After the chemicals have dissolved the solution is made up to 2 litres, allowed to stand for 24 hours, filtered and stored in a dark bottle.

4. *Dilute alcoholic zinc uranyl acetate wash solution*
   Reagent 3 .. .. 120 ml
   95% alcohol .. .. 40 ml

   This solution is placed in an ice-bath for 1 hour and then saturated with 32 mg of solid manganous triple salt, prepared as described below, and, after at least an hour, is filtered through a No. 4 sintered glass crucible using suction.

   This reagent remains servicable for three weeks if stored in a dark bottle.

   Sodium manganous uranyl acetate salt is prepared by treating 125 ml of Reagent 1 with 2 ml of a 5% sodium chloride solution. After 30 minutes the liquid is centrifuged. The supernatant liquid is removed and the precipitate washed 3 times with 95% alcohol, twice with ether, and dried at ambient temperature.

5. *Oxidising solution*
   Potassium periodate (KIO_4) .. .. 0.75 g
   Distilled water .. .. 150 ml
   Phosphoric acid (syrupy) (H_3PO_4) .. .. 25 ml

   The periodate is dissolved in the water, the acid added, and the solution diluted to 200 ml. This solution is stable for 4 weeks in a cool place.

6. *Sulphuric acid* (H_2SO_4) $2N$

The Standard Lovibond Comparator Disc 3/42

The disc covers the range 50 ppm to 250 ppm of sodium (Na) in the following steps:—
50, 75, 100, 125, 150, 175, 200, 225, 250 ppm.

Technique

Place 1 ml of the test solution, containing 50-250 ppm of sodium, in a 30 ml beaker. Add 10 ml of reagent 2 and mix. Stand for at least 4 hours at room temperature. After this time filter through a No. 4 sintered glass crucible using suction. Wash the beaker and the precipitate with five separate 5 ml portions of ice-cold wash solution (reagent 4). Wash the outside of the crucible with water. Dissolve the precipitate with three successive 5 ml portions of warm $2N$ sulphuric acid (reagent 6), using slight suction. Wash the crucible twice with distilled water. Collect the filtrate and washings in a 100 ml beaker, remove the alcohol by boiling for 5-7 minutes using a low flame. Add 25 ml of oxidising solution (reagent 5). Continue boiling for a further 5-7 minutes to ensure maximum colour development. Cool and make up to 50 ml in a volumetric flask. Fill a comparator tube with this solution and place the tube in the right-hand compartment of the comparator. Match the colour against the standards in the disc using a standard source of white light such as the Lovibond White Light Cabinet, or failing this, north daylight.

Note

For use with solutions containing less than 50 ppm 10 ml of the test solution is taken and reduced to 1 ml by evaporation before proceeding with the above technique. Make due correction when reporting the result.

The Determination of Sulphate

Introduction

Sulphur is an essential element for plant growth and is obtained from the soil in the form of sulphate ions. The estimation of the available sulphate in soils is therefore a necessary part of agricultural and horticultural control analysis. Sulphur is also essential in the body and the determination of sulphate in body fluids and tissues is important.

Methods of sample preparation
Biological materials[1-3]

Biological fluids[1,2] To a 2 ml aliquot of protein-free fluid add 5 ml of a 1% solution of benzidene ($C_{12}H_{12}N_2$) in ethanol (C_2H_5OH). Allow the solution to stand for 1-2 hours at 4°C, and then centrifuge. Dissolve the precipitate in 2 ml of hot water, add 3 ml of 95% ethanol and 1 ml of *p*H 4 buffer and proceed by *Method 1*[1]. Alternatively[2] for the analysis of urine, remove the interfering cations with an ion exchange resin add excess barium chloranilate and proceed by *Method 1*.

Tissues[3] Store 0.5 ml of the tissue homogenate with 0.5 ml of 2*N* hydrochloric acid at 80°C overnight, to hydrolyse the conjugates. Adjust the *p*H to 4.0 with aqueous ammonia (NH_4OH) and proceed as for biological fluids above.

Metals, alloys and plating baths[4]

Nickel Dissolve the sample in nitric acid and make up to 200 ml with water. Pass this solution through a column of Amberlite IR-120 (20-50 mesh, H^+ form) to remove the nickel, neutralise the percolate and make up to a known volume. Determine sulphate on an aliquot part of this solution.

As any sulphide present in the sample would be oxidised to sulphate by this procedure, the result obtained is the total of sulphide and sulphate concentrations. If it is necessary to differentiate between these ions, dissolve another sample in hydrochloric acid, boil off any hydrogen sulphide resulting from sulphide present in the sample and then proceed as above. This will give the true sulphate concentration and sulphide, if required, can be obtained from the difference between the first and second results.

Miscellaneous[5]

General Remove cations by an ion exchange resin, adjust the concentration of the solution to 50% in ethanol (C_2H_5OH) and proceed by *Method 1*. This method can be applied to the determination of sulphur in fuel oils if the sulphur is first oxidised to sulphate.

Plastics and polymers[6]

Cellulose nitrate Weigh 4-8 g of dry cellulose nitrate into a tall beaker, add 80-160 ml of water, keeping the water to sample ratio at 20:1. Cover the beaker and place in an autoclave. Heat at 50-60 pounds per sq in for 20-30 minutes and then at 160-170 pounds per sq in for 10 minutes. Cut off the heat and rapidly reduce the pressure. After cooling, filter and determine sulphate in the filtrate.

Soil[7-10]

Mix the air-dried soil with Morgan's reagent[8] (100 g of sodium acetate (CH_3COONa) and 30 ml of acetic acid (CH_3COOH) in 1 litre) in the proportion 1 part soil to 2 of reagent, shake and filter. Pass the extract through a 20 × 1 cm column of ZeoKarb 225 (H^+ form), reject the first 15 ml of percolate and then collect the next 5 ml for analysis[7]. Alternatively[9] take 10 g of air-dried soil, which has been ground to pass through a 2 mm sieve, and shake the soil with 20 ml of water for 15 minutes. Add 0.04 g of purified animal charcoal and shake for a further 15 minutes. Filter and determine sulphate on an aliquot part of the filtrate. These methods are only suitable for the determination of "available" or "soluble" sulphate.

Nemeth[10] recommends the following indirect method for the determination of sulphate in soil extracts:—Place 1 ml of the extract, which should contain 100—550 µg of sulphate, in a 15 ml centrifuge tube and add 5 ml of the reagent solution, containing 12.66 g of barium chromate ($BaCrO_4$) in 150 ml of 25% hydrochloric acid and 850 ml of water. Add 2 ml of 25% aqueous ammonia (NH_4OH), dilute to 15 ml with water, centrifuge for 5 minutes at 3000 r.p.m. and determine the concentration of chromium in the supernatant liquid using one of the standard methods for chromium (pp 155,156). The chromium ions released into solution by the precipitation of barium sulphate are equivalent to the sulphate ions originally present.

Water[11]

Adjust the *p*H of the water sample to 6.0. Pass through a column of Wofatit-K ion exchange resin and discard the first 50 ml. Use the next 10 ml for analysis (5 ml for the sample and 5 ml for the blank).

The individual tests

Method 1 is a modification of the original barium chloranilate method[5] which was specifically designed[7] to provide a simple field test for sulphate.

Method 2 is an indirect method[12] involving the precipitation of lead sulphate by lead nitrate solution and the determination of lead in the precipitate.

Method 3 is another indirect method[13] in which the sulphate is reduced to sulphur dioxide which is trapped and measured in the normal manner using *p*-rosaniline.

References
1. I. P. T. Hakkinen and L. M. Hakkinene, *Scand.J.Clin.Lab.Invest.*, 1959, **11**, 294; *Analyt.Abs.*, 1961, **8**, 3397
2. A. Wainer and A. L. Koch, *Anal.Biochem.*, 1962, **3**, 457; *Analyt.Abs.*, 1963, **10**, 267
3. I. P. T. Hakkinen, *Scand.J.Clin.Lab.Invest.*, 1959, **11**, 298; *Analyt.Abs.*, 1961, **8**, 3397
4. R. D. Srivastava and H. Gesser, *J.Prakt.Chem.*, 1968, **38** (5-6), 262; *Analyt.Abs.*, 1970, **18**, 1673
5. R. J. Bertolacini and J. E. Barney, *Analyt.Chem.*, 1957, **29**, 281
6. A. F. Dawoud and A. A. Gadalla, *Analyst*, 1970, **95** (9), 823
7. W. Y. Magar and A. G. Pollard, *Chem. and Ind.*, 1961, 505; *Analyt.Abs.*, 1961, **8**, 4857
8. M. F. Morgan, Chemical Soil Diagnosis by the Universal Soil Testing System, *Bull.Conn.Agric.Expt.Sta.*, 1941, 372
9. A. Massoumi and A. H. Cornfield, *Analyst*, 1963, **88**, 321
10. K. Nemeth, *Z.PflErnahr.Dung.*, 1963, **103** (3), 193; *Analyt.Abs.*, 1965, **12**, 3069
11. L. Prochazkova, *Z.Analyt.Chem.*, 1961, **182**, 103; *Analyt.Abs.*, 1962, **9**, 885
12. R. Baronowski, J. Ciba, J. Czerniec and Z. Gregorowicz, *Chemia analit.*, 1965, **10** (3), 499
13. S. Nunez Cubero and J. Ortega Abellan, *An.R.Soc.esp.Fis.Quim.B.*, 1965, **61**, 1097; *Analyt.Abs.*, 1967, **14**, 1392

Sulphate Method 1

using barium chloranilate

Principle of the method

The chloranilic acid liberated from barium chloranilate by a portion of the test solution under conditions of controlled pH, is estimated by comparison of the colour produced with the colours of a range of Lovibond permanent glass standards.

Reagents required

1. *Barium chloranilate* $(BaC_6Cl_2O_4)$
2. *Ethyl alcohol* $(EtOH)$ absolute
3. *Potassium hydrogen phthalate* 0.05 M. $(C_6H_4(COOH)COOK)$ pH 4.0 buffer solution

The Standard Lovibond Comparator Disc 3/49

The disc covers the range 5 to 80 ppm (5, 10, 20, 30, 40, 50, 60, 70, 80) of sulphate (SO_4^{2-}).

Technique

Transfer 2 ml of the sample solution to a centrifuge tube together with 1 ml of phthalate solution (reagent 3), 2 ml of water and 5 ml of ethyl alcohol. Mix, add approximately 0.03 g of barium chloranilate (reagent 1), shake the mixture for 5 minutes and then centrifuge at 2,500 r.p.m. for 5 minutes (Note 3).

Transfer the supernatant liquid to a standard test tube and place this in the right-hand compartment of the Comparator. Fill an identical tube with water and place in the left-hand compartment. Compare the colour of the test solution with the permanent glass standards in the disc, using a standard source of white light such as the Lovibond White Light Cabinet or, failing this, north daylight.

Notes

1. The potassium hydrogen phthalate is added to the sample solution *to bring it to a final value between 3.0 and 3.5 pH, which value is necessary in order to produce the correct colours in the test*.
2. 0.03 g of barium chloranilate is conveniently measured by a spatula made of a piece of glass rod of suitable size, which may be found by experiment.
3. Small hand-operated centrifuges for field use are readily available, and may be obtained from The Tintometer Ltd.
4. Calcium, magnesium, chloride, and nitrate do not interfere with this test.

Sulphate Method 2

using lead nitrate

Principle of the method

Sulphate ions, in 1:1 acetone water solution or, in the presence of a large excess of chloride, in 1:1 water-dioxan, are precipitated by means of lead nitrate. The precipitated lead sulphate is collected, dissolved in sodium acetate solution and the lead is extracted with a solution of dithizone in chloroform and determined by Lead *Method 1* (p. 227).

Reagents required

1. *Lead nitrate solution* 0.02N   Dissolve 3.312 g of lead nitrate ($Pb(NO_3)_2$,) in 1 litre of water.
2. *Acetone* (CH_3COCH_3)
3. *1,4-Dioxan* ($CH_2.CH_2.O.CH_2.CH_2.O$)
4. *Sodium acetate 30%*   Dissolve 30.0 g of sodium acetate trihydrate ($CH_3COONa.3H_2O$) in 100 ml of water.

All chemicals used in the preparation of reagents should be of analytical reagent quality.

In addition to the above reagents, the reagents needed for Lead *Method 1*, listed on page 227, will also be required.

The Standard Lovibond Comparator Disc 5/17

This disc includes standards corresponding to 2, 4, 6, 8, 10, 12, 14, 16 and 20 μg of lead.

These are equivalent to 0.9, 1.8, 2.8, 3.7, 4.6, 5.5, 6.4, 7.4, 8.3 and 9.2 μg of sulphate (SO_4). This is equivalent to the range 0.09—0.9 ppm of sulphate if a 10 ml sample is used.

Technique

Take 10 ml, or other suitable volume, of sample and add an equal volume of acetone. If the sample is known to contain a large excess of chloride ions use dioxan in place of acetone. Add 1 ml of lead nitrate solution (reagent 1) and set aside for 30 minutes. After this time filter the solution and wash the filter paper with three successive 10 ml portions of 1:1 water:acetone (or dioxan in the presence of excess chloride). Discard filtrate and washings. Dissolve the precipitate on the paper by washing with five 2 ml portions of 30% sodium acetate solution (reagent 4). Collect the 10 ml of sodium acetate solution and use this as the sample for the determination of lead by *Method 1* (p. 227).

Sulphate Method 3

using *p*-rosaniline

Principle of the method

The sulphate ions in the sample solution are reduced to sulphur dioxide which is distilled out of solution in a stream of nitrogen, trapped in potassium chloromercuriate and determined with *p*-rosaniline by the standard procedure (Sulphur dioxide *Method 2* page 549).

Reagents required

1. *Reducing mixture* Heat 5 g of powdered copper with 10 ml of 85% phosphoric acid (H_3PO_4) until white fumes of phosphorus pentoxide (P_2O_5) are evolved. This reagent should be freshly prepared immediately before use.
2. *Absorbing reagent (potassium chloromercuriate)* Dissolve 10.9 g of mercuric chloride and 5.9 g of potassium chloride (KCl) in 1 litre of distilled water. *This reagent is highly poisonous!* If it is spilled on the skin it must be flushed off *immediately* with water.

All the chemicals used in the preparation of these reagents should be of analytical reagent quality.

In addition to the above reagents, those necessary for the determination of sulphur dioxide by *Method 2*, which are listed on page 549, will also be required.

The Standard Lovibond Comparator Disc 6/32

This disc contains standards corresponding to 2.5, 5, 7.5, 10, 12.5, 15, 17.5, 20 and 25 µg of sulphur dioxide (SO_2). These standards correspond to 4, 7.5, 11, 15, 19, 22.5, 26, 30 and 37.5 µg of sulphate (SO_4). If a 10 ml sample is used this is equivalent to the range 0.4—3.75 ppm of sulphate.

Technique

Place 10 ml of sample solution in a 50 ml distillation flask fitted with inlet and outlet tubes and a delivery funnel which is fitted with a stop-cock. The inlet tube should dip below the surface of the sample solution. To the outlet tube connect a midget impinger bubbler containing 10 ml of the absorbing reagent (reagent 2). Connect a cylinder of nitrogen, through a pressure reducing valve, to the inlet tube of the distillation flask and pass a stream of nitrogen through the equipment at about 100 ml a minute. Place 1 ml of the freshly prepared reducing mixture (reagent 1) in the delivery funnel and drop this slowly into the flask. After 15 minutes disconnect the bubbler and determine the amount of sulphur dioxide which has been trapped, using *Method 2* page 549.

The Determination of Sulphite
using potassium iodide-iodate

Introduction
Sulphites are encountered in industrial wastes, in boiler water, and in water exposed to atmospheres containing sulphur dioxide. Solutions of sulphur dioxide are also used in certain bleaching processes, and in preserving foodstuffs.

Principle of the method
The test is based on the quantitative reaction between sulphites and excess iodine in acid conditions, and the standards are matched against the residual colour obtained when a constant amount of iodine is reacted with different amounts of sulphite. All substances reacting with iodine under the conditions of the test will interfere, and hence *e.g.* nitrites and sulphides must be absent. The sample must be oxygen free.

Reagents required
1. *Potassium iodide solution* $N/10$ 16.6 g of KI per litre
2. *Potassium iodate solution* $N/40$ 0.892 g of KIO_3 per litre
3. *Hydrochloric acid* (*HCl*)   Analytical reagent grade, (wt. per ml at 20°C at 1.18 g)

The Standard Lovibond Nessleriser Disc NOB
The disc covers the range from 2 to 50 parts per million sodium sulphite (Na_2SO_3) (colour standards for 2, 5, 10, 15, 20, 25, 30, 40, 50) using a 25 ml sample.
To obtain the amount of SO_2, divide the answer by 2. The range for SO_2 is therefore 1 to 25 ppm on a 25 ml sample.

Technique
Place 1 ml of potassium iodate solution, 1 ml of potassium iodide solution and 0.5 ml of hydrochloric acid in a Nessleriser glass in that order; then add 25 ml of the sample and, after mixing, adjust the volume to 50 ml with distilled water. Stand for 15 minutes. Place the Nessler tube in the right hand compartment of a Lovibond Nessleriser and a blank of distilled water in the left hand compartment. Stand the Nessleriser facing a standard source of white light, such as the Lovibond White Light Cabinet, or failing this, north daylight, and compare the colour of the test solution against the disc colour standards, selecting the nearest match. The figures represent ppm Na_2SO_3 when a 25 ml sample is used.

Note
It must be emphasized that the readings obtained by means of the Lovibond Nessleriser are only accurate provided that Nessleriser glasses are used which conform to the specification employed when the discs were calibrated, namely that the 50 ml calibration mark shall fall at a height of 113 mm ± 3 mm measured internally.

The Determination of Sulphur

Introduction

Sulphur occurs widely in nature both in the elemental form and, more commonly, in combination with other elements. For example it is found in beds of gypsum and limestone in association with hydrocarbons, carbonates and sulphates, in sulphide minerals and in many materials of biological origin.

Sulphur has many uses in addition to the uses of its compounds and derivatives. Large quantities are used in the vulcanisation of natural rubber and as a fungicide. From the earliest times burning sulphur has been used as a fumigant, for bleaching fabrics and the ancients used it medicinally. It is still used in medicine as a mild antiseptic. Impregnation of wood and paper with sulphur is used to improve durability by preventing the ravages of micro organisms.

Because of its ready convertibility to either its oxides or to sulphides sulphur is rarely, if ever, determined directly. As the three most common forms in which it is determined, hydrogen sulphide, sulphur dioxide and sulphate, are all covered individually elsewhere in this book it would be pointless to repeat the individual methods here. In considering methods which have been described for the determination of sulphur in various materials only the details of the techniques used for sample preparation will be reported, the technique used for the final determination will be found in its appropriate place, as also will the descriptions of sample preparation methods devoted to the individual compounds rather than to sulphur in general.

Methods of sample preparation
Chemicals[1-3]

Organic compounds Either[1] treat the sample in a micro-Kjeldahl flask with 1 ml of water-washed Raney nickel, boil and stir for 20 minutes then add dilute sulphuric acid to decompose the nickel sulphide formed, sweep out the hydrogen sulphide with a stream of carbon dioxide and determine by *Method 1* or *2* (pp. 198, 199); or[2] pyrolyse the sample in the presence of copper powder in a stream of nitrogen at 700°C, and absorb and determine the sulphur dioxide which results from the oxidised forms of sulphur, then heat the residue in a stream of air to oxidise sulphides and determine the sulphur dioxide as before (*Method 2* p. 549). Alternatively[3] the same author suggests that one should pyrolyse the sample at 950°C, hydrogenate the products in a stream of hydrogen and nitrogen, obtained from the thermal decomposition of ammonia in the presence of reduced copper and platinum, and determine the hydrogen sulphide formed (*Method 1* p.198).

Fuels and lubricants[4-7]

In addition to the S.T.P.T.C. method (*Method 1*) the following alternative methods have been described.

React the sample with activated Raney nickel catalyst in an atmosphere of oxygen-free nitrogen to form nickel sulphide from elemental and organic sulphur. Add hydrochloric acid and determine the amount of hydrogen sulphide which is liberated[4,5]. Dokladalova[5] recommends adding 5 ml of isopropanol (($CH_3)_2CHOH$) to not more than 10 ml of sample and carrying out the reduction at a temperature below the boiling point of isopropanol (83°C). Alternatively[6,7] pyrolyse the sample either in a stream of oxygen at 800°C to give sulphur dioxide[6] or in a stream of hydrogen at 1200°C to give hydrogen sulphide[7].

*Metals, alloys and plating baths*s[8-15]

Copper-, iron-, and nickel-based alloys[8] Separate the sulphur from the metal matrix by combustion in oxygen in a high-frequency induction furnace, in the presence of pre-treated vanadium pentoxide (V_2O_5) as flux-accelerator, and determine the sulphur dioxide formed.

Ferrovanadium[9] Dissolve 0.5 g of sample by heating with 10 ml of nitric acid, evaporate twice with hydrochloric acid, dissolve the residue in water, dilute to 120 ml with hot water, reduce the iron and vanadium by the addition of 15 ml of 10% hydroxylammonium chloride ($HONH_3Cl$) solution, and adjust the concentration of hydrochloric acid to be $0.3N$. Pass this solution through a paper-pulp filter followed by a column containing 25 g of cationite

KU-2. Wash the filter and column with 50 ml of $0.2N$ hydrochloric acid and determine sulphur in the percolate as sulphate (*Methods 1, 2, 3* pp. 350, 351, 352).

General[10,11] Dissolve the sample in nitric acid and convert the sulphates in the sample solution to hydrogen sulphide by heating with a mixture of sodium hypophosphite (NaH_2PO_2), hydrochloric acid and hydriodic acid (HI). Entrain the hydrogen sulphide in a stream of nitrogen and determine by *Method 1* (page 199)[10]. Alternatively[11] burn 0.2 g of sample in a current of 1 litre a minute of oxygen in a high-frequency furnace at 1300°C in the presence of sulphur-free copper beads. Trap the sulphur oxides with 80 mesh silver gauze at 550°C. Pass nitrogen at 1 litre a minute for 1 minute and then reduce the silver sulphate formed in a current of 300 ml a minute of hydrogen at 550°C. Absorb the hydrogen sulphide and determine by *Method 1* (page 198).

Iron[12,13] Dissolve a 1 g sample in hydrochloric acid, wash the gases evolved with water, to remove hydrochloric acid, and then absorb and determine the hydrogen sulphide[12]. Alternatively[13] dissolve the sample in aqua regia (HNO_3+3HCl), reduce ferric iron to ferrous with stannous chloride ($SnCl_2$) and sulphates to hydrogen sulphide with hypophosphorous (H_3PO_2) and hydriodic (HI) acids. Distill off the hydrogen sulphide and determine by *Method 1* (page 198).

Nickel[14] Dissolve up to 1 g of sample in nitric acid containing bromine, eliminate the nitric acid by repeated evaporation with hydrochloric and formic (HCOOH) acids, reduce the sulphates with titanous phosphate ($TiPO_4$) and pass the hydrogen sulphide into zinc acetate using a stream of nitrogen. Proceed by *Method 1* (page 198).

Tin[15] Combust 1 g of degreased fine tin turnings in a stream of oxygen at 1350°C and measure the sulphur dioxide evolved (*Method 2* page 549).

Minerals, ores and rocks[16,17]

Mix the sample with vanadium pentoxide (V_2O_5) and heat in a stream of pure nitrogen at 900—950°C. Reduce the sulphur trioxide evolved to sulphur dioxide by passage over heated copper turnings in the same combustion tube and determine the sulphur dioxide[16]. Alternatively[17] sublime 10 mg of ore under vacuum at 200°C, then ignite the sublimate at 300—350°C for 10 minutes in a stream of oxygen and determine the sulphur dioxide formed.

Miscellaneous[10,18,19]

Fertilizers[18] Reduce sulphates to hydrogen sulphide with a boiling mixture of 59% hydriodic acid, $6N$ hydrochloric acid and 50% hypophosphorous acid (H_3PO_2) and determine by *Method 1* or *2* (pages 198,199).

General[19] Dissolve the sample in aqua regia (HNO_3+3HCl) and add bromine. Dilute until the solution is $0.3M$ in hydrochloric acid and pass it at 3 ml a minute through a column of cationite K.U-2 (H^+ form). Determine sulphate in the percolate (*Methods 1, 2* and *3* pages 350, 351, 352).

Semi-conductor compounds[10] Use the first general method described above for the analysis of metallurgical samples.

Paper[20,21]

Heat 1 g of sample to 225°C with 1 g of N-bis-(4-methoxyphenyl)methylenebenzylamine and determine the hydrogen sulphide formed[20]. To determine the different forms of sulphur[21] determine sulphide by treating the sample with $6N$ hydrochloric acid and measuring the hydrogen sulphide evolved; free sulphur is reduced to hydrogen sulphide with N-bis-(4-methoxyphenyl)methylenebenzylamine as described earlier; sulphite is treated with oxalic acid $((COOH)_2.2H_2O)$ and the sulphur dioxide formed is estimated; sulphate is reduced with a mixture of sodium hypophosphite (NaH_2PO_2) and hydriodic acid and determined as hydrogen sulphide.

Plants[22,23]

Combust the sample in a stream of oxygen at 1280°C, pass the gases through two absorbers in series each containing 30 ml of 3% hydrogen peroxide (H_2O_2), combine the contents of these bubblers, dilute to 100 ml and determine sulphate (pages 350, 351, 352) on an aliquot[22]. Alternatively[23] digest 1 g of sample in a mixture of nitric and perchloric ($HClO_4$) acids, evaporate to dryness, dissolve the residue in hydrochloric acid, adjust the volume to 50 ml with water and determine sulphate on an aliquot.

The individual test

Method 1 is the standard S.T.P.T.C. Method[24-27] for the determination of sulphur in benzole.

References

1. S. D. Iordanov and Ch. P. Ivanov, *C.r.Acad.bulg.Sci.*, 1968, **21** (5), 451; *Analyt.Abs.*, 1969, **17**, 2147
2. M. A. Volodina, M. Abdukarimova and L. V. Kozlovskaya, *Vest.mosk.gos.Univ., Ser.khim.*, 1968, (6), 109; *Analyt.Abs.*, 1970, **18**, 1697
3. M. A. Volodina, M. Abdukarimova and A. P. Terent'ev, *Zh.analit. Khim.*, 1968, **23** (9), 1420; *Analyt.Abs.*, 1970, **18**, 2471
4. E. E. H. Pitt and W. E. Ruppelrecht, *Fuel*, 1964, **43** (6), 417; *Analyt.Abs.*, 1966, **13**, 1365
5. J. Dokladalova, *Chemicky.Prum.*, 1965, **15** (3), 175; *Analyt.Abs.*, 1966, **13**, 3612
6. J. Dokladalova, *Ropa.Uhlie.*, 1966, **8** (2), 53; *Analyt.Abs.*, 1967, **14**, 4796
7. L. L. Farley and R. A. Winkler, *Analyt.Chem.*, 1968, **40** (6), 962; *Analyt.Abs.*, 1969, **17**, 1491
8. K. E. Burke, *Analyt.Chem.*, 1967, **39**, (14), 1727; *Analyt.Abs.*, 1969, **16**, 1228
9. V. V. Stepin and I. A. Onorina, *Trudy.vses.nauchno.–issled.Inst.standart.Obraztsov.*, 1968, **4**, 136; *Analyt.Abs.*, 1970, **18**, 943
10. V. G. Goryushina and E. Ya. Biryukova, *Zavodskaya.Lab.*, 1965, **31** (11), 1303; *Analyt.Abs.*, 1967, **14**, 1391
11. T. Takeuchi, I. Fujishima and M. Yamada, *Japan Analyst*, 1967, **16** (12), 1354; *Analyt.Abs.*, 1969, **16**, 1227
12. S. Turina and V. Marjanovic-Krajovan, *Z.analyt.Chem.*, 1963, **196**, 32; *Analyt.Abs.*, 1964, **11**, 2615
13. K. Polyakova and A. Bocek, *Hutn.Listy.*, 1969, **24** (5), 360; *Analyt.Abs.*, 1970, **19**, 1285
14. A. Parker, C. G. Wallace and T. J. Webber, *Rep.U.K.atom.Energy Auth.*, A.E.R.E.—R6009, 1969; *Analyt.Abs.*, 1970, **18**, 4021
15. E. Pell, H. Malissa, N. A. Murphy and B. R. Chamberlain, *Analyt.chim.Acta*, 1968, **43** (3), 423; *Analyt.Abs.*, 1970, **18**, 2324
16. J. G. Sen Gupta, *Analyt.Chem.*, 1963, **35**, 1971
17. L. S. Nadezhina, E. K. Bespalenkova and E. L. Grinzaid, *Zh.analit.Khim.*, 1968, **23** (5), 787; *Analyt.Abs.*, 1970, **18**, 142
18. R. Millet, *Chim.Anal.*, 1963, **45**, 174; *Analyt.Abs.*, 1964, **11**, 1987
19. L. D. Zinov'eva, K. F. Gladysheva and T. P. Zelenina, *Sb.nauch.Trudy.vses.nauchno.–issled.gornometallurg.Inst.tsvet.Metall.*, 1965, (9), 118; *Analyt.Abs.*, 1967, **14**, 125
20. O. Huber, H. Kolb and J. Weigl, *Z.analyt.Chem.*, 1967, **227** (6), 416; *Analyt.Abs.*, 1968, **15**, 4036
21. O. Huber, H. Kolb and J. Weigl, *Z.analyt.Chem.*, 1967, **227** (6), 420; *Analyt.Abs.*, 1968, **15**, 4037
22. M. Buck, *Z.analyt.Chem.*, 1963, **194**, 116; *Analyt.Abs.*, 1964, **11**, 678
23. C. Egoumenides, *Fruits.Paris*, 1963, **18**, 69; *Analyt.Abs.*, 1964, **11**, 1388
24. G. Claxton and K. H. V. French, *J.Inst.Petroleum*, 1949, **35**, 496
25. Standardisation of Tar Products Tests Cttee. (S.T.P.T.C.), *"Standard Methods for Testing Tar and its Products"*, 6th Edn. 1967, Method RLB20—67
26. Brit. Standards Inst., *"Benzenes and Benzoles,"* B.S. 135:1963, London 1963
27. Brit. Standards Inst., *"Xylenes"*, B.S. 458:1963, London 1963

Sulphur 1
using lead acetate

Introduction

This disc has been produced in collaboration with the Technical Department of the National Benzole Company and is specified in the S.T.P.T.C. Method, Serial R.L.B.20–67. The details of the technique are published with the permission of that Committee.

Principle of the method

A strip of clean copper foil is immersed in the sample under examination and held for 2 hours at 50°C. The stain (if any) of sulphide deposited on the copper is converted into hydrogen sulphide, and this is estimated colorimetrically by means of lead acetate paper.

The Standard Lovibond Comparator Disc 3/22

The disc covers the range 1 to 8 micrograms of sulphur, in 8 steps. This disc must be used with the Lovibond Comparator and Gas Paper Test stand DB418.

Technique

For full details for producing the stain on a copper strip, reference should be made to the S.T.P.T.C. Handbook, "Standard Methods for Testing Tar and its Products."

The test papers are prepared by soaking No. 1 Whatman filter paper for 1 minute in a solution of 10 g of lead acetate in 90 ml distilled water, to which are added 5 ml of glacial acetic acid and 10 ml pure glycerol. The superfluous liquid is allowed to drain off and the papers are dried in an oven at 90°—100°C. until they are just dry (10—15 minutes). Such a paper is placed in position in an aluminium holder in a streaming apparatus, (see reference), and pure hydrogen gas streamed through for 10 minutes to clear out all other gas. Hydrochloric acid 18% w/w is then allowed to react with the stained copper strip, and the gas stream continued for 30 minutes, to ensure that all the hydrogen sulphide produced is swept through the test paper.

The stained paper is removed from the apparatus with tweezers, allowed to dry for one minute, and then placed on the right-hand side of the base of the test paper stand, while a clean blank of similar filter paper is placed on the left-hand side underneath the colour standards of the disc. The apparatus is held facing a standard source of white light or placed in the standard lighting cabinet so that both papers are evenly illuminated, and the disc rotated until a match is obtained, when the value is read from the indicator. A second paper is inserted in the hydrogen stream, and if any further stain is obtained, the value of this is added to the first result.

The sample of benzole is reported as causing a corrosion of the copper strip that will produce hydrogen sulphide equal to the number of micrograms of sulphur (as hydrogen sulphide) as read from the disc.

Note

The test may also be employed to obtain a figure for the hydrogen sulphide present. A suitable volume of the benzole itself is inserted in the streaming apparatus and the gas streamed out by means of the hydrogen. The reading obtained represents the amount of hydrogen sulphide actually present in the volume taken. If too great a concentration of hydrogen sulphide is present, the sample is diluted with pure benzole free from hydrogen sulphide.

The Determination of Tin

Introduction

Tin (Sn) is widely used as an alloying constituent in metallurgy and as a surface plating material for food containers. The determination of tin is of importance in metals, ores, foodstuffs and also in animal and plant tissues.

Methods of sample preparation

Biological materials[1,2]

Ash 30 – 40 g of the sample at 450 – 500°C in a muffle furnace. Dissolve the residue in 20 ml of $6N$ hydrochloric acid, cover and heat at 100°C for 15 minutes. Add 30 ml of water and filter into a 100 ml flask. Ash the washed filter paper at 450°C, dissolve the ash in 10 ml of $6N$ hydrochloric acid, add 20 ml of water and filter. Remove any silica in the ash with hydrofluoric acid (HF), evaporate with hydrochloric acid, dilute and filter. Combine all the filtrates and make up to 100 ml. Take a 50 ml aliquot in a centrifuge tube. Add 2 ml of 2% ferric ammonium sulphate $((NH_4)_2SO_4Fe(SO_4)_3)$ solution and 1 ml of 5% ammonium hydrogen phosphate $((NH_4)_2HPO_4)$ solution. Add aqueous ammonia (NH_4OH) until the solution is cloudy, clear with $2N$ hydrochloric acid and adjust the pH to 4 – 5 with sodium acetate (CH_3COONa) solution. Heat at 100°C for 20 minutes, centrifuge for 20 minutes and decant the supernatant liquid. Wash the precipitate with 20 ml of 1% sodium acetate solution, then dissolve it in 2 ml of $6N$ hydrochloric acid and dilute the solution to 25 ml. Add 5 ml of 10% potassium thiocyanate (KSCN) solution and then add ascorbic acid $(C_6H_8O_6)$ until a pale pink colour persists. Add 20 ml of ethyl acetate $(CH_3COOC_2H_5)$, shake for one minute, separate the layers and repeat with two further 10 ml portions of ethyl acetate. Combine the organic layers, evaporate to dryness, dissolve the residue in 1 ml of sulphuric acid. Add 20 drops of 30% hydrogen peroxide (H_2O_2), cover and heat at 120°C for 30 minutes. Add 0.1 g of hydroxyammonium sulphate $((NH_2OH)_2H_2SO_4)$ and heat again for 30 minutes. Proceed by *Method 1*.

Alternatively[2] destroy the organic matter by wet oxidation with nitric and sulphuric acids, or with nitric perchloric $(HClO_4)$ and sulphuric acids, or with 50% aqueous hydrogen peroxide (H_2O_2) and sulphuric acid. Dilute the residue with 10 ml of water and evaporate until white fumes appear and repeat this process until the solution is clear. Dilute the final clear solution to a known volume, sufficient to ensure that the sulphuric acid concentration does not exceed 4% by volume. Take a 10 ml aliquot of this solution, which should contain 30 – 150 μg of tin, shake with successive 5 ml portions of a 0.02% solution of dithizone $(C_6H_5.N:N.CSNH.NH.C_6H_5)$ in chloroform $(CHCl_3)$ until the extracts remain green. Discard the organic layers, which contain any copper present in the sample, wash the aqueous layer with two 5 ml portions of carbon tetrachloride and then treat proceed by *Method 1*.

The method[3] described below for the analysis of plants may also be used for biological material.

Foods and edible oils[2,4-9]

Canned fruit, fruit juices and vegetables[4,5] Wet ash the sample with sulphuric and nitric acids, with the addition of hydrogen peroxide (H_2O_2) if necessary to produce a final colourless solution. Evaporate this solution twice with 5 ml of water, take up the residue in hydrochloric acid, dilute to 50 ml with water and determine tin on a 10 ml aliquot.

Corned beef[6] Ignite the sample overnight at 500°C. Extract the tin from the ash by boiling with 50% caustic soda (NaOH) solution, neutralise the extract, filter, make up to a known volume and determine tin on an aliquot.

General[2,7-9] Ash the sample with magnesium oxide (MgO) and then fuse the ash with a mixture of potassium cyanide (KCN) and sodium carbonate (Na_2CO_3). Extract the melt with hydrochloric acid and determine tin on this solution[7]. Alternatively[8] char a 5 – 10 g sample and ignite the residue at 600°C. Fuse the ash with sodium carbonate + potassium cyanide (3+1), boil the melt with hydrochloric acid and filter. Remove copper from the filtrate by extraction with diethylammonium diethyldithiocarbamate $((C_2H_5)_2N.CS.S.NH_2(C_2H_5)_2)$ in chloroform $(CHCl_3)$. Treat an aliquot of the copper-free solution with hydrochloric acid and proceed by *Method 1*.

The second method[2] described for biological materials is also generally applicable to the analysis of foodstuffs. Gajek[9] recommends ashing at 450–500°C followed by solution of the ash in hydrochloric acid. He claims that dry ashing is more rapid than wet ashing yet gives results within 4% of those obtained by wet ashing.

Metals, alloys and plating baths[10-21]

Copper alloys[10-12] Dissolve 0.3–0.4 g of sample by heating with 5 ml of 1:1 nitric acid, to the solution add 2 ml of sulphuric acid and then evaporate until white fumes are evolved. Transfer the residue to a distillation flask by means of 5 ml of water and 10 ml of sulphuric acid, add a few drops of hydrochloric acid, heat to 200°C, cool, add 10 ml of hydrobromic acid (HBr sp.gr. 1.4) one drop at a time and distil off the stannic bromide ($SnBr_4$) at 200–220°C, for 1 hour. Collect the distillate in three 5 ml receivers connected in series, combine the distillates and dilute to 50 ml[10]. Alternatively[11] dissolve the sample in aqua regia ($HNO_3 + 3HCl$), add sulphuric acid and evaporate until white fumes appear, dilute and filter. Pass the filtrate through a column (0.5 sq cm cross section) containing 3 g of Dowex 1-X8-resin (SCN^- form) which has been pre-treated with 1.5M hydrochloric acid, M ammonium thiocyanate (NH_4SCN) washed with water and dried at 50°C for 3 hours. Elute the tin with 15 ml of a solution 0.5M in caustic soda (NaOH) and 0.5M in sodium chloride (NaCl). Kurbatova[12] recommends the separation of tin from the acid solution by passing the solution through a column of an anionite (Cl^- form) previously washed with 3N hydrochloric acid, removing the other metals by washing the column with 60 ml of 3N hydrochloric acid at 0.8–1.0 ml a minute and then desorbing the tin with 150 ml of 0.5N hydrochloric acid at 1 ml a minute.

Ferrotungsten[13] Fuse 0.5 g of powdered sample with 2 g of sodium carbonate (Na_2CO_3) and 9 g of sodium peroxide (Na_2O_2) in an iron crucible. Dissolve the melt in 150 ml of water, add 5 ml of ethanol (C_2H_5OH) and boil and stir for 5 minutes. Make the volume up to 250 ml with water and filter. Discard the first portion of the filtrate and then take a 25 ml aliquot. Add 5 ml of 10% tartaric acid ($(CHOH.COOH)_2$) solution and 1 drop of phenolphthalein indicator solution, neutralise with 10N sulphuric acid, add 0.5 ml excess of the acid and boil. Cool, neutralise with 10% caustic soda (NaOH) solution, make acid with 1 drop of 10N sulphuric acid and add 10 ml of acetate buffer (pH 5.5). Add 5 ml of 1% diethyldithiocarbamate (($C_2H_5)_2N.CS.S.NH_2$) solution and, after 5 minutes, extract for 1 minute with 25 ml of chloroform ($CHCl_3$). Repeat the procedure with a further 2 ml of diethyldithiocarbamate solution and 10 ml of chloroform. Combine the organic extracts, add 5 ml of 10 N sulphuric acid, 8 ml of water and 2 ml of 2% potassium permanganate ($KMnO_4$) solution, shake for 1 minute and then add 2 ml of 3% hydrogen peroxide (H_2O_2) and shake until the solution becomes colourless. Add 5 ml of 10% tartaric acid ($(CHOH.COOH)_2$) solution, shake and discard the chloroform layer. Filter the aqueous layer, wash the filter with two 5 ml portions of water, make up to a known volume and determine tin on this solution.

General[14,15] Completely dissolve the sample in a mixture of nitric and hydrofluoric acids. Pass the acid solution through a column of Zerolit 225 (equivalent to Amberlite IR 120 or Dowex 50) to remove interfering metals and determine tin on the eluate[14]. Alternatively[15] dissolve the sample in about 45 ml of 9N sulphuric acid, add 5 ml of 5M sodium iodide (NaI) solution and sufficient 7.8% sulphurous acid (H_2SO_3) to remove the colour of the liberated iodine. Then extract with benzene (C_6H_6), wash the extract with a 10:1 mixture of 9N sulphuric acid and 5M sodium iodide, remove the benzene and iodine by warming the extract in a stream of air, destroy organic matter with perchloric and sulphuric acids, evaporate to dryness and dissolve residue in 1:3 hydrochloric acid. Make the solution up to a known volume.

Iron and steel[16-19] Dissolve 0.1–0.5 g of sample in a mixture of sulphuric and perchloric ($HClO_4$) acids, treat this solution with aqueous titanous sulphate ($Ti_2(SO_4)_3$) solution and then with aqueous sodium iodide (NaI) to give a solution 7–9N in sulphuric acid and 0.5M in sodium iodide, add 1 drop of phosphoric acid (H_3PO_4), the use of more than 1 drop will precipitate titanium, and extract with 10 ml of benzene (C_6H_6). Determine tin on the organic layer[16]. Alternatively[17] dissolve 0.2–0.5 g of sample in either 1:4 sulphuric acid with the subsequent addition of a small volume of nitric acid, or in 25 ml of 1:4 sulphuric acid and 10 ml of 70% perchloric acid. Evaporate the solution until copious fumes are evolved, dissolve the residue by boiling with 60–70 ml of water and 10 ml of hydrochloric acid and make up to a known volume.

Among other methods which have been suggested for the separation of tin from interfering ions are precipitation of the tin as its sulphide, solution of the sulphide in 1:9 sulphuric acid, extraction of the tin into diethylammonium diethyldithiocarbamate ((C_2H_5)$_2$N.CS.S.NH$_2$(C_2H_5)$_2$), evaporation of the extract and solution of the residue in dilute hydrochloric acid[18]; and extraction from 37% hydrochloric acid solution into $2.5M$ tributylphosphate ((C_4H_9)$_3$PO$_4$) in benzene, washing the organic phase twice with hydrochloric acid and determining tin on the combined aqueous phase and washings[19].

Lead and lead alloys[20] Dissolve 0.25–2.5 g of sample in 25 ml of 1:4 nitric acid and 10 g of citric acid (C(OH)(COOH) (CH$_2$COOH)$_2$.H$_2$O), add 2 ml. of aqueous ammonia (NH$_4$OH sp. gr. 0.880), 10 ml of $0.1M$ ammonium chloride (NH$_4$Cl) and 2 g of E.D.T.A. (ethylenediamine-tetra-acetic acid, ((CH$_2$N(CH$_2$COOH)$_2$)$_2$) and adjust the pH to 5.5. with aqueous ammonia or citric acid as required. Pass this solution through a column of silica gel previously washed with hydrochloric acid, water and citrate solution of pH 5.5. Elute tin with 1:1 hydrochloric acid and make volume up to 50 ml.

High purity nickel[21] Dissolve 1 g of the nickel, containing 0.001–0.005% of tin, in 15–20 ml of 1:1 nitric acid, evaporate to dryness and heat to decompose nitrates. Dissolve the residue in 50 ml of 1:1 hydrochloric acid and extract with two 25 ml portions of a 10% solution of Amberlite LA-2 in xylene (C_6H_4(CH$_3$)$_2$). Wash the combined extracts with three 10 ml portions of 1:1 hydrochloric acid and re-extract the tin into three 30 ml portions of $0.5N$ nitric acid. Wash the combined aqueous extracts with 25 ml of xylene. To the aqueous phase add 0.5 ml of 1:1 sulphuric acid, evaporate to dryness, add sulphuric acid and 30% hydrogen peroxide (H$_2$O$_2$), evaporate to point where copious white fumes are evolved, dissolve the residue in dilute hydrochloric acid and proceed by *Method 1*.

Minerals, ores and rocks[22-24]

Grind 0.5g of sample to a fine powder and heat the powder to red heat, in a silica test tube, with 1.0 g of ammonium iodide (NH$_4$I) until all the iodine has volatilised (15-30 minutes). Cool, add 4 ml of $10N$ sulphuric acid, 5 ml of water and 1 ml of alcohol. Shake until the sublimate has dissolved. Filter into 0.5 ml of thioglycollic acid (HSCH$_2$COOH), dilute to 50 ml and determine tin on an aliquot[22]. Alternatively[23] decompose 0.2- 2 g of sample by heating with 20-25 ml of hydrofluoric acid and 10 ml of 1:1 sulphuric acid in a platinum crucible. Evaporate to the point when fumes appear, dissolve the residue in 100 ml of water and precipitate tin with other hydroxides by adding aqueous ammonia (NH$_4$OH) and heating. Filter, wash the precipitate with a 5% solution of ammonium chloride (NH$_4$Cl), ignite the paper and precipitate at 500-600°C and then fuse the ash in an iron crucible with 5 g of caustic soda (NaOH) and 0.5 g of sodium peroxide (Na$_2$O$_2$). Extract the cooled melt with hot water, cool, neutralise to Congo red with 1:1 sulphuric acid, and add about 4 drops of the acid in excess. Filter, to the filtrate add 2 ml of 10% manganous sulphate (MnSO$_4$.4H$_2$O) solution and 1 ml of 4% potassium permanganate (KMnO$_4$) solution, boil for 10-15 minutes, collect the precipitate and wash it with hot water acidified with sulphuric acid. Heat the precipitate and filter paper with 10 ml of 1:1 hydrochloric acid, filter, wash the residue with hot water and make up the filtrate to 25 ml.

Ruzinova[24] recommends fusion of the sample with sodium peroxide, solution of melt in hot dilute hydrochloric acid, precipitation of hydroxides in presence of 5 ml of 1% gelatin, and solution in hydrochloric acid.

Miscellaneous[25]

General Destroy organic matter by wet ashing. Remove arsenic, antimony etc, by distillation of their chlorides and then distil out tin from a mixture of hydrochloric and hydrobromic acids (1:3) at 137-147°C. Treat the tin solution with sulphuric acid and hydrogen peroxide (H$_2$O$_2$) to remove bromine and determine tin on the resulting solution.

Plants[3]

Wet ash sufficient sample to produce 5-40 μg of tin using sulphuric acid+nitric acid. Eliminate residual nitric acid by boiling with water, add 1-2 g of hydrazine sulphate ((NH$_2$)$_2$H$_2$SO$_4$) and boil until the temperature reaches 160°C. Carefully add a mixture of 15 ml of hydrochloric acid and 7 ml of hydrobromic acid (HBr) and then distil for 15-20 minutes at 145-160°C. Collect the distillate in a flask containing water. Evaporate the distillate to fumes with sulphuric acid+nitric acid, add 7 ml of water to the residual acid and determine tin on this solution.

Tin Method 1
using dithiol

Principle of the method

In the presence of an acid solution of divalent tin, dithiol (1:2-dimercapto-4-methyl-benzene) forms a red lake. In this test the tin is reduced to the divalent state by means of thioglycollic acid, and the red lake is held in suspension by sodium lauryl sulphate.

Reagents required

1. *Hydrochloric acid concentrated.*
2. *Thioglycollic acid* ($HSCH_2COOH$)
3. *Sodium lauryl sulphate* 1.0% w/v aqueous solution
4. *Dithiol (1:2-dimercapto-4-methyl-benzene)* 0.125% w/v solution in 1.0% w/v aqueous sodium hydroxide, freshly made up as required.
   (N.B.—This reagent seldom maintains its reactive strength for more than two days).
5. *0.5N hydrochloric acid* Approximately 60g concentrated hydrochloric acid solution per litre.

All chemicals used in the preparation of these reagents should be of analytical reagent quality.

The Standard Lovibond Comparator Disc 3/35

The disc contains colour standards for 1, 2, 3, 4, 5, 6, 8, 10 and 12 parts per million of tin, present in the final 20 ml of solution.

Technique

Place in a 20 ml volumetric flask 10 ml of the solution under test, which should be $0.5N$ in terms of hydrochloric acid (i.e. containing about 5% HCl) and should contain not more than 0.24 mg of tin. If a preliminary test shows that there is a higher concentration of tin present, a suitable dilution with $0.5N$ HCl should be carried out before a determination is undertaken.

Add to the solution in the flask in the following order:—

    1 drop of thioglycollic acid (reagent 2)
    2.0 ml of hydrochloric acid, concentrated (reagent 1)
    0.5 ml of 1% sodium lauryl sulphate solution (reagent 3)
    1 ml of dithiol (reagent 4)

Mix thoroughly after the addition of each of these reagents. At the same time prepare a "blank" solution containing 10 ml of $0.5N$ hydrochloric acid (reagent 5) in place of the 10 ml of test sample and the other reagents exactly as above.

Stopper the flasks and place in a water bath at 60°C for ten minutes. Cool to 20°C and make up the volumes in each case to the 20 ml mark with distilled water. Pour the test solution into a 13.5 mm comparator cell or test tube in the right-hand compartment of the comparator and the "blank" in a similar cell or test tube in the left-hand compartment so that it comes behind the glass colour standards.

Hold the comparator facing a standard source of white light such as the Lovibond White Light Cabinet or, failing this, north daylight, and compare the colour produced in the test solution with the colours in the standard disc, rotating the latter until a colour match is obtained. The figure shown at the indicator window represents parts per million of tin in the final 20 ml solution. Due correction must therefore be made to relate this to the amount of sample originally taken.

Note

The determination of tin is disturbed by the presence of copper, nickel and bismuth, and less so by cobalt, since in this case tin is preferentially precipitated. Compounds produced by silver, mercury, cadmium, arsenic, and lead are yellow, and therefore only slightly reduce the accuracy of the results. Nitrites and phosphates can interfere.

The Determination of Titanium

Introduction
Titanium is used in the metallurgical, paint, textile and plastics industries. Methods have been published for its determination in a wide range of materials. The following are some of the methods which have been suggested for sample preparation.

Methods of sample preparation
Chemicals[1]
Organometallic compounds Dissolve about 0.2 g of the sample by warming it with 3 ml of of sulphuric acid in a platinum crucible, cool, add 0.2 g of ammonium nitrate (NH_4NO_3) and shake gently. Repeat the addition of ammonium nitrate followed by warming the mixture until a colourless solution is obtained. Make this solution up to a known volume and determine titanium on an aliquot.

Fuels and lubricants[2-4]
Coal ash Ignite 75 – 300 mg of sample at 700°C for 3 hours, first in air and then finally in oxygen. Cool, add 10 – 15 ml of hydrofluoric acid (HF) and heat to fuming on a water bath for 10 minutes. Repeat with three 5 ml portions of $6N$ hydrochloric acid. Dissolve the residue in 5 ml of $6N$ hydrochloric acid and pass the solution through a 20×20 cm column of Lewatit KSN ion exchange resin at a flow rate of 4 ml per minute. Wash the column with 150 ml of $0.15N$ hydrochloric acid. Elute the cations retained on the column with 200 ml of $4N$ hydrochloric acid. Reduce the volume of the eluate to 10 ml by evaporation. Precipitate iron, aluminium and titanium by boiling with the addition of aqueous ammonia (NH_4OH). Separate the precipitate by centrifuging, dissolve it in 10 ml of $8N$ hydrochloric acid and pass the solution through a column of Amberlite IRA-400 resin which has been previously washed with $8N$ hydrochloric acid. Collect the eluate, make about $4N$ with respect to sulphuric acid and determine titanium by *Method 1*[2,3]. Alternatively[4], grind the sample to pass through a 250 mesh sieve. Mix 0.5 g of the finely ground sample with 2 g of lithium metaborate ($LiBO_2$) in a platinum crucible and heat in a muffle furnace at 900°C until the melt is clear. If the sample contains more than 15% of iron add 30 – 50 mg of ammonium vanadate (NH_4VO_3) to act as a flux. Cool the melt and transfer it to a beaker containing 8 ml of nitric acid and 150 ml of water. Stir vigorously until dissolution is complete. Add 2.5 g of tartaric acid ($(CHOH.COOH)_2$) and make the volume up to 250 ml.

Metals alloys and plating baths[5-24]
Alloys[5-12] Dissolve 1 g of sample in hydrochloric acid in the presence of an oxidising agent. Remove any precipitated silicon or tungsten by filtration. Make the filtrate $4N$ with respect to sulphuric acid and make up to 50 ml with $4N$ sulphuric acid and proceed by *Method 1*[5,6]. Alternatively[8,10] dissolve the alloy in aqua regia (hydrochloric acid + nitric acid, 3 + 1), boil off nitrogen oxides, cool, dilute and pass the solution through a column of Dowex-1 anion exchange resin. Elute aluminium, nickel and titanium with $7M$ hydrochloric acid. Pass the eluate through two columns of anionic and cationic exchange resins in series. Elute with $0.8M$ hydrofluoric acid (HF) + $0.06M$ hydrochloric acid, when titanium will be retained on the anionic resin. Separate the two columns and elute the anionic column with $3M$ hydrochloric acid. Make the eluate $4N$ with respect to sulphuric acid and proceed by *Method 1*. Golovatyi[11] and Kenna[12] also advocate the use of ion exchange resins to separate titanium from other elements which could interfere with its determination. It is also suggested[7,9] that interfering metals can be removed by electrolysis.

Aluminium[6,13-14] Dissolve 0.5 – 2.0 g of sample in sulphuric, nitric, hydrochloric acids and water 5:4:7, remove bismuth, tin, lead, chromium, iron, nickel and cobalt electrolytically at a mercury cathode and determine titanium on the residual solution after this has been adjusted to about $4N$ with respect to sulphuric acid[6].

Abramov[13] claims that only reduction of iron to the ferrous state, by means of ascorbic acid (O.CO.C(OH):C(OH).CH.CH(OH).CH_2OH) is required before proceeding with the determination of titanium. Volkova[14] uses no separation of interfering metals, neither does Pilipenko[15].

Aluminium-chromium catalysts[16] Fuse 0.15–0.25 g of the sample with 7 g of potassium pyrosulphate ($K_2S_2O_7$) at 600–700°C for 1–1.5 hours. Extract the melt with hot water, dilute the solution to 100 ml, filter if necessary, and determine titanium on an aliquot of the filtrate.

Iron and Steel[15,17-24] Dissolve 0.25 g of sample in 50 ml of 1:4 sulphuric acid and add a few drops of nitric acid. Boil off the oxides of nitrogen, cool, dilute to 100 ml and determine titanium on an aliquot[17]. Alternatively[19] dissolve 1 g of sample in nitric acid, add 3 ml of phosphoric acid (H_3PO_4) and reduce the volume by evaporation. Add 10 ml of nitric acid and 3 ml of phosphoric acid and repeat these additions and evaporation until the carbides dissolve. Reduce any oxidised manganese with hydrogen peroxide (H_2O_2) and remove excess hydrogen peroxide by boiling. Filter, make the filtrate up to known volume with $4N$ sulphuric acid and determine titanium on an aliquot. Nadkarni et al[18] recommend removal of iron by extraction of its chloride with ether ($C_2H_5OC_2H_5$) and removal of titanium by co-precipitation with zirconium as phosphates. Kawahata et al[20] on the other hand removes titanium with E.D.T.A. Asmus[21], Tananaiko[22], Tribalat[23] and Scneiderman[24], however, do not find it necessary to recommend any separation procedures.

Minerals, ores and rocks[4,5,15,25-33]

Clays[25,26,30] Decompose 1 g of sample in a platinum crucible with 25 ml of 40% hydrofluoric acid (HF) and 2.5 ml of sulphuric acid and heat on a sandbath until all the hydrofluoric acid has evaporated. Fuse the residue with 3 g of potassium pyrosulphate ($K_2S_2O_7$) until the melt begins to crystalise. Dissolve the residue in warm $0.1N$ sulphuric acid, transfer the solution to a 250 ml volumetric flask and make up to the mark with $0.1N$ sulphuric acid. Determine titanium on an aliquot[25]. Alternatively[26,30] calcine 1 g of sample, fuse the ash with potassium hydrogen sulphate ($KHSO_4$) in a porcelain crucible. Dissolve the melt in a mixture of sulphuric acid and hydrogen peroxide (H_2O_2) and proceed by *Method 1*.

General[4,5,15,27-29,31,32] Mix 20 g of finely ground sample with 0.2–0.5 g of ammonium fluoride (NH_4F) in a platinum crucible and heat at 350–450°C, until the sample has completely decomposed. Add 0.5 g of oxalic acid ($(COOH)_2.2H_2O$), heat and ignite to remove excess oxalate. Cool, moisten the walls and bottom of the crucible with 2–3 ml of 1:3 sulphuric acid, heat until no more fumes are evolved and then add a further 2–3 ml of 1:3 sulphuric acid and warm until a clear solution is produced. Make up to a known volume and determine titanium on an aliquot[29]. Alternatively[22,28] dissolve the ore in hydrochloric acid, fuse any residue with potassium hydrogen sulphate ($KHSO_4$) and take up the melt in water. Filter and fuse any solid matter still remaining with fusion mixture (an equimolar mixture of potassium and sodium carbonates, $K_2CO_3+Na_2CO_3$). Dissolve the melt in sulphuric acid, add to the filtrate and proceed by *Method 1*. Schoffmann and Malissa[5] suggest that the heavy metals should be removed by complexing with tetramethylenedithiocarbamic acid, extracting the complexes with chloroform ($CHCl_3$) and determining titanium on the aqueous phase. The final procedure, described above, for the analysis of coal ash[4] can also be used for analysis of minerals. Pilipenko[15] and Klassova[31] both recommend solution of the sample in sulphuric and hydrofluoric acids. Van Loon[32] recommends the following fusion technique:— Prepare a graphite crucible by igniting it for 30 minutes at 950°C and then cool it without disturbing the powdery inner surface. Mix 0.2 g of rock powder with 1.4 g of lithium metaborate ($LiBO_2$) in a porcelain crucible and then transfer the mixture to the prepared graphite crucible. Fuse the mixture at 900°C for exactly 15 minutes, then pour the melt into a flat bottomed plastic beaker containing 100 ml of nitric acid diluted with 24 times its volume of water. Stir with a Teflon-coated glass rod, wash the solution into a 250 ml volumetric flask and dilute to the mark with 1:24 nitric acid. Determine titanium on an aliquot of this solution.

Gypsum or calcite[30] Treat 0.5–2.0 g of sample with 20 ml of 1:1 hydrochloric acid and 20 ml of water. Filter, ignite the insoluble residue with 2 g of potassium pyrosulphate ($K_2S_2O_7$), dissolve the melt in 10 ml of 1:1 hydrochloric acid and 20 ml of water, filter, combine the two filtrates and make up to 200 ml.

Ilmenite[33] Decompose 0.2–0.3 g of the sample with sulphuric and hydrofluoric acids and evaporate to dryness to remove both silicon and fluorine. Fuse the residue with potassium pyrosulphate ($K_2S_2O_7$), dissolve the melt in dilute sulphuric acid and dilute to a known volume.

COLORIMETRIC CHEMICAL ANALYTICAL METHODS — Titanium

Miscellaneous[2,3,25,26,30,34-36]

Cement[30,34] Dissolve 0.1–0.5 g of sample in 5 ml of hydrochloric acid containing 0.5 g of ammonium chloride (NH_4Cl), filter and dilute the filtrate to 100 ml[30]. Or[34], fuse the sample with sodium carbonate (Na_2CO_3) and caustic soda (NaOH), dissolve the melt in dilute hydrochloric acid, filter, make up to known volume and determine titanium on an aliquot.

Dusts[2,3] Use the first method described above for coal ash.

Fertilizers and liming materials[35] Grind the sample in an agate mortar until it will pass through a No. 100 sieve. Take 0.5 g (limestone) or 0.2 g (silicate) of the sample in a 75 ml nickel crucible, remove any organic matter which may be present by heating at 900°C for 15 minutes with the crucible uncovered, and then cool. Mix 0.3 g of potassium nitrate (KNO_3) with the sample and then cover the mixture with 1.5 g of caustic soda (NaOH) pellets. Cover the crucible with a nickel cover and heat to a dull red heat over a gas flame, not in a furnace, for 5 minutes. Remove the flame and swirl the melt around the sides of the crucible. Cool, add about 50 ml of water and then warm to disintegrate the fused cake. Transfer the contents pf the crucible to a 150 ml beaker containing 15 ml of $5N$ perchloric acid ($HClO_4$). Scrub the crucible and its lid with a plastic coated glass rod and wash any resulting residues into the beaker. Transfer the contents of the beaker to a 100 ml volumetric flask and dilute to the mark.

Refractories[25,26] Use either of the methods described above for the analysis of clays.

Textiles[36] Destroy the organic matter by wet or dry ashing. Acidify the solution to 2–$3N$ with sulphuric acid, add sodium sulphate (Na_2SO_4), and proceed by *Method 1*.

Plastics and polymers[37-39]

Ignite 1.5 g of sample in a Parr oxygen bomb. Fuse the ashed residue with not more than 0.3 g of potassium hydrogen sulphate ($KHSO_4$) and extract the melt with 5% sulphuric acid. Remove interfering elements by co-precipitating the titanium with zirconium as the arsenate. Dissolve the mixed arsenates in 10% sulphuric acid and proceed by *Method 1*[37]. Alternatively[38] place 2 g of plastic in a 250 ml flask, add 20 ml of sulphuric acid, char by heating and when the sample is completely charred add 20 ml of nitric acid slowly through a dropping funnel. If any black particles remain add further nitric acid. When the solution is clear, carefully add 5 ml of 70% perchloric acid ($HClO_4$) drop by drop. Heat until white fumes are evolved keeping the volume at about 5 ml by adding further sulphuric acid of necessary. Cool, add 50 ml of water and re-heat. Cool, make up to 100 ml and determine titanium on an aliqiot. Novak[39] uses hydrogen peroxide (H_2O_2) in place of nitric acid for the oxidation of the sample in his procedure for the analysis of polypropylene.

Water[40]

To the filtered sample add ascorbic acid ($C_6H_8O_6$), adjust the pH to 4–4.5 and pass the solution through a column of Dowex 1–X8 ion exchange resin. Convert the absorbed titanium-ascorbic acid complex to the titanium-fluoride complex by adding $0.01N$ sulphuric acid containing sodium fluoride (NaF). Elute the titanium with $0.1N$ sulphuric acid containing hydrogen peroxide (H_2O_2) and proceed by *Method 1*.

Separation from other ions[41-45]

In addition to the separation procedures recommended for specific applications above, the following general procedures have been described:—

From iron, aluminium, chromium, indium, beryllium and uranium[45] Precipitate titanium from solution at pH 2 with cinnamic acid ($C_6H_5.CH:CH.COOH$) to which ammonium chloride (NH_4Cl) has been added to speed coagulation. This precipitation completely separates titanium from aluminium, indium, beryllium and uranium. If iron or chromium is present, ignite the precipitate, fuse with potassium pyrosulphate ($K_2S_2O_7$), dissolve the cooled melt in acid, add sodium dithionite ($Na_2S_2O_4$) to reduce iron and then re-precipitate titanium as above.

From vanadium[44] Pass the test solution at pH 1 through a column of cationite KY-3 (H^+ form). Pass 10% caustic soda (NaOH) solution through the column whereupon vanadium is eluted. Titanium can then be eluted with $4N$ hydrochloric acid.

General[41-43] Both Korkisch[41] and Strelow[42] use ion exchange columns to separate titanium from acid solutions. Lewandowski[43] recommends the use of the following column which he claims is specific for titanium:— Mix a solution of 9.6 g of chromotropic acid, (1, 8 dihydroxy-naphthalene–3,6–disulphonic acid, $C_{10}H_8O_8S_2$) and 4.4 g of resorcinol ($C_6H_4(OH)_2$) in 25 ml of water with 15 ml of 30% formaldehyde (HCHO) solution. Set the mixture aside for 1 hour at room temperature and then heat on a water bath for 20 minutes. Wash the product with 0.5N caustic soda (NaOH), N hydrochloric acid and water and dry at room temperature. Sodium, potassium, ammonium, thallium, silver, magnesium, calcium, strontium, barium, copper, zinc, cadmium, tin, uranium oxide, manganese, nickel, cobalt, iron, chromium, aluminium, lanthenum, thorium and zirconium are easily eluted with 0.5–1.0N hydrochloric acid. To elute titanium 2N or preferably 6N hydrochloric acid is needed.

The individual test

Method 1 is an adaptation of the reaction between titanium sulphate and hydrogen peroxide originally reported by Weller[46] and more recently discussed by Neale[47].

References
1. S. Kohama, *Bull. Chem. Soc. Japan*, 1963, **36**, 830; *Analyt. Abs.*, 1964, **11**, 4365
2. F. M. Kessler and L. Dockalova, *Sbirka. Praci. Vyzkumm. Ust.*, 1957, **A8**, (17-26), 188; *Analyt. Abs.*, 1958, **5**, 4044
3. J. Fitzek and H. Stegemann, *Beitr. Silikose-Forsch.*, 1957, (47), 41; *Analyt. Abs.*, 1958, **5**, 2515
4. P. L. Boar and L. K. Ingram, *Analyst*, 1970, **95**, 124
5. E. Schoffmann and H. Malissa, *Arch. Eisenhuttenw.*, 1957, **28**, 623; *Analyt. Abs.*, 1958, **5**, 2156
6. G. Matelli and L. Monti, *Alluminio*, 1960, **28**, 553; *Analyt. Abs.*, 1960, **7**, 3188
7. Z. I. Kardokova, *Sb. Statei, Mosk. Vyssh. Teckhn. Uch.*, 1955, **36**, 30; *Analyt. Abs.*, 1957, **4**, 1142
8. L. E. Hibbs and D. H. Wilkins, *Talanta*, 1958, **2**, 16; *Analyt. Abs.*, 1959, **6**, 4672
9. V. Duriez and J. Barboni, *Rev. Nickel*, 1958, **24**, 11; *Analyt. Abs.*, 1960, **7**, 155
10. D. H. Wilkins, *Talanta*, 1959, **2**, 355; *Analyt. Abs.*, 1960, **7**, 2056
11. R. N. Golovatyi and V. V. Oshchapovskii, *Ukr. Khim. Zhur.*, 1963, **29**, 187; *Analyt. Abs.*, 1964, **11**, 1293
12. B. T. Kenna and F. J. Conrad, *Analyt. Chem.*, 1963, **35**, 1255
13. M. I. Abromov, *Uch. Zap. Azerb, Univ. Ser. Khim. Nauk.*, 1963, (3), 31; *Analyt. Abs.*, 1965, **12**, 1654
14. A. I. Volkova, T. E. Getman and N. A. Emtsova, *Ukr. Khim. Zhur.*, 1964, **30** (1), 102; *Analyt. Abs.*, 1965, **12**, 1655
15. A. T. Pilpenko, E. A. Shpak and Yu. P. Boiko, *Zavodskaya Lab.*, 1965, **31** (2), 151; *Analyt. Abs.*, 1966, **13**, 2917
16. T. A. Cheremukhina, *Prom. sintet. Kauchuka*, 1966, (2), 40; *Analyt. Abs.*, 1967, **14**, 6709
17. A. I. Lazarev and V. I. Lazareva, *Zavodskaya Lab.*, 1958, **24**, 145; *Analyt. Abs.*, 1958, **5**, 3750
18. M. N. Nadkarni, G. G. Nair and C. Venkateswarlu, *Anal. Chim. Acta.*, 1951, **21**, 511; *Analyt. Abs.*, 1960, **7**, 3753
19. M. Freegarde and B. Jones, *Analyst*, 1959, **84**, 393
20. M. Kawahata, H. Mochizuki and T. Misaki, *Japan Analyst*, 1960, **9**, 1019; *Analyt. Abs.*, 1962, **9**, 3713
21. E. Asmus, W. Richly and H. Wunderlich, *Z. analyt. Chem.*, 1963, **199** (4), 249; *Analyt. Abs.*, 1965, **12**, 1739
22. M. M. Tananaiko and G. N. Vinokurova, *Zh. analit. Khim.*, 1964, **19** (3), 316; *Analyt. Abs.*, 1965, **12**, 3312
23. S. Tribalat and J. M. Caldero, *Bull. Soc. chim. France*, 1964, (12), 3187; *Analyt. Abs.*, 1966, **13**, 1762
24. S. Ya. Scneiderman and E. N. Knyazeva, *Zh. analit. Khim.*, 1966, **21** (4), 419; *Analyt. Abs.*, 1967, **14**, 7485
25. O. Glemser, E. Raulf and K. Geisen, *Z. analyt. Chem.*, 1954, **141**, 86; *Analyt. Abs.*, 1954, **1**, 1218
26. J. Gottfried, *Chem. Prumysl.*, 1958, **8**, 176; *Analyt. Abs.*, 1959, **6**, 1251
27. D. N. Grindley, E. H. W. Burden and A. H. Zaki, *Analyst*, 1954, **79**, 95
28. J. A. Corbett and D. H. Pankhurst, *Proc. Aust. Inst. Min. Metall.*, 1957, (182), 55; *Analyt. Abs.*, 1958, **5**, 1161

29. S. I. Smyshlaev, *Razvedkai Okhrana Nedr.*, 1959, (2), 48; *Analyt. Abs.*, 1960, **7**, 1318
30. H. Ishii, *Japan Analyst,* 1967, **16** (2), 110; *Analyt. Abs.,* 1968, **15**, 6692
31. N. S. Klassova and L. L. Leonova, *Zh. analit. Khim.*, 1964, **19** (1), 131; *Analyt. Abs.*, 1965, **12**, 2175
32. J. C. Van Loon and C. M. Parissis, *Analyst,* 1969, **94**, 1057
33. G. E. Lunina and E. G. Romanenko, *Zavodskaya Lab.*, 1968, **34**, (5), 538; *Analyt. Abs.*, 1969, **17**, 1418
34. K. Kawagaki, M. Saito and K. Hirokawa, *J. chem. Soc. Japan, ind. Chem. Sect.*, 1965, **68** (3), 465; *Analyt. Abs.*, 1967, **14**, 5434
35. P. Chichilo, *J. Assoc. offic. agric. Chem.*, 1964, **47**, 1019; *Analyt. Abs.*, 1965, **12**, 6778
36. Shirley Inst. Test Leaflet 1954, No. Chem. 21; *Analyt. Abs.*, 1954, **1**, 2725
37. R. A. Anduze, *Analyt. Chem.*, 1957, **29**, 90
38. W. T. Bolleter, *Analyt. Chem.*, 1959, **31**, 201
39. K. Novak and V. Mika, *Chem. Prumysl.*, 1963, **13**, 360; *Analyt. Abs.*, 1964, **11**, 4414
40. J. Korkisch, *Z. analyt. Chem.*, 1960, **178**, 39; *Analyt. Abs.*, 1961, **8**, 3047
41. J. Korkisch, G. Arrhenius and D. P. Kharker, *Analyt. chim. Acta*, 1963, **28**, 270; *Analyt. Abs.*, 1964, **11**, 1252
42. F. W. E. Strelow, *Analyt. Chem.*, 1963, **35**, 1279; *Analyt. Abs.*, 1964, **11**, 3046
43. A. Lewandowski and W. Szczepaniak, *Chem. Stosow*, 1963, 603; *Analyt. Abs.*, 1964, **11**, 3598
44. D. Shiskov and L. Shiskova, *Compt. rend. Acad. bulg. Sci.*, 1963, **16**, 833; *Analyt. Abs.*, 1964, **11**, 5440
45. I. I. Volkov and E. A. Ostroumov, *Zh. analit. Khim.*, 1964, **19** (10), 1223; *Analyt. Abs.*, 1966, **13**, 1227
46. A. Weller, *Ber.*, 1882, **15**, 2592
47. W. T. L. Neale, *Analyst.*, 1954, **79**, 403

Titanium Method 1
using hydrogen peroxide

Principle of the method

The test is based on the yellow colours produced by the reaction between titanium sulphate and hydrogen peroxide in acid solution.

Reagent required

Hydrogen peroxide (H_2O_2) 20 volumes analytical reagent grade.

The Standard Lovibond Discs, Comparator disc 3/26, and Nessleriser discs NRA and NRB

Disc 3/26 covers the range 1.0 to 10 per cent. titanium (Ti). This disc is designed for use with a 5 mm cell.

Disc NRA covers the range 0.025 to 0.225 per cent. of TiO_2, in steps of .025%.

Disc NRB covers the range 0.25 to 2.25 per cent. of TiO_2, in steps of 0.25%.

The above percentages are based on 1 g of sample dissolved in 250 ml of test solution.

Technique

To 50 ml of test solution which should be $4N$ with respect to sulphuric acid, (i.e approximately 20% H_2SO_4), add 10 ml of hydrogen peroxide, dilute with distilled water to 100 ml and mix thoroughly.

Comparator

Pour into a 5 mm cell. Place this cell in the right-hand compartment of the comparator, and leave the left-hand compartment blank. Hold the comparator facing a standard source of white light, and rotate the disc until the nearest match is obtained. The figure shown in the indicator window represents percentage titanium in the original gram of sample taken.

Nessleriser

Transfer 30 ml of the yellow solution to a Nessleriser glass and place it in the right-hand compartment of the Nessleriser. In the other compartment of the Nessleriser place a Nessleriser glass containing 30 ml of distilled water. Stand the Nessleriser before a standard source of white light such as the Lovibond White Light Cabinet and compare the colour of the test solution with the colours in one of the standard discs, rotating the disc until a colour match is obtained. The markings on the discs indicate the percentage of TiO_2, assuming that 1 g of the sample has been dissolved in the original 250 ml of test solution. If the original material contains more than 2.5 per cent. TiO_2 but not more than 4 per cent., 25 ml of the original solution should be taken to make up to 100 ml instead of 50 ml. The results obtained should be corrected accordingly.

Notes

1. The application of the Lovibond Nessleriser to the determination of titanium was suggested by the British Refractories Research Association. The colour standards against which the discs are matched were prepared in the Laboratories of the Association.

2. It must be emphasised that the readings obtained with the Lovibond Nessleriser and disc are accurate only if Nessleriser glasses are used which conform with the specification used at the time the discs were calibrated, namely that the 50 ml mark shall fall at a height of 113 ± 3 mm measured internally.

3. 30 ml, instead of the usual 50 ml as in most Nessleriser tests, is used with discs NRA and NRB.

The Determination of Vanadium

Introduction

Vanadium is used as an alloying constituent in steels, where it imparts hardness and toughness. It is also frequently found in oils and water and is an ubiquitous trace element in plants.

Methods of sample preparation

Air[1]

Pass 100–200 litres of air through a filter paper at the rate of not more than 28 litres a minute. Digest the filter with 5 ml of nitric acid in a Kjeldahl flask, evaporate nearly to dryness, add 10 ml of water and again evaporate. Dissolve the residue in 10 ml of water acidified with 2–3 drops of nitric acid and make up to 25 ml.

Beverages[2,3]

Mineral water[2] Take sufficient sample to contain 2–20 μg of vanadium and not more than 170 milliequivalent of total cations, add enough of a 1:1 mixture of $6N$ hydrochloric acid and anhydrous acetic acid (CH_3COOH) to adjust the pH to lie between 1.0 and 2.0, remove carbon dioxide by boiling, cool and apply the solution to a column of Dowex 50-X8 resin (H^+ form) at a rate of 1.0–1.5 ml per square cm per minute. Wash the column with six 50 ml portions of water and elute the vanadium with 100 ml of 0.6% hydrogen peroxide (H_2O_2) solution at a rate of 0.4–0.5 ml per square cm per minute. Discard the first 25 ml of eluate, evaporate the rest in a Teflon crucible with 2 ml of 2% sodium carbonate (Na_2CO_3) solution and then dissolve the residue in a few ml of hot water.

Wine[3] Ash 50 ml of wine and fuse the ash with a mixture of sodium and potassium carbonate ($Na_2CO_3 + K_2CO_3$). Dissolve the melt in water, filter, adjust the filtrate to pH 4 with phthalate buffer and extract vanadium and iron by shaking with 0.5% 8-hydroxyquinoline in chloroform ($CHCl_3$). Add water and raise the pH to 9.4 (ammonium nitrate (NH_4NO_3) buffer). Shake, separate the aqueous layer. Wash with chloroform, neutralise with nitric acid and determine vanadium by *Method 1*.

Biological Materials[1,4,5]

Urine Evaporate a 100 ml sample with 10 ml of nitric acid in a Kjeldahl flask until nearly dry, add nitric acid a drop at a time until the residue is white, then add 25 ml of water, 1 ml of nitric acid, evaporate to about 20 ml, add aqueous ammonia (NH_4OH) until pH is 4.0 and then make the solution up to a known volume[1]. Alternatively[4], to 50 ml sample add 10 ml of nitric acid and evaporate to 3 ml. Transfer the concentrate to a crucible with two 2 ml portions of water, evaporate to dryness, ash the residue at 500°C, cool, dissolve the ash in 2 ml of nitric acid and 5 ml of water and wash the solution into a flask with 20 ml of water. Evaporate to dryness, add 50 ml of water and 1 ml of nitric acid and then evaporate to about 25 ml. Make up to 50 ml with water, add 2 drops of methyl orange indicator solution and neutralise the solution with $4N$ aqueous ammonia. Extract with three 5 ml portions of a 0.5% solution of 8-hydroxyquinoline ($N:CH.CH:CH.C_6H_3OH$) in chloroform ($CHCl_3$) by shaking for 2.5 minutes for each extraction. Collect the organic layers in a flask containing 50 ml of buffer solution of pH 9.4, shake for 5 minutes, discard the organic phase and make the aqueous phase up to a known volume. Chan[5] recommends that the vanadium in the sample be co-precipitated with 10 mg of ferric hydroxide ($Fe(OH)_3$) at pH 5–6 and then separated from other elements by means of a cation exchange resin, from which vanadium can be eluted with 0.3% aqueous hydrogen peroxide (H_2O_2).

Fuels and lubricants[6-9]

Dry ash the sample and dissolve the ash in acid. Complex the vanadium with 8-hydroxyquinoline and extract the complex into chloroform ($CHCl_3$) at pH 5.5 in the presence of the calcium salt of E.D.T.A. Determine vanadium on this extract[6]. Rayner[7] suggests that the ashing is best carried out in a calorimetric bomb. Other methods suggested for ashing are evaporation to dryness with sulphuric acid followed by ignition in the presence of magnesium carbonate ($MgCO_3$)[8]; and digestion with sulphuric and nitric acids to which 30% aqueous hydrogen peroxide has been added[9].

COLORIMETRIC CHEMICAL ANALYTICAL METHODS — Vanadium

Metals, alloys and plating baths[10-15]

Magnesium[10] Dissolve 0.1 g of sample in nitric acid, evaporate excess acid and make the solution up to a known volume with water.

Steel and refractory alloys[11-15] Dissolve 0.05–0.2 g of sample in 1:4 sulphuric acid then add hydrogen peroxide solution a drop at a time until a clear solution is obtained, heat until any carbides which may be present have dissolved, transfer to a flask and dilute to 50 ml. Add 2.5% potassium permanganate ($KMnO_4$) solution until a pink colouration just persists, add 2 ml of aqueous hydrogen peroxide (pH 1) and pass the solution through a column (15 × 2 cm) of cationite KU-2 at 3 ml a minute. Wash the column with 75–100 ml of 1% hydrogen peroxide in $0.1N$ sulphuric acid, evaporate the percolate until white fumes are evolved, add 1–2 ml of nitric acid to destroy any organic matter, again evaporate to fumes, dissolve the residue in water and dilute to volume.

Minerals, ores and rocks[14-22]

Dissolve the ore by treatment with hydrochloric acid followed by the fusion of any residual material with potassium hydrogen sulphate ($KHSO_4$). Take up the melt in hydrochloric acid. Re-fuse any residue with a mixture of sodium and potassium carbonates (fusion mixture, $Na_2CO_3 + K_2CO_3$ in equimolar proportions). Dissolve the melt in hydrochloric acid and combine all the solutions. Extract iron from this strongly acid solution with ether ($C_2H_5OC_2H_5$), make up to a known volume and determine vanadium on an aliquot[17]. Alternatively[16] the interference from iron can be prevented by masking with E.D.T.A. which is prevented from complexing with the vanadium by the addition of thorium nitrate ($Th(NO_3)_4$). Extract the vanadium complex with carbon tetrachloride (CCl_4), evaporate to dryness, take up the residue in nitric acid, dilute, and proceed by *Method 1*.

Other methods which have been suggested for the solution of samples of rocks are that described above for steel[14,15]; evaporation with a mixture of sulphuric and hydrofluoric acids[18,19]; fusion with sodium carbonate[20]; with a 4:1 mixture of sodium carbonate and magnesium oxide (MgO)[21]; and evaporation with mixed nitric, sulphuric and hydrofluoric acids[22].

Alternative methods suggested for the removal of interfering elements, especially iron, are extraction of the iron into isopentyl acetate ($CH_3COOCH_2CH_2CH(CH_3)_2$)[18]; and extraction into 8-hydroxyquinoline[20].

Miscellaneous[14]

Slag Use the method described for steel above.

Plastics and Polymers[23]

Ethylene-propylene rubbers Dry ash the sample in a platinum crucible, dissolve the ash in a mixture of nitric and phosphoric (H_3PO_4) acids and dilute the solution to a known volume.

Soil[24]

Clay Fuse the sample with sodium carbonate, extract the melt with water and filter. Adjust the pH of the filtrate to about 4.5 and extract with 8-hydroxyquinoline in chloroform. Evaporate the chloroform extract and calcine the vanadium complex. Fuse the residue with sodium carbonate, dissolve the melt in water, filter and make the filtrate up to a known volume.

Water[5,25-28]

No sample preparation is normally required unless it is necessary to concentrate the sample. The amount of iron present in the test solution should not exceed 2 ppm as effective suppression by means of pyrophosphate is impossible above this concentration. Fluorides, chlorides, bromides, iodides, sulphates, sulphites, nitrates and nitrites do not interfere, nor do moderate amounts of calcium, magnesium, sodium, potassium, manganese, lead or zinc. Copper interferes when present in concentrations comparable to the vanadium content. Interference by titanium may be prevented by the addition of sodium fluoride (NaF) which prevents the formation of the titanium complex[25,26].

In those cases where it is necessary to concentrate the vanadium one of the following methods can be used:- Acidify 0.5–1.0 litres of sample with 1–2 ml of 1:1 hydrochloric acid and add 1 ml of normal potassium hypochlorite solution (KOCl—mix 600 g of bleaching powder with 2 l of water add 1 l of 56% potassium carbonate solution and 250 ml of 20% caustic potash (KOH) solution and after 3–4 days syphon off the clear solution), boil the solution for 5 minutes, make alkaline to bromocresol purple with 5% aqueous ammonia (NH_4OH), collect the precipitated hydroxides, wash the precipitate with a 1% solution of ammonium nitrate (NH_4NO_3) make alkaline to bromocresol purple with aqueous ammonia, dry the precipitate and ignite in a platinum crucible. Fuse the residue with 0.5 g of sodium carbonate, extract the melt with 30–40 ml of hot water, filter, wash any residue with two portions of a 1% sodium carbonate solution and neutralise the combined filtrates to methyl orange with $4N$ sulphuric acid. Add 1.2 ml excess of $4N$ sulphuric acid and dilute to 50 ml[27]. Chan[5] recommends co-precipitation of vanadium with ferric hydroxide followed by cation exchange using 0.3% aqueous hydrogen peroxide as the eluting liquid. For sea water Riley[28] recommends the following method:- Filter 3 l of sample and adjust its pH to 5.0 ± 0.2 with $0.2N$ nitric acid. Prepare a 6×1 cm column of Chelex-100 resin (50–100 mesh) which has previously been digested with $2N$ nitric acid. Wash the column with 20 ml of $2N$ nitric acid and then with water until the pH of the percolate is 4.5. Pass the sample through the column at not more than 5 ml a minute, wash the column with 200 ml of water and then elute the vanadium with 24 ml of $2N$ aqueous ammonia (NH_4OH).

Separation from other ions[29]

From titanium Pass the sample solution at pH 1.0 through a column of cationite KY-3 (H^+ form). Pass a 10% solution of caustic soda (NaOH) through the column, when the vanadium is eluted. Titanium can then be eluted with $4N$ hydrochloric acid.

The individual test

Method 1 utilises the formation of the vanadium-8-hydroxyquinoline complex in dilute acetic acid solution. The presence of up to 100 μg of iron in the test solution can be tolerated, but if more than this is liable to be present then one of the procedures for the exclusion of iron must be used as mentioned above.

Tanaka[30] has shown that the colour of the complex increases with increasing concentration of the 8-hydroxyquinoline. It is therefore important to keep this concentration constant.

Kurmaiah[31] has reported that the complex can be extracted into a 4:1 mixture of benzene (C_6H_6) and butanol (C_4H_9OH) provided that the aqueous phase is $0.05M$ in either sulphuric or phosphoric acid. Beer's Law holds provided the vanadium concentration does not exceed 14 μg/ml. Iron can be masked by the addition of pyrophosphate ions.

References

1. W. Jaraczewska and M. Jakubowski, *Chemia analit.,* 1964, **9** (5), 969; *Analyt. Abs.,* 1966, **13**, 999
2. V. Nevoral and A. Okac, *Cslka. Farm.,* 1966, **15** (5), 229; *Analyt. Abs.,* 1967, **14**, 5796
3. H. Eschnauer, *Z. Lebensmitt.Untersuch,* 1959, **110**, 121; *Analyt. Abs.,* 1960, **7**, 2469
4. E. Arato Sugar and V. K. Falus, *Munkavedelem,* 1967, **13** (1–3), 35; *Analyt. Abs.,* 1969, **15**, 3464
5. K. M. Chan and J. P. Riley, *Analyt. chim. Acta,* 1966, **34** (3), 337; *Analyt. Abs.,* 1967, **14**, 4333
6. R. J. Nadalin and W. B. Brozda, *Analyt. Chem.,* 1960, **32**, 1141
7. J. E. Rayner, *J. Inst. Fuel,* 1964, **37**, 30; *Analyt. Abs.,* 1965, **12**, 2289
8. Z. Skorko-Trybula, *Nafta Krakow,* 1966, **22** (5), 141; *Analyt. Abs.,* 1967, **14**, 5505
9. I. Buzas, *Magy. Kem. Lap.,* 1968, **23** (12), 716; *Analyt. Abs.,* 1970, **18**, 2537
10. S. U. Kreingol'd and E. A. Bozhevol'nov, *Zavodskaya Lab.,* 1965, **31** (7), 784; *Analyt. Abs.,* 1966, **13**, 6180
11. T. S. Studenskaya, N. D. Fedorova, V. V. Stepin and V. L. Zolotavin, *Trudy. vses. nauchno-issled. Inst. Standartn. Obraztsov. Spektr. Etalonov.,* 1964, **1**, 22; *Analyt. Abs.,* 1966, **13**, 3393
12. V. L. Zolotavin and N. D. Fedorova, *Trudy. vses. nauchno-issled. Inst. Standartn. Obraztsov. Spektr. Etalonov.,* 1964, **1**, 31; *Analyt. Abs.,* 1966, **13**, 6178
13. V. A. Podchainova, A. V. Dolgorev and V. Ya. Dergachev, *Zavodskaya Lab.,* 1965, **31** (7), 790; *Analyt. Abs.,* 1966, **13**, 6223
14. I. M. Gol'tsberg, G. L. Koval' and G. A. Klemshov, *Sb. Trudy. ukr. nauchno-issled. Inst. Metall.,* 1965, (11), 387; *Analyt. Abs.,* 1967, **14**, 161

15. A. T. Pilipenko, E. A. Shpak and G. T. Kurbatova, *Zh. analit. Khim.*, 1967, **22** (7), 1014; *Analyt. Abs.*, 1969, **16,** 1874
16. A. W. Ashbrook and K. Conn. *Chemist Analyst,* 1961, **50** (2), 47; *Analyt. Abs.*, 1962, **9,** 667
17. D. N. Grindly, E. H. W. J. Burden and A. H. Zaki, *Analyst,* 1954, **79,** 95
18. K. P. Stolyarov and F. B. Agrest, *Zhur. analit. Khim.*, 1964, **19** (4), 457; *Analyt. Abs.*, 1965, **12,** 3727
19. V. Patrovsky, *Chemicke Listy,* 1966, **60** (11), 1545; *Analyt. Abs.*, 1968, **15,** 739
20. N. T. Tarakhanova, R. Kh. Dzhiyanbaeva and Sh. T. Talipov, *Trudy tashkentok. gos. Univ.*, 1967, (288), 58; *Analyt. Abs.*, 1968, **15,** 3295
21. R. Fuge, *Analyt. chim. Acta,* 1967, **37** (3), 310; *Analyt. Abs.*, 1968, **15,** 3296
22. P. G. Jeffrey and G. O. Kerr, *Analyst,* 1967, **92,** 763
23. A. J. Smith, *Analyt. Chem.*, 1964, **36** (4), 1944; *Analyt. Abs.*, 1965, **12,** 4035
24. F. Burriel-Marti, C. A. Herrero and F. Noriega, *Revta. port Quim.*, 1967, **9** (1), 1; *Analyt. Abs.*, 1969, **16,** 2432
25. R. Montequi and M. Gallego, *Anal. Soc. Esp. Fisiconium,* 1934, **32,** 134
26. J. M. Bach and R. A. Trelles, *Anal. Soc. Quim. Argentina,* 1940, **28,** 111
27. E. P. Mulikovskaya, *Trudy. vses. nauchno-issled. geol. Inst.*, 1964, **117,** 79; *Analyt. Abs.*, 1965, **12,** 4887
28. J. P. Riley and D. Taylor, *Analyt. chim. Acta,* 1968, **41** (1), 175; *Analyt. Abs.*, 1969, **17,** 1212
29. D. Shishkov and L. Shishkova, *Compt. rend. Acad. bulg. Sci.*, 1963, **16,** 833; *Analyt. Abs.*, 1964, **11,** 5440
30. M. Tanaka and I. Kojima, *Analyt. chim. Acta,* 1966, **36** (4), 522; *Analyt. Abs.*, 1968, **15,** 1360
31. N. Kurmaiah, D. Satyanarayana and V. Pandu Ranga Rao, *Talanta,* 1967, **14** (4), 495; *Analyt. Abs.*, 1968, **15,** 3894

Vanadium Method 1
using 8-hydroxyquinoline

Principle of the method

The procedure is based upon a qualitative test which consists in forming the reddish-brown, 8-hydroxyquinoline complex in dilute acetic acid solution, followed by extraction with amyl alcohol. For the purpose of the comparator disc, the published method has been modified (by the addition of sodium pyrophosphate), with a view to preventing the interference of iron.

Reagent required

1. *Sodium pyrophosphate* ($Na_4P_2O_7.10H_2O$) A 5 per cent. w/v aqueous solution
2. *Glacial acetic acid* (CH_3COOH)
3. *8-Hydroxyquinoline* A 2.5 per cent. w/v. solution in a 10 per cent. aqueous solution of acetic acid
4. *Amyl alcohol* ($C_5H_{11}OH$)

Analytical reagent grade chemicals should be used throughout.

The Standard Lovibond Comparator Disc 3/20

The disc covers the range from 10 μg to 100 μg of Vanadium (V) in steps of 10 μg omitting 90, (i.e. 0.2 to 2.0 parts per million if a 50 ml sample is taken).

Technique

To 50 ml of solution to be tested, contained in a separator, add 10 ml of a 5 per cent. aqueous solution of sodium pyrophosphate (reagent 1), followed by 0.2 ml of glacial acetic acid (reagent 2), and 0.2 ml of 8-hydroxyquinoline (reagent 3). Add 10 ml of amyl alcohol (reagent 4), and shake vigorously for one minute. Allow to separate, reject the aqueous layer, and transfer the organic layer to a graduated cylinder. Dilute to 10 ml with more amyl alcohol, and transfer to a 13.5 mm cell or test tube and place in the right-hand compartment of the comparator. In the left-hand compartment place a similar cell containing distilled water, and, with the comparator facing a standard source of white light such as the Lovibond White Light Cabinet or, failing this, north daylight, rotate the disc until a match is obtained. The figure shown indicates the amount of vanadium present in the 50 ml of sample taken; thus, if a colour equivalent to 50 μg is produced, the amount of vanadium present is 0.05 mg per 50 ml, or 1 part per million.

The Determination of Zinc

Introduction

Zinc is, after copper and aluminium, the most widely used non-ferrous metal. Its most common use is in the form of brass, but of almost equal importance are zinc protective coatings of steel. In atmospheric exposure zinc coatings corrode only one tenth as fast as iron. Zinc is also used in pigments; in rubber, where zinc oxide plays an active part in the vulcanization process; in pharmaceuticals: and zinc is also an essential trace element in most living matter.

The determination of zinc is, therefore, of importance in a wide variety of industries from metallurgy to the rubber industry, from pigments to pharmaceuticals.

Methods of sample preparation

Beverages[1,2]

Whisky[1] Shake a 5 ml sample with 2 ml of a 25% solution of sodium thiosulphate ($Na_2S_2O_3$) and 15 ml of acetate buffer (pH 4), followed by 10 ml of a 0.001% solution of dithizone in carbon tetrachloride (CCl_4). Separate the carbon tetrachloride layer, evaporate off the solvent, remove organic matter by wet ashing, evaporate the acid solution to dryness and take up the residue in a known volume of dilute sulphuric acid.

Wine and vinegar[2] Take 10 ml of sample or an amount containing about 10–50 μg of zinc, destroy the organic matter by digestion with nitric and perchloric ($HClO_4$) acids in the presence of sodium chloride (NaCl), evaporate to dryness, take up the residue in hydrochloric acid, re-evaporate and dissolve the residue in distilled water. Filter, make up to 25 ml and then determine zinc on this solution.

Biological materials[3-14]

Blood and serum[3-6] Wet ash the sample with nitric and perchloric ($HClO_4$) acids, and either[3] evaporate to dryness with hydrochloric acid and then take up the residue in sulphuric acid or[4] extract the zinc into dithizone and then back extract into dilute sulphuric acid. Alternatively[5] take 3 ml of serum, add 1.5 ml of N hydrochloric acid, mix and heat in a boiling water bath for 5 minutes. Cool, add 1.5 ml of 10% trichloracetic acid (CCl_3COOH) mix and centrifuge. Remove the supernatant liquid to a 10 ml volumetric flask, add 2 ml of N hydrochloric acid to the residue in the centrifuge tube, mix and re-centrifuge. Add the supernatant liquid to the original liquid in the volumetric flask, make up to 10 ml with N hydrochloric acid and determine zinc on this solution.

Cernikova[6] recommends Kjeldahl digestion, evaporation of the digest to dryness, solution of the residue in water, precipitation of heavy metals with sodium dihydrogen phosphate (NaH_2PO_4) at 100°C, solution of the precipitate in hydrochloric acid, evaporation to dryness, solution of the residue in butanol (C_4H_9OH) saturated with hydrochloric acid, ascending chromatography of this solution on Whatman No. 1 paper, location of zinc by dipping the paper in 0.5% 8-hydroxyquinoline solution and then exposing the paper to ammonia vapour, drying, elution of the zinc spot with 0.001% dithizone in chloroform ($CHCl_3$) and then back extraction into dilute sulphuric acid.

Tissue[7-13] Mix 10 g of sample with 3–4 g of calcium dihydrogen phosphate ($Ca(H_2PO_4)_2 \cdot H_2O$) and dry for 1 hour at 140°C followed by 1 hour at 400°C and 1 hour at 600°C. Finally continue heating at 900°C until ashing is complete. The zinc forms a stable polymer with the calcium phosphate and this prevents loss of zinc even at 900°C. Dissolve the cooled residue in 100 ml of $3N$ hydrochloric acid, make up to 250 ml and filter. Determine zinc on an aliquot of the filtrate[10]. Alternatively[9,11] digest sufficient sample to contain 10–50 μg of zinc with nitric acid–perchloric acid. Dilute and extract the zinc from solution at pH 5 with dithizone in chloroform ($CHCl_3$). Re-extract zinc from the chloroform layer into 0.1N hydrochloric acid and determine the zinc on this acid solution. Other extractants which have been suggested are diphenylthiocarbazone[7] and diethylthiocarbamate[8].

Kooi[12] recommends separation of the zinc from a concentrated hydrochloric acid solution of the tissue by means of Dowex-1 ion exchange resin, while D'yachenko[13] ashes the sample, at a temperature not exceeding 450°C and dissolves the ash in sulphuric acid.

Urine[14] Buffer the urine to pH 5.7 and extract four times with carbon tetrachloride (CCl_4) to remove any emulsifying substances. Extract the zinc with five 5 ml portions of 0.003% solution of dithizone in carbon tetrachloride. Combine the carbon tetrachloride layers in another separating funnel and extract with four 5 ml portions of $0.1N$ hydrochloric acid. Determine the zinc on this aqueous extract.

Chemicals[15]

Evaporate the sample with 20 ml of 57% perchloric acid ($HClO_4$) until copious white fumes are evolved, dilute the residue with a little water, evaporate and repeat this treatment three or four times. Finally evaporate nearly to dryness, dissolve the residue in 100 ml of water, filter to remove any silica and pass the solution through a column of cationite KU-2. Wash the column with water and desorb the zinc with 100 ml of $2N$ hydrochloric acid, evaporate to small volume and then dilute to 50 ml.

Dairy Produce

Cheese[16] Take a 10 g sample, dry overnight at 105°C, then gradually increase the heat until no more fumes are evolved. Moisten the black residue with a few drops of water and crush with a glass rod. Evaporate off the water at 105°C and then heat in a muffle furnace at 500–600°C for 1 hour. Add 10 ml of 25% phosphoric acid (H_3PO_4) and 5 ml of water and heat on a water bath. Filter off the carbon, wash the residue, combine the filtrate and washings and make up to 100 ml.

Food[17-20]

Fish[17] Clean the fish with distilled water, dry at 80°C for 48 hours, grind the dried sample and place 0.2 g in a size 0 crucible. Char the sample at 200°C then raise the temperature to 550°C and combust for 16 hours. Cool, dissolve the ash in 1:1 nitric acid and dilute to a known volume.

General[18,19] Ash a 5 g sample with 0.5 ml of sulphuric acid at 550°C. Dissolve the ash in 2.5 ml of N hydrochloric acid and dilute to 100 ml. Determine zinc on an aliquot of this solution.

Meat[20] Heat 2–3 g of homogenised sample with nitric acid in a Kjeldahl flask, add small amounts of hydrogen peroxide (H_2O_2), heating the mixture between successive additions, until the solution is colourless. Cool and dilute to a known volume.

Vegetables[20] Calcine 2–5 g of dried and ground sample with ammonium nitrate (NH_4NO_3) at 550–650°C. Dissolve the residue in 0.5–1.0 ml of $2M$ hydrochloric acid and dilute the solution to a known volume.

Fuels and lubricants[21-24]

Oils and oil additives Take 10 g of oil and dissolve it in 50 ml of ether ($C_2H_5OC_2H_5$). Shake with 30 ml of 4% hydrochloric acid, separate the aqueous layer and then extract the organic layer with three further 30 ml portions of 4% hydrochloric acid. Combine the aqueous extracts, evaporate to 100 ml to remove all the ether, dilute to 250 ml with water and determine zinc on an aliquot[21,22].

Alternatively either[23] carbonise 0.5–1.0 g of oil in a platinum crucible over a very low flame, moisten the residue with 2–3 drops of hydrochloric acid, remove excess hydrochloric acid by careful heating and then oxidise the carbon by heating at not more than 600°C, dissolve the ash in $6N$ hydrochloric acid and dilute the solution to 100 ml; or[24] wet ash 3 g of sample with sulphuric and nitric acids to which hydrogen peroxide (H_2O_2) is added as required until the solution is colourless, dilute to about 50 ml, add 2 g of tartaric acid (($CHOH.COOH)_2$) and then add aqueous ammonia (NH_4OH) until the solution is faintly coloured to phenolphthalein indicator, add 10 ml of a 20% solution of ammonium citrate (($NH_4)_3C_3H_5O_7$) and 5 ml of a 2% solution of sodium diethyldithiocarbamate (($C_2H_5)_2NCSSNa.3H_2O$) and extract with chloroform ($CHCl_3$), re-extract the zinc from the organic layer into $0.1N$ hydrochloric acid and determine by *Method 1*.

Metals, alloys and plating baths[25-41]

Alloys in general[25-29] Dissolve the sample in acid and then evaporate to dryness. Dissolve the residue in $0.12N$ hydrochloric acid containing 10% of sodium chloride (NaCl). Pass the solution through a column of De-acidite FF ion exchange resin, elute any copper with $2N$ hydrochloric acid and then elute the zinc with N nitric acid[25-28]. Alternatively for anti-friction alloys[29] dissolve a 1 g sample in nitric acid + hydrochloric acid, remove antimony and tin by precipitation with sodium sulphide (Na_2S) in tartaric acid–caustic soda solution at 100°C. Adjust the filtrate to contain 5% nitric acid and electrolyse to remove most of the copper and lead. Remove cadmium and bismuth, together with any residual traces of copper, lead and tin by precipitation with hydrogen sulphide (H_2S) in dilute sulphuric acid. Heat the final filtrate to fuming, dilute to 50 ml with water and determine zinc on an aliquot of this solution.

Aluminium and its alloys[30-33] Dissolve 0.5 g of the sample in warm hydrochloric acid, add 1 ml of $2M$ ammonium citrate (($NH_4)_3C_6H_5O_7$), adjust the pH to 8.5 with $6.5N$ aqueous ammonia (NH_4OH) and extract with 10 ml of a 0.01% dithizone solution in chloroform ($CHCl_3$). Save the organic layer and re-extract the aqueous layer with a further 5 ml of the dithizone solution. Combine the organic extracts, re-extract the zinc into $0.1N$ hydrochloric acid and determine by *Method 1*[31]. Alternatively[30] dissolve 0.3 g of sample in 10 ml of $6N$ hydrochloric acid and 2 ml of hydrogen peroxide (H_2O_2), add a further 20 ml of $6N$ hydrochloric acid, extract any iron which is present by shaking the solution with 20 ml of isobutyl methyl ketone (($CH_3)_2CHCH_2COCH_3$) and then shake the aqueous layer with two successive 20 ml portions of 15% Amberlite LA-1 solution in kerosine. Extract zinc from the combined organic phases with two 20 ml portions of $0.3N$ nitric acid.

Kharin[32] recommends complexing iron, aluminium, magnesium, manganese, chromium and silicon impurities with sodium pyrophosphate ($Na_4P_2O_7$), converting copper and zinc into complexes by the addition of aqueous ammonia and then passing the complexed solution through a column of KU-2 resin (NH_4^+ form). The zinc and copper complexes are retained on the resin while the pyrophosphate complexes pass through. The zinc can then be eluted with a 5% solution of caustic soda (NaOH) at 4 ml a minute until the eluate gives no reaction with dithizone solution. Rozycki[33] does not consider any prior separation to be necessary.

Chromium plating baths[34] Take a 10 ml sample of the plating solution and dilute to 100 ml. Mix a 10 ml portion of the dilute solution with 5 ml of 10% ammonium thiocyanate (NH_4SCN) solution and 1 g of ammonium fluoride (NH_4F). Adjust the pH to 1.5–2.5 by the addition of aqueous ammonia and then shake with 20 ml of isobutyl methyl ketone. Separate the organic phase and wash with three 5 ml portions of 5% ammonium thiocyanate containing 0.5 g of ammonium fluoride, and then back extract the zinc with aqueous ammonia–$0.04M$ ammonium chloride (NH_4Cl) buffer at pH 10 (2 × 25 ml).

Electrolytic copper[35] Dissolve 1–3 g of the sample in 20 ml of aqua regia (3 HCl + 1 HNO_3), evaporate the solution to a small volume and remove any remaining nitric acid by twice evaporating to dryness after the addition of hydrochloric acid. Dissolve the residue in 20–30 ml of $2M$ hydrochloric acid and pass the solution through a column of anionite EDE-10P at 1 ml a minute. Elute the copper with six or seven 10 ml portions of $2M$ hydrochloric acid and then elute the zinc with 100 ml of $0.02M$ hydrochloric acid at 1.5 ml a minute.

Iron and steel[36,37] Dissolve 1 g of sample in 50 ml of $2M$ hydrochloric acid, pass the solution through a column of Dowex 1-X8 resin, wash the column with 10–25 ml of $0.5M$ hydrochloric acid and then elute the zinc with 25 ml of $0.005M$ nitric acid[36]. Alternatively[37] dissolve 2 g of the sample in aqua regia (3 HCl + 1 HNO_3), evaporate the solution with hydrochloric acid and dissolve the residue in 50 ml of 3:7 hydrochloric acid:water. Filter and pass the solution through a column of EDE-10P resin (H^+ form), wash free from nickel, chromium, vanadium, iron, aluminium and manganese with 200–250 ml of $2N$ hydrochloric acid, elute the zinc with 200 ml of $0.02N$ hydrochloric acid and dilute the eluate to 250 ml.

Lead[38] Dissolve a 5 g sample in 25 ml of 1:4 nitric acid and dilute the solution to 200 ml. Take a 50 ml aliquot and heat this with $4M$ aqueous ammonia (NH_4OH). Adjust the pH to about 4 with acetic acid (CH_3COOH) and then boil with 50 ml of a $0.5M$ solution of sodium thiosulphate ($Na_2S_2O_3$) until the precipitate turns brown. Cool, filter, pass the filtrate through Amberlite IR-120 ion exchange resin. Wash the column with $0.05M$ sodium thiosulphate until the eluate is free from lead (test with Eriochrome black T) and then elute the zinc with 90 ml of $2M$ sodium chloride (NaCl) solution.

Nickel, its alloys and electrolyte solutions[39,40] Dissolve 1–5 g of sample in 20 ml of 1:1 nitric acid and evaporate to dryness with 20 ml of hydrochloric acid. Dissolve the residue in 50 ml of

$2M$ hydrochloric acid and pass the solution through a 20 x 1 cm column of ANEX-L resin which has been washed before use with 100 ml of $2M$ hydrochloric acid at a rate of 30 drops a minute. After passing the sample solution, wash the column with $2M$ hydrochloric acid until the percolate gives a negative reaction for nickel and then eluate zinc with $0.1M$ hydrochloric acid or, in the absence of cadmium, with water[39]. For the analysis of nickel electrolytes adjust the acidity to be $2M$ in hydrochloric acid before passing the sample through the column. Alternatively[40] dissolve up to 300 mg of sample, which should contain less than 15 μg of zinc, in 2 ml of 50% nitric acid, evaporate with 2 ml of hydrochloric acid and dissolve the residue in 5 ml of water.

Add 20 mg of ascorbic acid (O.CO.C(OH):C(OH).C H.CH(OH).CH$_2$OH) and wash the solution into a separating funnel with 5 ml of water. Add 10 ml of 50% hydrochloric acid and 5 ml of a 5% solution of dioctylmethylamine ((CH$_3$(CH$_2$)$_7$)$_2$NCH$_3$) in xylene ((CH$_3$)$_2$C$_6$H$_4$), shake the funnel, separate the organic layer, wash this layer with two 10 ml portions of 25% hydrochloric acid and then with 10 ml of a 4% aqueous solution of caustic potash (KOH). Run the aqueous layer into a beaker containing about 0.5 g of ammonium chloride (NH$_4$Cl) and dilute to a known volume with water.

Silver[41] Dissolve sufficient sample to contain 50 – 150 μg of zinc in 5 ml of 1:1 nitric acid and dilute with water to 100 ml. To a 10 ml aliquot of this solution add 25 ml of a mixture of sodium thiosulphate (Na$_2$S$_2$O$_3$) and potassium cyanide (KCN) solutions, to mask other ions, extract with dithizone solution in chloroform (CHCl$_3$), re-extract the zinc into $0.1N$ hydrochloric acid and proceed by *Method 1*.

Minerals, ores and rocks[31,42-46]

Dissolve 1 g of powdered sample into 20 ml of 1:1 nitric acid and then heat to white fumes with 10 ml of 1:1 sulphuric acid and 10 ml of hydrobromic acid (HBr.) Cool and dissolve the residue in 50 ml of water. Filter and evaporate the filtrate to dryness. Dissolve the residue in 50 ml of $0.12N$ hydrochloric acid containing 10% sodium chloride (NaCl) and pass this solution through a column of Amberlite IRA-140 ion exchange resin at 5 ml per minute. Wash the column with the same hydrochloric acid – sodium chloride solution and then elute zinc with N aqueous ammonia (NH$_4$OH) containing 2% of ammonium chloride (NH$_4$Cl)[42-45]. Alternatively either use the technique described earlier for the analysis of aluminium[31] or decompose 0.5 g of a finely ground sample with 30 ml of a 1:1:1 mixture of water and hydrochloric and hydrofluoric acids, add 10 ml of perchloric acid (HClO$_4$) and 5 ml of sulphuric acid, evaporate to 3 ml on a hot plate, cool, repeat the heating with 5 ml of perchloric acid, add 10 ml of hydrochloric acid and 25 ml of water, digest on a steam bath for 30 minutes, cool and dilute to 100 ml, pass the solution through a column of Dowex 1X-8 resin (Cl$^-$ form), pre-conditioned with $1.2M$ hydrochloric acid, and elute the zinc with 45 ml of $0.01M$ hydrochloric acid[46].

Miscellaneous[9,47-56]

Fertilizers[9] Wet ash the sample and then use the dithizone method to extract the zinc, as described for plants above.

Meteorites[47] Take a 1 g sample and dissolve it in either hydrochloric acid – nitric acid (for iron meteorites) or sulphuric acid – hydrofluoric acid (for silicon meteorites). Evaporate this solution to dryness and fuse the residue with sodium carbonate (Na$_2$CO$_3$). Dissolve the melt in $2N$ hydrochloric acid and pass this solution through a column of Dowex 1X-8 ion exchange resin. Elute the column with 50 ml of $2N$ hydrochloric acid and discard this eluate, unless it is required to determine metals other than zinc, then elute the zinc with 25 ml of $0.001N$ hydrochloric acid.

Pharmaceuticals[48-52] Heat a 2.5 g sample with 20 ml of chloroform (CHCl$_3$) and 30 ml of 12.5% hydrochloric acid until solution is complete. Add 20 ml of hot water, separate the chloroform layer and filter the aqueous layer into a 100 ml volumetric flask. Dilute to 100 ml and determine zinc on this solution.[48-51] If any dyes are present in the pharmaceutical preparation these can be removed by adsorption onto active charcoal[50]. Alternatively[52] wet oxidise the sample with a mixture of sulphuric acid, perchloric acid (HClO$_4$) and hydrogen peroxide (H$_2$O$_2$).

Phosphors[53] Dissolve about 50 mg of sample in hydrochloric acid and evaporate to dryness. Repeat to destroy all the sulphide, dissolve the residue in a small amount of water, treat the solution with caustic soda (NaOH) until a faint turbidity appears and then add hydrochloric acid until the turbidity just disappears. Dilute to 50 ml and pass this solution through a 16 mm diameter column of Dowex 50W-X8 (H^+ form). Elute cadmium with 100 ml of $0.5N$ hydrochloric acid and then elute zinc with 60 ml of $2.5N$ hydrochloric acid.

Photographic gelatin[54] Dissolve the gelatin in water, pass the solution through a column of Dowex 50W resin and elute the metals with $3N$ hydrochloric acid. Extract copper into a 0.005% solution of dithizone in carbon tetrachloride (CCl_4), then add sodium acetate (CH_3COONa) and sodium thiosulphate ($Na_2S_2O_3$) and extract the zinc into 0.005% dithizone in carbon tetrachloride. Separate the organic layer, evaporate off the solvent, destroy organic matter by digestion with sulphuric and nitric acids, evaporate to dryness, take up the residue in water and make up to a known volume.

Wood preserving solutions[55,56] Remove chromium from the sample by extraction with isobutyl methyl ketone (($CH_3)_2CHCH_2COCH_3$) in hydrochloric acid (1 + 1) precipitate copper with cupferron, filter, add glycine to the filtrate, adjust the pH to 7.0 ± 0.5 and precipitate the zinc as phosphate. Filter off the precipitate and dissolve in dilute sulphuric acid. Determine zinc on this solution.

Plants[9,57-60]

Remove organic matter by wet or dry ashing, taking care to keep the temperature below 600°C to prevent loss of zinc. Take up the residue in $2-6M$ hydrochloric acid and pass the solution through a column of Dowex-1 ion exchange resin. Elute the zinc with $0.005M$ hydrochloric acid [57,58]. Alternatively[9] extract the zinc from the acid solution, buffered to pH 5 with acetate buffer (sodium acetate 300 g, glacial acetic acid 10 ml and citric acid 30 g to 1 litre to water), with 30 ml of a 0.05% solution of dithizone in toluene ($CH_3.C_6H_5$). Separate the toluene layer, wash with 20 ml of a 1% sodium sulphide (Na_2S) solution and then three times with water. Extract the zinc into the aqueous phase with $0.08N$ hydrochloric acid and determine. Bradfield[59] and Page[60] both recommend that wet ashing is used for the removal of organic matter.

Plastics and polymers[61,65]

Destroy the organic matter by wet or dry ashing and dissolve the residue in hydrochloric acid. The zinc may be determined either directly on this ash solution[63] or the zinc may be separated by passing the ash solution through a column of ion exchange resin[61,62,65]. Either use Wofatit L150 resin and elute zinc with $0.05M$ hydrochloric acid[33,34] or use EDE-10P resin, elute aluminium, iron, calcium and magnesium with $4N$ hydrochloric acid and then elute zinc with water[37]. For vulcanized rubber[64] ignite a 1-2 g sample below 550°C, dissolve the residue in 30 ml of $5N$ hydrochloric acid, filter off silicon, barium, titanium and lead; boil with nitric acid and precipitate iron and aluminium with aqueous ammonia (NH_4OH). Filter and make the filtrate $2N$ in hydrochloric acid and then pass this solution through a column of Amberlite IRA-140 ion exchange resin. Elute the zinc with $0.1N$ nitric acid.

Sewage and industrial wastes[66-68]

Remove organic matter from the sample by wet oxidation. Then either[66,68] evaporate to dryness, ignite at 300°C and dissolve the residue in hydrochloric acid, or[67] extract the solution at pH 5.5 with a chloroform ($CHCl_3$) solution of dithizone, and extract the zinc from the chloroform layer into dilute hydrochloric acid.

Soil[9,69-76]

For "acid extractable zinc"[72] shake 2 g of air-dried soil in a glass stoppered centrifuge tube with 20 ml of $0.1N$ hydrochloric acid for 5 minutes. Centrifuge, and extract the residual soil with two further 20 ml portions of acid. Combine the extracts, make up to 100 ml with distilled water and determine zinc on this solution. For "total zinc"[71] take a 1 g sample of soil, ground to pass through a 100 mesh sieve, and fuse with 3 g of sodium carbonate (Na_2CO_3). Dissolve the melt in 100 ml of N hydrochloric acid. Evaporate to dryness and then add 20 ml of $3N$ hydrochloric acid and repeat the evaporation. Heat the residue with a further 20 ml of $3N$ hydrochloric acid, filter the solution into a 50 ml volumetric flask and dilute to the mark. Determine zinc on an aliquot of this solution. Verigina[73] uses sulphuric and hydrofluoric (HF) acids to decompose the sample. Koter[74] extracts the soil with N-potassium chloride

(KCl), Stewart[75] prefers $2M$ magnesium chloride ($MgCl_2$) as the extractant while Mizuno[76] recommends fusion with sodium carbonate as above.

Water[69,77-83]

In many cases no sample preparation will be necessary. Where the concentration of zinc is too low for it to be determined directly, concentration may be effected either by co-precipitation with calcium carbonate ($CaCO_3$), filtering off the precipitate and dissolving it in acid[78] or by extraction with diethyldithiocarbamate[69,79] or dithizone[80] followed by extraction of the zinc into acid solution. The determination of zinc in condensate[80] is typical of these extraction methods. Take 1 litre of condensate in a 1,200 ml separating funnel, add 25 ml of 4% ammonium citrate (($NH_4)_3C_6H_5O_7$) solution, 8 ml of aqueous ammonia (NH_4OH) and 15 ml of 0.01% dithizone in chloroform ($CHCl_3$). Shake for 5 minutes. Decant the chloroform layer, add a further 5 ml of dithizone solution and re-extract. Combine the extracts and shake with 10 ml of $0.02N$ hydrochloric acid for 5 minutes. Filter off the aqueous layer, make up to 25 ml and determine zinc on an aliquot of this solution.

Gurkina[81], and Sadilkova[83] also recommend extraction into dithizone[81] or diethyldithiocarbamate[83] to concentrate the zinc.

Fel'dman[82] recommends that water be sampled only into glass vessels, that analysis be carried out immediately after sampling, that filter papers are *not* used for filtration, and that all reagents be purified by extraction with dithizone.

Separation from other ions[31,84,85]

Separation from aluminium, iron, gallium, copper, silicon, titanium, calcium, magnesium and sodium has already been described under the analysis of aluminium[31].

From iron, cadmium and uranium[84] Apply solution containing between 0.1–0.3 milli-equivalents of the ions to a column (10×0.8 cm) of zirionium phosphate resin (NH_4^+ form), and elute cadmium with $2.5N$ ammonium chloride (NH_4Cl) followed by zinc with $5N$ ammonium chloride.

From manganese, cobalt and nickel[85] Pass the solution in $5N$ hydrochloric acid in 60% methanol (CH_3OH) through a 10×1.2 cm column of Amberlite CG 400 (Cl^- form), when nickel is not absorbed, manganese is eluted with $10N$ hydrochloric acid, cobalt with $5N$ hydrochloric acid and finally zinc with $10N$ hydrochloric acid.

From all other metals[86] It is claimed[86] that after masking other metals with a mixture of sodium thiosulphate ($Na_2S_2O_3$) and potassium cyanide (KCN) extraction of zinc with dithizone effects complete separation.

The individual method

Method 1 This method is based on Houghton's modification[77] of the Brilliant Green procedure originally developed by Hermanowicz and Sikorowska[87]. It covers the range 0–50 µg of zinc.

References

1. R. A. Nelson and M. P. Pro, *J. Ass. off. agric. Chem.*, 1957, **40**, 1100; *Analyt. Abs.*, 1958, **5**, 2790
2. J. Deshusses and J. Vogel, *Pharm. Acta Helv.*, 1962, **37**, 401; *Analyt. Abs.*, 1963, **10**, 799
3. B. Tvaroha and O. Mala, *Mikrochim. Acta*, 1962, (4), 634; *Analyt. Abs.*, 1962, **9**, 4819
4. J. Vogel and D. Monnier, *Mitt. Lebensmitt. Hyg. Bern.*, 1961, **52**, 539; *Analyt. Abs.*, 1963, **10**, 705
5. L. A. Williams, J. S. Cohen and B. Zak, *Clin. Chem.*, 1962, **8**, 502; *Analyt. Abs.*, 1963, **10**, 1897
6. M. Cernikova and B. Konrad, *Biochim. Biophys. Acta*, 1963, **71** (1), 190; *Analyt. Abs.*, 1964, **11**, 246
7. G. Weitzel and A. M. Fretzdorff, *Hoppe-Seyl. Z.*, 1953, **292**, 212; *Analyt. Abs.*, 1954, **1**, 354
8. J. A. Stewart and J. C. Bartlet, *Analyt. Chem.*, 1958, **30**, 404
9. K. Scharrer and H. Munk, *Z. Pfl Ernähr. Dung.*, 1956, **74**, 24; *Analyt. Abs.*, 1959, **6**, 378
10. W. A. C. Campen and H. Dumoulin, *Chem. Weekbl.*, 1959, **55**, 623; *Analyt. Abs.*, 1960, **7**, 3647

11. K. O. Baker, *Z. analyt. Chem.*, 1960, **173**, 57; *Analyt. Abs.*, 1960, **7**, 4400
12. P. V. Kooi, B. Heagan and J. T. Lowman *Proc. Soc. Exp. Biol. Med.*, 1963, **113**, 772; *Analyt. Abs.*, 1964, **11**, 4440
13. R. A. D'yachenko and N. P. D'yachenko, *Ukr. Biokhim. Zhur.*, 1964, **36** (5), 791; *Analyt. Abs.*, 1965, **12**, 6586
14. J. H. R. Kagi and B. L. Vallee, *Analyt. Chem.*, 1958, **30**, 1951
15. V. V. Stepin, V. I. Ponosov, G. N. Emasheva and N. A. Zobnina, *Trudy. vses. nauchno. -issled. Inst. standart. Obraztsov*, 1968, **4**, 100; *Analyt. Ans.*, 1970, **18**, 900
16. W. Schwabe, *Dtsch. Levensmitt Rdsch.*, 1955, **51**, 245; *Analyt. Abs.*, 1957, **4**, 1345
17. G. A. Knauer, *Analyst*, 1970, **95**, 476
18. R. Kalinowska and J. Siedlecka, *Roczn. Panstw. Zakl. Hig.*, 1955, **6**, 93; *Analyt. Abs.*, 1957, **4**, 403
19. A. C. Francis and A. J. Pilgrim, *Analyst*, 1957, **82**, 289
20. G. Balogh and E. Felszeghy, *Studia, Univ. Babes-Bolyai, Ser. Chem.*, 1967, **12** (1), 97; *Analyt. Abs.*, 1968, **15**, 6930
21. W. Fisher, *J. Inst. Pet.*, 1962, **48**, 290; *Analyt. Abs.*, 1963, **10**, 2781
22. M. Palovcikova, *Ropa Uhlie*, 1962, **4** (4), 114; *Analyt. Abs.*, 1963, **10**, 3758
23. K. S. Anand, P. Dayal and O. N. Anand, *Z. analyt. Chem.*, 1968, **239** (1), 33; *Analyt. Abs.*, 1969, **17**, 3567
24. S. Hypta, *Chemia analit.*, 1968, **13** (5), 1141; *Analyt. Abs.*, 1970, **18**, 306
25. A. M. Amim, *Chemist Analyst*, 1956, **45**, 95, 101; *Analyt. Abs.*, 1957, **4**, 2116
26. L. M. Budanova and T. V. Matrosova, *Zavodskaya Lab.*, 1961, **27**, 661; *Analyt. Abs.*, 1962, **9**, 569
27. M. Freegarde, *Metallurg. Manchr.*, 1958, **58**, 261; *Analyt. Abs.*, 1959, **6**, 4301
28. A. I. Lazarev and V. I. Lazareva, *Zavodskaya Lab.*, 1959, **25**, 542; *Analyt. Abs.*, 1960, **7**, 905
29. P. Reboul, *Chim. Anal.*, 1961, **53**, 408; *Analyt. Abs.*, 1962, **9**, 1383
30. M. Ishibashi and H. Komaki, *Japan Analyst*, 1962, **11**, 43; *Analyt. Abs.*, 1964, **11**, 55
31. D. Monnier and G. Prod'hom, *Analyt. chim. Acta*, 1964, **30** (4), 358; *Analyt. Abs.*, 1965, **12**, 3243
32. A. N. Kharin and N. N. Soroka, *Zhur. Prikl. Khim.*, 1964, **37**, 672; *Analyt. Abs.*, 1965, **12**, 3244
33. C. Rozycki, *Chemia analit.*, 1969, **14** (3), 459; *Analyt. Abs.*, 1970, **19**, 1093
34. N. Tajima and M. Kurobe, *Japan Analyst*, 1960, **9**, 612; *Analyt. Abs.*, 1962, **9**, 3090
35. R. G. Pats, N. V. Lukashenkova, E. D. Shuvalova and T. V. Zaglodina, *Zavodskaya Lab.*, 1968, **34** (10), 1173; *Analyt. Abs.*, 1970, **18**, 780
36. T. Okubo and F. Uehara, *Japan Analyst*, 1962, **11**, 761; *Analyt. Abs.*, 1964, **11**, 2613
37. A. P. Savranskaya and L. A. Fartushnaya, *Zavodskaya Lab.*, 1969, **35** (5), 556; *Analyt. Abs.*, 1970, **19**, 238
38. T. Katsura, *Japan Analyst*, 1961, **10**, 1323; *Analyt. Abs.*, 1963, **10**, 4110
39. V. Lehecka, *Hutn. Listy.*, 1964, **19** (7), 516; *Analyt. Abs.*, 1965, **12**, 5677
40. T. R. Andrews and P. N. R. Nichols, *Analyst*, 1965, **90**, 161
41. Z. Skorko-Trybala and J. Chivastowska, *Chem. Anal. Warsaw*, 1963, **8**, 859; *Analyt. Abs.*, 1965, **12**, 1091
42. L. F. Rader, W. C. Swadley and H. H. Lipp, *U.S. Geol. Survey Profess. Paper No. 400B*, 1960, 477; *Analyt. Abs.*, 1961, **8**, 2313
43. D. H. Wilkins, *Analyt. Chim. Acta*, 1959, **20**, 271; *Analyt. Abs.*, 1959, **6**, 4420
44. M. Hisada and K. Kashikawa, *Japan Analyst*, 1959, **8**, 235; *Analyt. Abs.*, 1960, **7**, 1680
45. K. Itsuki and M. Nagao, *Japan Analyst*, 1960, **9**, 836; *Analyt. Abs.*, 1962, **9**, 3091
46. C. Huffman, H. H. Lipp and L. F. Rader, *Geochim. Cosmoch. Acta*, 1963, **27**, 209; *Analyt. Abs.*, 1964, **11**, 1223
47. M. Nishimura and E. B. Sandell, *Analyt. Chim. Acta*, 1962, **26**, 242; *Analyt. Abs.*, 1962, **9**, 4098
48. B. Schmitz, *Dtsch. Apoth. Ztg.*, 1957, **97**, 399; *Analyt. Abs.*, 1958, **5**, 695
49. M. B. Shchigol and N. B. Burchinskaya, *Aptechnoe Delo*, 1958, **7**, 48; *Analyt. Abs.*, 1959, **6**, 721
50. P. M. Parikh, D. J. Vadodaria and S. P. Mukherji, *Indian J. Pharm.*, 1960, **22**, 229; *Analyt. Abs.*, 1961, **8**, 1218

51. B. Schmitz, *Dtsch. Apoth. Ztg.*, 1961, **101**, 1673; *Analyt. Abs.*, 1962, **9**, 2917
52. J. Hollos and I. Horvath-Gabai, *Pharmazie*, 1965, **20** (4), 207; *Analyt. Abs.*, 1966, **13**, 4367
53. H. G. Meyer, *Z. analyt. Chem.*, 1969, **244** (6), 394; *Analyt. Abs.*, 1970, **18**, 3781
54. J. Saulnier, *Sci. ind. Photogr.*, 1964, **35**, 1; *Analyt. Abs.*, 1965, **12**, 4037
55. W. J. Wilson, *Analyt. Chim. Acta*, 1956, **15**, 508; *Analyt. Abs.*, 1957, **4**, 1140
56. W. J. Wilson, *Analyt. Chim. Acta*, 1957, **16**, 419; *Analyt. Abs.*, 1957, **4**, 3868
57. R. K. Jackson and J. G. Brown, *Proc. Amer. Soc. Hort. Sci.*, 1956, **68**, 1; *Analyt. Abs.*, 1958, **5**, 282
58. T. L. Yuan and J. G. A. Fiskell, *J. Assoc. offic. agric. Chem.*, 1958, **51**, 424; *Analyt. Abs.*, 1959, **6**, 1003
59. E. G. Bradfield, *J. Sci. Food Agric.*, 1964, **15**, 469; *Analyt. Abs.*, 1965, **12**, 5957
60. E. R. Page, *Analyst*, 1965, **90**, 435
61. F. Oehlmann, *Plaste u. Kautsch*, 1957, **4**, 183; *Analyt. Abs.*, 1958, **5**, 3420
62. F. Oehlmann, *Chem. Prumysl.*, 1958, **8**, 44; *Analyt. Abs.*, 1958, **5**, 3814
63. M. Piazzi, *Ann. Chim. Roma*, 1960, **50**, 1176; *Analyt .Abs.*, 1961, **8**, 2533
64. F. Fujita, H. Matsushita and H. Omori. *J. Soc. Rubber Ind. Japan*, 1960, **33**, 661; *Analyt. Abs.*, 1961, **8**, 3845
65. N. M. Vitalskaya and N. F. Pantaeva, *Kauchuk. i. Rezina*, 1962, (6), 53; *Analyt. Abs.*, 1963, **10**, 2811
66. A. A. Christie, J. R. W. Kerr, G. Knowles and G. F. Lowden, *Analyst*, 1957, **82**, 336
67. Joint A.B.C.M. -S.A.C. Ctte. on Methods for Anal. of Trade Effluents, *Analyst*, 1957, **82**, 443
68. E. V. Mills and B. L. Brown, *Analyst*, 1964, **89**, 551
69. A. D. Miller and R. I. Libina, *Zh. analyt. Khim.*, 1958, **13**, 664; *Analyt. Abs.*, 1959, **6**, 2792
70. A. E. Martin, *Analyt. Chem.*, 1953, **25**, 1853
71. P. F. Pratt and G. R. Bradford, *Proc. Soil Sci. Soc. Amer.*, 1958, **22**, 399; *Analyt. Abs.*, 1960, **7**, 286
72. J. L. Nelson, L. C. Boawn and F. G. Viets, *Soil Sci.*, 1959, **88**, 275; *Analyt. Abs.*, 1961, **8**, 2195
73. K. V. Verigina, *Mikroelementy v. S.S.S.R.*, 1963, (5), 50; *Analyt. Abs.*, 1965, **12**, 2024
74. M. Koter, A. Krauze and B. Bardzicka, *Chemia analit.*, 1965, **10** (6), 1247; *Analyt. Abs.*, 1967, **14**, 1745
75. J. A. Stewart and K. C. Berger, *Soil Sci.*, 1965, **100**, 244; *Analyt. Abs.*, 1967, **14**, 4344
76. N. Mizuno, T. Ogata and K. Takao, *Japan Analyst*, 1969, **18** (9), 1077; *Analyt. Abs.*, 1970, **19**, 5216
77. G. U. Houghton, *Proc. Soc. Water Treatment Exam.*, 1957, **6**, 60; *Analyt. Abs.*, 1958, **5**, 3538
78. V. B. Aleshovskii, A. D. Miller and E. A. Sergiev, *Trudy Komiss. Anal. Khim. Akad. Nauk. S.S.S.R.*, 1958, **8**, 217; *Analyt. Abs.*, 1959, **6**, 2369
79. T. D. Kolesnikova, *Gidrokhim. Materialy*, 1961, **32**, 165; *Analyt. Abs.*, 1962, **9**, 2942
80. N. S. Litvinova, *Elektr. Stantsii*, 1961, (9), 81; *Analyt. Abs.*, 1962, **9**, 4596
81. T. V. Kurkina and A. M. Igoshin, *Zh. analit. Khim.*, 1965, **20** (7), 778; *Analyt. Abs.*, 1967, **14**, 1730
82. M. B. Fel'dman and E. P. Nakhshina, *Gidrobiol. Zh.*, 1967, **3** (1), 86; *Analyt. Abs.*, 1968, **15**, 484
83. M. Sadilkova, *Mikrochim. Acta*, 1968, (5), 934; *Analyt. Abs.*, 1970, **18**, 624
84. I. Gal and N. Peric, *Mikrochim. ichnoanalyt. Acta*, 1965, (2), 251; *Analyt. Abs.*, 1966, **13**, 2838
85. L. Mazza, T. Frache, A. Dadone and R. Ravasio, *Gazz. chim. ital.*, 1967, **97**, 1551; *Analyt. Abs.*, 1968, **15**, 5224
86. H. Fischer and G. Leopoldi, *Z. analyt. Chem.*, 1937, **107**, 241; *Chem. Abs.*, 1937, **31**, 969
87. W. Hermanowicz and C. Sikorowska, *C. Przem. Chem.*, 1951, **7**, 353; *Water Pollution Res. Abst.*, 1955, 841

Zinc Method 1
using Brilliant Green

Principle of the method

The method, using brilliant green, is based on Houghton's modification of the method of Hermanowicz and Sikorowska. It was developed specifically for the determination of zinc in water. Any ferric iron is inhibited with ammonium fluoride, and at pH 1.7 the interference from copper is slight and is almost entirely prevented by the addition of sodium diethyldithiocarbamate. The colour produced is measured by comparison with a series of Lovibond permanent glass colour standards.

Reagents required

1. *Sulphuric acid* 3.75N Add 100 ml of analytical reagent quality H_2SO_4 slowly, with stirring, to 800 ml of distilled water. When cool dilute with distilled water to 1 litre.
2. *Fluoride-thiocyanate reagent* Dissolve 30 g of ammonium thiocyanate (NH_4SCN), of analytical reagent quality, and 5 g of similar quality ammonium fluoride (NH_4F) in sufficient distilled water to make 100 ml. Store in a polythene bottle.
3. *Acacia solution* Add 3 g of finely crushed or powdered acacia to 100 ml of hot distilled water and dissolve the powder by boiling. Cool, filter, and add 0.1 ml of toluene ($CH_3.C_6H_5$) as a preservative.
4. *Carbamate solution* Dissolve 0.1 g of sodium diethyldithiocarbamate, analytical reagent quality, in 100 ml of distilled water. **This reagent must be freshly prepared each day.**
5. *Brilliant green solution* Dissolve 0.1 g of brilliant green C.I.42040 in 100 ml of cold anhydrous industrial methylated spirits.

The Standard Lovibond Comparator Disc 3/69

This disc covers the range 0–50 µg. (0–0.05 mg) of zinc (Zn) in steps of 0, 2.5, 5.0, 10.0, 15, 20, 30, 40, 50 µg. This represents for example, a range 0–2 ppm if 25 ml sample is used, or 0–10 ppm with a 5 ml sample.

Technique

Measure a suitable volume, not exceeding 30 ml, of the sample solution, which should be approximately neutral, into a Nessler cylinder and make up the volume to 30 ml with distilled water if necessary. Add 2 ml of sulphuric acid (reagent 1) and mix well. Add 2 ml of fluoride thiocyanate (reagent 2), mix again and then allow to stand for 3 minutes. Cool to $20\pm1°C$ and maintain the solution at this temperature for the remainder of the test. Add 5 ml of acacia solution (reagent 3) mix and then add 2 ml of carbamate solution (reagent 4), mix again and immediately add 1.0 ml of brilliant green solution (reagent 5). Mix well, dilute to the 50 ml mark with distilled water, mix again and allow to stand for 12 minutes. Transfer 10 ml of this solution to a comparator test tube and place this in the right-hand compartment of the comparator. Place an identical tube filled with distilled water in the left-hand compartment. Place the comparator before a standard source of white light, such as the Lovibond White Light Cabinet or, failing this, north daylight, and compare the colour of the test solution with the standards in the disc. The comparison should be made exactly 15 minutes after the addition of the brilliant green solution. The figure shown in the indicator window when a match is found represents micrograms of Zinc in the volume of sample used for the test, e.g. 10 µg on a 25 ml sample represents a concentration of 0.4 ppm of Zinc.

A blank should be determined on each batch of chemicals used in preparing the reagents, by repeating the above procedure using 30 ml of distilled water as the sample. The blank figure should be recorded and subtracted from the readings obtained on samples under test.

Notes

1. It is essential to carry out the test at the temperature stated.
2. The colours produced fade on standing, and readings should be taken exactly at the times prescribed above.

Biochemistry

Pathology

and Pharmacology

Foreword

This section contains details of tests which can be carried out with a minimum of apparatus to be found in the smaller laboratory, and in many cases these are simple enough to be used as side-room methods when the main laboratory in a large hospital is not manned.

It is not suggested that the Lovibond Comparator will take the place of the highly sophisticated automatic equipment available in large laboratories, but even here there are occasions when it is uneconomic to set up and calibrate such apparatus for one or two tests, and the Lovibond Comparator is reliable and ready for instant use.

For smaller hospitals, in private practice, and for field use, the comparator is the method of choice. A much simplified handbook, covering the procedure for the more frequently required medical tests, is published by The Tintometer Limited under the title "Simple Colorimetric Analysis for Small Medical Laboratories" by Dr. D. Stansfield; this is specially written with the less highly skilled technician in mind.

The Determination of α-Amylase in Blood or Urine
(Amyloclastic method)

Introduction

The pancreas contains a high concentration of α-amylase. When it suffers damage this is released into the interstitial tissue, from there into the blood, to be finally excreted by the kidney.

The determination of serum α-amylase is used mainly in the diagnosis of acute pancreatitis. In this condition, there is a rapid increase in serum amylase activity, levels in excess of 1,000 Somogyi units per 100 ml of serum being seen occasionally. In the presence of normal renal function this is rapidly cleared by the kidney, which results in greatly increased levels of amylase (diastase) activity in the urine.

Raised levels of amylase are seen sometimes in other abdominal conditions, e.g. intestinal obstruction and acute peritonitis but the increase rarely exceeds 400 units per 100 ml and the increase is only seen after several days. Mumps may also produce an increased amylase activity, but this may be easily differentiated on clinical grounds. The normal serum level is 15–110 Somogyi units per 100 ml.

A number of units are in use to express amylase activity, derived from the varying methods used in its determination. A Somogyi unit is that amount of enzyme activity required to liberate 1 mg of reducing substance. Since the method employed here measures the degradation of starch, it is assumed that 1 mg of reducing substance is released by the breakdown of 10 mg of starch.

One Somogyi unit = 20.7 International units.

Principle of the method

α-Amylase present in blood is allowed to act on a buffered starch substrate which splits it into dextrins and sugars. Iodine is then added which produces a blue colour with the residual starch but not with the degradation products. This colour is then compared with that produced by the original starch solution. The diminution in colour indicates the amount of starch degraded and therefore the amylase activity present.

Reagents required

1. *Buffer pH 7.2* Dissolve 2.922 g of sodium chloride 6.135 g of anyhdrous disodium hydrogen phosphate and 2.286 g of anhydrous potassium dihydrogen phosphate in distilled water in a 1 litre volumetric flask and make up to the mark. Stable for 1 month.
2. *Sorbic-Starch solution* Prepare a saturated solution of sorbic acid by adding 2 g to 1,200 ml of water. Heat, allow to cool and filter off excess acid.
    Add 1 g of soluble starch to approximately 50 ml of sorbic acid solution in a litre volumetric flask. Add three 50 ml portions of boiling sorbic acid solution, mixing well after each addition. Cool and make to mark with cold sorbic acid solution. Store in polythene bottle. Stable for one year at room temperature.
3. *Buffer-Substrate* Mix equal quantities of sorbic-starch solution and phosphate buffer pH 7.2. Stable for one week at 4°C.
4. *Iodine Reagent* Add 1.0 ml of potassium iodate solution containing 492 mg per litre to 10 ml of 0.1N hydrochloric acid. Add 200 mg of potassium iodide and make up to 100 ml with distilled water. Stable for one week at 4°C.
5. *Hydrochloric acid 0.1N*

    All chemicals used should be of reagent grade and distilled water must be used.

The Standard Lovibond Comparator Disc 5/46

This covers the range 25, 100, 150, 175, 200, 225, 250, 300, 350 Somogyi units per 100 ml.

Technique

1. Take two test tubes, labelling one 'Blank' and one 'Test.'
2. Add 2 ml of buffer substrate to each tube.
3. To 'Blank' tube add 0.5 ml of 0.1 N HCl.
4. Allow to warm to 37°C by standing in water bath for 10 minutes.
5. To both tubes add 0.02 ml of serum (or urine).
6. Place in 37°C water bath for 30 minutes.
7. Add 0.5 ml of 0.1 N HCl to 'Test' tube.
8. Add 10 ml of iodine reagent to each tube.
9. Transfer contents of both tubes to 13.5 mm comparator cells, placing the blank solution in the RIGHT HAND compartment of the comparator and the test solution in the LEFT HAND compartment i.e. the reverse of the normal procedure.
10. With disc 5/46 in position in the Comparator, match the two colours in north daylight or in White Light Cabinet.
11. Read off the amylase activity directly in Somogyi units per 100 ml.

Notes

1. Some difficulty may be experienced at first in matching the rather deep colours. However, with a little practice a match can be obtained without much difficulty. A high degree of precision should not be attempted and in fact is not required clinically.

2. If the result is higher than 350 units repeat test after diluting the serum 1 in 2 or 1 in 4 with distilled water, and multiply by the appropriate factor.

References

1. B. W. Smith and J. H. Roe, *J. biol. Chem.* 1949, **179**, 53
2. B. W. Smith and J. H. Roe, *J. biol. Chem.* 1957, **227**, 357
3. E. W. Rice, *Clin. Chem.* 1959, **5**, 592

The Determination of Ammonia
field method for estimation of ammonia in urine

Principle of the method

Urine contains substances (e.g. creatine) which interfere with direct Nesslerisation for determining the ammonia content. The ammonia is therefore removed by "Permutit-Decalso" (sodium aluminium silicate). The supernatant urine is discarded, and the "Permutit" washed. The "Permutit" is then made alkaline to liberate the ammonia, which is estimated colorimetrically by means of the Lovibond Comparator and the standard blood urea discs.

Reagents required

1. *Sodium hydroxide* ($NaOH$) 10% w/v.
2. *Nessler's solution* Dissolve 150 g potassium iodide (KI) in 100 ml distilled water. Transfer this solution to a litre flask and add 200 g mercuric iodide (HgI_2) and allow to dissolve. When solution is complete, make up to the litre mark with distilled water and filter. Add 1 litre distilled water to this filtrate. Dilute 15 ml of this stock solution immediately before use with 85 ml of distilled water.
3. *"Permutit"* A convenient fineness is that which will pass through a 60 mesh but not an 80 mesh sieve. If the "Permutit" as obtained is too coarse, it may be ground gently in a mortar. Fine particles should then be removed by repeated shaking with water, and decantation of the turbid supernatant liquid. The "Permutit" can be used repeatedly. The "Permutit" collected from a day's work should be freed from any traces of ammonia by washing with 10% w/v sodium hydroxide. The "Permutit" is then washed in 20% v/v acetic and finally with water. It is dried in air without heat; if it is oven-dried the activity is greatly reduced.

The activity of each new batch of "Permutit" must be established before use, and determined against the Lovibond Blood Urea discs. The method is as follows. Prepare three standard ammonia-nitrogen solutions:—
   (a) 10 mg ammonia-nitrogen per 100 ml. Weigh out 94.4 mg of ammonium sulphate (($NH_4)_2.SO_4$) and make up to 200 ml in a graduated flask with distilled water.
   (b) 5 mg ammonia-nitrogen per 100 ml. Solution (*a*) diluted 1 in 2 with distilled water.
   (c) 2 mg ammonia-nitrogen per 100 ml. 20 ml of solution (*a*) diluted to 100 ml with distilled water.

Take 1 ml of each of these standard solutions through the procedure described below for urine, and draw a graph showing Lovibond Comparator readings plotted against mg ammonia-nitrogen per 100 ml.

The Standard Lovibond Comparator Discs 5/9A and 5/9B

Disc A covers the range 20 mg to 100 mg urea per 100 ml of blood, in 9 steps of 10 mg
Disc B covers the range 110 mg to 220 mg urea per 100 ml of blood, in 9 steps of 10 mg except for the last 3, which are of 20 mg.

Technique

In a test tube place 100 mg "Permutit." Add 1 ml urine (diluted if necessary) and shake for 5 minutes. Add approximately 10 ml distilled water, shake, and allow to settle. Pour off the supernatant fluid carefully, and add another 10 ml of water. After shaking, again allow to settle and decant. Add 1 ml of reagent 1 to the "Permutit," shake, and allow to stand for 5 minutes. Add 8 ml of water and 1 ml of Nessler's solution. Shake, place in a comparator test tube or 13.5 mm cell in the right-hand compartment of the comparator, a "blank" of distilled water in the left-hand compartment behind the glass standards, and compare the colour using a standard source of white light, such as the Lovibond White Light Cabinet or, failing this, north daylight, revolving the disc until a match is obtained. Estimate the ammonia present by reference to the graph prepared for that batch of "Permutit".

With dilute urines the estimation can be done directly on 1 ml of the urine, but with more concentrated specimens the urine will have to be diluted. The usual dilution is 1 in 2, but specimens have been met in which dilutions of 1 in 5 or even 1 in 10 have had to be made. If a dilution is used, the ammonia-nitrogen concentration as read from the graph must, of course, be multiplied by the dilution factor.

Note.—"Permutit" is the Registered Trade Mark of The Permutit Co. Ltd., London.
"Permutit-Decalso" is specially manufactured and graded for biochemical purposes.

COLORIMETRIC CHEMICAL ANALYTICAL METHODS

The Determination of Serum Bilirubin
(Direct and Total)
(Lathe and Ruthven — modified)

Introduction

Bilirubin is produced in the reticulo-endothelial system from the destruction of hæmoglobin. It circulates in the blood mainly attached to the albumin fraction. In this form it is not soluble in water but, in the liver cell, it is rendered soluble by conjugation with glucuronic acid to form the diglucuronide and in this form it is excreted in the bile.

The conjugated soluble form of bilirubin is able to react with the colour developing reagent used in its estimation, and is referred to as the direct component. The insoluble form of bilirubin prior to its passage through the liver cell requires the presence of a coupling reagent before the colour reaction can take place, and is therefore referred to as the indirect fraction.

The normal total serum bilirubin is less than 1 mg/100 ml with insufficient direct reacting pigment to give a colour reaction. In hæmolytic anæmia the increase in bilirubin is mainly in the direct fraction. Bilirubin levels of up to 6 mg/100 ml may be found in chronic hæmolytic anæmia with higher levels being reached in acute hæmolytic anæmia. In obstructive jaundice the bilirubin is mainly direct reacting and total serum bilirubin values of up to 25 mg or more may be found in cases of long standing jaundice. When the jaundice is due to liver cell damage (e.g. hepatitis) both direct and indirect pigments are found, the proportion of each varying with the stage of the disease, though the total bilirubin levels rarely reach those found in obstructive jaundice.

Principle of the method

Bilirubin is allowed to react with a modified Van Den Berghs' diazo reagent with the formation of a purple complex, azobilirubin. If the reaction is allowed to take place in aqueous solution the direct acting pigment may be estimated from the colour produced. Methanol is then added as a coupling agent in a concentration less than that required to precipitate the plasma protein and the total bilirubin measured. The indirect component may then be determined by difference.

Reagents required

1. *Diazo reagent A* Dissolve 10 g of sulphanilic acid in 18 ml of concentrated HCl and make to 1 litre with water. Stable for at least one month.
2. *Diazo reagent B* Dissolve 0.5 g of sodium nitrite in water and make to 100 ml. Prepare a fresh solution weekly.
3. *Absolute methanol.*

The Standard Lovibond Comparator Disc 5/42.

This disc covers the range 1-9 mg/100 ml of bilirubin in steps of one milligram.

Technique

1. Add 0.2 ml of serum to 2.2 ml of water.
2. Prepare fresh diazo by mixing 10 ml of diazo reagent A with 0.3 ml of diazo reagent B.
3. Add 0.7 ml of freshly mixed diazo reagent to test and mix.
4. Add 2.9 ml of absolute methanol.
5. Mix and stand in the dark for five minutes.
6. Transfer to 13.5 mm cell. Place cell in right-hand compartment of the comparator, and place a blank cell containing water in the left-hand compartment. Match colours using White Light Cabinet or north daylight and read off total bilirubin concentration from the indicator window.

Notes

1. If it is required to estimate the direct bilirubin level, the method is repeated substituting 2.9 ml of water for absolute mehtanol in step four of the above scheme.

2. If the bilirubin concentration exceeds 9 mg/100 ml, repeat the test using 0.1 ml of serum in 2.3 ml of water in step one, and multiply the result by two. If the level exceed 18 mg/100 ml, use 0.05 ml of serum in 2.35 ml of water and multiply by four.

3. Though the method given here is less affected by the presence of hæmaglobin than other methods, every effort should be made to obtain serum free from hæmaglobin for this estimation.

4. Protect the test from excessive sunlight during the colour development.

Reference

Lathe and Ruthven. *J. Clin. Pathol.* 1958, **11**, 155

The Determination of Bromide
using gold chloride

Introduction

Estimation of the amount of bromide in blood is of value in confirming the clinical diagnosis of bromism as manifest by a skin eruption or by mental changes. Formerly patients with mental disease were often heavily sedated with bromides and on rare occasions bromide intoxication, characterised by disorientation and hallucinations, developed. In such cases it is important to distinguish between the symptoms of bromide intoxication and those due to the underlying mental disease itself.

Blood normally contains 0.25 to 2 mg sodium bromide per 100 ml. In patients with mental symptoms due to bromide intoxication, levels of 200 mg or more per 100 ml are found. For the estimation 5 ml venous blood, collected in a clean tube containing a pinch of potassium oxalate powder, are required.

Principle of the method

The proteins of serum, or plasma, are precipitated by trichloracetic acid. The filtrate is treated with gold chloride solution, which combines with bromides to give a double salt, the colour of which is a deeper yellow-brown than that of the added gold solution: the colour is compared with that of a series of Lovibond permanent glass standards prepared to match bromide standard solutions similarly treated.

Reagents required

1. *Trichloracetic acid* (CCl_3COOH) 20% w/v.
2. *Gold chloride* Dissolve 0.5 g of gold chloride ($AuCl_3HCl.3H_2O$), in distilled water, and make up to 100 ml.

The Standard Lovibond Comparator Disc 5/23

The disc covers the range 0, 10, 25, 50, 75, 100, 125, 150, 175 mg of sodium bromide per 100 ml of serum or plasma.

Technique

Mix in a test tube:—

    Serum or plasma   ..   2 ml.
    Distilled water   ..   6 ml.
    20% trichloracetic acid ..   2 ml.

Filter. To 5 ml filtrate (=1 ml serum) add 1 ml of 0.5% gold chloride solution.

Mix, transfer to a comparator test tube and place in the right-hand compartment of the comparator. Hold facing a standard source of white light such as the Lovibond White Light Cabinet or, failing this, north daylight, and rotate the disc until the colour of the liquid is matched by one of the glass standards. The value (expressed in mg sodium bromide per 100 ml serum or plasma) is then read from the indicator recess at the bottom right-hand corner of the comparator.

Notes

1. It is important to check each batch of reagents against the 0 of the disc. Prepare a blank from:—

    4 ml distilled water
    1 ml trichloracetic acid solution
    1 ml gold chloride solution

and mix well.

If this does not match the 0 of the disc, it is useless to proceed; the gold chloride may be different from that used in standardising the disc.

2. Iodides, as well as bromides, give double salts with gold chloride. It is therefore essential to make sure that the patient is not taking iodides.

Confirmatory test

If the patient is suffering from skin lesions thought to be due to bromism, and the bromide found by the above test is low (25 mg or less), the fluorescein confirmatory test should be performed, to ascertain that increased bromide is in fact present.

Bromides are oxidised to bromine by chloramine T. The bromine is taken up by fluorescein to yield red tetrabromofluorescein (eosin).

Reagents required for confirmatory test

1. *Chloramine-T 0.4%.*
2. *Fluorescein* Dissolve 125 mg of fluorescein in 25 ml of $0.1\,N$ NaOH and add distilled water to 100 ml.
3. *Buffer solution pH 5.3 to 5.4* Dissolve 6.6 g of anhydrous sodium acetate (or 10.9 g of $CH_3COONa.3H_2O$), and 1.2 ml of glacial acetic acid in water to 100 ml.

Technique

Mix in a small test tube :—

Serum filtrate, as prepared for gold chloride test	1 ml
Buffer solution	0.5 ml
Fluorescein solution	1 drop
Chloramine-T	1 drop

Treat 1 ml of distilled water, or 1 ml of protein-free filtrate from normal serum, in the same way in parallel.

Observe the colour of the contents of the two tubes: the filtrate from the patient's serum will turn orange-red or red, in a few seconds, if more than 5 mg NaBr is present. Observe the tubes for five minutes before declaring the reaction negative. If bromides were not in excess of normal, the treated filtrate of the patient's serum will not alter in colour, but will remain like the treated water or treated normal filtrate.

Remarks

Iodide, in this test, will give tetraiodofluorescein, which is similar in colour. The iodide can be removed by treating the protein-free filtrate with sodium nitrite solution and extracting with chloroform, but it is simpler to ask whether the patient is taking iodide.

Reference

G. A. Harrison, *"Chemical Methods in Clinical Medicine"*, 4th Edition, J. & A. Churchill, London, 1957, page 361

The Determination of Cholesterol
using ferric chloride and sulphuric acid

Principle of the method

Cholesterol and its esters are extracted with warm ferric chloride and acetic acid following protein precipitation. Concentrated sulphuric acid is added and the resulting colour compared with a series of coloured glass standards.

Reagents required

1. *Ferric chloride – acetic acid reagent* Dissolve 50 mg of $FeCl_3.6H_2O$ in 100 ml of glacial acetic acid. Store at room temperature. Stable for one month.
2. *Sulphuric acid concentrated*.

All reagents must be of analytical reagent quality.

The Standard Lovibond Comparator Disc 5/45

The disc covers the range 150 to 550 mg per 100 ml in steps of 50 mg.

Technique

1. Add 0.1 ml of plasma or serum to 9.9 ml of ferric chloride - acetic acid reagent.
2. Mix and stand for 15 minutes.
3. Centrifuge.
4. Take 5 ml of supernatant and place in large boiling tube.
5. Add 3 ml of concentrated sulphuric acid.
6. Mix well and stand for 15 minutes.
7. Prepare blank by adding 3 ml of concentrated sulphuric acid to 5 ml of ferric chloride – acetic acid reagent in a similar boiling tube. Mix and stand for 15 minutes.
8. Using 13.5 mm comparator cells place blank in left-hand compartment of comparator and the test in the right-hand compartment.
9. Match the colour to that of the standard disc using north daylight or White Light Cabinet.
10. Read off result directly in mg per 100 ml.

Notes

1. As traces of water interfere with the reaction, always replace cap on sulphuric acid and do not use any bottle which has been repeatedly opened.
2. Oxalated plasma should not be used for this method as it produces significantly low results.
3. As haemoglobin interferes with the estimation, grossly haemolysed samples should not be used.
4. The colour should be matched as soon as it has developed.

References

1. A. Zlatkis, B. Zak, A. J. Boyle, *J. Lab. Clin. Med.* 1953, **41,** 486
2. A. A. Henly, *Analyst* 1957, **82,** 286

The Rapid Field Determination of Cholinesterase

Introduction

Farmers, spray operators, and research workers using organic phosphorus insecticides may run some risk of significant absorption of these toxic chemicals despite the adoption of protective measures. Over-absorption, and unsafe working methods are detectable by depression of the normal cholinesterase enzyme activity of the blood. See also Note 11. Several methods exist for the accurate measurement in the laboratory of cholinesterase activity, but a rapid 'field' method is frequently needed for use in emergencies or away from the laboratory. The method of Limperos and Ranta[1] has been modified by Edson[2] to give a combination of speed, convenience and reasonable precision for emergency or routine cholinesterase determination with simple equipment.

The cholinesterase activity in the blood from the subject under test is expressed as a percentage of the activity in normal blood. Depending on the result obtained, the following action is recommended.

100%—75% of normal	No action, but retest in near future.
75%—50% of normal	Over-exposure probable; repeat test. If confirmed, suspend from further work with organic phosphorus insecticides for 2 weeks; then retest to assess recovery.
50%—25% of normal	Serious over-exposure: repeat test. If confirmed, suspend from all work with insecticides. If indisposed or ill arrange medical examination.
25%—0% of normal	Very serious and dangerous over-exposure. Repeat test: if confirmed, suspend from all work pending medical examination.

Principle of the method

Blood contains an enzyme, cholinesterase, which liberates acetic acid from acetylcholine thereby changing the pH. A mixture of blood, indicator and acetylcholine perchlorate is prepared and allowed to stand for a fixed time. The change of pH in this time is a measure of the cholinesterase activity.

It has been established by Watson & Edson[10] that, as suggested by Davies & Nicholls[7], it is possible to dispense with a timed control sample and to substitute a temperature/time relationship provided that the temperature range involved is not excessive and that the modified instructions are followed exactly.

Reagents required

1. *Indicator solution*

    Sodium salt of bromo-thymol blue (B.D.H. Ltd. water soluble) 0.112 g
    Distilled water (CO_2 free) 250 ml

    This solution is stable for several months and should be kept tightly stoppered to prevent absorption of CO_2. The concentration is critical, the permissible limits being ± 0.01 g.

    NOTE re Reagent 1 (Indicator solution)
    The concentration of Bromo-thymol blue indicator solution is critical, the **permissible limits being 107-117 mg/250 ml distilled** water (CO_2 free). Outside these limits colour matching becomes unreliable. While every attempt is made to ensure the ampoules are precisely filled, it is recommended that accurate weighing be carried out where a very critical study is proposed. Accurate weighing is of course necessary where the indicator has been obtained in bulk.

2. *Substrate solution*

    Acetylcholine perchlorate 0.5 ± 0.2 g
    Distilled water (CO_2 free) 100 ml

    This substrate solution should be freshly prepared each day. The exact concentration is not critical, and slight turbidity can be ignored.

The Standard Lovibond Comparator Disc 5/30

The disc covers the range 0—100% normal activity in steps of 12½%, and is calibrated for use with 2.5 mm cells (see note 12).

Technique

(a) *Determination of blood cholinesterase in human beings*

1. Test the reagents by mixing 0.5 ml of indicator, 0.01 ml of finger-prick blood from a normal "control" subject, and 0.5 ml of substrate solution. Mix well and place in a 2.5 mm cell in the right-hand compartment of the comparator. Hold the comparator facing a source of uniform white light, such as the north sky in the northern hemisphere, and revolve the disc until a close colour match with the test solution is obtained. It should be no more yellow (acid) than the 12.5% activity colour. If the indicator solution is too acid, due to the absorption, of carbon dioxide, it can be restored to normality by bringing it momentarily to the boil. If this does not reduce acidity to the 12.5% colour the substrate is at fault and it must be discarded, and a fresh solution prepared.

2. Make a blood "blank" by adding 0.01 ml of finger-prick blood to 1 ml distilled water in a 2.5 mm cell and place in the left-hand compartment of the comparator.

3. Set up a reaction tube for each patient, and pipette 0.5 ml of indicator solution into each tube. Up to 15-20 tests can be carried out at the same time.

4. From each patient pipette 0.01 ml blood, obtained by finger or thumb prick, into a reaction tube, rinsing the pipette two or three times with the indicator in the tube. A "control" blood sample (i.e. that of a normally healthy person unexposed to organic phosphorus insecticides) should be put into tube No. 1.

5. Pipette 0.5 ml substrate solution into tube No. 1 ("control" tube). Note the time ("zero time"), immediately transfer the reaction mixture to a 2.5 mm cell and note its colour in the comparator. It should be no more acid than the 12.5% colour.

6. Add 0.5 ml substrate solution to the other reaction tubes at 1 minute intervals from zero time, stopper them and shake to mix.

7. Wait for the contents of the control sample cell to reach 100% activity colour (which will take 20—30 minutes, depending largely on temperature) decant control sample and at 1 minute intervals from this time transfer the contents of the other tubes in turn to the cell. Match each sample against the disc, by placing the cell in the right-hand compartment of the comparator and revolving the disc until the nearest match is found, and record the activity. The discs are graduated in % of normal (i.e. of control) activity.

(b) *Simplified Technique*

Test the reagents, make up a blood blank, and set up the reaction tubes as directed in paragraphs 1-3 of the original instructions. From each patient pipette into a reaction tube 0.01 ml of blood, obtained by finger or thumb prick, rinsing the pipette two or three times with the indicator in the tube. Note the shade temperature. Add 0.5 ml of substrate to the reaction tubes at 1 minute intervals, stopper the tubes and shake to mix. Allow each tube to stand for exactly the time corresponding to the shade temperature shown in the Table. Immediately this time has expired transfer the contents of the tube to the 2.5 mm sample cell, place this in the right-hand compartment of the comparator and match against the standards in the disc.

Time-Temperature Table

Use time column corresponding most closely to colour value in reagent test

Shade Temp. °C	Reagent = 0%	Reagent = 12½%
10	41	36
15	33	29
20	27	24
25	24	21
30	21	18.5
35	18	16
40	16.5	14.5
45	16	14

Time minutes

For temperatures above 45°C the original procedure must be used.

(c) *Determination of blood cholinesterase in animals*[6]

Blood is collected from the ear, except in the case of mice when the tail is used, 0.01 ml being used for guinea-pigs and mice, 0.02 ml for other animals. The blood is immediately transferred to a tube, and mixed with 0.5 ml of indicator solution, (reagent 1). 0.5 ml of substrate solution (reagent 2) is added and the time noted. After the elapse of the appropriate time, given in the table below, the mixture is transferred to a 2.5 mm comparator cell and the percentage determined.

Cattle	80 minutes
Dogs	40 ,,
Sheep	120 ,,
Mice	20 ,,
Guinea-pigs	80 ,,

A conversion of 75 per cent. or more within the stated time —blood normal
    50–75 ,, ,, —moderate depression
    35–50 ,, ,, —no clinical symptoms but caution is required
    below 35 ,, ,, —severe exposure requiring palliative measures

It has not been possible to obtain satisfactory results using this method with goats and rats.

Notes

1. Good agreement has been found with results obtained by the electrometric method of Michel[4].

2. Acid, alkali or insecticide contamination should be avoided by carefully cleaning the skin before taking the blood sample, first with soap and water, then with surgical spirit, and drying the area with clean cotton wool. The needle must also be carefully cleaned each time.

3. The control blood sample must be obtained from a normally healthy person, who has had no probability of exposure to organic phosphorus insecticides in the past three months. **Alternatively,** a 'stock' sample of normal blood may be used, but only heparin is suitable for use as the anticoagulant. Citrate or oxalate cause interfering pH changes and should not be used.

4. The reaction tubes and bungs must be carefully washed with distilled water between tests, because, as this test depends on a change of pH, there must be no contamination from any acid or alkali.

5. All pipetting must be done with teats to avoid producing acidity and false results due to contamination with CO_2 in the breath.

6. The 100% normal chlolinesterase value of the cholinesterase disc is based on average values for healthy European males. Non-Europeans of tropical countries may show lower normal values, in the range of 75—87%. Females normally have slightly lower ChE activity than males.

7. As there is considerable personal variation in normal blood ChE activity, it is very desirable wherever possible, to determine the normal activity of all workers before they begin a campaign with organic phosphorus insecticides.

8. The indicator-blood dilution is stable only for up to four hours in cool climates, and two hours in hot climates. Individual 0.01 ml blood samples from isolated workers may thus be collected and retained in stoppered reaction tubes, as dilutions in 0.5 ml indicator solution, but should be analysed within the above times.

9. An almost immediate colour change from green-blue to orange suggests acid contamination. Repeat the test after washing out tubes with distilled water.

Failure to develop any colour change in the reaction tube may be due to a complete absence of cholinesterase in the blood sample from a grossly over-exposed worker, but may also be due to failure to add the blood or substrate, or to alkaline contamination. All such instances therefore necessitate a repeat test for confirmation.

10. A portable field set is available for use with this test, details of which may be obtained from The Tintometer Ltd., Salisbury.

11. Cholinesterase is involved in the metabolism of the muscle relaxant succinylcholine. Deficiency of the enzyme results in prolonged action of this normally short acting relaxant, leading to prolonged apnoea. Apart from low enzyme levels that are sometimes found following liver disease and exposure to organophosphorus insecticides, a small number of people are found to have a genetically determined deficiency. The finding of a prolonged apnoea following succinylcholine administration, due to cholinesterase deficiency, is a strong indication for screening the remaining members of the family for the presence of a similar defect.

12. A special 2.5 mm comparator cell, W727, is made for this test.

References

1. G. Limperos and K. E. Ranta, *Science,* 1953, **117**, 453
2. E. F. Edson, *Brit. Med. J.,* 1955, i, 841
3. *idem., ibid.,* 1955, i, 1218
4. H. D. Michel, *J. Lab. Clin. Med.,* 1949, **34**, 1954
5. J. H. Wolfsie and G. D. Winter, *Arch. Ind. Hyg.,* 1954, **9**, 396
6. D. W. Jolly and B. D. Ratcliffe, *The Vet. Record,* 1958, **70**, 289
7. D. R. Davies and J. D. Nicholls, *Brit. Med. J.,* 1955, i, 1373
8. E. F. Edson, *World Crops,* 1956, **8**, 240
9. E. F. Edson, *World Crops,* 1958, **10**, 49
10. W. A. Watson and E. F. Edson, *personal communication,* 1964
11. D. M. Forsyth & Carole Rashid, *Lancet* 1967, 909 (28 October)

The Determination of Hæmoglobin

Introduction

The estimation of hæmoglobin is probably the most frequently requested laboratory investigation, and is used in establishing a diagnosis of anæmia and to a lesser extent polycythemia. The hæmoglobin level when considered with the packed cell volume and the microscopical examination of a well-stained blood film will often indicate the type of anæmia and can be used to follow the response to therapy. It may also be helpful in assessing the degree of hæmoconcentration in patients suffering from shock or dehydration.

Techniques

Due to the importance of the investigation many techniques have been devised to determine the hæmoglobin concentration. Three only are presented here. *Method 1* is that devised by Harrison and is included because of its simplicity, since it employs undiluted blood and does not use any reagents. In *Method 2* the hæmoglobin is converted to oxyhæmoglobin. This method is widely used, is simple, uses only a very simple reagent and gives satisfactory results in the majority of cases. Its main disadvantage, that of finding a suitable standard, has been solved by the use of Lovibond permanent glass standards. *Method 3* measures the hæmoglobin as cyanmethæmoglobin (official name hemiglobincyanide). The method has been recommended by the International Committee for Standardization in Hæmatology as the standard method.

When hæmoglobinometry was first introduced the results were expressed as a percentage of normal. This was derived from an arbitrary normal population with the result that several widely differing standards came into use. In 1961 The Protein Commission of the International Union of Pure and Applied Chemistry ruled that 100% should be equivalent to 14.6 g per 100 ml. This is incorporated in B.S. 3420/1961 which now replaces the earlier B.S. 1079/1956 (which equated 100% with 14.8 g per 100 ml).

In health the average hæmoglobin concentration in men is 13—18 g per 100 ml and 11.5—16.4 in women. Children, except during the neonatal period, have a lower level than adults.

Clinical Feature

A low hæmoglobin level indicates anæmia and is due to a reduced number of red cells or a reduced amount of hæmoglobin in each cell or to both factors in combination. The most common causes are blood loss; failure to produce sufficient normal red cells due to deficiency of iron, vitamin B_{12} or folic acid; failure to produce red cells due to leukæmia or other malignancy and failure to produce red cells due to primary marrow failure e.g. aplastic anæmia.

An increased hæmoglobin level is seen less commonly, and is due to an increase in circulating red cells which may be primary, as in polycythæmia vera, or secondary to long standing anoxia due to pulmonary disease, heart disease or even altitude. A transient increase can also be seen due to dehydration.

Capillary Blood Specimen

Capillary blood can be obtained from the finger, the ear lobe or in infants the heel. To ensure a free flow of blood the site must be warm and this can be achieved either by gently massaging the area or in the case of babies' heel by placing it in a bowl of warm water for a few minutes. The site is dried, cleaned with 70% ethyl alcohol and allowed to dry. A straight cutting needle is used to prick the patient to a depth of 3—4 mm. It is possible to obtain sterile disposable blood lancets which have a stop to prevent excessive penetration and also avoid the risk of hepatitis.

The first drop of blood is wiped away and the pipette filled from the next drop. Squeezing should be avoided. If the pipette is held almost horizontally the blood flows in more easily and the operator can see easily when it is filled to just above the mark. Still holding the pipette almost horizontal wipe the outside clean with a paper tissue or cotton wool and gently touch the end on a piece of filter paper until the blood column is just up to the mark.

The contents are blown into the diluting fluid, which is then sucked in and out of the pipette several times before being finally blown out.

The Determination of Hæmoglobin (1)
using undiluted blood
(Harrison's method.)

Principle of the method

Blood, obtained by pricking the finger, is run directly into a special cell without any preliminary pipetting or diluting. The colour is compared with a series of glasses in a Lovibond Comparator. Errors due to inaccuracies in the use of, or calibration of, blood pipettes is avoided, and the method is particularly suited for use by the clinician at the bed-side or in the consulting room.

The Standard Lovibond Comparator Discs 5/8A and 5/8B

Disc A covers the range 20, 24, 28, 32, 36, 40, 46, 52 and 58% hæmoglobin.

Disc B covers the range 64, 70, 76, 84, 92, 100, 110, 120 and 130%.

To bring these values into conformity with British Standard 3420/1961 they must be multiplied by a factor of 1.14 as shown in the accompanying table.

To establish the colour value of these hæmoglobin standards, the glasses were matched against bloods of known hæmoglobin content in the special cell in the Comparator; the hæmoglobin contents of the bloods were calculated from their oxygen combining power, which had been determined by Van Slyke's technique.

The discs have been calibrated for use with the special 0.004″ cell. The blood cell consists of a base-plate and a cover-plate constructed from plain white glass. Fused on the cover-plate are three small studs of glass which are ground until they project 0.004″. By means of this three point separation the two plates, when in apposition, create a cell of 0.004″ thickness. For convenience whilst filling the cell, the two parts may be held together by a metal clip as shown in figure A. After filling the cell by running the blood directly into the cell from the side, as shown in figure C, the clip may be removed if desired, as the two parts will adhere by capillary attraction. Each cell is checked with blood of known hæmoglobin content on final inspection.

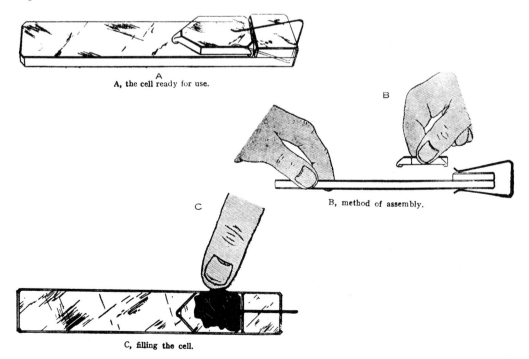

A, the cell ready for use.

B, method of assembly.

C, filling the cell.

Technique

The base-plate and cover-plate are cleaned thoroughly, first with water and then with alcohol, and dried with a clean non-fluffy rag. If the cleaning is not thorough, air bubbles form when the blood is run into the cell and spoil the colour matching. The two plates are placed in apposition and clipped (Figures A and B). The cell is filled from the side with capillary blood in the usual way (Fig. C.). Oxalated venous blood may be used, but is liable to appear bluer than the standards, which renders estimation of the intensity of colour difficult, and if it has been allowed to stand, it may, in spite of thorough mixing, give a film which is speckled owing to rouleaux formation, and so cause an error. For these reasons venous blood is often unsatisfactory. Another condition giving rise to rouleaux formation is a very cold glass cell. In cold conditions, therefore, the cell may be slightly warmed before use.

After filling, the cell is lowered carefully into the right-hand slot of the comparator. The film must completely fill the right-hand field of view and be free from air bubbles. Holding the apparatus about 18 in. away, and facing a white light (Note), the operator revolves the disc until the unknown matches a standard, and reads the hæmoglobin percentage at the bottom right-hand corner. With a little experience, it is easy to interpolate between the standards, and particularly between those in the lower part of the scale. All those who use this method should obtain closely comparable results. The reading should be taken within a few minutes of filling the cell. Wash the cell as soon after use as possible.

Note

Observations must be made with the comparator facing a standard source of white light, such as the Lovibond White Light Cabinet or, failing this, north daylight. Direct sunlight or uncorrected artificial light gives incorrect results.

Reference

G. A. Harrison, *Lancet*, 1938, ii, 621

Conversion Table to relate disc readings to B.S. 3420:1961

Reading on disc 5/8A or B	B.S. equivalent value	g./100 ml
20	22.8	3.3
24	27.5	4.0
28	32.0	4.7
32	36.5	5.3
36	41.0	6.0
40	45.5	6.7
46	52.5	7.3
52	59.3	8.7
58	66.0	9.7
64	73.0	10.7
70	80.0	11.7
76	86.6	12.7
84	95.8	14.0
92	105.0	15.3
100	114.0	16.7
110	125.5	18.3
120	137.0	20.0
130	148.0	21.7

The Determination of Hæmoglobin (2)
using oxyhæmoglobin

Introduction

Hæmoglobin can exist in the blood in several forms, but the proportion that cannot readily be transformed into oxyhæmoglobin is very small. For clinical purposes therefore, an oxyhæmoglobin method of hæmoglobin measurement is completely adequate. The advantages of this oxyhæmoglobin method[1,2] are simplicity, speed in execution, minimum equipment requirements, and the absence of liquid standards.

For hæmatological purposes, the best method of expressing values is as grams of hæmoglobin per 100 ml of blood (g/100 ml). The iron content of pure hæmoglobin is accepted internationally as the basis for calibrating hæmoglobin, and blood is reckoned as 100% when it contains 14.6 g Hb per 100 ml (see ref. 3). The disc for the present test has been calibrated accordingly.

Principle of the method

The colour of an alkaline solution of blood is compared with a series of Lovibond permanent glass standards calibrated directly in grams of hæmoglobin per 100 ml of blood.

Reagent required

Ammonia solution. 4 ml of ammonium hydroxide (NH_4OH sp. gr. 0.880) in 1 litre of distilled water.

The Standard Lovibond Comparator Discs 5/37 and 5/37X

Disc 5/37 contains 9 standard covering the range 8–16 g/100 ml HbO_2, in steps of 1 g except for the last three, which are 14, 14.6, 16.

Disc 5/37X contains 9 standards, with values 3, 4, 5, 6, 7, 8, 10, 12, 14 g/100 ml HbO_2.

Technique

Fill the standard test tube, or 13.5 mm comparator cell, to the 10 ml mark with the ammonia solution. Measure 0.05 ml of blood in a blood pipette and carefully remove all surplus blood from the outside of the pipette. Deliver the blood into the ammonia solution and wash out the pipette by repeatedly drawing up and expelling the ammonia solution. Mix well. Place the tube in the right-hand compartment of the comparator and match against the standards in the disc, using a standard source of white light, such as the Lovibond White Light Cabinet or, failing this, north daylight. Read off the hæmoglobin content in the indicator window in the bottom right-hand corner of the comparator.

Note

A complete field kit for this test is available. Details from Tintometer Ltd.

References

1. I. D. P. Wootton, *"Microanalysis in Medical Biochemistry"* 4th editn., J. & A. Churchill London, 1964
2. H. J. Woodliff, P. Onesti and D. W. Goodall, *Med. J. Australia* 1966 (Aug.) 410
3. *J. Clin. Path.* 1965, **18,** 353

The Determination of Hæmoglobin (3)
using cyanmethæmoglobin

Introduction

The accurate estimation of hæmoglobin in the blood has, in the past, been made difficult by the absence of reliable hæmoglobin reference standards. The use of stable cyanmethæmoglobin solutions as reference standards has now been officially accepted[1,2] and the present test[3] is based on the use of this new standard.

Principle of the method

The blood sample is converted into a cyanmethæmoglobin solution by accurate dilution with a reagent containing potassium cyanide. The cyanmethmogælobin colour thus formed is measured by comparison with a series of Lovibond permanent glass colour standards.

Reagent required

Diluent solution Dissolve 200 mg of potassium ferricyanide [$K_3Fe(CN)_6$], 50 mg of potassium cyanide (KCN) and 140 mg of potassium dihydrogen phosphate (KH_2PO_4) in distilled water and make up to 1 litre. Add 0.5 ml of Sterox SE (Hartman-Leddon Co., of Philadelphia, U.S.A.) or 1 ml of Nonidet P40 (Shell Chemical Co. of London). Preserve in a brown bottle and discard if the reagent becomes cloudy. All chemicals used must be of analytical reagent quality. Alternatively, Drabkins's Reagent supplied by BDH Chemicals Ltd. (reference 22040) may be used.

The Standard Lovibond Comparator Discs 5/40 and 5/40X

Disc 5/40 contains standards corresponding to 7, 9, 10, 11, 12, 13, 14, 15, and 17 g of hæmoglobin per 100 ml of blood. Disc 5/40X contains standards for 5, 6, 7, 8, 9, 10, 11, 12, 13 g/100 ml.

Technique

Wash 0.04 ml of blood into a standard comparator tube with diluent solution and make up to the 10 ml mark. Mix thoroughly and, after 3 minutes, place this tube in the right-hand compartment of the comparator. Place an identical tube filled with the diluent in the left-hand compartment. Match the colour of the diluted sample with the standards in the disc using a standard source of white light, such as the Lovibond White Light Cabinet or, failing this, north daylight. When a match has been obtained read off the hæmoglobin content of the blood from the indicator window of the comparator.

Note

The collaboration of Professor I. D. P. Wootton in the preparation and checking of the colour standards is gratefully acknowledged.

References

1. Brit. Standard *"Cyanmethæmoglobin Solution for Photometric Hæmoglobinometry,"* 3985: 1966
2. J. V. Dacie and S. M. Lewis, *"Practical Hæmatology"* 4th edition (J. & A. Churchill Ltd.), 1968
3. I. D. P. Wooton, *"Micro-analysis in Medical Biochemistry"* (Earl J. King), J. & A. Churchill Ltd., 4th editn., London, 1964

The Determination of Iron in Serum, and Total Iron-Binding Capacity

using bathophenanthroline

Introduction

The iron in serum is present coupled to a transport protein transferrin (siderophilin). The normal level for men is 80–180 μg/100 ml and for women 60–160 μg/100 ml, though there is an appreciable diurnal variation. It is, therefore, advisable to collect blood for iron determination when the patient is fasting in the morning. Serum iron levels are low in iron deficiency anæmia and normal or increased in megaloblastic anæmias. They are also increased in infective hepatitis and hæmachromatosis.

Transferrin is normally only partially saturated with iron and is capable of binding more iron. The total amount of iron that can be carried is termed the total iron-bonding capacity and is 250–400 μg/100 ml. The serum iron level expressed as a percentage of the total iron-binding capacity is termed the saturation index.

This is low in the iron deficiency anæmias and high in megaloblastic anæmias compared with a normal level of 25–50%.

Principle of the method

Iron is separated from its protein carrier by hydrochloric acid. The proteins are then precipitated with trichloracetic acid. Ferric iron is reduced to the ferrous state with thioglycollic acid and bathophenanthroline is then added, which gives a purple colour proportional to the iron present.

In order to determine the total iron-binding capacity, excess iron in the form of ferric chloride is added to the serum. Surplus unbound iron is then removed by adsorption onto magnesium carbonate. The total serum iron content is now determined to give the iron-binding capacity.

The Standard Lovibond Comparator Disc 5/50

This disc covers the range 25–225 μg/100 ml iron in steps of 25 μg. (Micrograms per 100 ml).

Reagents required for serum iron determination

1. *30% trichloracetic acid, $CCl_3.COOH$* Dissolve 30 g of trichloracetic acid in double glass distilled water and make up to 100 ml. Stable for one year.
2. *Concentrated hydrochloric acid, HCl* Analytical grade. Stable indefinitely.
3. *Thioglycollic acid $CH_2(SH).COOH$* Stable indefinitely.
4. *50% sodium acetate ($CH_3COONa\ 3H_2O$)* Dissolve 50 g of sodium acetate in double glass distilled water and make up to 100 ml. Stable for one month.
5. *0.02% Bathophenanthroline in isopropanol* Dissolve 20 mg of 4,7–diphenyl 1,10–phenanthroline ($C_{24}H_{16}N_2$) in 100 ml of isopropyl alcohol analytical grade. Stable for one month.

Technique for Serum iron determination

1. To 2 ml of serum add 3 ml of double glass distilled water, one drop of concentrated hydrochloric acid and one drop of thioglycollic acid.
2. Mix well and stand for thirty minutes.
3. Add 1 ml of 30% trichloracetic acid.
4. Mix well and stand for 10 minutes.
5. Centrifuge.
6. To 3 ml of supernatant in a tube labelled "test" add 0.4 ml of 50% sodium acetate and 2 ml of bathophenanthroline. Mix and stand for 10 minutes.
7. To 3 ml of glass distilled water in a tube labelled "blank" add 0.4 ml of 50% sodium acetate and 2 ml of bathophenanthroline. Mix and stand for 10 minutes.

8. Pour "test" liquid into 13.5 mm comparator cell and place in right hand compartment of comparator. Pour "blank" into a similar cell and place in left hand comparator compartment. Match colours using white light equipment or north daylight and read off serum iron concentration in μg/100 ml in indicator window.

Reagents required for total iron binding capacity

1. *Ferric chloride solution* Dissolve 145 mg of $FeCl_3$ in $0.5\,N$ hydrochloric acid and make up to 100 ml. This stock solution is stable for 1 year. Make "working solution" by adding 1 ml to 99 ml of water. Stable for one month.
2. *Magnesium Carbonate* "Light" quality as used for absorption
3. All reagents mentioned above for serum iron determination

Technique for total iron-binding capacity

1. To 2 ml of serum add 4 ml of working ferric chloride solution.
2. Mix well and stand for 10 minutes.
3. Add 400 mg of magnesium carbonate.
4. Shake vigorously every five minutes for 30 minutes or place stoppered tube on rotary mixer for 60 minutes.
5. Centrifuge.
6. Take 3 ml of supernatant fluid and add 2 ml of glass distilled water, one drop of concentrated hydrochloric acid, and one drop of thioglycollic acid.
7. Follow the technique given for serum iron determination from step (2) and multiply the value observed in the comparator indicator window by 2 to give the total iron binding capacity. The saturation index may be calculated by expressing the serum iron concentration as a percentage of the iron binding capacity.

Notes

1. Due to the sensitivity of the method, care should be taken to ensure that specimen tubes and pipettes and test tubes are not contaminated with iron. All glass ware should be washed in hot 50% HCl and rinsed in glass-distilled water. It is advisable to keep a set of glassware solely for iron determination and to collect specimens in disposable plastic tubes.
2. Special care should be taken to ensure that specimens taken for analysis show no hæmolysis.
3. High quality deionised water may be used as a substitute for glass distilled water.
4. Place a small cork or cotton wool spacer beneath each comparator cell in the comparator so that the meniscus of the test and blank solution is raised above the viewing window, or a small metal clip may be obtained from Tintometer Ltd.

References

1. T. J. Giovanniello, T. Peters Jr., *"Standard Methods of Clinical Chemistry"* edited David Seligson. Academic Press New York. Vol. 4, 139 (1963)
2. W. N. M. Ramsay, *"Advances in Clinical Chemistry"* edited H. Sobotka & C. P. Stewart. Academic Press New York Vol. II. (1958)

The Determination of Serum Lactate Dehydrogenase Activity

using lactate and 2:4-dinitrophenylhydrazine

Introduction

Lactate dehydrogenase is an intracellular enzyme and is found in many tissues where it is responsible for the inter-conversion of pyruvic to lactate acid. Cellular damage to these tissues results in a release of enzyme into the blood. Both liver and heart muscle are particularly rich in the enzyme and its assay is used as an aid to diagnosis of conditions affecting these organs.

The normal level of lactate dehydrogenase is 70–240 milli I.U./ml with values of up to twice this level being found during early infancy. Serum L.D.H. activity is raised following myocardial infarction and elevated values can still be found 7–10 days later after the transaminases have returned to normal. Its estimation is therefore often of help in the delayed diagnosis of cardiac infarction. Moderately raised values are also found in acute hepatitis and when the liver contains secondary carcinoma deposits.

Since red blood cells also contain high levels of the enzyme it should be noted that increased levels will be found in the hæmolytic anæmias and those anæmias that are associated with a reduced cell life e.g. the megaloblastic anæmias.

Principle of the method

Lactate dehydrogenase in the presence of nicotinamide adenine dinucleotide (NAD) catalyzes the conversion of L-lactate to pyruvate and reduced NAD ($NADH_2$). The amount of pyruvate produced is determined by the brown colour of the pyruvic dinitrophenylhydrazone which is formed on the addition of alkaline 2:4-dinitrophenylhydrazine.

Reagents required

This method is designed to be used with the LDH Assay kit* marketed by BDH Chemicals Ltd. of Poole, Dorset, England for the determination of Serum Lactate dehydrogenase, but details are given to make up the reagents required, if it is so desired, or should the kits not be available.

*Code number 25010

1. *Buffered substrate*
   Dissolve the contents of the vial in distilled water and make up to 250 ml. Filter if necessary and add one drop of chloroform to prevent bacterial growth. Store at 4°C.
   (Alternatively: 1. Dissolve 1.501 g of glycine and 1.17 g of sodium chloride and make up to 200 ml.
   2. To 125 ml of glycine buffer prepared as above, add 75 ml of 0.1 N sodium hydroxide and 5 ml of 70% sodium lactate solution. This constitutes the buffer substrate solution.)

2. *Nicotinamide adenine dinucleotide (NAD)*
   Dissolve the contents of the vial in 5 ml distilled water. Store at 4°C. Stable for six weeks.
   (Alternatively: Dissolve 25 g of nicotinamide adrenine dinucleotide (NAD) in 5 ml of distilled water.)

3. *2:4-Dinitrophenylhydrazine reagent*
   Dilute the contents of the bottle with 4 volumes of N hydrochloric acid. Store in the dark at room temperature.
   (Alternatively: Dissolve 200 mg of 2:4-dinitrophenylhydrazine in hot N hydrochloric acid, cool, and make up to 100 ml with cold acid. Store in a brown bottle in the dark).

4. *0.4 N Sodium hydroxide*
   Dilute 10 ml 4 N sodium hydroxide to 100 ml with carbon dioxide free distilled water

The Standard Lovibond Comparator Disc 5/56

This disc contains standards representing 83, 167, 250, 333, 416, 500, 583, 667, and 883 milli-International Units/ml.

Technique

1. Hæmolysis-free serum is used for the test and is diluted 1 in 5 before use.
2. Add 1 ml of buffered substrate to two tubes, one labelled 'test' and the other 'blank'.
3. Add 0.1 ml of dilute serum to each tube. Place tubes in water bath at 37°C for 5 minutes to warm.
4. Add 0.2 ml of NAD to the 'Test' tube and 0.2 ml of water to 'Blank' tube. Mix and incubate for 15 minutes exactly.
5. Add 1 ml of 2:4-dinitrophenylhydrazine to each tube. Mix well and remove from water bath.
6. Add 5 ml of 0.4 N sodium hydroxide to each tube and mix.
7. Pour 'test' solution in 13.5 mm comparator cell and place in the right hand compartment of the Lovibond comparator. Fill a second 13.5 mm cell with the 'blank' solution and place in left hand compartment of comparator. Match the colour using Lovibond White Light Cabinet or north daylight and read off L.D.H. activity in m I.U./ml from the indicator window of the Comparator.

Reference

J. King, *J. Med. Lab. Tech.*, 1959, **16**, 265-272

Lange's Colloidal Gold Reaction in Cerebrospinal Fluid

Introduction

Examination of the cerebro-spinal fluid is an essential step in the diagnosis of many diseases of the central nervous system, and in syphilitic involvement of the brain and meninges and in disseminated sclerosis characteristic changes may occur in the proteins of the cerebro-spinal fluid. The globulin content of the fluid is increased and the albumin:globulin ratio lowered. These changes are readily detected by Lange's colloidal gold reaction.

Principle of the method

The test is carried out by adding cerebro-spinal fluid in falling dilutions to ten test-tubes each containing the same amount of gold sol. An increased globulin content and a low albumin:globulin ratio cause precipitation of the gold sol and a change in the colour of the supernatant fluid. The colour change in each of the tubes is compared with a glass standard and the result recorded numerically as described below.

Normal cerebro-spinal fluid causes little or no precipitate of the gold sol and therefore there is no change in the colour of any of the ten tubes. The normal result is 0000000000 or 0000110000. In syphilis of the central nervous system the result obtained depends on the stage of the disease. In general paralysis of the insane, a paretic curve is obtained: complete precipitation of the gold sol occurs in the first four or five tubes where the concentration of cerebro-spinal fluid is greatest, and the result is recorded 5555432100. A similar response is obtained in some cases of tabes dorsalis and meningo-vascular syphilis, and also in some 50 per cent of cases of disseminated sclerosis, but in this last condition the Wassermann reaction is negative. In most cases of tabes dorsalis and meningo-vascular syphilis a luetic curve is obtained, the precipitation being most marked in the middle tubes where there is a medium concentration of cerebro-spinal fluid, e.g. 0123210000. In cases of pyogenic meningitis a meningitic type of response is obtained: this resembles the luetic curve but is shifted to the right, the precipitation occurring in tubes with a lower cerebro-spinal fluid concentration, e.g. 0001244310.

Taken in conjunction with the history, physical findings and other changes in the cerebro-spinal fluid, the colloidal gold reaction is of value in diagnosing syphilis of the central nervous system and disseminated sclerosis, although a positive result is obtained in only half the patients with the latter disease. The reaction is also valuable in distinguishing general paralysis of the insane, in which a paretic curve is obtained and not readily altered by treatment, from other forms of neuro-syphilis showing a paretic curve which as a rule is more readily altered by anti-syphilitic treatment.

Formerly the changes produced by the cerebro-spinal fluid in the colour of the gold sol were determined by the unaided eye. This was difficult for observers with a poor colour memory and led to considerable differences in the results obtained from different laboratories. The development of a Lovibond disc containing standards with which the colour changes can be compared has not only made the results obtained by different workers more uniform, but has allowed the colour of the gold sol used in the test to be standardised.

Reagents required

1. *Saline* 0.4 g pure sodium chloride (NaCl) in redistilled water to 100 ml. This will keep for several weeks if sterilised after withdrawing what is required for immediate use, but otherwise should be made fresh weekly.
2. *Sodium citrate* ($Na_3C_6H_5O_7.2H_2O$) 5% w/v in redistilled water.
3. *Brown gold chloride* 1 per cent solution w/v ($H.AuCl_4 3H_2O$) in redistilled water. Solutions 2 and 3 should be prepared fresh each time the gold sol is prepared.
4. *Gold Sol* Add 4 ml of the gold chloride solution to about 450 ml of distilled water in a 500·ml conical Hysil flask and bring to the boil. When just boiling add 5 ml of citrate solution and continue boiling until volume is reduced to 400 ml. (A mark scratched on the flask is adequate.)
   This gold sol should not be kept for longer than one month.

Note.—All water used should be redistilled from glass, without rubber connections.

The Standard Lovibond Comparator Disc 5/16

The disc is fitted with nine glass colour standards, three to check the gold sol before commencing the test, and six, numbered 0, 1, 2, 3, 4, 5, to determine the degree of precipitation of the gold in the diluted sol.

The three standard colours for the gold sol are labelled
G   representing an average colour;
GY   representing the yellowest sol permissible for satisfactory results;
GB   representing the bluest sol permissible for satisfactory results.

Technique

To check that the gold sol is satisfactory, pour about 5 ml into a clean comparator tube, and place it in the right-hand compartment of the comparator. In the left-hand compartment place another tube containing distilled water. Hold the comparator facing a standard source of white light such as the Lovibond White Light Cabinet or, failing this, north daylight, rotate the disc and note whether the gold sol is matched by one of the three lettered glass standards.

If the gold sol matches G, or falls within the limits shown by GY and GB, it is suitable for the test. Otherwise, it should be rejected.

For each test ten clean dry comparator test tubes are set out in a rack. Put 0.9 ml of the saline in the first tube and 0.5 ml into each of the remaining nine tubes. With a clean dry pipette, 0.1 ml of cerebrospinal fluid is mixed with the 0.9 ml saline in the first tube, 0.5 ml of this is transferred to the 2nd, and after mixing, 0.5 ml of this is transferred to the 3rd tube, and so on until the tenth tube is reached; from this 0.5 ml fluid is taken and thrown away. The dilutions of cerebrospinal fluid in the ten tubes are thus 1 in 10, 20, 40, 80, 160, 320, 640, 1280 2560 and 5120. Add to every tube 2.5 ml of colloidal gold solution. The racks are shaken and put aside for 12 to 24 hours after which each tube in turn is transferred to the right-hand compartment of the Lovibond Comparator, a tube of distilled water being placed in the left-hand compartment and the disc is revolved until the colour standard most nearly approaching the colour of the solution appears in the left-hand aperture. The figure visible at the indicator aperture is recorded.

0   Denotes no change in colour of the diluted sol
1   Bluer red, scarcely purple
2   Purple to lilac
3   Violet to deep blue
4   Light blue to light purple with purplish precipitate, (See Notes below)
5   Complete decolorisation of top fluid with a heavy bluish precipitate

Notes

1. Owing to the sensitivity of the colloidal solution of gold, it is necessary to use special care in the preparation of all the glassware used in the test. The test tubes, in particular, should be cleaned, boiled, washed in weak hydrochloric acid, rinsed thoroughly in running tap water, and dried in an oven at 150°–160°C.

2. Occasionally it is found that a deposit of metallic gold occurs in tubes preceding a 5 or 4, with a resultant loss of colour in the supernatant fluid, which is reddish or purplish. Such tubes are not easy to read, but may be gauged by taking into account the degree of loss of colour and the amount of precipitate, rather than the colour of the upper fluid. They will usually then be found to fall not more than one or two degrees of colour change below those immediately succeeding them.

3. The success of this test is very dependent on the *meticulous* observance of every detail of the technique as set out above. Any departure is likely to lead to failure.

References

1. J. G. Greenfield and E. A. Carmichael, *"The Cerebro-spinal Fluid in Clinical Diagnosis"*, MacMillan & Co., London, (Significance and interpretation of results)
2. P. N. Panton and J. R. Marrack, *"Clinical Pathology"*, J. & A. Churchill, London, 1945, (Technique)

The Determination of Lead
in urine and fæces, using dithizone

Introduction
This method has been developed for the estimation of lead in urine and fæces, as an aid in assessing the degree of exposure to lead workers in certain industries who have not developed evidence of plumbism.

The average excretion of lead in urine is between .003 and .005 mg per 100 ml, while up to .01 mg per 100 ml may be considered to be within normal limits. In fæces an average of .25 mg per 24 hours has been stated. Workers with lead may excrete as much as 10 times these amounts, while in cases of acute lead poisoning even these amounts are greatly exceeded.

Principle of the method
The lead is precipitated by entrainment with calcium oxalate. Interfering substances remain in solution in the supernatant fluid, which is discarded. The precipitate is digested with perchloric acid and the solution thus obtained is treated with ammonia and sodium cyanide. When the mixture is extracted with a chloroform solution of dithizone, a red colour is obtained which is proportional to the amount of lead present.

Reagents required, all of analytical reagent grade
1. *Dilute acetic acid* Glacial acetic acid (CH_3COOH) 10 ml made up to 100 ml with water.
2. *Nitric acid (HNO_3)* Fuming, sp gr 1.5
3. *Ammonia solution (NH_4OH)* .880
4. *Dilute ammonia solution* 10 ml of .880 ammonia made up to 100 ml with water.
5. *Ammonium oxalate [$(COO\ NH_4)_2.H_2O$]* Saturated solution
6. *Calcium chloride ($CaCl_2$) solution* 2.5 g in 100 ml water
7. *Perchloric acid ($HClO_4$) 60%*
8. *Citric acid ($C_6H_8O_7.H_2O$)* 10 g in 100 ml water
9. *Dithizone solution.* 4 mg dithizone ($C_6H_5.N:N.CS.NH.NH.C_6H_5$) in 100 ml chloroform. This must be freshly made.
10. *Sodium cyanide ($NaCN$) solution* 5 g recrystallised in 100 ml water
11. *Chloroform ($CHCl_3$)*
12. *Indicators*—Bromo-thymol blue and Bromo-cresol green

The Standard Lovibond Comparator Disc 5/17
The disc includes standards for the range 0.002 to 0.02 mg of Lead (Pb) in steps of .002, which will cover subjects exposed to lead hazards, and patients actually suffering from lead poisoning. These values are equivalent to 0.2 to 2.0 ppm if a 10 ml sample is taken.

Technique
Lead in Urine

Adjust 10 ml of the urine to approximately pH 4.5 by the addition of dilute acetic acid or dilute ammonia solution, using bromo-cresol green as indicator. Place in a centrifuge tube, add 1 ml saturated ammonium oxalate solution and 0.5 ml calcium chloride solution, stand for 20 minutes and then centrifuge. Decant the supernatant fluid, allow the tube to drain on a filter paper for 2—3 minutes, and digest the precipitate with 0.5 ml perchloric acid until the solution is colourless. To the cooled contents of the tube add 1 ml of water, 1 ml dilute citric acid, 0.1 ml of bromo-thymol blue indicator, and concentrated ammonia drop by drop until the mixture is just blue. Transfer to a 25 ml separating funnel containing 1 ml of the dithizone solution and 1 ml of sodium cyanide solution, and shake until the red colour in the chloroform reaches a maximum. The chloroform layer should then be run off and the extraction repeated with a further 1 ml of dithizone solution. If the second chloroform extract shows an appreciable pink colour, a third or even a fourth extraction with 1 ml portions of the dithizone solution must be made. The combined chloroform extracts are then placed in a separating funnel and shaken with 10 ml of water, 0.1 ml of sodium cyanide solution and 0.1 ml of concentrated ammonia, when the green colour due to excess of reagent goes into the aqueous layer. Transfer the chloroform layer to a graduated vessel, make up to 5 ml with chloroform, and place in a test tube or 13.5 mm cell in the right-hand compartment of the comparator. Parallel with the above work, a "blank" test is carried out, using 10 ml of distilled water in place of urine: this blank is placed in the left-hand compartment of the comparator.

Hold the comparator facing a standard source of white light such as the Lovibond White Light Cabinet, or failing this, north daylight, and rotate the disc until the colour of the solution is matched by one of the Lovibond permanent glass standards. The value, expressed as mg of lead (Pb) in the original volume of sample taken, is read from the indicator at the bottom right-hand corner. If the colour is deeper than the darkest standard, the test must be repeated using a smaller quantity of the original sample.

Lead in Fæces

Place 1 g of dried fæces in a 500 ml beaker. Carefully cover with pure fuming nitric acid and gently warm until the vigorous reaction has ceased. Evaporate the solution, with the addition of more nitric acid if necessary, until it consists of a dark brown clear liquid containing a little insoluble matter and a little fat. After cooling, the solution is diluted to 100 ml and filtered, and 10 ml of this solution are treated as above described for urine. The reading obtained from the comparator disc, multiplied by 10, will represent the amount of lead in the 1 gram sample taken, if the above quantities have been adhered to throughout. Appropriate adjustments must be made in the calculations according to any alterations made in the amount taken.

Note

All glassware used for this test must be scrupulously clean; it should be boiled with nitric acid and thoroughly rinsed with distilled water before use.

References

1. Aub *et al.*, *"Lead Poisoning,"* Medicine Monograph No. 7, London & Baltimore, 1926
2. C. C. Lucas and J. R. Ross, *J. Biol. Chem.*, 1935, **111**, 285
3. R. A. Kehoe, F. Thamann and J. Cholak, *J. Amer. Med. Assn.*, 1935, **104**, 90

The Determination of Nicotinic Acid
using phosphomolybdic acid

Introduction

Nicotinic acid has been identified as one of the constituents of the Vitamin B complex. It is used in medicine for the treatment of deficiency diseases, such as pellagra, and of other nutritional disorders. The estimation of the nictonic acid content of foodstuffs is therefore important, particularly in the case of some special dietary preparations. Its estimation in urine affords a method of control in cases where nicotinic acid is being administered for medicinal purposes. The present method was developed by Daroga[1], for use with the Tintometer but has been adapted for use with the Lovibond Comparator.

Principle of the method

The nicotinic acid is precipitated with phosphomolybdic acid. After removal of excess reagent the nicotinic acid-phosphomolybdate complex is reduced with stannous chloride to produce the molybdenum blue colour, which is compared with permanent glass standards in the Lovibond Comparator.

Reagents required

1. *Phosphomolybdic acid solution*,
Phosphomolybdic acid ($H_3PO_4.12MoO_3.24H_2O$)	2.5 g
Hydrochloric acid (HCl 20%)	20 ml
Distilled water	20 ml

   The phosphomolybdic acid is dissolved, by warming, in the hydrochloric acid and water.
   The solution is cooled, filtered, and the filtrate diluted to 50 ml.
   This reagent is stable for 4 weeks if stored in darkness.

2. *Wash liquid*
Glacial acetic acid (CH_3COOH)	5 ml
Distilled water	95 ml

3. *Sodium hydroxide*   0.1 N approx.

4. *Reducing solution*
Stannous chloride ($SnCl_2.2H_2O$)	10 g
Hydrochloric acid (HCl conc.)	25 ml

   This reagent deteriorates on keeping and should be freshly prepared on the day it is to be used.

The Standard Lovibond Comparator Disc 5/33

This disc covers the range 0.4 to 1.2 mg of nicotinic acid, in steps of 0.1.

Technique

To 1 ml of the sample solution (which should contain 0.1–1.0 mg of nicotinic acid) in a 30 ml beaker, add 0.2–0.3 ml of reagent 1. Warm the solution until the precipitate has dissolved, adding a further 0.1 ml of reagent 1 if required. Cool the solution to room temperature and allow to stand for at least 2 hours. Collect the precipitate, using a sintered glass crucible of porosity G4, and wash the crucible and beaker with three successive portions of 1 ml of reagent 2, shutting off the suction during each addition of wash liquid. After the final washing suck the crucible dry for 2 minutes and then empty and wash the filter flask.

Pipette 1–2 ml of reagent 3 into the crucible and allow to stand without suction for 2–3 minutes to dissolve the precipitate. Then suck the solution into the flask and wash the filter with 15 ml of water at 25°C. Transfer the liquid and washings from the filter flask to the beaker used for the precipitation and add 1 ml of reagent 4. Transfer the solution to a 50 ml flask and dilute to the mark with distilled water. Maintain the flask at 25°C for 5 minutes to develop the full colour. 10 minutes after the addition of reagent 4 fill a 13.5 mm test tube to the 10 ml mark with the solution from the flask. Place this tube in the right-hand compartment of the comparator and immediately compare the colour of the solution with the standards in the disc using a standard source of white light such as the Lovibond White Light Cabinet or, failing this, north daylight.

Notes

1. Maximum colour is obtained within 5 minutes at 25°C and the colour commences to fade after 15 minutes. Matching of the colour with the standards should, therefore, always be carried out 10 minutes after the reducing agent (reagent 4) is added.

2. If the concentration of nicotinic acid falls outside the limits of the standard colours the concentration of the test solution should be adjusted by concentration, or dilution, as required. It has been demonstrated[1] that a solution of nicotinic acid in excess of dilute hydrochloric acid may be evaporated without loss.

3. Extraction of nicotinic acid from foodstuffs

To a suspension of the foodstuff in 250 ml of distilled water add 3 g of calcium oxide and 20 g of sodium chloride. Pass a current of steam through the heated suspension and collect 400 ml of distillate in 20 ml of 20% hydrochloric acid. Evaporate the distillate to about 2 ml and determine the nicotinic acid in an aliquot as described.

4. Extraction of nicotinic acid from urine

Two compounds known to occur in normal human urine, trigonelline and methylpyridinum hydroxide, are likely to be precipitated by reagent 1. These compounds must, therefore, be removed from the urine before releasing the nicotinic acid.

Dilute 100 ml of urine to 250 ml with distilled water. Steam distil for 20 minutes and reject the distillate. Allow the residual mixture of urine and water to cool, add 3 g of calcium oxide and 20 g of sodium chloride and proceed as in Note 3.

Reference

1. R. P. Daroga, *J.S.C.I.,* 1941, **60,** 263

The Determination of pH of Urine
Using Universal Indicator

Introduction

One of the main functions of the kidneys is to maintain the acid-base equilibrium of the blood. Acids are constantly being formed as carbon-dioxide is produced by the oxidation of foodstuffs, and as phosphoric and sulphuric acids are produced by the oxidation of phosphorus and sulphur contained in ingested protein. On the other hand vegetable foods contain basic ions such as sodium, potassium and calcium which tend to make the reaction of the blood more alkaline. By the appropriate excretion of acidic or basic radicles the kidneys maintain the pH of the blood within narrow limits. When alkalæmia tends to occur, alkaline urine is excreted; in acidæmia, acid urine is excreted. The extreme range of pH changes in the urine which healthy kidneys can achieve is from 4.8 to 7.9.

Acid Urine

Freshly voided urine is usually slightly acid if the subject is eating an ordinary mixed diet. The degree of acidity is increased by a high protein intake. Marked acidity of the urine occurs in uncontrolled diabetes mellitus when aceto-acetic and hydroxybutyric acids are formed and eliminated in the urine. Very acid urines are also formed in such conditions as emphysema and chronic bronchitis, when the inadequate ventilation gives rise to a respiratory acidosis.

In the treatment of urinary infections, the urine may be made acid by giving ammonium chloride, or calcium or ammonium mandelate. The optimal degree of acidity in a freshly voided urine sample is pH 5.0.

Alkaline Urine

The reaction of urine may become alkaline after a meal—the so-called alkaline tide caused by the secretion of hydrochloric acid into the stomach. The reaction of the urine also tends to be alkaline on a vegetarian diet. A strongly alkaline urine is often found when there is infection of the renal tract with urea-fermenting organisms, such as *B. Proteus*, which produce ammonia. An alkaline urine is excreted in respiratory alkalosis such as occurs in hysterical hyperventilation, in which the patient exhales large amounts of carbon-dioxide. The urine is also alkaline in conditions which cause metabolic alkalosis such as vomiting in pyloric stenosis and the ingestion of large amounts of alkali.

Urine is deliberately made alkaline in the treatment of certain urinary infections and during sulphonamide therapy by giving sodium or potassium citrate or bicarbonate. The night urine is the most difficult to make alkaline and when the pH is greater than 7.0 sufficient alkalis are being given.

Principle of the method

The pH of the freshly voided urine is determined with Universal Indicator, the colour produced being compared with a series of Lovibond permanent glass standards representing the colours produced by solutions of known pH value.

Reagent required

Universal Indicator.

The Standard Lovibond Comparator Disc 2/1 P

This covers the pH range from 4.0 to 11.0 in steps of 1.0.

Technique

Fill both test tubes or cells to the 10 ml mark with the freshly voided urine. To the right-hand tube only add 0.2 ml Universal Indicator. Do not immerse the tip of the pipette beneath the surface of the urine. Carefully mix by stirring with a clean glass rod or by pouring into another tube and back again. Hold the comparator about 18 inches in front of the eye facing a standard source of white light such as the Lovibond White Light Cabinet or, failing this, north daylight and rotate the disc until the colours match. The pH of the specimen is shown in the indicator recess at the bottom of the right-hand corner.

Phenolsulphonphthalein (Phenol Red) Excretion Test

Introduction
Phenolsulphonphthalein (P.S.P.) is excreted mainly by the kidney, up to 94% being cleared by tubular excretion. The clearance of P.S.P., which is determined by the renal plasma flow, lies between that of p-amino hippuric acid and insulin. While a clinical plasma clearance rate may be determined for P.S.P. it is possible by measuring its excretion during a specific time period to obtain an estimate of renal efficiency without resorting to the time consuming and more complicated standard clearance methods. Prior to the administration of potentially toxic antibiotics, this test may be used as a check on adequate renal function for their excretion.

Principle of the method
After ingestion of a quantity of water to induce an adequate urine flow, P.S.P. is injected intravenously and the urine collected over set time periods. From the red colour developed following the addition of alkali to the urine, the amount of P.S.P. excreted is estimated and from this amount, expressed as a percentage of the total dose, the percentage excretion during a particular time interval is calculated.

Several different time periods have been suggested for the collection of urine, the more common ones being 0–15 mins., 15–30 mins., 0–60 mins. and 0–120 mins. The use of a short collection period increases the sensitivity of the test but introduces the disadvantage that the patient requires catheterization to ensure accurate urine collection.

The normal values found in the varying time periods are

0–15 mins.	25–50%	15–30 mins.	12–28%
0–60 mins.	40–80%	0–120 mins.	60–85%

These values are reduced when there is impairment of renal function and may indicate renal damage before there has been an increase in blood urea.

Reagents required
1. *25% Sodium hydroxide (NaOH)* Dissolve approximately 25 g of NaOH in 100 ml of water. Stable for at least 1 year.
2. *Phenolsulphonphthalein* Injection of 6 mg per ml.

The Standard Lovibond Comparator Disc 5/44
This disc covers the range 5 to 45% excretion in 9 steps of 5%.

Technique
1. Patient given 300–400 ml of water to drink.
2. 30 minutes later patient empties bladder and then 1 ml of P.S.P. solution containing 6 mg per ml is injected intravenously.
3. Urine collected at 15 mins., 30 mins., 60 mins. or 120 mins. as required.
4. Entire urine collection for a particular time period transferred to a 1,000 ml volumetric flask.
5. 5 ml of 25% NaOH added, mixed, and the flask made up to volume with water.
6. 13.5 mm comparator cell filled and placed in right-hand compartment with a similar cell filled with water in left-hand compartment.
7. The colour is matched with those of disc 5/44 using White Light Cabinet or north daylight.

Notes

1. Urine may be collected for any or all of the time periods suggested. Where the 0–15, 15–30 or 0–30 minute periods are used the patient should be catheterized. In addition, if urine is only collected for the 0–15 minute period the bladder should be rinsed with 100 ml of sterile water or saline and the washings added to the urine collection.

2. The P.S.P. may be given by intramuscular injection. This is not recommended where urine is only collected for 15 or 30 minutes, due to the uncertain period of absorption. When intramuscular injection is used, urine should be collected at 70 mins. or 130 mins. to allow for adsorption. The normal excretion values then equate to those when intravenous injection is used.

3. There may be a variation of up to 17% in the concentration of the P.S.P. injection so that it is advisable to standardise each batch obtained and compute a correction factor if necessary.
    (a) Wash contents of one ampoule into a 1,000 ml volumetric flask.
    (b) Add 5 ml of 25% NaOH and make to mark with water.
    (c) Add 2 ml of dilute P.S.P. solution to a graduated 13.5 ml comparator cell (Vol. A), fill to the 7 ml mark and mix well.
    (d) Compare colour with the 25% excretion colour glass in the comparator.
    (e) If necessary add water to the comparator cell, mixing well after each addition until the colour matches that of the 25% standard.
    (f) Note the final colume (Vol. B).

Calculation

$$\text{Correction factor} = \frac{\text{Vol. A}}{\text{Vol. B}} \times \frac{100}{25}$$

True excretion = Observed reading × Correction Factor.

References

1. L. G. Rowntree, & J. T. Geraughty, *Arch. Internal. Med.* 1912, **9**, 284
2. R. J. Henry, S. L. Jacobs, & S. Berkman, *Clin. Chem.* 1961, **7**, 231
3. E. J. Storen, *Acta Chir. Scand.* 1963, **125**, 177

The Determination of Alkaline or Acid Phosphatase (1)

in serum and plasma
using phosphotungstic-phosphomolybdic acid

Phosphatase is an enzyme which hydrolyses phosphoric esters with the liberation of inorganic phosphate. Two types, present in plasma, are of importance in clinical medicine—alkaline phosphatase, so-called because its activity is greatest at pH 9.3, and acid phosphatase which is most active at pH 5.0.

ALKALINE PHOSPHATASE

Introduction

Alkaline phosphatase is present in plasma, leucocytes and many tissues, especially bone and kidney. Its estimation is of value in certain disorders of bone and in liver disease.

Bone Disease

Osteoblasts in bone secrete alkaline phosphatase, which hydrolyses phosphoric esters brought in the blood-stream and thereby increases locally the concentration of phosphate ions. The increased amount of phosphate combines with calcium and the resulting product is deposited in and around the osteoblastic cells. Bone diseases in which there is increased osteoblastic activity are associated with an increase in plasma alkaline phosphatase and the level rises above normal range of 3 to 12 King-Armstrong units per 100 ml. High concentrations of alkaline phosphatase are found in osteomalacia, a condition due to several causes, but in which the underlying abnormality is either too little inorganic phosphate or too little calcium to allow the precipitation of calcium phosphate in the bone tissue. The main causes of osteomalacia are rickets and steatorrhœa, and the severity of the bone lesion is reflected in the height of the alkaline phosphatase level. The estimation is not only of particular value in the diagnosis of osteomalacia but also in following the response to treatment: the alkaline phosphatase level falls but remains above normal for some months during active bone repair. Very high phosphatase levels (over 50 King-Armstrong units) are found in Paget's disease (osteitis deformans), especially when the disease is active and widespread, and in extensive carcinomatous metastases in the bone. Moderately high levels occur in hyperparathyroidism when bone involvement is present (osteitis fibrosa cystica), and slightly increased activity is found in some cases of healing fractures and osteogenic sarcoma. As a rule normal values are found in multiple myelomatosis.

Liver Disease

Phosphatase is excreted by the liver into the bile. An increase in serum alkaline phosphatase occurs almost invariably in obstruction of the biliary passages, and is of great diagnostic value in distinguishing between obstructive and infective jaundice. In infective hepatocellular disease the level seldom exceeds 30 units, whereas in obstructive jaundice the level is raised above 30 units as a rule, and if the obstruction is complete this increase is present from an early stage.

Principle of the method

The amount of alkaline phosphatase present is measured by determining the amount of phenol liberated from phenylphosphate when the enzyme and substrate are incubated for thirty minutes at 38°C[1]. One King-Armstrong unit is defined as the amount of phosphatase per 100 ml plasma that will liberate 1 mg phenol in the given time under the conditions of the test. (See Note).

COLORIMETRIC CHEMICAL ANALYTICAL METHODS — Alkaline or Acid Phosphatase (1)

Reagents required

1. *Alkaline Buffer Substrate* M/200 disodium mono-phenylphosphate in M/20 sodium barbitone—1.09 gm disodium mono-phenylphosphate and 10.3 gm sodium barbitone dissolved in water and made up to 1 litre. This should be preserved with a few drops of chloroform and should be kept in a refrigerator.

2. *Folin and Ciocalteu dilute phenol reagent* Dissolve 100 g sodium tungstate ($Na_2WO_4.2H_2O$) and 25 g sodium molybdate ($Na_2MoO_4.2H_2O$) in 700 ml water in a 1,500 ml flask. Add 50 ml syrupy (89 per cent w/w) phosphoric acid (H_3PO_4) and 100 ml of concentrated hydrochloric acid (HCl). Boil gently under a reflux condenser for 10 hours. Then add 150 g lithium sulphate, 50 ml water and a few drops of bromine. Cool and dilute with water to 1,000 ml. Store in a glass-stoppered bottle with a glass—or paper—cover to exclude dust. The finished reagent must not have a green tint.
Before use dilute 1 part with 2 parts of water.

3. *Sodium carbonate solution* 20% Na_2CO_3 (w/v). This should be kept in a warm place to prevent the carbonate from crystallizing out.

4. *Physiological (Normal) saline* 0.85% sodium chloride solution. This is required for diluting the serum when it has a content of more than 60 units.

The Standard Lovibond Comparator Disc 5/12

The disc covers the range from 8 to 60 units of phosphatase per 100 ml of serum or plasma, the steps being 8, 12, 16, 20, 25, 30, 40, 50, 60 King-Armstrong units.

Technique

Test Measure 4 ml of buffer substrate into a test tube and place in a water-bath at 38° for 5 minutes. Without removing the tube from the bath, add exactly 0.2 ml of serum or plasma (which must be cell-free, as otherwise phosphate from red cells is added to the result). Stopper, mix, and allow to remain in the bath exactly 30 minutes. At the end of this time add at once 1.8 ml of the dilute phenol reagent, mix and centrifuge of filter through a small (5.5 cm.) filter paper.

Blank At the same time prepare a "blank" solution by mixing 4 ml of buffer substrate and 0.2 ml of serum or plasma. Add *immediately* 1.8 ml of diluted phenol reagent, without heating, and filter or centrifuge.

Pipette 4 ml of filtrate from the test and "blank" solutions into test tubes, and 1.0 ml of the sodium carbonate solution, mix, and place the test tubes in the water-bath (38°) for 10 minutes to develop the colour.

Transfer to comparator test tubes or 13.5 mm cells and place the vessel containing the test solution in the right-hand compartment of the comparator, and the "blank" in the left-hand compartment, so that it comes behind the colour standards in the disc. Hold the comparator facing a standard source of white light such as the Lovibond White Light Cabinet or, failing this, north daylight and rotate the disc until the colour of the test solution is matched by one of the glass standards. The value, expressed in King-Armstrong units per 100 ml of serum or plasma, (equal to mg of phenol liberated in 30 minutes), is then read from the indicator window at the bottom right-hand corner of the comparator.

Technique when the enzyme content is very high

So long as the enzyme content of the serum is less than 60 units, the colorimetric reading is carried out as described above. When it is very high it is customary to dilute the serum or plasma with normal saline, so that the number of units per 100 ml of diluted serum will not be in excess of 60.

Note: The modified phosphatase procedure described by King, Haslewood and Delory[2] can also be carried out with this Standard Disc.

ACID PHOSPHATASE

Introduction

Acid phosphatase is present in plasma and many tissues, especially prostatic epithelium. The normal level in plasma is 1 to 4 King-Armstrong units per 100 ml. In obtaining the blood sample it is most important to avoid hæmolysis, because red corpuscles contain considerable amounts of acid phosphatase and falsely high values will be obtained if the red cells are lysed. High plasma levels of acid phosphatase (over 10 units per 100 ml) are found in patients with a prostatic carcinoma when there are metastases with invasion of lymphatics and blood vessels. Thus estimation of the plasma acid phosphatase level is of value in confirming the diagnosis of carcinoma of the prostate with metastases, and is essential in controlling the treatment of this condition with stilbœstrol, sufficient oestrogens being given to reduce the plasma acid phosphatase to normal levels.

A slight increase in acid phosphatase (less than 10 units) may occur in advanced Paget's disease, hyperparathyroidism, and carcinoma of the breast with skeletal metastases.

Principle of the method

This is essentially the same as that described above for estimating alkaline phosphatase except that the reaction is carried out in an acid buffer at pH 4.9.

Reagents required

1. *Acid Buffer Substrate* Dissolve 21.0 g of crystalline citric acid ($C_6H_8O_7.H_2O$) in water, add 188 ml of N sodium hydroxide (NaOH), and make up to 500 ml. Adjust the pH value to 4.9 by dropwise addition of N NaOH or N hydrochloric acid (HCl). Dissolve 1.09 g of disodium mono-phenylphosphate ($Na_2C_6H_5PO_4$) in water, add to the above, and finally dilute to 1 litre.
   Preserve with a few drops of chloroform, and keep in a refrigerator.
2, 3, and 4 as for alkaline phosphatase

The Standard Lovibond Comparator Disc 5/12

The same disc is used as for alkaline phosphatase.

Technique

Exactly as for alkaline phosphatase, but using the acid instead of the alkaline buffer substrate. Incubate for one hour instead of 30 minutes. The result is expressed in King-Armstrong units per 100 ml of serum or plasma (equal to mg of phenol liberated in 1 hour.)

The colours are rather pale for normal sera, and also for prostate cases in which the acid phosphatase is not greatly increased. In such instances it may be preferable to use a longer incubation period, e.g. 3 hours. If this is done, the mg phenol liberated must be divided (by e.g. 3) to reduce the figure to 1 hour.

Note

There are a number of different "units" of phosphatase, and techniques for estimating phosphatase, in the literature, and each has a different normal range. It is essential, therefore, to state the method used when quoting results. The Joint Sub-Commission on Clinical Enzyme Units of the International Union of Biochemistry has recommended that a unit of enzyme activity be defined as that which under specified conditions utilises one micromole of substrate per minute per litre of specimen. The position with regard to phosphatase is made more complicated by the fact that the activity of the enzyme varies according to the substrate used. For the King and Armstrong method given here, one King-Armstrong unit/100 ml, is equivalent to 7.1 mI.U's for alkaline phosphatase and 1.8 mI.U's for acid phosphatase.

References

1. E. J. King and A. R. Armstrong, *Canad. Med. Ass. J.,* 1934, **31,** 376
2. E. J. King, G. A. D. Haslewood and G. E. Delory, *Lancet,* 1937, **i,** 886
3. A. B. Gutman and E. B. Gutman, *J. Clin. Invest.,* 1938, **17,** 473
4. N. F. MacLagan, *Brit. Med. J.,* 1947, **ii,** 197

The Determination of Alkaline and Acid Phosphatase (2)

in serum, using 4–amino antipyrine

Principle

Phosphatase present in serum is used to split phenol from phenylphosphate. In alkaline solutions the phenol liberated reacts with 4–amino antipyrine to produce a red coloured complex. Serum is therefore incubated at 37°C with disodium phenyl phosphate at either an acid or alkaline pH according to the type of enzyme being determined with the liberation of phenol. The colour developed on addition of 4–amino antipyrine is proportional to the phenol liberated. The enzyme activity is expressed in King-Armstrong units, one unit of which is defined as that amount of activity which will liberate 1 mg of phenol under the conditions of the experiment.

Reagents—Alkaline phosphatase

1. *Alkaline buffer pH 10* Dissolve 6.36 g of anhydrous sodium carbonate and 3.36 g of sodium bicarbonate in a litre of water. Store at 4°C.
2. *Phosphatase substrate* Dissolve 2.18 g disodium phenylphosphate in a litre of water. Preserve with a few drops of chloroform. Store at 4°C.
3. *Sodium hydroxide 0.5 N* 20 g sodium hydroxide per litre.
4. *Sodium bicarbonate* 42 g sodium bicarbonate per litre.
5. *4–Amino antipyrine* Dissolve 6 g of 4–amino antipyrine in a litre of water. Keep in a brown bottle.
6. *Potassium ferricyanide* Dissolve 24 g of potassium ferricyanide in a litre of water.

Standard Alkaline Phosphatase Lovibond Comparator Disc 5/53

This covers the range 2—18 King-Armstrong units/100 ml in steps of 2 K.A. units.

Technique—Alkaline phosphatase

1. Into two test tubes marked 'Blank' and 'Test' respectively add 1 ml of acid buffer (reagent 1) and 1 ml of phosphatase substrate (reagent 2).
2. Warm the 'Test' tube and then add 0.2 ml of serum.
3. Incubate in 37°C water bath for 15 minutes.
4. Add 0.8 ml of $0.5N$ sodium hydroxide (reagent 3) and 1.2 ml of $0.5N$ sodium bicarbonate (reagent 4) to each tube.
5. Add 0.2 ml of serum to 'Blank' tube.
6. Add 1 ml of 4–amino antipyrine (reagent 5) and 1 ml of potassium ferricyanide (reagent 6) to each tube.
7. Pour 'Test' into 13.5 mm comparator cell and place in right-hand compartment of Lovibond comparator. Pour 'Blank' into a similar cell and place in left-hand compartment. Match colours using White Light Cabinet or north daylight and read off alkaline phosphatase activity in King-Armstrong (K.A.) units per 100 ml from indicator window.

Reagents—Acid phosphatase

The reagents are as for alkaline phosphatase except for the buffer:—
Acid buffer pH 4.9

Dissolve 29.41 grams sodium citrate in $0.2N$ hydrochloric acid and make up to 500 ml with more acid. Preserve with a few drops of chloroform and store at 4°C.

Standard Lovibond Comparator Disc 5/53 as for alkaline phosphatase

Technique—acid phosphatase

1. Into two test tubes marked 'Blank' and 'Test' respectively add 1 ml of alkaline buffer and 1 ml of phosphatase substrate (reagent 2).
2. Warm the 'Test' tube and then add 0.4 ml of serum.
3. Incubate in 37°C water bath for one hour.
4. Add 1 ml of 0.5N sodium hydroxide (reagent 3) and 1 ml of sodium bicarbonate (reagent 4) to each tube.
5. Add 0.4 ml of serum to 'Blank' tube.
6. Add 1 ml of 4-amino antipyrine (reagent 5) and 1 ml of potassium ferricyanide (reagent 6) to each tube.
7. Pour 'Test' into 13.5 mm comparator cell and place in right-hand compartment of comparator. Pour 'Blank' into a similar cell and place in left-hand compartment. Match colour using White Light Cabinet or north daylight. Divide the figure that appears in the indicator window by 2 to give the acid phosphatase activity in King-Armstrong (K.A.) units per 100 ml of serum.

Note

Should the enzyme activity be higher than the range on the disc, repeat the test having first diluted the serum with an equal volume of saline and then read the indicated value.

References

1. P. R. N. Kind and E. J. King, *J. clin. Path.* 1954, **7**, 322
2. E. J. King and K. A. Jegatheesan, *J. clin. Path.* 1959, **12**, 85

The Determination of Formol Stable Acid Phosphatase

Introduction
Acid phosphatases are present in a number of tissues of the body. Several methods have been used to attempt to differentiate prostatic acid phosphatase from those derived from other tissues. It has been found that formalin inhibits many other acid phosphatases with the exception of that derived from the prostate, and the determination of formol stable acid phosphatase can be helpful in investigating suspected cases of carcinoma of the prostate.

Principle
Neutral formalin is added to the standard acid phosphatase reaction mixture to inhibit all but the prostatic enzyme.

Reagents required
The reagents are as for acid phosphatase (Method 2) using 4-amino antipyrine with the addition of:—

Neutral formaldehyde To 100 ml of 40% formalin add one drop of phenolphthalein solution. To this add 0.5 N sodium hydroxide drop by drop until the solution is just pink.

Standard Lovibond Comparator Disc 5/55
This disc covers the range 1 to 9 King-Armstrong units/100 ml in steps of 1 K.A. unit for formol stable acid phosphatase.

Technique
This is exactly the same as for acid phosphatase determination using 4-amino antipyrine except that 0.5 ml of neutral formalin is added to the 'Blank' and 'Test' tubes prior to incubation for one hour.

Reference
M. A. M. Abdul Fadl, and E. J. King, *J. clin. Path*. 1948, **1**, 80

The Determination of Phosphorus
using amino-naphthol-sulphonic acid
IN BLOOD

Introduction

Normal adult blood contains 2 to 3 mg of inorganic phosphate (expressed as phosphorus) per 100 ml, whereas the blood of children contains somewhat more, about 4 to 5 mg per 100 ml plasma. Phosphorus is present in blood as inorganic phosphate and also as phosphate bound to other substances (so-called ester or organic phosphates). In clinical work only the inorganic phosphorus is usually of importance, and after the blood has been taken by venepuncture the estimation should be made at once with the freshly-drawn blood, or the plasma should be separated as soon as possible in order to avoid the formation of extra inorganic phosphate by enzymic hydrolysis of the ester phosphates.

Estimation of the plasma inorganic phosphate level is of value in the diagnosis of certain bone disorders, especially rickets and diseases of the parathyroid glands.

Hyperphosphatæmia

Increased amounts of inorganic phosphate are found in the plasma of patients taking excessive amounts of vitamin D; in patients with hypoparathyroidism, because the parathyroid hormone regulates the amount eliminated in the urine, and if parathyroid function is diminished the serum calcium falls and the plasma phosphate rises in proportion; and in patients with renal insufficiency in its later stages.

Hypophosphatæmia

A low plasma phosphate level is often found in rickets and other causes of osteomalacia, such as steatorrhœa, in which the absorption of phosphates or calcium is defective, due to a deficiency of vitamin D. In rickets a level of 1 to 2 mg per 100 ml is commonly found, and usually associated with a high serum concentration of alkaline phosphatase. In other cases of rickets the phosphate level is normal but the serum calcium is reduced.

In hyperparathyroidism excessive secretion of the parathyroid hormone causes increased urinary excretion of phosphorus and increased mobilisaiton of calcium from the bones. As a result low levels of phosphorus and high levels of calcium are found in the blood. In long-standing cases of hyperparathyroidism, calcium phosphate salts may be deposited in the renal tissues and cause renal insufficiency. The resulting retention of phosphorus may then mask the usual biochemical findings, because the serum phosphate level is raised to within normal limits, or above, and the serum calcium level is reduced accordingly to 10 to 12 mg per 100 ml.

Principle of the method

The method is based on King's modification[3] of the Fisk and Subbarrow procedure.[1] The inorganic phosphate (i.e. the phosphoric ion in the protein-free filtrate of blood) is combined with molybdic acid, and reduced with amino-naphthol-sulphonic acid to give a blue colour, which is proportional to the amount of phosphate ion.

The total phosphate (inorganic plus ester phosphates) can be measured in the same way after destroying the organic matter and liberating free phosphate, perchloric acid being used for this purpose.

Reagents required

1. *Trichloracetic acid* (CCl_3COOH) 10%
2. *Perchloric acid* $(HClO_4)$ 60%
3. *Ammonium molybdate* $[(NH_4)_6Mo_7O_{24}.4H_2O]$ 5%
4. *Amino-naphthol-sulphonic acid solution*
   0.2 g 1:2:4-amino-naphthol-sulphonic acid, 12 g sodium metabisulphite $(Na_2S_2O_5)$, and 2.4 g sodium sulphite (Na_2SO_3) dissolved in 100 ml distilled water.
   Keep in a dark bottle, and make fresh every two weeks
5. "Porous pot," small pieces about $\frac{1}{8}''$ diameter

The Standard Lovibond Comparator Disc 5/14

The disc covers the range 0.02 mg to 0.16 mg of Phosphorus (as P) per ml, in steps. 02, .03, .04, .05, .06, .08, .1, .12, .16 mg.

Technique

Inorganic Phosphate

Add 2 ml of freshly-drawn blood or oxalated plasma to 8 ml of 10% trichloracetic acid, shake and filter. Place 5 ml of filtrate (equivalent of 1 ml of blood or plasma) in a 10 ml test tube, and add 0.7 ml. perchloric acid, 0.7 ml ammonium molybdate, and 0.5 ml amino-naphthol-sulphonic acid solution, and finally distilled water to the 10 ml mark. Mix, and leave for 10 minutes for the colour to develop. At the end of this time, place the solution in a comparator test tube or 13.5 mm cell in the right-hand compartment of the comparator, and a "blank" of distilled water in the left-hand compartment so that it comes behind the colour standards in the disc. Hold the comparator facing a standard source of white light such as the Lovibond White Light Cabinet or, failing this, north daylight and rotate the disc until the colour of the liquid is matched by one of the glass standards. The value expressed in mg of phosphorous in the test solution (i.e. in 1 ml of the blood or plasma) is then read from the indicator window at the bottom right-hand corner of the comparator. Multiplied by 100, this result gives mg inorganic phosphate (as P) per 100 ml of blood or plasma.

Total Phosphate

Place 5 ml of the trichloracetic acid filtrate from plasma, or 0.5 ml of the filtrate from whole blood, in a test tube, add 1 ml of perchloric acid and a small piece of porous pot. Heat the tube carefully with a micro-burner or on an electric heater at such a rate that the liquid boils gently. When this has evaporated to a small volume the solution turns black or brown, and white fumes of perchloric acid are evolved. Continue the heating until the hot mixture turns yellow and then colourless. Remove from heater and allow to cool, then dilute to about 5 ml with distilled water. Add 0.7 ml ammonium molybdate, 0.5 ml amino-naphthol-sulphonic acid solution, and distilled water to 10 ml. The solution is then transferred to the comparator test tube or 13.5 mm cell and the reading taken as described above.

When plasma is used, the result is multiplied by 100, and when whole blood is used the result is multiplied by 1,000 to obtain mg per 100 ml of total trichloracetic acid-soluble phosphate. Ester phosphate = total phosphate—inorganic phosphate.

IN URINE

The amount of inorganic phosphate in a 24-hour collection of urine depends largely on the phosphate dietary intake and usually ranges from 1.5 to 3 g per day. Estimation of the urinary excretion of phosphate is of little clinical value except in balance studies carried out to investigate cases of metabolic bone disease.

Technique

Accurately measure 0.2 ml of urine into a test tube. Add water, perchloric acid, molybdate and amino-napthol-sulphonic acid solution as in the estimation of inorganic phosphate in the trichloracetic acid filtrate from blood.

The figure shown at the indicator window gives the mg P contained in the measured volume of urine. If the colour is too dark a lesser volume of urine is taken.

References

1. C. H. Fiske and Y. Subbarrow, *J. Biol. Chem.*, 1925, **66**, 375
2. H. D. Kay, *J. Biol Chem.*, 1932, **99**, 85
3. E. J. King, *Biochem. J.*, 1932, **26**, 292
4. G. Stearns and E. Warweg, *Amer. J. Dis. of Child.*, 1935, **29**, 79

The Determination of Proteins (1)
using copper sulphate
(Biuret method)

Introduction

Urine In health there is no appreciable amount of protein in urine, but in many diseases proteinuria occurs, and this is revealed by ordinary qualitative tests. Unfortunately these give only a rough indication of the amount of protein present. Accurate quantitative estimation of the amount of proteinuria is of value in following the course of glomerulo-nephritis and also in determining the significance of proteinuria in otherwise healthy people who show no other signs of renal disease.

In glomerulo-nephritis there is a rough correlation between the severity of the disease and the degree of proteinuria. Estimations, repeated at weekly or monthly intervals, indicate whether the condition is progressing and give some indication of the rate of progress. In acute nephritis the degree of proteinuria in the early stages is often considerable, but with healing of the renal lesion, progressively small amounts of protein are excreted, and in some 80 per cent of cases the urine is protein-free in four to six weeks. In the subacute nephrotic stage enormous amounts of protein may appear in the urine, and the loss may exceed 20 g in a twenty-four hour period. To maintain protein balance these patients require a high protein intake and the amount of extra protein required in the diet can be decided by estimating the daily loss in the urine.

Healthy subjects and some pregnant women, who are otherwise well, sometimes excrete protein in the urine after they have been laying down or taking exercise but not when they are standing or walking. This orthostatic proteinuria is of no significance if it is found that the amount of protein excreted is less than about 0.2 g per 100 ml and there are no abnormalities in the urinary deposit.

Cerebrospinal Fluid Accurate determination of the amount of protein present is an essential determination in the examination of any specimen of cerebrospinal fluid. Normal fluid taken from the lumbar sac contains less than 40 mg protein per 100 ml. Apart from meningismus, an increase in the protein content is found in a large number of conditions affecting the central and peripheral nervous systems, and is always indicative of some abnormality. The degree of increase taken in conjunction with other abnormalities such as the cellular, chloride, and glucose contents and the serological reactions of the fluid is often helpful in determining the underlying pathology. A moderate increase in the protein content up to 200 mg per 100 ml is seen in many conditions causing irritation and inflammation of the meninges, especially syphilitic and viral infections, peripheral neuritis and diphtheritic paralysis. Disseminated sclerosis may be associated with increased protein as an isolated abnormality, and this also occurs in intracranial tumours, cerebral abcess and prolapsed intervertebral disc. A greater increase in the protein content, up to 500 mg per 100 ml, occurs in infective polyneuritis (often without any increase in cells) and when there is a block in the subarachnoid space, giving rise to Froin's syndrome, comprising xanthrochromic fluid, high protein content and a positive Queckenstedt's test.

Normal cerebrospinal fluid contains a trace of globulin which is detected by qualitative tests such as the Nonne-Apelt and Pandy reactions. In most conditions which cause an increase in the protein content of the cerebrospinal fluid there is a parallel increase in the globulin content. However in general paralysis of the insane, subarachnoid block and some cases of meningitis and tabes dorsalis, there is a disproportionate rise in the globulin fraction. This is indicated by a moderate rise in total protein and a strongly positive Pandy or Nonne-Apelt reaction. Where a quantitative estimation of the amount of globulin present is required, the estimation may be conveniently be made by the method described overleaf.

Plasma Changes in the total amount of protein in the plasma and also in the amount of its two chief constituents, albumin and globulin, occur in a large number of chronic infective and neoplastic conditions.

The normal values in health are:—

Total protein	6.3 to 8.2 g per 100 ml
Albumin	3.5 to 5.7 g per 100 ml
Globulin	1.5 to 3.0 g per 100 ml

An increase in the total plasma proteins may be due to a rise in albumin or the globulin fraction or both. In dehydration the total protein content is raised because of the loss of plasma water, and there is a proportionate rise in both albumin and globulin. In most other conditions hyperproteinæmia is due to a disproportionate increase in the globulin fraction. This occurs in severe hepatitis and hepatic cirrhosis, rheumatoid arthritis, advanced pulmonary tuberculosis and other chronic infective conditions, in nephrotic nephritis, sarcoidosis, multiple myelomatosis, malaria, kala-azar and trypanosomiasis. In liver disease an increase in the globulin fraction is of diagnostic value in distinguishing jaundice due to hepatocellular disease from that due to obstruction of the biliary passages. In the other conditions detection of hyperglobulinæmia may be of diagnostic help and also of value in following the response to treatment.

A decrease in the total plasma protein content is usually due to hypoalbuminæmia and occurs in starvation and malnutrition from any cause, and when there is loss of plasma proteins from hæmorrhage, burns and heavy proteinuria in the nephrotic syndrome. In acute hepatitis and in cirrhosis the hypoalbuminæmia is usually accompanied by a compensatory increase in globulin. The degree of reduction in plasma albumin may indicate the severity of the liver condition, and values below 2.5 g per 100 ml usually indicate severe damage, and if persistent on repeated testing, a bad prognosis.

Fibrinogen is normally present in plasma in amounts ranging from 0.2 to 0.4 g per 100 ml In many infective and non-specific conditions the level is raised and this rise is in part responsible for the increase in erythrocyte sedimentation rate. Accurate determination of the fibrogen content of plasma is of practical value only when there is a reduction in the fibrinogen content and this occurs very rarely—in congenital fibrinogenopenia in which there is delayed coagulation of blood and a clinical condition resembling hæmophilia, and in severe liver disease when fibrinogen synthesis in the liver is defective.

The biuret method of determination

The biuret method of determining proteins has been known for many years, but has not been employed much in clinical work owing to the labour or difficulty in preparing suitable protein or biuret standards. The application of coloured glasses, permanent in hue, saturation and brightness, as the standards makes the method simple and suitable for routine clinical investigations.

The results in determining the total protein in urine or cerebrospinal fluid are read directly from the figures on the disc containing the glass standards, and are expressed in mg per 100 ml. Multiplying factors are required in the other techniques, details of which are given below.

Some clinical methods for urinary and cerebrospinal fluid proteins are admittedly unsatisfactory, and the comparator-biuret method has proved reliable. A similar technique provides a very useful clinical method for serum or plasma proteins. The results for the globulin of serum or plasma, being calculated by difference, must be acknowledged as only approximate.

Principle of the method

The proteins are precipitated by trichloracetic acid. The precipitate is dissolved in caustic soda solution and copper sulphate is added. The resulting purple solution is centrifuged to throw down the suspended copper hydroxide, after which it is matched against the glass standards and the result noted.

In the differential estimations, the globulins are precipitated by half-saturation with ammonium sulphate, and the albumin is estimated in the filtrate by the biuret technique;

in plasma, globulins and fibrinogen are thrown down together. Howe's method of fractionation with 22 per cent. sodium sulphate may be used instead of the ammonium sulphate, but it is more tedious, and so is not described below (see Notes).

The glass standards were prepared by matching in the Lovibond comparator a series of solutions of known protein content, which had been prepared from normal human serum and treated by the same biuret technique. The protein content of the diluted serum was estimated by the Kjeldahl method, due allowance being made for the non-protein nitrogen.

Reagents required

1. *Trichloracetic acid* (CCl_3COOH) 10% solution
2. *Sodium hydroxide* ($NaOH$) 30% solution
3. *Copper sulphate solution* 5 g of crystalline $CuSO_4.5H_2O$ in water to 100 ml
4. *Saturated ammonium sulphate solution* 52.8 g of analytical reagent grade $(NH_4)_2SO_4$ in water to 100 ml. If necessary warm slightly to dissolve and make up to volume at room temperature.
5. *Ammonium sulphate solution (analytical reagent grade)* 27.79%
6. *Sodium chloride* ($NaCl$) *solution* 0.85%
7. *Calcium chloride solution* 2.5 g of anhydrous $CaCl_2$, or 5 g of crystalline $CaCl_2.6H_2O$ in water to 100 ml

The Standard Lovibond Comparator Discs 5/5A and 5/5B

Disc A covers the range 20 mg to 180 mg protein per 100 ml, in nine steps of 20 mg each. Disc B covers the range 200 mg to 360 mg protein per 100 ml, in nine steps of 20 mg each. It is therefore possible to read approximately to 10 mg. In dealing with serum or plasma a calculation factor is applied (see text).

Technique

(a) *Total Proteins in Urine*

Check the reaction with litmus paper; if acid, neutral or very slightly alkaline do not modify it, but if markedly alkaline (e.g. due to gross bacterial decomposition) add glacial acetic acid drop by drop to some 20 to 50 ml urine till the reaction is neutral or slightly acid (this method introduces a negligible dilution as a rule; in exceptional cases a fresh specimen must be substituted, or the dilution must be done quantitatively).

Mix in a graduated centrifuge tube (Note 4) by inverting repeatedly, but without vigorous shaking lest some precipitate become entangled in froth, and not be thrown down when centrifuged

Urine	5 ml
10% trichloracetic acid	5 ml

Allow to stand for a few minutes until the precipitate clumps. Centrifuge thoroughly, and decant the supernatant fluid, as completely as possible, by inverting the tube carefully, and wiping the mouth with filter paper.

Add to the precipitate 1 or 2 ml of water and 1 ml of 30% $NaOH$ (reagent 2); shake till the protein has dissolved.

Add 1 ml of 5% copper sulphate solution (reagent 3), water to exactly 10 ml, and then mix thoroughly for at least 1 minute. Do not mix after the addition of the copper sulphate and before the water is added, because frothing may be troublesome and a higher result may be obtained. Centrifuge well until all the precipitate of cupric hydroxide has been thrown down. Transfer the clear supernatant fluid to a comparator tube, or 13.5 mm cell, and place in the right-hand compartment of the comparator. Hold the comparator facing a standard source of white light such as the Lovibond White Light Cabinet or, failing this, north daylight and rotate the disc until the colour of the liquid is matched by one of the glass standards. The value (expressed in mg per 100 ml) is then read from the indicator recess at the bottom right-hand corner of the comparator. If the unknown exceeds 180 mg substitute the second disc (200 to 360 mg). If the unknown is above 360 repeat the test from the beginning using less urine, read, and multiply by the appropriate factor. Thus if 2 ml of urine, with 2 ml of trichloracetic acid, be employed, multiply by 2.5.

COLORIMETRIC CHEMICAL ANALYTICAL METHODS — Proteins (1)

Differential Determination of Urinary Proteins

The urine must first be made slightly alkaline to litmus, e.g. by adjustment of a relatively large bulk with glacial acetic acid or 30 to 40% NaOH. The correct reaction is pH 7.4 and this may be secured by adding phenol red and titrating with acid or alkali to that pH by means of a Lovibond Comparator and appropriate pH disc, but with a little experience litmus paper may be employed instead.

Mix, in a boiling tube or flask, 10 ml of the adjusted urine and 10 ml of saturated ammonium sulphate solution (52.8%). Allow to stand till the precipitate of globulin (plus mucus, if present), flocculates, and filter.

Mix in a graduated centrifuge tube 5 ml of filtrate and 5 ml of 10% trichloracetic acid, centrifuge and decant completely. Add another 5 ml of filtrate and 5 ml of the acid, centrifuge and decant completely. Treat the combined precipitate (which is derived from 5 ml of urine) with NaOH and $CuSO_4$ as described under "Total Proteins". The reading gives directly the mg of albumin per 100 ml.

Calculate the globulin (or globulin plus mucus) by difference. (Total Protein minus Albumin).

The above technique will be found satisfactory in the majority of cases, but the details may be modified to suit special circumstance or conditions. Thus if a big centrifuge and tube is available 10 ml of filtrate may be treated with 10 ml of trichloracetic acid in one stage; if the total protein is high, less than 10 ml, or, if the proportion of albumin is low, more than 10 ml of filtrate may be better, in which case the appropriate calculation factors must be introduced.

(b) *Total Proteins in Cerebrospinal Fluid*

Mix in a graduated centrifuge tube:—
- Cerebrospinal fluid .. .. 2 ml
- 10% trichloracetic acid .. .. .. 2 ml

Centrifuge and decant.

Add to the precipitate
- Water, about .. .. 1 ml
- 30 per cent. NaOH .. .. 0.5 ml

and shake till the protein has dissolved. Add 0.5 ml of 5% copper sulphate solution, water to exactly 4 ml and then mix thoroughly for at least 1 minute, and centrifuge down completely the precipitate of cupric hydroxide. Do not mix after the addition of the copper sulphate and before the water is added.

Transfer the supernatant fluid to the comparator and read the answer directly.

Fluids yielding a colour less intense than the 40 mg glass are of normal protein content, those yielding a deeper colour are pathological. The glasses in the two discs cover the pathological range very satisfactorily. As before, the technique may easily be modified to suit special conditions.

Differential Determination of Proteins in Cerebrospinal Fluid

In pathological fluids a reasonably close estimate of the proportions of albumin and globulin may usually be made. In brief, centrifuge or filter a mixture of:—
- Cerebrospinal fluid .. .. 2 ml
- 52.8% $(NH_4)_2SO_4$ .. .. 2 ml

In a graduated centrifuge tube mix:—
- Supernatant fluid (or filtrate) .. .. .. 2 ml
- 10% trichloracetic acid .. .. .. 2 ml

Centrifuge, decant completely, and add to the precipitate the NaOH solution (0.5 ml), the $CuSO_4$ solution (0.5 ml) and water to a total of 4 ml, exactly as described under Cerebrospinal Total Protein. Read in the comparator and multiply by 2 to obtain the albumin in mg per 100 ml.

Calculate the globulin by difference. (Total Protein minus Albumin).

(c) *Total Protein in Serum (Albumin+Globulin), or in Plasma (Albumin+Globulin+Fibrinogen)*

0.2 ml of serum or plasma are substituted for 5 ml of urine; otherwise the technique is the same as for urinary total protein. For convenience a brief summary is given.

Serum or plasma	0.2 ml	Centrifuge in a
Water	4.8 ml	graduated tube,
10% trichloracetic acid	5.0 ml	and decant.

Add to the precipitate 1 or 2 ml of water and 1 ml of 30% NaOH and shake till dissolved. Add 1 ml of 5% $CuSO_4$ solution, water to exactly 10 ml and then mix for 1 minute, centrifuge, transfer the supernatant fluid to the comparator and read. Do not mix after the addition of the copper suphate and before the water is added.

Total protein = (reading × 25) mg per 100 ml.

$$= \left(\text{reading} \times \frac{25}{1,000}\right) \text{grams} = \frac{\text{reading}}{40} \text{grams per 100 ml.}$$

(d) *Separate Determination of Albumin in Serum or Plasma*

Mix
Serum or plasma	0.5 ml
27.79% ammonium sulphate	9.5 ml

Filter through two thicknesses of No. 44 Whatman filter-paper, and, if necessary, refilter till clear.

Place 5 ml of the filtrate (=0.25 ml serum or plasma) in a graduated centrifuge tube, continue exactly as for total protein and read in the comparator.

Albumin = (reading × 20) mg per 100 ml

$$= \left(\text{reading} \times \frac{20}{1,000}\right) \text{grams} = \frac{\text{reading}}{50} \text{grams per 100 ml.}$$

(e) *Fibrinogen of Plasma*

Mix in a boiling tube or small beaker:

0.85% NaCl	28 ml
Plasma	1 ml
Calcium chloride solution (reagent 7)	1 ml

Incubate at 37°C for 30 minutes.* Insert a capillary pipette with sealed tip, or a slender glass rod with a pointed end, and whip gently; the fibrin, initially a jelly, contracts down and adheres to the rod. Filter, and just before completion of filtration inspect carefully and if necessary pick up on the rod any bits of fibrin which became detached in the whipping process. Grip the fibrin in a dry folded filter paper and slip it off the rod. Then press the clot thoroughly in the paper to absorb adherent fluid. Transfer the clot to a graduated centrifuge tube.

Add 1 or 2 ml of water and 1 ml of 30% NaOH and place in a boiling water bath until solution is complete. Cool. Add 1 ml of 5% $CuSO_4$ solution, water to exactly 10 ml and then mix for 1 minute, centrifuge, transfer the supernatant fluid to the comparator, and read.

Fibrin(ogen) = (reading × 5) mg per 100 ml

$$= \left(\text{reading} \times \frac{5}{1,000}\right) \text{grams} = \frac{\text{reading}}{200} \text{grams per 100 ml}$$

Calculation of Globulin in Serum or Plasma

For serum, globulin = total protein — albumin.
For plasma, globulin = total protein — albumin — fibrin(ogen).

* Occasionally 30 minutes is not enough, and the mixture may have to be left for an hour, or even overnight, for a gel to be formed.

Notes

1. It is important that the reagents do not become contaminated by ammonia absorbed from the laboratory atmosphere. If they do, the final colour will be too blue and too high results will be obtained. This is easily checked by setting up a blank from time to time using water instead of urine, cerebrospinal fluid or serum. The blank of course should be colourless.

2. In the differential estimations it is essential that after precipitation with trichloracetic acid the supernatant fluid shall be decanted as completely as possible; otherwise sufficient ammonium sulphate is left behind to yield, on the subsequent addition of NaOH, enough ammonia to keep some of the cupric hydroxide in solution. It has been shown by experiment that if the decantation is done carefully, only a very slight positive error is introduced. This can be avoided by using sodium sulphate instead, but as noted previously this is more tedious.

3. It is important to add the water after the copper sulphate, before any mixing takes place. If the copper sulphate is added to the protein dissolved in the NaOH and this mixture is shaken **before** the water is added, high results may be obtained.

4. Centrifuge tubes preferably should be tapered and of about 15 ml total capacity, graduated from 1 to 10 ml in steps of at least 0.5 ml. Though not so satisfactory, it is possible to work with non-graduated tubes, the volume of each reagent added being accurately measured so that the total volume is as prescribed (10 or 4 ml).

References

1. E. Reigler, *Z. anal. Chem.*, 1914, **53**, 242
2. W. Autenrieth, *Munch, med. Woch.*, 1915, **62**, 1418 and 1917. **64**, 241
3. A. Hiller, *Proc. Soc. Exp. Biol. Med.*, 1927, **24**, 385
4. A. Hiller, J. F. McIntosh and D. D. Van Slyke, *J, Clin. Investig.*, 1927, **4**, 235
5. J. P. Peters and D. D. Van Slyke, *"Quantitative Clinical Chemistry,"* Bailliere Tindall & Cox, London ,1932
6. J. Fine, *Biochem. J.*, 1935, **29**, 799 and *J. Lab, Clin. Med.*, 1936, **21**, 1084

The Determination of Proteins (2)
using the Biuret reaction

Introduction

The violet colour produced by the reaction of copper present in the Biuret reagent with protein, forms the basis of a number of methods for the determination of serum protein. The main difference between them lies in the methods used to separate the various protein fractions. The method described here, which uses a mixed sodium sulphate-sulphite salting-out reagent, produces less denaturing of the protein than *Method 1* described in this book.

Principle of the method

The total protein concentration of serum is determined by measuring the colour produced with Biuret reagent. The globulin fraction of serum is then removed by salting out with a sodium sulphate-sulphite mixture, and the remaining albumin determined by the colour produced on the addition of Biuret reagent. The globulin concentration is then obtained by difference, by subtracting the albumen from the total protein concentrate.

Reagents

1. *Sodium sulphate-sulphite solution 27.8%*
   Dissolve 208 g of anhydrous sodium sulphate and 70 g of anhydrous sodium sulphite in about 900 ml of warm water and add 2 ml of concentrated sulphuric acid. Make up to one litre. The reagent should be stored in a 37°C water bath or incubator to prevent it from recrystalizing.

2. *Biuret reagent*
   Dissolve 45 g of potassium sodium tartrate (Rochelle salt) in about 500 ml of *0.2N* sodium hydroxide. While stirring continuously add 15 g of hydrated copper sulphate and then 5 g of potassium iodide. Make up to one litre with *0.2N* sodium hydroxide.
   Before use, this reagent should be diluted 1 in 5 with *0.2N* sodium hydroxide containing 5 g of potassium iodide per litre.

3. *Ether (analytical grade)*

Standard Lovibond Comparator Discs 5/54 A & B

Disc A covers the range 0.5 to 4.5 g/100 ml in steps of 0.5 g protein.
Disc B covers the range 4.5 to 8.5 g/100 ml in steps of 0.5 g protein.

Technique

1. Add 5 ml of dilute Biuret reagent to each of two tubes labelled 'Total Protein' and 'Albumin' respectively and place on one side.
2. Layer 0.4 ml of serum on to 6 ml of sodium sulphate-sulphite solution in a test tube.
3. Cover the tube and invert it to mix. Immediately remove 2 ml of the mixture and add it to the tube marked 'Total Protein'.
4. Add 3 ml of ether to the sulphate-sulphite serum mixture, stopper the tube and invert twice a second for twenty seconds, evenly without excessive shaking.
5. Cap tube with a metal cap and centrifuge. Take 2 ml of the lower solution, taking care not to disturb the globulin clot present in the middle of the tube, and add it to the tube marked 'Albumin'.
6. Shake both 'Total protein' and 'Albumin' tubes and place in a 37°C water bath for 10 minutes.
7. Cool and pour 'Total Protein' tube into a 13.5 mm comparator cell and place in right-hand compartment of the Lovibond comparator. Fill a second 13.5 mm comparator cell with 2 ml sodium sulphate/sulphite and 5 ml of the diluted Biuret reagent and place in the left-hand compartment. Match colour using a white light equipment, or north daylight and read off total protein concentration in grams/100 ml from the indicator window.
8. Repeat for the albumin tube. Subtracting this value from the total protein value will give the globulin concentration.

Reference

J. G. Reinhold, *Standard methods in clinical chemistry*. Volume I, p. 88. Academic Press, New York, 1957

COLORIMETRIC CHEMICAL ANALYTICAL METHODS

The Determination of Salicylate in Blood
using ferric nitrate (Trinder's method)

Introduction

The number of patients admitted to hospital suffering from poisoning is steadily increasing[1]. About 15% of these admissions, in adults, are cases of acute salicylate poisoning, where a knowledge of the plasma salicylate level and of its rate of change during the period immediately following admission provides a valuable indication of the severity of poisoning, and a very useful guide to the type of treatment required[2]. As a high proportion of these admissions occurs between the hours of 5 p.m. and 8 a.m., it is desirable that a qualitative and quantitative test for salicylate in blood should be simple, and reliable enough to be carried out by physicians in the ward side-room, as well as by the staff of the chemical toxicology laboratory.

The present method was developed to meet this requirement[2]. It is an adaptation[3] of Caraway's modification[4] of the method originally devised by Trinder[5].

Plasma (or serum) should be used in preference to whole blood since salicylate is largely excluded from red cells.

Principle of the method

Salicylate reacts with ferric ion to form a purple coloured complex. The intensity of this colour, which is proportional to the salicylate concentration up to approximately 100 mg per 100 ml, is measured by comparison with a series of Lovibond permanent glass colour standards. Mercuric chloride and hydrochloric acid are used to precipitate the plasma proteins.

Reagent required

Trinder's reagent. Transfer 40 g of ferric nitrate ($Fe(NO_3)_3.9H_2O$) and 40 g of mercuric chloride ($HgCl_2$) into a 1 litre volumetric flask. Add 250 ml of 0.5 N hydrochloric acid (HCl), dissolve and dilute to the mark with water. Filter. This reagent is stable indefinitely at room temperature. It should be stored in a dark bottle clearly marked "TOXIC, CORROSIVE—HANDLE WITH CARE".

The Standard Lovibond Comparator Disc 5/51

This disc contains nine standards corresponding to 15, 25, 35, 45, 55, 65, 80, 95 and 110 mg of salicylic acid per 100 ml of plasma (Note 1).

Technique

To 0.5 ml of plasma (Note 2) in a test tube add 5 ml of Trinder's reagent. Stopper and mix well by shaking vigorously; allow the tube to stand for 5 minutes. Centrifuge for 5 minutes and then place the supernatant fluid in a Lovibond Comparator test tube in the right-hand compartment of the comparator with a blank prepared from 0.5 ml of water and 5 ml of the reagent in a tube in the left-hand compartment. Stand the comparator in front of a standard source of white light, such as the Lovibond White Light Cabinet or, failing this, north daylight and match the colour of the supernatant liquid in the tube with the standards in the disc.

Notes

1. The common symptoms of salicylate poisoning are generally considered to be evident when plasma levels exceed 30 mg per 100 ml. It is exceptional[2] to find patients with levels higher than 80 mg per 100 ml; in such cases, the assay should be repeated with 0.25 ml of plasma, 0.25 ml of water and 5 ml of reagent, and the result multiplied by two to give the plasma level.

2. Heparin should be used as the anticoagulant for the original blood sample; oxalate or citrate is not suitable. High levels of bilirubin, of glucose or of urea, or a slight degree of haemolysis, do not affect the salicylate reading.

References
1. General Register Office, "Report on Hospital in-patient Enquiry for 1962", H.M.S.O., London 1966
2. S. S. Brown, Jean C. Cameron and H. Matthew, *Brit. Med. J.,* 1967, **2,** 738
3. S. S. Brown and A. C. A. Smith, *Brit. Med. J.* 1968, **4,** 327
4. W. T. Caraway, "Microchemical methods for Blood Analysis" Thomas, Springfield, Illinois, 1960, p. 105
5. P. Trinder, *Biochem. J.* 1954, **57,** 301

The Determination of Sugar (1)
in blood, using phosphomolybdic acid
(Folin and Wu's method)

Introduction

Determination of the blood sugar level is essential in elucidating the significance of glycosuria discovered on routine examination of urine, and also in the diagnosis and treatment of diabetes mellitus and of other conditions associated with disturbance of carbohydrate metabolism. The significance of glycosuria depends upon the patient's symptoms and upon tests designed to detect disordered carbohydrate metabolism such as the fasting blood sugar level and the glucose tolerance test. A heavy glycosuria in a patient with symptoms of diabetes mellitus, ketonuria and a raised blood sugar level under fasting conditions is diagnostic of diabetes. In other cases the significance of glycosuria may not be so clear, and it is essential to study the blood sugar level after the ingestion of 50 g glucose. Glycosuria may occur when the renal threshold for sugar is reduced, and glucose passes into the urine without the blood sugar level being abnormally raised. Renal glycosuria is a benign condition which may be discovered on routine urine analysis and be confused with diabetes mellitus unless blood sugar estimations are carried out. Glycosuria due to a lowered renal threshold is not uncommon during the later months of pregnancy and may occur transiently in association with intracranial lesions, especially subarachnoid hæmorrhage.

The normal fasting blood sugar as determined by the method described below is 80 to 120 mg per 100 ml blood. After ingesting 50 g glucose by mouth the level rises within the first hour 40 to 50 mg above the fasting value and returns to the fasting level within two hours.

High Blood Sugar In diabetes mellitus the fasting blood sugar level is raised above normal in all but very mild cases. Following the ingestion of glucose the level rises gradually for an hour or more to a height 80 mg or more above the fasting value and fails to return to the fasting level after two hours: it may remain elevated for three hours or longer. In some pyogenic infections, particularly carbuncles and boils, a diabetic type of glucose tolerance curve is found but often the disturbance in carbohydrate metabolism is only temporary, and the curve returns to normal as the infection subsides. Most cases of Cushing's syndrome have a normal fasting blood sugar value but after ingesting glucose a diabetic type of curve is obtained.

In some otherwise normal subjects, and in some patients after partial gastrectomy or gastro-enterostomy, a normal fasting blood sugar level is followed after 50 g glucose by an excessive rise above 180 to 200 mg per 100 ml and sugar appears in the urine. This type of response differs from the diabetic type of curve in that the blood sugar level falls rapidly within two hours. This "lag type" of curve does not indicate diabetes mellitus.

Low Blood Sugar A low blood sugar level may occur after overdosage with insulin and in tumours of the islet tissue of the pancreas, which secrete insulin. A flat type of glucose tolerance curve is found in myxoedema and Addison's disease, in renal glycosuria when excessive amounts of sugar are lost in the urine, and in conditions, such as steatorrhœa, when there is poor or delayed absorption from the gastro-intestinal tract.

Principle of the method

The method described below measures the total reducing substances in blood. The chief of these is glucose but other non-glucose reducing substances mainly present in the corpuscles are included. Thus the values obtained are slightly higher than those for "true" blood sugar. This is of no clinical importance except in the diagnosis of hypoglycæmia when the levels obtained by the Folin and Wu method may be somewhat higher than those obtained by "true" blood sugar methods.

The proteins are precipitated by tungstic acid.

The protein-free filtrate is heated with an alkaline cupric sulphate solution under standard conditions. It is then treated with a solution of phosphomolybdic acid, which is reduced in proportion to the amount of cuprous salt and, therefore, in proportion to the quantity of sugar.

COLORIMETRIC CHEMICAL ANALYTICAL METHODS — Sugar (1)

The compound formed by reduction of phosphomolybdic acid is blue, and the intensity of this colour is compared in the comparator with that of a series of standard glasses which have been prepared by matching against known solutions of pure dextrose similarly treated.

Reagents required

1. *Sodium tungstate* ($Na_2WO_4.2H_2O$) 10%
2. *Sulphuric acid* (H_2SO_4) 2/3 N
3. *Folin & Wu's alkaline copper solution*

    Dissolve 40 g of anhydrous sodium carbonate (Na_2CO_3) in about 400 ml of water and transfer to a 1,000 ml flask. Add 7.5 g of tartaric acid ($C_4H_6O_6$) and wait till this has dissolved. Then transfer quantitatively to the flask 4.5 g of crystalline copper sulphate ($CuSO_4.5H_2O$) which has been dissolved in about 100 ml of water. Mix and make up to volume. A sediment often forms in time, in which case decant the clear supernatant solution.

4. *Folin & Wu's phosphate-molybdate solution,* formerly called "phosphomolybdic acid solution."

    Dissolve 35 g of molybdic acid (H_2MoO_4). and 5 g of sodium tungstate ($Na_2WO_4.2H_2O$) in 200 ml of 10 per cent. sodium hydroxide (NaOH) plus 200 ml of water in a litre beaker. Boil vigorously for twenty to forty minutes so as to remove as completely as possible the ammonia present in the molybdic acid. Cool and transfer to a 500 ml volumetric flask, washing in with sufficient water to make the volume about 350 ml. Add 125 ml of 90 per cent. w/w phosphoric acid (H_3PO_4 S.G. 1.75) and make up to 500 ml.

The Standard Lovibond Comparator Discs 5/2A, 5/2B and 5/2C

Disc A covers the range 60 mg to 220 mg glucose per 100 ml of blood, in nine steps of 20 mg each.

Disc B covers the range 240 mg to 400 mg glucose per 100 ml of blood, in nine steps of 20 mg each.

It is therefore possible to read approximately to 10 mg with the above discs.

Disc C covers the range 20 mg to 100 mg glucose in steps of 10 mg each, and may be used for C.S.F. and other body fluids.

Technique

In a small test tube (about 20 × 75 mm externally) place
- 3.5 ml of distilled water
- 0.1 ml of blood
- 0.2 ml of 10% sodium tungstate
- 0.2 ml of 2/3 N sulphuric acid

Mix. Stand for ten minutes or until protein precipitate clumps. Filter through acid-washed filter paper (7 cm Whatman No. 41).

In a Folin's tube (see Figure) place
- 2 ml of blood filtrate.
- 2 ml of alkaline copper solution.

Mix and place in a boiling water bath for exactly six minutes. Cool for one or two minutes only, and without shaking. (If cooling is prolonged there is a risk of oxidation of cuprous oxide by air).

Add 2 ml of reagent 4, dilute to the 12.5 ml mark with water, and mix thoroughly.

Transfer at once to a comparator tube or 13.5 mm cell which should be rotated briskly to make the CO_2 bubbles rise, and place in the right-hand compartment of the comparator. Hold the comparator facing a standard source of white light such as the Lovibond White Light Cabinet or, failing this, north daylight and rotate the disc until the colour of the solution is matched by one of the glass colour standards. The result is then read from the indicator recess at the bottom right-hand corner of the comparator, expressed as mg glucose per 100 ml of blood.

If the colour developed exceeds the highest value on Disc A (220), substitute Disc B, which covers the range up to 400.

Alternatively the test may be repeated using 1 ml of protein-free filtrate plus 1 ml of water; after completion to the stage of colour development and matching, the reading is multiplied by 2.

Reference

O. Folin and H. Wu, *J. Biol. Chem.*, 1920, **41,** 367

The Determination of Sugar (2)

in blood, using phosphomolybdic acid

Introduction

This method, a modification of the Folin and Wu method[1], has been developed[2,3] to provide an acurate method for the estimation of glucose in solution and also in blood. The results obtained are identical with those obtained by the titration method of Harding[4] and are 'true sugar values' as opposed to total reducing substances. The latter, which occur mainly in the corpuscles, are prevented from entering the solution by the use of isotonic sodium sulphate solution.

Principle of the method

Proteins are precipitated by sodium tungstate and copper sulphate[5]. The filtrate is treated with a modified Harding sugar reagent and the cuprous oxide so formed is estimated by the blue colour formed with a phosphomolybdic acid solution. The intensity of this colour, which is a measure of the sugar content, is estimated by comparison with Lovibond permanent glass standards.

Reagents required

1. *Modified Harding reagent*

Sodium bicarbonate ($NaHCO_3$)	50 g
Anhydrous sodium carbonate (Na_2CO_3)	40 g
Potassium oxalate (($COOK)_2.H_2O$)	36.8 g
Sodium potassium tartrate ($KNaC_4 H_4O_6.4H_2O$)	24 g
Distilled water	2,000 ml

    Dissolve the bicarbonate in 700 ml of distilled water and, when completely dissolved, add the carbonate. When this has also dissolved add a solution of the oxalate in 120 ml of warm water. Finally add a solution of the tartrate in 100 ml of distilled water. Transfer the mixture to a 2 litre volumetric flask and dilute to the mark.

2. *Phosphomolybdic acid reagent*

Molybdic acid (H_2MoO_4)	35 g
Sodium tungstate ($Na_2WO_4.2H_2O$)	5 g
Sodium hydroxide solution (NaOH 2N)	20 g NaOH in 250 ml
Phosphoric acid (H_3PO_4 sp. gr. 1.75)	125 ml
Distilled water	to 500 ml

    Dissolve the molybdic acid and tungstate in the 250 ml of sodium hydroxide and boil for 30 minutes. Make volume up to approximately 350 ml with distilled water, add the phosphoric acid, cool, transfer to a 500 ml volumetric flask and dilute to the mark with distilled water.

3. *Isotonic sodium sulphate solution*

Sodium sulphate ($Na_2SO_4.10H_2O$)	3% w/v solution
Copper sulphate ($CuSO_4.5H_2O$)	7% w/v solution

    Mix 320 ml of the sodium sulphate solution with 30 ml of the copper sulphate solution.

4. *Sodium tungstate ($Na_2WO_4.2H_2O$) solution* .... 10 g per 100 ml
    Store in a waxed bottle.

The Standard Lovibond Comparator Discs 5/36A and 5/36B

The disc 5/36A covers the range 40 to 200 mg of glucose per 100 ml (based on a 0.05 ml blood sample) in 9 steps each of 20. The disc 5/36B covers the range 220 to 400 mg per 100 ml in steps of 20 except for the last two, which are 30. The normal range for fasting individuals is 68-96 mg per 100 ml of blood[3]. The extreme values of the normal range are 55 and 109 mg per 100 ml[3].

Technique

Pipette 0.05 ml of whole blood into 3.9 ml of isotonic sodium sulphate (reagent 3), in a conical centrifuge tube. Add 0.05 ml of sodium tungstate (reagent 4) and shake throughly. Spin down the precipitated proteins and copper tungstate in a centrifuge. Transfer 2 ml of the supernatant fluid and 2ml of the Harding reagent (reagent 1) to a standard Lovibond comparator tube. Prepare a blank from 2 ml isotonic sodium sulphate (reagent 3) and 2 ml of Harding reagent (reagent 1) in an identical tube. Stopper both tubes with cotton wool and place in a boiling water-bath for exactly 10 minutes. Cool immediately, add 6 ml of the phosphomolybdic acid (reagent 2) and 2.5 ml of distilled water to each tube and mix well. Place the sample tube in the right-hand compartment of a Lovibond Comparator and the blank tube in the left-hand compartment. Compare the colour of the sample with the colours of the permanent glass standards in the disc, using a standard source of white light, such as the Lovibond White Light Cabinet or, failing this, north daylight.

References

1. O. Folin and H. Wu, *J. Biol. Chem.*, 1920, **41**, 367
2. A. Asatoor and E. J. King, *Biochem. J.*, 1954, **56**, xliv
3. E. J. King and I. D. P. Wootton, "*Micro-analysis in Medical Biochemistry*," J. & A. Churchill Ltd., London, 1959
4. V. J. Harding and C. E. Downs, *J. Biol. Chem.*, 1933, **101**, 487
5. M. Somogyi, *J. Biol. Chem.*, 1931, **90**, 725

The Determination of Sugar (3)
in blood, using glucose oxidase (Trinder's method)

Introduction

Glucose oxidase has been used to determine blood glucose levels in the hope that this method would give a true blood sugar level. It has been found though that glutathione, uric acid, bilirubin and hæmoglobin inhibit the enzyme systems used to varying degrees. In addition Vitamin C and fluoride also interfere. However, with careful choice of protein precipitant and blood preservative these errors can be minimised, and this method is presented for those who wish to utilize the advantages of glucose oxidase for the determination of true glucose level.

The original methods devised, employed o-tolidine or o-dienisidine in the colour reagent. Recently, attention has been focused on the fact that these substances are potential carcinogens[2]. 4-aminophenozone has been used as oxygen acceptor in the present method to avoid risk.

Principle

Glucose is enzymically oxidised to gluconic acid by glucose oxidase. Oxygen from the hydrogen peroxide formed is then transferred to an oxygen acceptor. Phenol in the presence of this oxygen reacts with 4-aminophenozone to produce a red colour.

Reagents required

1. *Protein precipitant* Dissolve 10 g of sodium tungstate ($Na_2WO_4.2H_2O$), 10 g of disodium hydrogen phosphate (Na_2HPO_4) 10 g of phenol and 9 g sodium chloride in about 800 ml of distilled water. Add normal hydrochloric acid until the pH is 3.0. About 125 ml will be needed. Mix and make up to 1 litre with water. Stable for six months.

2. *Colour reagent* Dissolve 3 g of disodium hydrogen phosphate (Na_2HPO_4), 0.3 g of sodium azide (NaN_3) and 90 mg 4-amino-phenozone in 300 ml of distilled water. To this add 5 ml of Fermcozyme 653AM* and 5 ml of 1% peroxidase (R.Z.O.6).* Mix well and store in refrigerator. Stable for 4 weeks.
   *See Note 2

The Standard Lovibond Comparator Disc 5/48

The disc covers the range 50 to 250 mg glucose/100 ml in steps of 25 mg.

Technique

1. To 5.8 ml of protein precipitant add 0.2 ml of blood. Mix well and centrifuge.
2. Into a tube labelled 'Test' put 2 ml of supernatant fluid and 6 ml of colour reagent.
3. Into a second tube labelled 'Blank' place 2 ml of protein precipitant and 6 ml of colour reagent.
4. Mix both tubes well and place in 37°C water bath for ten minutes.
5. Pour 'Test' into 13.5 mm comparator cell and place in right-hand compartment of Lovibond comparator. Pour 'Blank' into a similar cell and place in left-hand compartment. Match colour using white light equipment or north daylight read off glucose concentration in milligrams/100 ml from the indicator window.

Notes

1. Should the colour exceed the maximum on the disc, repeat using 1 ml of supernatant and multiply the value indicated on the disc by 2.
2. Fermcozyme 653AM and Peroxidase R.Z.O.6. are obtainable from: Hughes & Hughes (Enzymes) Ltd., 12a High Street, Brentwood, Essex, England.

References

1. P. Trinder *J. clin. Path.* 1969, **22**, 246
2. Chester Beatty Research Institute. *Precautions for Laboratory Workers who handle Carcinogenic Aromatic Amines.* Institute of Cancer Research, London (1966)

The Determination of Sulphetrone
(and Drugs allied to Diaminodiphenylsulphone)
using N-(1-naphthyl) ethylenediamine dihydrochloride

Introduction

A number of derivatives of diaminodiphenylsulphone have been developed in recent years which have proved to be treatment of choice in leprosy. Sulphetrone is one of these derivatives and is given orally in three doses of 1 to 3 g daily. To obtain the best therapeutic effect the blood concentration should be maintained between 7.5 and 10 mg per 100 ml. Toxic symptoms are likely to occur if the concentration exceeds 12 mg per 100 ml.

Principle of the method

Sulphanilyl drugs and allied sulphones owe their therapeutic activity to free amino groups which, however, are to a substantial degree acetylated or glycuronated in the body. It is a remarkable fact, supported by pharmacological evidence, that Sulphetrone, although active, does not appear to be conjugated in this way. This appears to imply that the substituent groupings of the amino nitrogens are not removed by hydrolysis; the fact that they "unmask" the amino group sufficiently to allow diazotisation is the basis of the colour reaction. The method differs from that usually employed for sulphanilyl drugs in that Sulphetrone is first treated with N HCl, without heating and before deproteinising.

Reagents required

1. *Sodium nitrite* ($NaNO_2$)   0.3% (w/v) aqueous solution, prepared weekly
2. *Ammonium sulphamate* ($NH_4SO_3NH_2$)   1.5% (w/v) aqueous solution
3. *N-(1-naphthyl) ethylenediamine dihydrochloride*   0.1% aqueous solution
4. *N hydrochloric acid* (HCl)
5. *Trichloracetic acid* (CCl_3COOH)   12% (w/v) aqueous solution

All reagents must be of analytical reagent grade, and the sodium nitrite solution should be freshly prepared each week. The ethylenediamine reagent should be freshly prepared each month and these two solutions, together with the solution of ammonium sulphamate, are best kept in amber glass-stoppered bottles.

The Standard Lovibond Comparator Disc 5/25

The disc covers the range 1 mg to 9 mg of Sulphetrone per 100 ml of body fluid in 9 steps of 1 mg.

Technique

(a) *Determination of Sulphetrone in Blood*

To 0.5 ml of blood, or other body fluid, add 5 ml of N HCl, mix, and then add 2 ml of trichloracetic acid solution. Filter through a 9 cm Whatman grade 5 paper, and refilter the filtrate, if necessary, to give a bright solution; alternatively the material may be centrifuged. To 3 ml of the filtrate, placed in a comparator tube, add 0.05 ml (1 drop) of 0.3% sodium nitrite solution; shake and allow to stand for 3 minutes. Add 0.05 ml (1 drop) of the 1.5% ammonium sulphamate solution, mix well and allow to stand 2 minutes. Finally add 0.05 ml of 0.1% N-(1-naphthyl) ethylenediamine dihydrochloride reagent. The colour is allowed to develop for 20 minutes after the addition of the last reagent before being compared. A "blank" solution is prepared at the same time in another comparator tube, substituting 0.5 ml of distilled water for the body fluid. Hold the comparator facing a uniform source of white light, and match the solution against the disc.

If the first estimation falls outside the upper range of the disc (9 mg per 100 ml), the determination should be repeated, the 0.5 ml of blood being diluted with more than 5 ml N HCl and appropriate correction made in reading the result. Comparison by dilution of the colour should not be attempted, since it leads to inaccurate results.

(b) *Determination of Sulphetrone in Urine*

A preliminary dilution of the urine with distilled water is necessary before estimation by the above method. A dilution of 1 in 100 will usually be suitable, but there is considerable variation. The value read from the disc (expressed as mg per 100 ml) must, of course, be multiplied by the dilution.

(c) *Determination of Sulphetrone in Skin*

Cut a piece of skin weighing 0.1 g into small pieces with scissors, add 1 g of silver sand which has been previously washed with N HCl, and thoroughly grind until smooth. Crystalline sugar, washed quartz and coarse silicon carbide are equally suitable. To the mass add 9 ml of N HCl, and triturate the suspension for 10 minutes; 4 ml of trichloracetic acid solution are now added, and the volume is adjusted to 15 ml, with N HCl. Treat 2 ml of the solution with the colour reagents as previously described. The answer derived from the disc must be multiplied by ten to give an answer in mg Sulphetrone per 100 g of tissue.

Determination of Allied Drugs

Although the visible spectra of the coloured derivatives of diaminodiphenylsulphone and its derivatives, such as Sulphetrone, are similar enough to warrant estimation by the method described, a correction factor must be applied for each drug other than Sulphetrone.

References

1. G. Brownlee, A. F. Green and M. Woodbine, *Brit. J. Pharmacol.,* 1948, **3,** 15
2. G. Brownlee, *Lancet,* 1948, **ii,** 131
3. M. G. Clay and A. C. Clay, *Lancet,* 1948, **ii,** 180
4. D. G. Madigan and G. Brownlee, *Lancet,* 1948, **ii,** 791

The Sulphobromophthalein Test

Introduction
The sulphobromophthalein retention test is considered to be one of the most sensitive indices of liver function. It is usually performed to establish the presence of liver disease in the non-jaundiced patient or it may be repeated at intervals to assess progress during, for example, acute hepatitis. The test is of no value in differentiating types of jaundice since abnormal results may be caused by biliary obstruction, liver cell disorder or intereference with hepatic blood flow from extra-hepatic causes.

Principle of the method
Sulphobromophthalein solution is injected intravenously, the amount of solution being calculated to provide 5 mg of solid dye per kg body weight. Clearance of the dye from the circulation takes place *via* the liver. The concentration of the dye remaining in the serum after 45 minutes is measured by comparison with the standard glasses in the Comparator disc, which are calibrated directly in terms of "% retention." In normal persons, there is less than 5% retention: in cases of liver disease, values of 10—100% may be found.

Reagents required
1. *Sterile sulphobromophthalein solution for injection (50 mg/ml)*.
2. *Dilute sodium hydroxide (NaOH) solution* Approximately 0.05 N (2 g/l)
3. *Dilute hydrochloric acid (HCl) solution* Approximately 0.05 N (6 g concentrated acid solution/l.)

The Standard Lovibond Comparator Disc 5/32
The disc covers the range 0—50% retention, in 9 steps of 5% up to 30% and then two steps of 10%.

Technique
The test should be carried out on the fasting patient. A dose of the sterile solution amounting to 0.1 ml per kg body weight is injected slowly into an arm vein with care to avoid any leakage into the tissues. Forty-five minutes later, blood is withdrawn from the other arm (to avoid contamination) into a clean dry syringe and is allowed to clot in a clean tube.

Mix 3 ml of serum with 4.5 ml of dilute sodium hydroxide solution to develop the colour of the dye, and place in a 13.5 mm test tube or cell in the right aperture of the comparator. Mix another sample of 3 ml of serum with 4.5 ml of dilute hydrochloric acid to make a "blank" solution to compensate for inherent colour, and place in a similar 13.5 mm test tube or cell in the left hand compartment of the comparator so that it comes behind the colour standards in the disc. Hold the comparator facing a standard source of white light such as the Lovibond White Light Cabinet or, failing this, north daylight and rotate the disc until the colour of the test solution is matched by one of the glass standards. The value, expressed as % retention, is then read off from the indicator window at the bottom right-hand corner of the comparator.

Technique when the retention is very high
If the value found is more than 50%, test and blank solutions should each be diluted with an equal volume of water. The reading is then repeated and the result found is multiplied by two.

Notes
1. Since different batches of dye may vary slightly in their colour strength, it may be considered proper to check the dye against the colour standard. To do this, the sulphobromophthalein solution for injection is accurately diluted 1 : 1000. The resulting solution, containing 5 mg/100 ml, is treated exactly like a blood serum, and should match the disc at 50% retention.

2. Urine passed after the test may be stained red if it is alkaline. This has occasionally caused alarm.

3. Alternative technique using a lower dosage

Some workers prefer to use a smaller dose of the dye, i.e. 2 mg/kg body weight instead of 5 mg/kg, since they consider that reactions are less frequent, especially in heavy patients. The following modifications to the technique are required.

1. The injected dose is one twenty-fifth ml per kg body weight.
2. After 45 minutes, a large amount of blood must be withdrawn, sufficient to yield at least 15 ml of serum. The clear serum is placed in equal amounts into two 13.5 mm cells or test tubes; to one is added two or three drops of $2N$ NaOH, making the test solution for the right-hand tube. The other cell, containing the blank serum, is treated with two or three drops of $0.5N$ HCl, and placed in the left-hand tube behind the glass colour standards.

The answer is read exactly as described above.

References

1. C. W. Wirts and A. Cantarow, *Amer, J. Digest Dis.,* 1942, **9,** 101
2. J. G. Mateer, J. I. Baltz, D. F. Marion and J. M. McMillan, *J, Amer. Med. Ass.* 1943, **121,** 723
3. N. Zamcheck, T. C. Chalmers, F. W. White and C. S. Davidson, *Gastroenterology,* **14,** 343
4. S. Sherlock, "*Diseases of the liver and biliary system,*" Blackwell, Oxford, 1955
5. E. J. King and I. D. P. Wootton, "*Microanalysis in medical biochemistry,*" Churchill, London, 1956

The Determination of Sulphonamides
using N-(1-naphthyl) ethylenediamine dihydrochloride
(Bratton and Marshall's method)

Introduction

Sulphonamides, particularly sulphadiazine, are the treatment of choice in patients with meningococcal infections, and are still widely used in the treatment of infections due to *pneumococci, streptococci* and *Esch. coli*. In body fluids some of the sulphonamide is in the free form and has antibacterial activity; some is in a conjugated inactive form. A satisfactory therapeutic response is only to be expected when the concentration of free sulphonamides reaches and is maintained at a level of 8 to 15 mg per 100 ml. The total sulphonamide (free plus conjugated) is usually 20 to 50 per cent. greater than the free.

Some 90 per cent of the ingested dose of sulphonamide is excreted in the urine, and if the urine output amounts to 1,000 ml a day, and the intake of sulphonamide, given six hourly, amounts to 2 g, the urinary concentration will reach a level of the order of 200 mg per 100 ml. Thus it is easy to obtain satisfactorily high concentrations in the urine to treat renal tract infections. Only rarely is it necessary to determine the urinary concentration of sulphonamides.

Principle of the method

The sulphonamide is diazotised with sodium nitrite, after precipitation of proteins with trichloracetic acid, and coupled with N-(1-naphthyl) ethylenediamine dihydrochloride. The red colour formed is compared with Lovibond permanent glass standards.

Reagents required

1. *Trichloracetic acid* (CCl_3COOH)   15% w/v
2. *Sodium nitrite* ($NaNO_2$)   0.1% solution w/v This should be freshly made.
3. *Ammonium sulphamate* ($NH_4SO_3NH_2$)   solution 0.5% w/v
4. *N-(1-naphthyl) ethylenediamine dihydrochloride*   0.1% aqueous solution w/v. Keep in a dark bottle.
5. *4N hydrochloric acid* (HCl)

The Standard Lovibond Comparator Disc 5/10A

The disc covers the range 0.5 to 8.0 mg sulphanilamide per 100 ml, in 9 steps of one mg from 1.0 onwards.

Technique

(a) *Determination of free sulphanilamide in blood*

Mix in a boiling tube: serum or plasma 2 ml, distilled water 30 ml, 15% trichloracetic acid 8 ml (if whole blood is used, it must be laked thoroughly with the water before adding the trichloracetic acid). Filter. Take 10 ml of the filtrate, add 1 ml of sodium nitrite solution and mix well. Stand for 3 minutes. Add 1 ml of ammonium sulphamate solution, mix and stand for another 2 minutes. Add 1 ml of the naphthyl ethylenediamine dihyrochloride, and pour into the test tube or 13.5 mm cell in the right-hand compartment of the comparator. A "blank" solution is prepared at the same time, substituting 2 ml of distilled water for the serum, and this is poured into a test-tube and placed in the left-hand compartment of the comparator. Hold the comparator facing a standard source of white light such as the Lovibond White Light Cabinet or, failing this, north daylight, and rotate the disc until the colour of the test solution is matched by one of the glass standards. The value, expressed as mg per 100 ml of blood, is then read from the indicator window at the bottom right-hand corner of the comparator. If there is a shortage of sample, the test may be performed on 1 ml of serum, using half quantities throughout.

(b) *Determination of total sulphanilamide in blood*

To 10 ml of filtrate after trichloracetic-acid precipitation, add 0.5 ml of 4N HCl, heat in a boiling-water bath for 1 hour, cool, and adjust the volume to 10 ml. Then proceed as above.

(c) *Determination of sulphanilamide in urine*

Test a sample for protein.

Add sufficient water to a measured volume of the urine to reduce the sulphonamide content to between 1 and 4 mg per 100 ml, and note the degree of dilution. A dilution of 1 in 5 or 1 in 10 may be tried in the rough preliminary test.

Protein-containing urine To 2 ml of the diluted urine add 30 ml of water and 8 ml of trichloracetic acid solution. Filter, and treat 10 ml of filtrate as a blood-filtrate.

Protein-free urine Place 5 ml of the diluted urine in a 100 ml volumeteric flask. Add 5 ml of $4N$ HCl, and water to the mark. For free sulphonamide, treat 10 ml of the mixture as a blood filtrate: for total sulphonamide, heat 10 ml without further addition of acid.

The figure shown at the indicator window must be multiplied by the initial dilution of the urine.

(d) *Determination of allied drugs*

Most drugs of the sulphanilamide group can be estimated by the above technique, as they all give almost identical colours. A correction factor must be applied for each drug; these factors can be determined by preliminary estimations of known concentrations of the particular compound being studied.

Note

To check reagent

8 ml of water and 2 ml of trichloracetic acid are taken through the test as for 10 ml of filtrate. A colourless blank should result. This should be repeated with each fresh batch of reagent.

Reference

A. C. Bratton and E. K. Marshall, Junr., *J. Biol. Chem.*, 1939, **128,** 537

The Determination of Serum Transaminase
using α-ketoglutaric acid

Introduction

The use of the test for serum transaminase in diagnostic and prognostic investigations outside hospitals has been limited by the difficulty of preparing and storing the necessary substrates. This difficulty has been overcome by the commercial availability of stable substrate concentrates which require only the addition of water before use. The original tests have been slightly modified, and a standard disc has been prepared for use with the B.D.H. substrates.

Principle of the method

Serum glutamic-oxaloacetic transaminase (Aspartate transaminase) transaminates L-aspartic acid and α-ketoglutaric acid to oxaloacetic acid and glutamic acid respectively. The oxaloacetic acid is decomposed to pyruvic acid by the action of aniline citrate solution. The addition of 2,4-dinitro-phenyl-hydrazine converts the pyruvic acid to the corresponding hydrazone and this, in alkaline solution, gives an intense brown colour[1-4]. The intensity of this colour, which is proportional to the pyruvic acid concentration and hence to the SGO-T concentration, is measured by comparison with a series of Lovibond permanent glass colour standards in a comparator.

Serum glutamic-pyruvic transaminase (Alanine transaminase) transaminates L-alanine and α-ketoglutaric acid to pyruvic acid and glutamic acid respectively. The pyruvic acid is determined as for SGO-T.

Reagents required

1. *SGO-T buffered substrate* Each B.D.H. tube contains sufficient substrate for 25 tests and blanks. Dissolve the contents in water and adjust the volume to 60 ml. This solution is stable for at least 6 weeks if stored in a deep freeze and preserved with 2 drops of chloroform.
2. *SGP-T buffered substrate* Each B.D.H. tube contains sufficient substrate for 25 tests and blanks. Dissolve the contents in water and adjust the volume to 60 ml. This solution is stable for at least 6 weeks if stored in a deep freeze and preserved with 2 drops of chloroform.
3. *2,4-Dinitro-phenyl-hydrazine reagent* Dilute one volume of the concentrated ($5mM$) solution in the B.D.H. ampoule with four volumes of N hydrochloric acid before use. It is advisable to dilute only the amount required for each series of tests.
4. *Sodium hydroxide 0.4N (free from carbonate)* Prepare the solution from $4N$ sodium hydroxide (free from carbonate) by dilution with freshly de-ionised or carbon dioxide free water.
5. *Aniline citrate reagent* Dissolve 5 g of citric acid in 5 ml of water and add 5 ml of aniline.

All chemicals used in the preparation of reagents should be of analytical reagent quality.

The Standard Lovibond Comparator Disc 5/39

This disc covers the range 0–50 international units (in steps 0, 5, 10, 20, 30, 40, 50) of either SGO-T or SGP-T according to which substrate is used in the test.

The master disc against which all reproductions are checked was tested and approved under the supervision of Professor I. D. P. Wootton, Postgraduate Medical School, London.

Technique

Blood specimens should be separated immediately. The serum may be stored at 4°C for a short period. Citrate, oxalate, heparin and Sequestrene are said to be satisfactory as anticoagulants, but fluoride is not suitable. Hæmolysed blood serum should not be used for these estimations.

SGO-T Label two tubes "Sample" and "Control" respectively. Into each pipette 1 ml of SGO-T substrate and warm to 37°C by placing the tubes in a water-bath. The temperature of this water-bath should be regulated to $37\pm0.5°C$. After noting the time add 0.2 ml of serum to the "Sample" tube. Incubate the tubes at 37°C for 60 minutes and, with the tubes still in the water-bath, add 1 drop of the aniline citrate reagent to each. After 5 minutes add 1 ml of 2,4-dinitro-phenyl-hydrazine reagent to each tube. Add 0.2 ml of serum to the "Control" tube and again note the time. Continue incubation for a further 20 minutes, remove the tubes from the water-bath and add 10 ml of 0.4 N sodium hydroxide solution to each. Mix by inversion and allow the tubes to stand for 10 minutes.

SGP-T Proceed exactly as for SGO-T except that SGP-T substrate is used and that the first incubation period is 30 minutes instead of 60 minutes, and the addition of aniline citrate is omitted.

After standing for 10 minutes transfer the solutions to 13.5 mm comparator tubes.

Place a tube of water in the left-hand compartment of the comparator and the "Sample" tube in the right-hand compartment. Compare the colour of the tube with the standard in the disc using a standard source of white light, such as the Lovibond White Light Cabinet. In the absence of a standard source, north daylight should be used. Read off the unit figure from the indicator window in the bottom right-hand corner of the comparator when the colours are matched.

Repeat this procedure with the "Control" tube substituted for the "Sample" tube.

The transaminase activity is then given by the difference between the readings of "Sample" and "Control" tubes (in International Units).

If the colour of the "Sample" is deeper than that of the highest standard on the disc, dilute the serum with water (a dilution of 1 in 5 should be adequate) and repeat the determination using 0.2 ml of the diluted serum. Alternatively shorten the incubation time and make the appropriate adjustment to the calculation.

Notes

1. The collaboration of Professor I.D.P. Wootton in the preparation and checking of the standards used in this test is gratefully acknowledged.

2. The units used in the calibration of the disc are the international units determined spectrophotometrically at 25°C. For conversion to other units reference should be made to the appropriate literature[4-6].

3. In view of the unstable nature of the reagents used for these tests, each batch should be tested by the following procedure :—
   (a) A $2mM$ pyruvate standard is prepared by dissolving the contents of the appropriate tube in distilled water and making up to 100 ml.
   (b) A series of tubes is set up as shown.

Tube	Water	2mM standard	SGOT or SGPT substrate	Colour equivalent to International Unit
1	0.2 ml	0 ml.	1.0 ml.	0
2	0.2 ml	0.10 ml	0.90 ml	13
3	0.2 ml	0.15 ml	0.85 ml	21
4	0.2 ml	0.20 ml	0.80 ml	30
5	0.2 ml	0.25 ml	0.75 ml	40
6	0.2 ml	0.30 ml	0.70 ml	54

Incubate the standards at 37°C for 30 minutes, add 1 ml. of dinitro-phenyl-hydrazine reagent to each, and incubate for a further 20 minutes. Remove the tubes from the water-bath add to each 10 ml $0.4N$ sodium hydroxide, mix and transfer the coloured solutions to 13.5 mm comparator tubes.

Place the tubes in turn in the right-hand compartment of the comparator. With a tube of water in the left-hand compartment, the coloured solutions should have an intensity as shown in the right-hand column of the table.

References

1. S. Reitman and S. Frankel, *Amer. J. Clin. Path.,* 1957, **28,** 56
2. J. King, *J. Med. Lab. Tech.,* 1958, **15,** 17
3. H. Varley, "*Practical Clinical Biochemistry*", Heinemann, London, 3rd editn., 1962, p. 217
4. I. D. P. Wootton, "*Micro Analysis in Medical Biochemistry* ", (King), Churchill, London, 4th editn., 1964
5. H. U. Bergmeyer, "*Methods of Enzymatic Analysis*", Academic Press, New York, 1963, p. 841
6. J. Daly and A. Jordan, *Lancet,* 1959, **1,** 256

The Determination of Trichloracetic Acid
using pyridine

Introduction
Trichlorethylene is used as a solvent for degreasing in the engineering and textile industries, for the extraction of fats and oils, for dry-cleaning, and in the paint trade. Although originally thought non-toxic, there has been a growing realisation that prolonged exposure to trichlorethylene may cause symptoms of irritation of the mucous membranes, headache, dizziness, fatigue and the clinical picture of drunkenness. Although these symptoms are usually transient, in severe cases death may occur.

In the body a proportion of trichlorethylene is converted into trichloracetic acid, and the amount of this excreted in the urine is related to the degree of exposure to fumes of trichlorethylene. When the excretion of trichloracetic acid is less than 20 mg per litre of urine no ill-effects are encountered. When the excretion is 40 to 75 mg per litre some of the subjects have symptoms of nausea, dryness in the mouth, headache, flatulence, somnolence, and a constricting sensation in the chest. These symptoms are frequently present in subjects excreting 100 mg of trichloracetic acid per litre. Above 100 mg these symptoms are usually pronounced.

Principle of the method
Trichloracetic acid is decomposed by sodium hydroxide to chloroform and carbon dioxide. In alkaline solution chloroform combines with pyridine to form a pink colour, the intensity of which is proportional to the amount of chloroform present.

Reagents required
1. *Sodium hydroxide (NaOH) solution* 25%
2. *Pyridine (C_5H_5N), redistilled*
3. *Ethyl alcohol (C_2H_5OH)* 95%

The Standard Lovibond Comparator 5/31
The disc covers the range 10 to 200 mg (10, 20, 30, 40, 50, 75, 100, 150, 200 mg) trichloracetic acid per litre.

Technique
To obtain reliable results the details of the technique must be closely followed, the duration of heating in the water-bath and thorough shaking at 3 or 4 minute intervals being critical.

Place 16 ml sodium hydroxide solution in a tube, add 4 ml pyridine and 1 ml of the urine sample to be tested. Shake well and stand in a water-bath at 60°C for exactly 20 minutes, shaking the tube thoroughly every 3 to 4 minutes to ensure mixing. Cool the tube, and transfer the contents to a separating funnel. Drain off the clear layer completely. To the coloured layer in the funnel add 7 ml ethyl alcohol and shake; this will remove any turbidity present. Transfer the coloured solution to a 13.5 mm Lovibond test tube or cell and place in the right-hand compartment of the comparator. Hold the comparator facing a standard source of white light such as the Lovibond White Light Cabinet or, failing this, north daylight and match the colour against the disc immediately. The value shown in the indicator window represents mg trichloracetic acid per litre urine. A more accurate result is obtained[5] if a correction is made for the specific gravity of the urine as follows:—

$$\frac{\text{T.C.A. reading} \times 16}{(\text{Sp. Gr} - 1.000) \times 1000}$$

References
1. R. K. Waldman and L. A. Krause, *Occupational Health*, 1952, **12**, 110
2. A. Ahlmark and S. Forssman, *Arch. ind. Hyg.*, 1951, **3**, 386
3. K. Fujiwara, *Chem. Abs.*, 1917, **11**, 3201
4. H. O. Engel, *Trans. Assn. Indus. Med. Officers*, 1956, **6**, 96
5. S. Buchwald, *Am. Occupational Hygiene*, 1964, **7**, 125

The Determination of Urea (1)

using urease

(Archer and Robb's Urease-Nesslerisation method)

Introduction

Urea is a waste-product of protein metabolism which is formed in the liver and excreted via the kidneys into the urine. The blood concentration of urea largely depends upon the amount excreted and this in turn depends upon the number of glomeruli functioning and on the volume and pressure of the blood flowing through the glomerular capillaries. Estimation of the blood urea level is of great value in the study of renal disease and in a large number of other conditions in which the kidneys are secondarily involved. For the test about 5 ml venous blood are collected in a clean tube containing a pinch of potassium oxalate powder.

Normally the blood urea concentration varies from 20 to 40 mg per 100 ml, but values up to 50 mg cannot be considered abnormal in the elderly. Since the urea level is influenced by disease of the kidneys and by conditions in which the kidneys are not primarily involved, the causes of a raised blood urea are conveniently considered under two main headings—renal and extra-renal.

Renal Causes of Urea Retention The degree of renal damage that must be present before there is urea retention depends to some extent upon the amount of protein in the diet. On an average intake, adequate elimination of urea can be effected by as little as two-thirds of one kidney, and therefore wide-spread bilateral kidney damage must be present before there is nitrogen retention of purely renal origin.

In acute glomerulo-nephritis the blood urea is often raised but rarely exceeds 100 mg per 100 ml. In 90 per cent of cases the symptoms and signs of the acute phase subside and the blood urea returns to normal, although in a small proportion proteinuria may persist and chronic glomerulo-nephritis develop many years later. Some five per cent. of patients die in the acute attack with severe oliguria and a blood urea that climbs steadily to levels of 300 to 400 mg per 100 ml. In another five per cent the disease is a little more protracted; these patients proceed over a year or eighteen months to progressive renal failure with a blood urea that rises steadily until the time of death. In the early stages of nephrotic nephritis, characterised by gross oedema, a normal blood-pressure and heavy proteinuria, the blood urea is normal, but if the condition progresses, urea retention due to renal failure occurs in the later stages. During the early phases of chronic glomerulo-nephritis there is no rise in the blood urea. The daily urine volume is normal and renal disease may only be revealed by proteinuria, casts, a raised blood-pressure or a poor water-concentration test. As the disease progresses more glomeruli become damaged. Filtration of fluid through the glomeruli is decreased and less urea is filtered into the tubules. The blood urea concentration therefore rises but this automatically results in more urea being filtered and despite the diminished glomerular filtration rate the same amount of urea is eliminated but at the price of a higher level in the blood. Thus a balance is struck, and a blood urea which rises from 30 mg to 100 mg per 100 ml and remains stationary at this abnormally high level is compatible with many years of happy and useful life especially if the patient's blood pressure is normal. As the disease progresses, more glomeruli are involved and the blood urea rises higher. The tempo with which the disease advances varies from patient to patient but is usually more rapid in those with hypertension. Estimations of the blood urea concentration repeated at monthly or bimonthly intervals are helpful in deciding the speed of progression, and a urea level which rises in a month from 60 to 100 mg and then to 120 mg a month later usually indicates a worse prognosis than a level which remains stationary at 100 mg per 100 ml.

The blood urea level often remains within normal limits during the entire course of essential hypertension but may be raised during episodes of cardiac failure and in the malignant phase of the disease. A raised blood urea level may occur in chronic infective, neoplastic or polycystic renal disease and indicates bilateral involvement unless there are additional extra-renal causes of urea retention.

Extra-Renal Causes of Urea Retention Any condition which reduces the volume or pressure of the blood flowing through the glomeruli may cause a rise in the blood urea level, and this occurs in dehydration and shock due to severe and protracted vomiting, profuse diarrhoea, haemorrhage (particularly bleeding from a peptic ulcer), uncontrolled diabetes and circulatory failure. In these pre-renal conditions which cause urea retention the raised blood level indicates a disturbance of renal circulation rather than intrinsic renal disease, and provided the underlying cause can be corrected promptly, does not in itself signify a bad prognosis.

Any condition which obstructs the outflow of urine will cause back-pressure on the kidneys and reduce the glomerular filtration rate. These post-renal causes of urea retention include prostatic hypertrophy, bilateral calculi and carcinoma of the bladder involving the ureteric orifices. If the urinary obstruction can be relieved the blood urea returns rapidly to normal unless permanent renal damage has occurred

Principle of the method

Blood is incubated with urease to convert the urea into ammonium carbonate. The proteins are precipitated by tungstic acid, and an aliquot part of the supernatant fluid is treated with Nessler's reagent. The intensity of the resulting brown colour is compared in the comparator with the standard colour glasses. The preformed ammonia in blood is negligible.

Reagents required

1. *Urease suspension* Grind one tablet of B.D.H. urease (one tablet = 50 mg urea) in 5.0 ml of 30 per cent alcohol. This suspension usually keeps for at least 4 or 5 days at room temperature.
2. *10% sodium tungstate ($Na_2WO_4.2H_2O$)*
3. *2/3 N sulphuric acid (H_2SO_4)*
4. *Nessler's Reagent*
    (a) *Double Iodide Solution* Dissolve 150 g of potassium iodide (KI) in 100 ml distilled water. Add 200 g of mercuric iodide (HgI_2) and wait till solution is complete. Then dilute to 1,000 ml with distilled water and filter. Dilute the filtrate to 2,000 ml.
    (b) *10% Sodium Hydroxide* Prepare a saturated solution of sodium hydroxide (NaOH) (about 55%) by adding an excess of NaOH to about 200 ml water and stopper securely. After two or three days decant the clear supernatant fluid, and dilute with distilled water to 10%. (Add 45 ml of water to each 10 ml of supernatant fluid.) Check the concentration by further diluting 10 ml to 25 ml with distilled water, and titrating 10 ml of the supposed 4% NaOH with N acid. If the concentration differs from the theoretical by more than $\pm$ 5% (i.e. if in the titration 10 ml of sodium hydroxide require more than 10.5 ml or less than 9.5 ml of N acid) it must be adjusted.

    Preparation of Nessler's Reagent
10% Sodium hydroxide	700 ml
Double iodide solution	150 ml
Distilled water	150 ml

The Standard Lovibond Comparator Discs 5/9A and 5/9B

Disc A covers the range 20 mg to 100 mg urea per 100 ml of blood in nine steps of 10 mg each.

Disc B covers the range 110 mg to 220 mg; the first six steps are of 10 mg, the last three steps are of 20 mg each.

It is therefore possible to read approximately to 5 mg, except at the top of the scale.

Technique

In a non-tapered centrifuge tube place 2 ml of distilled water and 0.2 ml of blood. The latter is measured with a blood pipette calibrated "to contain". The blood is delivered beneath the 2 ml of water. The pipette is then raised and washed out with the water two or three times. Add 0.2 ml of the urease suspension and shake well to mix.

Place the centrifuge-tube in a water-bath at 55°C for fifteen minutes (or in a beaker of water which has been heated to 60°C).

Remove the tube and add 0.3 ml of 10 per cent sodium tungstate, 0.3 ml of two-thirds normal sulphuric acid and 5 ml of distilled water. Shake well to mix, stand a few minutes till the protein precipitate flocculates, and centrifuge till the supernatant fluid is quite clear. Pipette off 5 ml of the supernatant fluid into a clean test tube and add 5 ml of water and 2 ml of Nessler's reagent. Mix and transfer to a comparator tube or 13.5 mm cell and place in the right-hand compartment of the comparator.

Hold the comparator facing a standard source of white light such as the Lovibond White Light Cabinet or, failing this, north daylight, and rotate the disc until the colour of the solution is matched by one of the glass colour standards. The result is then read from the indicator recess at the bottom right-hand corner of the comparator, expressed as mg urea per 100 ml of blood.

If the colour developed exceeds the highest value on Disc A (100), substitute Disc B, which covers the range up to 220.

If the blood urea exceeds 220 mg, the test must be repeated from the beginning but using 0.1 ml of blood plus 0.1 ml of water instead of 0.2 ml of blood; the final reading, of course, must then be multiplied by 2.

Notes

1. With each new set of reagents, including each new batch of urease tablets, a blank test should be performed using 0.2 ml of water instead of blood. There should be no perceptible colour.

2. The protein precipitate must be removed by centrifuging and not by filtration, because filter papers, unless specially washed, contain ammonium salts, and give a positive colour reaction with Nessler's reagent.

3. All water used in the test and for making solutions must, of course, be free from ammonia. Freshly distilled water is usually safe, but each supply should be tested with Nessler's reagent.

Reference

H. E. Archer and G. D. Robb, *Quart. J. Med,* 1925, **18,** 274

The Determination of Urea (2)
using urease and King's modification

Introduction

This test[1], which measures serum or plasma urea, has been developed as a simpler modification of the urease nesslerisation methods [2,3]. It is designed for emergency use in large hospitals, for routine use in smaller hospitals and for use in private medical practice.

Principle of the method

Blood serum or plasma is treated with urease and the ammonium ions so formed are reacted with Nessler's reagent. The intensity of the brown colour formed, which is proportional to the urea concentration, is measured by comparison with a series of Lovibond permanent glass colour standards.

Reagents required

1. *Nessler's reagent*

    (a) *Double iodide solution* Dissolve 150 g of potassium iodide (KI) in 100 ml of distilled water. Add 200 g of mercuric iodide (HgI_2) and wait until solution is complete. Then dilute to 1 litre with distilled water, and filter. Dilute the filtrate to 2 litres.

    (b) 10% *sodium hydroxide* Prepare a saturated solution of sodium hydroxide (NaOH) by adding an excess (about 120 g) of NaOH to 200 ml of distilled water. Stopper securely. After two or three days decant the clear supernatant fluid to give a 10% solution by adding 45 ml of water to each 10 ml of supernatant fluid. Check the concentration by titrating with standard acid by diluting 10 ml of the 10% NaOH to 25 ml and titrating 10 ml of this supposedly 4% NaOH with N hydrochloric acid (HCl). If more than 10.5 or less than 9.5 ml of N HCl is required, then adjust the concentration of the 10% NaOH to bring the titration within these limits.

    Preparation of Nessler's reagent Mix 700 ml of 10% NaOH solution and 150 ml of the double iodide solution and then make up to 1 litre with distilled water.

2. *Glycerol-urease reagent* Take 5 g of jack beans, or potent jack bean meal, pulverise in a mortar, add 100 ml of 70% glycerol, mix well, leave overnight, centrifuge hard, and use the supernatant fluid. Store the extract in a dark bottle in a refrigerator.

The Standard Lovibond Comparator Discs 5/9A and 5/9B

Disc 5/9A covers the range 20—100 mg of urea per 100 ml of sample in 9 steps of 10 mg each.

Disc 5/9B covers the range 110—220 mg in six steps of 10 mg and three steps of 20 mg.

Technique

Measure 0.1 ml of serum or plasma into a conical centrifuge tube graduated to 10 ml in steps of 0.1 ml. Citrated or sequestrinated plasma may be used but **NOT** oxalated plasma. Add two drops of glycerol urease reagent and mix well by flicking the bottom of the tube. Leave to incubate at room temperature for 20 minutes (Note 1). Add distilled water to the tube up to the 8 ml mark. Mix by inversion with the thumb held over the end of the tube. Add Nessler's reagent to exactly the 9.6 ml mark and mix again. Transfer the solution to a 13.5 mm comparator cell or test tube and place the tube in the right-hand compartment of the comparator. Illuminate the comparator with a standard source of white light, such as the Lovibond White Light Cabinet or, failing this, north daylight. Compare the colour of the sample with the colour of the standards in the disc. Read off the corresponding urea concentration from the indicator window in the comparator, interpolating between adjacent standards if necessary.

If the urea concentration exceeds 220 mg per 100 ml repeat the test using 0.1 ml of a diluted sample and multiplying the final reading by the dilution factor.

Notes

1. This period of incubation has been made deliberately excessive as a safeguard against the use of a weak enzyme preparation.

2. All water used in the test and in the preparation of reagents must be free from ammonia. Freshly distilled water is usually safe but each supply should be tested with Nessler's reagent and should give no perceptible colour.

References

1. M. H. King "*Medical Care in Developing Countries.*" Oxford University Press, Nairobi, 1966
2. H. E. Archer and G. D. Robb, *Quart. J. Med.,* 1925, **18,** 274
3. E. C. Lile, M. F. Villamil, M. C. Rhees and B. H. Scribner, *J. Amer. Med. Assoc.,* 1957, **164,** 277

The Determination of Urea (3)
(using diacetyl monoxime)

Principle of the method

Proteins are precipitated from the blood sample and removed by centrifugation. The supernatant liquid is heated with acidic diacetyl monoxime, when the urea reacts with the diacetyl to produce a yellow colour which is compared with a series of standard glasses.

Reagents required

1. *Sodium tungstate* 10%
2. *Sulphuric acid* 2/3 N
3. *2% diacetyl monoxime in 2% acetic acid* To 60 ml of water add 2 g of diacetyl monoxime. Add 2 ml of glacial acetic acid. Warm gently and shake to dissolve. When dissolved add water to make up to 100 ml.
4. *Sulphuric acid—Phosphoric acid reagent* To 140 ml of water add 150 ml of 85% phosphoric acid. Mix well and continue mixing while 50 ml of concentrated sulphuric acid are added slowly. This solution keeps for 1 year.

The Standard Lovibond Comparator Disc 5/43

This covers the range 20—180 mg per 100 ml in steps of 20 mg

Technique

1. Dilute 0.1 ml of blood in 3.3. ml of water.
2. Add 0.3 ml of 10% sodium tungstate and 0.3 ml of 2/3 N sulphuric acid.
3. Mix and centrifuge.
4. To 2 ml of supernatant add 0.8 ml 2% diacetyl monoxime, 3.2 ml of sulphuric-phosphoric acid mixture, and 4 ml of water.
5. Place in boiling water bath for 30 minutes.
6. Cool and fill a 13.5 cm comparator cell, which is placed in the right-hand comparator compartment.
7. Compare colour with Comparator disc, using water in left-hand comparator cell as a blank.

Note

Should the blood urea be higher than 180 mg per 100 ml repeat the method using 1 ml of supernatant and one ml of water. The result is then multiplied by 2.

References

1. S. Natelson, M. L. Scott and C. Beffa. *Amer. J. Chem. Path* 1951, **21,** 275
2. S. Natelson, *Microtechniques of Clinical Chemistry for the Routine Laboratory,* C. C. Thomas, Springfield, Illinois. p. 381

The Determination of Uric Acid in Serum
using alkaline phosphotungstate

Introduction

Uric acid is produced as an end product of purine metabolism which is derived from nucleic acid. It is therefore formed from the catabolism of endogenous nucleo-protein and also that taken in with the diet.

The normal level varies to some extent with the method used for its determination. The normal levels for the phosphotungstic acid method are 1.5—6.0 mg/100 ml for women and 2.5—7.0 mg/100 ml for men.

The serum levels are increased in a number of conditions, the principal ones being:—
   (a) Gout, where there is probably an increased renal tubular absorption of uric acid.
   (b) Tissue destruction such as is seen in severe acute infections and sometimes in eclampsia.
   (c) Increased nucleo-protein metabolisms as seen in leukaemia (particularly when under treatment with cytotoxic drugs), polycythaemia and haemolytic anaemias.
   (d) Renal failure.

Principle of the method

Following protein precipitation, uric acid is allowed to reduce alkaline phosphotungstate to produce a blue colour. This is a modification of the Henry, Sobel and Kim method.

Reagents required

1. *Phosphotungstic acid* 40 g of analytical reagent grade sodium phosphotungstate (sodium dodeca-tungstophosphate) are dissolved in 300 ml of distilled water. 32 ml of 85% ortho-phosphoric acid are added and the whole refluxed gently for 2 hours. The solution is cooled and made up to one litre with water. 32 g of lithium sulphate ($Li_2SO_4H_2O$) are added and the reagent well mixed. Store in the refrigerator. Stable for one year.
2. *Sodium Carbonate 14%* Dissolve 14 g of anhydrous sodium carbonate (Na_2CO_3) in distilled water and make up to 100 ml. Store in a polythene bottle. Stable for one month.
3. *Sulphuric Acid 2/3N* To 800 ml of distilled water add slowly 18 ml of concentrated sulphuric acid. Make up to one litre and mix.
4. *Sodium Tungstate 10%* Dissolve 10 g of sodium tungstate ($Na_2WO_4.2H_2O$) in water and make up to 100 ml. Stable for one year.

The Standard Lovibond Comparator Disc 5/47

The disc covers the range 1 to 9 mg/100 ml in steps of 1 mg.

Technique

1. To 6 ml of water add 2 ml of serum.
2. Mix, add 1 ml of 2/3N Sulphuric acid and finally 1 ml of 10% sodium tungstate.
3. Mix, stand for 10 minutes and then centrifuge.
4. Take 5 ml of supernatant fluid and add 1 ml of phosphotungstic acid reagent and 1 ml of 14% sodium carbonate.
5. Mix and stand for 15 minutes to allow the colour to develop.
6. Pour the test solution into a 13.5 comparator cell and place in right-hand compartment of the Lovibond Comparator, and place a blank cell containing water into left-hand compartment. Match colours using White Light Cabinet or north daylight, and read off uric acid concentration from the indicator window.

Notes

1. The test should be read within 30 minutes.
2. Serum is recommended for this analysis. If plasma is used potassium oxalate must not be used as an anti-coagulant.

Reference

R. J. Henry, C. Sobel, Kim, *J. Am. Clin. Path.* 1957, **28**, 152, 645

Toxic Substances

in Air

COLORIMETRIC CHEMICAL ANALYTICAL METHODS

The Determination of Toxic Gases

in low concentrations in air

Introduction

Section 63 of the 1961 Factories Act requires that all practicable measures must be taken to protect industrial workers against the inhalation of any dust, or fume, or other impurity, of such a character, and to such an extent, as to be likely to be injurious. With the ever increasing complexity of chemical industry there is a growing list of recognized hazards and maximum permissible concentrations have now been laid down[1,2] for over 400 vapours, fumes and dusts commonly encountered in industry. Of these, in over 200 cases the toxic hazard is estimated as being at least as great as that from hydrogen cyanide (prussic acid), which has long been regarded by the general public as the most toxic vapour. However some of the vapours now encountered in industry are estimated to present hazards a hundred to a thousand times as great as that presented by prussic acid vapour.

General advice on the protection of workers against toxic hazards has been issued by the Department of Employment, who are also responsible for the development of a series of tests[3], listed below, for the detection and determination of noxious fumes and vapours at concentrations below the level at which they are dangerous to life. These tests were originally developed by the Chemical Defence Establishment of the Ministry of Defence, for the Department of Scientific and Industrial Research, acting with the financial and technical co-operation of the Association of British Chemical Manufacturers. The tests are now the responsibility of H.M. Factory Inspectorate of the Department of Employment and are being revised and extended by the Government Chemist's Laboratory[4-8]. Permission to quote from those official tests[3] is gratefully acknowledged.

In addition, a number of other similar tests, not included in this official series, but developed by the Tintometer Ltd. in collaboration with other workers, are described in the following section.

Principle of the method

The methods recommended are colorimetric. A known volume of the air under examination is drawn through a filter paper (or in some cases a solution) containing a reagent, the colour change of which is used as a measure of the concentration of the toxic gas. This concentration is evaluated by comparison of the colour of the stain on the filter paper (or the colour of the solution) against either the standard colour charts issued with the Stationery Office series of booklets, or against standard coloured solutions.

As neither the colour charts nor the standard solutions provide permanent standards, discs for use with the Lovibond Comparator have been made available for the interpretation of the colours obtained in the tests. These permanent Lovibond glass standards have been officially tested and approved as an alternative method of comparison. Details of the gases for which the official discs are available are given in the Table.

COLORIMETRIC CHEMICAL ANALYTICAL METHODS — Toxic Gases

Technique

The critical requirement, in all the tests for noxious substances in air, is to sample accurately a known volume of air at a specified flow rate. There are several alternative methods of meeting this requirement, and the method adopted must depend on personal preference in the light of the facilities which are available. Some simple methods are described below.

(a) *Aspirator*

By the use of a 'constant head' aspirator, as shown in Figure 1, a uniform sampling rate may be maintained and the total volume of air sampled may be determined directly by measuring the amount of water run off from the aspirator. By running the air inlet tube down to the approximate level of the outlet tube the effective head of water at the outlet, and thus the rate of flow, remains constant irrespective of the height of water in the aspirator. The sampling rate can be adjusted by the use of a capillary attached to the outlet, a suitable capillary being used for each sampling rate required. Small adjustments of flow rate can be made by adjusting the height of the inlet tube above the outlet. Provided that the volume of water above the level of the bottom of the inlet tube is greater than the volume of gas to be sampled, then the requirements of constant flow will be maintained. If an aspirator with an outlet at the bottom is not available, then a suitable alternative may be constructed using a normal Winchester Quart bottle, as shown in Figure 2. Provided that the syphon tube is filled with water, and that the same conditions regarding the relative levels of inlet and outlet tube are maintained as before, then this arrangement will also ensure a constant flow.

The aspirator offers the simplest, and possibly the most certain, means of sampling a known volume of air at a fixed sampling rate. It is not however always convenient to carry liquids about, and, where the sample to be taken is large, the equipment is correspondingly bulky. It has however the advantage of flexibility in its application and of being readily constructed from materials available in the normal laboratory.

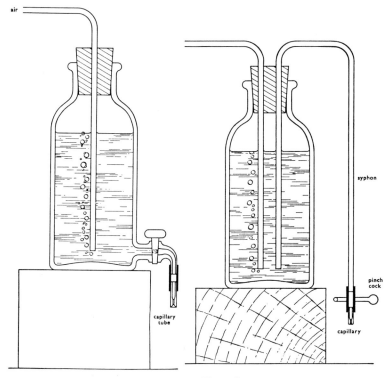

Figure 1
"Constant head" Aspirator

Figure 2
Alternative arrangement for 'Constant head' aspirator.

(b) *The 'critical orifice'*

It is well known that, if a constriction is introduced into a pipe-line, then a pressure difference is set up across the constriction. The magnitude of this pressure difference depends *inter al.*, on the flow rate through the constriction. This effect is used as a basis for many laboratory types of flowmeter. However it is not so well known that if the pressure on the suction side of such a constriction is reduced to such a level that its ratio to the inlet pressure is 0.527, or less, then there is a maximum rate of flow through the constriction[9].

This effect is the basis of an extremely simple device for producing constant sampling rates. A piece of thick-walled glass tubing is heated until its end is nearly sealed and a narrow passage is produced by means of a wire of known diameter. After cooling, this constricted tube (Figure 3) is connected to a source of suction capable of producing a pressure of less than half an atmosphere, such as a water-pump, and the rate of flow through the orifice is measured. By varying the diameter of the wire it is possible to produce orifices to give any desired rate of flow.

Having prepared and calibrated a suitable orifice, the test equipment is set up as follows. The orifice is connected to the *inlet* side of the bubbler or test-paper holder, as the case may be, and the outlet side is connected to a suction source. This may be a water-pump, an electric suction pump, or a vacuum line. The suction is turned on and the volume of air sampled is determined by multiplying the flow rate for the orifice by the time during which the equipment is running.

Figure 3   Critical orifice

(c) *Battery Operated Pump*

Several small battery operated pumps are now available commercially. One of these is that manufactured by Rotheroe and Mitchell Ltd, Victoria Rd., Ruislip, Middlesex, England. This is illustrated in figure 4. The completely self-contained unit comprises an electrically driven pump, a monitoring flowmeter with flow control valve and a red L.E.D. which flashes when the batteries need recharging. This unit is capable of sampling over the range 5 to 500 ccs/minute and its internal sealed rechargeable batteries will run for over ten hours on one charge.

Other battery operated pumps with greater flow rates are also available.

Figure 4   Battery operated pump, type C500

Notes

1. In certain cases complete test outfits for a particular application are available. These contain the necessary pump, filter papers or bubbler, comparator and disc etc. Details from Tintometer Ltd.

2. For full details of the chemical procedure for the tests given in the Table, reference should be made to the appropriate booklet in the series *"Methods for the detection of toxic substances in air,"*[3] as much of the matter is Crown copyright.

3. The advantage to be gained from the use of the comparator rather than a printed colour chart is the permanence of the Lovibond glass colour standards and the fact that these can reproduce exactly the colour developed in the test. The convenience resulting from the use of glass standards rather than coloured solutions is self evident.

4. Reference is made in the instructions to the use of test-tubes for the comparison of the colours. The use of rectangular cells, of 13.5 mm optical path, in the place of test-tubes, has the advantage of presenting a uniform field of observation and thus facilitating comparison.

References

1. Department of Employment *"Dust and fumes in factory atmospheres"*, 1968, H.M. Stationery Office, London, New Series No. 8. 4th Edn. 1969
2. International Union of Pure and Applied Chemistry, *"Methods for the Determination of Toxic Substances in the Air,"* Butterworths, London, 1959
3. Department of Employment, *"Methods for the Detection of Toxic Substances in Air,"* H.M.S.O. London
4. B. E. Dixon, G. C. Hands and A. F. F. Bartlett, *Analyst,* 1958, **83,** 199
5. B. E. Dixon and P. Metson, *Analyst,* 1960, **85,** 122
6. G. A. Sergeant, B. E. Dixon and R. G. Lidzey, *Analyst,* 1957, **82,** 27
7. B. E. Dixon and G. C. Hands, *Analyst,* 1959, **84,** 463
8. G. C. Hands and A. F. F. Bartlett, *Analyst,* 1960, **85,** 147
9. E. Ower, *"The Measurement of Air Flow,"* 3rd. edn., Chapman & Hall Ltd., London, 1949, pp 143—4

Reagents required and the Standard Lovibond Comparator Discs for Official Tests.

Gas	Reagents used	Type of test	Range of concentrations	Standard Disc No.	Number of Standards	H.M.S.O. Booklet No.
Acetone	Sodium nitroprusside	Solution	500 to 2000 ppm	6/48	7	23
Acrylonitrile	Alkaline permanganate	Solution	10, 20, 40 ppm	6/29	6	16
Aniline	Bleaching powder Phenol	Solution	5 to 200 ppm	6/11	5	11
Benzene	Paraformaldehyde	Solution	12 to 50 ppm	6/24	3	4
Carbon Disulphide	Diethylamine Copper acetate	Solution	0, 5, 10, 20 and 40 ppm	6/56	4	6
Chlorine	3,3'-dimethylnaphthidine	Solution	0.5 to 2.0 ppm	6/23	3	10
Chromic Acid Mist	Diphenylcarbazide	Solution	0.05 to 0.2 ppm	6/30	3	17
Copper oxide fumes	Cuprizone	Solution	0 to 0.2 mg/m³	6/51	4	22
Cyclohexanone	H-acid diazonium salt	Solution	Cyclohexanone 0 to 100 ppm Methylcyclohexanone 0 to 100 ppm	6/53	4	26
Fluorides, Inorganic	Lanthanum-alizarin	Solution	0 to 5 mg/m³	6/47	4	19
Hydrogen Cyanide	Ferrous hydroxide	Paper	2.5 to 50 ppm	6/35	5	2
Hydrogen Fluoride	Zirconium-Solochrome cyanine	Solution	0-6 ppm	6/46	4	19
Hydrogen Sulphide	Lead acetate	Paper	5 to 40 ppm	6/38	4	1
Isocyanates, Aromatic	Naphthylethylene diamine	Solution	0.01 to 0.04 TDI MDI NDI PAPI	6/42 6/43 6/44 6/45	3 3 3 3	20
Iron oxide fumes	Diphenylphenanthroline disulphonic acid	Paper	1 to 20 mg/m³	6/52	4	21
Isophorone	Dodecamolybdophosphoric acid	Solution	0, 5, 10, 20, 25 and 50 ppm	6/58	3	24
Ketone vapour	Dinitrophenylhydrazine	Solution	From 12.5 to 200 ppm according to which ketone being tested	6/55	4	
Lead (total)	Dithizone	Solution	0 to 0.8 mg/m³	6/49	5	14

Reagents required and the Standard Lovibond Comparator Discs for Official Tests—*continued*

Gas	Reagents used	Type of test	Range of concentrations	Standard Disc No.	Number of Standards	H.M.S.O. Booklet No.
Mercury	Selenium sulphide	Paper	0 to 200 $\mu g/m^3$	6/37	5	13
Nitrous Fumes	p-Anisidine Alkaline arsenite/ naphthylethylene diamine	Paper Solution	1.3 to 10 ppm 2.5 to 10 ppm	6/39 6/33	4 3	5 5
Ozone	Potassium iodide	Solution	0.05 to 0.4 ppm	6/34	9	18
Phosgene	4-p-nitrobenzylpyridine N-benzylaniline	Paper	0.05 to 0.4 ppm	6/36	4	8
Styrene	Sulphuric Acid	Solution	50 to 200 ppm	6/26	3	4
Sulphur Dioxide	Ammoniacal zinc nitroprusside	Paper	1 to 20 ppm	6/40	5	3
Toluene and/or Xylene	Potassium Iodate	Solution	100 to 300 ppm	6/25	3	4
Zinc oxide fumes	T.A.R. (Thiazolylazo resorcinol)	Solution	0 to 10 mg/m^3	6/50	4	25

The Determination of Acetone
using sodium nitroprusside

Introduction

Acetone is widely used in industry. In 1965 in the U.S.A. alone, 520,000 tons were produced, of which 14% was used in plastics production, 41% as an intermediate in chemical syntheses, 28% as a solvent and the remaining 17% for miscellaneous purposes.

The vapour of acetone has a strongly irritant and narcotic effect on man, and can also form flammable and explosive mixtures with air. Whereas the concentration limits for flammability are 2.5 to 12.8%[1], the recommended threshold limit value[2] is 1,000 ppm (2,400 mg/m^3). Enlightened industrial hygiene practice inclines towards controlling concentrations at limits below the threshold limit value and this test[3,5], which is a modification of the earlier work of von Bitto[4], was designed to determine the concentration of acetone vapour in air at about the threshold limit value.

Principle of the method

Acetone reacts with sodium nitroprusside, in the presence of ammonium salts, to produce a complex which, on the addition of alkali, develops a purple colour. The intensity of the colour, which is proportional to the acetone concentration, is measured by comparison with a series of Lovibond permanent glass colour standards. As it has been shown[3] that the colour intensity varies with the temperature at which the colour development is carried out, two sets of colour standards, corresponding to two different working temperatures, are provided in the comparator disc.

Reagents required

The chemicals used in the preparation of these solutions should all be of analytical reagent quality.

1. *Sodium nitroprusside reagent* Dissolve 1.00 g of sodium nitroprusside and 25.0 g of ammonium acetate in distilled water and dilute to 100 ml. This solution should be stored in the dark, at a temperature not exceeding 27°C, and must be used between 16 and 24 hours after preparation.

2. *Ammonia solution* Dilute 25 ml of concentrated ammonia solution (sp.gr. 0.880) to 50 ml with distilled water (Note 1). This solution should be prepared freshly each day.

All solutions should be stored in an acetone-free atmosphere and kept tightly stoppered when not in use.

The Standard Lovibond Comparator Disc 6/48

This disc contains seven standards labelled 2000A, 1000A, 500A, 0, 500B, 1000B, 2000B, ppm of acetone (CH_3COCH_3), based on a 25 ml air sample.

The 'A' standards are used when the test is carried out at temperatures between 19° and 22°C.

The 'B' standards are used when the test is carried out at temperatures between 22° and 26°C.

Apparatus

Bubbler as illustrated. Receiver – Overall length 155 mm, external diameter 15 mm, internal diameter 12.8 ± 0.4 mm. Distance from bottom of side arm to bottom of bubbler 110 mm. Inlet Tube – Approximate overall length 185 mm, external diameter 6 mm, internal diameter 4.5 ± 0.1 mm. Distance from bottom of inlet tube to bottom of bubbler 7 mm. Cone and Socket B10.

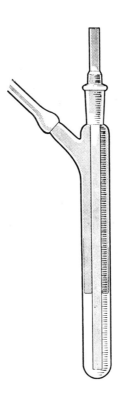

Technique

In an atmosphere free from acetone vapour, transfer (Note 1) 3 ml of the sodium nitroprusside reagent (reagent 1) into an absorption tube of the type illustrated in the Figure (Note 2). Insert the inlet tube and connect the bubbler to a suitable source of suction (Note 3). Transfer the apparatus to the sampling site and collect 250 ml of the suspect atmosphere. Remove the apparatus to an acetone-free atmosphere, disconnect the source of suction, remove the inlet tube and with a safety pipette transfer 1 ml of the ammonia reagent (reagent 2) into the absorption tube. Stopper the tube and mix the contents well by inverting the tube several times. Take care that no liquid escapes through the side arm. Place the tube in a beaker of water at a temperature within the range 19° to 26°C and note the temperature. Best results will be obtained with the temperature of the water adjusted to either 20° or 25°C, as these were the temperatures at which the two ranges of standards were calibrated.

After 10 ± 0.5 minutes transfer the absorption tube bubbler to the right-hand compartment of the comparator. Place the comparator before a standard source of white light such as the Lovibond White Light Cabinet or, failing this, north daylight, and match the colour of the sample with the standards in the disc, using the range of standards appropriate to the temperature of the water in the beaker during colour development.

Notes

1. For reasons of safety, pipette fillers should be used to dispense the reagents used in this test.
2. Bubblers manufactured to these dimensions may be obtained from Tintometer Ltd.,
3. Sources of suction are described in pages 462 to 464.
4. Formaldehyde gives a faint red colour with these reagents, but this colour disappears within the 10 minute colour development period and will not therefore interfere.

Ethyl methyl ketone, isobutyl methyl ketone and diacetone alcohol would all interfere, but only at concentrations above their respective threshold limit values[2]. They should, therefore, not be present in the atmosphere at concentrations which would interfere with this test.

Variations in relative humidity within the limits encountered in the United Kingdom have no effect on this test.

References

1. "*Handbook of Chemistry and Physics*", The Chemical Rubber Publishing Co., Cleveland, Ohio, U.S.A., 44th editn. 1962, p. 1941
2. Department of Employment and Productivity, H.M. Factory Inspectorate, Technical Data Note 2/69 "Threshold Limit Values for 1969"
3. A. F. Smith and R. Wood, *Analyst,* 1970, **95,** 683
4. B. von Bitto, *Justus Liebigs Annln. Chem.,* 1892, **267,** 372
5. "*Methods for the detection of toxic substances in air*. Booklet 23, *Acetone*" H.M. Stationery Office 1971

The Determination of Acrylonitrile (vinyl cyanide)
using alkaline permanganate

Introduction

Acrylonitrile, which is widely used, particularly in the plastics industry, is highly toxic either by absorption of the liquid through the skin, or by inhalation of the vapour. Its effects are similar to those of cyanide. The Threshold Limit Value for a continuous eight hour exposure has been set[1] at 20 ppm, by volume, of air.

The present test[2] has been developed to determine the concentration of acrylonitrile vapour in factory atmospheres using only simple equipment. It covers the range 10—40 ppm of acrylonitrile.

Principle of the method

The air is sampled through an alkaline solution of potassium permanganate which converts the acrylonitrile to cyanide. Excess permanganate is removed by means of sodium arsenite and the cyanide is determined by means of the modified Aldridge procedure[3,4] using pyridine-aniline reagent. The intensity of the resulting reddish-yellow colour, which is proportional to the concentration of acrylonitrile, is measured by comparison with a series of Lovibond permanent glass colour standards.

Reagents required

1. *Alkaline permanganate solution* Dilute 3.5 ml of $0.1N$ potassium permanganate ($KMnO_4$) solution (5.267 g $KMnO_4$ per litre) to 100 ml with $0.1N$ sodium hydroxide (NaOH) solution (4 g NaOH per litre). This reagent is stable for only 3 days.
2. *Sodium arsenite solution* 0.75% w/v aqueous solution of $NaAsO_2$.
3. *Hydrochloric acid* Approximately $2N$ HCl, dilute 240 g of concentrated HCl to 1 litre.
4. *Bromine water* Saturated aqueous solution.
5. *Pyridine-aniline solution* Add 40 ml of pyridine to 60 ml of water, mix, cool and add 2 ml of aniline, mix thoroughly. This reagent is stable for 7 days.

All chemicals used should be of analytical reagent grade.

The Standard Lovibond Comparator Disc 6/29

This disc contains standards equivalent to 10, 20 and 40 ppm of acrylonitrile, based on a 120 ml air sample. As the colours produced in the test vary slightly with temperature, two sets of standards are included in the disc. These correspond to test temperatures of 12°C and 20—25°C respectively.

Technique

Pipette 2 ml of alkaline permanganate solution (reagent 1) into a bubbler of the type shown in the Figure, having an internal diameter of 12.8 ± 0.4 mm (Note 1). Insert the inlet tube and attach the chosen source of suction (Note 2) to the side arm. Adjust the sampling rate to about 40 ml per minute and sample exactly 120 ml of the atmosphere through the bubbler. Disconnect the suction and remove the inlet tube, draining as much liquid as possible into the receiver. Add 1 ml of sodium arsenite solution (reagent 2) washing down the walls of the bubbler to ensure complete reduction of the excess permanganate. Then add 0.5 ml of $2N$ hydrochloric acid (reagent 3). Shake and allow to stand for about 30 seconds until the contents of the bubbler becomes colourless. Add 0.5 ml of bromine water (reagent 4) mix thoroughly and then add 1.0 ml of sodium arsenite solution (reagent 2) to discharge the bromine colour. Add rapidly 1.0 ml of pyridine-aniline solution (reagent 5) and shake thoroughly. Allow the colour to develop for 4 minutes and then place the tube in the right-hand compartment of the comparator. Hold the comparator facing a standard source of white light, such as the Lovibond White Light Cabinet or, failing this, north daylight, and match the colour of the sample with the standards in the disc. Use the set of standards which correspond most nearly to the temperature at which the test was carried out.

COLORIMETRIC CHEMICAL ANALYTICAL METHODS — Acrylonitrile (vinyl cyanide)

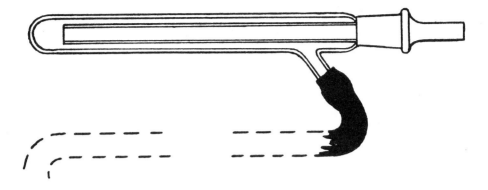

Notes

1. The bubbler tube may be obtained from Tintometer Ltd. Any similar tube is suitable provided that its internal diameter falls within the limits specified i.e. 12.8 ± 0.4 mm.

2. Methods of taking gas samples are described in the Introduction to this Section.

References

1. Min. of Labour, *"Toxic Substances in Factory Atmospheres"*, H.M. Stationery Office, London, 1960
2. Min. of Labour, *"Methods for the Detection of Toxic Substances in Air, Booklet No. 16 Acrylonitrile"* H.M. Stationery Office, London, 1966
3. W. N. Aldridge, *Analyst*, 1945, **70**, 474
4. Cyanide Method 2

The Determination of Aniline
using bleaching powder and phenol

Introduction

Aniline exists in small quantities in coal tar, but most of the industrial aniline is produced synthetically. It is used as a starting material for the preparation of synthetic dyestuffs. In chemical works where aniline is used, dangerous concentrations may occur in the atmosphere unless proper precautions are taken. The maximum permissible concentration of aniline in an industrial atmosphere has been specified as 5 ppm by volume[1]. The present test[2] has been designed to enable factory atmospheres to be tested quickly and easily.

Principle of the method

In the presence of bleaching powder aniline forms a purplish colour which, on treatment with ammonia and phenol, turns a deep blue. The intensity of this colour, which is proportional to the aniline concentration, is measured by comparison with a series of Lovibond permanent glass colour standards.

Reagents required

1. *Hydrochloric acid* 1% Dilute 25 ml of concentrated hydrochloric acid (HCl) to 1 litre with distilled water.

2. *Bleaching powder solution* Warm 5 g of bleaching powder [$Ca(OCl)_2$] in 100 ml of distilled water to 50-60°C. and shake continuously. Filter while hot. The bleaching powder used should contain not less than 25% of available chlorine.

3. *Phenol reagent* Dissolve 5 g of phenol (C_6H_5OH) in 100 ml of an ammonia solution containing 5 ml of 0.880 ammonia (HN_4OH)

All chemicals should be of analytical reagent quality.

The Standard Lovibond Disc 6/11

This disc contains five standards corresponding to the standards specified in the official test for aniline[2]. See table on next page.

Technique

Sample the contaminated atmosphere through a bubbler containing 10 ml of hydrochloric acid (reagent 1). A suitable bubbler consists of a side-arm test tube of about 0.75 in. internal diameter and 6 in. long, fitted with a rubber bung through which passes a delivery tube (approximately 0.125 in. bore and terminating in a fine jet which will just pass a wire of 18–25 S.W.G.), reaching nearly to the bottom of the test tube. Sample the air at 750 ml per minute by a suitable standard method[3] and record the volume of air sampled. For a preliminary test a volume of 630 ml is convenient. When the desired volume of air has been sampled, disconnect the bubbler and transfer the contents to an ordinary test tube. Add two drops of bleaching powder solution (reagent 2), allow the mixture to stand for 5 minutes and then boil. Add 5 ml of the phenol reagent (reagent 3) and allow the mixture to stand for 15 minutes. Transfer the liquid to a 13.5 mm comparator cell or standard test tube and place this in the right-hand compartment of the comparator. Place the comparator before a standard source of white light, such as the Lovibond White Light Cabinet or, failing this, north daylight, and compare the colour of the solution with the standards in the disc. Read off the concentration, corresponding to the depth of colour and the volume of air sampled, from the following table:—

Volume of air (ml.)	126	378	630	1260
Concentration ppm by volume	Standard No.	Standard No.	Standard No.	Standard No.
200	3–4	–	–	–
100	2	4	–	–
66	–	3–4	5	–
50	1	3	4	–
33	–	2	3	5
20	–	–	2	3–4
10	–	–	1	2
5	–	–	–	1

Notes

1. If a colour less intense than Standard No. 1 is obtained in a test, repeat the test taking a larger volume of air then calculate the concentration of aniline from the Table by proportion.

2. Care must be taken to ensure that the test tube and the delivery tube are thoroughly clean before starting the test.

References

1. Min. of Labour, "*Toxic Substances in Factory Atmospheres*", H.M.S.O., London, 1960
2. Min. of Labour, H.M. Factory Inspectorate, "*Methods for the detection of Toxic Substances in Air, Booklet No. 11, Aniline Vapour*", H.M.S.O., London, 1968
3. See pages 462 to 464

The Determination of Arsine
using mercuric chloride

Introduction

Arsine is an extremely poisonous gas, constituting one of the major hazards in industry. The maximum permissible concentration in an industrial atmosphere is 0.05 parts per million (0.2 mg per cubic metre of air)[1]. On this basis it is twenty times as toxic as phosgene and two hundred times as toxic as hydrogen cyanide. One of its main dangers results from the cumulative effect of repeated exposure to very low concentrations.

Arsine is formed during many industrial processes, particularly during electrolytic processes and the smelting of ores containing arsenic. It is readily formed in the presence of nascent hydrogen, even from trace amounts of arsenic in other metals. An example of this is the production of arsine, from traces of arsenic in the lead plates, in storage batteries during charging.

The present test was adopted for the detection of arsine in industrial atmospheres, despite its lack of sensitivity compared with alternative methods. It has the advantage that the test papers used are stable, if stored in the dark and kept dry, and the stains produced in the presence of arsine are reproducible and adequately stable. The test is sufficiently sensitive to detect the maximum permissible concentration of arsine provided that a sufficient volume of air is sampled.

Principle of the method

Air is drawn through a paper impregnated with mercuric chloride. In the presence of arsine a yellow to orange coloured stain of $As(HgCl)_3$ is produced. The amount of arsine in the sample is determined by comparing the colour of the stain with Lovibond permanent glass standards.

Reagent required

Mercuric chloride solution 5 g of analytical reagent grade mercuric chloride ($HgCl_2$) in 100 ml of water.

Preparation of Test Papers

Cut a sheet of Postlip 633 Extra Thick White filter paper into strips 2 inches wide. Immerse these strips for 2 minutes in cold mercuric chloride solution. Drain off superfluous liquid and dry the strips in warm air which is free of traces of arsine or hydrogen sulphide. When dry cut off, and discard, the bottom two inches from each strip. Cut the remainder into 3 inch lengths and store in a well stoppered bottle.

The Standard Lovibond Comparator Disc 6/8

The disc contains 4 standards, corresponding to 0.1, 0.2, 0.5 and 1.0 mg arsine.

Technique

Place a piece of the mercuric chloride paper in a suitable paper holder[2] and place a piece of dry lead acetate paper on the inlet side of the test paper. This will absorb any hydrogen sulphide, which may be present in the air, before the air reaches the mercuric chloride test paper. Connect to a suitable source of suction[3] and sample until a stain is obtained on the paper. Place the paper on the right-hand side of the Lovibond Comparator Test-Paper Viewing Stand. Place a piece of blank test paper on the left-hand side and compare the colour of the stain with that of the permanent glass standards in the disc, using a standard source of white light, such as the Lovibond White Light Cabinet, or failing this north daylight. Read off the mg of arsine, to which the appropriate standard corresponds, from the following Table.

Standard No.	mg arsine
1	0.1
2	0.2
3	0.5
4	1.0

Calculate the arsine concentration by means of the following formula:—

$$\text{Arsine conc. (ppm)} = \frac{\text{mg of arsine} \times 250}{\text{air sample volume in litres}}$$

References

1. Ministry of Labour, *"Toxic Substances in Factory Atmospheres,"* London, H.M.S.O., 1960
2. B. E. Dixon, G. C. Hands and A. F. F. Bartlett, *Analyst,* 1958, **83,** 199
3. See pages 462 to 464

The Determination of Benzene
in the presence of toluene, xylene, and styrene, using paraformaldehyde

Introduction

Benzene is a narcotic agent with insidious chronic effects, described as metaplastic anaemia with a varied pattern of pathology, which often progress to a fatal termination. There is also a growing body of evidence that benzene may be a carcinogen, causing leukaemia. The recommended maximum concentration of benzene in factory atmospheres is 25 ppm v/v (80 mg/m^3).[1] The present test[2] has been designed to enable the benzene concentration in a contaminated atmosphere to be measured quickly and simply.

Principle of the method

Toluene and xylene are removed from the air sample by absorption in a solution of selenous acid in sulphuric acid. The benzene is absorbed in a sulphuric acid solution of paraformadehyde with which it forms a yellowish brown colour. The intensity of this colour, which is proportional to the benzene concentration, is measured by comparison with a series of Lovibond permanent glass colour-standards.

Reagents required

1. *Selenous acid solution* Dissolve 2 g of selenous acid (H_2SeO_3) in diluted sulphuric acid prepared by adding 90 ml of H_2SO_4 (Sp gr 1.84) to 10 ml of water, slowly and with cooling (Note 1). This reagent is stable for 1 week after which time it must be discarded.
2. *Paraformaldehyde solution* Add 100 ml of H_2SO_4 (sp gr 1.84) to 2 g paraformaldehyde. Shake until the solid is completely dissolved. This reagent is stable for 3 days after which time it must be discarded.

The chemicals used in the preparation of reagents should be of analytical reagent quality. It is especially important that the selenous acid should contain not more than 0.002% of nitrate (as NO_3). If suitably pure sample of selenous acid is not readily obtainable it may be possible to use the available selenous acid provided that 0.1% of benzoic acid (C_6H_5COOH) is added to reagent 1.

The Standard Lovibond Comparator Disc 6/24

The disc contains the official standards corresponding to 12, 25 and 50 ppm of benzene based on a sample of 360 ml of air.

Technique

Place 5 ml of paraformaldehyde solution (reagent 2) in a dry bubbler of the type and size shown in the figure. Attach a suitable form of suction[3] to the side arm and insert the inlet tube. Place 5 ml of selenous acid solution (reagent 1) in another dry bubbler, insert the inlet tube and connect the two bubblers in series by means of glass tubing, using the minimum amount of plastic tubing to make the joints. Adjust the air sampling rate to about 25 ml per minute and sample exactly 360 ml of air. During sampling, at 5 minute intervals, momentarily interrupt the air flow, by squeezing the plastic tubing connected to the side arm of the paraformaldehyde bubbler, so that the reagent rises about 2 cm above the ring of any benzene colour formed inside the inlet tube. After the 360 ml of air has been sampled disconnect the paraformaldehyde bubbler and dissolve the resin remaining on the inside of the inlet tube by raising and lowering the tube several times into the liquid and, if necessary, by rubbing the inside of the tube with a thin glass rod. Transfer the bubbler to the right-hand compartment of the comparator, place the comparator before a standard source of white light, such as the Lovibond White Light Cabinet or, failing this, north daylight, and compare the colour of the liquid with the colour standards in the disc.

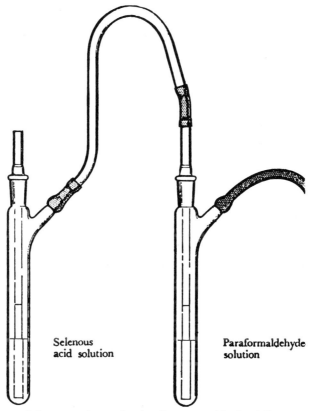

Selenous acid solution

Paraformaldehyde solution

Apparatus

Bubbler – 2 bubblers of the type shown in the diagram with the following dimensions:—

Receiver—Overall length 155 mm, external diameter 15 mm, internal diameter 13 mm. Distance from bottom of side arm to bottom of bubbler 110 mm.

Inlet Tube—Approximate overall length 185 mm, external diameter 6 mm, internal diameter 4.5 mm. Distance from bottom of inlet tube to bottom of bubbler 7 mm. Cone and Socket B10.

Notes

1. The concentrated sulphuric acid used in this test is dangerous and must be handled with great care. In no circumstances should water be added to the acid. When diluting the acid, always add acid to water allowing adequate time between successive additions for the acid to mix and for heat to dissipate. Take adequate precautions to prevent the acid or reagents 1 and 2 from coming into contact with the bare skin and especially with the eyes. Always use safety pipettes for measuring out the reagents.

2. Toluene, xylene and ethyl benzene up to at least 300 ppm do not interfere with this determination, nor does styrene up to at least 170 ppm.

3. If it is necessary to sample from outside the contaminated area, use a glass tube attached to the inlet of the first bubbler. Keep the amount of plastic tubing used to connect the glass down to a minimum. On no account must rubber tubing be used at any point between the contaminated atmosphere and the paraformaldehyde bubbler.

References

1. Min. of Labour, *"Toxic Substances in Factory Atmospheres"*, H.M.S.O., London, 1960
2. Min. of Labour, H.M. Factory Inspectorate, *"Methods for the Detection of Toxic Substances in Air, Booklet No. 4, Benzene, Toluene and Xylene, Styrene"*, H.M.S.O., London, 1966
3. See pages 462 to 464

The Determination of Carbon Dioxide
using mixed indicator solution

Introduction

Enrichment of glass-house atmospheres with carbon dioxide (CO_2), as a means of encouraging photosynthesis and promoting plant growth, is receiving increasing attention by commercial growers. Normal atmospheric CO_2 concentrations are about 300 volumes per million (v.p.m.). To ensure the economic employment of techniques to enrich the atmosphere in glass-houses by the addition of CO_2, there is a requirement for a simple method of estimating the CO_2 concentration in the atmosphere. In order to cover the important aspects of deficiency and enrichment it is advisable to be able to cover the concentration range 200—4,000 v.p.m. (0.02—0.4%) CO_2. The present test [1,6] has been developed to meet this requirement and a kit incorporating the method (Note 1) has been tested in comparison with kits based on alternative methods and shown to be the most reliable [1,7].

Principle of the method

This test is based on Slavik and Catsky's [2] modification of the method suggested by Claypole and Keefer [3]. This method estimates the concentration of CO_2 by measuring the pH obtained by equilibrating the atmosphere with a dilute sodium bicarbonate-potassium chloride solution. The equilibrium pH of this solution has been shown [3] to be directly proportional to the logarithm of the partial pressure of the CO_2 in the atmosphere [4] provided that sufficient bicarbonate ions are present in the solution prior to establishing the equilibrium. The pH range corresponding to CO_2 concentrations of 0.02 to 0.5% has been shown by Zeller [5] to be 7.0—8.3. The pH value of the solution, which is a direct measure of the CO_2 concentration, is determined by comparing the colour of the mixed indicator solution with a range of Lovibond permanent glass colour standards, which are calibrated in % CO_2.

Reagent required

Tablets specially prepared for Tintometer Ltd., containing a carefully adjusted mixture of Thymol Blue and Cresol Red indicators in sodium bicarbonate.

The Standard Lovibond Comparator Disc 6/27

This disc covers the range 0.02 to 0.4% CO_2 (vol./vol. 0.1% =1,000 ppm v/v) in 9 steps (0.02, 0.03, 0.04, 0.05, 0.075, 0.1, 0.15, 0.2, 0.4) and is designed to be used with a 25 mm cell.

Technique

Dissolve one tablet in 10 ml freshly boiled distilled (not deionised) water (Note 2), and transfer this into the clean wide-necked jar provided in the test kit (Note 3). Place the jar in the position to be sampled and leave exposed to the atmosphere being tested for 2 hours to come to equilibrium with the environment, note the temperature of the reagent, and then carefully decant the clear liquid into the 25 mm cell, taking care not to disturb any sediment.

Place the cell in the right-hand compartment of the comparator and compare the colour of the reagent with the Lovibond permanent glass colour standards in the disc using a standard source of white light to illuminate the comparator (Note 4). The value of the standard nearest in colour to that of the reagent is read off from the window in the bottom right-hand corner of the comparator face and represents % by volume CO_2. If the temperature of the reagent differs from 20°C the true CO_2 concentration may be obtained from the following Table. The values for temperatures between those quoted may be obtained by interpolation.

It is not advisable to use this test for temperatures above 30°C.

The solution may be poured back into the jar and left exposed as before, and examined at intervals to follow the variations in the CO_2 concentration.

Make up the solution fresh daily.

Approximate CO₂ concentrations (% by volume) at various temperatures

Actual disc reading	Temperature corrected reading					
	0°C	10°C	15°C	20°C	25°C	30°C
0.02	0.015	0.016	0.018	0.02	0.022	0.024
0.03	0.021	0.024	0.027	0.03	0.034	0.039
0.04	0.027	0.032	0.036	0.04	0.048	0.056
0.05	0.034	0.04	0.044	0.05	0.062	0.075
0.075	0.046	0.055	0.063	0.075	0.095	0.12
0.1	0.059	0.072	0.082	0.1	0.13	0.17
0.15	0.078	0.1	0.115	0.15	0.2	0.275
0.2	0.1	0.13	0.15	0.2	0.28	0.38
0.4	0.16	0.22	0.27	0.4	0.59	0.88

Notes

1. A complete kit for this test, containing a Lovibond Comparator, Lovibond Disc No. 6/27, a 25 mm cell, stoppered sampling jar, a thermometer, 2 calibrated test tubes and 250 reagent tablets complete in a fitted case, may be obtained from Tintometer Ltd.

2. Cleanliness of the cells, test tubes and sampling jars is of the utmost importance. It is recommended that they be rinsed with several changes of distilled water before use.

3. Care must be taken not to exhale above the jar during sampling, whilst reading temperatures, before the jar is stoppered, or whilst transferring the reagent to the cell. Exhaled air contains up to 5.0% CO_2 and if this comes in contact with the reagent it will lead to high results. Sampling jars must be stoppered during transportation from the sampling area to the comparator.

4. A standard source of white light, such as the Lovibond White Light Cabinet or natural north daylight must be used to illuminate the comparator if accurate colour-matching is to be achieved. Matching against a background of a blue sky has been shown to lead to high results.

5. Replacement of the reagent must be ordered from Tintometer Ltd.

6. This test was developed by Mr. R. B. Sharp, Instrumentation Department, National Institute of Agricultural Engineering, Wrest Park, Silsoe, Bedford.

References

1. R. B. Sharp, *J. Agric. Eng. Res.* 1964, **9**, 87
2. J. Catsky and B. Slavik, *Biologia Pl.,* 1960, **2**, 107
3. L. L. Claypool and R. M. Keefer, *Proc. Am. Soc. hort. Sci.* 1942, **40**, 177
4. M. H. Martin and C. D. Piggott, *J. Ecol.* 1965, **53**, 153
5. O. Zeller, *Planta,* 1951, **39**, 500
6. G. E. Bowman, *The Grower,* 1966, 558
7. D. A. Browne, *Parks and Sports Grounds,* 1968, **33**, No. 6, 458

The Determination of Carbon Disulphide (2)
using Diethylamine and copper acetate

Introduction

The existing test[1] for carbon disulphide vapour in air (Carbon Disulphide 1) has been re-examined for three reasons: the mixed reagents occasionally become turbid: the volume of air (2.5 litres) required was inconvenient as it required twenty strokes of the standard pump but was too small to justify the use of an electric pump: and it did not provide a standard at the threshold limit concentration. No alternative chemical method has been found which is more suitable for the determination of carbon disulphide so the existing method has been improved[2] to overcome the objections outlined above.

Principle of the method

Carbon disulphide vapour reacts with an alcoholic solution of diethylamine and copper acetate to produce copper diethyldithio-carbamate, the intensity of the yellow-brown colour of which is directly proportional to the carbon disulphide concentration, and is measured by comparison with a series of Lovibond permanent glass colour standards.

Traces of hydrogen sulphide will interfere with this test. Concentrations of hydrogen sulphide up to 40 ppm v/v can be removed by drawing the air sample through a filter paper impregnated with lead acetate solution, prior to passing through the absorption bubbler.

Reagents required

1. *Absorption solution* Dissolve 0.01 g of copper acetate monohydrate $((CH_3COO)_2Cu.H_2O)$ in a small amount of alcohol (industrial methylated spirit is suitable) and transfer the solution to a 500 ml flask using a further 100 ml of alcohol. Add 5 ml of triethanolamine $((HOC_2H_4)_3N)$ from a measuring cylinder and wash any residue from the cylinder into the flask with a little more alcohol. Agitate the contents of the flask until all the triethanolamine has dissolved and then add 2 ml of diethylamine $((C_2H_5)_2NH)$ and dilute contents to 500 ml with more alcohol.

    This reagent can be used for up to four weeks if it is stored in tightly stoppered bottles at a temperature below 30° C.

2. *Lead acetate solution* Dissolve 10 g of lead acetate trihydrate $((CH_3COO)_2Pb.3H_2O)$ in 90 ml of distilled water. Add 5 ml of glacial acetic acid (CH_3COOH) and 10 ml glycerol $(CH_2OH.CHOH.CH_2OH)$ to the solution and mix thoroughly.

3. *Lead acetate papers* Suspend strips of Whatman 3MM chromatographic paper, 20 mm wide and 100 mm long, for one minute in the lead acetate solution contained in a 100 ml measuring cylinder. Remove the papers, allow the excess liquid to drain off, suspend the papers vertically in an atmosphere free from hydrogen sulphide, and allow them to dry at room temperature. When dry cut off and discard 25 mm strips from each end of the papers and store the centre 50 mm portions in stoppered, wide-necked, dark-glass bottles. Use within 14 days of preparation.

    All chemicals used for the preparation of reagents should be of analytical reagent quality where possible.

The Standard Lovibond Comparator Disc 6/56

This disc contains standards corresponding to 0, 5, 10, 20 and 40 ppm v/v of carbon disulphide, based on a 500 ml sample of the atmosphere. This disc is designed to be used with a Lovibond 1000 Comparator with a Large Cells Attachment and special holder (DB 430) and 10 mm i.d. tubes calibrated at 50 mm, and the colour is viewed down through the depth of the tube.

Technique

In a carbon disulphide-free atmosphere pipette 10 ml of the absorption solution (reagent 1) into a standard glass bubbler, the dimensions of which are shown in Figure 1.

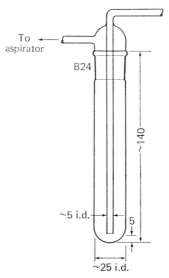

Fig. 1 Diagram of bubbler.
(All measurements are in millimetres)

Insert the inlet tube and attach the exit tube to an aspirator, or pump, capable of drawing air through the bubbler at 125 ml a minute. If the atmosphere to be tested is suspected of containing hydrogen sulphide, connect a suitable paper-holder containing a lead acetate paper to the inlet of the bubbler. Readjust the aspirator, if necessary, so as to sample 125 ml/min since the resistance of the paper may reduce the volume of air sampled. Transfer the sampling equipment to the test site and draw 500 ml of the atmosphere through the bubbler at 125 ml a minute. Remove the equipment to a clean atmosphere and, exactly 10 minutes after collecting the sample, transfer the solution from the bubbler into the 60 mm comparator tube, filling the tube exactly to the 50 mm mark. Fit the Large Cells Attachment complete with a Special Tube Holder (DB 430) onto the back of a Lovibond 1000 Comparator. Hold the Comparator horizontally (Figure 2) and insert the 60 mm tube into the aperture in the centre of the disc as shown. Fill a similar tube 55 mm long with distilled water and insert this into the other aperture to act as a blank. If a 60 mm tube is inserted into this aperture it will foul the disc. View the liquid through its depth, either by holding the comparator over a white background or by placing it in an inverted Lovibond White Light Cabinet. Compare the colour of the solution with the standards in the disc.

Fig. 2
Comparator and DB430

Note

This modification[2] of the original test[1] was developed by members of the staff of the Laboratory of the Government Chemist. Permission to quote details of the test from their published paper is gratefully acknowledged.

References

1. Department of Employment, *"Methods for the Detection of Toxic Substances in Air, Booklet No. 6, Carbon Disulphide Vapour"*. H.M. Stationery Office, London, 1968.
2. E. C. Hunt, W. A. McNally and A. F. Smith, *Analyst*, 1973, **98,** (8), 585.

The Determination of Chlorine
in atmospheres, using 3,3′-dimethylnaphthidine

Introduction

Chlorine is a highly toxic gas which is extensively used in the chemical industry and may also be produced during the decomposition of some commercial disinfectants and bleaches, e.g. when hypochlorite reacts with acids. The maximum permitted concentration of chlorine in industrial atmospheres is 1 ppm (3 mg/m^3 at 20°C.)[1]. At this concentration the presence of chlorine is not readily detected by the sense and this test[2] has been developed for the testing of factory atmospheres. It is somewhat more sensitive than the earlier test using *ortho*-tolidine, which it supersedes as the official test for chlorine in factory atmospheres.

Principle of the method

The atmosphere being tested is drawn through an acetic acid solution of 3,3′-dimethylnaphthidine. If chlorine is present in the air a mauve coloration is produced, the intensity of which is proportional to the chlorine concentration. The intensity of the mauve colour is measured by comparison with a series of Lovibond permanent glass colour standards.

Reagent required

0.005% *solution of 3,3′-dimethylnapthidine* Dissolve 0.01 g of finely ground 3,3′-dimethylnaphthidine in 5 ml of glacial acetic acid and immediately dilute to 200 ml with distilled water. Store this reagent in the dark. When freshly prepared this solution is supersaturated and the reagent slowly crystallises out. However the supernatant solution is satisfactory for use.

This reagent should be freshly prepared every 2–3 weeks.

All chemicals used in the preparation of the reagent should be of analytical reagent grade.

The Standard Lovibond Comparator Disc 6/23

This disc contains three standards corresponding to 0.5, 1.0 and 2.0 ppm of chlorine (Cl_2) in 360 ml of air.

Technique

Place 4 ml of the reagent solution in a bubbler, insert the inlet tube and connect a suitable source of suction (Note 1) to the side arm. Aspirate exactly 360 ml of air through the bubbler at a rate of about 3 ml per second. After sampling is complete, transfer the contents of the bubbler to a 13.5 mm cell or standard comparator tube and place this tube in the right-hand compartment of the comparator. Make certain that the liquid in the tube covers the aperture in the comparator, raising the tube if necessary. Compare the colour of the reagent with the colours of the standards in the disc (Note 2) using a standard source of white light such as the Lovibond White Light Cabinet or, failing this, north daylight.

Notes

1. A variety of convenient methods for taking air samples are described in the Introduction to this Section.
2. The colour comparison must be carried out within two minutes of the completion of the sampling.
3. This test is not specific for chlorine; other oxidising agents such as bromine, chlorine dioxide and nitrogen dioxide may also produce colours with this reagent.

References

1. Dept. of Employment, *"Dust and fumes in factory atmospheres"* 1968 H.M. Stationery Office, London, 4th edn, 1969
2. H.M. Factory Inspectorate, Min. of Labour, *"Methods for the Detection of Toxic Substances in Air,"* Booklet No. 10 *"Chlorine"* (1966), H.M.S.O., London, 1966

The Determination of Chloroform in Air
using pyridine
(Fujiwara reaction)

Introduction

The widespread use of chloroform in anaesthesia has resulted in the search for a method of determining this substance in air, in body fluids and in tissue. The methods so far devised fall into two groups involving either decomposition of the chloroform, followed by estimation of chlorine either gravimetrically or volumetrically, or the direct colorimetric determination of chlorine. In general the methods involving decomposition of the chloroform are laborious without being sensitive, and colorimetric methods are therefore preferred for the determination of trace amounts. Daroga and Pollard[1] investigated the available colorimetric reactions and concluded that none of the existing methods was sufficiently reliable for routine use. They have therefore developed the following modification of the Fujiwara reaction.[2]

Principle of the method

Chlorinated hydrocarbons react with pyridine, in the presence of strong alkali, giving a red colour. By careful control of the experimental conditions this colour can be stabilised and compared with Lovibond permanent glass standards.

Reagents required

1. *20% Sodium hydroxide* (NaOH)
2. *Pyridine* (C_5H_5N pure, colourless)
3. *Hydrochloric acid* $0.05N$ (HCl)

The Standard Lovibond Comparator Disc 3/45

This disc covers the range 0.05 to 0.45 mg of chloroform ($CHCl_3$), in steps of 0.05.

Technique

The atmosphere to be sampled is drawn, at 1 litre per minute, through a Schott[3] apparatus containing 50 ml of $0.05N$ hydrochloric acid (reagent 3), and the chloroform estimated on an aliquot of the resulting solution.

Measure 10 ml of sodium hydroxide (reagent 1) into a 50 ml graduated flask (see Note 1). Add exactly 20 ml of pyridine (reagent 2) by means of a burette, and then add the prepared test solution (see above). Loosely cork the flask to avoid evaporation of pyridine, and immerse the flask in boiling water for 5 minutes, shaking continually. Cool under running water for 2 minutes. The coloured pyridine layer now separates on top of the alkali. Transfer 10 ml of this supernatant liquid to a standard 13.5 mm comparator cell and place it in the right hand compartment of the Comparator. Using a standard source of white light such as the Lovibond White Light Equipment or, failing this, north daylight, compare the colour with the permanent glass standards. The figures represent the amount in mg in the volume of the test solution taken.

Notes

1. It is advisable to use a Pyrex volumetric flask to avoid breakage from thermal shock during heating and cooling.
2. The amount of pyridine added must be accurately measured as the colour developed is a function of the pyridine concentration.

References

1. R. P. Daroga and A. G. Pollard, *J.S.C.I.*, 1941, 60, 218.
2. K. Fujiwara, *Sitzungber Abhandl. Naturforsch. Ges Rostock*, 1914, 6, 1.
3. P. H. Prausnitz, *Ind. Eng. Chem. (Anal. Ed.)*, 1932, 4, 432.

The Determination of Chromic Acid Mist
using diphenylcarbazide

Introduction

Chromic acid is used on a large scale in connection with chromium plating and with anodic oxidation. In those processes where gas bubbles are released at the surface of the liquid, either as a result of the passage of an electric current or as the result of operating the bath at or near boiling point, the danger of inhalation of spray exists. Contact of this spray with the skin or with the mucous membranes of the nose and throat usually results in a chronic type of ulceration which is slow to heal. A sensitization type of dermatitis may also occur.

The maximum permissible concentration of chromic acid mist in an industrial atmosphere has been set at 0.1 mg per cubic metre[1,3]. To enable suspected atmospheres to be tested rapidly this simple test has been devised[2].

Principle of the method

The chromic acid mist is collected on filter paper, which is immediately extracted with acid diphenylcarbazide reagent. This forms a red complex with chromium, and the intensity of this red colour, which is proportional to the chromium concentration, is measured by comparison with a series of Lovibond permanent glass colour standards.

Reagents required

1. *Diphenylcarbazide solution* Dissolve 0.3 g of diphenylcarbazide in a mixture of 5 ml of glacial acetic acid (CH_3COOH) and 95 ml of isopropyl alcohol (($CH_3)_2CHOH$).
2. *Sulphuric acid 0.5N* Add, slowly, 4.0 ml of concentrated sulphuric acid (H_2SO_4) to 300 ml of distilled water, taking care to avoid any contact between the acid and the skin.
3. *Diphenylcarbazide-sulphuric acid reagent* Mix one volume of reagent 1 with four volumes of reagent 2.

   All chemicals used in the preparation of reagents should be of analytical reagent quality.

The Standard Lovibond Comparator Disc 6/30

This disc contains three standards corresponding to 0.05, 0.10 and 0.20 mg of chromic oxide (CrO_3) per cubic metre of air, based on a 40 litre air sample.

Technique

Place a 2.5 cm Whatman No. 2 filter paper in a stainless steel holder. (A suitable design for such a holder is illustrated in Reference 2. See also Note 1). Connect the holder to a convenient source of suction (Note 2) and draw 40 litres of air through the filter paper at a rate of 2 to 10 litres per minute. The air should be drawn from a point a few inches above the exhaust plane of the bath if lip exhaust ventilation is in operation; from a point about 12 inches above the surface of the bath during plating if a spray suppressing material is in use; or at the operator's breathing level at the side of the plating bath. Immediately sampling is finished transfer the filter paper to a 500 ml plastic beaker (polyethylene or polypropylene) and add 6 ml of the diphenylcarbazide-sulphuric acid reagent (reagent 3). Swirl the beaker gently for two to three minutes until all colour has been extracted from the paper. Decant the liquid into a standard 13.5 mm test-tube and place this in the right-hand compartment of the comparator. If necessary raise the level of the test-tube by means of a plug of cotton wool until the liquid in the tube completely covers the viewing aperture in the comparator. Place the comparator in front of a standard source of white light, such as the Lovibond White Light Cabinet or, failing this, north daylight, and compare the colour of the solution with the standards in the disc.

Notes

1. A suitable 25 mm holder for filter paper can be obtained from Tintometer Ltd., (Code No. AF401).

2. Convenient methods of taking air samples, of known volume and at constant rate, are described on pages 462 to 464.

3. A complete test kit is available for this test. Details from Tintometer Ltd.

References

1. Ministry of Labour, *"Safety Health and Welfare, New Series Booklet No. 8 Dust and Fumes in Factory Atmospheres"* H.M. Stationery Office, London, 1960

2. Ministry of Labour, H.M. Factory Inspectorate, *"Methods for the Detection of Toxic Substances in Air, Booklet No. 17, Chromic Acid Mist"* H.M. Stationery Office, London, 1967

3. *Chromium Plating Statutory Regulations, 1931 and 1973*

… COLORIMETRIC CHEMICAL ANALYTICAL METHODS

The Determination of Copper Fume and Dust
using cuprizone

Introduction

Copper fume may arise in any industrial situation in which copper or a copper alloy is heated to a temperature high enough to volatilize the copper. Typical situations in which this may occur are the extraction of copper from its ores, the preparation of alloys such as brasses and bronzes, and the welding of copper and its alloys. Copper dust can arise in the course of any operation involving copper, its alloys, or its salts, particularly in processes such as grinding the metals or in the use of aerosols containing copper where the copper is likely to be projected into the air in the form of fine particles. Such situations are likely to arise in the spraying of copper-containing paints and fungicides.

In trace quantities copper is a normal constituent of the human body, being ingested in food and any excess excreted in faeces and urine. If, however, more copper is ingested than can be excreted then excessive retention may cause disease of the liver and of the central nervous system. An acute reaction, very similar to that caused by the inhalation of zinc oxide fume, has occasionally arisen due to inhalation of copper oxide fume. The symptoms are fever and shivering and no appropriate first aid treatment is known. The maximum concentrations of copper fume and copper dust which may be permitted in an industrial atmosphere have been set at 0.1 mg m^{-3} for copper fume and 1.0 mg m^{-3} for copper dust[1,2].

The test described below has been developed[3] to provide a simple method for the determination of copper fume and dust in industrial atmosphere. This test is intended to give a rapid indication of whether or not the atmosphere is dangerous. It is not designed to give an extremely accurate result, and a result which approaches the statutory limit value of 0.1 mg m^{-3} (1.0 mg m^{-3} for dust) should be regarded as an indication of unsatisfactory conditions.

Principle of the method

The copper fume from a known volume of air is collected on a filter paper and then dissolved in acid. This acid solution is reacted with a solution of cuprizone (oxalyldi(cyclohexylidenehydrazide)) and the intensity of the colour of the blue complex formed, which is proportional to the copper concentration, is determined by comparison with a series of Lovibond permanent glass colour standards.

Reagents required

1. *Hydrochloric acid* Dilute concentrated hydrochloric acid with an equal volume of water.
2. *Buffer solution* Dissolve 50 g of citric acid ($C(OH)(COOH)(CH_2COOH)_2.H_2O$) in 300 ml of water. Add 100 ml of aqueous ammonia (NH_4OH, d.0.880), 25 ml at a time, mixing thoroughly and cooling after each addition. Finally dilute to 500 ml with water. (Note 1).
3. *Cuprizone (oxalyldi(cyclohexylidenehydrazide)),* $(C_6H_{10}:N.NH.CO)_2)$ Dissolve 0.25 g of cuprizone in a hot mixture of 50 ml of ethanol (C_2H_5OH) and 25 ml of water. Cool and dilute to 100 ml with water. This reagent is only stable for 4 weeks and should be renewed after that time.

All chemicals used in the preparation of reagents should be of analytical reagent quality and only distilled or deionized water should be used.

The Standard Lovibond Comparator Disc 6/51

This disc contains standards corresponding to 0, 0.05, 0.1 and 0.2 mg m^{-3} of copper, when a 50 litre air sample is used. If a 5 litre sample is used then the standards correspond to 0, 0.5, 1.0 and 2.0 mg m^{-3} of copper.

Technique

(a) *Copper fume* Place a 25 mm diameter Millipore AA (0.8 micron) filter paper in a suitable filter paper holder (Note 2) and connect this to an appropriate source of suction (Note 3). Take a 50 litre sample of the contaminated atmosphere at a rate of 5 litres a minute for 10 minutes, disconnect the holder, remove the filter paper and place this in a 25 ml beaker having a diameter large enough to take the paper without having to fold it. Add 1 ml of hydrochloric acid (reagent 1) and after 5 minutes transfer the acid solution to a 25 ml graduated flask using 10 ml of water to wash the beaker and paper. Add 5 ml of the buffer solution (reagent 2) and make sure that the solution temperature is below 25°C, cooling the flask if this proves to be necessary. Add 2.5 ml of the cuprizone solution (reagent 3), dilute to volume with water and set aside for 5 minutes to allow the maximum colour to develop. After this time pour sufficient of the test solution into a 60×10 mm flat bottomed tube to fill it to a depth of 50 mm. Fit the large cells attachment complete with the special tube holder onto the back of a Lovibond 1000 Comparator, and while holding the comparator in a horizontal position insert the 60 mm tube into the central aperture, as shown in Figure. Fill a similar 55 mm tube to a depth of 50 mm with distilled water and place this in the other aperture, as shown in the Figure, to act as a blank. Place the comparator over a standard source of white light such as the Lovibond White Light Cabinet tipped onto its back or, failing this, a white surface illuminated by north daylight, and match the colour of the sample with the standards in the disc viewing down the depth of the liquid.

(b) *Copper dust* Use exactly the same technique as that described above for copper fume but only take a 5 litre sample of the atmosphere using a sampling rate between 0.5 and 1.0 litres a minute. Multiply the disc reading by 10.

Notes

1. The buffer solution (reagent 2) tends to lose ammonia by evaporation. It should therefore be tested at frequent intervals by taking a 25 ml flask, adding 1 ml of hydrochloric acid (reagent 1), 10 ml of water, 5 ml of the buffer solution (reagent 2), 4 drops of thymol blue indicator solution and then diluting to volume. The colour should be strongly blue. If it is green the buffer should be discarded and a fresh batch prepared.

2. A suitable filter paper holder is obtainable from Tintometer Ltd., Code No. AF401.

3. Suitable suction methods for sampling contaminated atmospheres are described in the Introduction to this Section. In view of the large sample required for the determination of copper fume, a hand pump would not be a convenient source of suction for this test and one of the alternative sources is recommended.

4. It has been reported[4] that 5–100 µg of copper can be determined without interference in the presence of 1000 µg of aluminium, calcium, chromium, iron, magnesium, manganese, nickel and zinc, while other workers[3] have shown that insignificant interference is caused to the determination of 5 µg of copper by the presence of 1000 µg of antimony, arsenic, cadmium, lead and tin.

References

1. Department of Employment, 'Safety Health and Welfare New Series No. 8 *"Dust and Fumes in Factory Atmospheres* 1968"', H.M. Stationery Office, London, 1968
2. Department of Employment, Technical Data Note 2, *"Threshold Limit Values"*, H.M. Stationery Office, London, 1970
3. Department of Employment, *Methods for the Detection of Toxic Substances in Air,* Booklet No. 22 *"Copper Dust and Fume"*, H.M. Stationery Office, London, 1971
4. Brit. Standards Inst., B.S. 2690 *"Methods of Testing Water Used in Industry, Part* 1 *Copper and Iron"*, London, 1964

The Determination of (Methyl) Cyclohexanone
using H-acid diazonium salt

Introduction

Both cyclohexanone and methylcyclohexanone are used in industry as solvents. Their vapours have a strong irritant and narcotic action, and limits have been proposed[1] for the maximum concentration of these vapours which are permissible in industrial atmospheres. The current threshold limit for cyclohexanone in air is 200 mg m^{-3}. The limit assigned to 2-methylcyclohexanone is 460 mg m^{-3} and it is reasonable to assume a similar threshold limit for commercial methylcyclohexanone, which is a mixture of the isomeric methylcyclohexanones of indeterminate composition.

The present test[2,6] has been developed to give a simple field method for the determination of these alicyclic ketones in the presence of the aliphatic ketones which might also be present in industrial atmospheres. This test is a modification of the test originally described by Maslennikov[3,4] and further developed by Adamiak[5].

Principle of the method

Cyclohexanone is coupled with the diazonium salt of H-acid (4-amino-5-hydroxynaphthalene-2, 7-disulphonic acid), in the presence of sodium sulphite and sodium hydroxide, to give a red azo dye. The intensity of the red colour produced, which is proportional to the concentration of cyclohexanone, is measured by comparison with a series of Lovibond permanent glass colour-standards.

Methylcyclohexanones give an identical colour and, providing that it is known which of the two solvents is present in the atmosphere, which is usual in an industrial process where solvents are used, then the same test may be used for determining either of the solvents.

Reagents required

1. *H-acid solution* Dissolve 1.25 g of the monosodium salt of H-acid (4-amino-5-hydroxynaphthalene-2,7-disulphonic acid) in 250 ml of 0.1N sulphuric acid (reagent 2). This reagent is stable for up to 2 months.
2. *0.1N sulphuric acid* Take 3 ml of pure concentrated sulphuric acid (H_2SO_4 d 1.84) and pour it cautiously into 10–20 ml of water, cool, mix thoroughly and dilute to 1 litre.
3. *Sodium nitrite solution* Dissolve 1 g of sodium nitrite ($NaNO_2$) in 100 ml of water. This solution is stable for at least two months.
4. *Sodium sulphite-sodium hydroxide solution* Dissolve 10 g of anhydrous sodium sulphite (Na_2SO_3) and 5 g of sodium hydroxide (NaOH) in water and dilute to 100 ml. This reagent is stable for up to 1 month.

All chemicals used in the preparation of these reagents should be of analytical reagent quality and solutions should be prepared with distilled or de-ionised water. Solutions should be kept tightly stoppered when not in use and stored, at a temperature below 25°C, in dark glass bottles in an uncontaminated atmosphere.

Reagents should be dispensed with pipette fillers.

The Standard Lovibond Comparator Disc 6/53

This disc contains standards 0, 25, 50 and 100 ppm of cyclohexanone in air based on a 500 ml air sample. These standards correspond to 0, 100, 200 and 400 mg m^{-3}.

The same standards may also be used for the determination of methylcyclohexanone. In this case the standards correspond to 0, 115, 230 and 460 mg m^{-3}, of methylcyclohexanone, based on a 500 ml sample of air.

Technique

For cyclohexanone atmospheres In an uncontaminated atmosphere well away from the suspected source of cyclohexanone vapour pipette 3 ml of water into a glass bubbler (Figure 1, Note 1), insert the inlet tube and connect a suitable source of suction to the side arm (Note 2). At the sampling site collect a 500 ml sample of the atmosphere at a rate of 125 ml a minute. Remove the apparatus from the contaminated atmosphere, disconnect the suction, remove the inlet tube and, to the receiver, add successively, by means of safety pipettes, 2 ml of H-acid solution (reagent 1), 1 ml of sodium nitrite solution (reagent 3) and 1 ml of alkaline sodium sulphite solution (reagent 4). Stopper the receiver and mix the contents well by inverting several times. Take care to ensure that no liquid is lost through the side arm. Allow the solution to stand for 20 minutes at a temperature within the range 15–30°C. Transfer 1 ml of the resulting solution to a standard 13.5 mm comparator tube, dilute to the 10 ml graduation mark with water and mix thoroughly. Place this tube in the right-hand compartment of the comparator. Fill an identical tube to the 10 ml mark with water and place this tube in the left-hand compartment to act as a blank. Place the comparator before a standard source of white light, such as the Lovibond White Light Cabinet or, failing this, north daylight, and match the colour of the sample against the standards in the disc.

For methylcyclohexanone atmospheres Carry out the sampling as described above for cyclohexanone and sample 500 ml of the atmosphere into 2 ml of water. Then proceed as for cyclohexanone.

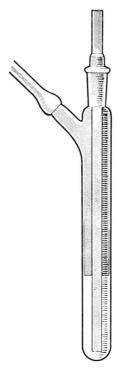

Figure 1

Notes

1. *Dimensions of Receiver.* Overall length 155 mm external diameter 15 mm internal diameter 13 mm, distance from bottom of side arm to bottom of bubbler 110 mm. B10 socket. Inlet tube—overall length 185 mm external diameter 6 mm internal diameter 4.5 mm, distance from bottom of inlet tube to bottom of bubbler 7 mm B10 cone.

This bubbler may be obtained from Tintometer Limited or Glass of Mark Limited, Northwich, Cheshire.

2. Suitable sources of suction for taking gas samples are described in the Introduction to this Section.

References

1. Dept. of Employment and Productivity, Technical Data Note 2/69, *"Threshold Limit Values for 1969"*, H.M. Stationery Office, London, 1969
2. P. Andrew, A. F. Smith and R. Wood, *Analyst,* 1971, **96** (7), 528
3. A. S. Maslennikov, *Zh. prikl. Khim. Leningr.,* 1958, **31,** 1277
4. A. S. Maslennikov, *Zh. analit. Khim.,* 1958, **13,** 599
5. J. Adamiak, *Chemia analit.,* 1968, **13,** 895
6. *"Methods for the determination of toxic substances in air"* Booklet 26. *"Cyclohexanone and Methylcyclohexanone"* H.M. Stationery Office, London, 1972

The Determination of Fluoride (Total)
using lanthanum-alizarin fluorine blue

Introduction

The toxic hazard arising from fluoride-bearing dusts or fumes is considered[1] to be as great as that from hydrogen fluoride vapour.

Atmospheric contamination by inorganic fluorides occurs in a variety of industrial processes such as the manufacture of phosphates, phosphoric acid and aluminium – where the hazard arises from the dusts of fluoride-bearing rocks; metal welding – where fumes resulting from the use of electrode flux containing fluoride are the danger, and chemical processes using or producing hydrogen fluoride.

The field test proposed[2,3], for the determination of hydrogen fluoride vapour in air is unsuitable for atmospheres polluted with fluoride-containing dusts, as these dusts also contain aluminium or phosphate which would interfere with the method used. The present method[4] was developed to provide a simple and versatile method for the determination of total inorganic fluoride in air.

Principle of the method

The fluoride is collected on an alkali impregnated filter paper. The fluoride is released by treatment with perchloric acid, in a micro diffusion chamber, and trapped by an alkaline coating on the lid of the diffusion vessel. The alkali is quantitatively transferred to a volumetric flask containing lanthanum-alizarin fluoride blue reagent and the intensity of the colour developed, which is proportional to the concentration of fluoride, is measued by comparison with a series of Lovibond permanent glass colour standards.

Reagents required

1. *Alkali impregnated filter papers* Drop 40 microlitres of 0.7% sodium hydroxide solution onto the centre of a Whatman No. 30 filter paper 25 mm in diameter. Dry in a fluoride free atmosphere and store in closed containers.
2. *Perchloric acid* Dilute 750 ml of 60% w/w perchloric acid to 1 litre with distilled water.
3. *Methanolic sodium hydroxide solution* Dissolve 2 g of sodium hydroxide in the minimum volume of water and dilute to 100 ml with methanol. Store in a well stoppered bottle and discard when sediment appears.
4. *Lanthanum-alizarin fluorine blue reagent* Dissolve 34 g of hydrated sodium acetate ($CH_3COO\,Na.3H_2O$) in about 150 ml of water and transfer to a 500 ml graduated flask. Dissolve 120 mg of alizarin fluorine blue in 0.25 ml of concentrated ammonia solution (Sp. gr. 0.880) and 2.5 ml of 20% w/v ammonium acetate solution and transfer to the graduated flask, using the minimum amount of water. Then add 15 ml of glacial acetic acid and 250 ml of 2-ethoxyethanol. Dissolve 270 mg of lanthanum nitrate ($La(NO_3)_3.6H_2O$) in 5 ml of $2M$ hydrochloric acid, transfer this solution to the mixture in the graduated flask and mix well. Allow to stand for 1 hour then dilute to 500 ml with water and again mix well. Store in a stoppered darkened bottle.

   For use, to 12.5 ml of this solution add 10 ml of 2-ethoxyethanol and dilute to 100 ml with water.

All chemicals used in the preparation of reagents should be of analytical reagent grade.

The Standard Lovibond Comparator Disc 6/47

This disc contains standards corresponding to 0, 1.25, 2.5 and 5.0 mg of fluoride per cubic metre of air, based on a 2.5 litre sample.

COLORIMETRIC CHEMICAL ANALYTICAL METHODS — Fluoride (Total)

Technique

Place an impregnated filter paper in the filter holder, (Note 1), attach the assembly to a convenient source of suction[5] and draw a sample of the atmosphere through the paper at a fixed rate of 0.5 litre per minute for 5 minutes. Remove the paper from the holder and place it in the bottom section of a microdiffusion vessel (Note 2). Add 0.1 ml of the methanolic sodium hydroxide solution (reagent 3), dropwise onto the underside of the lid of the microdiffusion vessel. Ensure uniform wetting of the lid surface and dry in a fluoride-free atmosphere. When dry add 3 ml of the perchloric acid (reagent 2) to the bottom section of the vessel, quickly cover with the lid and place in an oven at $60\pm1°C$. Leave for 16 hours i.e., overnight. Remove the vessel from the oven, and allow to cool. By means of distilled water quantitatively transfer the alkaline coating on the underside of the lid to a 25 ml graduated flask containing 10 ml of lanthanum-alizarin fluorine blue reagent. Make up to 25 ml with distilled water and wait for 10 minutes to allow the colour to develop. Pour sufficient of the test solution into a 60×10 mm flat-bottomed tube to fill it to a depth of 50 mm. Fit the Large Cells Attachment (DB411) complete with Special Tube Holder (DB430) on to the back of the Lovibond 1000 Comparator, and holding the Comparator horizontally insert the 60 mm tube into the disc centre aperture as shown in figure. Fill a similar tube 55 mm long to a depth of 50 mm with distilled water and place in the other compartment, as shown, to act as a "blank" behind the disc colours.

Insert the appropriate disc and match the colour of the test solution with the standards in the disc, viewing the liquid through its depth either by holding the comparator in the hand over a white background or placing it in a Lovibond White Light Cabinet tipped over on its back.

Notes

1. A suitable filter-paper holder to take papers 25 mm in diameter may be obtained from Tintometer Ltd., (Code No. AF401).

2. A suitable microdiffusion vessel is a disposable polystyrene Petri dish 50 mm diameter. These dishes are obtainable from Sterilin Ltd., 43–45 Broad St., Teddington, Middlesex.

3. It is recommended that a "Blank" determination be carried out simultaneously with any batch of samples. This is done by placing an unused impregnated filter paper in a microdiffusion vessel and treating it in the same way as the sample papers.

References

1. Ministry of Labour, *Safety, Health and Welfare, New Series No. 8, Dusts and fumes in Factory Atmospheres* Fourth Edition, H.M. Stationery Office, London, 1969
2. B. S. Marshall and R. Wood, *Analyst*, 1968, **93**, 821
3. See page 501
4. B. S. Marshall and R. Wood, *Analyst*, 1969, **94**, 493
5. See pages 462 to 464
6. "*Methods for the Detection of Toxic Substances in Air*" Booklet 19 *Hydrogen Fluoride and other inorganic fluorides*. H.M. Stationery Office, London, 1970

The Determination of Formaldehyde
using MBTH (3-methylbenzothiazol-2-one hydrazone hydrochloride)

Introduction

Formaldehyde is one of the most important industrial chemicals, being used in the manufacture of phenolic resins, artificial silks, and cellulose esters, dyestuffs, glass mirrors and explosives. It is a preservative and coagulant for rubber latex and in photography it is used as a hardener for gelatin plates and papers.

Formaldehyde is a toxic chemical. The 1973 Threshold Limit Value for its vapour in air is 2 ppm by volume (3 mg per cubic metre).

Most of the colorimetric methods which exist for the determination of formaldehyde are unsuitable for adaptation for use as a simple field test for the vapour in air because of the slowness of reaction, insensitivity or the potentially hazardous nature of the reagents used. However, a method[1] based on the reaction of aliphatic aldehydes with 3-methylbenzothiazol-2-one hydrazone hydrochloride (MBTH) in the presence of ferric chloride to form a blue cationic dye in acidic media, did appear to be capable of adaptation.

Principle of method

In the procedure[2] described below, formaldehyde vapour is collected in a solution of MBTH, the solution heated and then reacted with ferric chloride. After 10 minutes acetone is added to quench the reaction and the resultant blue solution compared against the standard Comparator disc.

In the presence of furfuryl alcohol an additional yellow complex is formed which imparts a green colour to the solution. In certain industrial processes, e.g., the preparation of resin-bonded mould liners in foundries, furfuryl alcohol is used with formaldehyde and may therefore occur as a co-contaminant. A simple procedure is described, for use where necessary, whereby the yellow complex is extracted with tetrachloroethylene leaving the blue complex formed from formaldehyde in the aqueous phase.

Reagents required

All reagents should be of analytical quality where possible and the water distilled or deionised.

1. *3-Methylbenzothiazol-2-one hydrazone hydrochloride* (MBTH) Reagent. ($S.C_6H_4.N(CH_3).C:N.NH_2.HCl$) Molecular weight 215.70.

    Dissolve 0.05 grams of MBTH in 30 ml of water and, in a separate vessel, 10 grams of sodium chloride. (NaCl) in 40 ml of water. Mix the two solutions thoroughly in a 100 ml volumetric flask and dilute to the mark with water. The solution should be stored in dark glass bottles below 25 °C and freshly prepared after two days. If stored in a refrigerator the solution remains satisfactory for seven days.

2. *Ferric chloride solution.*
    Dissolve 2.0 grams of ferric chloride hexahydrate ($FeCl_3.6H_2O$) in water, add 20 ml of concentrated hydrochloric acid (HCl SG. 1.18) and dilute to 1 litre with water. Prepare a fresh solution after one month.

    Note: The above solutions must be kept in well stoppered, glass bottles when not in use and stored in an aldehyde-free atmosphere. In the interest of safety, pipette fillers should be used for dispensing reagents used in this test.

The Standard Lovibond Comparator Disc No. 26 591 8

This disc contains 4 standards corresponding to 0, 1, 2 and 4 ppm by volume of formaldehyde (HCHO) based on a 250 ml air sample.

COLORIMETRIC CHEMICAL ANALYTICAL METHODS — Formaldehyde

Apparatus

Bubbler as illustrated. Receiver — Overall length 155 mm, external diameter 15 mm, internal diameter 13 mm. Distance from bottom of side arm to bottom of bubbler 110 mm. Inlet Tube — Approximate overall length 185 mm, external diameter 6 mm, internal diameter 4.5 mm. Distance from bottom of inlet tube to bottom of bubbler 7 mm. Cone and Socket B 10.

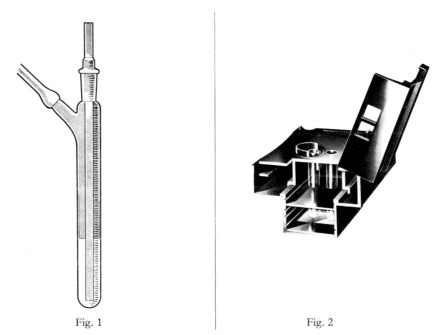

Fig. 1   Fig. 2

Technique

In an uncontaminated atmosphere, well away from the suspected source of formaldehyde, pipette 5 ml of the MBTH reagent into the receiver and assemble the apparatus as shown in the diagram, Fig. 1. At the sampling site, draw a 250 ml sample of the atmosphere through the reagent at the rate of 125 ml per minute. Remove the apparatus from the test atmosphere, disconnect the source of suction, remove the inlet tube and stopper the receiver. Heat in a boiling water bath for 2 mins., then transfer to a bath containing water at a temperature between 15 and 30°C. After a further 5 mins. add 0.5 ml of the ferric chloride solution from a pipette, replace the stopper and mix by inverting the tube several times ensuring that no liquid is lost through the side arm. Ten $\pm$ 0.5 mins. after the addition of ferric chloride add from a pipette 4 ml of acetone and mix thoroughly by inversion as before. Pour the contents of the tube into the special 60×10 mm flat bottomed comparator tube until the 50 mm calibration mark is reached, and place this tube in the right-hand compartment of the Comparator in the manner shown in Fig. 2.

This is done by fitting a Large Cell Attachment with Special Tube Holder (DB430) on to the back of the Lovibond Comparator and holding the Comparator horizontally: insert the 60 mm tube into the disc centre aperture as shown. The "blank" tube is only 55 mm high to avoid fouling the disc, and this is placed in the left hand aperture filled to the 50 mm mark with distilled water. Place the Comparator over a standard source of white light such as a white tile reflecting north daylight or in a Lovibond White Light Cabinet tipped on its back and compare the colour of the sample solution with the standards in the disc.

Atmospheres containing furfuryl alcohol in addition to formaldehyde

If the presence of furfuryl alcohol is expected or suspected in addition to formaldehyde, proceed as above as far as the addition of 4 ml of acetone. Then add 2 ml of tetrachloroethylene, re-stopper the receiver and mix well by repeated inversion of the tube for 2 mins., ensuring that no liquid is lost through the side arm. Allow the tube to stand until the two layers separate and the upper (aqueous layer) is free from turbidity. (NOTE: It may be necessary to tap the tube gently to dislodge droplets of tetrachloroethylene from the side or the top of the aqueous layer). Transfer the upper aqueous phase, by means of a suitable pipette, to the 60 mm tube and compare the standards as above.

Interferences

Aliphatic aldehydes give a positive response to this test producing similar blue colours. It is unlikely, however, that these will be present where the test is commonly used.

References

1. E. Sawiki, T. R. Hauser, T. W. Stanley and W. Elbert, *Analyt. Chem.*, 1961, **33**, 93.
2. I. A. Carmichael, *et al.* – *In preparation.*

The Determination of Hydrogen Cyanide
using ferrous hydroxide

Introduction

Hydrogen cyanide (prussic acid) vapour is popularly regarded as one of the most toxic vapours likely to be met in industry. While not as toxic as some other industrial gases, it does present a serious hazard in some circumstances. A concentration of 1 part in 10,000 by volume is stated[1] to be very dangerous while a concentration of 1 part in 500 would almost certainly be lethal. Such concentrations are liable to be encountered in some circumstances consequent upon the use of hydrogen cyanide as a fumigant or the use of metal cyanides in industrial processes.

The present test[2] replaces the two previous tests described in previous editions of "Methods for the Detection of Toxic Gases in Industry—Booklet No. 2". It was developed by Dixon, Hands and Bartlett[3], to overcome certain shortcomings in the original official tests, from the earlier work of Gettler and Goldbaum.[4] The papers developed by Dixon et al[3] have proved to be stable, under suitable storage conditions, for many months and are sensitive to slightly less than 1 ppm of hydrogen cyanide in air.

Principle of the method

Paper is impregnated with a solution of ferrous sulphate which is converted into ferrous hydroxide by treatment with sodium hydroxide solution. The prepared test paper, after exposure to hydrogen cyanide vapour, gives a blue colour due to the formation of sodium ferric ferrocyanide (Prussian blue) when immersed in diluted sulphuric acid.

Reagents required

1. *Ferrous sulphate solution* 10 g $FeSO_4.7H_2O$ in 100 ml distilled water
2. *Sodium hydroxide solution* 20 g NaOH in 100 ml distilled water
3. *Sulphuric acid solution* 30 ml of concentrated H_2SO_4 added to 70 ml of distilled water.

All chemicals must be of analytical reagent grade.

Preparation of test papers

Immerse a sheet of Whatman No. 50 filter paper in a developing dish containing 100 ml of ferrous sulphate solution. After 5 minutes remove the paper and dry in a current of warm air. Cut off, and discard, an inch strip from the bottom of the paper and then cut the remainder of the sheet into strips $1\frac{1}{2}'' \times 1''$. Immerse these strips singly in the sodium hydroxide solution for about 15 seconds, blot with absorbent paper and dry under vacuum in a desiccator. Satisfactory papers are grey-green to pale brown in colour and are dry, but not brittle, to the touch. For storage the papers should be individually vacuum sealed in glass tubing.

The Standard Lovibond Comparator Disc 6/35

The disc, containing 5 colour standards, covers the range of 2.5 to 50 ppm of hydrogen cyanide (2.5, 5.0, 10, 20, 50), based on a 360 ml air sample, and is designed for use with the Lovibond Comparator and Test Paper Viewing Stand with reducing aperture.

Technique

Break the sealed glass tube and remove the test paper. Fix this in a suitable holder which will expose a 1 cm circle of the paper[3] and draw exactly 360 ml of air through the paper by means of a suitable pump or aspirator. This air sample should be taken at a rate not exceeding 6 ml per second. After the completion of sampling remove the paper from the holder and immerse in the sulphuric acid solution (Reagent 3). If hydrogen cyanide was present in the air sampled a blue colour will develop on the paper in 30 seconds to 1 minute. Remove the paper from the dish, wash well with water and dry. After drying (see Note 1) place the test paper on the right-hand side of the Lovibond Comparator test-paper viewing stand. Place a piece of filter paper on the left-hand side and compare the colour of the test paper with that of the permanent glass standards in the disc, using a standard source of white light, such as north daylight or the Lovibond Lighting Cabinet. The special field reducing diaphragm should always be used for making the comparison with the spot on the test paper.

Notes

1. For an approximate rapid estimation of the hydrogen cyanide concentration the comparison may be carried out before drying the test paper. Alternatively, drying may be expedited by rinsing the paper with acetone and waving the paper in the air.

2. This test is specific for hydrogen cyanide and blue stains are only produced by this compound or by substances which could react to produce it e.g. cyanide dust or cyanogen. Dust may be removed by means of a filter placed before the test paper.

3. Hydrogen chloride and ammonia in concentrations below 400 ppm do not interfere, neither do sulphur dioxide and hydrogen sulphide in concentrations below 50 ppm.

4. Chlorine inhibits the reaction of the test papers with hydrogen cyanide.

5. If it is necessary to estimate concentrations of hydrogen cyanide outside the range of standards on the disc, this can be accomplished by suitably adjusting the volume of air sampled. For example if only 180 ml of air are sampled then the concentration is twice that of the equivalent standard on the disc.

References

1. Y. Henderson and H. W. Haggard, *"Noxious Gases and the Principles of Respiration influencing their Action,"* American Chemical Society Monograph No. 35, 2nd edition, Reinhold, New York, 1943
2. D.S.I.R., *"Methods for the Detection of Toxic Substances in Air,* Booklet 2, *Hydrogen Cyanide"*, H.M. Stationery Office, London, 1973
3. B. E. Dixon, G. C. Hands and A. F. F. Bartlett, *Analyst,* 1958, **83,** 199
4. A. O. Gettler and L. Goldbaum, *Anal. Chem.,* 1947, **19,** 270

The Determination of Hydrogen Fluoride
using zirconium — Solochrome cyanine R reagent

Introduction

Hydrogen fluoride is widely used in the chemical industry. It is a highly toxic gas with a threshold limit value of 3 ppm v/v in air[1]. There is therefore a need for a simple and rapid test for the determination of its concentrations in industrial atmospheres. The present test has been developed[2,3,6] to meet this requirement.

Principle of the method

In this test, which is a development of earlier tests described by Megregian[4] for the determination of fluorine, the atmosphere is sampled into an acid solution of the zirconium-Solochrome cyanine R reagent. The degree to which the hydrogen fluoride bleaches the colour of this reagent is measured by comparison with a series of Lovibond permanent colour glass standards.

Reagents required

1. *Zirconium solution* Dissolve 36 mg of zirconium oxychloride octahydrate ($ZrOCl_2 \cdot 8H_2O$) in water, add 120 ml of concentrated hydrochloric acid and dilute to 500 ml.

2. *Solochrome cyanine (C.I. 43820) solution* Dissolve 60 mg of the purified reagent in water, add 2.5 ml of N hydrochloric acid and dilute to 250 ml. The normal chemical commercially available can be purified by extraction with methanol. After filtering the solution the extract is evaporated to dryness under reduced pressure, and the resulting purified material is used to prepare the reagent solution.

3. *Zirconium-Solochrome cyanine R reagent* Combine 10 ml of each of reagents 1 and 2 and dilute to 200 ml with water. Although this reagent solution appears to be stable for a considerable time it is recommended[2] that a fresh batch be prepared for each series of tests.

All chemicals used in the preparation of these reagents should be of analytical reagent quality.

The Standard Lovibond Comparator Disc 6/46

This disc contains four standards corresponding to 0, 1.5, 3.0, and 6.0 ppm of hydrogen fluoride (HF) based on a 500 ml air sample.

Technique

Prepare an absorbing tube as shown in figure 1, place 5 ml of the mixed reagent (reagent 3) in the tube, insert the inlet tube and connect a suitable source of suction[5] to the side-arm. Sample the suspect atmosphere at 125 ml per minute for four minutes. Disconnect the suction and, if the sampling has been carried out in a humid atmosphere, raise the level of the reagent in the inlet tube to within about 10 cm of the top of the tube by blowing gently into the side-arm. Hold the reagent in that position for a few seconds by placing a finger over the end of the inlet tube. Transfer the contents of the absorber to a 60×10 mm flat-bottomed tube to fill it to a depth of 50 mm. Fit the Large Cells Attachment (DB411) complete with Special Tube Holder (DB430) on to the back of the Lovibond 1000 Comparator, and holding the Comparator horizontally insert the 60 mm tube into the Disc centre aperture as shown in figure 2. Fill a similar tube 55 mm long to a depth of 50mm with distilled water and place in the other compartment as shown to act as a "blank" behind the disc colours.

Insert the appropriate disc and match the colour of the test solution with the standards in the disc, viewing the liquid through its depth either by holding the comparator in the hand over a white background or placing it in a Lovibond White Light Cabinet tipped over on its back.

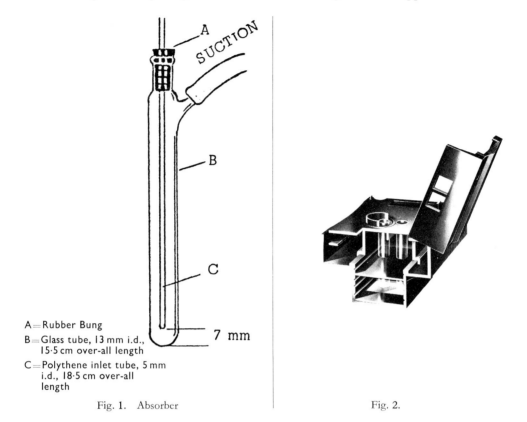

A = Rubber Bung
B = Glass tube, 13 mm i.d., 15·5 cm over-all length
C = Polythene inlet tube, 5 mm i.d., 18·5 cm over-all length

Fig. 1. Absorber

Fig. 2.

Notes

1. The setting up and dismantling of the apparatus should be carried out in a clean atmosphere well away from the source of suspected hydrogen fluoride contamination.

2. To avoid cross-contamination between samples each absorber, both test tube and inlet tube, should be rinsed with a few ml of the mixed reagent and the inlet tube dried before re-use.

References

1. "*Ministry of Labour, Safety, Health and Welfare. New Series No. 8, Dusts and Fumes in Factory Atmospheres.*" 4th edition, H.M. Stationery office, London, 1969
2. B. S. Marshall and R. Wood, *Analyst,* 1968, **93,** 821
3. B. S. Marshall and R. Wood, *Analyst,* 1969, **94,** 493
4. S. Megregian, *Anal. Chem.* 1954, **26,** 1161
5. See pages 462 to 464
6. "*Methods for the Detection of Toxic Substances in Air*" Booklet 19 *Hydrogen Fluoride and other inorganic fluorides.* H.M. Stationery Office, London, 1970

The Determination of Hydrogen Sulphide (1)
using dimethyl-*p*-phenylene diamine dihydrochloride

Introduction

This technique is especially suited to the quantitative determination of very small quantities of hydrogen sulphide and its ammoniacal derivatives in fuel gases, using only a small volume of gas.

Principle of the method

The most sensitive of all the reactions of hydrogen sulphide is used. Hydrogen sulphide and its mineral derivatives in a hydrochloric acid medium and in the presence of ferric chloride react with dimethyl-*p*-phenylene diamine to give methylene blue, and the intense blue colour of this compound is compared with Lovibond permanent glass standards representing different concentrations.

Reagents required

1. *Cadmium acetate* 1% solution, rendered slightly acidic (about *p*H 5.0) with acetic acid.
2. *Dimethyl-p-phenylene diamine dihydrochloride* 0,5 g in 100 ml 1:1 hydrochloric acid. The solution should be practically colourless. Decolorise with animal charcoal if necessary. It is advisable to check each fresh batch of reagent against known H_2S standards and the disc.
3. *Ferric chloride* ($FeCl_3.6H_2O$) 54 g in 1 litre concentrated hydrochloric acid (HCl). Dilute 1 volume with 4 volumes of distilled water immediately before use.

The Standard Lovibond Comparator Disc 6/14

The disc covers the range 0.2—1.6 ppm hydrogen sulphide (H_2S) in steps of 0.2 ppm w/v when 1 cu. ft. gas is taken for the test.

The disc is designed for use with 40 mm cells.

Technique

Place 10 ml of Reagent No. 1 in a Dreschel gas wash bottle with sintered glass (G1 or G2 porosity) distribution tube and connect this to the supply of gas, connecting either a gas meter or aspirator in the train after the wash bottle. Adjust the gas rate to 1—2 cu. ft./hr. and pass 1 cu. ft. of gas or such a volume as will contain about 30 micro-litres of H_2S through the wash bottle.

Disconnect the wash bottle (sealing the inlet and outlet tubes for transporting), and then add 4 ml of Reagent No. 2 followed by 1 ml of Reagent No. 3 (diluted), to the solution in the wash bottle, and shake to ensure that all the precipitate of cadmium sulphide is dissolved. (Gentle intermittent blowing down the sintered glass tube will ensure that all the precipitate retained in the sintered glass is dissolved). Transfer the liquid from the wash bottle to a 50 ml calibrated flask or Nessler glass, washing with a minimum of distilled water and make up the solution to 50 ml. Maintain at 40°C in a water bath for 10 minutes to develop the colour. Cool and transfer to a 40 mm comparator cell. Place this in the right-hand compartment of the Lovibond Comparator.

At the same time as Reagent Nos. 2 and 3 are added to the solution in the wash bottle, a blank is prepared by adding 4 ml of Reagent No. 2 and 1 ml of Reagent No. 3 to 10 ml of Reagent No. 1 in a standard flask or Nessler glass, and diluting to 50 ml. This blank is placed in the water bath at the same time as the test solution, and after 10 minutes is cooled and placed in a similar cell in the left-hand compartment of the comparator so that it comes behind the glass colour standards.

Hold the comparator facing a standard source of white light such as the Lovibond White Light Cabinet or, failing this, north daylight and compare the colour produced in the test solution with the colours in the standard disc, rotating the latter until a colour match is obtained. The answer is read from the indicator recess as ppm H_2S in the gas tested if 1 cu. ft. of gas is passed. If a greater or smaller volume is used, due allowance must be made.

Note

The following table allows conversion of the readings to give micro-litres of H_2S present, and it is possible by means of a simple calculation to extend the range of the disc by passing different volumes of gas.

ppm H_2S 1 cu. ft. gas passed	micro-litres H_2S
0.2	5.7
0.4	11.4
0.6	17.1
0.8	22.8
1.0	28.5
1.2	34.2
1.4	39.9
1.6	45.6

References

1. J. F. Fogo and M. Popowsky, *Anal. Chem.*, 1949, **21**, 732/4
2. A. E. Sands, M. A. Grafius, H. W. Wainwright and M. W. Wilson, *U.S. Bur. Mines Rept. Invest.,* No. 4547, **19**, (1949)

The Determination of Hydrogen Sulphide (2)
using lead acetate

Introduction

Hydrogen sulphide (or sulphuretted hydrogen) is a very poisonous gas and is of very widespread occurrence. In addition to its formation during the decomposition of organic matter containing sulphur, it is encountered in many important industries among which are the following:

    Artificial silk manufacture
    Chemical industry
    Dye-making and dyeing
    Coke oven and by-product plants
    Gas production
    Grease refining
    Petroleum refining
    Tar distillation
    Sewage purification

It is also encountered in sewers and cesspools.

Hydrogen sulphide is one of the more toxic gases encountered in industry and in moderately high concentrations, by its action on the respiratory centre in the brain and without premonitory symptoms, it can cause sudden unconsciousness and possibly death.

The characteristic smell of hydrogen sulphide is not a reliable guide to its presence. At concentrations in excess of 150 ppm its systemic toxicity increases rapidly and its odour is no longer appreciated.

Prolonged exposure to concentrations of the order of 50 ppm or less can cause irritation of the eyes, conjunctivitis and, in the worst cases keratitis. Concentrations of about 200 ppm will cause irritation of the respiratory tract progressing, on prolonged exposure, to pulmonary oedema. The action on the central nervous system may occur with dramatic suddenness and becomes increasingly important at concentrations in excess of 500 ppm. Concentrations much in excess of 1,000 ppm are likely to cause immediate death.

The recommended threshold limit value for an eight-hour day of continued exposure is 10 ppm (v/v) in air, i.e. 15 mg per cubic metre.

Principle of the method

The concentration of hydrogen sulphide is measured by drawing 125 ml of the air through a test paper impregnated with lead acetate. The complete absorption of the hydrogen sulphide is ensured by the incorporation of glycerol into the test paper.

Reagents required

1. *Lead acetate* $(CH_3COO)_2Pb.7H_2O$ analytical reagent quality
2. *Acetic acid* CH_3COOH glacial
3. *Glycerol* $CH_2(OH)CH(OH)CH_2OH$ pure

Preparation of test papers

Dissolve 10 g of lead acetate in 90 ml of distilled water. Add 5 ml of glacial acetic acid and 10 ml of glycerol to the solution and mix well. This solution should be freshly prepared each time a batch of test papers is made.

Place the solution in a 100 ml measuring cylinder. Immerse strips of Whatman No. 3 MM chromatographic paper 20 mm wide and 100 mm long, vertically in this solution for one minute. Allow the excess liquid to drain off, suspend the papers vertically and allow to dry as completely as possible at room temperature in an atmosphere free from hydrogen sulphide. When dry, cut off the top and bottom 25 mm of the prepared papers and store the remaining 50 mm in stoppered wide-necked, dark glass bottles. Use the test papers within 14 days of preparation.

The Standard Lovibond Comparator Disc 6/38

This disc, which replaces Disc 6/2, contains four standards corresponding to 5, 10, 20 and 40 parts per million v/v of hydrogen sulphide in a 125 ml sample. This disc should be used in conjunction with the Lovibond Comparator and a Test Paper Viewing Stand.

Technique

Fit the test-paper centrally between the gaskets of the test-paper holder and screw up the holder until it is finger-tight. Then attach the holder to the source of suction by means of a piece of rubber tubing, and draw air steadily through the paper at a rate of 125 ml per minute.

Remove the test paper and place this in the right hand aperture of the Test Paper Viewing Stand. Place the comparator and stand before a source of white light, such as the Lovibond White Light Cabinet or, failing this, north daylight and compare the colour of the stain with the standards in the disc.

The stain should be only on that side of the test paper exposed to the gas entering the holder. If the stain is visible on both sides of the paper this indicates that the hydrogen sulphide has not been fully absorbed by the paper. Fresh test papers should be prepared and the test repeated with these.

Reference

"*Methods for the detection of toxic substances in air*" Booklet 1 *Hydrogen Sulphide*, H.M. Stationery Office, London, 1969

The Determination of Iron Oxide Fume
using bathophenanthrolinedisulphonate

Introduction

Iron oxide fume is found in many industrial atmospheres where processes involving the melting of iron or steel are carried out. Although the fume is generally regarded as a nuisance rather than a health hazard a threshold limit has been assigned[1] for the concentration of iron oxide fume in industrial atmospheres. The present test[2,4] has been developed to provide a simple and rapid field test for the identification and determination of iron oxide fume in air at its threshold limit concentration of 10 mg m^{-3}. It is based on the original test described by Reynolds[3] which uses the specific reaction between ferrous iron and bathophenanthroline. In the present test the water soluble disodium salt of bathophenanthrolinedisulphonic acid is used to obviate the necessity for solvent extraction of the coloured complex formed.[5]

Principle of the method

The iron oxide fume is collected on a filter paper. This deposit is then dissolved in a hot solution of hydroxylammonium chloride in hydrochloric acid, which ensures that all the iron present is in the ferrous form, and complexed with a solution of bathophenanthroline-disulphonate. The intensity of the red colour of the complex, which is proportional to the iron oxide concentration, is measured by comparison with a series of Lovibond permanent glass colour standards.

Reagents required

1. *Hydrochloric acid* (1+1 v/v) Dilute concentrated hydrochloric acid (HCl) with an equal volume of distilled water.
2. *Hydroxylammonium chloride solution* Dissolve 20 g of hydroxylammonium chloride (HONH$_3$Cl) in 100 ml of distilled water.
3. *Buffer solution* Dissolve 33.3 g of hydroxylammonium chloride and 60 g of sodium acetate trihydrate (CH$_3$COO.Na.3H$_2$O) in 360 ml of distilled water.
4. *Bathophenanthrolinedisulphonate solution* Dissolve 100 mg of disodium 4,7-diphenyl-1,10-phenanthroline-disulphonate in 100 ml of distilled water.

All chemicals used in the preparation of reagents should be of analytical reagent quality. Solutions should be stored in iron-free glass or polythene bottles.

The Standard Lovibond Comparator Disc 6/52

This disc contains 4 standards corresponding to 1, 5, 10 and 20 mg m^{-3} of iron oxide fume, based on a 2.5 litre sample of air.

Technique

Place a Millipore Type AA filter paper, 25 mm diameter, in a suitable holder (Note 1), attach this to a suitable source of suction (Note 2) and draw the polluted atmosphere through the paper at 0.5 litre a minute for 5 minutes. Disconnect the holder from the source of suction, remove the filter paper and place the paper in a 25 ml beaker of sufficient diameter to avoid having to fold the paper. Add 1 ml of hydrochloric acid (reagent 1) and 1 ml of hydroxyl-ammonium chloride solution (reagent 2), cover the beaker with a watch-glass and place it on a steam bath. After 10 minutes remove the beaker, transfer the solution to a 25 ml volumetric flask, wash the filter paper thoroughly with 10 ml of distilled water and add the washings to the solution in the flask, add 2 ml of bathophenanthrolinedisulphonate solution (reagent 4) followed by 10 ml of the buffer solution (reagent 3). Dilute to 25 ml with distilled water, mix well, set aside for 5 minutes and then transfer a portion of the solution to a standard comparator tube. Place this tube in the right-hand compartment of the comparator, fill an identical tube with distilled water and place this tube in the left-hand compartment to act as a blank. Place the comparator before a standard source of white light such as the Lovibond White Light Cabinet or, failing this, north daylight and match the colour of the sample with the standards in the disc.

Notes

1. Suitable filter holders can be obtained from Tintometer Ltd., (Code No. AF401).
2. Suitable methods for taking air samples at controlled flow rates are described in the Introduction to this Section.
3. Up to 1000 μg of fluoride, manganese, phosphate, vanadate and zinc, 200 μg of tin, 50 μg of nickel and tungstate and 20 μg of chromium and titanium will not interfere with the determination of 10 μg of iron.
4. This method is versatile and can be adapted to the determination of other iron containing materials including metallic iron fume or dust. It can also be used to determine the concentration of water-soluble iron salts in air. The threshold limit for these is[1] 1 mg m^{-3}. Provided that an air sample of exactly 17.5 l is taken, at a flow rate of 0.5 to 2.5 l a minute, then the colour standards in the disc will represent 0.1, 0.5, 1.0 and 2.0 mg m^{-3} of soluble iron salts (as iron).

References

1. Department of Employment and Productivity, *"Technical Data Note 2/69, Threshold Limit Values* 1969", H.M. Stationery Office, London, 1969
2. D. W. Meddle and R. Wood, *Analyst,* 1971, **96,** 62
3. R. G. Reynolds and J. L. Monkman, *Amer.Ind.Hyg.Assoc.J.,* 1962, **23,** 415
4. *"Methods for the detection of toxic substances in air"* Booklet 21. *"Iron oxide fume"*, H.M. Stationery Office, London, 1971
5. D. Blair and H. Diehl, *Talanta,* 1961, **7,** 163

The Determination of Aromatic Isocyanates
using N-1-naphthylethylenediamine

Introduction

The production of polyurethane plastics and adhesives was originally based on the use of tolylene-2,4-di-isocyanate (TDI). In view of the toxicity of TDI and its volatility, the use of other isocyanates of lower volatility was investigated and 4,4′-di-isocyanatodiphenyl-methane (MDI), polymethylene polyphenyl isocyanate (PAPI), naphthylene-1,5-di-isocyanate (NDI) are now used in place of TDI wherever practicable.

Despite the lower vapour pressures of MDI, PAPI and NDI it is still possible for hazardous concentrations of these materials to be generated in the form of droplets or aerosols in industrial atmospheres. The recommended maximum permissible concentration for both TDI and MDI is 0.02 ppm v/v in air[1] and PAPI and NDI are probably equally toxic. It is therefore necessary to monitor the atmospheres of all working areas where these isocyanates are used, and the present test[2,7] has been developed for this purpose from the methods proposed by earlier workers[3,4,5].

Although the sampling method and subsequent chemical manipulations are the same for all the aromatic isocyanates, the colours produced by these compounds vary somewhat one from the other and a separate disc of colour standards must be used for each compound

Principle of the method

The isocyanate is absorbed in a mixture of dimethylformamide and hydrochloric acid, which hydrolyses the isocyanate to the corresponding amine. This amine is then coupled with N-1-naphthylethylenediamine to form a coloured complex. The intensity of the colour formed, which is proportional to the concentration of the isocyanate, is measured by comparison with a series of Lovibond permanent glass colour standards.

Reagents required

1. *Dilute hydrochloric acid*   Dilute 15 ml of concentrated acid to 100 ml with distilled water.
2. *Dimethylformamide*
3. *Diazotisation solution*   Dissolve 3 g of sodium nitrite and 5 g of sodium bromide in distilled water and dilute to 100 ml.
4. *Sulphamic acid solution*   10% w/v solution in distilled water.
5. *N-1-naphthylethylenediamine*   Dissolve 0.75 g of N-1-naphthylethylenediamine dihydrochloride in distilled water, add 2 ml of concentrated hydrochloric acid and dilute to 100 ml with distilled water. Prepare a fresh solution after two days.

All the chemicals used for the preparation of reagents should be of analytical reagent quality.

Apparatus required

1. All-glass absorbers—see Fig. 1.
2. Suction system. This should be capable of drawing 10 litres of air though the absorber at a rate of 1 litre a minute[6].

The Standard Lovibond Comparator Discs 6/42, 6/43, 6/44 and 6/45

Each of these discs contains standards corresponding to 0.01, 0.02 and 0.04 ppm v/v of the appropriate isocyanate in air, based on a 10 litre air sample.

    Disc 6/42 is for use with TDI
    Disc 6/43 is for use with MDI
    Disc 6/44 is for use with NDI
    Disc 6/45 is for use with PAPI

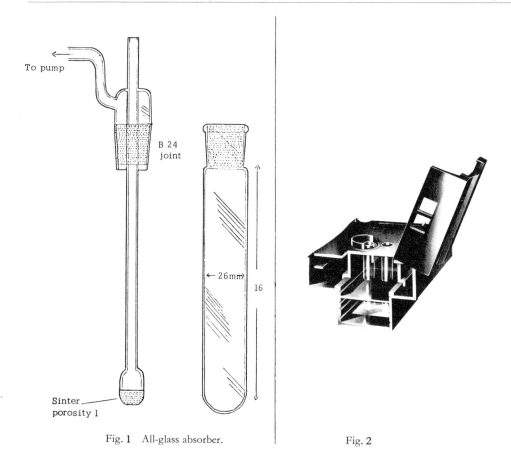

Fig. 1 All-glass absorber.   Fig. 2

Technique

Place 3 ml of dimethylformamide (reagent 2) and 2 ml of dilute hydrochloric acid (reagent 1) in the absorber (Fig. 1) and insert the inlet tube. Mount and fix the absorber at the sampling site in a vertical position (this is especially important when sampling an isocyanate atmosphere in the form of an aerosol). Attach the pump to the absorber and draw 10 litres of the atmosphere through the absorber at a rate of one litre a minute. Allow the absorbing solution to stand for 10 minutes to ensure the complete hydrolysis of any isocyanate which has been collected. (This waiting period is not required when TDI is being determined). Lift the inlet tube so that the sinter is clear of the liquid and expel the liquid trapped in the domed sinter as completely as possible. Ensure that the temperature of the absorbing solution is not above 20°C, then add 0.5 ml of the diazotisation solution (reagent 3). Shake the absorber to mix the contents and allow to stand for 2 minutes. Add 0.5 ml of the sulphamic acid solution (reagent 4) and shake the mixture until effervescence has ceased; 2 minutes after the addition of the sulphamic acid solution add 0.5 ml of N-1-naphthylethylenediamine solution (reagent 5) and mix well. Pour sufficient of the test solution into a 60×10 mm flat-bottomed tube to fill it to a depth of 50 mm. Fit the Large Cells Attachment complete with Special Tube Holder (DB430) on to the back of the Lovibond 1000 Comparator, and holding the comparator horizontally insert the 60 mm tube into the disc centre aperture as shown in figure 2. Fill a similar tube 55 mm long to a depth of 50 mm with distilled water and place in the other compartment as shown, to act as a "blank" behind the disc colours.

Insert the disc appropriate to the isocyanate to be determined and after 10 minutes match the colour of the test solution with the standards in the disc, viewing the liquid through its depth either by holding the comparator in the hand over a white background or placing it in a Lovibond White Light Cabinet tipped over on its back.

References

1. Ministry of Labour, *"Safety Health and Welfare Booklets, New Series, No. 8, Dusts and Fumes in Factory Atmospheres,"* 4th Editn., H.M. Stationery Office, London, 1969
2. D. W. Meddle, D. W. Radford and R. Wood, *Analyst,* 1969, **94,** 369
3. K. Marcali, *Anal. Chem.,* 1957, **29,** 552
4. K. E. Grim and A. L. Linch, *Amer. Ind. Hyg. Ass. J.,* 1964, **25,** 285
5. D. A. Reilly, *Analyst,* 1963, **88,** 732
6. See pages 462 to 464
7. Dept. of Employment. H.M. Factory Inspectorate *"Methods for the detection of toxic substances in air"* Booklet 20 '*Aromatic Isocyanates*', H.M. Stationery Office, London, 1970

The Determination of Isophorone
using dodecamolybdophosphoric acid

Introduction

Isophorone is extensively used as a solvent for many synthetic and natural materials especially resins and plastics. It is particularly outstanding as a solvent for polyvinylchloride (PVC) and vinyl copolymers, and it is increasingly used in the manufacture of vinyl lacquers. Even when used in small quantities isophorone will improve the gloss, flow and adhesion of many stoving systems; it improves the emulsifiability, and emulsion stability in aqueous dilution, of insecticide and herbicide concentrates, and it is increasingly used as an intermediate in the synthesis of peroxides, caprolactam, polyamides, polyesters, diisocyanates, epoxy hardener and other chemicals.[1]

Isophorone is a relatively toxic chemical. The 1973 proposed Threshold Limit Value[2] for its vapour in air is 10 ppm (55 mg per cubic metre). This test[3,5] has been developed as a rapid field method for determining concentrations of isophorone vapour at about its threshold limit value, and is a modification of the earlier test developed by Kacy and Cope.[4]

Principle of the method

Isophorone vapour is trapped in water. This aqueous solution reduces dodecamolybdophosphoric acid to molybdenum blue in the presence of perchloric acid. The yellow colour of any excess dodecamolybdophosphoric acid is removed by reaction with sodium citrate solution. The residual molybdenum blue colour, the intensity of which is proportional to the isophorone concentration, is measured by comparison with a series of Lovibond permanent glass colour standards.

Reagents required

All the chemicals used in the preparation of these solutions should be of analytical reagent quality.

1. *Dodecamolybdophosphoric acid reagent* Dissolve 4.0 g of sodium hydroxide and 12.5 g of dodecamolybdophosphoric acid ($H_3PO_4.12MoO_3.24H_2O$) in distilled water and dilute to 50 ml. Use the solution within three days of its preparation. The suitability or otherwise of any batch of dodecamolybdophosphoric acid must be assessed prior to its use. A suitable sorting test is given below.

2. *Dilute perchloric acid solution* Dilute 75 ml of 60% w/w perchloric acid ($HClO_4$) to 90 ml with distilled water.

3. *Trisodium citrate solution* Dissolve 30 g of trisodium citrate dihydrate ($Na_3C_6H_5O_7.2H_2O$) in distilled water and make up to 100 ml.
   These reagents should be stored in well stoppered glass bottles in an isophorone-free atmosphere.

The Standard Lovibond Comparator Disc 6/58

This disc contains standards corresponding to 0, 5, 10, 20, 25, 50 ppm of isophorone based on a 500 ml air sample. This disc is designed to be used with special comparator tubes graduated at a depth of 50 mm and viewed through the depth of the tube.

Apparatus

Bubbler as illustrated (see over page). Receiver – Overall length 155 mm, external diameter 15 mm, internal diameter 13 mm. Distance from bottom of side arm to bottom of bubbler 110 mm. Inlet Tube – Approximate overall length 185 mm, external diameter 6 mm, internal diameter 4.5 mm. Distance from bottom of inlet tube to bottom of bubbler 7 mm. Cone and Socket B10.

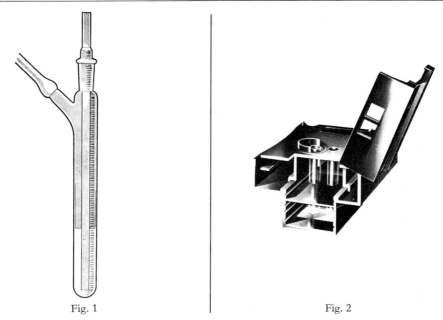

Fig. 1   Fig. 2

Technique

Pipette 2 ml of distilled water into a bubbler, of the type shown in Fig. 1 (Note 1), in an uncontaminated atmosphere. Insert the inlet tube and connect the side-arm to a suitable source of suction. At the sampling site draw a 500 ml sample of the atmosphere through the water. Remove the apparatus to an uncontaminated atmosphere, disconnect the source of suction, remove the inlet tube and add, by means of a safety pipette, 1 ml of the dodecamolybdophosphoric acid solution (reagent 1) and 2 ml of the perchloric acid solution (reagent 2). Stopper the tube and mix the contents well by inverting it several times. Take care that no liquid is lost through the side-arm.

Suspend the bubbler vertically in a boiling water bath in such a manner that it does not touch the bottom of the bath. Heat the tube and its contents for 15 minutes, then remove the tube from the bath and cool for 5 minutes in a water bath at a temperature below 27°C. Add 3 ml of the citrate solution (reagent 3) using a safety pipette, remove the tube from the cooling bath and mix the contents by inversion as before. Pour the contents of the tube into the special 60×10 mm flat bottomed Comparator tube until the 50 mm calibration mark is reached, and place this in the right-hand compartment of the Comparator in the manner shown in Fig. 2. This is done by fitting a Large Cell Attachment with Special Tube Holder (DB430) on to the back of the Lovibond 1000 Comparator and holding the Comparator horizontally: insert the 60 mm tube into the disc centre aperature as shown. The "blank" tube is only 55 mm high to avoid fouling the disc, and this is placed in the left-hand aperture filled to the 50 mm mark with distilled water. Exactly 5 minutes after adding reagent 3 place the Comparator over a standard source of white light such as a white tile reflecting north daylight or in a Lovibond White Cabinet tipped on its back and compare the colour of the sample solution with the standards in the disc.

Sorting test for dodecamolybdophosphoric acid

Pipette 2 ml of water, 1 ml of dodecamolybdophosphoric acid reagent and 2 ml of perchloric acid solution into a receiver. Proceed as above from the stage at which the receiver is stoppered. If any blue colour obtained is more intense than that of the "O" standard, discard the reagent and prepare again using a different batch of dodecamolybdophosphoric acid.

Notes

1. Bubblers, manufactured to the dimensions shown in Fig. 1, may be obtained from Glass of Mark Ltd., Jubilee Street, Northwich, Cheshire, England, or from Tintometer Ltd.

2. The solvents most likely to be present together with isophorone in an industrial atmosphere are 3-methylcyclohexanone and cyclohexanone. The colour produced by an atmosphere containing 200 ppm of 3-methylcyclohexanone, i.e. twice the threshold limit value, is equivalent to that produced by 12 ppm of isophorone. Similarly a concentration of 100 ppm of cyclohexanone, i.e. twice the threshold limit value produces a colour equivalent to a concentration of 7 ppm of isophorone. These compounds should not therefore interfere with the determination of isophorone at the concentrations at which they should be encountered in industrial atmospheres.

The maximum levels tested at which other solvents showed no interference were:

Compound	Concentration ppm
Acetone	15,000
Butan-2-one	3,000
4-Methylpentan-2-one	1,500
5-Methylheptan-3-one	250
Benzene	375
Toluene	3,000
o-Xylene	1,500
m-Xylene	1,500
Styrene	500
α-Methylstyrene	500
Vinyltoluene	500
Methanol	12,300
Ethanol	15,000
Propanol	3,000
Pentanol	1,000
Butanol	1,500
Cyclohexanol	750
3-Methylcyclohexanol	1,500
2-Ethoxyethanol	3,000
2-Methoxyethanol	375

References

1. K. Schmitt, *Chemische Industrie Internat.*, 1966, 3.
2. Department of Employment, H.M. Factory Inspectorate, Technical Data Note 2/73 *"Threshold limit values for 1973"*.
3. P. Andrew and R. Wood, *Analyst Lond.*, 1970, 95, 691.
4. H. W. Kacy and R. W. Cope, *Amer. Ind. Hyg. Ass. Q.*, 1955, 16, 55.
5. "Methods for the detection of toxic substances in air" Booklet 24. *"Isophorone"*, H.M. Stationery Office, London, 1974.

COLORIMETRIC CHEMICAL ANALYTICAL METHODS

The Determination of Ketone Vapour
using 2.4 - dinitrophenylhydrazine

Introduction

Ketone vapours are frequently encountered in industrial atmospheres, where they can occur as a result of their widespread use as solvents and as intermediates in chemical syntheses. Their toxicities vary widely depending on the individual chemical structures, the threshold limit values[1] for individual ketones falling within the ranges 25 – 1,000 ppm. Even when relatively non-toxic, however, ketone vapours can be strongly irritant, can have a narcotic effect on workers exposed to them, and in some cases can form flammable and explosive mixtures with air.

In order to keep the concentration of ketone vapour in an industrial atmosphere within safe limits a simple field test was required which would be applicable to a variety of ketones and to a range of concentrations. The present test[2] has been developed to meet this need. In view of the widely differing threshold limit values for the various ketones it is not possible to use exactly the same procedure for all and to obtain a measurable range of colours. Accordingly conditions have been established for each of the seven common ketones to which this test applies, which will result in the colour produced, by the threshold limit concentrations, matching a single colour standard. Additional standards are provided which correspond to concentrations of twice and half the threshold limit value. The sampling conditions which are varied to ensure this constant colour are the volume of the absorbent liquid and the volume of the atmosphere which is sampled.

This test, therefore, requires that the ketone which is being determined is known beforehand. This limitation is not as restrictive as it might appear at first sight, as in industrial situations the materials being used in a process are known. The Ketones to which the test applies are specified in Table 1.

Principle of the method

Ketone vapour, from a specified volume of the suspect atmosphere, is collected in a specified volume of water. The resulting solution is reacted with acidic 2.4-dinitrophenylhydrazine. On the addition of methanolic potassium hydroxide a red colour is formed. The intensity of this colour, which is proportional to the ketone concentration, is measured by comparison with Lovibond permanent glass colour standards.

Reagents required

1. *Methanol* (CH_3OH) It is essential that the methanol used in the preparation of the 2.4-dinitrophenylhydrazine solution should be free from carbonyl compounds. This can be checked by carrying out this test for ketone vapours by using 1 ml of distilled water instead of the ketone solution. The resulting colour should not be more intense than that of the blank standard. Some samples of methanol, such as the spectroscopic grade, will meet this requirement. If the methanol which it is proposed to use does not meet this standard then it should be purified in the following manner:

   Reflux, in an all glass apparatus, 1 litre of the methanol with 5 g of 2.4-dinitrophenylhydrazine and 5 drops of concentrated hydrochloric acid for 2 hours. Then distil the methanol twice, rejecting the first 100 and final 150 ml from each distillation. Store in a well stoppered, dark-glass bottle. Re-purify after two weeks.

2. *2.4-dinitrophenylhydrazine reagent* Dissolve 0.1 g of 2.4-dinitrophenylhydrazine ((NO_2)$_2$ $C_6H_3.NH.NH_2$) in about 75 ml of the purified methanol (reagent 1) to which 0.4 ml of hydrochloric acid (sp. gr. 1.18) has been added and dilute to 100 ml with the purified methanol.

3. *Potassium hydroxide solution* Dissolve 100 g of potassium hydroxide in 200 ml of distilled or de-ionised water, cool and dilute to 1 litre with commercial methanol. Store in a polythene bottle fitted with a well fitting screw top. This reagent is stable for several weeks.

   All chemicals used in the preparation of reagents should be of analytical reagent quality. All reagents, and the distilled or de-ionised water used for the collection of samples, should be stored and dispensed in a ketone-free atmosphere. Pipette fillers should be used to dispense the reagents.

The Standard Lovibond Comparator Disc 6/55

This disc contains four standards labelled Blank, A, B and C. The B standard corresponds to the Threshold Limit Value[1] concentration for the particular ketone being determined. Standards A and C correspond to half and twice that concentration respectively. The concentrations of the individual ketones to which these standards correspond, under the sampling conditions specified in Table I, will be found in Table II.

This disc has been designed for use with the Lovibond 1000 Comparator fitted with the special holder and miniature tubes (Catalogue No. DB430) illustrated in Figure 1.

Fig. 1
Lovibond Comparator with holder and miniature tubes.

Technique

This test uses a special absorption tube illustrated in Figure 2.

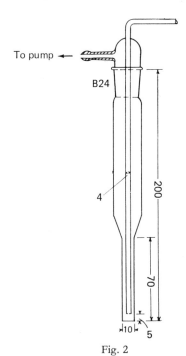

Fig. 2
All-glass absorber. (All measurements are in millimetres)

From Table 1 ascertain the sampling conditions to be used for the ketone to be determined. Place the designated volume of distilled or de-ionised water into the bubbler by means of a pipette. Insert the inlet tube and connect the bubbler to the aspiration equipment (Note 1). Transfer the apparatus to the sampling site and aspirate the required volume of air, as designated in Table 1, at a rate of 125 ± 5 ml a minute. Keep the temperature of the bubbler below 30°C during sampling by cooling if necessary (Note 2). After the completion of sampling, transfer the equipment to a Ketone-free atmosphere, remove the inlet tube and, by means of a pipette, place 1 ml of the water from the bubbler in a 20 ml test tube, fitted with a glass stopper. Add 1 ml of the 2.4-dinitrophenylhydrazine solution (reagent 2), stopper the tube and mix the solutions by gentle swirling.

Maintain the mixed solution at a temperature between 18 and 25°C (Note 2) for 10 ± 0.5 minutes then add 5 ml of potassium hydroxide solution (reagent 3) replace the stopper and mix the solutions by inverting the tube several times. (The dense black precipitate which forms on the addition of reagent 3 will slowly dissolve to give a clear solution). After about 8 minutes pour sufficient of this test solution into a 60×10 mm flat bottomed tube (supplied with holder DB430) to fill it to the 50 mm height mark. Fit the miniature tube holder DB430 on to the back of the Comparator and holding the Comparator horizontally insert the sample tube into the disc centre aperture as shown in Figure 1.

Fill a similar 55×10 mm tube (supplied with the holder) to the 50 mm mark with distilled water and place in the other compartment as shown, to act as a "blank" behind the disc colours.

Exactly 10 minutes after adding reagent 3 to the solution, compare the colour of the solution with the standards in the disc, viewing through the depth of the liquid either by holding the Comparator in the hand over a well lit white background, or placing it in a Lovibond Comparator Lighting Cabinet tipped on to its back. Read off the concentration, corresponding to the nearest matching standard, from Table II. As the red colour is unstable, the matching procedure must be completed within one minute or the test will have to be repeated.

TABLE I
Sampling Conditions for Various Ketones

Ketone	Volume of water in absorption tube	Volume of air Sample	Sampling Time at 125 ml min^{-1}
	ml	ml	min
Propan-2-one	25	125	1
Butan-2-one	20	625	5
4-Methylpentan-2-one	5	750	6
4-Hydroxy-4-methylpentan-2-one	5	1625	13
4-Methylpent-3-en-2-one	5	1375	11
Cyclohexanone	15	1125	9
Methylcyclohexanone	15	500	4

TABLE II
Ketone Concentrations Corresponding to Standards

Ketone	Concentrations (ppm) Corresponding to Standard		
	A	B	C
Propan-2-one	500	1000	2000
Butan-2-one	100	200	400
4-Methylpentan-2-one	50	100	200
4-Hydroxy-4-methylpentan-2-one	25	50	100
4-Methylpent-3-en-2-one	12.5	25	50
Cyclohexanone	25	50	100
Methylcyclohexanone	50	100	200

Standard B corresponds to the current[1] Threshold Limit Values.

Notes

1. Suitable methods of aspirating the sample are described in the Introduction to this Section.

2. The permissible temperature range can normally be maintained by immersing the absorption tubes and the reaction tubes in water whose temperature has been adjusted to 22°C.

3. This test was devised by the staff of the Laboratory of the Government Chemist. Their collaboration in the calibration of the disc and their permission to quote from their published results[2] are gratefully acknowledged.

4. *Interferences* Of the compounds studied[2], which included hydrocarbons, alcohols, aldehydes, halogenated hydrocarbons, esters and ethers, only aldehydes showed any positive interference. The concentrations of formaldehyde and acetaldehyde which significantly interfered with the test for ketones were above the threshold limit values for these aldehydes and, therefore, themselves constituted hazards.

References

1. "*Threshold Limit Values for 1970*" Technical Data Note 2/70, Department of Employment, H.M. Factory Inspectorate, H.M. Stationery Office, London, 1970.
2. A. F. Smith and R. Wood, *Analyst*, 1972, **97,** 363.

The Determination of Total Airborne Lead
using dithizone

Introduction

The original field method for the determination of total airborne lead[1,2] was introduced in 1960. Experience with the use of the test over the ensuing years has revealed some disadvantages and the present test[3,4] is a modification of the original test which has been developed to overcome these disadvantages.

In particular the use of Millipore filters is recommended in the interests of standardisation with other tests for metallic dust or fume in air. This change also enables the sample dissolution procedure to be simplified. Highly toxic carbon tetrachloride has been replaced as solvent by the less toxic 1,1,1-trichloroethane, and this has improved the stability of the dithizone reagent solution. The range of colour standards has also been modified; the highest standard has been dropped and an intermediary standard, corresponding to the half threshold limit value, has been introduced in its place.

Principle of the method

The airborne lead is collected on a filter, dissolved in acid and complexed with dithizone. The lead-dithizonate is extracted into 1,1,1-trichloroethane and the intensity of the colour of this solution, which is proportional to the lead concentration, is determined by comparison with a series of Lovibond permanent glass colour standards.

Reagents required

1. *Nitric acid hydrogen peroxide solution* Dilute 5 ml of lead-free nitric acid (HNO_3, sp.gr. 1.42) to 100 ml with distilled water and add 0.2 ml of 30% w/v hydrogen peroxide (H_2O_2).
2. *Buffer solution* Dissolve 3 g of potassium cyanide (KCN) 6 g of sodium metabisulphite ($Na_2S_2O_5$) and 5 g of ammonium citrate ((NH_4)$_3C_6H_5O_7$) in about 200 ml of distilled water, add 325 ml of aqueous ammonia (NH_4OH, sp.gr. 0.880) and dilute to 1 litre with distilled water. This buffer solution is stable for several months and should be stored in a polythene screw-topped bottle.
3. *1,1,1-Trichloroethane (CCl_3CH_3)* This solvent should be inhibitor free. Solvents supplied by Hopkin & Williams Ltd., and by Fisons Scientific Apparatus Ltd., have been recommended[3] for use in this determination.
4. *Dithizone stock solution* Dissolve 40 mg of dithizone ($C_6H_5.N:N.CS.NH.NH.C_6H_5$) (Note 1) in 100 ml of 1,1,1-trichloroethane (reagent 3). This solution must not be stored for more than seven days.
5. *Dithizone working solution* Dilute 5 ml of reagent 4 to 50 ml with reagent 3. This solution must not be stored for more than 8 hours.

All chemicals used in the preparation of reagents should be of special lead-free analytical quality.

The Standard Lovibond Comparator Disc 6/49

This disc contains standards corresponding to 0.0, 0.1, 0.2, 0.4 and 0.8 mg m^{-3} of lead (Pb) based on a 15 litre air sample.

Technique

Place a 25 mm diameter Millipore type AA filter paper in the filter-paper holder (Note 2), attach the holder to a suitable source of suction (Note 3) and draw a 15 litre sample of the atmosphere through the holder at a rate of 3 to 5 litres a minute. Disconnect the holder, remove the filter paper and place the paper in a small beaker which is wide enough to hold the paper without folding. Add 2.5 ml of reagent 1 and set aside for 5 minutes.

During this 5 minutes prepare a 100 ml separating funnel by adding 15 ml of reagent 2 to the funnel followed by 5 ml of reagent 5. Stopper the funnel and shake vigorously for about 15 seconds. Allow the layers to separate and then run off, and discard, the lower layer. Dry the inside of the stem of the funnel with a spill of filter paper and then insert into the stem a rolled-up strip (20 × 80 mm) of Whatman No. 1 chromatographic paper.

Transfer the acid solution from the beaker to the funnel, wash the beaker with two 1 ml aliquots of water and add the washings to the solution in the funnel. Add 5 ml of reagent 3 to the mixture, stopper the funnel and shake vigorously for about 15 seconds. Allow the layers to separate and run the lower layer into a standard 13.5 mm comparator tube. Place this tube in the right-hand compartment of the comparator, place the comparator before a standard source of white light such as the Lovibond White Light Cabinet or, failing this, north daylight, and match the colour of the sample with the standards in the disc.

Notes

1. The suitability of each new batch of dithizone reagent should be checked before use in the following manner:—Prepare a 100 ml separating funnel as described above but instead of adding the acid solution of the sample add 2 ml of a dilute lead solution (3 µg of lead per ml of solution, which can be prepared by dissolving 0.192 g of lead nitrate ($Pb(NO_3)_2$) in 1 litre of $0.1N$ nitric acid and then diluting 5 ml of this solution to 200 ml with $0.1N$ nitric acid). Add 5 ml of reagent 3 to the mixture in the funnel, stopper the funnel and shake for about 15 seconds. Allow the layers to separate and then run the lower layer into a 13.5 mm comparator tube, place this in the right-hand compartment of the comparator and compare the colour with that of the 0.4 mg m^{-3} standard. Unless a good match in both colour and intensity is obtained reagents 4 and 5 should be rejected and fresh solutions prepared from a fresh batch of dithizone.

2. Suitable filter paper holders may be obtained from Tintometer Ltd., (Code No. AF401).

3. Suitable sources of suction for sampling industrial atmospheres are described in the Introduction to this Section.

4. Up to 3 µg of copper, 15 µg of antimony, 6 µg of cadmium, 50 µg of iron and 50 µg of zinc do not interfere with the test. Tin does interfere at the 15 µg level but experiments have shown[3] that tin trapped on the filter paper is not extracted in sufficient quantity to cause serious error, up to 60 µg of tin spotted on the filter paper having failed to interfere.

5. As less than 5 ml of the lower layer will be available, it is desirable to raise the tube in the comparator so that the liquid covers the viewing aperture. A pad of cotton wool is suitable, or a special clip may be obtained for the purpose from Tintometer Ltd.

References

1. B. E. Dixon and P. Metson, *Analyst* 1960, **85**, 122
2. Ministry of Labour, *"Methods for the detection of toxic substances in air, Booklet No. 14, Lead and compounds of lead"*, H.M. Stationery Office, London, 1962
3. D. M. Groffman and R. Wood, *Analyst,* 1971, **96**, 140
4. Department of Employment, *"Methods for the detection of toxic substances in air"* Booklet 14. *"Lead and Compounds of Lead"* H.M. Stationery Office, London, 1971

The Determination of Lead (Fume)
using tetrahydroxy-*p*-benzoquinone

Introduction

Lead and its compounds are highly poisonous and, as the poison is cumulative in the body, serious poisoning may result from prolonged ingestion or inhalation of even small amounts of these materials. The metal and its compounds are encountered in more than 200 industries, and from recorded cases of lead poisoning it appears that lead dust and lead fumes are the chief sources of danger[1-9]. Lead was the first industrial poison for which quantitative limits were established, and it has been suggested that 2 mg a day is the maximum amount of lead which can safely be ingested for long periods.

The present test has been developed by Dixon and Metson[10] as a field test for determining small amounts of lead fume in industrial atmospheres. It is based on the earlier work of Amdur and Silvermann[11]. After the determination of lead fume, another sample may be used for the determination of total lead by the method described in another test in this section of this book. Tests designed to check possible lead poisoning by the determination of lead in fæces and urine are described in another part of this volume, as also are tests for the determination of lead in waters and foodstuffs.

Principle of the method

Air is sampled through filter paper impregnated with tetrahydroxy-*p*-benzoquinone. On spraying with a mixture of acetic acid and acetone the particles of lead fume dissolve producing a uniform purple stain. The intensity of this stain is proportional to the lead concentration and is compared with Lovibond permanent glass standards. Particles of greater diameter than those in lead fume—which is usually defined as consisting of particles of 1μ diameter or less—produce isolated spots of more intense colour which do not interfere with the comparison of the general background stain with the standards.

Reagents required

1. *Tetrahydroxy-p-benzoquinone solution* 0.2% w/v solution in ethanol or methanol (See Note 1)
2. *Spraying solution* 1 volume lead-free glacial acetic acid mixed with 3 volumes acetone. Both chemicals must be of analytical reagent grade.

The Standard Lovibond Comparator Disc 6/17

This disc covers the range 0.1 to 0.6 mg (0.1, 0.2, 0.4, 0.6 mg) of lead fume (as Pb) per cubic metre of air, based on a 15 litre sample. The disc also contains a range of standards for the determination of total lead by the method of Dixon and Metson[13].

Technique

Cut Munktell No. 00 or acetic acid washed Whatman No. 2 filter paper into strips $2\frac{1}{2}$ inches by 1 inch. Dip the strips in the tetrahydroxy-*p*-benzoquinone solution to a depth of 1 inch, drain, and allow to dry, taking care not to allow the fingers to touch the impregnated portion of the paper. These test papers should be freshly prepared each day and must be protected from contact with alkaline vapours.

Place the impregnated portion of the paper in a suitable holder[12] and draw 15 litres of the suspect air through the paper by means of a suitable pump or aspirator (see Note 2). The flow rate through the paper should be between 3 and 5 litres per minute. Remove the test paper from the holder and spray the exposed side with the spraying solution by means of a suitable insufflator (see Note 3), spraying for 5 seconds at a distance of 6 inches. Allow the colour to develop for 1 minute. Place the developed paper on the right-hand side of the Lovibond Comparator Test Paper Viewing Stand with reducing aperture and place a piece of blank filter paper on the left-hand side. Compare the colour of the test paper with that of the permanent glass standards using a standard source of white light, such as north daylight or the Lovibond White Light Cabinet.

Notes

1. The extinction coefficient of an ethanolic solution of the tetrahydroxy-p-benzoquinone crystals, measured at 310 nm, should be not less than 15,000.

2. Suitable pumps may be obtained from laboratory equipment suppliers and should be capable of withdrawing a 15-litre sample against a pressure of 3.5 cm of mercury per litre per minute.

3. A suitable laboratory spray gun is available from Townson and Mercer Ltd., Beddington Lane, Croydon, Surrey.

4. Cadmium zinc and antimony fumes also form stains under the conditions of this test. Stains caused by cadmium and zinc may be removed by immersing the developed test paper in 10% v/v aqueous acetic acid for 30 seconds, washing with distilled water slightly acidified with acetic acid, and drying with absorbent paper. An antimony stain can be similarly removed together with any cadmium or zinc by substituting a 5% aqueous solution of sodium hydrogen tartrate for the dilute acetic acid. After removal of the stain of the interfering metal, the uncontaminated lead stain is compared with the standards in the normal manner.

5. The permission of the authors to quote from their paper[10] is gratefully acknowledged.

References

1. L. T. Fairhall, *"Industrial Toxicology"* 2nd edit., Williams & Wilkins Co., Baltimore, 1957, p.69
2. L. Dautrebande, H. Beckmann and W. Walkenhorst, *A. M. A. Arch. Ind. Health,* 1957, **16,** 179
3. N. I. Sax, *"Dangerous Properties of Industrial Materials,"* Chapman & Hall Ltd., London, 1957, p.817
4. P. E. Palm, J. M. McNerny and T. Hatch, *A. M. A. Arch. Ind. Health,* 1956, **13,** 355
5. T. Hatch, *ibid,* 1955, **11,** 212
6. M. Eisenbud, *A. M. A. Arch. Ind. Hyg.* 1952, **6,** 214
7. H. D. Landahl, T. N. Tracewell and W. H. Lassen, *ibid,* 1951, **3,** 359; 1952, **6,** 508
8. R. E. Lane, *Brit. J. Ind. Med.* 1949, **6,** 129
9. L. T. Fairhall and R. R. Sayers, *Public Health Bull. No.* 253, U.S. Public Health Service, 1940
10. B. E. Dixon and P. Metson, *Analyst,* 1959, **82** (1), 46
11. M. O. Amdur and L. Silverman, *A. M. A. Arch. Ind. Health,* 1954, **10,** 152
12. B. E. Dixon, G. C. Hands and A. F. Bartlett, *Analyst,* 1958, **83,** 199
13. B. E. Dixon and P. Metson, *Analyst,* 1960, **85,** 122

COLORIMETRIC CHEMICAL ANALYTICAL METHODS

The Determination of Mercury and Compounds of Mercury

using selenium sulphide

Introduction

Mercury is used in industry in the form of the metal and its compounds. The metal and organo-mercurial compounds are volatile at room temperature and inhalation of vapour is the main cause of poisoning. Inhalation of mercury or of mercury compounds in the form of dust and spray, and absorption through the skin or gastro-intestinal tract may also give rise to poisoning.

Metallic mercury is used mainly in the manufacture, maintenance and repair of electrical apparatus such as DC meters and mercury arc rectifiers, in the manufacture of thermometers, barometers and other scientific apparatus, and in electrolytic processes in chemical works. Other sources of exposure include the making of amalgams, the manufacture of mercury compounds, the use of mercury in laboratories, fire gilding and when mercury with chalk is used as a fingerprint powder.

Mercury and its oxides and salts are used in the manufacture of certain types of dry cell batteries. The breaking down of these batteries and other processes designed for the recovery of mercury are important sources of exposure.

Organo-mercurial compounds are used for seed dressing for the prevention of fungus-disease of plants, for the control of paper mill slimes and for antifouling paints.

Mercury fulminate is used in making certain types of detonators and percussion caps. The toxic effects are discussed in full in the official booklet "Mercury and compounds of mercury"[1].

Principle of the method

Two methods of sampling the contaminated atmosphere are described. The first is for the determination of all forms of mercury but is principally used when mercury vapour is the major hazard. The second, simpler, method of sampling is for use when organo-mercurial dusts or vapours are the main hazard and mercury vapour is not retained to any significant degree. In both cases, the mercury and mercury compounds are volatilised and reacted with selenium sulphide test papers to give a coloured stain.

Sampling

(a) *Total Mercury* The arrangement of the sampling tube is shown in figure 1. The inlet side consists of a piece of glass tubing (10.5 to 11 mm internal diameter) which contains 0.8 g of mineral wool filter. The inlet tube is attached by a short piece of rubber tubing to a glass bulb (25 mm external diameter) which is partly filled with granular calcium chloride (3 g of 8–14 mesh) held in place by cotton wool. The narrow outlet tube (5.5 to 6 mm internal diameter) holds the iodised activated carbon (0.35 g) which is retained by thimbles of fine steel gauze. This sampling tube is connected to a source of suction capable of drawing 2.5 litres of air per minute against a back pressure of 2 to 3 lbs/sq. in.

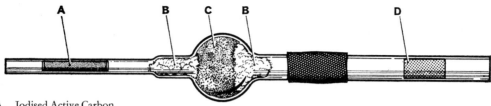

A. Iodised Active Carbon
B. Cotton Wool
C. Calcium Chloride
D. Mineral Wool Filter

Figure 1

The packing should be prepared as follows:—

Mineral wool—ignite a dust-free sample at 450–500°C until all carbonaceous matter has burned away.

Calcium chloride—anhydrous granular 8 to 14 mesh.

Iodised active carbon—sift a quantity of 18–30 mesh carbon to remove dust and weigh out 50 g of the sifted material (active carbon 270 C, 18 to 30 mesh, obtainable from Sutcliffe, Speakman and Co. Ltd., Leigh, Lancs., has been found to be a suitable base material). Add 1 g of iodoform powder, mix well, transfer to a covered porcelain basin and heat to 800°C in a furnace. Cool and store in a desicator.

To take the sample connect the narrow end of the sampling tube to the source of suction and draw a 50 litre sample through the tube in not less than twenty minutes. For sampling diethylmercury at least 125 litres should be sampled.

(b) *Organo-mercurial vapours and dusts* This method of sampling is applicable to ethylmercury chloride, diphenylmercury, methyl mercury dicyandiamide and ethylmercury phosphate but not to diethylmercury or mercury vapours, for both of which the previous method should be used.

This method uses a 7 cm diameter cadmium sulphide treated glass fibre pad, supported in a stainless steel filter paper holder (A suitable holder. is obtainable from B.S.A. Sintered Components Limited, Montgomery Street, Birmingham.) An airflow of 33.3 litres per minute is sucked through the pad by means of a pump and monitored by a flowmeter.

The cadmium pads are prepared by sliding a 7 cm Whatman GF/A glass fibre pad into a shallow dish, such as a petri dish, containing a 2% aqueous solution of cadmium acetate. After 2 minutes the pad is removed by means of forceps, blotted between two pieces of filter paper to remove excess liquid, turned over, then immersed in a second dish containing a 2% aqueous solution of sodium sulphide for 2 minutes. The excess solution is again blotted off with filter paper, the pad is once more turned over and it is slid into the original cadmium acetate solution. After 2 minutes it is removed, washed with water and dried at 100°C for one hour and stored in an air tight jar.

To take a sample place a cadmium sulphide pad in the holder, connect to the flowmeter and pump, and draw air through the pad at 33.3 litres per minute for 15 minutes i.e. sample 500 litres air (see Note 1).

Preparation of test papers

Dissolve 2 g of selenous acid in 100 ml of distilled water. Dip sheets of Whatman No. 1 filter paper in this solution, drain for a few minutes and then dry at 100°C. Trim 1 cm off the edges of the dried paper and store the remainder of the impregnated sheets in a stoppered bottle. Dissolve 1 g of thioacetamide in 20 ml of alcohol and store in a dropping bottle. Prepare each test paper immediately before use by adding two drops of the thioacetamide solution to a piece of the impregnated paper 2 cm square and allow to dry for a few minutes during which time an orange colour develops. Both solutions keep indefinitely if stored in well-stoppered bottles, but the papers will darken in time, making colour matching difficult.

The Standard Lovibond Comparator Disc 6/37

This disc[3], which replaced Disc 6/16[2], contains five standards corresponding to 0, 25, 50, 100, and 200 $\mu g/m^{-3}$ of mercury,

The Comparator should be used in conjunction with a Test Paper Viewing Stand. (DB418)

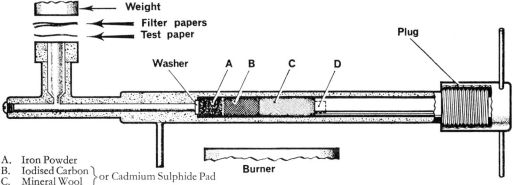

A. Iron Powder
B. Iodised Carbon ⎱ or Cadmium Sulphide Pad
C. Mineral Wool ⎰
D. Sodium Oxalate

Figure 2

Technique

Having sampled the mercury contaminant by the appropriate technique described above, remove the plug from the ignition tube (Figure 2 and Note 3) and charge the tube by introducing the following items in the order given.

1. 2 g of electrolytic iron powder 40–60 mesh
2. The iodised carbon from the sampling tube. This must be transferred quickly to avoid moisture being picked up from the atmosphere.
   The mineral wool filter from the sampling tube. This must be pressed well in.
   or
   The cadmium sulphide glass fibre pad from the airline filter.
3. Pack the cavity in the plug nearly to the brim with sodium oxalate.
4. Add a little synthetic graphite powder to the recess containing the washer, sufficient to maintain a sealing layer of powder, and screw the plug firmly into the tube.

Adjust an Amal Major laboratory Bunsen burner to consume about 7 litres of gas per minute, using a flowmeter. The central cone of the flame should be reduced to lie flat on the grid of the burner.

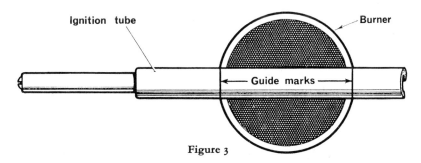

Figure 3

Lay the ignition tube on the burner rests and align by means of the guide marks (see Figure 3).

Without delay place a piece of freshly prepared test paper on the head of the ignition tube followed by two squares of Whatman No. 1 filter paper and finally the brass weight. Heat for eight minutes. Remove the test paper and place in the right-hand side of the Test Paper Viewing Stand fitted with the Lovibond Comparator and compare the colour of the stain with the standards in the disc.

Notes

1. If a 500 litre sample was taken (i.e. method for organo mercurials) divide the result obtained from the disc by 10 to obtain micrograms of organo mercurial, calculated as mercury, per cubic metre of air.

2. Before each test the ignition tube should be heated sufficiently to remove traces of condensed moisture. A blank test should be carried out with a charge of iron powder and sodium oxalate only, to ensure that the ignition tube and materials are free from contamination by mercury. A blank stain equal to or lighter than the zero standard is satisfactory.

References

1. *"Methods for the Determination of Toxic Substances in Air"* Booklet No. 13 *"Mercury and Compounds of Mercury"*. H.M. Stationery Office, London, 1968
2. G. A. Sergeant, B. E. Dixon and R. G. Lidzey, *Analyst* 1957 **82,** 27
3. A. A. Christie, A. J. Dunsdon, and B. S. Marshall, *Analyst* 1967, **92,** 185

The Determination of Methyl α-chloro-acrylate
using potassium permanganate

Introduction

The determination of methyl α-chloro-acrylate in the atmosphere is important in the control of working conditions in the manufacture of poly-methyl α-chloro-acrylate polymer.

Principle of the method

Methyl α-chloro-acrylate reacts with dilute aqueous solutions of potassium permanganate, producing a colour which ranges from red through orange-red to orange-yellow as the concentration of methyl α-chloro-acrylate increases.

Reagent required

0.001 N potassium permanganate solution

The Standard Lovibond Comparator Disc APMC

The disc covers the range 0–100 μg (0–0.1 mg) of methyl α-chloro-acrylate in the following steps: 0, .01, .02, .03, .04, .05, .06, .08, .1 mg.

Technique

The methyl α-chloro-acrylate is absorbed in 'jet' bubblers[1] containing 10 ml of 0.001 N potassium permanganate solution. Two of these bubblers are used in series, the purpose of the second being to trap any acrylate which slips through the first bubbler. The dimensions of these bubblers are:—

Length 260 mm, internal diameter 22 mm, tube diameter 6.5 mm drawn out to 1–1.5 mm at the tip.

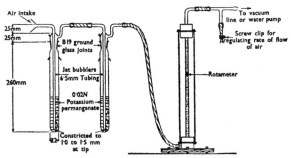

Air is drawn through the bubblers at 1.5 litres a minute until the colour of the permanganate solution in the first bubbler appears to be suitable for colour comparison. The airflow is then discontinued and the duration of sampling is recorded.

The contents of the first bubbler are drained into a 13.5 mm cell which is placed in the right-hand compartment of the comparator, and a similar cell, filled with distilled water, is placed in the left-hand compartment. The colour of the sample is then matched with the permanent glass standards in the disc, using a standard source of white light, such as the Lovibond White Light Cabinet or, failing this, north daylight. The reading in the indicator window gives the amount, in mg, of methyl α-chloro-acrylate contained in the solution.

The contents of the second bubbler are treated in the same manner and the amount of acrylate found is added to that from the first bubbler. The concentration of methyl α-chloro-acrylate in the air is calculated as follows:—

$$\text{Concentration (mg/m}^3) = \frac{\text{Total reading in mg}}{\text{Sampling time in min.}} \times \frac{1000}{1.5}$$

This concentration is converted to parts per million v/v at 20°C and 760 mm pressure by dividing by 5.

Notes

1. Other acrylates, such as methyl methacrylate interfere with this method, but are unlikely to be present in the same atmosphere, particularly in a plant manufacturing polymethyl α-chloro-acrylate.

2. This method was developed by the staff of I.C.I. (Plastics) Ltd.

Reference

1. J. Haslam, S. M. A. Whettem and W. W. Soppet, *Analyst,* 1951, **76,** 628

The Determination of Nitrobenzene
using disodium 2-naphthol-3:6-disulphonic acid (R salt)

Introduction

Nitrobenzene is widely used in perfumery and flavouring, in the preparation of aniline and otherwise in the dye industry. It is one of the most toxic vapours encountered in industry, the maximum permissible concentration in working atmospheres being 1 part per million (approx. 5 mg per cubic metre of air).[1]

In order to control atmospheres to this limit a sensitive method of determining the nitrobenzene concentration is required, and methods depending on the further nitration of the compound followed by colorimetric determination of the resulting dinitrobenzene[2], or on reduction to aniline followed by diazotisation and coupling[3,4], have been recommended for this purpose. None of these methods is completely satisfactory as a field test however, as all require heat or other laboratory facilities.

The present method[5] has been developed to overcome these drawbacks and to provide a test of adequate sensitivity which can be carried out in the field.

Principle of the method

The nitrobenzene vapour is trapped in Cellosolve (2-ethoxyethanol) and is subsequently reduced to aniline by shaking with liquid zinc amalgam[6] and acid, at ambient temperature. The aniline thus produced is diazotised with sodium nitrite and then coupled with disodium 2-naphthol-3:6 disulphonate (R salt). The resulting colour, the intensity of which is proportional to the concentration of nitrobenzene in the original solution, is compared with Lovibond permanent glass standards.

Reagents required

All reagents should be of analytical reagent grade.
1. *"Cellosolve"* (2-ethoxyethanol)
2. *Hydrochloric acid* Dilute 5 ml of concentrated hydrochloric acid (HCl) to 100 ml with water.
3. *Sodium nitrite solution* Dissolve 3.5 g of sodium nitrite ($NaNO_2$) in 100 ml of water. This solution should not be kept for longer than 1 month.
4. *Sodium carbonate solution* Dissolve 10 g of anhydrous sodium carbonate (Na_2CO_3) in 100 ml of water.
5. *R salt solution* Dissolve 0.8 g of purified R salt* in 100 ml of boiling water, adjust the *p*H to about 8 (between 7.5 and 8.5) by adding sodium carbonate (reagent 4), cool to room temperature, and filter. Store in darkness. This solution should not be kept for longer than one month.
6. *Ammonia solution* Dilute 20 ml of ammonia solution ((NH_4OH) sp. gr. 0.880) to 100 ml with water.
7. *Liquid zinc amalgam* Add 6 g of powdered zinc to 300 g of mercury and 5 ml of 20% v/v sulphuric acid (H_2SO_4). Stir, set aside for 2 hours, transfer to a separating funnel, and wash three times with dilute sulphuric acid. This reagent should be kept under a layer of dilute sulphuric acid and can be conveniently stored in, and dispensed from, a small polythene wash-bottle.

*For full chemical name, see page heading.

The Standard Lovibond Comparator Disc 6/22

This disc has 4 standards corresponding to 0, 0.5, 1, 2 ppm of nitrobenzene in air, based on a 6 litre air sample.

Technique

Place 2 ml of Cellosolve (reagent 1) in a clean glass bubbler. This bubbler (Fig. 1) should be of all glass construction and of the dimensions shown. Connect the bubbler, via a trap (Figure 2) to a suitable source of suction (Note 1). Take a 6 litre sample of the atmosphere to be tested at a flow rate of 1.5 litres per minute (Note 2). Detach the bubbler and add to its contents 1 ml of liquid zinc amalgam (reagent 7) and then 4 ml of hydrochloric acid solution (reagent 2). Close the bubbler with a stopper and shake vigorously for 1 minute. By means of a 5 ml teat pipette transfer 5 ml of the aqueous layer to a clean, dry, comparator tube or 13.5 mm cell, taking care not to transfer any of the zinc amalgam. Add 0.5 ml of sodium nitrite solution (reagent 3), shake gently to mix, and allow to stand for 2 minutes. Add 2 ml of sodium carbonate solution (reagent 4), followed immediately by 0.5 ml of R salt solution (reagent 5). Shake until mixed, add 2 ml of ammonia solution (reagent 6), shake again and then place the tube in the right-hand compartment of the comparator. Place an identical tube filled with distilled water in the left-hand compartment and compare the colour of the sample with the standards in the disc, using a standard source of white light such as the Lovibond White Light Cabinet or, failing this, north daylight.

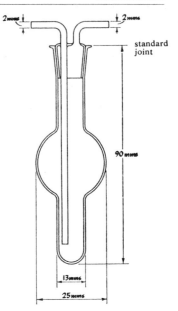

Figure 1 Gas absorption bubbler

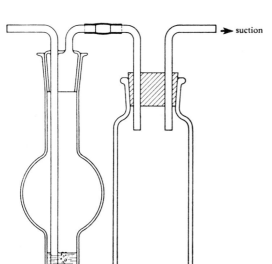

Figure 2 Absorption train

Notes

1. Several alternative methods of taking air samples at standard flow-rates are described in the Introduction to this Section.
2. The sampling efficiency of the bubbler is about 85% at a flow-rate of 1.5 litres per minute. The standards have been calibrated to allow for this efficiency and the readings obtained are the concentration of nitrobenzene in the atmosphere as tested.

References

1. Ministry of Labour, *"Toxic Substances in Factory Atmospheres"*, H.M.S.O., London, 1960
2. M. B. Jacobs, *"Analytical Chemistry of Industrial Poisons, Hazards and Solvents,"* 2nd editn., Interscience Publishers Inc., London and New York, 1949
3. N. Strafford and D. A. Harper, *J. Soc. Chem. Ind.,* 1939, **58**, 169T
4. Internat. Union of Pure and Applied Chem., *"Methods for the Determination of Toxic Substances in Air"*, Butterworth, London, 1959
5. G. C. Hands, *Analyst,* 1960, **85**, 843
6. B. Gordon, *Neft. Khoz.,* 1938, **19**, 52; *Chem. Abst.,* 1939, **33**, 6576

The Determination of Nitrogen Dioxide and Nitric Oxide

using N-(1-naphthyl)-ethylenediamine dihydrochloride (Saltzman Method)

Introduction

Nitrogen dioxide is the most toxic of the various oxides of nitrogen. The maximum permitted concentration in industrial atmospheres is 5 ppm[1]. Oxides of nitrogen, containing nitrogen dioxide, may be released during the use of explosives, in chemical processes involving nitration or the use of nitric acid, in welding and in the exhaust gases from internal combustion engines. The present test[2,3,4] was devised to give a simple means of determining both nitric oxide and nitrogen dioxide concentrations in polluted atmospheres.

Principle of the method

The atmosphere to be tested is bubbled, at a controlled low flow rate, through a reagent of sulphanilic acid and N-(1-naphthyl)-ethylenediamine dihydrochloride in acetic acid. If nitrogen dioxide is present in the air a stable red-violet colour is formed. The intensity of this colour, which is proportional to the nitrogen dioxide concentration, is measured by comparison with a series of Lovibond permanent glass colour standards.

Nitric oxide is determined after prior oxidation to nitrogen dioxide by means of acid potassium permanganate solution[5].

Reagents required

1. *0.1% N-(1-naphthyl)-ethylenediamine dihydrochloride solution* Dissolve 0.1 g of N-(1-naphthyl)-ethylenediamine dihydrochloride in 100 ml of distilled water.
2. *Absorbing reagent* Dissolve 5 g of sulphanilic acid in 800 ml of water and 140 ml of glacial acetic acid. The dilute acetic acid may be warmed gently if required to speed up the dissolution of the sulphanilic acid. To the cooled mixture add 20 ml of reagent 1 and dilute to 1 litre.
3. *Acid permanganate solution* Dissolve 2.5 g of potassium permanganate in about 90 ml of water, add 2.5 g of concentrated sulphuric acid (or 5.2 ml of 1:3 H_2SO_4) and dilute to 100 ml with distilled water. This reagent should be discarded when an appreciable precipitate of brown manganese dioxide is noticed.

All chemicals used in the preparation of these reagents should be of analytical reagent quality.

Apparatus required

The special bubbler illustrated below must be used.

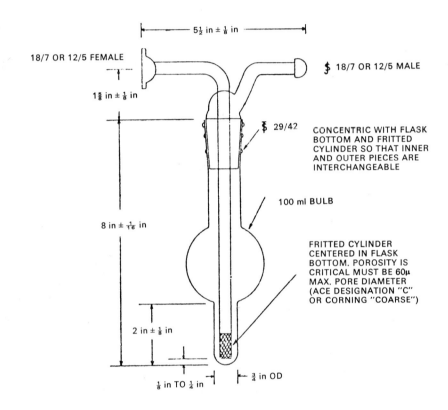

Figure 1 Fritted bubbler for sampling nitrogen dioxide

The Standard Lovibond Comparator Disc 6/31

This disc contains nine colour standards corresponding to 0.5, 1.0, 1.5, 2.0, 3.0, 4.0, 5.0, 6.0, and 7.0 µg of nitrogen dioxide (NO_2) per 10 ml of absorbing reagent.

Technique

a) Determination of nitrogen dioxide Pipette 10 ml of absorbing reagent (reagent 2) into the fritted bubbler (see Figure). Draw an air sample through the bubbler at a rate not exceeding 400 ml per minute (Notes 1 & 2), until sufficient colour has developed in the bubbler. Note the total volume of air sampled. Transfer the contents of the bubbler to a 13.5 mm comparator cell or standard test tube and place this tube in the right-hand compartment of the comparator. Fill an identical cell, or tube, with fresh absorbing reagent (reagent 2) and place this in the left-hand compartment of the comparator. Allow 15 minutes for complete colour development and then compare the colour of the sample with the colours in the disc using a standard source of white light such as the Lovibond White Light Cabinet, or failing this, north daylight. Read off the corresponding weight of NO_2 in the bubbler from the indicator window when a colour match is achieved. Calculate the NO_2 concentration by means of the following formula:—

$$\text{ppm } NO_2 \text{ (at } 25°C \text{ and } 760 \text{ mm Hg)} = \frac{\text{disc reading} \times 530}{\text{volume of air sample in ml}}$$

volume of air sample = rate of sampling in ml per minute × the time of sampling in minutes

b) Determination of nitric oxide (NO) Assemble an absorbing train consisting of, in order, a fritted absorber, an acid permanganate bubbler and a second fritted absorber. Pipette 10 ml of absorbing reagent (reagent 2) into each end of the fritted absorbers and pipette 10 ml of acid permanganate solution (reagent 3) into the central bubbler. Proceed exactly as for the determination of NO_2. Measure the colour of the liquid in the third bubbler. If a simultaneous determination of NO_2 is required it may be obtained by measuring the solution from the first bubbler also. Calculate the NO concentration by the use of the following formula (Note 3):—

$$\text{ppm } NO \text{ (at } 25°C \text{ and } 760 \text{ mm Hg)} = \frac{\text{disc reading (bubbler 3)} \times 350}{\text{volume of air sampled in ml.}}$$

Notes

1. Suitable methods for air sampling are described in pages 462 to 464.
2. The sampling efficiency of the fritted bubbler at flow rates not exceeding 400 ml per minute is at least 95%.
3. The conversion efficiency of NO to NO_2 by the permanganate bubbler may be as low as 70%. If more accurate results for the NO concentration are required then the method of Ripley et al[6] should be used for the oxidation. This is claimed to give conversion efficiencies of 95–100%. Instead of the permanganate bubbler use a 17 mm O.D. glass U tube, containing one 7 cm dia. sheet of impregnated glass-fibre paper cut into $\frac{1}{4}''$ strips, and a flow rate of 290 ml per minute. As the paper deteriorates if it is used downstream from a bubbler, discard the first fritted absorber, and use the paper packed tube followed by a single fritted absorber. The colour developed in this absorber then corresponds to the total of NO_2+NO and a separate absorber must be used to measure the NO_2 concentration. If exactly the same volume of air is sampled through each absorbing train then the disc reading for NO_2 can be subtracted from that for $NO+NO_2$ and the NO formula used to calculate the NO concentration from the difference in readings.

Prepare the impregnated glass-fibre paper as follows:—impregnate a stack of 25, 7 cm dia. discs of the paper with a solution containing 2.5% of sodium dichromate and 2.5% sulphuric acid. Dry the sheets in a vacuum oven at 70°C, or on a hot-plate at 95°C. Discard the top and bottom sheets and store in a dessicator.

4. A fivefold ratio of ozone to NO_2 causes little interference. A tenfold ratio of sulphur dioxide to NO_2 has no effect. Interference from other nitrogen oxides and from other gases found in polluted air is negligible.

References

1. Min. of Labour, *"Toxic Substances in Factory Atmosphere"*, H.M.S.O., London, 1960
2. B. E. Saltzman, *Anal. Chem*, 1954, **26**, 1949
3. U.S. Dept. Health Educ. and Welfare, *"Selected Methods for Measurement of Air Pollutants"*, 1965
4. S. Hochheiser & G. A. Rogers, *Environmental Science and Technology*, 1967, Vol. I, 75
5. M. D. Thomas, J. A. MacLeod, R. C. Robbins, R. C. Goettelman, R. W. Eldridge and L. H. Rogers, *Anal. Chem.*, 1956, **28**, 1810
6. D. L. Ripley, J. M. Clingenpeel and R. W. Hurn, *J. Air and Water Poll. (London)*, 1964, **8**, 455

The Determination of Nitrous Fumes
using 1, alkaline arsenite sulphanilic acid and naphthyl ethylene diamine dihydrochloride
or, 2, *p*-anisidine

Introduction

Nitrous fumes are usually reddish-brown in colour and consist of a mixture of nitrogen dioxide (NO_2) and dinitrogen tetroxide (N_2O_4) of variable composition (usually referred to as nitrogen peroxide), although traces of nitric oxide (NO) may occur under certain conditions. They are evolved when nitric acid acts on metals and organic material, during the burning of nitrated materials such as celluloid or as the result of the detonation of certain explosives, such as dynamite. They are encountered in concentrations which may be dangerous in many important industries, for example in the manufacture and use of nitric acid and nitrates, the manufacture of sulphuric acid by the chamber process and where nitrations are carried out.

They are also encountered where nitric acid is used in electro-plating, engraving, metal cleaning and photogravure processes. They are formed during gas welding, particularly when the flame plays on cold steel in a confined space as may occur in shipbuilding and repair works.

The toxic effects are dealt with in the official booklet[1]. The threshold limit value is 5 ppm v/v for an eight hour day.

Principle of Method 1 (liquid bubbler method)

Nitrogen dioxide is collected in an alkaline arsenite solution of sulphanilic acid, an acid solution of N-(1-naphthyl)-ethylene-diamine dihydrochloride (see Note 1) is added, and any colour produced is compared with Lovibond permanent glass colour standards.

Apparatus required

1. *Bubbler* of type shown in Figure. Dimensions: Receiver length 155 mm, external diameter 15 mm internal diameter 12.8 mm. Distance from bottom of side arm to bottom of bubbler 110 mm.
    Inlet tube Overall length 185 mm, external diameter 6 mm, internal diameter 4.5 mm. Distance from bottom of inlet tube to bottom of bubbler 7 mm. Cone and socket B10.
2. *Aspirator* Rubber hand-bulb aspirator of capacity 120 ml as shown in Figure.

Reagents required

1. *Absorbing solution* Dissolve 1 g of sodium hydroxide, 0.1 g sodium arsenite (laboratory grade) and 0.75 g sulphanilic acid in water and dilute to 100 ml. Prepare fresh solution each week.
2. *Colour forming solution* Dissolve 0.02 g of N-(1-naphthyl)-ethylene-diamine dihydrochloride (laboratory grade—Note 1) and 6.0 g oxalic acid in water and dilute to 100 ml. Prepare fresh solution each week.

All chemicals used should be of analytical reagent quality unless otherwise stated.

The Standard Lovibond Comparator Disc 6/33

This disc contains three standards corresponding to 2.5, 5 and 10 parts per million v/v of nitrogen dioxide in a 120 ml sample.

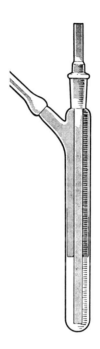

Technique

Place 5 ml of the absorbing solution in a bubbler and connect the source of suction to the side arm, (see Note 1). Transfer the apparatus to the atmosphere to be tested and collect 120 ml in one minute. Disconnect the aspirator, add 3 ml of colour forming solution, shake thoroughly, and allow the colour to develop for at least five minutes. Remove the connection from the side arm, take out the inlet tube and place the bubbler in the right-hand compartment of the comparator. Place the comparator before a standard source of white light, such as the Lovibond White Light Cabinet, or failing this, north daylight and match the colour in the bubbler against the standards in the disc.

Notes

1. Suitable sources of suction are described in the Introduction to this Section.
2. N-(1-naphthyl)—ethylenediamine dihydrochloride should be handled with care. Inhalation of the dust and skin contamination should be avoided.
3. Hydrochloric acid at 5 ppm does not interfere. Nitric oxide produces a slight colour at 5 ppm which is indistinguishable in hue from the colour produced by the dioxide, but is very much less intense.

Sulphur dioxide interferes but is removed by passing the atmosphere through a tube containing cotton wool impregnated with lead acetate.

Sulphur trioxide at 3000 ppm slightly enhances the colour produced by nitrogen dioxide at 5 ppm. Ozone at 20 ppm in the absence of nitrogen dioxide gives a slight colour equivalent to 1 ppm of nitrogen dioxide. Ozone at 20 ppm in the presence of 5 ppm of nitrogen dioxide enhances the colour by 40 per cent.

Principle of Method 2 (test paper method alternative)

Air is drawn through a test paper treated with p-anisidine and any brown colour produced is compared with Lovibond permanent glass colour standards.

Preparation of test papers Dissolve 5.0 g of p-anisidine in about 70 ml of absolute alcohol. Add 12 ml of glycerol, dilute to 100 ml with absolute alcohol and mix well. This solution should be freshly prepared each time a batch of papers is made and should be no more than pale yellow in colour. Transfer the solution to a 100 ml measuring cylinder. Immerse strips of Whatman chromatography paper No. 3MM, 12 cm long and 2 cm wide, vertically in the solution for 3 minutes. Remove the paper, suspend freely in air away from chemical fumes and allow to dry for 15 minutes. Discard the top and bottom 3 cm of the treated papers and store the remainder in stoppered wide-necked dark glass bottles. Renew the papers after 2 months.

The Standard Lovibond Comparator Disc 6/39

This disc contains four standards corresponding to 1.3, 2.5, 5, and 10 parts million per v/v of nitrogen dioxide in a 240 ml sample, and must be used with a Lovibond 1000 Comparator and Test Paper Viewing Stand.

Technique

Place a strip of impregnated test paper in the paper holder, attach the source of suction and draw 240 ml of air through the paper in two minutes. Remove the paper from the holder and place it in the right hand compartment of the Test Paper Viewing Stand. Place the Viewing Stand before a standard source of white light such as the Lovibond White Light Cabinet, or failing this, north daylight and compare the colour of the stain with the standard colours in the disc.

Notes

Hydrogen chloride at 5 ppm does not interfere. Nitric oxide does not interfere. Sulphur dioxide does interfere but is removed by passing the atmosphere through a tube containing cotton wool impregnated with lead acetate. Sulphur trioxide at 30 ppm in a moist atmosphere produces no interference. Ozone at 0.1 ppm (the threshold limit value) has little effect either in the presence or absence of nitrogen dioxide.

References

1. H.M. Factory Inspectorate. *"Methods for the detection of toxic substances in air"*. Booklet 5 *Nitrous Fumes*. H.M. Stationery Office, London, 1969
2. H. A. Christie, R. G. Lidsey, and D. W. Radford. *Analyst*, 1970, **95,** 519

The Determination of Atmospheric Oxidants
using phenolphthalin

Introduction

The role of oxidising compounds in the atmosphere and their importance in atmospheric pollution has been recognised in recent years,[1,2,3,4] especially in the United States. The photochemical oxidation of organic material, in the presence of oxides of nitrogen, which was first noted in Los Angeles smog[1], is stated to result in eye-irritating haze, objectionable odours and plant damage. These effects are due to the presence in the air of a number of compounds, the most important of which are:—

(a) Ozone
(b) Nitrogen compounds such as nitrogen dioxide, alkyl nitrites and peroxy alkyl nitrites
(c) Peroxides formed in the oxidation of organic material.

The peroxides are claimed[5] to be the compounds chiefly responsible for the physiologically irritating action of smog.

This test was developed by Haagen-Smit and Brunelle[5], from a test originally suggested by Schales[6] for the determination of hydrogen peroxide in biological preparations, to determine the concentration of oxidising substances in the atmosphere. It has been applied by these authors to measure the concentration of oxidising material in Los Angeles air over a period of five years using a continuous sampling device.

As a result of this extensive investigation it has been shown that when the total oxidant reached a level of 0.15 ppm, calculated as hydrogen peroxide, smog odour was apparent; at a level of 0.25 ppm eye irritation was generally experienced and at 0.5 ppm the smog was generally evaluated as severe.

Principle of the method

The method is based on the oxidation of phenolphthalin to phenolphthalein giving an intense reddish-purple colour. The intensity of this colour is measured by comparison with Lovibond permanent glass colour standards.

Reagents required

1. *Phenolphthalin* Take 1 g of phenolphthalein, 10 g of sodium hydroxide (NaOH) 5 g of zinc dust and 20 ml of distilled water. Warm, in a flask equipped with a reflux condenser, on a water bath for approximately 2 hours, until colourless, filter through a sintered glass filter and dilute with 50 ml of distilled water. Store over granulated zinc in a closed bottle in the dark preferably in a refrigerator.
2. *Copper sulphate 0.01M in distilled water* Dissolve 2.497 g of $CuSO_4.5H_2O$ in 1000 ml of distilled water.

The Standard Lovibond Comparator Disc 6/21

The disc covers (in steps of 0.05) the range 0.05 to 0.50 ppm of oxidants, calculated as hydrogen peroxide, in 10 litres of air.

Technique

Dilute 1 ml of the concentrated phenolphthalin solution (reagent 1) with 3 ml of distilled water. To 100 ml of distilled water add 1 ml of the diluted phenolphthalin solution and 0.5 ml of the copper sulphate solution (reagent 2), and mix thoroughly. This is the solution used for the test.

Place 10 ml of this solution in a bubbler and bubble air through the solution at 1 litre per minute for 10 minutes. Transfer the solution to the comparator tube and place in the right-hand compartment of the comparator. Fill an identical comparator tube with unreacted reagent and place it in the left-hand compartment of the comparator. Allow to stand for 15 minutes to ensure that colour development is complete. At the end of this time compare the colour of the sample with those of the permanent glass standards, using a standard source of white light, such as the Lovibond White Light Cabinet, or failing this, north daylight.

Notes

1. This disc was prepared at the request of the Bay Area Pollution Control District, San Francisco, California, U.S.A.
2. Glass to glass joints should be used on the inlet side of the bubbler, as rubber and other connecting tubes have been shown to lead to erroneous results[5].
3. If a flowmeter is used in the sampling system this must be connected on the **outlet** side of the bubbler.
4. Atmospheric oxygen does not interfere with this test.

References

1. A. J. Haagen-Smit, *Ind. Eng. Chem.*, 1952, **44**, 1342
2. E. R. Stephens, W. E. Scott, P. L. Hanst and R. C. Doerr, *"Recent Development in the Study of Organic Chemistry of the Atmosphere,"* paper presented at the 21st Mid-Year Meeting of the American Petroleum Institute, Montreal, May, 1956
3. L. H. Rogers, *J. Chem. Ed.*, 1958, **35**, 310
4. W. L. Faith, N. A. Renzetti and L. H. Rogers, *Fifth Technical Progress Report*, Report No. 27, Air Pollution Foundation, San Marino, California, U.S.A., March, 1959
5. A. J. Haagen-Smit and Margaret F. Brunelle, *Internat. J. Air. Pollution*, 1958 **1**, 51
6. O. Schales, *Ber. dtsch. chem. Ges.*, 1938, **71b**, 447

The Determination of Ozone
using potassium iodide

Introduction

Ozone is one of the most toxic gases which may be present in industrial atmospheres. In addition to its direct uses in chemical synthesis, in the sterilization of water, and as a bleaching agent for textiles, it also occurs in the vicinity of high voltage electrical discharges, ultra violet lamps and electric arc-welding. In the latter cases it usually occurs together with nitrogen dioxide, another toxic gas.

The maximum permissible concentration (MPC) of ozone in an industrial atmosphere[1] is 0.1 ppm (v/v) whereas the M.P.C. for nitrogen dioxide is 5 ppm. It is therefore necessary to be able to measure low concentrations of ozone in the presence of nitrogen dioxide and the present test[2,3] has been developed to meet this requirement.

Principle of the method

Ozone reacts with a solution of potassium iodide to liberate iodine. Nitrogen dioxide also liberates iodine from potassium iodide. Two samples are therefore drawn, in parallel, through buffered potassium iodide solutions at 2 litres a minute for 20 minutes. One sample, prior to its passage through the iodide solution, is drawn through a plug of cotton wool. The ozone is completely removed by this plug without any loss of nitrogen dioxide. The iodine liberated in the sample taken through the plug is thus a measure of the nitrogen dioxide concentration while that liberated in the other sample is a measure of ozone plus nitrogen dioxide. The concentration of iodine is measured by the addition of starch solution. The intensity of the colour of the blue starch iodine complex is measured by comparison with a series of Lovibond glass permanent colour standards. The difference between the colours in the two samples is a measure of the ozone concentration and the colour standards are calibrated in ozone equivalents.

Reagents required

1. *Neutral, buffered potassium iodide (KI) solution* Dissolve 20g of potassium iodide, 14.2 g of disodium hydrogen phosphate (Na_2HPO_4) and 13.6 g of potassium dihydrogen phosphate (KH_2PO_4) in water and dilute to 1 litre.
2. *Starch solution* Dissolve 0.25 g of soluble starch in about 70 ml of boiling water, cool and dilute to 100 ml. This reagent must be prepared on the day on which it is to be used.
3. *Cotton wool, absorbent, B.P.C.* Wash successively with tap water and then distilled or de-ionised water and dry at 105°C. Store in air-tight glass bottles.

All chemicals used for the preparation of reagents should be of analytical reagent quality.

The Standard Lovibond Comparator Disc 6/34

This disc contains nine standards corresponding to 0.05, 0.075, 0.1, 0.15, 0.2, 0.25, 0.3, 0.35 and 0.4 ppm of ozone (O_3), based on a 40 litre sample of air (Note 1).

Technique

The apparatus required is illustrated in the Figure (Note 2).

Pack 1 g of the prepared dry cotton wool (reagent 3) into the glass tube to form a plug about 7 cm long. Pipette 10 ml of potassium iodide solution (reagent 1) into each absorber. Before entering the suspect atmosphere assemble the appratus as shown in the figure (Notes 3 and 4). Draw the test atmosphere from a single sampling point through each absorber at a rate of 2 litres a minute for 20 minutes (Note 5). Return to an uncontaminated atmosphere before continuing the test. Remove the absorbers from the assembly. Remove the absorber heads, taking care to allow any liquid in the inlet tubes to drain into the bulk of the solution. Pipette 5 ml of starch solution (reagent 2) into each absorber. Mix well and transfer the contents of the absorber which was not preceded by the cotton wool plug into a comparator test tube. Place this tube in the right-hand compartment of the comparator. Place the comparator before a uniform source of white light such as the Lovibond White Light Cabinet or failing this, north daylight and measure the colour against the standards in the disc. Record

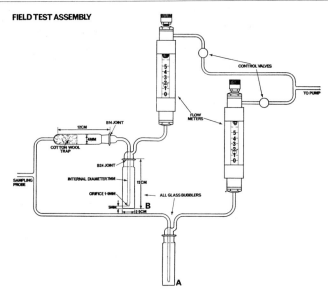

Field test assembly for the determination of ozone in the presence of nitrogen dioxide.
Reproduced by permission of the Society for Analytical Chemistry.

the value, as shown in the indicator window, of the matching standard. Repeat the measurement with the solution from the absorber which was preceded by the wool plug and record the value of the matching standard. The difference between the two recorded values is the ozone concentration. It is important that both readings are completed within 15 minutes of adding starch solution to the absorbers.

Notes

1. A single absorber of the type specified, has been shown[2] to trap 87% of the ozone over a range of 0–0.6 ppm of ozone in the air. The callibration has been adjusted to allow for this incomplete absorption. It is important therefore that the same type of absorber is used for field tests, otherwise the accuracy of the calibration cannot be guaranteed.

2. The all-glass absorbers as shown in the figure, may be obtained from:—Glass of Mark Ltd., Jubilee Street, Northwich, Cheshire, England. Suitable flowmeters are obtainable from Rotameter Manufacturing Company Ltd., Croydon, or from G. A. Platon Ltd., 281 Davidson Road, Croydon, Surrey, England.

3. No grease should be used in any of the joints of this apparatus. The preferred method of jointing components is ungreased glass ball-and-socket joints. Alternatively make sure that all glass connections are butted together and held by PVC, **not rubber,** tubing. If a flexible connection is required between the inlet and the contaminated atmosphere, then this should be of nylon tubing 5 mm internal diameter. **It is important that a single sampling point is used for both absorbers.**

4. The flowmeters should be calibrated, in the laboratory, before use in the field to indicate a flow of 2 litres a minute in the assembled apparatus. This can be done with either an additional flowmeter having one end open to the atmosphere or with a soap-bubble flowmeter. When used in the assembled apparatus these flowmeters are working under reduced pressure and this renders the normal calibration invalid.

5. Suitable methods of taking gas samples of known volume are described on pages 462 to 464.

References

1. Ministry of Labour, *"Safety, Health and Welfare Booklets, New Series No. 8—Toxic Substances in Factory Atmospheres"* H.M.S.O., London, 1960
2. I. C. Cohen, A. F. Smith and R. Wood, *Analyst,* 1968, **93,** 507
3. H. M. Factory Inspectorate *"Methods for the detection of toxic substances in air"* Booklet No. 18, *Ozone.* H.M. Stationery Office, London, 1969

The Determination of Phosgene
using 4-*p*-nitrobenzylpyridine and N-benzylaniline

Introduction

The extensive use of phosgene, in the pharmaceutical, organic chemical and dye-stuffs industries, coupled with its high toxicity has led to the development of this test[8] for its detection and estimation in industrial atmospheres. Although phosgene may be produced by the decomposition of chlorinated hydrocarbons, hazards arising from this source are not normally appreciable, with the possible exception of phosgene produced by the pyrolysis of carbon tetrachloride in fire-fighting with some types of fire extinguishers. The main hazard undoubtedly arises where phosgene itself is being used as part of a manufacturing process.

The chief danger of phosgene as an industrial poison is its high toxicity coupled with its insidious nature. Atmospheres containing lethal concentrations of phosgene may be relatively non-irritant. The maximum permissible concentration for prolonged exposure is laid down[1] as 0.1 part per million. Previous tests[2,3,4] have proved either too insensitive or to have other drawbacks, and the present test was developed by Dixon and Hands[5], on the basis of earlier work by Brown, Wilzbach and Ballweber[6], in order to overcome these drawbacks in the tests previously used.

Principle of the method

Phosgene reacts with 4-*p*-nitrobenzylpyridine in the presence of N-benzylaniline to produce a red colour. The intensity of this red colour is compared with Lovibond permanent glass standards.

Reagent required

Solution containing 2% w/v of 4-*p*-nitrobenzylpyridine and 4% w/v of N-benzylaniline in benzene.

Preparation of test papers

Immerse strips of Whatman No. 1 paper, chromatographic grade, 2 cm wide, in the above solution for about 10 seconds. Drain and allow to air-dry vertically in a fume cupboard free from other chemical fumes. Discard the top and bottom 2 cm of the paper strip. Store the papers in dark, stoppered glass bottles. Renew after 2 months.

The Standard Lovibond Comparator Disc 6/36

The disc contains 4 standards corresponding to 0.05, 0.1, 0.2 and 0.4 ppm of phosgene when a 600 ml sample of air is taken. This disc is designed to be used with a Lovibond Comparator and Test Paper Viewing Stand (DB418).

Technique

Place a piece of the prepared test paper in a holder which exposes a circle of paper 1 cm in diameter[7] (see Note 1) and draw 600 ml of air through the paper by means of a suitable pump or aspirator. The air flow should not exceed 2 ml per second. Transfer the paper to the right-hand side of the Lovibond Comparator Test Paper Viewing Stand and place a piece of unexposed test paper on the left-hand side. Compare the stain with the standards in the disc, making certain that the special reducing diaphragm is in position. A standard source of white light such as north daylight or the Lovibond White Light Cabinet should be used for illumination.

Notes

1. Normal humidity variations have no effect. Sulphur dioxide up to at least 20 ppm has little effect. Chlorine at 5 ppm has no effect; at 20 ppm chlorine, phosgene colour is reduced by half. Hydrogen chloride interferes but can be removed up to 20 ppm by inserting a paper treated with sodium iodide and sodium thiosulphate in front of the test paper. Acetyl chloride also

interferes, but the effect can be suppressed up to 160 ppm by using a test paper containing sodium carbonate in addition to the other reagents. Benzoyl chloride produces a transient orange stain which fades in 15 minutes leaving the phosgene colour unimpaired. Tolylene-2,4-di-isocyanate at high concentrations slightly enhances the phosgene colour: at the threshold limit value of 0.02 ppm its effect is negligible. Trichloroethylene (175 ppm) slightly reduces phosgene colour. Toluene (150 ppm), chloroform (200 ppm) and benzyl chloride (275 ppm) have little effect.

2. For accurate results it is important that exactly 600 ml of air are drawn through the paper.

3. This test replaces the previous test and Disc No. 6/18. This change has become necessary owing to the lower value now stated[1] for the maximum permissible concentration of phosgene. Disc 6/18 can still be used with this test if a sample of 600 ml is taken and the old disc readings are converted to the new values by dividing by five, as follows:—

Reading on Disc 6/18	Revised reading
0.25	0.05
0.5	0.1
1.0	0.2
2.0	0.4
5.0	1.0
10.0	2.0

References

1. Department of Employment, *"Safety Health and Welfare, New Series No. 8, Dusts and Fumes in Factory Atmospheres,"* 4th Edn. H.M. Stationery Office, London, 1969
2. D.S.I.R., *"Methods for the Detection of Toxic Gases in Industry,* Leaflet No. 8 *Phosgene"*, H.M. Stationery Office, London, 1939
3. H. Maureu, P. Chovin and L. Truffert, *Compt. Rend.,* 1949, **228,** 1954
4. M. Hayashi, M. Okazaki and Z. Shinohara, *J. Soc. Org. Syn. Chem. Japan,* 1954, **12,** 273
5. B. E. Dixon and G. C. Hands, *Analyst,* 1959, **84,** 463
6. W. G. Brown, K. E. Wilzbach and E. G. Ballweber, *Library of Congress, P.B. No.* 5945, September, 1945
7. B. E. Dixon, G. C. Hands and A. F. F. Bartlett, *Analyst,* 1958, **83,** 199
8. Department of Employment H.M. Factory Inspectorate, *"Methods for the Detection of Toxic Substances in Air"*, Booklet No. 8 *Phosgene.* H.M.S.O. 1967

The Determination of Pyridine
using copper silicomolybdic acid

Introduction

Pyridine is one of the more toxic vapours encountered in industry. The maximum permissible concentration for an industrial atmosphere has been set at 10 ppm (v/v)[1,2]. It is thus as toxic as hydrogen cyanide (prussic acid) and ten times as toxic as carbon monoxide. As the toxic concentrations are readily tolerated, indeed workers quickly become oblivious of the presence of the vapour, the danger of toxic effects is greater than is the case with more irritant vapours. Great care must therefore be taken to ensure that, when there is a possibility of an atmosphere becoming contaminated with pyridine, adequate measures are taken to keep the pyridine concentration below the danger level. This test was devised[3] to enable the concentration of pyridine in the atmosphere to be determined.

Principle of the method

Pyridine vapour is trapped in hydrochloric acid and the solution is subsequently concentrated by evaporation. The pyridine is then precipitated as a complex with copper-silicomolybdic acid reagent. This complex is then reduced to the blue molybdenum compound by means of sodium sulphite and glycine. The intensity of the blue colour is proportional to the pyridine concentration which is estimated by comparison with Lovibond permanent glass standards.

Reagents required

1. *Copper-silicomolybdic acid*

Molybdic anhydride (MoO_3)	14.4 g
Sodium hydroxide (NaOH Normal solution)	100 ml
Silica (as sodium silicate solution)	0.7 g
Hydrochloric acid (HCl) 10% solution	
Copper chloride ($CuCl_2$)	2.5 g

    Dissolve the molybdic anhydride in the caustic soda, by warming, and add the sodium silicate. Next add hydrochloric acid, a little at a time, until the solution becomes green. Dilute this green solution to 900 ml and heat on a water bath for 3 hours. Cool and stand for 24 hours at ambient temperature. Filter off any excess silica and add the copper chloride dissolved in a little water. Make up to one litre and store in a dark bottle. The reagent is stable for three months.

2. *Wash liquid*
    10% sodium chloride (NaCl) in 0.1 N hydrochloric acid

3. *Reducing solution*

Glycine	1 g
Sodium sulphite (Na_2SO_3 20% aqueous solution)	15 ml
Ammonium hydroxide (NH_4OH sp. gr. 0.880)	5 ml

    Dissolve the glycine in the sodium sulphite solution and add the ammonium hydroxide slowly, while stirring. Dilute to 100 ml.

    This reagent is only stable for 24 hours and must be freshly prepared each day.

4. *Hydrochloric acid (HCl) 0.1 N.*

The Standard Lovibond Nessleriser Disc NOG

This disc covers the range 0.6 to 4.5 mg of pyridine (0.6, 0.8, 1.0, 1.5, 2.0, 2.5, 3.0, 4.0, 4.5 mg). This is equivalent to the range 9—64 ppm of pyridine based on a 20 litre air, sample. (Note 4).

Technique

Place 40 ml of 0.1 N hydrochloric acid in a Schott bubbler[4] and connect to a convenient source of suction. Draw the contaminated atmosphere through the bubbler at exactly a litre per minute for 20 minutes. Transfer the contents of the bubbler to an evaporating basin and heat on a water bath until the volume is reduced to 1—2 ml. (Note 1). Add 1 ml of copper-silicomolybdic acid solution (reagent 1) and 1 ml of 0·1N hydrochloric acid. Warm and agitate until the precipitate coagulates. Allow to stand for at least an hour at ambient temperature and then collect the precipitate on a disc of Whatman No. 42 filter paper in a small (No. 00) Gooch crucible. Wash the evaporating basin and the precipitate twice with successive 0.5 ml portions of wash liquid (reagent 2). Empty and wash the filter flask.

Pipette 1—2 ml of reducing solution (reagent 3) into the evaporating basin to dissolve any residual precipitate. Pour the solution into the Gooch crucible and suck through into the filter flask. Wash the basin and crucible with more reducing solution. Transfer the liquid from the filter flask to a 50 ml graduated flask and make up to the mark with distilled water. Heat on a water bath at 45°C for 25 minutes. Cool, transfer the solution to a standard Nessleriser tube and place in the right-hand compartment of a Lovibond Nessleriser. Compare the colour of the solution with the permanent glass standards in the disc, using a standard source of white light, such as the Lovibond White Light Cabinet, or failing this, north daylight.

Notes

1. It has been proved[3] that solutions of pyridine in hydrochloric acid can be concentrated, without loss, by evaporation.

2. This procedure must be followed exactly, as it has been shown[3] that the temperature of colour development, the acid concentration during precipitation, and the time elapsing between precipitation and filtration all affect the intensity of the final colour.

3. The readings obtained by means of the Lovibond Nessleriser and disc are only accurate provided that standard Nessleriser tubes are used. These are tubes in which the 50 ml calibration mark falls at a height of 113 ± 3 mm measured internally.

4. To convert mg to ppm use the following formula—

$$\text{ppm (volume/volume at N.T.P.)} = \frac{\text{mg} \times 1000}{3.5 \times \text{volume of sample in litres}}$$

References

1. Ministry of Labour, "*Dusts and Fumes in Factory Atmosphere*", H.M.S.O., London, 1969
2. American Conference of Governmental Industrial Hygienists, 1958
3. R. P. Daroga and A. G. Pollard, *J.S.C.I.*, 1941, **60**, 207
4. P. H. Prausnitz, *Ind. Eng. Chem. (Anal. Ed.)*, 1932, **4**, 432

The Determination of Styrene
using sulphuric acid

Introduction

Styrene is a narcotic agent whose vapour is especially irritating to the eyes, nose and throat. Higher concentrations will irritate the deeper respiratory passages causing fits of coughing. The maximum permissible concentration of styrene vapour in factory atmospheres has been stated as 100 ppm v/v (420 mg/m^3)[1]. The present test[2] has been developed to provide a simple and rapid method for checking the styrene concentration in contaminated atmospheres.

Principle of the method

Styrene reacts with concentrated sulphuric acid to produce a yellow colour. The intensity of this colour, which is proportional to the styrene concentration, is measured by comparison with a series of Lovibond permanent glass colour standards.

Apparatus

Bubbler—bubbler of the type shown in the diagram with the following dimensions:—

Receiver—Overall length 155 mm, external diameter 15 mm, internal diameter 13 mm. Distance from bottom of side arm to bottom of bubbler 110 mm.

Inlet Tube—Approximate overall length 185 mm, external diameter 6 mm, internal diameter 4.5 mm. Distance from bottom of inlet tube to bottom of bubbler 7 mm. Cone and Socket B.10

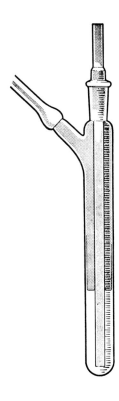

Reagent required

Concentrated sulphuric acid H_2SO_4 sp.gr. 1.84 of analytical reagent quality

The Standard Lovibond Comparator Disc 6/26

This disc contains three standards which correspond to concentrations of 50, 100, and 200 ppm of styrene when a 120 ml air sample is used.

Technique

Place 5 ml of sulphuric acid in a dry bubbler of the type shown in the Figure. Attach a convenient source of suction to the side arm and insert the inlet tube. Sample 120 ml of the contaminated atmosphere through the bubbler at 25 ml per minute. Disconnect the bubbler and dissolve any resin on the inside of the inlet tube by raising and lowering the tube into the liquid several times and, if necessary, by rubbing it with a thin glass rod. Place the bubbler in the right-hand compartment of the comparator and stand the comparator in front of a standard source of white light such as the Lovibond White Light Cabinet, or failing this, north daylight. Compare the colour of the sample with the standards in the disc within 5 minutes of completing sampling.

Notes

1. Concentrated sulphuric acid is dangerous. Care should be taken to prevent the acid from coming into contact with the skin, especially the eyes. The acid should be measured out and transferred to the bubbler by means of a safety pipette.

2. Alpha methyl styrene and phenyl acetylene produce very little colour at 200 ppm. Vinyl styrene reacts like styrene. Acetone (500 ppm) and ethyl methyl ketone (200 ppm) reduce the colour produced by 100 ppm of styrene by about 20% and 10% respectively.

3. Suitable bubblers can be obtained from Tintometer Ltd.

References

1. Min. of Labour, *Dusts and Fumes in Factory Atmospheres,"* H.M.S.O., London, 1969
2. Min. of Labour, H.M. Factory Inspectorate, *"Methods for the Detection of Toxic Substances in Air, Booklet No. 4, Benzene, Toluene and Xylene, Styrene"*. H.M. Stationery Office, London, 1966

COLORIMETRIC CHEMICAL ANALYTICAL METHODS

The Determination of Sulphur Dioxide (1)
using ammoniacal zinc nitroprusside

Introduction

Sulphur dioxide is produced during the burning of any substance containing sulphur. It is also widely used in industry and in fumigation and disinfection.

In high concentrations sulphur dioxide gas is irrespirable and causes asphyxiation. Lower concentrations produce irritation to eyes, nose throat and lungs. Although its presence in an atmosphere is thus obvious to the senses, and it may therefore not be regarded as being as dangerous as the more insidious toxic gases, nevertheless prolonged exposure to tolerable concentrations may cause inflamation of the nose and throat and set up bronchitis. The maximum permissible concentration for an exposure of several hours has been stated[1] to be 10 ppm and a concentration of 500 ppm is dangerous for even short exposures.

The method previously recommended for the determination of sulphur dioxide in air has been found to be unsatisfactory as it is not specific for sulphur dioxide; and further, unless great care is taken in the production of the test papers, uneven stains are produced which are difficult to match against a standard. The present method[2,3] based on that of Gandolfo[4], has been developed to overcome these difficulties.

Principle of the method

On reaction of sulphur dioxide with ammoniacal zinc nitroprusside a brick-red colour is formed, the intensity of which is proportioned to the concentration of sulphur dioxide. The colour obtained by drawing a known volume of air, at a controlled rate, through filter paper impregnated with this reagent is compared with Lovibond permanent glass standards.

Reagents required

1. *Sodium nitroprusside* ($Na_2Fe(CN)_5NO.2H_2O$ *Reagent grade*) 10% w/v solution in distilled water
2. *Zinc sulphate heptahydrate* ($ZnSO_4.7H_2O$ *Reagent grade*) 6% w/v solution in distilled water
7. *Ammonium acetate* (CH_3COONH_4)
4. *Glycerol*

All chemicals used in the preparation of reagents should be of anlytical reagent quality.

Preparation of test papers

Add, while stirring, 50 ml of the zinc sulphate solution (reagent 2) to 50 ml of the sodium nitroprusside solution (reagent 1). Add solid ammonium acetate (reagent 3) gradually, with stirring, to the mixture until the precipitated zinc nitroprusside has dissolved. Finally add 20 ml of glycerol (reagent 4) to the mixture while stirring. Soak 1-inch wide strips of Whatman Grade 3 MM Chr filter paper in this solution, drain off the superfluous liquid, and dry the papers at a temperature not exceeding 40°C in an atmosphere free from chemical fumes. Store the dried papers in a light-tight stoppered container.

The test papers if stored in a stoppered light-tight container are stable for about one month. Any papers which on drying or after storage are found to have developed a pink colouration of an intensity approaching that of the 1 ppm colour standard should be discarded.

The Standard Lovibond Comparator Disc 6/40

The disc contains 5 standards covering the range 1 to 20 ppm (1, 2.5, 5, 10, 20) of sulphur dioxide, based on a 360 ml air sample, and is designed to be used in conjunction with a field-reducing aperture in the Lovibond Test Paper Viewing Stand.

Technique

Fix the test-paper in a suitable holder[5] and attach this to an aspirator (Note 1) with a piece of rubber tubing. Draw a 360 ml sample through the paper at a rate of 90 to 180 ml per minute. Remove the test paper from the holder and place on the right-hand side of the Comparator test-paper viewing stand. Place a piece of unimpregnated test paper on the left-hand side. Compare the stain with the standards in the disc making sure that the aperture-reducing diaphragm is in position. A standard source of white light such as north daylight or the Lovibond White Light Cabinet should be used for illumination.

Notes

1. The air sample may be taken by any means, such as water aspirator, bulb aspirator or pump, which is capable of giving a regulated rate of flow.
2. The permission of the authors to quote from their paper[3] is gratefully acknowledged.
3. Hydrogen sulphide in excess of 15 ppm will interfere.

References

1. Y. Henderson and H. W. Haggard, *"Noxious Gases and the Principles of Respiration influencing their Action"*
   American Chemical Society Monograph No. 35, 2nd edition, Reinhold, New York, 1943
2. H.M. Factory Inspectorate, *"Methods for the Detection of Toxic Substances in Air, Booklet No. 3: Sulphur Dioxide"*, H.M. Stationery Office, London, 1961
3. G. C. Hands and A. F. F. Bartlett, *Analyst,* 1960, **85**, 147
4. N. Gandolfo, *Rend. Ist. Super. Sanità, Rome,* 1948, **11**, 1268
5. B. E. Dixon, G. C. Hands and A. F. F. Bartlett, *Analyst,* 1958, **83**, 199

The Determination of Sulphur Dioxide (2)
using *p*-rosaniline

Introduction

This modification[1,2,5] of the West-Gaeke[3] method for the determination of sulphur dioxide in air, has been developed to overcome variability in the original technique. The *p*-rosaniline reagent is specially purified, phosphoric acid is used to control the final *p*H and to assist the liberation of sulphur dioxide from its mercury complex. The interference of nitrogen dioxide is removed by the addition of sulphamic acid[4] to the absorbing reagent before analysis. These alterations result in greater sensitivity, increased reproducibility and adherence to Beer's Law over a greater working range (0 to 35 µg SO_2).

Principle of the method

The reaction of sulphur dioxide with acid-bleached *p*-rosaniline and formaldehyde produces an intense orange-red colour. The intensity of this colour, which is proportional to the sulphur dioxide concentration, is measured by comparison with a series of Lovibond permanent glass colour standards.

Reagents required

1. *Absorbing reagent* Dissolve 10.9 g of mercuric chloride ($HgCl_2$) and 5.9 g of potassium chloride (KCl) in 1 litre of distilled water. **This reagent is highly poisonous.** If spilled on the skin, immediately flush off with water.

2. *Acetate buffer solution* Dissolve 34.02 g of sodium acetate trihydrate ($CH_3COONa.3H_2O$) in 100 ml of water. Add 14.3 ml of glacial acetic acid (CH_3COOH) and dilute to exactly 250 ml with water.

3. *Purified p-rosaniline stock solution* Dissolve 100 mg of purified *p*-rosaniline hydrochloride (Note 1) in normal hydrochloric acid (HCl) and make up to 50 ml with N HCl.

4. *p-Rosaniline reagent* To 20 ml of the stock solution, in a 250 ml glass-stoppered volumetric flask, add 25 ml of $3M$ phosphoric acid (H_3PO_4). Dilute to 250 ml with distilled water. This reagent is stable for at least a month.

5. *Sulphamic acid 0.6%* Dissolve 0.6 g of sulphamic acid (NH_2SO_3H) in 100 ml of distilled water. This reagent can be stored for a few days if protected from air.

6. *Formaldehyde 0.2%* Dissolve 5 ml of 40% formaldehyde (HCHO) in distilled water and make up to 1 litre. This reagent must be prepared on the day of use.

7. *Standard sulphite solution* Dissolve 400 mg of sodium sulphite (Na_2SO_3) in 500 ml of distilled water. This produces a solution which contains 360 to 400 µg per ml as SO_2. Standardise by the usual iodine-thiosulphate titration just before use and dilute further as required to give a final concentration of 10 µg per ml (Note 3).

All chemicals used in the preparation of reagents must be of analytical reagent quality.

The Standard Lovibond Comparator Disc 6/32

This disc contains standards corresponding to 2.5, 5, 7.5, 10, 12.5, 15, 17.5, 20 and 22.5 µg. of sulphur dioxide (SO_2).

Technique

Draw the air sample through a midget impinger bubbler containing 10 ml of the absorbing reagent (reagent 1). The flow rate is not critical provided that it is low enough to prevent entraintment of the liquid. Take a sufficient volume of air to collect 3 – 20 µg of sulphur dioxide, and measure the volume of air sampled (Note 2). If the sample must be stored for more than a few days before analysis, store at 5°C.

After collection transfer the sample quantitatively to a 25 ml volumetric flask (filter if any particles are visible in the sample). Use about 5 ml of distilled water for rinsing the bubbler and the filter and add the washings to the sample. Prepare a blank by adding 10 ml of unexposed absorbing reagent to another 25 ml volumetric flask. To each flask add 1 ml of sulphamic acid solution (reagent 5) and allow to react for 10 minutes to destroy any nitrite resulting from the absorbtion of oxides of nitrogen from the atmosphere. Pipette 2 ml of formaldehyde (reagent 6) and then 5 ml of p-rosaniline reagent (reagent 4) into each flask and start a laboratory timer which has been set for 30 minutes. Make up all the flasks to volume with freshly boiled distilled water. After 30 minutes transfer 10 ml of the sample solution to a standard comparator 13.5 mm cell and place this in the right-hand compartment of the comparator. Transfer 10 ml of the blank solution into an identical cell and place this in the left-hand compartment. Place the comparator before a standard source of white light, such as the Lovibond White Light Cabinet, or failing this, north daylight, and compare the colour of the sample with the standards in the disc (Note 3). Calculate the concentration of sulphur dioxide in the atmosphere by the following formula:—

$$\text{ppm} = \frac{\text{disc reading} \times 24.47}{V \times 64.0}$$

where V is the sample volume in litres corrected to 25°C: and 760 mm Hg pressure.

Notes

1. Suitable purified p-Rosaniline reagent may be obtained from Harleco, 60th and Woodland Avenue, Philadelphia, Penn. 19147, U.S.A. or from S.E.A.C., 33 rue Carnot, Levallois-Perrett (Seine) France, or prepared according to reference 2.

2. Suitable methods for taking air samples are described on pages 462 to 464.

3. It is important that the analysis of the sample be carried out at 23°C. If the laboratory temperature is more than a few degrees different from this temperature then a correction factor must be applied to the results obtained. To determine this correction factor add an amount of the standard sulphite solution (reagent 7), equivalent to 10 μg of SO_2, to 10 ml of absorbing solution and proceed with the determination as above. Multiply all results by the result obtained from this standard divided by 10.

4. Losses of sulphur dioxide from the absorbing reagent on storage of collected samples vary linearly with time and are temperature dependent. Loss rates are approximately 1.6% per day at 25°C and are negligible at 5°C.

5. Strong oxidants interfere with colour development. Interference from ordinary atmospheric concentrations of ferric iron is negligible.

References

1. F. P. Scaringelli, B. E. Saltzman and S. A. Frey, *"The effects of various parameters on the spectrophotometric determination of sulphur dioxide with p-rosaniline"*, Paper presented to Division of Water, Air and Waste Chemistry, 150th. National Meeting of the American Chemical Society, Atlantic City, Sept. 13th, 1965

2. F. P. Scaringelli, B. E. Saltzman and S. A. Frey, *"Spectrophotometric determination of sulphur dioxide in the atmosphere with p-rosaniline"*, Paper presented to the Division of Water, Air and Waste Chemistry, American Chemical Society, Pittsburgh, March 23, 1966

3. P. W. West and G. C. Gaeke, *Anal Chem.*, 1956, **28**, 1816

4. J. B. Pate, B. E. Ammons, G. A. Swanson and J. P. Lodge, *Anal. Chem.*, 1965, **37**, 942

5. S. Hochheiser, F. P. Scaringelli and L. A. Elfers *"Estimation of atmospheric SO_2 concentration by use of a visual color comparator"*, U.S., D.H.E.W., Public Health Service, National Center for Air Pollution Control, Cincinnati, Ohio. April 1967

The Determination of Toluene and Xylene
using potassium iodate

Introduction

Toluene and xylene are narcotic agents whose toxicity is thought to be a function of the amount of benzene present as an impurity. However maximum permissible concentrations have been specified for these chemicals in factory atmospheres[1].

The maximum permissible concentration of toluene is 200 ppm v/v (750 mg/m^3), and that for xylene has been tentatively specified as 100 ppm v/v (435 mg/m^3). The present test[2] has been developed to enable the concentration of toluene and xylene to be determined quickly and easily.

Principle of the method

Toluene and xylene react with an acid solution of potassium iodate to produce a yellow colour. The intensity of this colour which is proportional to the toluene and xylene concentrations is measured by comparison with a series of Lovibond permanent glass colour standards.

Reagent required

Potassium iodate solution Add 85 ml of sulphuric acid (H$_2$SO$_4$ sp gr 1.84) to 15 ml of water slowly and with cooling (Note 1). Dissolve 1 g of finely ground potassium iodate (KIO$_3$) in this diluted acid. This reagent is stable for 1 week after which time it must be discarded.

Chemicals used in the preparation of reagents should be of analytical reagent quality.

The Standard Lovibond Comparator Disc 6/25

This disc contains three standards which correspond to 100, 200, and 300 ppm of toluene and/or xylene when a 120 ml sample of air is used.

Technique

Place 5 ml of the potassium iodate solution in a dry bubbler of the type illustrated in the Figure. Attach a suitable source of suction[3] to the side arm and insert the inlet tube. Suck 120 ml of the contaminated atmosphere through the bubbler at 25 ml per minute. Disconnect the bubbler and dissolve any resin on the inside of the inlet tube by raising and lowering the tube into the liquid several times and, if necessary, by rubbing it with a thin glass rod. Place the tube in the right-hand compartment of the comparator and then place the comparator in front of a standard source of white light, such as the Lovibond White Light Cabinet, or failing this, north daylight. Compare the colour of the sample with the standards in the disc, 2 to 4 minutes after the completion of sampling. Read off the toluene/xylene concentration from the indicator window in the comparator.

To estimate xylene concentrations of 50 ppm in the absence of toluene take 240 ml of sample. The 100 ppm standard will then represent 50 ppm. When toluene and xylene are present together, the total contamination should be quoted as xylene.

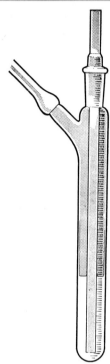

Apparatus

Bubbler—bubbler of the type shown in the diagram with the following dimensions:—
  Receiver—Overall length 155 mm, external diameter 15 mm, internal diameter 13 mm. Distance from bottom of side arm to bottom of bubbler 110 mm.
  Inlet Tube—Approximate overall length 185 mm, external diameter 6 mm, internal diameter 4.5 mm. Distance from bottom of inlet tube to bottom of bubbler 7 mm. Cone and Socket B10.

Notes

  1. The concentrated sulphuric acid used in this test is dangerous and must be handled with care. In no circumstances should water be added to the concentrated acid. When diluting the acid always add acid to water, allowing adequate time between successive additions for the acid to mix and for heat to dissipate. Take adequate precautions to prevent the acid or the mixed reagent, from coming into contact with the bare skin, and especially with the eyes, and always use a safety pipette for measuring out the reagent.

  2. Styrene and ethyl benzene react like toluene and xylene. Benzene also reacts but at 25 ppm or less the colour produced is insignificant. Ortho and para xylene and some commercial xylenes give results which may be 10% low.

  3. Suitable bubblers may be obtained from Tintometer Ltd.

References

  1. Min. of Labour, *"Dust and Fumes in Factory Atmospheres"*, H.M.S.O., London, 1966
  2. Min. of Labour, H.M. Factory Inspectorate, *"Methods for the Detection of Toxic Substances in the Air, Booklet No. 4, Benzene, Toluene and Xylene, Styrene"*, H.M.S.O., London 1966
  3. *See pages* 462 to 464.

COLORIMETRIC CHEMICAL ANALYTICAL METHODS

The Determination of Zinc Oxide Fume

using 4-(2'-thiazolylazo) resorcinol (TAR)

Introduction

The main industrial health hazard from zinc is from the inhalation of particles of freshly formed zinc oxide. Over-exposure to fume of this nature is claimed to cause metal-fume fever[1]. Zinc oxide fume can arise in processes such as the casting of zinc-based alloys and the welding of galvanised steel, and a threshold limit of 5 mg m^{-3} is currently recommended[2] for industrial atmospheres.

This test has been developed [3,5] to enable the concentration of zinc oxide fume present in the air to be determined quickly and reliably without the aid of sophisticated equipment. It is realised that this test may determine some zinc not originally present as the oxide, but this over-estimation will err on the side of safety and is, therefore, considered to be acceptable for the application for which the test is designed.

Principle of the method

Zinc oxide film is collected on a Millipore filter, dissolved in acid and the zinc concentration of the resulting solution is determined by reaction with 4-(2'-thiazolylazo) resorcinol (TAR) as originally suggested by Kawase[4]. TAR forms a red water-soluble complex with zinc and the intensity of the red colour, which is proportional to the zinc concentration, is determined by comparison with a series of Lovibond permanent glass colour standards. Colour differentiation is improved by screening the yellow background colour of the TAR reagent by the addition of Pontamine sky blue dye solution.

The only metal which may be present together with zinc in the industrial atmosphere and which will interfere significantly with the zinc determination, when present in a ratio of twice the threshold limit value of the metal to one half the threshold limit value of zinc, is iron. This is removed from the test solution by means of a column of ion exchange resin.

Reagents required

1. *Hydrochloric acid, 5M*   Dilute 43.5 ml of concentrated hydrochloric acid (sp.gr. 1.19 at 20°C) to 100 ml with water.
2. *Triammonium citrate solution*   Dissolve 10 g of triammonium citrate ((NH_4)$_3C_6H_5O_7$) in 100 ml of water.
3. *Nitric acid (1 + 1)*   Dilute concentrated nitric acid with an equal volume of water.
4. *Aqueous acetone*   Dilute 6 volumes of acetone (CH_3COCH_3) with 4 volumes of water.
5. *Acetone-hydrochloric acid solution*   Add 10 ml of 5M hydrochloric acid (reagent 1) to 60 ml of acetone, and dilute to 100 ml with water.
6. *Ion exchange resin*   Place about 50 g of Zeo-Karb 225 resin (52 – 100 mesh, Na^+ form) in a large glass column about 500 mm in length and 30 mm in internal diameter, fitted with a No. 1 porosity sintered glass disc in the lower part and terminating in a tap (Note 1). Back-wash the column with distilled water to remove any fines, then calculate the bed volume of the resin (height x cross-sectional area of the wet resin). Drain off the water and wash with four bed volumes of the triammonium citrate solution (reagent 2) followed by a similar volume of 5M hydrochloric acid (reagent 1). Finally wash the resin with water until the eluate gives no cloudiness with silver nitrate. Remove excess water with a filter pump and allow the resin to dry in air. Store in a screw topped bottle.
7. *4-(2'-thiazolylazo) resorcinol (TAR) solution*   Dissolve 0.1 g of TAR ($\underline{S.CH.CH.N.C.}$ $N{:}N.C_6H_3(OH_2)$) in 100 ml of methanol (CH_3OH).
8. *Pontamine sky blue solution*   Dissolve 0.168 g of Pontamine sky blue (C.I.24410) in 100 ml of water.

9. *TAR reagent* Weigh 15 g of triethanolamine ($N(CH_2CH_2OH)_3$) into a 100 ml volumetric flask, add 60 ml of N sodium hydroxide (NaOH), 1.5 ml of TAR solution (reagent 7) and 1.5 ml of Pontamine sky blue solution (reagent 8). Dilute to 100 ml with water. Store at a temperature below 30°C and renew after 24 hours.

The chemicals used in the preparation of these reagents should be of analytical reagent quality wherever possible, and all water used should be either distilled or de-ionised.

The Standard Lovibond Comparator Disc 6/50

This disc contains standards corresponding to 0, 2.5, 5 and 10 mg m^{-3} of zinc oxide (ZnO) based on a 5 litre sample of air.

Technique

Place a 25 mm Millipore AA filter in a filter paper holder (Note 2), attach this holder to a suitable source of suction (Note 3) and draw 5 litres of air through the filter at a constant rate of 1 litre a minute. Disconnect the holder, remove the filter and place it in a small beaker (25 - 30 mm diameter).

Samples containing no iron Add 2 ml of 5M hydrochloric acid (reagent 1) to the filter in the beaker. After 5 minutes transfer the acid solution to a 50 ml volumetric flask. Wash out the beaker with a little water and add the washings to the liquid in the flask. Add 20 ml of aqueous acetone (reagent 4), 10 ml of TAR reagent (reagent 9) and make up to 50 ml with water. Fill a 60 x 10 mm flat bottomed comparator tube to a depth of 50 mm with the solution and place this in the right hand compartment of the comparator in the manner shown in Figure 1, i.e. fit a "Large Cells Attachment" with Special Tube Holder (DB430) onto the back of the Lovibond 1000 Comparator, holding the comparator horizontally, and insert the tube into the disc central aperture as illustrated. Hold the comparator over a standard source of white light, such as a white tile reflecting north daylight, or the Lovibond White Light Cabinet tipped on to its back, and compare the colour of the sample solution with the standards in the disc.

Samples containing iron Prepare an ion exchange column in the following manner:– Place a small plug of cotton wool in the narrow end of the column holder (Figure 2) and add a slurry of 1 g of the prepared resin in 20 ml of the acetone-hydrochloric acid solution (reagent 5). Allow nearly all of the liquid to pass through the resin and then add a further 20 ml of reagent 5. Stopper the lower end of the column holder with a piece of plugged rubber tubing just as the liquid meniscus enters the resin. Transport the prepared columns to the sampling site in a vertical position.

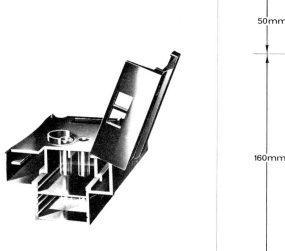

Figure 1

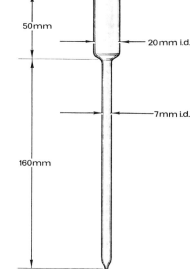

Figure 2

To the filter in the beaker add 1 ml of 1 + 1 nitric acid (reagent 3) and then, after 5 minutes, add 8 ml of the aqueous acetone solution (reagent 4). Transfer the combined solutions onto the ion exchange column, wash the beaker with a further 2 ml of reagent 4 and add these washings to the column. Remove the stopper from the bottom of the column and allow the solution to pass through the resin until just before the meniscus reaches the top of the resin. Discard this eluate, add 20 ml of the acetone-hydrochloric acid solution (reagent 5) and collect all the subsequent elute from the column in a 50 ml volumetric flask. Proceed as described above, for solutions not containing iron.

Notes

1. A suitable column and tap may be obtained from Quickfit and Quartz Ltd., catalogue Nos CR 51½/50 and CR 3T/23.
2. A suitable filter holder is obtainable from Tintometer Ltd., (Code No. AF401).
3. Suitable sources of suction for taking samples of polluted atmospheres are described in the Introduction to this Section.

References

1. E. Browning, *"Toxicology of Industrial Metals"*, 2nd Editn., Butterworths, London, 1969, p. 352
2. Department of Employment and Productivity, *"Threshold Limit Values* 1969", Technical Data Note 2/69, H.M. Stationery Office, London, 1969
3. B. S. Marshall, I. Telford and R. Wood, *Analyst*, 1971, **96**, 569
4. A. Kawase, *Talanta*, 1965, **12**, 195
5. Dept. of Employment, H.M. Factory Inspectorate *"Methods for the detection of toxic substances in Air"* Booklet No. 25 *"Zinc Oxide Fume"* HMSO London 1972

Colour Grading and Quality Tests

The Colour-grading of Commodities

Introduction

The careful standardisation of the colour of commodities is an important factor in assuring the confidence of the purchasers. This applies not only to goods—such as paints or dyes—which are brought primarily for their colour, but also to such items as oils, foodstuffs, beverages, confectionery and many others. Colour is often a direct indication of quality, in which case colour-grading is also grading according to quality. In other cases, colour is added to improve the appearance or to simulate a natural product: in this latter case, standardisation of colour indicates care and uniformity in preparation, and is taken as a criterion of reliability.

The simplest method of accurate colour grading (as distinct from actually measuring the colour) is by means of the Lovibond Comparator and a suitable disc containing permanent glass colour standards.

Official and unofficial standards

Two distinct types of grading discs are made. Firstly those which represent an agreed range of standard colours laid down by some Government or similar authority, or by an accepted standardising body, such as National Standards Institutions or trade organisations, and secondly those which are made especially to meet the needs of an individual case. Examples of the former are colour standards for Whale Oil and Brewing colours (International agreement) rosin (Government decree), lubricating oils (Standardising Institution) and varnish (trade organisation). Examples of the latter are discs for brewers, wine blenders, honey producers, and manufacturers of pharmaceutical preparations.

The value of unofficial standards

No standard disc can be prepared for these special and similar commodities, as each user will have different requirements, but The Tintometer Ltd. are able to prepare comparator discs containing glass colour standards exactly matching the purchaser's own samples. For producers of proprietary lines which are sold in sealed containers, such discs are of value during the manufacturing processes and in ensuring reproducibility of colour in the final product. For those who prepare commodities for sale in bulk, such as brewers or vinegar makers, there is the added value that subsequent dilution or other tampering with the product can easily be detected by means of this instrument.

How to order

Samples of the commodity to be tested should be sent to The Tintometer Ltd., Salisbury, England, with as much information as possible regarding requirements. For example, instructions may be given to match exactly the colour of the samples sent, or one sample may be sent with instructions to dilute in various proportions for other standards: another possibility is to send a few samples, with instructions to match these and also to insert intermediate points on a linear scale. Up to 10 standards can be fitted to a disc, and of course several discs can be supplied if more colours comprise the required scale.

It must be emphasised that only a range of fixed colours can be dealt with in this way. If varying combinations of colours are likely to be required, only a Lovibond Tintometer is applicable to the work.

On receipt of the samples and instructions, the makers will advise what sized cell is appropriate for use with the particular colour, and quote a price, which will be according to the preparatory work involved. Unless specific instructions are given to the contrary, all matchings are carried out using the standard Lovibond White Light Cabinet.

Acid Wash Test for the Quality Grading of Refined Lower Boiling Products of Coal Tar

Introduction

The products falling within the scope of this test are defined by the Standardisation of Tar Products Test Committee (S.T.P.T.C.) as "Refined Products (of Coal Tar) consisting essentially of hydrocarbons boiling below 200°C and substantially free from phenols and pyridine bases". Heavy Naphtha is regarded as falling within this definition.

The S.T.P.T.C. official "Standard Methods 1967"[1] Acid Wash Test uses colour standards of potassium dichromate solutions, but specifies the discs herein described as an alternative. Similar standards are adopted by the British Standards Institution[2], the National Benzole Association[3], and by Indian Standards[4].

Principle of the method

Equal volumes of the sample and sulphuric acid of stated strength are shaken together. The colour of the acid layer is then compared with specified standards to give some indication of the degree to which acid refining has been carried.

The Standard Lovibond Comparator Discs 4/7A, 4/7B and 4/7C

The range is covered by 15 permanent glass colour standards, which correspond to the colour of the stated weight (in grams) of potassium dichromate in one litre of diluted sulphuric acid, viz.—

Disc A	**0.3**	0.4	**1.0**	1.3	1.5	**2.0**	2.6	**3.0**	4.0
Disc B	**5.0**	6.5	**7.0**	9.0	**10.0**	**13.0**			

Those values in heavy type correspond to the 8 official grades, and the next darker standard corresponds to the permitted upper limit for each.

For dealing with the purer grades of lower-boiling coal-tar products, such as benzenes and toluenes covered by N.B.A. specifications, a special disc is available covering the paler colours as follows:—

Disc C	0.1	0.2	0.3	0.4	0.5	0.6	0.7	0.8	0.9

These discs are designed for use with special stoppered shaking tubes which may be obtained from Tintometer Ltd.

Technique

"The sample shall be passed through a filter paper (e.g. Whatman No. 5), the first 10 ml of the filtrate being rejected. One of the stoppered shaking tubes (dry) obtainable with the comparator shall then be filled to the lower mark with the filtered sample, the stopper inserted, and the tube placed in a water-bath at 20°C plus or minus 1° for at least 15 minutes. A quantity of sulphuric acid 95% w/w (plus or minus 0.5% w/w H_2SO_4) shall also be immersed in the water bath for at least 15 minutes to attain the same temperature. The shaking tube containing the sample shall then be filled to the upper mark with the acid, stoppered, vigorously shaken for exactly two minutes, and returned to the water-bath for exactly ten minutes. The tube shall then be placed in the right-hand compartment of the comparator (the left-hand compartment remaining empty) and the colour immediately compared with that of the glass standards of the discs, the latter being revolved until a match is obtained. The comparison shall be made with the case held about 18 inches from the eye and facing a north daylight. The number designating the standard effective in the left-hand aperture is shown in the hole at the bottom right-hand corner of the instrument.

According to the result of the comparison, the sample shall be reported as "equal to the required standard, better than the standard, 'darker than the standard but within the limits of error permitted,' or as failing."

For the full details of test, and permitted limits of error in the various grades, reference should be made to the official publication.

Note

Stringent requirements are laid down in the official publication as to the purity and conditions of storage of the acid, and also of the exact dimensions and graduations of the shaking tubes.

References

1. S.T.P.T.C., *"Standard Methods for Testing Tar and its Products"* 6th Edn., Leeds, 1967: —Test RLB 10–67
2. British Standards 135, 458, 479 and 805 all of 1963
3. National Benzole Assocn., *"Standard Specifications for Benzole and Allied Products"*, 4th Edn., 1960, Method D
4. Indian Standards 534 and 535: 1965

The Determination of α-amylase (1)

using the A.S.B.C. dextrinization procedure for malt

Introduction

The various methods which are available for the determination of α-amylase in malt have been studied by a combined committee of the American Society of Brewing Chemists, (A.S.B.C.), and the Association of Official Agricultural Chemists. Three general methods were considered: (a) the measurement of dextrinization time in the presence of excess β-amylase[1], (b) a liquefaction procedure[2], and (c) the determination of diastatic power after differential inactivation of β-amylase[3]. The committee reported that the dextrinization method was both the simplest and the most accurate.

The original dextrinization procedure[1] specified 30°C as the reaction temperature, but the procedure was subsequently modified[4] for use at 20°C. The modified method has been collaboratively tested and was officially accepted in 1949[5]. However it was subsequently demonstrated [6,7] that the specified extraction procedure was unsatisfactory, and the present method[8] has been adopted as the official method[9] after further investigation and collaborative testing.

Principle of the method

In the presence of excess β-amylase, α-amylase converts 'soluble starch' (α-amylodextrin) to maltose. The unit of α-amylase activity is defined[8] as that quantity of α-amylase which will dextrinize 'soluble starch', in the presence of excess β-amylase, to the specified end-point at the rate of 1 g per hour at 20°C.

When reacted with iodine, undextrinized starch produces a deep red-purple colour, the intensity of which is proportional to the amount of starch present. Incubation of the starch-α-amylase extract is continued until the starch content is reduced to a predetermined level as indicated, after reaction with iodine, by the colour of the Lovibond permanent glass standard.

Reagents required

1. *β-amylase* This enzyme preparation must be standardised to a Diastatic Power of 2000° Lintner, and must conform to the following specification:—

    (a) At the addition level used, the variation in the dextrinization of a standard malt infusion, when 1- and 3-day old substrates are compared, shall not exceed 5%.

    (b) The dextrinization by a standard malt infusion of a substrate prepared by adding 500 mg of β-amylase instead of the standard 250 mg shall not vary more than 5% from that obtained with the standard substrate, when both substrates are dextrinized 24 hours after their preparation.

    The β-amylase powder should be stored in a tightly closed bottle and kept in a refrigerator. The bottle should be allowed to warm up to room temperature before it is opened, in order to minimise the risk of water condensing on the cold enzyme.

2. *Buffered limit-dextrin (α-amylodextrin) substrate* Prepare a suspension of 10.00 g (dry weight) of Merck's soluble Lintner starch in 100 ml of cold water and pour slowly into 300 ml of boiling water. Boil with stirring for 1 to 2 minutes, cool, add 25 ml of buffer solution and 250 mg of β-amylase dissolved in a small amount of water. Make the total volume up to 500 ml with distilled water saturated with toluene, and store at 20°C for not less than 18 hours nor more than 72 hours before use.

3. *Buffer solution* Dissolve 164 g of anhydrous sodium acetate (CH_3COONa) in water. Add 120 ml of glacial acetic acid (CH_3COOH) and dilute the solution to 1 litre.

4. *Sodium chloride solution (0.5%)* Dissolve 5 g of analytical reagent grade sodium chloride (NaCl) in distilled water and make up to 1 litre.

5. *Stock iodine solution* Dissolve 5.50 g of iodine crystals and 11.0 g of potassium iodide (KI) in water and dilute the solution to 250 ml. Store in a dark bottle. This solution should not be stored for more than 4 weeks.

6. *Dilute iodine solution* Dissolve 20 g of potassium iodide (KI) in water, add 2 ml of stock iodine solution and dilute to 500 ml.

The Standard Lovibond Comparator Disc 4/29

This disc contains a single colour standard which corresponds to the concentration of starch remaining in solution at the end-point of the dextrinization process as defined by the American Society of Brewing Chemists.

Technique

Preparation of malt infusion

Extract 10 g of finely ground malt with 200 ml of 0.5% sodium chloride solution (reagent 4) in a 300 ml flask for 1 hour at 30°C. Filter. Adjust volume to exactly 200 ml with 0.5% sodium chloride solution. Dilute this infusion further with 0.5% sodium chloride to give a final infusion of suitable concentration for dextrinization, e.g. for malts with activities in the range 30–100 D.U.'s (dextrinizing units) dilute 4 ml to 100 ml.

Dextrinization

Allow the temperatures of the substrate (reagent 2) and the malt infusion to come to equilibrium at 20°C and ensure that the final dilution of the infusion is carried out at this temperature. Transfer 10 ml of the diluted malt infusion to a 50 ml conical flask or 25×150 mm test tube and place in the water-bath at 20 ± 0.05°C. After a few moments add 20 ml of substrate, noting zero time from the moment the first of the substrate touches the infusion in the flask. Mix well by blowing the substrate from the pipette. At zero plus ten minutes withdraw 2 ml of the mixture and add to 10 ml of dilute iodine solution (reagent 6) which is also at 20°C. Mix and immediately transfer to a 13.5 mm comparator cell and place this in the right-hand compartment of the comparator. With an identical cell filled with distilled water in position in the left-hand compartment, immediately compare the colour of the iodine solution with the standard end-point colour in the disc, using a Lovibond White Light Cabinet or north daylight. At appropriate intervals remove additional 2 ml aliquots from the reaction mixture, add to the dilute iodine solution and immediately measure the colour. Continue this process until the α-amylase end-point colour is reached.

During the initial stages of the reaction it is not necessary that the 2 ml sample be measured with great accuracy, as at this stage only an indication of the progress of the reaction is required. However as the end-point is approached a fast flowing 2 ml pipette, such as a bacteriological pipette, should be used. Blow the contents of the pipette into the dilute iodine solution to ensure the fastest possible mixing of the liquids. Near the end-point take readings every half minute, on the minute and half minute. If two readings half a minute apart give colours one darker than and one lighter than the standard, then record the end-point at the nearest quarter minute. Shake out the cell used for colour comparison between successive readings.

For accuracy and convenience it is desirable that dextrinization times fall between 10 and 30 minutes. With malts of high α-amylase activity it may be necessary to use only 5 ml of the diluted infusion. In this case add 5 ml of 0.5% sodium chloride solution to the 5 ml of malt infusion before adding the substrate. The final volume of the reaction mixture should always be 30 ml.

Calculation of α-amylase activity

From the time interval necessary for dextrinization and the weight of malt in the infusion aliquot taken, calculate the α-amylase dextrinization units (20° D.U.) from the following formula:—

$$20° \text{ D.U.} = \frac{24}{W \times T} \text{ (as is)} = \frac{24}{W \times T} \times \frac{100}{100-M} \text{ (dry basis)}$$

where
W = weight, in gm, of malt *in aliquot taken*
T = dextrinization time in minutes
M = % moisture in sample.

In this formula 24 represents the weight of starch taken (0.4 g in 20 ml of substrate) multiplied by 60 (minutes in 1 hour). Report Dextrinizing Units to nearest 0.1 unit.

References

1. R. M. Sanstedt, E. Kneen and M. J. Blish, *Cereal Chem,* 1939, **16,** 712
2. S. Redfern and Q. Landis, *ibid,* 1946, **23,** 1
3. F. R. Graesser and P. J. Dax, *Wallerstein Lab. Common,* 1946, **9,** 43
4. W. J. Olsen, R. Evans and A. D. Dickson, *Cereal Chem,* 1944, **21,** 533
5. *Amer. Soc. Brewing Chemists Proc.* 1949, 165
6. W. J. Olson, M. T. Lowry and A. D. Dickson, *ibid,* 1948, 13
7. P. R. Whitt and R. L. Ohle, *ibid,* 1947, 45
8. D. B. West, J. *Ass. Off. Agric. Chemists*, 1954, **37,** 655
9. *ibid,* 1954, **37,** 89

The Determination of α-Amylase (2) as a test for pasteurised liquid eggs
using Starch Iodine

Introduction
It has been shown[1] that the reduction of α-amylase activity by heat treatment is directly related to the quantity of heat applied and is not influenced by any factor other than heat. The measurement of α-amylase activity has therefore been proposed as a measure of the efficiency of heat pasteurisation, in an analogous manner to the phosphatase test for milk pasteurisation[2,3,4]. Later work[5] has shown that the reduction in α-amylase activity is an efficient method of measuring the efficiency of heat pasteurisation of liquid whole egg, as measured by the destruction of heat resistant strains of salmonellae. The present test has been developed[5] as a quick and simple method of measuring residual α-amylase activity in pasteurised liquid whole egg, and became official in the United Kingdom on 1st January 1964[6]. **This test is not suitable for dried egg.**

Principle of the method
The α-amylase present in whole egg, when incubated with a standard starch solution, will degrade the starch and prevent the formation of the normal blue starch-iodine complex on the addition of iodine. The intensity of the blue colour formed on the addition of iodine to the incubated solution thus varies inversely as the residual α-amylase activity. The intensity of the blue colour is measured by comparison with a series of Lovibond permanent glass standards.

Reagents required
1. *Starch solution* Take an amount of soluble starch (analytical reagent grade) which is equivalent to 0.70 g dry weight (previously determined on a 1 g sample by drying at 100°C for 16 hours or at 160°C for 1 hour), make into a paste with cold, distilled water, pour quantitatively into 50 ml of boiling distilled water, boil for one minute, cool by immersion in cold water, add 3 drops of toluene and make up to 100 ml with distilled water.
   Make up fresh solution monthly, and store at room temperature.
2. *Trichloroacetic acid solution* 15% (w/v) trichloroacetic acid (CCl_3COOH analytical reagent grade) in distilled water.
3. *Iodine solution* Dissolve 25 g of analytical reagent grade potassium iodide (KI) in 30 ml of distilled water, add 12.70 g of iodine and make up to 1 litre. This approximately $0.1N$ stock solution is stable for 6 months. Before use, dilute 1 ml of stock solution with 99 ml of distilled water and then add 0.25 g of potassium iodide.

The Standard Lovibond Comparator Disc 4/26
This disc, which is designed for use with 25 mm cells, contains 7 arbitrary standards.

Technique
Place 15 g of whole egg into a 25 ml conical flask, or into a 7×1 inch boiling tube. Add 2 ml of starch solution (reagent 1). Incubate for 30 mins. in a water-bath at $44 \pm 0.5°C$. Pipette 5 ml of trichloroacetic acid solution (reagent 2) into a 100 ml flask, add 5 ml of the incubated egg-starch mixture and shake thoroughly. Add 15 ml of distilled water, mix thoroughly and filter. Pipette 10 ml of the filtrate into 2 ml of the iodine solution (reagent 3) in a 25 mm comparator cell. Place the cell in the right-hand compartment of the Lovibond Comparator and place an identical cell filled with distilled water in the left-hand compartment. Compare the colour produced in the right-hand cell with the permanent glass standards in the disc, using a standard source of white light, such as the Lovibond White Light Cabinet or, failing this, north daylight. A colour matching standards 1–3 is indicative of inefficient pasteurisation; a reading above 3 is considered satisfactory.

Notes

1. The reagents and procedures can be checked by preparing two control tubes at the same time as the experimental material is tested. In the first tube replace the egg by an equal amount of distilled water. In the second tube replace the starch solution by an equal volume of distilled water. In all other respects follow the above technique exactly. After processing, the colour from the solution in the first tube should be darker than any standard in the disc, and the colour from the solution in the second tube should be lighter than any standard in the disc.

2. All glassware should be soaked in 'Chloros' or 'Lysol' after use. Adhering egg should be washed off with water and, if necessary, with $N/10$ sodium hydroxide. The glassware should then be washed with chromic acid, followed by thorough rinsing with water and then distilled water. Particular care is necessary to avoid contamination with saliva.

References

1. J. Brooks, *J. Hyg. Camb.*, 1962, **60**, 145
2. H. D. Kay and W. R. Graham Jnr., *J. Dairy Res.*, 1935, **6**, 191
3. Statutory Instrument 1960, No. 1542, "*Milk and Dairying, The Milk (Special Designation) Regulations,* 1960"
4. A. H. Tomlinson, *Monthly Bull. Min. of Health & Public Health Lab. Service,* 1956, **15**, 65
5. D. H. Shrimpton, J. B. Monsey, Betty C. Hobbs and Muriel E. Smith, *J. Hyg. Camb.*, 1962, **60**, 153
6. Statutory Instrument 1963, No.1503,"*The Liquid Egg (Pasteurisation) Regulations* (1963)"

COLORIMETRIC CHEMICAL ANALYTICAL METHODS

The Determination of Anti-icing Additives (1)
using potassium dichromate

Introduction

This method has been developed for the determination of anti-icing additives in aviation fuels. The additive normally used consists of $99.6 \pm 0.04\%$ by weight of ethylene glycol monomethyl ether and $0.4 \pm 0.04\%$ by weight of glycerol. The method is applicable over the range 0.04 to 0.18% by volume of the additive (Note 1).

Principle of the method

The additive is extracted from the fuel with distilled water. The aqueous extract is then oxidised by heating with a standard acid dichromate solution. During the oxidation some of the orange dichromate ions are reduced to green chromous ions and the resulting colour of the solution depends on the relative proportion of these two ions, which is in-turn proportional to the amount of additive present. This colour is compared with a series of Lovibond permanent glass colour standards, which are calibrated directly in concentration of the additive.

Reagent required

Potassium dichromate reagent Dry approximately 6 g of potassium dichromate (analytical reagent quality) at 110°C for one hour. Cool 500 ml of distilled water in a 1 litre volumetric flask in an ice-bath; carefully add 375 ml of concentrated sulphuric acid; mix well and allow to cool. Weigh out 4.9035 g of the dried dichromate and dissolve this in the diluted acid. When the solution is cool, dilute to the 1 litre mark with distilled water.

The Standard Lovibond Comparator Disc 4/33

This disc covers the range 0.04 to 0.18% by volume of additive, in steps of 0.02% with an extra step at 0.15 and is calibrated for use with 25 mm cells.

Technique

Transfer, by means of 10 ml measuring cylinders, 10 ml of fuel and 10 ml of distilled water to a suitable 2 oz bottle fitted with a screw cap (Note 3). Replace the cap securely and shake the bottle and its contents steadily for 2 minutes. Place the bottle aside and allow it to stand for 5 minutes for the two layers to separate. Withdraw 5 ml of the lower (aqueous) layer by means of a 5 ml safety syringe pipette. Transfer this aqueous solution to a 60 mm diameter × 90 mm high polypropylene bottle (Note 4), taking precautions to ensure that no fuel is transferred at the same time, e.g. from the outside of the pipette. Add 10 ml of the dichromate reagent from an automatic dispensing pipette. Place the polypropylene bottle and its contents in a boiling water bath for exactly 10 minutes and then remove and quickly cool it to room temperature. Transfer the coloured solution to a 25 mm comparator cell and place this cell in the right-hand compartment of a Lovibond Comparator. Fill an identical 25 mm cell with distilled water and place this in the left-hand compartment. Compare the colour of the solution with the permanent glass colour standards in the disc, using the Lovibond White Light Cabinet to illuminate the comparator. If a White Light Cabinet is not available, then north daylight should be used for matching the colours. The figure shown in the comparator window when a colour-match is achieved represents the per cent by volume of anti-icing additive in the fuel sample. Record this to the nearest 0.01%, by interpolation between standards if necessary.

Notes

1. This method is also applicable to anti-icing additive of composition 98% by volume of ethylene glycol monomethyl ether and 2% by volume of glycerol.

2. The standards used in this disc were prepared in collaboration with, and have been checked by, the Technical Sales Dept. of Esso Petroleum Co. Ltd.

3. Suitable bottles may be obtained from French, Flint & Ormco Ltd., Ibex House, Minories, London, E.C.3. Order Pot No. 20. In the laboratory suitable stoppered flasks, etc., may be used as an alternative.

4. 25 × 150 mm test tubes made of a heat-resistant glass, such as Pyrex, may be used as an alternative to the polypropylene bottles, especially in a laboratory. The polypropylene bottles were chosen for the field testing of fuels. Bottles made of polyethylene (Polythene, Alkathene, etc.) should not be used, as they do not withstand boiling water.

The Determination of Anti-icing Additives (2)
using hexanitrato ammonium cerate

Introduction

The prevention of fuel system icing in some aircraft depends on the presence of an anti-icing additive in the fuel. The additive commonly used is ethylene glycol monomethyl ether (EGME). For complete protection it is necessary for the quantity of EGME to exceed a certain minimum level when the fuel enters the aircraft. It is therefore necessary to check the EGME content of the fuel at various points between the blending stage and the aircraft fuel tanks.

The present method[1,2], which is a modification by the author of his original method, has been developed to provide a field test for the determination of EGME in aviation turbine fuel. As no heating of the solutions is required during the colour development stage, the present test is more convenient for field use than *Method* 1.

Principle of the method

Ethylene glycol monomethyl ether is extracted from the fuel into water. On reaction with hexanitrato ammonium cerate this aqueous extract forms an orange red colour, the intensity of which is proportional to the concentration of EGME. The colour intensity is measured by comparison with Lovibond permanent glass colour standards.

As only water soluble impurities can interfere with this colour reaction, only the accidental presence of alcohol in the fuel could interfere with this determination.

Reagent required

Hexanitrato ammonium cerate solution Dissolve 150 g of hexanitrato ammonium cerate $((NH_4)_2.Ce(NO_3)_6)$ (ceric ammonium nitrate) in one litre of 0.25 N nitric acid (HNO_3). Filter through a Whatman No. 1 filter paper. This reagent is quite stable.

The Standard Lovibond Comparator Disc 4/44

This disc contains standards corresponding to 0, 0.05, 0.08, 0.10 and 0.15% of EGME by volume in fuel, based on the extraction of 50 ml of fuel with 10 ml of water.

Technique

Pour 50 ml of fuel into a 150 ml separating funnel. Add 10 ml of water, shake vigorously, allow the two layers to separate and run 5 ml of the lower, aqueous, layer into a comparator tube. Add 5 ml of the reagent solution, mix, and place the tube in the right-hand compartment of the comparator. Place the comparator before a standard source of white light such as the Lovibond White Light Cabinet or, failing this, north daylight and match the colour of the sample with the standards in the disc within 30 minutes of colour development.

Notes

1. If the aqueous extract appears cloudy, due to the retention of minute fuel droplets, this cloudiness can be removed by filtration through a Whatman No. 1 filter paper.

2. This disc replaces disc 4/43 which did not have the 0.08% step.

Reference

1. L. Gardner, *J. Inst. Petroleum*, 1969, **55**, 418
2. L. Gardner, *J. Inst. Petroleum*, 1971, **57**, 294

Barrett Scale
for the Colour Grading of Liquids

Introduction

One of the grading scales used in the colour grading of resins, shellacs and tar products is a series of solutions made from varying proportions of cobalt chloride, ferric chloride and potassium chromate in hydrochloric acid solution. These solutions cover a range from colourless to brown, known as the Barrett scale, and are compared with the sample in square bottles.

To avoid the trouble of handling such liquid colour standards, a Lovibond Comparator disc has been prepared which matches the colours of this series of solutions.

The Standard Lovibond Comparator Disc 4/36

This Disc covers the values 1, $1\frac{1}{2}$, 2, $2\frac{1}{2}$, 3, $3\frac{1}{2}$, 4, 5 and 6 on the Barrett scale, and must be used in conjunction with a 5 mm optical depth cell.

Technique

Pour the sample into a 5 mm Lovibond Comparator cell and place in the right-hand compartment of the Lovibond Comparator with the disc in position. Hold facing a uniform source of daylight and compare the colour against those in the disc. Interpolate between the steps if necessary.

Reference

"Synthetic Resins and Allied Plastics" by R. S. Morrell (Oxford University Press) Page 241

The Colour Grading of Beers, Malt Worts and Caramel Solutions
according to the E.B.C. (European Brewery Convention) Colour Scale

Introduction

The original E.B.C. colour scale was devised by Dr. Bishop in 1950 (*Journal of the Institute of Brewing,* Vol. 56, 373) and was adopted by the E.B.C. in 1953 (*Analytica,* October, 1953, 304E). Since then, modifications have been made (*Analytica* October 1958, 2–400E and *Journal of the Institute of Brewing,* 1965, 71, 471), and as a result of further development, a very considerable tightening of the grading tolerances towards the theoretically ideal scale has been achieved by entirely new methods of manufacture and selection. In order to take advantage of this increasing accuracy, it became desirable to increase observer discrimination in carrying out the test. To this end, the E.B.C. Lovibond 3-aperture Comparator has been introduced. (See Note).

The Standard apparatus and discs

The Lovibond E.B.C. 3-aperture Comparator[6] incorporates a prism unit and compartment for cells up to 40 mm optical path, white light cabinet (correct for colour temperature) with cooling fan, optical glass cell 25 mm and set of 4 discs as follows:—

Disc 1. EBC values 2, 3, 4, 5, 6, 7, 8, 9, 10.
Disc 2.   2.5, 3.5, 4.5, 5.5, 6.5, 7.5, 8.5, 9.5, 11.
Disc 3.   10, 12, 14, 16, 18, 20, 22, 24, 26.
Disc 4.   11, 13, 15, 17, 19, 21, 23, 25, 27.

Discs 1 and 3 are used in the left-hand compartment and discs 2 and 4 in the right-hand compartment, so that alternate steps are on each side of the sample.

Cells of the following optical depths are available and should be ordered as required:—

Code W680/1 mm   Code W680/15 mm
Code W680/2.5 mm   Code W680/25 mm
Code W680/5 mm   Code W680/40 mm
Code W680/10 mm

Technique
Institute of Brewing

The following instructions are reprinted by permission of the Council of the Institute of Brewing.

The colour match of the glasses is with pale worts and beers at the lower end of the scale and with dark worts and beers at the upper end of the scale. For some beers, intermediate sizes of cell will be required, because it is important that dark beers should be always measured at the upper end of the scale, and pale beers at the lower end. Obviously the minimum number of cells should be used and they should be of distinct sizes so that any two cannot be confused.

Every person entrusted with the measurement of wort or beer colour must be known to be free from colour blindness. To ensure this it is necessary to test each one by means of the book of charts by Ishihara.

Standard malt mash worts.—These should be protected from strong light during mashing and filtration and the colour should be measured as soon as possible. The worts should be bright before measurement in the comparator with E.B.C. discs.

Colours up to 10 (25 mm cell) are measured in the 40 mm cell. Colours from 10 to 26 are measured in 25 mm cell. Colours from 26 to 650 are measured in a suitable cell thickness to give a reading of between 20 and 26 units. Worts of above 650 colour are diluted ten times and measured in a cell thickness to give a reading of 20 to 26 units.

All results are calculated to that for 10% wort in a 25 mm cell.

Beers.—These should be filtered or fined-and-filtered so that they are bright before measurement.

Cell thicknesses are chosen as stated above and the reading is calculated for reporting to the value of the 25 mm cell.

Sugar and caramels.—These are prepared in 10% (w/v) solution and this is measured in the appropriate cell thickness either directly or after dilution as for malt mash worts. The value for the 10% solution in a 25 mm cell is reported.

E.B.C. method

In the official method, quoted under "References" below, it is stated:—

"The colour is to be measured on the standard, fine-grind-mash wort as soon as possible, in order to avoid darkening.

Where necessary, the wort should be clarified by the addition of 0.1% pure kieselguhr and refiltering. The first runnings should be returned to the filter".

Pale malt worts should be measured in a 25 or 40 mm cell. Munich type malt worts should be measured in a 5 or 10 mm cell. As far as possible, the figure measured should lie between 10 and 20 units. The result in each case should be calculated to that for a 25 mm cell.

A tolerance of ± 10% for sampling and analysis is to be allowed in comparing colour measurements between different laboratories.

Use of 3-aperture Lovibond Comparator

This comparator front is hinged forward to change the discs or to insert the cell containing the sample.

Prepare the sample according to the official procedure, and fill a cell of the selected size. Open the front of the Comparator, place the cell in the centre compartment (clear face towards operator) and close the Comparator. Switch on, and rotate the discs, one with each hand, until the colour of the sample is judged to fall between that of the colour standards. The discs should be rotated alternately, so that consecutive numbers are in position (e.g. 4 showing on the left and 4.5 on the right). Assess the colour value and calculate the corrected value for the standard depth (e.g. if a 10 mm cell has been used, multiply by 2.5 to bring it to the required figure for a 25 mm optical depth).

Keep the cells and colour standards scrupulously clean.

Note

Discs are also available for use in the Lovibond Comparator

Disc EBC/A values 2.0 to 6.0

Disc EBC/B values 6.0 to 10.0

Disc EBC/C values 10 to 18

Disc EBC/D values 19 to 27

References

1. *Journal of the Institute of Brewing,* 1950, **56**, 373
2. ,,   ,,   ,,   ,,   ,,   ,,   1952, **58**, 247
3. *Analytica,* European Brewery Convention, 1st Otcober 1953, Page 304E, and 2–400E, October, 1958
4. *Journal of the Institute of Brewing,* 1965, **71**, 471
5. *ibid* 1969. **75** 164
6. Page xxxvii

The Colour Grading of Benzene and Other Refined Lower Boiling Products of Coal Tar

Introduction

The colour grading of the refined lower boiling products of coal tar, by reference to a series of solutions containing known amounts of potassium dichromate and cobalt sulphate, has been adopted by:—

The British Standards Institution

The Indian Standards Institution

The National Benzole Association (U.K.)

The Standardisation of Tar Products Tests Committee (U.K.) and all of the above allow the use of the Lovibond Nessleriser Disc NU as an approved alternative to the impermanent liquid standards.

Principle of the method

The colour of the sample, in a Nessler cylinder, is compared with the permanent glass colour standards prepared to match the reference solutions prepared as follows:—

Solution A $0.1N$ potassium dichromate (4.91 g $K_2Cr_2O_7$/litre)

Solution B $0.1N$ cobalt sulphate (14.05 g $CoSO_4.7H_2O$/litre)

mixed in the following proportions and made up to 1 litre with distilled water.

Standard	Solution A	Solution B
1	0.8 ml	12.0 ml
2	1.0 ml	15.0 ml
3	2.0 ml	10.0 ml
4	2.0 ml	30.0 ml
5	7.5 ml	10.0 ml

The Standard Lovibond Nessleriser Disc NU

The disc is fitted with five standards which match the above solution standards.

Technique

Filter the sample and reject the first 10 ml of filtrate. Fill one of the Nessleriser glasses to the 50 ml mark with the filtered sample and place this in the right-hand compartment of the Nessleriser. Fill the other Nessleriser glass to the 50 ml mark with distilled water and place this in the left-hand compartment. Stand the Nessleriser before a standard source of white light, such as the Lovibond White Light Cabinet or, failing this, north daylight, and compare the colour of the product with the standards in the disc. In each case the colour standard is the limit of colour permissible.

The various standard limits for colours are as follows:—

Standard Colour	Product	British Standard	N.B.A. Test A 1960	Indian Standard
1	Benzene	1963–135/1	1	534 & 535
	Benzene for nitration	1963–135/2	2	
	Benzene 90's	1963–135/3	4	
	Toluene	1963–805 Sec. 1 & 2	6A	
	Toluene for nitration	1963–805 Sec. 3	7	
	Toluene 90's	1963–805 Sec. 4 & 5	8A	
	Xylene 3°	1963–458 Sec. 1, 2 & 3	10A	
	Xylene 5°	1963–458 Sec. 4, 5 & 6	11A	
	Xylene 10°	1963–458 Sec. 7	12	
	Xylene industrial solvent			539
2	Coal Tar solvent naptha 96/160	1963–479 Sec. 1, 2, 4 & 5	13A	
3	Motor benzole	1963–135/4	5A	
	Motor benzole low gravity	1963–135/5		
4	Coal Tar solvent naphtha 90/160	1963–479 Sec. 3	14A	
5	Coal Tar heavy naphtha 90/190	1963–479 Sec. 5	15A	

Note

It must be emphasised that the readings obtained with the Nessleriser and disc are only accurate provided that Nessleriser glasses are used which conform to the specification employed when the discs were being calibrated, namely that the 50 ml calibration mark shall fall at a height of 113 ± 3 mm measured internally.

References

1. British Standards 135, 458, 479 and 895 all of 1963
2. National Benzole Association, *"Standard Specifications for Benzole and Allied Products"*, 4th. edtn., 1960, Test Method A
3. Standardisation of Tar Products Tests Committee, *"Standard Methods for Testing Tar and its Products"*, 6th edtn., 1967, Test RLB 2-67
4. Indian Standards 534, 535 and 539, all of 1965

The Colour Grading of Butter

Introduction

The assessment of the colour of butter is important in dealing with marketing problems and local preferences, and a number of arbitrary printed colour scales have been offered from time to time. It was considered desirable by the dairy industry to obtain a colour scale based on actual samples over a long period of time, so that all variations might be embraced.

The technical control division of the Milk Marketing Board co-operated with The Tintometer Limited over a period of three years to obtain representative samples of butter produced in England and Wales over all seasons of the year. These were measured and recorded in Lovibond colour units, and the results plotted graphically to ascertain the complete range of colours during the whole period. From the averaged results, a colour scale was produced in permanent Lovibond glass standards which is suitable for the assessment and valuation of all usual butter samples.

Apparatus

Lovibond Comparator, Surface-viewing Stand (DB414) and White Light Cabinet.

Disc 4/34 with 9 permanent glass colour standards covering the Milk Marketing Board selected colour range.

Technique

Place a **freshly-cut** sample of butter in one of the porcelain dishes provided, cut surface upwards. The surface should not be smoothed down, as this changes the colour. Cut the sample of such a size that it can be laid straight into the dish with a minimum of contact with the knife.

Lay the sample in the base of the cabinet, underneath the centre hole of the Comparator, on the white opal glass which acts as a background to the glass colour standards. Revolve the disc and report the number of the nearest colour match.

Reference

Dairy Industries 1964. June Page 461

Determination of the 'Chemical Oxygen Demand' (C.O.D.)
using dichromate oxidation

Introduction

When water is polluted with organic substances, such as are found in sewage, these substances provide food for bacteria, which break them down into simple organic salts. In the process of break-down, oxygen dissolved in the water is consumed by the bacteria. Water which has lost its dissolved oxygen cannot support fish, and any fish present in such water will rapidly die. As sewage contamination of water is also harmful to man, both from its lack of oxygen and more particularly from bacteria present in the sewage, the detection of this type of pollution is a matter of great importance.

The determination of the 'Biochemical Oxygen Demand' is a measure of the organic pollution present in water, and this is one of the most important tests applicable to sewage effluents[1], as well as to many trade wastes. B.O.D. is defined[2] as ". . . . the amount of oxygen, in mgm, taken up during a 5-day period by one litre of the sample." It is therefore a time-consuming test. Where a more rapid assessment is required, or where the presence of trade wastes cause inhibition of bacterial action in the B.O.D. test, the use of the 4-hour permanganate test has been recommended[3,11] for which the Lovibond Comparator Oxygen disc may be used[10].

This C.O.D. test has been developed[4] to reduce further the time taken. It is a modified version of the well-known Chemical Oxygen Demand test[5]. It obviates the use of ferrous salt solution which is used in the titration method, and this is a great advantage as this solution deteriorates rapidly and requires constant checking. It has been shown, when used on crude dairy effluents, (i.e. where the B.O.D. is likely to be of the order of 200 ppm or above), to give results which are in close agreement with the B.O.D. determined by the 5-day incubation method. It may be necessary to adopt a different correlation chart for other types of sewage.

This rapid method is a chemical valuation of the oxygen capacity of an effluent, whereas the 5 day B.O.D. test is a measure of the bacterial uptake of oxygen during the 5 days. The two tests will therefore produce different results if

(a) there is present in the effluent, or in the reagents or apparatus, a chemical reducing agent not attacked by bacterial oxidation (See Note 3),

or (b) if the effluent contains some material which inhibits bacterial oxidation.

A suggested cross-check in these circumstances is the permanganate 4-hour test.

In the dairy and malting industries treated effluents are required by local River Boards to have a B.O.D. of 10-20 ppm. To meet this requirement the sensitivity of this test has been increased in order to enable C.O.D. values down to 10 ppm to be determined[6,7,8]. In order to achieve this sensitivity the composition of the oxidation mixture is varied according to the expected C.O.D. value. For convenience, discs corresponding to four ranges of C.O.D. have been prepared, each of which must be used with the appropriate oxidation mixture.

Principle of the method

The organic material present in the sample is oxidised by a standard amount of a potassium dichromate oxidising mixture. The excess of this reagent, after oxidation is complete, is measured by comparison with Lovibond permanent glass colour standards. By means of an appropriate factor the amount of oxidising reagent which has been consumed may be correlated with the 5-day B.O.D.

Reagents required

1. *N potassium dichromate solution ($K_2Cr_2O_7$ 49.03 g/litre)*
2. *Sulphuric acid (H_2SO_4 conc.)*
3. *Silver sulphate (Ag_2SO_4)*
4. *Distilled water*

All chemicals should be of analytical reagent quality. These reagents should be combined, in the proportions shown in Table 1 for the appropriate C.O.D. range, in the following manner:—

Clean a 5 litre flask with chromic acid by rinsing the inside surfaces and allowing to stand overnight. Rinse out thoroughly with distilled water and allow to drain. Measure the potassium dichromate solution (reagent 1) and the water (reagent 4) into the flask by means of a measuring cylinder. Slowly and carefully add about 200 ml of the conc. sulphuric acid (reagent 2), mixing throughly after each successive addition. Add the silver sulphate (reagent 3), breaking up lumps if necessary. Mix thoroughly until all the solid has dissolved. Then add the remaining 1600 ml of acid (reagent 2). This must be added slowly and carefully and the temperature of the mixture must be kept down by continuously cooling the flask under running water.

When completely cool store in brown glass bottles.

Table 1

Oxidation Mixture Composition for Various C.O.D. ranges

C.O.D. Range ppm	Quantities of Reagent required			
	Reagent 1 (ml)	Reagent 2 (ml)	Reagent 3 (g)	Reagent 4 (ml)
200–600	200	1800	10	nil
100–300	100	1800	5	100
40–120	40	1800	5	160
10–40	20	1800	5	180

The Standard Lovibond Comparator Discs 4/21, 4/21A, 4/21B and 4/21C

These discs cover the following ranges:—

Disc 4/21   200–600 ppm (mg/l) C.O.D. in 5 steps
Disc 4/21A 100–300 ppm (mg/l) C.O.D. in 9 steps
Disc 4/21B 40–120 ppm (mg/l) C.O.D. in 9 steps
Disc 4/21C 10–40 ppm (mg/l) C.O.D. in 4 steps

and must be used with the appropriate oxidation mixture (Table 1).

These discs are used in the Lovibond Comparator with standard 13.5 mm test tubes or cells.

Technique

Clean all flasks for this test with chromic acid, as described under preparation of oxidation mixtures. After cleaning, rinse with tap water followed by distilled water. Measure 20 ml of well mixed effluent into a 250 ml flask by means of a measuring cylinder. Add 25 ml of the oxidation mixture appropriate to the C.O.D. range expected (Table 1) by means of a safety pipette (Note 2), add 2 glass beads and mix thoroughly. Boil under reflux for one hour and then allow the mixture to cool. This usually takes about 30 minutes. Carefully transfer the liquid to a standard 13.5 mm test tube or comparator cell. Place the filled cell in the right-hand compartment of the comparator and match against the standards in the appropriate disc using the Lovibond White Light Cabinet. If this is not available, north daylight should be used for colour matching.

If the colour in the tube is darker than that of the highest standard in the disc, then repeat the determination either using the oxidation mixture and disc appropriate to a higher C.O.D. range or repeat after diluting the effluent (Note 1). Match the sample colour to the nearest half step on the disc, by interpolation between standards if necessary, and read off the C.O.D. value from Table 2.

COLORIMETRIC CHEMICAL ANALYTICAL METHODS — 'Chemical Oxygen Demand' (C.O.D.)

Notes

1. The volume taken for oxidation must always be 20 ml. If diluted effluent is used the total volume must always be adjusted to 20 ml as shown in Table 2. If the dilution is such that less than 10 ml of effluent are to be measured, then accuracy can be increased by diluting a larger sample and then taking a 20 ml aliquot of the final solution, e.g. instead of taking 5 ml of effluent and adding 15 ml of water, take 25 ml of effluent, add 75 ml of water, mix thoroughly and then take a 20 ml aliquot of the diluted sample.

2. The oxidation mixture is virtually concentrated sulphuric acid and must be handled at all times with this fact in mind. On no account should a pipette be used which could result in acid reaching the mouth.

3. This test is not a particularly good index of the quality of effluents from biological treatment plants, which may contain a relatively large proportion of their organic content in a biologically undegradable form: C.O.D. gives a measure of total organic content, and cannot distinguish between biodegradable and undegradable organic matter. It is more suitable for assessing the strength of sewages and trade wastes[11].

However, as the C.O.D. figure will always be higher than the B.O.D. figure in such cases, a sample which meets River Board requirements on the C.O.D. test must necessarily meet the B.O.D. requirement.

Table 2

C.O.D. for Dairy and Malting Effluents allowing for dilution and using Discs 4/21, 4/21A, 4/21B and 4/21C

a) Disc 4/21

Effluent ml	Water ml	Disc reading ppm								
		200	250	300	350	400	450	500	550	600
20	0	200	250	300	350	400	450	500	550	600
15	5	250	350	400	450	550	600	650	750	800
10	10	400	500	600	700	800	900	1,000	1,100	1,200
8	12	500	650	750	900	1,000	1,150	1,250	1,400	1,500
5	15	800	1,000	1,200	1,400	1,600	1,800	2,000	2,200	2,400
2	18	2,000	2,500	3,000	3,500	4,000	4,500	5,000	5,500	6,000

b) Disc 4/21A

Effluent ml	Water ml	Disc reading ppm								
		100	125	150	175	200	225	250	275	300
20	0	100	125	150	175	200	225	250	275	300
15	5	125	175	200	225	275	300	325	375	400
10	10	200	250	300	350	400	450	500	550	600
8	12	250	325	375	450	500	550	625	700	750
5	15	400	500	600	700	800	900	1,000	1,100	1,200
2	18	1,000	1,250	1,500	1,750	2,000	2,250	2,500	2,750	3,000

c) Disc 4/21B

Effluent ml	Water ml	Disc reading ppm								
		40	50	60	70	80	90	100	110	120
20	0	40	50	60	70	80	90	100	110	120
15	5	50	70	80	90	100	120	130	150	160
10	10	80	100	120	140	160	180	200	220	240
8	12	100	130	160	180	200	230	250	280	300
5	15	160	200	240	280	320	360	400	440	480
2	18	400	500	600	700	800	900	1,000	1,100	1,200

d) Disc 4/21C

Effluent ml	Water ml	Disc reading ppm						
		10	15	20	25	30	35	40
20	0	10	15	20	25	30	35	40
15	5	15	20	25	35	40	45	55
10	10	20	30	40	50	60	70	80
8	12	25	40	50	65	75	90	100
5	15	40	60	80	100	120	140	160
2	18	100	150	200	250	300	350	400

References

1. E. W. Taylor, "*The Examination of Waters and Water Supplies,*" Churchill, London, 1958
2. L. Klein, "*River Pollution, Vol. 1, Chemical Analysis,*" Butterworth, London, 1959
3. Thresh, Beale and Suckling, "*Examination of Waters and Water Supplies,*" 6th editn., Churchill, London, 1949
4. G. T. Lloyd and J. H. L. Royal, *J. Soc. Dairy Tech.,* 1964, **17**, 11
5. American Public Health Association, "*Standard Methods for the Examination of Water and Wastewater,*" 12th Edn., 1965, New York
6. S. H. Baker-Munton, *Proc. Eur. Brew. Conv.* 1963, Elsevier Publ. Co. Amsterdam
7. W. Wolfner, *J. Inst. Brew.,* 1964, **70**, 446
8. W. Wolfner, *Brauwissenschaft,* 1964, **10**, 378
9. K. Vogl, G. Schumann and G. Peters, *Monatsschrift für Brauerei* 1965 June, 166
10. See page **601**
11. Ministry of Technology. Water Pollution Research Laboratory, "*Notes on Water Pollution No.44—Simple methods for testing sewage effluents,*" March 1969, Page 4

The Colour Grading of Coconut Oil
by means of the Ceylon Research Institute Scale

Introduction

The Coconut Research Institute of Sri Lanka (Ceylon) has recommended a colour scale for grading coconut oil which is reproduced as a series of 9 colours in Lovibond glass mounted in a Comparator disc. It is compared with the oil sample in a 1″ optical depth.

The Standard Lovibond Comparator Disc 4/25

The disc has 9 colours with the arbitrary numbers 1 to 9.
It is used with a 1 inch cell.

Technique

Pour the sample into the 1″ cell and place this in the right-hand compartment of the comparator, with the disc 4/25 in position. Hold the comparator facing a uniform source of white light. Revolve the disc until the nearest match is found and report the figure showing in the indicator window. Interpolate if necessary.

Conditions as to temperature, filtering etc, should be agreed between the parties concerned.

The Colour Grading of Cresylic Acid

Introduction

The colour of cresylic acid is specified by the Standardisation of Tar Products Tests Committee (S.T.P.T.C.)[1] in terms of colour standards composed of chemical solutions, with the alternative of using the Lovibond colour standards herein described. The British Standards listed in the references below[2] have adopted the same procedure.

The Standard Lovibond Comparator Disc 4/8

The disc contains seven colour standards, No. 1 being the palest. The odd numbered standards are of a yellow tint while the even numbered standards are of a red tint.

This disc is designed for use with a 1 inch cell.

Technique

The sample is poured into a 1 inch glass cell, and the cell is placed in the right-hand compartment of the Lovibond Comparator. The left-hand compartment is left empty.

The comparator is held about 18 inches from the eye, facing a standard source of white light, such as the Lovibond White Light Cabinet. If this equipment is not available then north daylight should be used. The disc is then rotated and the colour of the sample is compared with the permanent glass colour standards in the disc. The colour of the sample shall be reported as not darker than that of the appropriate standard, as shown in the window at the lower right-hand corner of the instrument.

Any material darker in colour than 6 or 7 shall be reported as Grade 8.

Note

The S.T.P.T.C. Method[1] allows the use of the Lovibond Tintometer as an alternative, using the formula

$$C(olour) = Y(ellow) \text{ units} + 3R(ed) \text{ units}$$

Use a 6 inch cell unless C exceeds 80, in which case use a 1 inch cell and record this in the report.

References

1. S.T.P.T.C. (U.K.), *"Standards Methods for Testing Tar and its Products"* 6th edtn., 1967, Test No. PC 1–67
2. British Standards 517, 521, 522, and 524, all of 1964

The Colour Grading of Fats and Tallows
using the F.A.C. Scale

Introduction

For many years the colour scale approved by the Fats Analysis Committee (F.A.C.) of the American Oil Chemists Society, and known universally as the F.A.C. colour scale, consisted of a series of sealed glass tubes of coloured liquid. This was not a scientifically developed uniformly spaced scale, but an arbitrary collection of 26 colours which was found to fulfil the requirements of the trade. However considerable difficulty was experienced in exactly reproducing the colours, and the liquid standards were found to be impermanent. In 1964 the Committee asked the Tintometer Ltd. to reproduce, in permanent glass standards, the colours of an agreed master set of the coloured solutions. The resulting set of glass standards was then adopted as the primary standard for the F.A.C. scale. No attempt was made to correct any of the existing uneven intervals in the scale, nor to unify the hues. The F.A.C. scale as it stood was reproduced in permanent unfadeable Lovibond glass and made available by Tintometer Ltd., in a 3-aperture comparator devised and approved by the Committee, under the designation A.O.C.S., Cc 13a–64.

The Standard Lovibond Comparator Discs 4/32A, B, C and D

These four discs are supplied as part of the outfit (AF 229 illustrated on page xxxvii) which consists of a three aperture comparator, 4 discs (interchangeable to use any two at a time), a sample tube, a prism unit (accessory), and a standardised source of white light to the A.O.C.S. requirements in a fan-cooled cabinet. The discs are numbered as follows:—

Disc A. 1, 5, 9, 11a, 11b, 11c, 13, 17, 21
Disc B. 3, 7, 11, 11a, 11b, 11c, 15, 19, 23
Disc C. 25, 29, 33, 37, 41, 45
Disc D. 27, 31, 35, 39, 43

Technique

Place the sample in the test tube, and place this tube between the discs in the sample compartment of the 3-aperture comparator. Discs 1 and 3 are always used in the left-hand position and discs 2 and 4 in the right-hand position. Revolve the discs until the sample colour is matched by one of the standards or falls between two of them. Alternatively, when working to a specification, set the discs at the limiting values to ascertain if the sample falls within the specified range.

Note

To change the discs remove the two thumb screws at the back of the instrument and take off the front plate.

Reference

American Oil Chemists Society, *"Official and Tentative Methods"*, Method Cc 13a–64

Gardner 1963 Scale
The Colour Grading of Varnishes, Oils, Laquers and Resins

Introduction

The Gardner Colour Scale, with various revisions, has been in use in the U.S.A. for many years as a simple designation, in a one dimensional way, of the colour of oils and varnishes; and it is frequently quoted in other countries as well. The version originally used was known as the 1933 Gardner Scale (A.S.T.M. D154), and this consisted of a series of sealed glass tubes containing solutions of ferric and cobalt chlorides, graduated in concentration in 18 steps. These were supplied by the Gardner Laboratory Inc. of U.S.A. in the standard Gardner-Holdt viscosity tubes having an internal diameter of 10.75 mm, and the sample was placed in a similar tube.

These tubes have been criticised from time to time on the grounds of the impermanence of the colour and lack of reproducibility, and in 1953 a committee of the Inter-Society Colour Council of America proposed that the first 8 standards, in which fading had been most troublesome, should be replaced by solutions of potassium chloroplatinate, and that the whole scale of 18 steps should be defined by the spectral transmission of a 10 mm layer of the liquid, rather than by the chemical composition.

This recommendation was adopted by the A.S.T.M. in June 1958 as A.S.T.M. Specification D1544–58T for the testing of the colour of transparent liquids. Subsequently the whole scale was reviewed both by the American Oil Chemists Society and by the American Society for Testing Materials, and certain parts of the scale were respaced and the specification stated in terms of trichromatic co-ordinates. The lower part of the scale, values 1–9, is now slightly redder and darker than the original scale, and in particular, numbers 7, 8 and 9 are darker.

This finally approved scale bears the designation A.S.T.M. D1544 63T and A.O.C.S. code Td 1a–64T.

The specification not only defines the colours, but the apparatus in which they must be used, which is a 3-aperture comparator with its own built-in standardised source of white light (See Note).

The Standard Lovibond Comparator Discs 4/30-1 and 4/30-2

These discs are only supplied as part of the complete outfit (AF 228 illustrated on page xxxvii), which consists of a 3-aperture comparator, 2 discs, a tube for the sample, and a standardised light source to A.S.T.M. specification.

The discs are numbered as follows Disc 1 is used on the left-hand side of the sample and disc 2 on the right-hand side.

Disc 1. 1, 3, 5. 7, 9, 11, 13, 15, 17
Disc 2. 2, 4, 6, 8, 10, 12, 14, 16, 18

Technique

Place the sample in the test tube and place the tube in the compartment between the discs in the comparator. Revolve the discs until the colour of the sample matches one of the standards or falls between two of them.

Alternatively, when working to a specification, set the discs at the two limiting values to ascertain whether the sample falls within the permitted tolerance.

Note

Discs are also available for use in the Lovibond Comparator

Disc 4/30AS Gardner 1 to 9 } 1963 Scale
Disc 4/30BS Gardner 10 to 18 }

These discs must be used with a 10 mm cell.

References

1. American Society for Testing Materials, Method A.S.T.M. D1544–63T
2. American Oil Chemists Society, Method Td. 1a–64T

The Determination of Hardness
using Palin Eriochrome Tablets

Introduction

The hardness of water is a measure of its soap-consuming capacity; in practice it corresponds to the calcium and magnesium ion content. The soap test originally used for its estimation has now been superseded by complexometric methods using dyes such as Eriochrome Black T, developed by Biedermann and Schwarzenbach.[1] Instead of the sharp colour change necessary to denote the end-point in titrimetric methods, a modification has been introduced by Palin[2] to produce a gradual colour change with increasing hardness, so permitting the use of a simple colorimetric procedure.

To overcome the deterioration on storage associated with solutions of Eriochrome Black T the indicator is used in tablet form, which as an additional advantage provides much greater convenience in practice.

Principle of the Method

At a suitable pH a solution of Eriochrome Black T has a blue colour. In the presence of calcium and magnesium ions the colour changes to wine-red. The addition of the indicator to the water sample thus produces a colour which enables the hardness to be determined by comparison with a series of Lovibond permanent glass standards.

Reagents required

Palin Hardness (Eriochrome) tablet
Supplementary Buffer Powder (for highly alkaline boiler water). See note 4.

The Standard Lovibond Comparator Disc 4/38

This disc covers the range 0 to 60 ppm (mg/l) of hardness, calculated as $CaCO_3$, in steps as follows — 0, 5, 10, 15, 20, 25, 30, 40, 60.

By a five-times dilution of sample the range may be extended to 300 ppm, so enabling waters to be placed in the officially adopted classification[3].

The master disc, against which all reproductions are checked, was tested and approved by Dr. A. T. Palin.

Technique

Place 10 ml of sample in the comparator tube or cell and add one tablet. Break the tablet in two first if required to assist disintegration. Crush the tablet with the flattened end of a clean glass or plastic stirring rod and mix until dissolved. (Alternatively crush the tablet first with only one or two ml of sample in the tube, and afterwards make up to the 10 ml mark with sample. Dissolve by continued mixing as before). Now place this cell in the right-hand side of the comparator. In the left-hand side place a similar cell containing sample only, to serve as a blank. The colour of the treated sample develops by the time the tablet has dissolved. Match against the standards in the disc, using a standardized source of white light as the Lovibond White Light Cabinet, or, failing this, north daylight. See Note 6.

Make certain that the tablet is *completely dissolved* before matching.

To extend the range of the disc by dilution proceed as follows. Place 2 ml (or other appropriate quantity) of sample in the cell, add one tablet and crush as before. Then make up to the 10 ml mark with distilled water. Continue mixing to dissolve tablet completely. Prepare a blank cell by dilution of the same quantity of sample to 10 ml with distilled water and place in the left-hand side of the comparator. (If the original sample is reasonably colourless, use distilled water in the blank cell). Match against the disc as before, finally multiplying the observed reading by the dilution factor (5 in the example quoted, where 2 ml of sample was taken).

Notes

1. Samples such as boiler waters which may be coloured or turbid should first be filtered through a Whatman No. 1 paper. Discard the first runnings so as to obtain as clear a filtrate as possible for testing.

2. The following classification is officially adopted[3] for waters of varying hardness.

Hardness ppm	Description of water
0–50	Soft
50–100	Moderately soft
100–150	Slightly hard
150–200	Moderately hard
200–300	Hard
over 300	Very hard

3. The following recommendations are made[4] for the hardness of boiler and boiler feed-water.

Low pressure boilers:	5 ppm or less (carbonate treated)
	2 ppm or less (phosphate treated)
	Feed water under 20 ppm
Medium pressure boilers:	2 ppm or less
	Feed water under 10 ppm
High pressure boilers	2 ppm or less
	Feed water under 1 ppm

When internal alkali treatment is used, the boiler water hardness should be under 10 and preferably under 5 ppm. When phosphate is used for internal treatment, the desired boiler water hardness is 2 ppm or less.

4. In testing boiler waters of high caustic alkalinity, there may be interference with the test due to the colours produced being too red. To check this, add a small quantity of supplementary buffer powder (about 0.2 g) to the sample cell and mix vigorously until there is no further change of colour. Allow undissolved crystals to settle and match against the disc as before.

5. If result is required in grains per gallon, the conversion is 7 gpg = 100 ppm. In the German system of Degrees of Hardness, $1°DH = 10$ mg/l CaO. Therefore, to convert ppm $CaCO_3$ to $°DH$, multiply by 0.056.

6. A reading at the limit of the disc scale (60) should always be checked by repeating with a sample diluted with distilled water, to make certain that the true value is not higher. This may not be apparent from the colour produced in the first test. Always dilute until a reading below the figure of 60 is obtained, and then multiply by the appropriate factor.

References

1. W. Biedermann and G. Schwarzenbach, *Chimia*, 1948, **2**, 56
2. A. T. Palin, "Colorimetric Determination of Water Hardness" *Water and Water Engineering*, 1967, **71**, 109
3. Ministry of Health, "*Water Softening*," H.M. Stationery Office, London, 1949
4. P. Hamer, J. Jackson and E. F. Thurston, "*Industrial Water Treatment Practice*," Imperial Chemical Industries Ltd., London, 1961

The Colour Grading of Liquids
using the chloroplatinate/cobaltous chloride scale
(also known as the HAZEN or the APHA scale)

Introduction

A method for measuring the colour of potable water in terms of some easily reproducible colour standard is essential to water engineers responsible for the distribution of supplies for public consumption.

Colour determinations are useful for:—
    Detecting any irregular contamination of the supply due to flood waters, etc.,
    Checking the efficiency of the filter beds and of the decolourising treatments
    Maintaining a standard of clarity.

Turbid or coloured waters are unappetising, and even slight variations in the colour of the supply result in complaints from consumers.

The American Public Health Association in *"Standard Methods for the Examination of Water, and Wastewater"*, 1965, specifies that the colour of water shall be expressed by comparison with a series of solutions containing known amounts of platinic chloride and cobalt chloride. These solutions are known as "Hazen Colour Standards" and are adopted in British Standard 2690: 1970 Part 9 and in ASTM D1209–62 and D1045–58.

These Hazen Colour Standards have been matched with Lovibond glasses to produce permanent colour standards, and a disc containing nine colour standards exactly matching the Hazen standards is available for use with the Nessleriser. The unit adopted in this method of measurement is the colour produced by 1 milligram of platinum (present as chloroplatinate in association with cobaltous chloride) per litre of water; this is equivalent to 1 ppm of platinum.

These same colours have been adopted in the higher ranges for Tar Bases and for Polyester Resins (see below).

The Standard Lovibond Nessleriser Disc NSA

Disc NSA covers the range corresponding to 5, 10, 15, 20, 30, 40, 50, 60, 70, units on the Hazen Colour Scale. Water samples giving a higher reading should be diluted with glass-distilled water to bring their colour within the range of the disc, and due correction should be made for the dilution when reporting the result.

Special Lovibond Hazen Nessleriser Discs and Comparator Disc 4/28

Although, as stated above, darker samples should be diluted, there is a need in other trades for the Hazen Scale in its higher ranges, and the following discs are available to meet this need:—

Nessleriser Disc NSB range 70–250 units with the following steps: 70, 85, 100, 125, 150, 175, 200, 225, 250.

Nessleriser Disc NSX range 50–300 units, corresponding to 50, 60, 70, 80, 100, 150, 200, 250, 300. This disc was specially made to conform to British Standard 3532:1962 — Unsaturated Polyester Resin Systems.

Discs CAA and CAB cover the range 0 to 70 using a depth of 250 mm to give greater discrimination. CAA covers 0–30 and CAB covers 30–70.
For use only with the Lovibond comparator with Nessler attachment.

Comparator Disc 4/28 range 50–500 units with standards for 50, 75, 100, 150, 200, 250, 300, 400, 500. This disc is designed for use with a Lovibond Comparator and a 40 mm cell.

Technique

Disc NSA. Fill a Nessleriser glass to the 50 ml mark with the water under examination and place it in the right-hand compartment of the Nessleriser. Leave the left-hand compartment empty. Place the Nessleriser in the Lovibond White Light Cabinet and compare the colour of the sample with the colours in the disc.

The numbers on the disc represent the Hazen number of the water under test: thus a colour numbered 50 is that of the standard solution containing 50 ppm of platinum. The values for colours falling between those in the disc may be estimated by interpolation. Samples giving a reading above 70 units should be diluted with glass-distilled water to bring them within range of the disc.

Discs NSB and NSX Proceed as above, but obey any special regulations laid down by a particular specification. For example B.S. 3532 states that samples with colours above 300 should be diluted with a colourless solvent.

Discs CAA and CAB As above but using 250 mm tubes with the Lovibond 1000 Comparator, Nessler attachment, and White Light Cabinet.

Disc 4/28 Place the sample in a 40 mm cell in the right-hand compartment of a Lovibond Comparator and leave the left-hand compartment empty. Place the comparator in the Lovibond White Light Cabinet and match the colour of the sample against the colours in the disc.

This disc was made at the request of firms in the plastics industry, where it is used for grading the colour of various resins and plasticisers.

Notes
1. The White Light Cabinet must always be used with these discs.
2. It must be emphasised that the readings obtained by means of the Nessleriser and discs NSA, NSB, NSX are only accurate provided that Nessleriser glasses are used which conform to the specification employed when the discs were calibrated, namely that the 50 ml calibration mark shall fall at a height of 113 ± 3 mm, measured internally.

References
1. American Public Health Association, *"Standard Methods for the Examination of Water and Wastewater"*, 12th edtn., 1965
2. American Society for Testing Materials, Standards D 1045–58 and D 1209–62
3. British Standards 2690:1970 Part 9, 3532:1962
4. Standardisation of Tar Products Tests Committee, *"Standard Methods for Testing Tar and its Products"*, 6th Edn, 1967, Test GB 3–67

The Colour Grading of Honey

Introduction

The colour grading of clear honey for marketing and exhibition purposes is well established, and among the scales used are the Lovibond Series 52 brown brewing glasses and the German Pfund scale, in which the figures quoted refer to distances in millimetres along a wedge of brown glass. To enable users of the Lovibond Comparator to compare results, a conversion table is appended.

The Standard Lovibond Comparator disc 295175

This disc has been made up using the Lovibond Series 52 Brown Scale and calibrated in terms of Pfund Equivalent values 8, 17, 34, 48, 83 and 114, and is used in conjunction with a 10 mm cell.

Technique

Clear the honey by minimal warming if necessary, and stand to clear bubbles as far as possible. Pour the sample very carefully into the cell to avoid entrapping air, and tap the cell with the finger to encourage any bubbles to rise. Open the comparator and place the cell in the right-hand compartment of the comparator. With the disc in position, close the comparator and hold it facing a good source of daylight or a corrected artificial source such as the Lovibond White Light Cabinet.

Find the nearest match by revolving the disc. The relationship between the disc and the International Scale is shown below:

Honey Grading according to the Pfund Colorimeter		Lovibond Honey Disc No. 295175
INTERNATIONAL SCALE:	PFUND	PFUND EQUIVALENTS
Water White	0–8 mm	8
Extra White	9–17 mm	17
White	18–34 mm	34
Extra Light Amber	35–48 mm	48
Light Amber	49–83 mm	83
Amber	84–114 mm	114
Dark Amber	114 and below	

COLORIMETRIC CHEMICAL ANALYTICAL METHODS

The Estimation of Humidity with Cobalt Thiocyanate Papers

Introduction

It is often desirable to be able to measure the local humidity in small spaces or crevasses which may not be in dynamic equilibrium with the atmosphere, or which are otherwise not suitable for normal hygrometric tests. Examples of this are found in the measurement of humidity in badly ventilated areas during the inspection of buildings; against surfaces, such as concrete floors, which are not in moisture equilibrium with the atmosphere; within stored grain; in the micro-meteorological environments which are of interest in ecological studies; and in gases other than air, e.g. the measurement of the moisture content of gas mains. In applications such as these the humidity may be conveniently estimated by means of specially prepared cobalt thiocyanate papers and permanent colour standards against which the colour changes in these papers may be assessed. The staff of The Tintometer Limited collaborated with Mr. M. E. Solomon of The Department of Scientific and Industrial Research, Pest Infestation Laboratory, to produce permanent colour standards for use with the Lovibond Comparator, and suitable papers impregnated with cobalt thiocyanate are produced by BDH Chemicals Ltd, Poole.

The following gives a brief summary of the work, sufficient to enable the apparatus to be used: a full account has been published[3]. A general account, not applicable to the method in its present form, was given in an earlier paper.[1]

Principle of the method

Specially impregnated papers are exposed, in the atmosphere to be measured, for two hours. The paper is then transferred with forceps and immersed in a vessel of liquid paraffin to prevent any further change in colour. The colour of the paper is then assessed against the permanent glass colour standards provided in the Lovibond Comparator Disc, the papers being viewed by reflected light.

Apparatus

The Paper

The paper is a thin cellulose "condenser tissue," impregnated especially for use with these discs by BDH Chemicals Ltd, Poole. The amount of salt in the paper has been standardised at 0.55 mg anhydrous $Co(CNS)_2$ per square cm.

The stock of paper should not be exposed to extreme temperatures or humidities, and it is preferable to keep it, at least some time before being used, in a desiccator or some similar piece of apparatus at a humidity between 60% and 70% relative humidity. The paper is provided in sheets, and in order to avoid contamination, clean forceps and scissors should be used in cutting it up. A suitable size for each piece is $1\frac{1}{2}$ cm $\times$ 3 cm.

The Comparator

A Lovibond Comparator with Surface Viewing Stand DB414 is used and it should be set up in a Lovibond White Light Cabinet DB416, or failing that facing a good source of north daylight (in the northern hemisphere). Pieces of opal glass and cover glasses are available for use with this test.

The Standard Lovibond Comparator Discs 4/16A and 4/16B

There are two discs available containing 18 standards, labelled with humidity values in steps of 5% from 100% down to 60% relative humidity (Disc B), and then in steps of 10% from 70% down to zero (Disc A). The colours have been matched against papers exposed at the specified humidities and then mounted in oil as described below.

Technique

Expose a small piece of the impregnated paper, about $1\frac{1}{2}$ cm $\times$ 3 cm., at the site where the humidity is to be estimated, for about 2 hours, then remove it with forceps and immediately immerse it in a small bath of laboratory grade liquid paraffin to prevent any further change in colour. To match the colour, remove the test paper from the oil, place it on a piece of the white opal glass, double it over, and cover it with a piece of clear glass. Usually, enough oil will have been transferred with the paper to fill the space which should be left between the edges of the paper, and the edges of the cover glass, thus sealing it off from the outside atmosphere; more oil may easily be added if necessary. Once immersed in the oil, the test paper may be kept for some hours or even days before matching unless the humidity being measured is below 20% relative humidity, in which case the matching should be done as soon as possible. During the waiting period the papers should not be subjected unnecessarily to extremes of temperature, as for instance, being left in direct sunlight.

Place the test paper, mounted as above described, on the floor of the comparator stand, vertically underneath the right-hand viewing aperture. Beside it on the left, vertically below the left-hand viewing aperture, is a white tile. Place the comparator in the White Light Cabinet or facing north daylight, and, standing squarely above it, look down and rotate the disc until the nearest match is obtained. For greater accuracy, it is advisable to make two matchings of each paper.

Since the colours of the paper vary somewhat with temperature, the matching should be done if possible at the same temperature as at the site where the paper was exposed, or else with the papers at approximately 20°C, which is the temperature at which the colours of the glass standards were matched. If the temperature at which the matching is done differs appreciably from 20°C, allowance should be made for this as shown in the attached table of corrections.

In matching a paper against the coloured glasses, the aim should be to assess the redness and blueness of the paper relative to that of the glasses. For example, a paper may be a little redder and less blue than the 70% standard, and a good deal bluer and less red than the 75% standard. In such a case, it would be accorded a value between the two, but nearer the 70%.

Notes

1. In a research paper[3] in the *Bulletin of Entomological Research,* M. E. Solomon gives an account of the variations which can contribute to errors in the final result, such as differences between batches of paper, differences in observer readings, the effect of exposing at one temperature and matching at another, and the effect of temperature on the colours produced, and this has all been tabulated. The general conclusion is that it can be claimed with certainty that from humidities of 100% down to 60% the result can be relied upon within an accuracy of plus or minus 5%, and lower humidities than this to about plus or minus 10%. When great care is taken and proper precautions observed, very much better figures than these can, of course, be obtained. In particular, relative differences in humidity can be estimated to quite a high order of accuracy. The method is simple and easy to use and does not require any elaborate equipment.

2. A complete kit for carrying out this test is available, details of which may be obtained from The Tintometer Ltd., Salisbury.

References

1. M. E. Solomon, *Ann. appl. Biol.,* 1945, **32**, 75
2. M. E. Solomon, *Bull. ent. Res.,* 1951, **42**, 543
3. M. E. Solomon, *Bull. ent. Res.,* 1957, **48**, 489
4. M. E. Solomon, *J. Inst. Heating & Ventilating Eng.,* 1958, **26**, Sept.

TABLE Corrections to be made when papers are matched at temperatures other than 20°C.

TEMP. °C	\multicolumn{19}{c}{Approx. relative humidity indicated by paper}
	103	100	95	90	85	80	75	70	65	60	55	50	45	40	35	30	25	20	18
40°	.	.	+3.5	+5	+6	+6.5	+7.5	+8.5	+9	+10	+11	+11.5	+12.5	+13.5	+14	+12.5	+11	+10	+2
35°	.	.	+3	+4	+4.5	+5	+6	+6.5	+7	+8	+8.5	+9	+10	+10.5	+10	+9	+8	+3.5	.
30°	.	.	+2	+2.5	+3	+3.5	+4	+4.5	+5	+5.5	+6	+6.5	+7	+6.5	+6	+5.5	+5	+1.5	.
25°	.	.	+1	+1.5	+1.5	+2	+2	+2.5	+2.5	+3	+3	+3.5	+3.5	+3	+3	+2.5	+2	+0.5	.
15°	.	−1	−1.5	−1.5	−2	−2	−2.5	−2.5	−3	−3	−3.5	−3.5	−3	−3	−2.5	−2	−1.5	.	.
10°	.	−2	−3	−3.5	−4	−4.5	−5	−5.5	−6	−6.5	−7	−6.5	−6	−5.5	−5	−3.5	−2	.	.
5°	.	−3	−4.5	−5.5	−6	−7	−8	−8	−9.5	−10.5	−10	−9	−8.5	−8	−6.5	−4.5	−3	.	.
0°	−3	−5	−6	−7.5	−9	−10	−11	−12.5	−13.5	−13.5	−12.5	−11.5	−11	−10	−8	−5.5	−3.5	.	.
−5°	−4.5	−6.5	−8.5	−10	−11.5	−13.5	−15	−16.5	−17	−16	−15	−14	−13	−11.5	−9	−6.5	−4	.	.

To exemplify the use of the Table: if a paper reads 75% R.H. at 35° the corrected value is 75 + 6 = 81% at 20°C.

The Colour Grading of Liquids by Comparison with Iodine Solution Colours

Introduction

Among the many inorganic solutions which have been suggested as colour standards for the grading of liquids, a series of different concentrations of iodine in aqueous potassium iodide solution is commonly used for grading brown liquids.

As an improvement on using liquid standards, permanent glass colour standards for the Lovibond Comparator have been made to match these iodine solutions.

The Standard Lovibond Comparator Discs 4/23 and 4/24

Disc 4/23 covers the range of colours represented by 10, 20, 30, 40, 50, 60, 70, 80, 90 parts per million iodine in potassium iodide solution.

Disc 4/24 covers the range of colours represented by 100, 150, 200, 250, 300, 350, 400, 450, 500 parts per million iodine in potassium iodide solution.

Technique

Pour the sample to be graded into a 13.5 mm Lovibond moulded cell or test tube and place this in the right-hand compartment of the Lovibond Comparator with the colour disc in position. Hold facing a uniform source of daylight or in a Lovibond White Light Cabinet and compare the colour against those of the disc. Interpolate as necessary between the steps.

The Determination of Developed Lactic Acid in Milk and Milk Products

Introduction

A rejection test based on the amount of lactic acid developed in raw milk was described by Taylor & Clegg[2], who established the relationship between developed lactic acid and keeping quality. A modification to give greater sensitivity over a more restricted range was subsequently described by Pickering & Clegg[3] for use as a rapid "field" test. This test, with the great advantage of the simplified procedure made possible by the use of the Lovibond Comparator and permanent glass colour standards, has many applications in the dairy world for testing the lactic acid content of milk and milk products, and will be of use to farmers, milk distributors, and milk processors.

This lactic acid test requires no controlled temperature water bath, and none of the apparatus needs to be sterile.

Principle of the method

Milk proteins are precipitated by means of barium chloride, sodium hydroxide, and zinc sulphate, and ferric chloride is added to the filtrate to produce a yellow ferric lactate complex. The intensity of this yellow colour, under the conditions of the test, is proportional to the concentration of lactic acid present.

Reagents required

1. *Barium chloride* ($BaCl_2.2H_2O$) 197.5 g per litre
2. *Sodium hydroxide* (NaOH) 1.32 N
3. *Zinc sulphate* ($ZnSO_4.7H_2O$) 225 g per litre
4. *Ferric chloride* ($FeCl_3.6H_2O$) 5 g in 100 ml of $N/8$ HCl. This should be diluted, one volume of solution to four volumes distilled water, immediately before use, to give a 1% solution.

The Standard Lovibond Comparator Disc 4/17

The 6-standard disc covers the range 0–0.05% lactic acid in 5 steps of 0.01%, with a first step labelled 0 to represent the colour of the ferric chloride in a "blank" solution.

Technique

Place 25 ml of the milk in a 1″ diameter stoppered boiling tube graduated at 25 ml. Add 5 ml each of reagents 1, 2 and 3 in that order. Stopper, shake thoroughly, and stand for at least 30 seconds. Filter through a fluted 12.5 cm Whatman No. 40 filter paper into a Lovibond Comparator test tube graduated at 10 ml already containing 0.5 ml of the 1% ferric chloride reagent, until the liquid reaches the 10 ml mark. Mix, and place in the right-hand compartment of the comparator. Use a "blank" of water in a similar tube in the left-hand compartment. Read against the standards in the disc, using a standard source of white light, such as the Lovibond White Light Cabinet or, failing this, north daylight.

Notes

1. This method has been advocated as a more suitable platform test than the Resazurin test because the latter may reject milks containing micro-organisms which have no effect on the keeping quality of pasteurised milk. On the other hand, tests for keeping quality of raw milk may have little interest when most milk is destined for pasteurisation. Pickering & Clegg state that this test, which detects poor quality in raw milk, may assist in raising the level of palatability and the keeping quality of pasteurised milk.

2. Titratable acidity must not be confused with lactic acid. Fresh milk when titrated with $N/9$ caustic soda may require, for example 1.6 ml of the soda solution to attain the end point. This is sometimes erroneously referred to as (in the example quoted) 0.16% lactic acid. In fact, this is merely a measure of the buffering power of the milk, due largely to the casein.

References
1. E. R. Ling, *J. Sci. Fd. Agric.,* 1951, **2,** 279
2. P. B. Taylor & L. F. Clegg, *J. Dairy Research,* 1958, **25,** 32
3. A. Pickering & L. F. Clegg, *Dairy Industries,* 1958, **23,** 325

Lubricating Oil and Petrolatum Colour
using the ASTM D1500 scale

Introduction

The American Society for Testing Materials issued a new specification ASTM. D.1500-64 to replace the old "Union" D.155 colour scale (also known as the NPA scale). This new scale is more uniformly spaced and more precisely defined. It is intended for the colour grading of petroleum products such as lubricating oils, heating oils, diesel fuel oils, and petroleum waxes, and has been adopted by other standardising bodies[1-7]. The apparatus in which the scale is to be used is also defined.

The Seta Lovibond Comparator is produced to conform to this specification.

The Standard Apparatus

The Seta Lovibond Comparator AF760, exactly in conformity with ASTM. D.1500, using cylindrical glass sample jars, and with 16 standards in one wheel. Illustrations on page xxxvi.

Technique

Place in the sample container enough sample to cover the viewing field. If the sample is not clear, heat to 6°C above its cloud point and measure the colour at that temperature. Petroleum waxes, including petrolatum, should be heated 11° to 17°C above the congealing point. If the sample is darker than 8, mix 15 volumes of sample with 85 volumes of solvent kerosine. Rotate the disc until the sample is matched or until it falls between two standards. Report the figure, if an exact match to one, or the number of the next darker standard if intermediate between two, with the addition of "Lighter than". If the sample has been diluted, add the abbreviation "DIL".

References

1. U.K. Institute of Petroleum I.P. 196/66
2. American Standard ASA Z11.109–1960
3. France. NF T60–104
4. Germany DIN 515–78
5. Belgium NBN T52–109
6. Japan JIS K2824
7. I.S.O. 2049 : 1972

The Colour Grading of Milk and Milk Products

Introduction

In conjunction with Dr J. G. Davis,[1,2,3] The Tintometer Ltd has evolved three sets of permanent glass colour standards for the routine grading and checking of the colour of processed milk.

It is not intended to suggest that a good quality product must match any particular standard colour in the disc, but the discs permit a very simple matching procedure to be applied in routine checking, so that undesirable colours may be avoided, and colour be correlated with methods of preparation and acceptability of the product. The discs also facilitate standardisation within the industry, by introducing a very simple, single-figure, record of colour.

Colour is an essential characteristic of all foodstuffs, and acceptability by the consumer is decided mainly by colour and flavour. A "good" colour is an essential selling point, for processed milk as for other items, and it therefore becomes important for the manufacturer to exercise close control in this direction.

From the manufacturing point of view, much information regarding correctness of technique, quality of materials, and adequacy of storage conditions may be obtained from a consideration of the colour, and these discs are therefore of value from both the manufacturing and the selling angle.

Apparatus required

Lovibond Comparator, Surface Viewing Stand DB414 and White Light Cabinet.

The Standard Lovibond Comparator Discs 4/18A, 4/18B and 4/18C

Disc A   Sweetened Condensed Skimmed Milk (9 arbitrary standards)
Disc B   Sterilised Milk (7 arbitrary standards)
Disc C   Evaporated Milk (6 arbitrary standards)

For each disc, a large number of samples of the product from various sources was collected, and glass colour standards prepared to match under the conditions of the test. As milks vary in composition and colour, it is impossible to prepare standards which exactly match every possible sample, but by taking a very large number of samples it was possible to arrive at a compromise scale which gave a satisfactory match to every product tested.

Technique

A sample of the product to be tested is placed, in the porcelain tray, beneath the right-hand aperture of the comparator. Beside it, on the left, is placed a standard block of magnesium carbonate. The appropriate disc is placed in the comparator and rotated until the nearest match is found.

References

1. J. G. Davis and J. S. Bell, *Food,* 1958, **27,** 139
2. J. G. Davis and P. J. Stubbs, *Food Manufacture,* 1958, **33,** 501
3. J. G. Davis and P. J. Stubbs, *Dairy Engineering,* 1958, **75,** 262

The Determination of Penicillin in Milk
using 2,3,5-triphenyltetrazolium chloride (TTC)

Introduction

The presence of antibiotics in milk can cause difficulties in yoghurt production and in cheese manufacture. Fears have also been expressed[1] that the ingestion of milk containing traces of antibiotics may result in the sensitisation of the consumer to the antibiotic. This could result in a severe reaction if the antibiotic is later prescribed in the relatively large amounts normally used medicinally. Such a reaction could, at worst, be fatal and, at best, denies the patient the benefit of the antibiotic treatment. A further danger is that the continued ingestion of low concentrations of antibiotics could result in the evolution of micro-organisms resistant to the drug. This contamination may be present in the milk as the result of treating mastitis by the infusion of one or more antibiotics. Antibiotic residues may be found in the milk from animals so treated for up to 5 days or even longer, after the last infusion.

Published evidence[2,3] suggests that penicillin is the antibiotic most frequently found in milk. The present test[4] has therefore been developed to provide a simple, yet sensitive, routine method for the estimation of penicillin in milk, and has been adopted in British Standard 4285[7].

Principle of the method

The milk sample, diluted as necessary with penicillin-free sterilised milk, is inoculated with *Str. thermophilus* 'B.C.,' a strain of this organism which is used in commercial yoghurt production (Note 1). After a short incubation at 44–45°C, 2,3,5-triphenyltetrazolium chloride (TTC) is added. In the presence of actively metabolising cells this is changed into red formazane, but if metabolism is inhibited by penicillin or other inhibitors no colour is produced. The concentration of antibiotic in the milk is thus inversely proportional to the intensity of the red colour produced on incubation, which is estimated by comparison with Lovibond permanent glass standards. The presence of penicillin in milk is established if its inhibitory effect can be removed by adding penicillinase.

Reagents required

1. *Str. thermophilus* 'B.C.' This organism is normally stored as a freeze-dried culture from which it can be used after only two sub-cultures in milk (Note 1). Dilute culture 50:50 with quarter strength sterile Ringer solution for use in test.

2. *Penicillin-free sterilised milk* This is best prepared by processing a pre-tested penicillin-free bulk raw milk supply. Alternatively sterilised milk which has been pre-tested for the absence of penicillin may be used.
3. *Penicillinase* AVM 'Penase' (Note 2), is diluted with distilled water to give a working solution of 1,000 units per ml. This working solution can be kept for up to two weeks in cold store without loss of activity.
4. *Penicillin* Solution-tablets of Benzylpenicillin, buffered, each containing 15,000 units of benzylpenicillin, (obtainable from BDH Chemicals Ltd.*) should be used in the test. These should be freshly dissolved and diluted with distilled water to give a final strength of 1 iu/ml, for calibration purposes.
5. *2, 3, 5-Triphenyltetrazolium Chloride (TTC)* 1% aqueous solution.

Apparatus required

1. The Lovibond Comparator fitted to Milk Test Stand DB415. (This same apparatus is used for the Aschaffenburg & Mullen Phosphatase test and for the Resazurin test), with White Light Cabinet.
2. Lovibond disc 4/22
3. A water bath at 44–45°C
4. Lovibond "Resazurin" test tubes (AF 215TT) calibrated at 10 ml.

The Standard Lovibond Comparator Disc 4/22

This disc[6] contains 4 standards, labelled 0, 1, 2, 3, respectively, corresponding to the colours of standards prepared according to Liska and Calbert[5].

Technique

(a) *Estimation and identification of antibiotics* Pipette 5 ml of the milk (note 3) into each of a pair of "Resazurin" tubes, and make up to the 10 ml mark with penicillin-free sterilised milk. To one tube add 0.2 ml of penicillinase working solution (reagent 3). Inoculate each tube with 1 ml of *Str. thermophilus* culture solution (reagent 1), stopper with sterile rubber bungs, mix by inverting twice and place in a water bath, at 44–45°C. After incubating for 1½ hours add to each tube 1 ml of TTC solution (reagent 5), mix by inverting, and incubate for a further hour. Remove tubes from incubator, again mix by inversion and place, in succession, in the right-hand compartment of the milk test stand. Place an identical tube filled with milk, diluted to the same degree with the same sterilised milk as the sample but not incubated, in the left-hand compartment as a blank. Compare the colour of the incubated milk with that of the permanent glass standards in the disc, using a standard source of white light, and interpolating between standards if necessary.

(b) *Interpretation of results* If both the sample and the penicillinase-treated control tube match standard 3 after incubation, then no penicillin is present in the sample. If the control matches standard 3 and the sample matches, or lies between, one of the standards 1–3, then the penicillin concentration can be obtained from the Table. If the sample has a colour below standard 1 then the test should be repeated using a dilution of 1:4, 1:19 or 1:99, as necessary, until the colour falls within the range covered by standards 1–3.

If the colour of the control tube falls below that of standard 3 after incubation, this indicates either that the activity of the penicillinase has decreased or that an antibiotic other than penicillin is present. Test the penicillinase activity by the method given in section (c) below. If this is normal, then another antibiotic is present and the test method is no longer applicable.

*"*Penicillin solution tablets BPC54 code 331592 D*"

TABLE
Colour intensities developed in the presence of various concentrations of penicillin

Penicillin iu/ml of milk	TTC colour value after $2\frac{1}{2}$ hours at 44-45°C Samples diluted with penicillin-free milk				
	1:1 (Penase)	1:1	1:4	1:19	1:99
Nil	3	3	3	3	3
0.01	3	$2\frac{1}{2}$	3	3	3
0.02	3	$1\frac{1}{2}$	3	3	3
0.03	3	<1	$2\frac{1}{2}$	3	3
0.05	3	0	$1\frac{1}{2}$	3	3
0.075	3	0	<1	$2\frac{1}{2}$	3
0.10	3	0	0	2	3
0.20	3	0	0	1	3
0.30	3	0	0	<1	$2\frac{1}{2}$
0.50	3	0	0	0	2
1.0	3	0	0	0	1
1.5	3	0	0	0	<1

(c) *Standardisation procedure* Either daily, or with each batch of milks tested, set up an extra tube containing pre-tested sterilised milk and inoculate with the test organism (reagent 1), and carry out the test as detailed above. If the colour of this tube after incubation is found to be below that of standard 3 it indicates that the activity of reagent 1 is below standard and that a fresh subculture must be prepared.

At weekly intervals, checks should be carried out on the activity of the test organism (reagent 1) and of the efficiency of the penicillinase solution (reagent 3). These are carried out in the following manner. To 25 ml of the sterilised penicillin-free milk add 1 ml of penicillin solution (reagent 4). Mix well. This milk will now contain approximately 0.04 iu/ml of penicillin. Take 3 sterile "resazurin" tubes and label these 1, 2 and 3 respectively. Add 0.2 ml of Peñase solution (reagent 3) to tube 2. Into tubes 1 and 2 pipette 5 ml and into tube 3, 2 ml of the sterilised milk containing 0.04 iu/ml of penicillin. Make up to the 10 ml mark with sterilised penicillin-free milk, inoculate and proceed as in section (a) above. After incubation, tube 1 should show no activity (0), tube 3 should give some activity (2 approx), while tube 2, which contains added penicillinase, should give good activity (3 approx). A low or no activity in tube 2 indicates that reagent 3 is unsatisfactory. High activity in all tubes shows that either reagent 4 and/or 1 are unsatisfactory. They are best renewed. A low activity in all tubes shows reagent 1 to be unsatisfactory, provided that the bath temperature was correct and the sterilised milk satisfactory.

Notes

1. Propagation of *Str. thermophilus* 'B.C.', is carried out daily by inoculating penicillin-free sterilised milk at the rate of 1 drop per 10 ml and incubating overnight at 37°C. If propagation cannot be carried out for several days, sterilised milk is best inoculated with 10% of the clotted culture and cold stored. On the day before the culture is required the inoculated milk is transferred from the cold store into a 44-45°C bath until clotted, and from it sterilised milk is inoculated at the usual rate, followed by overnight incubation at 37°C.

In case of loss of activity or sensitivity, as indicated by the tests in section (c) above, a new freeze-dried culture should be used.

2. AVM 'Penase' may be obtained from A.V.M. Laboratories, P.O. Box 79, Wrexham, Denbighshire, Wales.

3. Raw milk should be tested within 2 hours of sampling, or if that cannot be done, the samples must be cooled to 40°F within this time and then maintained at this temperature until tested. Tests should not be carried out later than the day after sampling.

References

1. *New Scientist,* 1964, **22**, 467
2. A. J. Overby, *Dairy Sci. Abs.* 1954, **16**, 2
3. F. C. Storrs & W. Hiett-Brown, *J. Dairy Res.* 1954, **21**, 337
4. R. C. Wright & J. Tramer. *J. Soc. Dairy Tech.* 1961, **14**. 85
5. B. J. Liska & H. E. Calbert. *J. Dairy Sci.* 1958, **41**, 776
6. J. Tramer *J. Soc. Dairy Tech.* 1964, **17**, 95
7. British Standard 4285:1968

COLORIMETRIC CHEMICAL ANALYTICAL METHODS

The Determination of Permanganate Value
using acid permanganate

Introduction

The action of bacteria on organic matter found in polluted waters is to convert the organic matter into simple organic salts. During this process oxygen dissolved in the water is consumed by the bacteria. The consequent reduction in the dissolved oxygen content of the water is harmful to other forms of life and may result in the rapid death of fish.

The standard test for organic pollution in water is the "Biochemical Oxygen Demand" (B.O.D.) test[1]. B.O.D. is defined[2] as "... the amount of oxygen, in mg, taken up during a 5-day period by one litre of the sample". This is obviously a time-consuming test and where a more rapid assessment is required, or where the presence of trade wastes causes inhibition of bacterial action in the B.O.D. test, the use of the 4-hour permanganate test has been recommended[3,4].

The present test has been developed[4] as a field test, using the standard disc prepared for the determination of dissolved oxygen in waters[5].

Principle of the method

The sample is incubated at 27°C for 4 hours with a measured volume of acid potassium permanganate solution. The loss of permanganate resulting from oxidation of the organic matter in the sample is estimated by reacting the permanganate with potassium iodide and estimating the iodine which is liberated, by comparison with a series of Lovibond permanent glass colour standards.

Reagents required

1. *Sulphuric acid solution* Add 1 volume of sulphuric acid (H_2SO_4) to three volumes of water.
2. *N/80 potassium permanganate solution* (0.395 g of $KMnO_4$ per litre)
3. *Potassium iodide (KI)*

Chemicals used in the preparation of reagents should be of analytical reagent quality.

The Standard Lovibond Comparator Disc 3/3

This disc covers the range 4 to 12 parts of dissolved oxygen per million by weight, in steps of 1. Its readings are converted into Permanganate Values (mg/l) by the formula given in the next section. For permanganate values above 66, the sample must be diluted and the answer multiplied by the appropriate factor.

Technique

Measure 10 ml of sulphuric acid solution (reagent 1) and exactly 50 ml of potassium permanganate solution (reagent 2) into a clean 12 oz stoppered bottle. Add 50 ml of the sample. Mix thoroughly and then heat in a water bath at 27°C for 4 hours. Mix again after 1 hour if the sample contains much suspended matter. At the end of 4 hours add about 0.5 g of potassium iodide (reagent 3). Mix thoroughly. Dilute the contents of the bottle to 420 ml. If the solution is cloudy filter out the suspended matter. Place the clear solution in a 13.5 mm comparator tube or cell and place this in the right-hand compartment of the comparator. Stand the comparator before a standard source of white light, such as the Lovibond White Light Cabinet or, failing this, north daylight, and match the colour of the liquid with the standards in the disc.

Calculate the Permanganate Value by means of the following formula:—

Permanganate value (mg/l) = 100 — (8.4 × disc reading)

e.g. if disc reading = 5 parts per million
permanganate value = 100 — (8.4 × 5) = 100 — 42
= 58

References

1. E. W. Taylor, *"The Examination of Waters and Water Supplies,"* Churchill, London, 1958
2. L. Klein, *"River Pollution, Vol. 1, Chemical Analysis"* Butterworth, London, 1959
3. Thresh, Beale and Suckling, *"Examination of Waters and Water Supplies"* 6th Editn, Churchill, London, 1959
4. Ministry of Technology, Water Pollution Research Laboratory, *"Notes on Water Pollution No. 44 – Simple Methods for Testing Sewage Effluents"*, p.5, March, 1969
5. *Oxygen (Method 2)* page 297

COLORIMETRIC CHEMICAL ANALYTICAL METHODS

The Determination of Phosphatase in Milk (1)
using the Aschaffenburg and Mullen Phosphatase Test

Introduction

The milk phosphatase test of Kay and Graham[1] was formerly widely used in Britain both as the official test for pasteurised milk and as a routine test in plant control. Owing to the long incubation time which the test requires, viz: 24 hours, its value lies in future remedial action rather than in preventing the distribution of faultily treated milk. Bessey et al[2], working with blood phosphatase, showed that colourless alkaline solutions of the substrate *p*-nitrophenylphosphate are rapidly hydrolysed, and that the product of hydrolysis, *p*-nitrophenol, being yellow in alkaline solution, serves as a direct indicator of enzyme activity. Upon this substrate Aschaffenburg and Mullen[3] based a new rapid test for milk phosphatase which, they state, will detect minor processing faults after an incubation period of only 30 minutes, and which can be made more searching by the simple expedient of incubating for a further 90 minutes. Tramer and Wight[4] developed permanent colour standards for this test, simplified the technique, and established the relationship between the results of the new procedure and those of the Kay and Graham test. The new test is said to be much less liable to be affected by interfering substances. Subsequently to the Tramer and Wight modification, Aschaffenburg introduced a simplified buffer[6], which is cheaper and has better keeping qualities. This does not alter the colours produced in the test, and has been adopted in this revised version of the Lovibond test. No change whatever has been made in the disc colours. This test was adopted[8] as the official test for phosphatase in milk in 1960, and became effective as from January 1961. This is also an official test in France[9], and is adopted as a British Standard[10].

This test is not applicable to milk products.

Apparatus required

1. The Lovibond 1000 Comparator, complete with Milk Test Stand DB415 and White Light Cabinet (this is the same apparatus as is used for the Resazurin test)
2. Disc APTW7, containing seven permanent yellow glass colour standards
3. Test tubes AF215TT made of colourless glass, 13.5 mm internal diameter, conforming to B.S. 625. Soft rubber stoppers to fit
4. A water bath at 37°C
5. 1 ml pipettes—one for each milk sample, 5 ml pipette for buffer substrate

All glassware must be scrupulously cleaned by prolonged treatment in chromic acid, followed by rinsings with tap water and distilled water. After drying, all glassware should be stored under dust free conditions. New rubber stoppers should be boiled repeatedly in distilled water and dried and stored free from dust. Stoppers should always be boiled in distilled water after use.

Reagents required

1. *Buffer solution*

   3.5 g anhydrous sodium carbonate (Na_2CO_3 analytical reagent grade) ⎫

   1.5 g sodium bicarbonate ($NaHCO_3$ analytical reagent grade) ⎬ per litre distilled water ⎭

2. *Substrate* Disodium *p*-nitrophenyl phosphate.

3. *Buffer substrate* Place 0.15g of the substrate in a 100 ml measuring vessel and make up to the mark with the buffer solution. This mixture will remain in good condition for up to one week in a refrigerator. It should not be used if the colour exceeds 10 on the APTW disc when viewed by transmitted light in a 25 mm cell in the comparator, using a blank of distilled water on the left-hand side.

The Standard Lovibond Comparator Disc APTW7

The standard disc (reference APTW7) contains seven glass colour standards, corresponding to the colour produced by 0, 6, 10, 14, 18, 25 and 42 micrograms of *p*-nitrophenol per ml of milk.

Technique

A blank of boiled milk is required, and this must be of the same type as that undergoing test. With highly coloured milk, such as that from Channel Island cows, it is advisable to prepare a separate blank.

Place 5 ml of the buffer substrate solution in the requisite number of test tubes, stopper, and bring to 37°C in the water bath. Then add 1 ml of the milk to be tested, close with a rubber stopper and mix well. Incubate for 30 minutes at 37°C. The blank prepared from boiled milk and buffer substrate solution in the above proportions is incubated at the same time. At the end of 30 minutes remove from the water bath, mix well, and place in the comparator stand. The blank is placed on the left-hand ramp, so that it comes under the colour standards, the test sample is placed on the right-hand ramp, so that it is viewed through the centre aperture of the disc.

The diffusing screen in front of the blue daylight filter in the lighting cabinet may be removed to increase the illumination. Both tubes must be equally illuminated and free from shadows.

Place the comparator with stand on a bench facing a good source of north daylight if a White Light Cabinet is not available, direct sunlight must be avoided—at such a height that the operator is able, standing above it, to look down direct on to the two apertures. The disc is then revolved until the sample is matched. This should be done as soon as possible after removal from the bath. Readings falling between two standards are estimated and recorded by affixing a plus or minus sign to the figure for the nearest standard.

Interpretation

The interpretation of results suggested by Tramer and Wight[4] is as follows:—

Disc reading after 30 minutes incubation	*Report*
0 or trace	Properly pasteurised
6	Doubtful
10 or over	Underpasteurised

Milk giving a "doubtful" reading should be replaced for a further 90 minutes incubation. The suggested interpretation of the two hour test is as follows:

Disc reading after two hours incubation	*Report*
0 to 10	Properly pasteurised
>10 to 18	Slightly underpasteurised
>18 to 42	Underpasteurised
>42	Grossly underpasteurised

The more searching two hour test should be applied whenever time permits, but examination of the tubes after 30 minutes is useful in that it will reveal any serious fault in processing, and permits, in cases of doubt, correct classification by the simple expedient of re-examination after incubation for a further 90 minutes.

References

1. H. D. Kay and W. R. Graham Junr., *J. Dairy Research,* 1935, **6,** 191
2. O. Bessey, O. H. Lowry, M. J. Brook, *J. Biol. Chem.,* 1946, **164,** 321
3. R. Aschaffenburg and J. E. C. Mullen, *J. Dairy Research,* 1949, **16,** 58
4. J. Tramer and J. Wight, *J. Dairy Research,* 1950, **17,** 194
5. J. Wight and J. Tramer, *Dairy Ind.,* 1952, **17,** 54
6. R. Aschaffenburg, *Dairy Ind.,* 1953, **18,** 316
7. A. H. Tomlinson, *Monthly Bull., Min. of Health and the Public Lab. Service,* 1956, **15,** 65
8. Statutory Instrument 1960 No. 1542, *"Milk and Dairies, The Milk (Special Designation) Regulations,* 1960"
9. *J. Offic. Repub. Francaise,* 1955, 1037, p. 28
10. British Standard 4285: 1968

The Determination of Phosphatase in Milk (2)
using disodium phenyl-phosphate
(usually known as the Kay and Graham method)

Introduction

The milk phosphatase test depends on the detection of the enzyme, phosphatase, which is always present in raw milk but is destroyed at the temperature necessary for efficient heat treatment or pasteurisation. The absence of phosphatase indicates that the milk has been adequately heat treated, while its presence points to insufficient heating or to contamination with raw milk.

Principle of the method

When the milk containing phosphatase is incubated with disodium phenyl-phosphate, free phenol is liberated, and the quantity so liberated, which may be determined by a colorimetric method, is an approximate measure of the phosphatase present in the milk.

When the test is carried out under standardised conditions, a pale blue colour is produced, which, in the absence of phosphatase (i.e. in the case of correctly pasteurised milk) does not exceed a certain value.

Reagents required

1. *A buffer-substrate solution* containing 1.09 g disodium phenyl-phosphate buffered with 11.54 g sodium diethyl barbiturate and made up to one litre. This solution should be prepared with distilled water previously saturated by shaking with pure chloroform. If kept in a refrigerator it will be stable for 48 hours, but if kept for a longer period should be tested before use. In some cases it may be more convenient always to prepare a fresh solution and for this purpose compressed tablets are supplied. Two tablets should be dissolved in 50 ml of distilled water previously saturated with pure chloroform.

2. *Folin and Ciocalteu's phenol reagent*[8] *(Stock)*

Sodium tungstate ($Na_2WO_4.2H_2O$)	100 g
Sodium molybdate ($Na_2MoO_4.2H_2O$)	25 g
Distilled water	700 ml
Phosphoric acid (H_3PO_4 syrupy, 85%)	50 ml
Hydrochloric acid (HCl conc)	100 ml
Lithium sulphate ($Li_2SO_4.H_2O$)	150 g

   Dissolve the sodium tungstate and sodium molybdate in the water in a 1500 ml flask fitted with a reflux condenser. Ground glass joints should be used. Add the phosphoric and hydrochloric acids and reflux gently for 10 hours. Cool, add the lithium sulphate, a further 50 ml of distilled water and 4-6 drops of liquid bromine. Leave for 2 hours, and then remove excess bromine by boiling the mixture under the fume hood without the condenser for 15 minutes. Cool, dilute to 1 litre with distilled water, and filter.

   The reagent should have a golden yellow colour. Any reagent with a greenish tint should be discarded. If stored in a refrigerator this reagent should be stable for at least four months.

   It is important that this reagent be carefully protected from contact with dust and any reducing substance.

3. *Sodium-hexametaphosphate* $((NaPO_3)_6)$ 5% solution w/v

4. *Folin and Ciocalteu's reagent (Test reagent)*
   Add 1 volume of reagent 2 to 2 volumes of reagent 3. This mixture is stable for several weeks.

5. *Sodium carbonate solution* A 14 per cent solution of analytical reagent grade anhydrous sodium carbonate (14 g Na_2CO_3 in 100 ml distilled water). It is advisable to standardise this solution by titration.

The Standard Lovibond Comparator Disc APP9

The original work was carried out by Kay and Graham in a standard Lovibond Tintometer using a cell of optical depth 13 mm in which to examine the liquid. When this instrument is used, red and yellow glasses are also used to obtain a match, but only the blue units are recorded.

For greater convenience, a special Lovibond disc has been prepared, for use in the Lovibond Comparator, which incorporates the necessary red and yellow glasses with the blue to produce a match to the average hue produced with milks. A 25 mm glass cell is used to hold the sample, but the values recorded on the disc have been calculated to relate to the work done originally on the 13 mm cell.

Disc APP9 contains 9 standards, equivalent to
0.5   1.5   1.8   2.1   2.3   2.5   3.0   4.0 and 6.0 blue units

Technique

Tests should be carried out in duplicate.

To 10 ml of the buffer substrate solution contained in a test-tube, add 0.5 ml of the milk and mix. Add 3 drops of chloroform, stopper the tube, mix and incubate at 37°C $\pm 1°$ for 24 ± 2 hours. At the end of this time, cool, add 4.5 ml of the Folin and Ciocalteu test reagent, mix, allow to stand for 3 minutes, and filter into another tube marked at 10 ml. To 10 ml of this filtrate add 2 ml of the sodium carbonate solution, mix, and place in boiling water for exactly two minutes. If a number of tubes are being tested, the two minutes must be timed from the moment the water has recommenced boiling. Cool, filter if necessary, and match the colour in Lovibond blue units in one of the instruments mentioned above, using a standard source of white light such as the Lovibond White Light Cabinet or, failing this, north daylight.

Milks which give readings of 2.3 or less are classified as "giving a negative phosphatase test" or "sufficiently heat treated". If a reading above 2.3 but below 6.0 is obtained, report as "insufficiently heat-treated". Readings above 6.0 are reported as "grossly undertreated". Raw milk usually gives more than 30 blue units. A reading above 2.3 may signify incorrect pasteurisation or the addition of raw milk, and it is not possible to differentiate on these findings alone.

Control tests

Keep in the refrigerator some milk from all the samples which are under test, and at the completion of the above 24 hour incubation, carry out control tests as below on fresh portions of all those samples which gave a reading above 2.3.

Mix thoroughly 10 ml of the buffer substrate with 4.5 ml of the diluted Folin and Ciocalteu's reagent, add 0.5 ml of the milk and again mix. Stand for 3 minutes and filter into one of the test tubes. To 10 ml of the filtrate add 2 ml of the sodium carbonate solution, mix, and place the tubes in boiling water for 2 minutes. If a number of tubes are being tested, the two minutes must be timed from the moment the water has recommenced boiling. Cool, filter if necessary, and read the colour.

The colour should not exceed 1.5 Lovibond blue units. A control higher than 1.5 indicates either the presence of phenolic substances in the milk (which would not occur if the milk were uncontaminated and correctly pasteurised) or faulty reagents or faulty technique. If two or more different milk samples examined at the same time show a "control value" of over 1.5, faulty reagents or technique may fairly safely be assumed.

Test of reagents

The reagents should be tested by incubating a tube containing buffer substrate and chloroform *only*, with each batch of samples. The test is then completed in the usual way. This colour should not exceed 0.5 Lovibond blue units.

Notes

Precautions

Samples kept at room temperature should preferably be examined within 18 hours of having been heat-treated. They may be kept for longer in a cold store at 32° to 40°F, but must be raised to room temperature before testing. The usefulness of the test is dependent upon the strict attention to all details as given. In particular, the Ministry of Health pamphlet calls attention to the following points:—

1. The reagents must be kept in a dark cool place, and protected from dust
2. Tests must not be carried out in direct sunlight
3. Freshly boiled distilled water must be used
4. A fresh pipette must be used for each milk sample
5. Glassware must be chemically clean, and pipettes must not be contaminated by saliva
6. The test reagents and apparatus must be kept apart from phenols, disinfectants and detergents containing phenols, and soap containing carbolic acid. Bottle caps and rubber stoppers must not be used until they have been shown to be free from phenolic impurities
7. A sample must not be tested if it shows a taint or clot on boiling

Discussion

In experiments conducted by Kay and Graham it was observed that when the time of pasteurisation was exactly half an hour and the temperature 142.5°F (2.5°F below the statutory minimum) the colours obtained in a series of experiments on commercial mixed milk varied from 4.4 to 5.0 Lovibond blue units. In a larger series of unpublished experiments conducted by Neave, the lowest figure obtained with commercial milk was over 3.8 blue units (with milk samples from individual cows rather larger variations may be obtained).

When the temperature was reduced to 140°F the intensity of the blue colour yielded by commercial mixed milks was in the neighbourhood of 20 units.

In a further series of experiments in which the temperature was maintained at 145°F the reduction of the heating time to 25 minutes resulted in the test yielding colours between 1.9 and 3.1 blue units: while, if the time of heating was reduced to 20 minutes, the values obtained were all above the standard limit and ranged from 2.5 to 9.0 Lovibond blue units.

The admixture of 0.2 per cent of raw milk to a sample of correctly pasteurised milk gave a product which, when examined by this method, gave a colour equal to 2.5 blue units; while the addition of 0.5 per cent of raw milk had the effect of raising the blue value to 3.7 units in one case and to 4.2 units in another experiment. From these results it may be inferred that contamination with 0.5 per cent of raw milk can readily be detected by this method.

More recent experiments have shown that, in practice, 0.2 per cent of raw milk in a sample of correctly pasteurised milk may readily be detected.

Application of the test to milk pasteurised under different conditions

1. The minimum statutory temperature of pasteurisation in the United States varies somewhat from State to State and may be as low as 140°F. In Canada also, the range allowed is 140° to 145°, and exposure of the milk to a temperature of 140° for a period of 30 minutes is generally accepted as being sufficient to destroy any pathogenic organisms present, but it is insufficient to destroy the phosphatase completely. By decreasing the time of incubation, however, the test described above can be adapted to suit American and Canadian standards. Thus, no milk which has been pasteurised at 140°F for 30 minutes will give a blue colour greater than 2.3 Lovibond units after one hour's incubation, but many samples will give colours greater than 2.3 when incubated for $1\frac{1}{4}$ hours.

In the adjacent figure is shown a graph indicating the duration of the incubation corresponding to the temperature at which the milk is claimed to have been pasteurised. In all cases, if the blue colour which is developed exceeds 2.3 Lovibond units, the milk has not been heated as claimed or has been improperly handled.

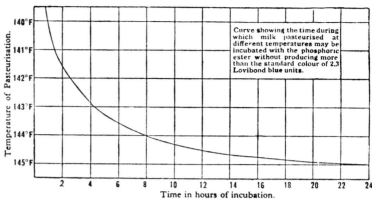

2. *High temperature short-time pasteurisation*

In two long series of controlled experiments, in one of which the apparatus of Stassano and in the other the A.V.P. apparatus for this type of heating was used correctly, no sample of milk gave a positive phosphatase test: that is, the phosphatase test may also be used for controlling the efficiency of pasteurisation by the high-temperature short-time process.

References

Among the extensive bibliography may be noted
1. H. D. Kay and W. R. Graham, *J. Dairy Research,* 1935, **6,** 191
2. H. D. Kay and F. K. Neave, *Lancet,* 1935, **1,** 1516
3. Ministry of Health circular 1533, London 24th April, 1936
4. H. D. Kay and F. K. Neave, *Dairy Ind.,* 1937, **2,** 5
5. E. B. Anderson, Z. Herchdorfer and F. K. Neave, *Analyst,* 1937, **62,** 86
6. H. D. Kay, R. Aschaffenburg and F. K. Neave, *Technical Comm. No. 1, Imp. Bureau of Dairy Sci.,* 1939
7. A. T. R. Mattick and E. R. Hiscox, *Medical Offr.,* 1939, **61,** 177
8. Ministry of Health pamphlet Memo 139/Foods Addendum March 1943
9. Norma Portuguesa Definitiva, NP458-1967

Picric Acid Test for the estimation of reducing sugars in potatoes

Introduction

The degree of darkening of a solution of potassium picrate is a measure of the reducing sugar content of a potato. This method is, basically, that of the Potato Chip Institute of America, with variations to reduce discrepancies between results from different operators and to reduce the time required for each test.

Principle of method

Picric acid reacts with reducing sugars to produce a red colour. In solutions containing excess picric acid the red colour will combine with the yellow of the picric acid to produce a series of shades from yellow, through orange to red. These colours are compared with a series of Lovibond permanent glass colour standards, which have been calibrated in terms of reducing sugar.

Apparatus required

Test tubes Pyrex or borosilicate 16 mm external x 150 mm
Test tube stand to hold 16 of above tubes in boiling water
Boiling water bath
No. 1 Cork borer to cut a core 4 mm diameter, with rod for pushing out the core.
Metal strip 25 mm x 150 mm
2 pipettes each to measure 1.5 ml

Reagents required

1. *Potassium carbonate anhydrous* (K_2CO_3) 20% solution w/v
2. *Picric acid* $((NO_2)_3 C_6H_2OH)$ Saturated solution. The clear solution is used for the test, but there must always be some picric acid crystals at the bottom of the stock bottle. Be careful not to allow these crystals to block the pipettes.

All chemicals used in the preparation of reagents should be of analytical reagent quality.

The Standard Lovibond Disc 4/45

This contains 9 colour standards, numbered arbitrarily 1 to 9, which are related to the reducing sugar content as shown in the table.

Technique

Take a representative sample of at least 16 potatoes from each load or bin it is desired to test.

Obtain a cylinder 4 mm in diameter from each of the 16 potatoes by pushing the cork borer through the centre of the potatoes, about halfway between the ends, and then pushing the core out of the borer. Lay the cores side by side and line up their ends with the metal strip. Lay the strip of metal on top and cut along each side to give specimens 25 mm long exactly. Drop each 25 mm × 4 mm specimen of potato into a test tube containing 1.5 ml of each reagent solution. Shake test tube to ensure potato specimen is in the test liquor. Rinse and wipe the borer, knife and strip between each test.

Put the rack, containing the 16 test tubes, into the boiling water bath and remove it after exactly 4 minutes, and shake the rack round by hand in such a way that the contents of all the test tubes are disturbed. The heat should be sufficient to prevent loss of boiling for more than a few seconds when the tubes are put in.

Place each of the 16 tubes in the comparator in turn, match against the disc, and note its colour value. Use either good north daylight or the Lovibond Comparator Lighting Cabinet for the illumination of the comparator.

Reporting the results

For every sample of potatoes tested, record the number of specimens in each colour and an average colour for the 16 pieces, e.g.

```
 9 at colour 1 =  9
 3   ,,    ,, 2 =  6
 2   ,,    ,, 3 =  6
 1   ,,    ,, 5 =  5
 1   ,,    ,, 6 =  6
───                ──
16   ,,    ,,      32
```

Average Colour 2.0

Relationship between Meredith and Drew Picric Acid Value of Potatoes and True Tuber Reducing Sugar Content

M and D Picric Acid Test Classification	Experimentally Determined Reducing Sugar Content on Whole Tubers
1	0.059%
2	0.12%
3	0.20%
4	0.28%
5	0.38%
6	0.49%
7	0.67%
8	0.95%
9	1.35%

Notes

1. The above relations hold good only so long as the Picric Acid Value is determined in strict accordance with Meredith & Drew's method as described above.

2. Acknowledgement is made to United Biscuits Ltd. for permission to publish this test which was devised in their laboratories.

3. If a glass vessel is used for the storage of the picric acid solution, care must be taken to keep the neck of the bottle scrupulously clean. If picric acid is allowed to crystallise inside the neck, then the friction of removing the stopper may lead to an explosion.

The Quality Grading of Milk
using the Resazurin Test

Introduction

In assessing the hygienic quality of milk, the estimation of bacterial content is of prime importance. The most accurate and informative method for good milk quality is the colony count on solid media. This is costly and slow, however, and is also inaccurate for high count milks. Indirect methods which are more rapid are now largely used for routine checking. The two best known indirect methods are the methylene blue test and the resazurin test[1]. Both tests measure the rate at which reducing systems are produced by growing bacteria such as coli, lactic streptococci and staphylococci. The resazurin test is the more sensitive to the weak static reducing systems of leucocytes and other cells present in the milk at the actual time of the test, and is thus responsive to high cell counts which are frequently caused by mastitis. Hence this test is of more value to those using milk for processing in any way, as udder disease is responsible for faults and taints in cheese, butter and other milk products. In the test, the resazurin dye changes colour from its original blue through pink to colourless, and the results are reported in arbitrary colour values (based on the Lovibond scale) as adopted by the Ministry of Agriculture and Fisheries[2,4], and by the British[9] and Portuguese[10] standards institutions.

For the regular periodic checking of milk, and for the checking of bulked raw milk at creameries, the standard (1 hour) resazurin test is recommended[2]. In some countries the older routine resazurin test[3] is still used. For a quick test to appraise producers' milk at the time of arrival at the point of first delivery, or for assessing any milk in Category C in the routine test, the 10 Minute (Rejection) resazurin test is recommended[4].

Johns and Howson[5] developed a "Triple-reading Resazuring Test" which they claim is superior to the 1 hour test for the detection of milk with high bacterial and leucocyte contents. It has been suggested[6] that the triple reading test at 37°C could be used for advisory purposes or to follow up failures on the 2-3 hour Hygiene Test. It has been reported[7,8] that this test is being extensively employed in North America and that it is more commonly used there than the 1 hour test.

Apparatus required

1. The Lovibond Comparator fitted to a Milk Viewing Stand DB415. (This same apparatus is used for the APTW phosphatase test to check efficiency of pasteurisation).
2. The resazurin disc 4/9
3. A water-bath controlled at $37.5 \pm 0.5°C$
4. Test tubes $6'' \times \frac{1}{2}''$ internal B.S.625, graduated at 10 ml (Tintometer reference AF215TT), and a 1 ml pipette

   All glassware must be washed and sterilised before use.

Reagent required

Resazurin may be obtained in tablet form, ready for dissolving in water.

Alternatively, a stock solution of approved resazurin, 0.05% in glass-distilled water, may be made up and kept in a refrigerator. For use this stock solution is diluted 1:10 to produce a 0.005% solution.

The Standard Lovibond Comparator Disc 4/9

The disc contains seven colour standards, ranging from blue through mauve and purple to pink and finally colourless.

Technique

(a) *Taking the sample*[2,3] When a composite sample is taken from all the cans in a consignment the sample from each can should be proportional to the volume of milk in the can. Otherwise the sample is best taken after the milk from all cans has been mixed.

The milk to be sampled should be vigorously mixed by stirring and plunging with a sterile dipper, and the sample should be taken from well below the surface. The sample should be poured into a sterile bottle and immediately stoppered.

(b) *1 Hour test*[2] Shake the sample bottle 25 times up and down, with an excursion of about one foot, and then pour the milk into a sterile test tube up to the 10 ml mark. Both the mouth of the sample bottle and the test tube should be flamed using a slightly yellow bunsen flame. Take 1 ml of the resazurin solution in a sterile pipette, inserting the pipette $\frac{1}{2}$ inch into the mouth of the tube and expelling the solution by blowing. Take care that the tip of the pipette does not touch the sides of the test tube. Replace the sterile stopper in the tube and mix the contents by inverting the tube twice. After adding resazurin solution to a batch of 10 tubes, place the tubes in a water-bath at $37.5 \pm 0.5°C$ and note the time. The level of the water in the bath must be maintained above the level of the milk in the tubes for the whole period of incubation.

After exactly 60 minutes remove the tube from the bath and examine immediately. Any tube showing no colour is recorded as 0. Very pale pink, pink and white mottling, or deep pink at the top with paler pink below record as $\frac{1}{2}$. Mix the contents of the other tubes by inverting twice, wipe off any water from the outside of the tube and place the tube in the right-hand slot at the back of the comparator stand so that it rests on the ramp and comes below the colourless centre of the disc. A "blank" of similar milk without resazurin is poured into an identical tube and placed on the left-hand ramp to act as a background for the coloured glass standards. When Guernsey or Jersey milk is being tested, especial care must be taken to use a blank from the same milk. The use of a blank in this manner compensates for the variable colour inherent in milk.

Place the comparator and stand on a bench facing a standard source of white light, such as the Lovibond White Light Cabinet or, failing this, north daylight. To increase the illumination, the opal diffusing screen may be removed from the blue correction filter in the Lighting Cabinet. Make sure that both tubes are equally illuminated. Fix the comparator at such a height that the operator can look down directly onto the two apertures. Revolve the disc until a colour match is obtained or until the sample colour falls between two successive standards. Read off the value of the matching colour from the indicator window in the comparator, interpolating half values between standards as required, i.e. record as $3\frac{1}{2}$ a sample with a colour falling between standard 3 and 4.

Any sample giving a disc reading of 4 or above on this test may be regarded as satisfactory.

(c) *Routine test*[3] Sample as in (a) above. After sampling, store the samples according to the instruction in Table 1 and take the minimum and maximum temperatures indicated.

Table I
Possible times of testing and temperature readings to be taken for various classes of samples

Sample	Time of Testing	Temperature Reading
Evening milk	(a) 4 p.m. day following date of production or (b) refrigerated at 4 p.m. on day following date of production until test at 9 a.m. next day	9 a.m. Min. and 4 p.m. Max. on day **following** date of production Ditto
Morning milk	9 a.m. day following date of production	4 p.m. Max. on day of production and 9 a.m. Min. on the following day
Mixed milk in which the evening milk is the older	(a) 4 p.m. day following date when **evening** milk was produced or (b) refrigerated at 4 p.m. on day following date when evening milk was produced until test at 9 a.m. on the following day	9 a.m. Min. and 4 p.m. Max. on day following date when evening milk was produced Ditto
Mixed milk in which morning milk is the older	9 a.m. day following date of production	4 p.m. Max. on day or production and 9 a.m. Min. on day following production

The times in all cases are B.B.C. times. Should samples freeze during storage they must be placed in water at 18°C *for not more than* 30 *minutes immediately before testing.*

Then proceed with the test as in (b) above, but incubate at 37.5 ± 0.5°C for the time prescribed in Table II.

Table II
Time of Incubation

Mean of Max. and Min. Shade Temperatures	Period of Incubation at 37.5 ± 0.5°C
	mins.
40°F and below	120
41–50°F	90
51–55°F	60
56–60°F	30
61°F and over	15

Tubes incubated for 90 or 120 minutes must be inverted at 60 minutes. The results are interpreted as follows:—

Disc reading	Category	Keeping Quality
4 or over	A	Satisfactory
1–3½	B	Doubtful
½ or 0	C	Unsatisfactory

(d) *10 minutes resazurin test*[4] This test must be started within 30 minutes of the arrival of the milk on the creamery platform. Sample as in (a) above and proceed as in (b) except that the batch of tubes must be removed from the water-bath and examined after 10 minutes ± 30 seconds.

The results are interpreted:—

4–6 accept

0–3½ reject

(e) *Triple reading resazurin test*[5,6] Sample and proceed as in (a) and (b) above but examine the tubes after incubation for 1, 2 and 3 hours, inverting the tubes before reading. The colour of the tubes should be reported as "over 3" or "3 or under". The results should be reported as:—

		Corresponding 1 hr test readings
Grade 1	over 3 in 3 hr	5–6
2	3 or under in 3 hr	3–4½
3	3 or under in 2 hr	1–2½
4	3 or under in 1 hr	0–½

Grade 4 is a good indication of a high bacterial content and activity and a colony count examination is recommended to detect if the milk has been handled in unsterile equipment.

Grade 3 is often an index of leucocyte activity and may be used as a preliminary screening method for sub-clinical mastitis. Incubation of Grade 3 samples could be extended to 4 or 5 hours to observe whether reduction is delayed at the pink stage, a sure indication of leucocyte induced reduction.

Grade 2 is considered fairly satisfactory.

Grade 1 is satisfactory.

Note

Resazurin powder and resazurin tablets must be from an approved source. A list of approved sources can be obtained from the National Institute for Research in Dairying, Shinfield, Nr. Reading, England.

All resazurin preparations must bear a label showing the following particulars:—

1. Name of manufacturer.
2. The words "this batch of resazurin powder (or tablets) has been tested and approved by the National Institute for Research in Dairying".

References

1. Pesch and Simmert, *Michl. Forsch.*, 1929, **8**, 551
2. Min. of Agric. Fisheries and Food, *"Bacteriological Techniques for Dairying Purposes"*, H.M. Stationery Office, London, 1962, Technique B.T. 13 – The 1 hr resazurin test
3. idem. Nat. Milk Testing Service Technique No. 9
4. idem., *"Bacteriological Techniques etc.,"* Technique B.T. 14 – The 10 minute (rejection) resazurin test
5. C. K. Johns and R. K. Howson, *J. Milk Tech.*, 1940, **3**, 320

6. S. B. Thomas and Pauline E. Makinson, *Dairy Ind.,* 1964, **29,** 432
7. C. K. Johns, *Proc.* 13*th. Int. Dairy Congr.* 1953, **2,** 241
8. Foster et al. *"Dairy Microbiology",* Prentice Hall, New Jersey, 1957
9. British Standard 4285: 1968
10. Norma Portuguesa Definitiva NP455 – 1967

The Quality Grading of Fish and Meat Products
using Resazurin

Introduction

The use of the resazurin test for the quality grading of milk is a standard procedure in many parts of the world. This test may also be used to advantage in the testing of raw or cooked meat and fish. This modification[1] of the standard procedure has been devised to enable the test to be applied to coloured products without interference, provided that no synthetic dyestuff has been used for colouring the sample.

Principle of the method

The sample is incubated with resazurin solution for 30 minutes at 37°C. The final solution is chromatographed on chromatography cellulose and the colour of the resazurin layer is compared with a series of Lovibond permanent glass colour standards.

Apparatus required

1. Incubator or water-bath at $37° \pm 0.5°C$.
2. McCartney bottles. These are 1 oz, wide mouthed bottles fitted with aluminium screw caps and rubber liners, as used for media, vaccines, etc.
3. Hirsch funnel – conical filter funnel with a perforated disc support.
4. Filter flask.
5. Water pump for suction.

Reagent required

Resazurin solution 0.005% Dissolve one standard resazurin tablet (Note 1) in 25 ml of sterile distilled water. Alternatively a stock solution of approved resazurin 0.05% in sterile distilled water may be made up and stored in a refrigerator. When required for use dilute this stock solution 1 in 10 with sterile distilled water.

The Standard Lovibond Grading Strip AF606 (illustrated on page xxxvi)

This strip contains seven colour standards ranging from blue through mauve and purple to pink and finally colourless. These colours correspond to the arbitrary standards adopted by the Ministry of Agriculture and Fisheries.[2]

Preparation of materials for Resazurin Test

Raw Meats (or fish)

Prepare the raw material by mincing, using a hand mincer which has been thoroughly washed and dried. The stationary plate of the mincer ($\frac{1}{4}$ inch dia holes approx.) is likely to become heavily contaminated and it is advisable to subject a number of these to a heat sterlisation process in addition to washing thoroughly. Mix the minced material and weigh two grams into a sterile wide mouth McCartney bottle with a cool stainless palette knife, which has been previously flamed to red heat on the tapering part of the blade. Apply the half hour Resazurin test.

Cooked Meats (and fish)

With cooked meats it is more important to ensure aseptic precautions throughout preparation. For most applications the meat can be mashed with a spatula (sterile Griffin Trulla) or a palette knife as above. Where the meat is not cooked to near disintegration, a sterile scapel or sharp knife used in conjunction with a sterile spatula is suitable for cutting $\frac{1}{8}$ inch slices from various parts of the meat. These slices may again need to be cut before weighing into a sterile bottle.

For meat products examined after a period of storage the test may be immediately applied, values being recorded at half hour and two hours respectively.

For quality control of freshly prepared meat products, it is necessary to incubate the product at 37°C for 18 hours before applying the test. It is desirable to incubate the whole product or a sufficiently large portion to ensure representative growth of bacteria, which may be sparsely distributed throughout the material. After incubation weigh a representative sample into a sterile McCartney bottle for the Resazurin test and examine after 30 minutes and two hours respectively.

Technique

Place 2 g of sample in a 1 oz McCartney bottle containing 20 ml of sterile distilled water at about 40°C. Add 1 ml of the resazurin solution, shake, and place in an incubator or water-bath at 37°C. Incubate for exactly 30 minutes. During this incubation period prepare a filter bed of chromatography cellulose approximately $\frac{3}{4}$ inch in depth in a Hirsch funnel. When the incubation period is completed, remove the bottle from the incubator, shake and pour half the contents onto the prepared filter bed. Draw the liquid through the bed by applying suction from a water pump. As the surface dries scrape away the top layer to reveal a brightly coloured layer immediately below. Hold the grading strip a few inches above a well-illuminated white surface, with the funnel in the other hand adjacent to the colour standards, and compare the colour of the layer with the standards. Record the number of the nearest standard.

For cooked meats return the bottle to the incubator for a further one and a half hours and examine the remaining half of the sample by the same method as above at the end of this time. It is found that the two results so obtained are of more value than the 30 minute test in this case.

For freshly cooked meat products incubate the sample alone for 18 hours, and then proceed with the resazurin test as above, using 2 g of the sample after thorough mixing.

The results obtained in these tests may be directly related to the bacteriological count of the sample. A raw meat having a value of 3 or less in the 30 minute test may be considered for rejection, since a good quality fresh meat will have a value of 6. Individual manufacturers can set their own quality control standards by means of this test. As a sorting test it may be used to pre-test samples from restaurants etc., only those samples in which the colour value falls below a predetermined level being submitted to a more complete bacteriological examination.

Note

Resazurin powder and resazurin tablets must be from an approved source. A list of approved sources can be obtained from the National Institute for Research in Dairying, Shinfield, Nr. Reading, England.

All resazurin preparations must bear a label showing the following particulars—

1. Name of manufacturer.
2. The words "this batch of resazurin powder (or tablets) has been tested and approved by the National Institute for Research in Dairying".

References

1. K. R. Baker, *Food Manufacture*, 1966, **41** (5), 49
2. Min. of Agric. Fisheries and Food, Nat. Milk Testing Service Technique No. 9

The Determination of Rubber Latex Film Colour

Introduction

This is one of a series of tests from British Standard 1672[1] designed to determine the chemical and physical condition of rubber latex. The tests are not necessarily suitable for latex from natural sources other than *Hevea brasiliensis* and are not intended for latices of synthetic rubber, compounded latex or artificial dispersions of rubber.

Principle of the method

A dried latex film prepared according to the method laid down in the British Standard[1] is compared with a series of Lovibond glass colour standards.

The Standard Lovibond Comparator Discs 4/19A and 4/19B

Disc 4/19A covers the range, 1.0, 1.5, 2.0, 2.5, 3.0, 3.5, 4.0, 4.5, 5.0 units
Disc 4/19B covers the range 5, 6, 7, 8, 9, 10, 12, 14, 16 units

The units are arbitrary steps specified by the British Standard[1], such that the Red content in Lovibond units is $\frac{1}{5}$ the numerical film colour value, and the Yellow value in Lovibond units is $(2.5R + 0.32R^2)$ where R is the Red content.

Technique

Prepare a film of the latex sample, in accordance with the British Standard method[1], to a thickness 1 ± 0.2 mm. Measure the film thickness at the time of the determination with a micrometer gauge to the nearest 0.1 mm, first peeling back the covering material so that only the film itself and not the covering material is included in the measurement.

Cut a piece of film of convenient size to fit into the special holder (which will be supplied on request), place this in position in the Lovibond Comparator in the right-hand compartment, and with the appropriate disc in position compare the colour against the glass colour standards, holding the comparator facing a standard source of white light. A Lovibond White Light Cabinet is recommended, but failing this, use north daylight.

Report the nearest match. If the film is judged to have a colour exactly between two adjacent steps, quote the higher value. Finally make a correction for the variation in film thickness from 1 mm, and report the film colour to the nearest half unit below a value of 5, and to the nearest whole unit above 5.

Examples

If the film sample is 1.1 mm thick and the colour is 3.0, the corrected colour is $\frac{3.0}{1.1}$ i.e. 2.73, or 2.5 to the nearest half unit. If the thickness is 0.9 mm and the colour 14, the corrected colour is $\frac{14}{0.9}$ i.e. 15.6, or 16 to the nearest unit.

Notes

1. The same Lovibond Comparator, with appropriate colour discs, may be used for the determination of copper and of iron as laid down by this same British Standard Specification. The code number of the Copper disc is 3/39 and for the Iron disc is 3/6.

2. This colour test was deleted in the 1972 edition of B.S. 1672, but these discs have been adopted in ASTM specification D3157–73.

Reference

1. British Standard 1672: Part 2: 1954 *"Methods of testing rubber latex"*. Amendment 3: 1961. Section 2.12

The Colour Grading of Sand and other Aggregate used in the manufacture of road-marking materials

Introduction

This test is specified in British Standard 3262: Part 1: 1960. (Hot applied thermoplastic materials—Superimposed type) and also in Part 2. (Hot applied thermoplastic materials—Inset type).

The specification says that "the aggregate shall consist of light coloured silica sand, calcite, quartz, calcined flint or other materials approved by the Engineer and that the colour when tested by Method A shall be at least as pale as Grade 6."

The Standard Lovibond Comparator Disc 4/20

This disc contains an arbitrary series of colour standards approved by the Road Research Laboratory, and labelled 1, 2, 3, 4, 5, 6, 7, 8, 9. It must be used in a Lovibond Comparator fitted to a Surface Viewing Stand (DB414).

Technique

A representative sample of the aggregate is obtained, preferably in accordance with the method laid down in British Standard 812 (which requires a minimum of 10 samplings, the bulk being then reduced by quartering). Wash and dry the sample and then obtain about 25 g, of the fraction graded between a 52 and a 100 B.S. sieve. Place in one of the Lovibond white porcelain dishes (AF733) and strike the surface off level. Place in position in the Lovibond Surface Viewing Stand (DB414) in a White Light Cabinet (DB416) and with the disc in position in the Comparator, compare the colours in the disc viewed against the magnesium carbonate block background.

Note

Other tests in this same British Standard include 'Colour of Rosin', for which the Lovibond Rosin Cube colour standard WG is required, and 'Colour of Oil' for which the Lovibond disc 4/11A is required, see Shellac (PRS scale). This disc is used with the Lovibond Comparator and a 10 mm, cell.

Reference

British Standard 3262 Part 1 1960, and British Standard 3262 Part 2 1960

COLORIMETRIC CHEMICAL ANALYTICAL METHODS

The Grading of Sand for Quality

Introduction

British Standard 812 : 1960 ("*The sampling and testing of mineral aggregates*") adopted an approximate test to indicate the amount of organic compounds present in a sample of sand, and this test (which was taken from a 1934 report of the Reinforced Concrete Structures Committee of the Building Research Board) is intended to show whether further, more detailed, tests are necessary.

Principle of the method

Fixed proportions of the sand under test and a caustic soda solution are shaken together and allowed to stand in contact. The amount of the discolouration of the caustic soda solution is taken as an assessment of the extent of organic impurity in the sand.

The Standard Lovibond Comparator Disc 4/6

This disc contains a single standard corresponding to the defined standard colour.

Reagent required

Sodium hydroxide ($NaOH$)   3% solution in water

Technique

Fill a 12 oz graduated clear-glass medicine bottle to the $4\frac{1}{2}$ oz mark with the sample of sand as received. Add the soda solution to a total final volume of 7 oz. Stopper, shake vigorously and allow to stand for 24 hours. Decant the supernatant solution into a Lovibond test tube, place in the right-hand compartment of the Lovibond Comparator and compare against the standard disc colour using a standard source of white light, such as the Lovibond White Light Cabinet, or failing this, north daylight. Report as lighter, or darker than, or equal to, the standard.

The Colour Grading of Shellac and Varnish
on the P.R.S. Scale

Introduction

Various liquids have been suggested as standards for the colour-grading of shellac, of which iodine has been most favoured, but investigators have from time to time pointed out that these liquids are in many cases not a good match, and in other respects leave much to be desired. In 1934 the Paint Research Station published a report[1], summarising the position and putting forward a suggested improved colour scale for varnishes which was based on Lovibond glass standards. Verman[3] later used this scale for shellac solutions. This scale has subsequently been reproduced in the form of standard discs for the Lovibond Comparator, quarter and eighth steps being added, where necessary, to give the required close steps at the most critical values.[2] This scale has also been adopted by the British Plastics Federation[6], and the British Standard for Road Paint[7] uses disc 4/11A for the colour of the oil.

The Standard Lovibond Comparator Discs 4/11, 4/11A and 4/11B

The numbers are the same as in the Paint Research Station Varnish Scale.

4/11 *Bleached Shellac* covers the range $\frac{3}{8}$; $\frac{1}{2}$; $\frac{5}{8}$; $\frac{3}{4}$; $\frac{7}{8}$; 1; $1\frac{1}{4}$; $1\frac{1}{2}$; $1\frac{3}{4}$;

4/11A *Shellac Disc A* covers the range 2; $2\frac{1}{2}$; 3; $3\frac{1}{2}$; 4; $4\frac{1}{2}$; 5; $5\frac{1}{2}$; 6

4/11B *Shellac Disc B* covers the range $6\frac{1}{4}$; $6\frac{1}{2}$; $6\frac{3}{4}$; 7; $7\frac{1}{4}$; $7\frac{1}{2}$; 8; $8\frac{1}{2}$; 9

These discs are designed for use with a 10 mm cell.

Technique

Pour the liquid into a 10 mm all-glass Lovibond Comparator cell, and place in the right-hand compartment of the Lovibond Comparator. Leave the left-hand compartment empty. Hold the comparator facing a standard source of white light, such as the Lovibond White Light Cabinet or, failing this, north daylight, and rotate the disc until the nearest match is found. Report the colour as matching a stated standard, or falling between two standards.

Note

The cells should be washed out as soon as possible after use, with methylated spirits or industrial alcohol, before the liquid dries. Owing to their fused method of construction, the cells will not be affected by soaking in any appropriate solvent.

References

1. D. L. Tilleard., *Research Assn. British Paint Colour & Varnish Manufacturers Technical Paper* 47, 1934
2. *idem., ibid., Research Memorandum* 125, Jan. 1946
3. L. C. Verman, *London Shellac Research Bureau Technical Paper,* 1936, 10
4. D. L. Tilleard, *J.O.C.C.A.,* 1937, **20**, 124
5. *Paint Research Station Memorandum,* March, 1950, No. 75
6. Surface Coating Synthetic Resin Manufacturers and the British Plastics Federation "*Standardisation of test methods for resin*"
7. British Standard 3262: 1960 "*Road Marking Materials*" Parts 1 & 2

The Control of Water Purification in Swimming Pools

using diethyl-*p*-phenylene diamine (Palin-DPD method) for chlorine
and
diphenol purple or phenol red for *p*H

Introduction

In the control of the Breakpoint method of chlorination the importance of determining both free and combined residual chlorine is emphasized in the Ministry's official publication.[1] Among the methods given therein appears the earlier Palin method of 1945. Subsequent improvements to this method have resulted in the development of an extremely simple procedure for these essential determinations.

In operating the Breakpoint process, maintenance of the free residual chlorine and the *p*H value between suitable limits ensures that the maximum bactericidal efficiency is obtained with the minimum discomfort to bathers. Above a *p*H value of 8.0 the bactericidal power of chlorine is reduced, whereas values of 7.0 or below increase the risk of producing objectionable chloramines and nitrogen trichloride.

Principle of the method

Free residual chlorine reacts with the DPD indicator to give a red colour. Combined residual chlorine is then caused to react by adding potassium iodide. A novel feature of the comparator method is the use of reagents in tablet form which, besides being far more convenient in use, also permits of further simplification to the technique.

Where differentiation is not required (as in ordinary or marginal chlorination) the total residual chlorine value is obtained by the use of one tablet only.

Reagents required

For free residual chlorine	—DPD tablets No. 1
For combined residual chlorine	—DPD tablets No. 3 after No. 1
For total residual chlorine	—DPD tablets No. 4 or Nos. 1 and 3
For *p*H	—Phenol red or diphenol purple tablets.

The Standard Lovibond Comparator Discs 3/40A, 3/40B, 3/40C, 3/40CZ and 3/40D.

Disc 3/40A (chlorine only) covers the range 0.1 to 1.0 parts per million chlorine, in steps of 0.1.

Disc 3/40B (chlorine only) covers the range 0.2 to 4.0 parts per million chlorine, (0.2, 0.4, 0.6, 1.0, 1.5, 2.0, 2.5, 3.0, 4.0).

Disc 3/40C (chlorine and *p*H) covers the range 0.25 to 3.0 parts per million chlorine (0.25, 0.5, 1.0, 1.5, 2.0, 3.0) and *p*H values 7.0, 7.5 and 8.0, using diphenol purple.

Disc 3/40CZ (Chlorine and *p*H) covers the range 0.5 to 4.0 parts per million chlorine (0.5, 1.0, 1.5, 2.0 and 4.0) and *p*H values 7.0, 7.4, 7.6 and 8.0 using phenol red.

Disc 3/40D Chlorine 0.5, 1, 2, 3, 4 and 6 ppm and *p*H (phenol red) 7.0, 7.5, 8.0.

The chlorine values obtained by the use of the discs are identical with those obtained by the FAS titration method of Palin. The master disc, against which all reproductions are checked, was tested and approved by Dr. Palin.

Technique

Free residual chlorine

Place in the left-hand compartment of the comparator a 13.5 mm cell or test tube containing the water sample only. Rinse a similar cell with the water sample, leaving in one or two drops, sufficient just to cover the tablet when added. Add one DPD tablet No. 1 and allow to disintegrate, (or alternatively, crush the tablet). Add the water sample up to the 10 ml mark, mix rapidly to dissolve the remains of the tablet and place in the right-hand compartment of the comparator. Match the colour at once by holding the comparator facing a standard source of white light, such as the Lovibond White Light Cabinet or, failing this, north daylight and revolve the disc. The figure then shown in the indicator window represents the parts per million of free residual chlorine present in the sample.

Combined residual chlorine

Continue the test by adding to the right-hand cell one DPD tablet No. 3, mix vigorously to dissolve and after allowing to stand for two minutes evaluate the colour as before. This gives total residual chlorine, from which deduct the free residual chlorine reading to obtain combined residual chlorine in parts per million chlorine.

Total residual chlorine (shorter method)

Carry out the test as above, but use one No. 4 tablet only (which is Nos. 1 and 3 combined together). Thus, rinse tube, leaving a little water in the bottom, Add the No. 4 tablet and allow to disintegrate. Fill up to the 10 ml mark, mix and match after two minutes.

For marginal chlorination, this is quicker and cheaper than using Nos. 1 and 3 tablets, and provides all the necessary information.

pH value (Discs 3/40C, 3/40CZ, or 3/40D)

After carefully rinsing the right-hand cell fill with the sample to the 10 ml mark. Then add one of the *p*H tablets, which contain sodium thiosulphate (to prevent the chlorine interfering with the indicator solution) dissolve and mix. Place this cell in the right-hand compartment of the comparator, and read the *p*H value on one of the glasses corresponding to values of 7.0, 7.5 or 8.0 or in the case of Disc 3/40CZ, 7.0, 7.4, 7.6 or 8.0. Rinse the cells carefully after use.

Notes

1. All glassware used must be very thoroughly rinsed after making residual chlorine tests, since only a trace of potassium iodide (as contained in DPD tablet No. 3) will cause the combined residual chlorine colour to develop. For the same reason handling the tablets, particularly DPD No. 1, should be avoided. With strip packing, one may be squeezed out direct into the cell after tearing the strip.

2. Samples containing more than six parts per million of residual chlorine must be diluted. If there is any doubt about the need for dilution, a simple check is to repeat the procedure using two DPD No. 1 tablets instead of one. A very decided increase in colour would indicate dilution to be necessary, in which case the requisite amount of distilled water is added first to the reagent, followed by the measured amount of sample. Concentrations of chlorine above 10 ppm will entirely bleach the colour and give an apparently zero reading, but at this concentration the smell of chlorine would be very apparent.

3. In performing the chlorine tests it is unnecessary to stand for any longer period than the two minutes specified, after which the tubes should be rinsed. If cells containing sample and reagent should be allowed to stand, a faint colour may develop even in the absence of residual chlorine. This is caused by dissolved oxygen and should, of course, be ignored.

Control of Swimming Pool Purification

The following is quoted from the Ministry's booklet[1] on *"The Purification of the Water of Swimming Baths."*

" . . . . . it is felt that the aim (in Breakpoint Chlorination) should be not to maintain as high a concentration of free residual as can be tolerated by the bathers, but as low a one consistent with the satisfactory operation of the process. This might be found to be in the region of 1 ppm under normal conditions, with increases to perhaps double that value when the bathing load is heavy and perhaps during the night, to allow its action to proceed as near to completion as possible."

". . . . . the pH value should be somewhat greater than 7. At the other extreme it is most inadvisable to allow it to exceed 8, as the operation of the filters might be interfered with and complaints of smarting eyes occur. This gives an operating range of say 7.2 to 8.0 Since the normal tendency of pH is to diminish, due to the addition of acid-forming substances in water treatment, most operators will probably prefer to work nearer the upper than the lower of these limits."

Operating standards recommended by Imperial Chemical Industries Limited in their pamphlet, "A Guide to the Breakpoint Chlorination of Swimming Bath Water," are as follows:—

A total chlorine residual of about 2 ppm of which at least 1.5 ppm is in the form of free chlorine.

A pH between 7.5 and 8.0

An alkalinity of about 200 ppm as $CaCO_3$.

The following recommendations are abstracted from official instructions used by swimming bath managers in the United Kingdom.

How to Maintain Breakpoint

To keep on breakpoint, free chlorine must be greater than combined chlorine, preferably by at least two to one. Failure to keep the proportions correct may lead to complaints. Aim at a total residual of around 2.0 ppm with at least 1.5 ppm as free chlorine. If the combined figure begins to rise, then increase the rate of chlorination despite the fact that this will cause the total residual to exceed 2.0 ppm. Return to the normal maximum of 2.0 ppm when peak conditions end. In slacker periods a free residual chlorine figure of 1.0 ppm is ample if the "combined" figure is only say, 0.1 or 0.2 or 0.3 ppm.

pH

pH must be kept over 7.5 at all times. Nearly 8.0 is much better than nearly 7.5. DO NOT EXCEED 8.4.

Bicarbonate Alkalinity

It is important to maintain the bicarbonate alkalinity of the water at over 200 ppm when operating the breakpoint system, as a reserve of alkalinity acts as a buffer against sudden depressions in the pH value. Tests should be carried out at intervals of at least one week and possibly even daily according to size of bathing loads. Unlike the Chlorine and pH tests great accuracy is not critical. If the result is below 200 ppm add sodium bicarbonate until further tests indicate that the level has risen above 200 ppm.

References

1. *"The Purification of the Water of Swimming Baths"*, Ministry of Housing and Local Govt., H.M. Stationery Office, London, 1951 (reprinted 1960)
2. A. T. Palin, *Analyst,* 1945, **70**, 203
3. A. T. Palin, *Jour. Amer. Water Wks. Ass.,* 1957, **49**, 873
4. A. T. Palin, *Baths Service,* 1958, **17**, 21
5. A. T. Palin, *Water and Water Engineering,* 1958, Jan.
6. W. H. Humphrey, *Baths Service,* 1970, **29**, 44

A note on "breakpoint" chlorination of swimming pool water
for those not familiar with this procedure

The discovery of "breakpoint" chlorination altered the entire approach to the sterilisation of swimming pool water. Although it was realised that, within reason, the more chlorine added the more sterile the water, a limit was set by the unpleasant smell and the irritation to eyes caused by increasing the chlorine content. The harmful nature of dirty swimming bath water arises from the presence of organic matter introduced by the bathers, or dust and dirt entering from the atmosphere: this matter is oxidised by the free chlorine and ultimately broken down and rendered harmless. It is now realised that the unpleasant smell and irritation is due, not primarily to free available chlorine, but to compounds of chlorine and organic matter known chemically as chloramines, of which nitrogen trichloride (trichloramine) is particularly objectionable.

The technique of "breakpoint" chlorination is to add more chlorine, beyond the point where one formerly stopped, and this has the effect of breaking down the chloramines into simpler elements which are inoffensive. A pH value of 7.5 to 7.8 covers the optimum range at which chlorine can oxidise the chloramines.

Chloramines answer to the chemical test for chlorine, and thus give a misleading appearance of adequate chlorine present in the water. (**A** in figure). With the continued application of more chlorine, a point is reached at which the chloramine breaks down and is removed, and the apparent chlorine content suddenly drops. This is the "breakpoint". (**B** in figure). Further addition of chlorine then causes the chlorine content to rise, but this is "free available chlorine," (**C** in figure) which can be tolerated in much heavier concentrations without discomfort to bathers, and this chlorine is available for immediate attack on fresh impurities which are introduced into the water.

Whereas combined chlorine (chloramine) can only be tolerated by bathers up to about 0.4 parts per million, the much more useful and active "free available chlorine" can be tolerated up to 2.0 ppm without bathers in fact being aware of its presence.

Breakpoint chlorination treatment requires more attention, and a test must be applied which easily distinguishes between free and combined chlorine.

Frequent tests must be applied to ensure that there is a continuous level of free available chlorine, and the DPD Chlorine tests has greatly simplified the task. If the presence of free available chlorine is lost, all the work of super-chlorinating to get beyond the "breakpoint" must be repeated. Thus, for large pools with a permanent attendant and a large turn-over of bathers, the breakpoint method and the necessary control is advocated, but for a small pool which does not carry a heavy load, and for which constant care is not available, the older "marginal" method will continue to be used. However, the small private pool (especially in the open air) could in fact benefit greatly by "breakpoint" chlorination, because sudden variations in load and contamination often occur, whereas in the large public bath the pattern of load usually follows a known and regular course.

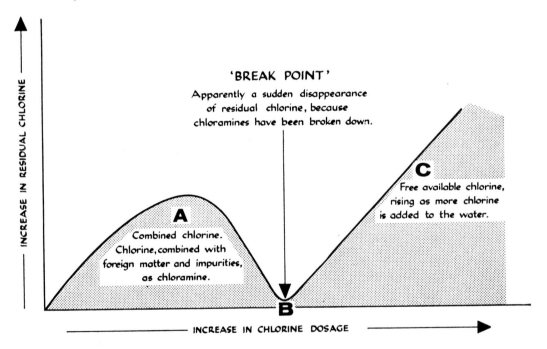